KV-414-645

WITHDRAWN

LIVERPOOL JMU LIBRARY
3 1111 01399 8792

Advances in **VIRUS RESEARCH**

VOLUME **84**

Viruses and Virus Diseases of Vegetables in the Mediterranean Basin

Advances in VIRUS RESEARCH

VOLUME **84**

Viruses and Virus Diseases of Vegetables in the Mediterranean Basin

Edited by

GAD LOEBENSTEIN
Agricultural Research Organization
Department of Virology
The Volcani Center
Bet Dagan, Israel

HERVÉ LECOQ
Institut National de la Recherche Agronomique (INRA)
Station de Pathologie Végétale
Montfavet, France

ELSEVIER

AMSTERDAM • BOSTON • HEIDELBERG • LONDON
NEW YORK • OXFORD • PARIS • SAN DIEGO
SAN FRANCISCO • SINGAPORE • SYDNEY • TOKYO
Academic Press is an imprint of Elsevier

Academic Press is an imprint of Elsevier

525 B Street, Suite 1900, San Diego, CA 92101-4495, USA
225 Wyman Street, Waltham, MA 02451, USA
32, Jamestown Road, London NW1 7BY, UK
Radarweg 29, PO Box 211, 1000 AE Amsterdam, The Netherlands

First edition 2012

Library of Congress Cataloging-in-Publication Data

A catalog record for this book is available from the Library of Congress

British Library Cataloguing-in-Publication Data

A catalogue record for this book is available from the British Library

ISBN: 978-0-12-394314-9
ISSN: 0065-3527

For information on all Academic Press publications
visit our website at store.elsevier.com

Printed and bound in United States in America
12 13 14 15   10 9 8 7 6 5 4 3 2 1

CONTENTS

CONTRIBUTORS

Yehezkel Antignus
Department of Phytopathology, Virology Unit, ARO, The Volcani Center, Bet Dagan, Israel

Miguel A. Aranda
Centro de Edafología y Biología Aplicada del Segura (CEBAS), Consejo Superior de Investigaciones Científicas (CSIC), Campus Universitario de Espinardo, Espinardo, Murcia, Spain

Marina Ciuffo
Istituto di Virologia Vegetale-CNR, Strada delle Cacce, Torino, Italy

Cécile Desbiez
INRA, UR407 Pathologie Végétale, Domaine Saint Maurice, Montfavet, France

Chrysostomos I. Dovas
Laboratory of Microbiology and Infectious Diseases, Faculty of Veterinary Medicine, Aristotle University of Thessaloniki, Thessaloniki, Greece

Alberto Fereres
Department of Plant Protection, Instituto de Ciencias Agrarias, ICA-CSIC, Madrid, Spain

Pedro Gómez*
Centro de Edafología y Biología Aplicada del Segura (CEBAS), Consejo Superior de Investigaciones Científicas (CSIC), Campus Universitario de Espinardo, Espinardo, Murcia, Spain

Victor Gaba
Department of Virology, Agricultural Research Organization, Bet Dagan, Israel

* Current address: School of Biosciences, University of Exeter, Penryn, United Kingdom

Donato Gallitelli
Dipartimento di Biologia e Chimica Agroforestale ed Ambientale, Università degli Studi di Bari "Aldo Moro"; and Istituto di Virologia Vegetale del CNR, U.O. di Bari, Bari, Italy

Inge M. Hanssen
Scientia Terrae Research Institute, Sint-Katelijne-Waver, Belgium

Mireille Jacquemond
INRA, UR407 Pathologie Végétale, Domaine Saint Maurice, Montfavet, France

Nikolaos I. Katis
Plant Pathology Laboratory, School of Agriculture, Aristotle University of Thessaloniki, Thessaloniki, Greece

Safaa G. Kumari
International Center for Agricultural Research in the Dry Areas (ICARDA), Aleppo, Syria

Moshe Lapidot
Department of Vegetable Research, Institute of Plant Sciences, Agricultural Research Organization, Volcani Center, Bet Dagan, Israel

Hervé Lecoq
INRA, UR407 Pathologie Végétale, Domaine Saint Maurice, Montfavet, France

Gad Loebenstein
Department of Virology, Agricultural Research Organization, Bet Dagan, Israel

Khaled Makkouk
National Council for Scientific Research, Beirut, Lebanon

Varvara I. Maliogka
Plant Pathology Laboratory, School of Agriculture, Aristotle University of Thessaloniki, Thessaloniki, Greece

Giovanni P. Martelli
Istituto di Virologia Vegetale del CNR, U.O. di Bari, Bari, Italy

Tiziana Mascia
Dipartimento di Biologia e Chimica Agroforestale ed Ambientale, Università degli Studi di Bari "Aldo Moro"; and Istituto di Virologia Vegetale del CNR, U.O. di Bari, Bari, Italy

Aranzazu Moreno
Department of Plant Protection, Instituto de Ciencias Agrarias, ICA-CSIC, Madrid, Spain

Benoît Moury
INRA, UR407 Pathologie Végétale, Domaine Saint Maurice, Montfavet, France

Hanu Pappu
Washington State University, Pullman, Washington, USA

Michel Pitrat
INRA, UR1052 Génétique et Amélioration des Fruits et Légumes, Montfavet, France

Raquel N. Sempere
Bioprodin SL, Campus de Espinardo s/n, Espinardo, Murcia, Spain

Luciana Tavella
DIVAPRA Entomologia e Zoologia applicate all'Ambiente, University of Torino, Via L. da Vinci, Grugliasco (TO), Italy

Antonio Tiberini
Plant Pathology Research Centre, Agricultural Research Council, Rome, Italy

Laura Tomassoli
Plant Pathology Research Centre, Agricultural Research Council, Rome, Italy

Massimo Turina
Istituto di Virologia Vegetale-CNR, Strada delle Cacce, Torino, Italy

Eric Verdin
INRA, UR407 Pathologie Végétale, Domaine Saint Maurice, Montfavet, France

Heinrich-Josef Vetten
Julius Kuehn Institute, Federal Research Centre for Cultivated Plants, Institute of Epidemiology and Pathogen Diagnostics, Braunschweig, Germany

PREFACE

Vegetables are one of the main ingredients in the Mediterranean cuisine, which is considered as one of the healthiest diets. The vegetables used include tomatoes, cucumbers, sweet peppers, onion and garlic, green salad, bean and peas, potatoes, etc. The climatic conditions and the use of covering the crops by plastic or glass in addition to irrigation enables the growth and supply of many of these vegetables continuously all the year around. This continuous planting also facilitates the establishment and perpetuation of pests and diseases, especially plant virus diseases and their vectors. In areas with a cold winter or dry and hot summers in contrast, there is a break in the vegetation, which reduces the vector population and the annual virus sources and thereby also virus dissemination.

In addition to the climate and cultural practices, emergence of new viruses brought into the area probably by import of contaminated seeds may cause severe damages to some vegetable crops. Thus, in the past decade, the outbreak of *Tomato torrado virus* was observed in Spain and several other Mediterranean countries. Some viruses as *Tomato yellow leaf curl virus* or *Tomato chlorosis virus* are now present in most tomato growing areas worldwide. This is probably associated with the spreading of their vector-*Bemisia tabacci.* Among other crops in the Mediterranean basin, cucurbits such as melon, watermelon, cucumbers, and squash also are often severely infected by viruses including *Zucchini yellow mosaic virus, Cucumber vein yellowing virus, Cucumber green mottle mosaic virus* and others.

Recent progress in the knowledge of the genome of plant viruses (or in molecular virology) has not only greatly increased the efficiency of diagnostic methods but also brought new light on the ecology and epidemiology of viruses infecting vegetable crops.

To overcome these diseases, breeding for resistance to viruses has to be intensified and efforts to exchange information between the various groups in the Mediterranean basin will help to locate new virus diseases. Vector control by novel methods especially against whiteflies will probably play a larger role in IPM to reduce virus diseases and their economic impact.

We hope that this book will contribute to this effort.

We thank Prof. Karl Maramorosch who encouraged this effort, all the authors who contributed to this volume, and the editorial staff who were instrumental in the production of this book.

Gad Loebenstein
Israel
Hervé Lecoq
France
Editors
January 2012

CHAPTER 1

Vegetable Crops in the Mediterranean Basin with an Overview of Virus Resistance

Michel Pitrat

Contents

Abstract

The Mediterranean area (MA) produces about 12% of the world vegetables both for local consumption and for export. With an average consumption of 242 kg per person and per year (and almost 400 kg in Turkey), vegetables are an important part of the Mediterranean diet. Vegetables are cultivated using different

INRA, UR1052 Génétique et Amélioration des Fruits et Légumes, Montfavet, France

Advances in Virus Research, Volume 84
ISSN 0065-3527, DOI: 10.1016/B978-0-12-394314-9.00001-4

cultivation techniques (for instance, open field or protected), and the importance of viruses varies greatly between these growing conditions. Breeding virus-resistant cultivars is a key component of an integrated pest management strategy. The origin and the diversity of the main vegetables are presented with the sources of virus resistance. The center of origin of most vegetables is not in the MA: for instance, tomato, potato, pepper, bean, squash and pumpkin, and sweetpotato have been introduced from the American continent. Very few original sources of resistance against viruses have been described in local landraces from the MA.

Vegetables are important crops in the Mediterranean area (MA) both for local consumption and for export. They are also socially and economically important for the jobs they need.

Vegetable crops are characterized not only by the great number of species but also by the diversity of cultivation techniques: open field, plastic tunnels, glasshouses with or without heating, allowing production in different seasons. The yield and the quality of the products are affected by these techniques. The importance of the diseases, including viruses, is highly variable according to the country, the season, the open field *versus* protected culture; the possibilities to control them vary accordingly. For instance, nets can be placed on the openings of the plastic tunnels or greenhouses to limit the entrance of aphids or whiteflies which are important virus vectors. Biological control through parasitoids is easier in protected crops than in the open field. Although there are no data on the relative importance of the different pests and diseases on vegetables in the MA, depending on the plant species and the cultivation techniques, viruses can be major problems and limiting factors for both yield and quality.

There is no single method to control viruses. Integrated pest management relies among other items on virus-free seeds or seedlings, control of the vectors, and plant resistance. For the classical resistances, the different steps are (i) the screening of the germplasm and the identification of resistance in the species or in related wild or cultivated species, (ii) the study of the inheritance of the traits, and (iii) the introgression of the resistance in elite material. Social and political constraints can be a limiting factor for the use of transgenic resistance, even if in some combinations plant × viruses, excellent field control has been demonstrated (Clough and Hamm, 1995; Dinant *et al.*, 1997; Fang and Grumet, 1993; Fuchs and Gonsalves, 1995; Fuchs *et al.*, 1997; Gubba *et al.*, 2002; Harrison *et al.*, 1987; Kaniewski *et al.*, 1990; Pang *et al.*, 1996; Yoshioka *et al.*, 1991).

I. SOME ECONOMIC DATA

Data on the production and the exchange between countries (export and import) are not always available and reliable. We have used the FAOSTAT data (FAOSTAT) which pool some products: for instance, lettuce and chicory, carrots and turnips, or summer and winter squash and pumpkin. Vegetables cultivated in private gardens for self-consumption are not included in the statistics of production and for some of them the production can be very important. For instance in France, it has been estimated that self-consumption of vegetables represents 14% in value of the total consumption of vegetables and 38% for the families with a vegetable garden (Caillavet and Nichele, 1999).

Production of the main vegetables in the MA is presented in Table I. Tomato is clearly the first vegetable with more than 40,000 Tmt[1] followed by potato (32,000 Tmt) and watermelon (11,000 Tmt). It is also interesting to compare the MA production with the world production and with the production of some large countries with important vegetable production like China, India, and USA (Table I). The MA produces 12% of the vegetables in the world. It appears that the MA is the leading area for artichoke production with circa 71% of the world production. The area is also significant for tomato and green bean (30% and 25%, respectively). For fresh pepper and melons, the MA produces about 20% of the world production. For onions, green peas, pumpkins and squash, carrots and turnips, lettuce and chicory, watermelons, cucumbers, and gherkins, the MA represents between 10% and 15% of the world production.

Vegetables being an important part of the diet in the Mediterranean countries, most of the production is for consumption within the countries. But there are also significant exports and imports between countries of the MA or to other countries, for instance northern Europe. For all the vegetables, 6.7% and 3.2% of the production are, respectively, exported and imported from countries of the MA. About 25% of the asparagus, cauliflowers and broccoli, dry peppers, and lettuce and chicory are exported. Between 11% and 15% of the carrots and turnips, chickpeas, fresh peppers, and cucumbers are exported. Spain is the main exporting country. Most of the dry beans (75%) consumed in the MA are imported, and there are also significant importations of asparagus, chickpeas, and dry peppers. France and Spain are the countries which import the most important quantities of vegetables.

A very rough estimate of the consumption of each vegetable per inhabitant can be calculated by the sum of the country production and

[1] Tmt = 10^3 metric tons.

TABLE I Percentage of the world production of some vegetables in China, India, USA, and the Mediterranean area (data FAOSTAT mean 2008–2009)

Item	China	India	USA	Mediterranean area (value in 10^3 metric tons)
Artichokes	4.5	0.0	3.4	71.4 (1065)
Tomatoes	24.5	7.7	10.0	30.0 (41,582)
Green beans	38.3	3.1	0.7	24.9 (1685)
Fresh chillies and peppers	51.0	0.1	3.6	20.0 (5657)
Melons	52.3	1.2	3.8	19.0 (5246)
Fresh onions and shallots	23.3	0.0	0.0	16.4 (601)
Green peas	29.4	30.2	5.4	14.4 (1280)
Pumpkins, squash, and gourds	30.5	8.4	3.7	13.4 (2816)
Carrots and turnips	35.2	0.7	5.1	13.1 (3605)
Lettuce and chicory	54.0	1.7	17.3	11.9 (2778)
Dry onions	29.0	9.4	4.7	11.8 (8516)
Watermelons	68.0	0.1	1.8	11.5 (11,398)
Cucumbers and gherkins	61.6	0.1	2.2	10.6 (4409)
Potatoes	21.0	10.5	5.9	9.9 (32,274)
Cauliflowers and broccoli	43.2	31.8	1.6	9.1 (1753)
Chickpeas	0.1	67.3	0.6	7.4 (675)
Cabbages and other brassicas	52.9	9.1	1.3	6.1 (4313)
Garlic	77.0	2.0	1.1	4.6 (759)
Dry chillies and peppers	8.7	21.6	0.0	2.9 (173)
Dry beans	8.1	13.2	5.8	2.1 (412)
Asparagus	89.0	0.0	0.7	1.8 (128)
sweetpotatoes	75.5	1.0	0.8	0.3 (352)
Total	39.6	7.5	4.8	12.2 (131,390)

the import minus the export divided by the number of inhabitants (Table II). For the vegetables listed in Table I, the main consumer is Turkey with 397 kg per person and per year, followed by Greece (362 kg/pers/year) and Romania (353 kg/pers/year). At the other end of the scale are Jordan (123 kg/pers/year), Slovenia (128 kg/pers/year), and Croatia (134 kg/pers/year).

II. CENTER OF ORIGIN AND BIOLOGICAL DATA

The center of origin of some vegetables is in the MA, but most of the economic important vegetables like potato, tomato, pepper, beans, squashes, and pumpkins were introduced only a few centuries ago

TABLE II Average consumption of vegetables in the Mediterranean area (kg/pers/ year) and in the three countries with the highest consumption

Item	Mediterranean area	First three countries
Artichokes	2.08	Italy (8.10), Spain (4.14), Malta (3.46)
Asparagus	0.25	Bulgaria (1.63), Spain (0.96), Italy (0.64)
Dry beans	1.69	Albania (8.58), Serbia (7.39), Bosnia-Herzegovina (5.62)
Green beans	3.52	Turkey (7.62), Serbia (6.06), Spain (5.98)
Cabbages and other brassicas	8.27	Romania (45.51), Serbia (41.81), Bosnia-Herzegovina (22.7)
Carrots and turnips	6.60	Portugal (18.67), Israel (15.39), Romania (12.31)
Cauliflowers and broccoli	2.96	Malta (15.12), Italy (6.57), Lebanon (5.87)
Chickpeas	1.52	Turkey (5.95), Jordan (4.02), Lebanon (2.82)
Dry chillies and peppers	0.34	Bosnia-Herzegovina (7.92), Romania (1.60), Slovenia (0.81)
Fresh chillies and peppers	10.24	Tunisia (27.76), Turkey (23.47), Serbia (20.72)
Cucumbers and gherkins	7.56	Lebanon (35.36), Turkey (21.69), Albania (17.53)
Garlic	1.64	Romania (3.52), Egypt (3.20), Serbia (3.01)
Lettuce and chicory	4.94	Italy (13.08), Slovenia (11.64), Spain (10.23)
Melons	10.70	Turkey (23.63), Albania (23.41), Morocco (18.80)
Fresh onions and shallots	1.20	Tunisia (17.72), Libya (8.59), Albania (8.17)
Dry onions	15.82	Libya (28.49), Turkey (24.37), Algeria (21.43)
Green peas	2.41	France (5.86), Serbia (5.76), Egypt (3.54)
Potatoes	60.56	Romania (172.10), Serbia (115.30), Bosnia-Herzegovina (114.48)
Pumpkins, squash, and gourds	5.45	Italy (8.45), Egypt (8.14), Jordan (7.62)
Sweetpotatoes	0.74	Israel (4.85), Albania (6.65), Egypt (3.16)
Tomatoes	74.63	Turkey (143.05), Greece (119.50), Egypt (114.89)
Watermelons	20.84	Turkey (53.77), Greece (47.39), Albania (42.50)
Total	242.07	Turkey (396.96), Greece (362.55), Romania (353.98)

(Table III). Some of these species have a great diversity, and only a part of them is cultivated in the MA.

Most data on the centers of origin and the diversity of the species have been taken from the following sources: Pitrat and Foury (2003), Zeven and de Wet (1982), and Zohary and Hopf (1993).

A. *Allium*

The center of origin of onion and garlic is probably in central Asia and of leek in the eastern Mediterranean basin (southwest Asia). Domestication occurred very early, and these plants were cultivated and consumed from at least the second millennium BCE in ancient Egypt and Greece. The roman emperor Nero was known as a "leek-eater" believing consuming leek in soup or in salad was good for his voice.

Garlic (*Allium sativum*) has always been used both as food and as medicine. Even if some fertile garlic plants have been found in central Asia, the cultivars are sterile and are vegetatively propagated (clones). As there is no sexual reproduction, there is little variability between cultivars; nevertheless, different types are available differing by the size and color of bulbs and cloves, the presence of a flowering stem, and the adaptation to climatic conditions. Resistances to *Onion yellow dwarf virus* (OYDV) and *Leek yellow stripe virus* (LYSV) have been described in a fertile clone of garlic (Lot *et al.*, 2001; Table IV).

China is the largest producer with more than 75% of the world production, and Egypt (217 Tmt) is the main producer in the MA followed by Spain (150 Tmt) and Turkey (105 Tmt). Spain is the main exporter among the Mediterranean countries.

Onion (*Allium cepa*) is one of the most important vegetables and is cultivated in almost all countries in the world. Two subspecies have been defined: *cepa* which is the true onion and *aggregatum* which is the shallot with multiple bulbs. A relative is the gray shallot (*Allium oschaninii*) cultivated in Italy and France. Many types have been selected varying by their pungency (from very pungent to sweet), the skin color (white, yellow, brown, red), the bulb size, and the adaptation to climatic conditions, mainly, the latitude for bulb formation. Scallions or green onions or green shallots usually refer to young onion shoots with no fully developed bulbs. Onion is propagated by seeds, except shallot which is usually vegetatively propagated. It is an allogamous plant with strong inbreeding effect. Traditional cultivars are populations, but modern cultivars are F_1 hybrids produced with a cytoplasmic male sterility. Resistance to OYDV is present in some cultivars (Vandijk, 1993).

China is the first producer with circa 30% of the world production. Turkey (1928 Tmt) is the main producer in the MA followed by Egypt

TABLE III Center of origin of the main Mediterranean vegetables

Family	Scientific name	Common English name	Origin
Alliaceae	*Allium cepa*	Onion, shallot	Central Asia
	Allium sativum	Garlic	Central Asia
	Allium ampeloprasum	Leek	North Africa and southwest Asia
Apiaceae	*Daucus carota*	Carrot	Europe, west and central Asia
Asteraceae	*Lactuca sativa*	Lettuce	Mediterranean
	Cynara cardunculus	Artichoke, cardoon	Mediterranean
Brassicaceae	*Brassica oleracea*	Cabbage, Cauliflower, Broccoli	Western and southern Europe
Convolvulaceae	*Ipomoea batatas*	Sweetpotato	Central America
Cucurbitaceae	*Cucumis melo*	Melon	Africa
	Cucumis sativus	Cucumber	Asia
	Citrullus lanatus	Watermelon	Africa
	Cucurbita pepo, C. moschata	Squash, pumpkin	Central America
Fabaceae	*Pisum sativum*	Pea	Mediterranean
	Phaseolus vulgaris	Bean	Central and South America
	Vicia faba	Broad bean	North Africa and southwest Asia
	Cicer arietinum	Chickpea	Mediterranean
Liliaceae	*Asparagus officinalis*	Asparagus	Europe
Solanaceae	*Solanum tuberosum*	Potato	South America
	Solanum lycopersicum	Tomato	South America
	Capsicum annuum, C. frutescens	Pepper	Central America

TABLE IV Genetic resistance to viruses in vegetables which have been described as possible sources of resistance and which are present in some commercial cultivars

Crops	Sources of resistance to	Commercial cultivars resistant to
Garlic	OYDV, LYSV	
Onion		OYDV
Leek		LYSV
Asparagus	Asparagus virus 2	
Pea	PEMV, ClYVV, WLMV	BYMV, PSbMV
Chickpea	PEMV	PSbMV
Bean	BYMV, BGMV, BGYMV	BCMV
Faba bean	PSbMV	
Cucumber		CMV, PRSV-W, WMV, MWMV, ZYMV, CVYV, CYSDV
Melon	CMV, CABYV, ZYMV, WMV, CYSDV, MWMV	PRSV-W, MNSV, CMV
Watermelon	ZYMV, WMV, PRSV	
Squash		CMV, WMV, ZYMV, PRSV
Cabbage, Cauliflower, Broccoli	TuMV, CaMV	
Potato	PVA, PVS, PLRV	PVY, PVX
Tomato	CMV	TMV, ToMV, TSWV, PVY, TYLCV, AMV
Pepper		PVY, TEV, PepMoV, PVMV, PMMoV, TMV, TSWV, CMV
Sweetpotato	SPFMV, SPCSV	
Carrot	CarVY, CRLV, CMoV	
Lettuce	BWYV, TSWV, LIYV	LMV, TuMV, MLBVV
Artichoke	ALV	

(1764 Tmt) and Spain (1147 Tmt); Spain followed by Turkey is the main exporter.

Leek (*Allium ampeloprasum*) is a tetraploid species. Traditional cultivars are populations and are being replaced by F_1 hybrids. The Egyptian leek or Kurrat belongs to the same species. Resistance to LYSV is present in commercial cultivars.

B. Carrot

Carrot (*Daucus carota*) is native from Europe and west and central Asia and was probably domesticated in Afghanistan. Cultivated carrots were introduced in the MA and in Europe around the eighth century CE. Ibn al-'Awwam mentions yellow and red carrots in southern Spain during the twelfth century. Orange carrots were selected in the Netherlands in the seventeenth century.

Carrot is an allogamous crop with strong inbreeding effect. Traditional cultivars are populations and modern varieties are more homogeneous F_1 or three-way cross hybrids produced with a cytoplasmic male sterility. Partial resistance to *Carrot virus Y* (Jones *et al.*, 2005) and to Carrot motley dwarf (a complex of *Carrot red leaf virus* (CRLV) and *Carrot mottle virus* (CMoV)) (Watson and Falk, 1994) has been described.

Carrots and turnips are mixed together in the FAO statistics. The main world producers of carrots and turnips are China (10,000 Tmt, 35% of the world) followed by Russia (1520 Tmt) and USA (1400 Tmt). In the MA, Italy (605 Tmt), Turkey (592 Tmt), and Spain (550 Tmt) are the main producers. Italy, Israel (107 Tmt), France (100 Tmt), and Spain (80 Tmt) are the main exporters, and France (122 Tmt) is the main importer.

C. Lettuce

The origin of lettuce (*Lactuca sativa*) is in Europe and southwest Asia. Lettuce was already cultivated by the ancient Egyptians, but it is not clear if it was for the production of oil from the seeds, for eating the leaves or as offers for religious purposes. Lettuce is cited in the first millennium BCE by Herodotus and Theophrastus.

Diversity of lettuce is large and is increasing in the past 20 years as butterhead, oak leaf, lollo, Cos (romaine and grasse), European Batavia, American Batavia (Iceberg).

Lettuce is strictly autogamous and only pure lines are cultivated. Lettuce is bolting in summer with long day length and is not well adapted to warm and dry conditions.

Resistance to several viruses has been described in *L. sativa* or in the closely related *Lactuca serriola*: *Lettuce mosaic virus* (LMV) ($mo1^1$, $mo1^2$) (Bannerot *et al.*, 1969; Ryder, 1970), *Tomato spotted wilt virus* (TSWV) (O'Malley and Hartmann, 1989), *Beet western yellows virus* (BWYV) (Pink *et al.*, 1991; Walkey and Pink, 1990), *Turnip mosaic virus* (TuMV) (Zink and Duffus, 1973), *Mirafiori lettuce big-vein virus* (MLBVV) (Ryder and Robinson, 1995), *Tomato bushy stunt virus* and *Lettuce necrotic stunt virus* (Grube and Ryder, 2003), and *Bidens mottle virus* (Zitter and Guzman, 1977). Other resistances have been described in related species like *Lactuca virosa*: LMV (*Mo3*) (Maisonneuve *et al.*, 1999), MLBVV (Hayes *et al.*, 2004),

BWYV (Maisonneuve *et al.*, 1991), or *Lactuca saligna*: CMV (Provvidenti *et al.*, 1980), TSWV (Wang and Cho, 1992), *Lettuce infectious yellows virus* (LIYV) (McCreight, 1987; McCreight *et al.*, 1986). Resistance to LMV and TSWV has also been described in the more distantly related *Lactuca perennis*.

Lettuce and chicory (*Cichorium endivia* and *Cichorium intybus*) are mixed together in the FAO statistics. China with 12,900 Tmt and USA (4000 Tmt) are the two main world producers followed by Spain (1000 Tmt) and Italy (850 Tmt). In the MA, the main exporters are Spain (558 Tmt) and Italy (106 Tmt) and the main importers are France (87.5 Tmt) and Italy (51.5 Tmt).

D. Artichoke

Artichoke (*Cynara cardunculus*) originated from the MA and was probably domesticated in the western part of the Mediterranean basin. Roman mosaics from the second century CE in Tunisia represent artichoke.

Globe artichoke is one of the few perennial vegetables and is mainly vegetatively propagated. Diversity is quite low but includes purple and green types more or less spiny. It is a typical Mediterranean plant which "disappears" during the summer. Cardoon is also cultivated since a long time for eating the petioles. It is an annual plant and seed propagated. Some traditional cultivars of globe artichokes are seed propagated; they are heterogeneous as they are populations. But new more homogeneous F_1 hybrids are finding their way to the markets.

Resistance to *Artichoke latent virus* (ALV) has been described (Manzanares *et al.*, 1995).

Artichoke is mainly cultivated in the MA, Italy being the first world producer (485 Tmt) followed by Spain (200 Tmt) and Egypt (180 Tmt). The main exporters are Spain (13.5 Tmt) and France (9.3 Tmt) and the main importers France (17.8 Tmt) and Italy (11 Tmt).

E. *Brassica*

Wild *Brassica oleracea* can be found along the Atlantic and Mediterranean coasts of Europe. There is no clear evidence that cabbage was cultivated before the Roman time. *B. oleracea* is one of the most diverse vegetables including cabbage and Savoy, kale, broccoli, cauliflower, Brussels sprout, kohlrabi, and some ornamental forms. Turnip and Chinese cabbage belong to the species *Brassica rapa*. Theophrastus, Cato the Elder, Columella, Ibn al-'Awwam (twelfth century in Andalusia) described different types of cabbage. It seems that the origin of broccoli and cauliflower is in the eastern MA and that they were introduced in Europe only during the

Middle Ages. In Italy, a large variability of cauliflower exists with different colors (white, orange, green, and purple) and shapes (Romanesco).

Cabbages and cauliflowers are allogamous crops with strong inbreeding effect. Traditional cultivars are populations. More homogeneous F_1 hybrids were developed by using self-incompatibility or cytoplasmic male sterility. Only partial resistance to TuMV has been described in *B. oleracea* (Pink *et al.*, 1986), but high level of resistance has been observed in *B. rapa* and *B. napus*. Partial resistances to *Cauliflower mosaic virus* (CaMV) (Pink and Walkey, 1988; Raybould *et al.*, 2000), to BWYV (Thomas *et al.*, 1990), and to *Turnip yellow mosaic virus* (Chod *et al.*, 1992) have been identified.

The FAO statistics on cabbage include the European cabbage (*B. oleracea*) and the Chinese cabbage (*B. rapa*). China produces 53% and 43%, respectively, of the world production of cabbages and cauliflowers. In the MA, Romania (986 Tmt), Egypt (691 Tmt), and Turkey (644 Tmt) are the main producers of cabbages; Italy (460 Tmt), Spain (435 Tmt), and France (390 Tmt) of cauliflowers and broccoli. Italy (71 Tmt) and Spain (60 Tmt) are the main exporters of cabbages; Spain (238 Tmt) and France (172 Tmt) of cauliflowers and broccoli.

F. Sweetpotato

Sweetpotato (*Ipomoea batatas*) is the only species of the Convolvulaceae family of economic importance. It is native from Central or South America and was largely cultivated when Europeans reached the American continent. Sweetpotato was introduced before 1000 CE in the Pacific islands (Polynesia). It is a minor crop in the MA, and its arrival and spread are not well documented.

Sweetpotato is vegetatively propagated, and diseases, including viruses, can be easily transmitted to the young plants from an infected mother plant.

Resistance to *Sweetpotato feathery mottle virus* (SPFMV) and *Sweetpotato chlorotic stunt virus* (SPCSV) are controlled by different genes (Mwanga *et al.*, 2002).

China (81,000 Tmt) produces 75% of the world production. Egypt (262 Tmt) is the largest producer in the MA followed by Israel (38 Tmt), Portugal (27 Tmt), and Spain (22 Tmt).

G. Cucumber, melon, and watermelon

Cucumber, melon, and watermelon are from the old world, and there is a lot of confusion between these three species in the old sources. It seems now well established that the center of origin and of domestication of cucumber is the foothills of the Himalaya (from northern India to southwest China) and that cucumber was unknown in the MA until a few centuries CE. Ancient Egyptians, Greeks, and Romans cultivated and ate a non-sweet elongated type of melon, known as *chate* for the shortest one and *flexuosus* for the longest one. It was used as a cucumber, raw in salad, or pickled, and is still popular under the name *faqqus* or *adjour* on the southern shore of the Mediterranean Sea or *carosello* in southern Italy. A sweet aromatic melon is described by Pliny the Elder. What is often described as "melon" in the Bible is probably watermelon and as "cucumber" is almost certainly a melon. The center of origin of melon is probably eastern Africa although recent studies indicated that the genus *Cucumis* is of Asian origin. Its center of diversification is Asia from the Mediterranean Sea to Far-East and centered on India where the highest diversity is observed. The center of origin and domestication of watermelon is southern Africa.

Cucumber, melon, and watermelon are allogamous crops pollinated by bees and bumble bees. However, there is little inbreeding effect, and pure lines can be cultivated. F_1 hybrids are more and more common.

Two main types of cucumber (*Cucumis sativus*) are cultivated according to their use: the pickling and the slicing. Among the slicing cucumber, the Beit Alpha group, adapted to the Middle East climatic conditions, is short with a smooth light green epidermis. The English or Dutch group has longer fruits with a neck and also a smooth light green epidermis. Other types with a tough dark green skin more or less spiny are also cultivated. Absence of bitterness in the fruit has been largely used by plant breeders. The parthenocarpic character originally found in the English type has been introduced in other horticultural groups, including the pickling cucumber. Traditional cultivars are monoecious; gynoecious (all female) cultivars are more and more common. The accession "Taichung Mou Gua" (TMG-1) from Taiwan has been extensively studied as it is resistant to several potyviruses: *Zucchini yellow mosaic virus* (ZYMV), *Zucchini yellow fleck virus*, *Papaya ringspot virus watermelon type* (PRSV-W), *Watermelon mosaic virus* (WMV), *Moroccan watermelon mosaic virus* (MWMV) (Gilbert-Albertini *et al.*, 1995; Kabelka and Grumet, 1997; Provvidenti, 1985; Wai and Grumet, 1991, 1995). Resistance to *Cucumber mosaic virus* (CMV) is usually observed in accessions from Far-East (Kawaide, 1975; Kooistra, 1969; Porter, 1928; Shifriss *et al.*, 1942; Wasuwat and Walker, 1961). Partial resistances to *Cucumber vein yellowing virus* (CVYV) (Picó *et al.*, 2008) and to *Cucurbit yellow stunting disorder virus* (CYSDV) have been described (Eid *et al.*, 2006).

China produces more than 60% of the world production of cucumber. In the MA, Turkey is the leading country (1707 Tmt) followed by Egypt (598 Tmt) and Spain (530 Tmt). Spain is the main exporter (45 Tmt) and France the main importer (71 Tmt).

Melon (*Cucumis melo*) has a large variability, and different types are cultivated in the Mediterranean basin. Fruits can be harvested before maturity in the *chate* and *flexuosus* types; they are not sweet and are consumed like pickling or slicing cucumber. Dessert types have sweet fruits at maturity and present a large diversity for the fruit shape and size, the skin color and aspect (netted, ribbed, or wrinkled), the flesh color (white, green or orange), the taste (aromatic or not), and the shelf life. Most of the traditional cultivars (except the *flexuosus* and the *chate* types) from the MA are andromonoecious. In order to facilitate the seed production of F_1 hybrids, breeders have developed monoecious cultivars for instance in the Charentais or the Galia types. Resistances to ZYMV (Pitrat and Lecoq, 1984), PRSV-W (Kaan, 1973; Webb, 1979), WMV (Diaz-Pendon *et al.*, 2005; Gilbert *et al.*, 1994), *Cucurbit aphid-borne yellows virus* (CABYV) (Dogimont *et al.*, 1996), CYSDV (López-Sesé and Gómez-Guillamón, 2000; McCreight and Wintermantel, 2008), CMV (Enzie, 1943; Karchi *et al.*, 1975; Kawaide, 1975; Risser *et al.*, 1977), *Melon necrotic spot virus* (MNSV) (Coudriet *et al.*, 1981), *Cucumber green mottle mosaic virus* (Rajamony *et al.*, 1987; Sugiyama *et al.*, 2006), *Kyuri green mottle mosaic virus* (Daryono *et al.*, 2005) have been described in accessions from India, Far-East or Africa. A resistance to colonization by the cotton-melon aphid *Aphis gossypii* and to the transmission of viruses has been found (Bohn *et al.*, 1973; Kishaba *et al.*, 1971; Lecoq *et al.*, 1979; Soria *et al.*, 2003). Resistance to MNSV has been introduced in many cultivars belonging to the "Galia" type of melon, and resistance to *A. gossypii* in some Charentais melons.

China produces 52% of the melon world production. In the MA, Turkey is the first producer (1715 Tmt) followed by Spain (1014 Tmt), Egypt (758 Tmt), and Morocco (733 Tmt). Spain is the leading country for export (337 Tmt) followed by Morocco (57 Tmt). France is the largest importer (142 Tmt).

Two subspecies of watermelon (*Citrullus lanatus*) are recognized: the subsp. *citroides* with white, firm, non-sweet flesh and the subsp. *lanatus* with sweet flesh. There is a large diversity for fruit size and shape (round or oval), skin color, flesh color (red, orange, or yellow), and size and color of the seeds. Almost all watermelon cultivars are monoecious, a few being andromonoecious. Seedless triploid watermelon cultivars are more and more common. Resistance to several potyviruses (WMV, PRSV, and ZYMV) has been described (Boyhan *et al.*, 1992; Gillaspie and Wright, 1993; Guner and Wehner, 2008; Strange *et al.*, 2002). A partial resistance to *Squash vein yellowing virus* (SqVYV) has also been described (Kousik *et al.*, 2009)

China is again the first producer with 68% of the world production. Turkey (3906 Tmt), Egypt (1493 Tmt), Algeria (940 Tmt), and Spain (774 Tmt) are the main producer in the MA. Spain (303 Tmt), Greece (127 Tmt), and Italy (111 Tmt) are the main exporters.

H. *Cucurbita*

The genus *Cucurbita* was endemic to the American continent. Before Columbus, the bottle gourd (*Lagenaria siceraria*) of African origin was commonly used as a vegetable and is one of the oldest domesticated plants in Africa, America, and Asia for food but also for utilitarian uses such as containers. After Columbus, the bottle gourd was more or less rapidly replaced by different species of *Cucurbita*. *C. pepo* originates from northern Mexico and southern United States, *Cucurbita moschata* from the lowland of Central America, and *Cucurbita maxima* from South America. The first two species were domesticated several millennium BCE, while *C. maxima* was domesticated more recently, around 700 CE.

Melon, watermelon, and *Cucurbita* (squash and pumpkin) were probably domesticated for eating the seeds which are rich in good-quality lipids and proteins. Indeed, the wild forms are very bitter due to the presence of cucurbitacins, some of them being toxic for humans.

The two most important cultivated species in the Mediterranean basin are as follows:

- *C. pepo* as summer squash of different colors and sizes: cylindrical or oval shape, more or less elongated, dark green to light green, uniform or striped. Most of the summer squashes are of bush type with very short internodes. The plants can be with or without ramifications.
- *C. moschata* as winter squash or pumpkin. Fruits can be of different shapes: flat/round or elongated. Plants are vigorous with long vines.

Cucurbita spp. are monoecious and allogamous and are pollinated by bees and bumble bees. There is little inbreeding effect, and pure line (homozygous) can be cultivated. Modern cultivars mainly of summer squash are F_1 hybrids.

Some *Cucurbita* are used as rootstocks in order to control some soil-borne diseases, as the interspecific crosses *C. maxima* × *C. moschata* for melon and *C. ficifolia* for cucumber.

Traditional *C. pepo* cultivars are susceptible to most diseases including viruses. Partial resistance to CMV has been described for instance in cv Cinderella (Pink, 1987), but higher levels of resistance have been introduced from interspecific crosses (Brown *et al.*, 2003). Resistances to potyviruses (ZYMV, WMV, PRSV-W, MWMV, *Clover yellow vein virus* (ClYVV)) have been identified in wild (*Cucurbita okeechobeensis*, *Cucurbita ecuadorensis*) or cultivated species (*C. moschata* "Menina" or "Nigerian local"), and some of these resistances have been introduced into *C. pepo* and *C. maxima* (Brown *et al.*, 2003; Gilbert-Albertini *et al.*, 1993; Herrington *et al.*, 2001; Paris and Cohen, 2000; Paris *et al.*, 1988). Partial resistance to *Squash leaf curl virus* (McCreight and Kishaba, 1991) has also been identified.

The FAO statistics on "gourds" include not only *Cucurbita* sp. (squashes, pumpkins, etc.) but also other genus like *Benincasa* which is not important in the MA but is of major importance in eastern Asia. China (6500 Tmt) followed by India (3500 Tmt) are the main producers. In the MA, Egypt (676 Tmt), Italy (417 Tmt), Turkey (395 Tmt), and Spain (375 Tmt) are the leading countries for production; Spain (222 Tmt) is the main exporter and France (148 Tmt) the main importer.

I. Peas

Archaeological records of pea date back to the Neolithic period in Syria and Turkey. Pea was cultivated by the ancient Egyptians circa 4500 BCE. In many civilizations, alimentation is based on an association of a legume and a cereal: soybean and rice in eastern Asia, bean and maize in Central America, and pea and wheat/barley in the Fertile Crescent. This area is also the center of origin of lentil, broad bean, and chickpea which are among the most ancient cultivated plants. Until the end of the Middle Ages, the dry seeds were mainly used and were an important source of proteins. At the Renaissance, the fresh seeds or "green peas" become more and more popular especially in England and France.

Different types of peas are cultivated. *Pisum sativum* subsp. *arvense* with very long stems, many leaves, and few seeds is grown for animal feed as well as the subsp. *sativum* with dry seeds (pulses). Dry seeds are commonly used as a food for instance in soups. The garden pea is harvested, while seeds are immature. *P. sativum* var. *macrocarpon* has large pods and is eaten as a vegetable while seeds are still very young. Cultivars with long vines are replaced by dwarf types for mechanical harvesting. The *afila* gene replacing the leaflets by tendrils is common in many modern cultivars. One of the traits studied by Mendel, the round *versus* wrinkled seeds, is of great horticultural importance as the wrinkled peas are sweeter and better than the round ones. Pea is a strictly autogamous plant, and only pure lines are cultivated, which can be produced by open pollination. Two recessive genes are involved in resistance to ClYVV (Andrade *et al.*, 2007, 2009; Nakahara *et al.*, 2010). Several pathotypes of *Pea seed borne mosaic virus* (PSbMV) and several genes for resistance have been described (Gao *et al.*, 2004; Kasimor *et al.*, 1997; Smykal *et al.*, 2010). Resistance to *Pea enation mosaic virus* (PEMV) (Larsen *et al.*, 2009; Lebeda *et al.*, 1999; Yu *et al.*, 1995), to *Bean yellow mosaic virus* (BYMV) (Bruun-Rasmussen *et al.*, 2007), and to *White lupin mosaic virus* (WLMV) (Provvidenti and Hampton, 1993) has been described. Cultivars with resistance to PSbMV or BYMV are available.

As the "dry peas" statistics include production for animal feed, we have presented only the "green peas" data. China and India provide each 30% of the world production. Within the MA, France (400 Tmt) is the first producer followed by Egypt (290 Tmt) and Morocco (118 Tmt). France is the main exporter.

Chickpeas (*Cicer arietinum*) are well adapted to Mediterranean or tropical dry conditions. Two main types are cultivated: the *desi* with smaller and darker seeds and the *kabuli* with larger and lighter seeds, the latter being the only type cultivated in the MA. Chickpea is an autogamous plant, and cultivars are pure lines. Resistance to PEMV (Larsen and Porter, 2010) and to PSbMV (Latham and Jones, 2001) has been described. Resistant cultivars to PSbMV are available.

India with two-thirds of the world production is the leading production country. Turkey is the world third producer (540 Tmt) and the first one in the MA followed by Syria (42 Tmt) and Morocco (38 Tmt). Turkey exports some of its production, while Spain, Algeria, and Jordan are the main importers.

J. Beans

The cowpea (*Vigna unguiculata*) is of African origin and was largely cultivated in the MA before the introduction of *Phaseolus* from America. Several species of *Phaseolus* were domesticated in Central America and the Andes. *Phaseolus vulgaris* is the most important although *Phaseolus coccineus* and *Phaseolus lunatus* are also cultivated in the MA as an ornamental or for eating the seeds. In pre-Columbian times, *P. vulgaris* was cultivated for the dry seeds, and today is still cultivated in the world more for the seeds than for the green beans (immature pods).

The color, the size, and the shape of the seeds are highly diverse. The color can be black, green, red, yellow, or white, uniform or mottled with two or three colors. The green beans can be of three types: the pod can be flat (snap bean) or with a more or less round section with or without a fibrous string along the pod. The color of the pod can be green, yellow, or purple, uniform or mottled. The plant can be bushy (20–60 cm) or running (2–3 m), the growth can be determinate or indeterminate, and the flowers can be inside or above the foliage. *P. vulgaris* is a strictly autogamous plant, and cultivars are pure lines which can be increased by open pollination. Resistance to ClYVV is controlled by one recessive gene, but strains overcoming this resistance have been identified (Sato *et al.*, 2003). Resistance to *Bean common mosaic virus* (BCMV) and *Bean common mosaic necrosis virus* is controlled by at least four recessive genes (Drijfhout, 1978). Many commercial cultivars are resistant to BCMV (Teran *et al.*, 2009). Resistance to BYMV is also controlled by a single recessive gene (Park and Tu, 1991). High levels of resistance to *Bean golden mosaic virus* (BGMV) have been described, and at least two recessive genes have been identified in different accessions (Bianchini, 1999; Velez *et al.*, 1998). Resistance to

Bean golden yellow mosaic virus (BGYMV) has been identified in *P. vulgaris* and *P. coccineus* (Osorno *et al.*, 2007).

Brazil is the first world producer of dry beans followed by India. Turkey is the leading producer in the MA (168 Tmt) followed by Egypt (98 Tmt) and Serbia (42 Tmt). There are no significant exports, and Italy (110 Tmt), Algeria, France, Spain, and Turkey (51 to 57 Tmt) are importers.

For green beans, China is the leading country (38% of world production). The MA produces about 25% of the world production of green beans. Turkey (583 Tmt), Egypt (274 Tmt), and Italy and Spain (185 Tmt) are the main producers. The main exporters are France (54 Tmt), Spain (29 Tmt), and Egypt (23 Tmt), and the main importers are Spain (97 Tmt) and France (58 Tmt).

The young or dry seeds of broad bean, *Vicia faba* var. *faba*, with many large and flat seeds per pod are used in human nutrition. The var. *equina* and *minor* have, respectively, medium size and small seeds and are used mainly to feed animals; the flour is sometimes added to wheat flour for bread. Broad bean seeds contain some alkaloids which can induce anemia in persons who have a deficiency in glucose-6-phosphate dehydrogenase. The *ful*, a broad bean preparation, is a kind of national dish in Egypt and Sudan. Resistance to PSbMV has been observed (Latham and Jones, 2001).

K. *Asparagus*

Asparagus officinalis is native to Europe with other species of *Asparagus* as *A. acutifolius* which is not cultivated but whose young shoots are gathered and eaten in the spring. Asparagus was cultivated as both a vegetable and a medicine (from which its species name *officinalis*) for its diuretic properties. Ancient Egyptians, Greeks, and Romans appreciated this vegetable. For instance, Cato the Elder in his *De agri cultura* describes the cultivation of asparagus. The king of France Louis XIV was particularly fond of asparagus and his gardener, J.B. de la Quintinie, developed methods for out of season production.

Asparagus is one of the few perennial vegetables, with artichokes. The young shoots are harvested in the spring. White asparagus is produced under a mound of earth, while green or purple asparagus grows in the light. It is a dioecious species with male and female plants. To replace the traditional populations which are quite heterogeneous, double cross hybrids were released in the 1970s. They were followed by F_1 hybrids between two parents that were vegetatively propagated. As the male plants are usually more productive than the female ones, all-male cultivars have then been developed. They are produced by using a "super-male" parent which is a doubled haploid line obtained after *in vitro* culture of anthers. A female parent crossed by the "super-male" line gives a 100% male progeny. A resistance to *Asparagus virus 2* (Elmer *et al.*, 1996) has been described, but it seems that this resistance has not been introduced in cultivars.

China is the first asparagus producer with almost 90% of the world production. Spain (48 Tmt) is the first producer in the MA, followed by Italy (35 Tmt) and France (19 Tmt). Greece and Spain are the two main exporting countries in the Mediterranean basin.

L. Potato

Within the Solanaceae family, potato, tomato, and pepper were unknown in the MA before Columbus as these species originated from America. Potato was domesticated between 3000 and 2000 BCE in the Andes (southern Peru?) and was introduced from the Americas in Spain around 1570 and in British Islands around 1590. It was only slowly adopted by European populations.

Potato is a tetraploid species. It is propagated vegetatively, and many diseases, including viruses, are transmitted through the tubers. Schemes for producing disease-free seed tubers have been implemented. *Solanum tuberosum* is the only species cultivated in the MA, but there are many related species which are cross fertile with *S. tuberosum*. Many genes for virus resistance (*Potato virus Y* (PVY), *Potato virus X* (PVX), *Potato virus A* (PVA), *Potato leafroll virus* (PLRV)) have been introduced from related species (for a review, see Barker and Dale, 2006).

The most important producers are China (69,000 Tmt), India (34,500 Tmt), Russia (30,000 Tmt), and Ukraine and USA (19,500 Tmt). In the MA, France (6990 Tmt), Turkey (4300 Tmt), and Romania and Egypt (3000 Tmt) are the main producers. France (1900 Tmt) and Egypt (400 Tmt) are the main exporters. Spain (760 Tmt), Italy (583 Tmt), and France (480 Tmt) are the main importers.

M. Tomato

Wild tomatoes from Peru were probably transported to Central America where they were domesticated at an unknown date. From Mexico, the Spanish took the tomato to Europe and the Philippines, at that time a Spanish colony. The first descriptions of tomato are from the mid-sixteenth century, and it seems that both yellow- (from which the Italian name *Pomi d'oro*) and red-fruited types were introduced. Tomato was slowly adopted and used cooked to prepare sauces. Tomato probably arrived in the eastern MA at the end of the eighteenth century. It is only in the twentieth century that tomato becomes one of the most important vegetables (second only to potato), both in gardens and for commercial production.

Tomatoes (*Solanum lycopersicum*) are classified according to the fruit traits: size, shape, and color. Cultivars with very firm flesh and determinate growth have been released for mechanical harvesting of processed tomatoes (juice, canning, and sauces). Both determinate and indeterminate cultivars are used for fresh tomato production. Adaptation to climatic conditions (winter/summer, open field/protected) is an important trait.

Tomato is an autogamous species and traditional cultivars are pure lines (homozygous). Modern cultivars are mostly F_1 hybrids even for processing tomatoes. In some conditions, for instance under greenhouses, self-pollination is not sufficient and the help of bumblebees is necessary to have a good harvest.

Resistances to viruses have been found only in related wild species, and several have been introduced in cultivated tomato. From *S. habrochaites*, resistance to *Tobacco mosaic virus* (TMV) (allele *Tm-1*) (Pelham, 1966), to *Alfalfa mosaic virus* (AMV) (allele *Am*) (Parrella *et al.*, 1998), to *Tobacco etch virus* (TEV) (Legnani *et al.*, 1996), and to PVY (Legnani *et al.*, 1995) has been identified. In *S. peruvianum*, resistance to *Tomato mosaic virus* (ToMV) (alleles *Tm-2* and $Tm\text{-}2^2$) (Alexander, 1963; Laterrot and Pecaut, 1969) and to TSWV (allele *Sw-5*) (Stevens *et al.*, 1991) was described. Resistance to CMV (Stamova and Chetelat, 2000) has been found in S. *chilense*. For resistance to *Tomato yellow leaf curl virus* (TYLCV), see the review by Lapidot and Polston (2006).

China (34,000 Tmt), USA (14,000 Tmt), and India (11,000 Tmt) are the three leading producers in the world. They are followed by Turkey (10,900 Tmt), Egypt (9600 Tmt), Italy (6180 Tmt), and Spain (4335 Tmt) in the MA. About 30% of the tomato in the world is produced in the MA. Spain (940 Tmt) and Turkey (440 Tmt) are the main exporters and France the main importer (482 Tmt).

N. Pepper

The genus *Capsicum* is also native from America, and contrarily to potato and tomato, pepper was immediately adopted by Europeans who were looking for spices and associated this new species with black pepper (*Piper nigrum*). Pepper was probably used several millennium BCE and was cultivated in Peru around 2500 BCE. The first introduced types were hot and small fruited. It is not clear if sweet and large fruited types were introduced later from America or were selected in Europe. *Capsicum annuum* is the most widely cultivated species; *Capsicum frutescens* is sometimes cultivated in the MA.

There is a large variability in pepper in the world and also in types cultivated in the MA. Pungency, shape, weight, color of the fruits varies greatly. Pepper for the fresh market can be harvested green (before maturity) or at maturity. Mature fruits can also been dried mainly to prepare chilli powder or sauces. Pepper is mostly autogamous, even if in some conditions allogamy can reach 50%, and pure lines can be cultivated. Most modern cultivars are F_1 hybrids.

Resistance to viruses has been found both in *C. annuum* and in wild or cultivated related species. Different alleles at the locus *L* from *C. annuum*, *C. frutescens*, *Capsicum chinense*, and *C. chacoense* control resistance to Tobamoviruses (TMV and *Pepper mild mottle virus* (PMMoV)) (Boukema, 1982; Rast, 1988). Resistance to another Tobamovirus, *Paprika mild mottle virus* (PaMMV), is controlled by another locus (Sawada *et al.*, 2005). Resistance to TSWV has been observed in *C. chinense* (Boiteux, 1995). Resistance to potyviruses (PVY, TEV, *Pepper mottle virus* (PepMoV), *Pepper veinal mottle virus* (PVMV)) has been found in *C. annuum* and *C. chinense* (Boiteux *et al.*, 1996; Caranta *et al.*, 1996; Kyle and Palloix, 1997). Partial resistance to CMV has been described in *C. annuum*, *C. frutescens*, and *C. baccatum* (Ben Chaim *et al.*, 2001; Grube *et al.*, 2000; Monma and Sakata, 1997; Suzuki *et al.*, 2003).

China is the largest producer of fresh pepper (14,300 Tmt) with more than 50% of the world production, followed by Mexico (2050 Tmt). Turkey is the third world producer (1800 Tmt) followed in the MA by Spain (1000 Tmt) and Egypt (752 Tmt). The MA produces 20% of the world harvest. Spain is the largest exporter (435 Tmt), and France is the main importer (135 Tmt).

India is the largest producer of dry chillies and peppers (1270 Tmt). In the MA, Egypt (46 Tmt), Romania (33 Tmt), and Bosnia and Herzegovina (30 Tmt) are the main producers. Spain (27 Tmt) and Tunisia (11 Tmt) are exporters. Spain (34 Tmt) is the main importer.

III. CONCLUSIONS

In the case of vegetables which are vegetatively propagated (potato, sweetpotato, garlic, artichoke), viruses are easily transmitted to the seedlings. It could also be the case with some seed-borne viruses such as PSbMV in pea, LMV in lettuce, or *Pepino mosaic virus* (PepMV) in tomato. The first step in protecting the crops against the viruses is to use virus-free seeds or seedlings. Production of virus-free seeds can be achieved in some areas with a low virus pressure or under protected conditions (greenhouses) and/or certified by control of the seed lots by ELISA for instance.

It is also important to avoid the transmission of viruses from one country to another or even between continents not only through the seeds or seedlings but also through the vegetables themselves which can act as a virus source for the vectors. For instance, aphids were able to recover PRSV and ZYMV from melon fruits imported to France from Central America and to contaminate melon and squash bait plants (Lecoq *et al.*, 2003).

Breeding virus-resistant cultivars is another method to control viruses (Table IV). As indicated in Table III, the centers of origin and/or diversity and/or domestication of most of the vegetables are not in the MA. Very few sources of resistance have been described in wild accessions or in local landraces from the MA. For instance, peas, lentils, and chickpeas originated from the Fertile Crescent (western Asia). In a survey of about 500 chickpea accessions for resistance to PEMV, the best sources of resistance were from India or Iran (Larsen and Porter, 2010). Similarly, resistance to this virus in lentils (Aydin *et al.*, 1987) and peas (Schroeder and Barton, 1958) has also been described in accessions from these two countries. Nevertheless, some virus resistances have been found in landraces from the MA. For instance, for lettuce whose the center of origin is Europe and western Asia, virus resistances have been found in related wild species or in *L. sativa* accessions from the MA (Section II.C). Some melon accessions from Spain are resistant to the melon/cotton aphid *A. gossypii* (Pitrat *et al.*, 1988). The cultivar "Menina" of *C. moschata* from Portugal is resistant to several potyviruses (Gilbert-Albertini *et al.*, 1993; Paris *et al.*, 1988).

When sources of resistance have been identified in the cultivated species or in related wild or cultivated species, the main challenges for the breeders are the following:

Time. Even by using biotechnological tools as marker assisted selection, breeding new cultivars requires between 5 and 10 years. Monogenic dominant resistances are easier to introduce in elite material than polygenic recessive ones.

Virus diversity. The virus situation is not stable. Among the emerging viruses, the PepMV, the *Tomato torrado virus* (ToTV), or whiteflies transmitted viruses are increasing, while other viruses are decreasing. Moreover, plants can be infected by several viruses with similar symptoms, for instance mosaic symptoms of potyviruses and cucumoviruses, or leaf yellowing by poleroviruses, criniviruses, or begomoviruses; resistance to only one of the viruses among the complex is of little value.

Virus importance. The release of commercial resistant cultivars is not always implemented according to the relative importance of virus resistance among all the traits: quality, yield, earliness, and other diseases. The importance of the diseases, including viruses, is highly variable according to the country, the season, the open field *versus* protected culture. Some viruses can be very frequent, but if the incidence on yield and quality is low, breeders will not put the resistance to this virus on the top of the list. This is the case for instance of CABYV on the Charentais cultigroup of melon. Conversely, epidemics of some viruses can be irregular, every 3 or 4 years for instance, but the incidence can be very high destroying completely a crop.

Plant diversity. Thousands of vegetable cultivars are used in the MA varying for their quality, climatic and soil adaptation, yield, consumer preferences, etc. Virus resistances are introduced only in the most important cultigroups of each species.

Stability or durability of the resistance. It can be defined by the nondevelopment of races or pathotypes of a virus overcoming a resistance which has been deployed over significant areas and time. For instance, the allele *Tm-2*^{2} controlling resistance of tomato to ToMV can be considered as stable: some strains have been isolated in laboratories that overcome this gene, but in the field, this allele is still effective. Similarly, one strain of MNSV overcoming the allele *nsv* for resistance of melon to this virus has been identified in Spain (Díaz *et al.*, 2002), but this strain does not seem to spread. The situation is more complex in the interaction between pepper and PVY. Two main loci are involved: at the locus *pvr2*, the allele *pvr2*^{3} is very frequently overcome, *pvr2*^{1} is quite stable, and *pvr2*^{2} is very stable; at the locus *Pvr4*, resistance is also very stable but for different reasons. At the locus *pvr2*, resistance-breaking strains occur according to the number of mutations in the virus genome, while the strains able to overcome the locus *Pvr4* have a very low fitness (Ayme *et al.*, 2007; Janzac *et al.*, 2010).

The abovementioned challenges could explain why there are not so many virus-resistant cultivars and that mainly monogenic resistances have been used. Nevertheless, breeding resistant cultivars is one of the key components of integrated pest management, and virus-resistant cultivars could be a factor which allows a farmer to earn one's living.

REFERENCES

Alexander, L. J. (1963). Transfer of a dominant type of resistance to the four known Ohio pathogenic strains of tobacco mosaic virus (TMV) from *Lycopersicon peruvianum* to *L. esculentum*. *Phytopathology* **53:**869.

Andrade, M., Sato, M., and Uyeda, I. (2007). Two resistance modes to *Clover yellow vein virus* in pea characterized by a green fluorescent protein-tagged virus. *Phytopathology* **97:**544–550.

Andrade, M., Abe, Y., Nakahara, K. S., and Uyeda, I. (2009). The *cyv-2* resistance to *Clover yellow vein virus* in pea is controlled by the eukaryotic initiation factor 4E. *J. Gen. Plant Pathol.* **75:**241–249.

Aydin, H., Muehlbauer, F. J., and Kaiser, W. J. (1987). Pea enation mosaic virus resistance in lentil (*Lens culinaris*). *Plant Dis.* **71:**635–638.

Ayme, V., Petit-Pierre, J., Souche, S., Palloix, A., and Moury, B. (2007). Molecular dissection of the *Potato virus Y* VPg virulence factor reveals complex adaptations to the *pvr2* resistance allelic series in pepper. *J. Gen. Virol.* **88:**1594–1601.

Bannerot, H., Boulidard, L., Marrou, J., and Duteil, M. (1969). Etude de la tolérance au virus de la mosaïque de la laitue chez la variété Gallega de Invierno. *Ann. Phytopathol.* **1:**219–226.

Barker, H., and Dale, M. F. B. (2006). Resistance to viruses in potato. *In* "Natural Resistance Mechanisms of Plants to Viruses" (G. Loebenstein and J. P. Carr, eds.), pp. 341–366. Springer, Dordrecht, NL.

Ben Chaim, A., Grube, R. C., Lapidot, M., Jahn, M., and Paran, I. (2001). Identification of quantitative trait loci associated with resistance to cucumber mosaic virus in *Capsicum annuum*. *Theor. Appl. Genet.* **102:**1213–1220.

Bianchini, A. (1999). Resistance to *Bean golden mosaic virus* in bean genotypes. *Plant Dis.* **83:**615–620.

Bohn, G. W., Kishaba, A. N., Principe, J. A., and Toba, H. H. (1973). Tolerance to melon aphid in *Cucumis melo* L. *J. Am. Soc. Hort. Sci.* **98:**37–40.

Boiteux, L. S. (1995). Allelic relationships between genes for resistance to tomato spotted wilt tospovirus in *Capsicum chinense*. *Theor. Appl. Genet.* **90:**146–149.

Boiteux, L. S., Cupertino, F. P., Silva, C., Dusi, A. N., MonteNeshich, D. C., vanderVlugt, R. A. A., and Fonseca, M. E. N. (1996). Resistance to potato virus Y (pathotype 1-2) in *Capsicum annuum* and *Capsicum chinense* is controlled by two independent major genes. *Euphytica* **87:**53–58.

Boukema, I. W. (1982). Resistance to a new strain of TMV in *Capsicum chacoense* Hunz. *Capsicum Newslett.* **1:**49–51.

Boyhan, G. E., Norton, J. D., Jacobsen, B. J., and Abrahams, B. R. (1992). Evaluation of watermelon and related germplasm for resistance to zucchini yellow mosaic virus. *Plant Dis.* **76:**251–252.

Brown, R. N., Bolanos-Herrera, A., Myers, C. H., and Jahn, M. M. (2003). Inheritance of resistance to four cucurbit viruses in *Cucurbita moschata*. *Euphytica* **129:**253–258.

Bruun-Rasmussen, M., Moller, I. S., Tulinius, G., Hansen, J. K. R., Lund, O. S., and Johansen, I. E. (2007). The same allele of translation initiation factor 4E mediates resistance against two Potyvirus spp. in *Pisum sativum*. *Mol. Plant Microbe Interact.* **20:**1075–1082.

Caillavet, F., and Nichele, V. (1999). Autoconsommation et jardin. Arbitrage entre production domestique et achats de légumes. *Econ. Rurale* **250:**11–20.

Caranta, C., Palloix, A., GebreSelassie, K., Lefebvre, V., Moury, B., and Daubeze, A. M. (1996). A complementation of two genes originating from susceptible *Capsicum annuum* lines confers a new and complete resistance to pepper veinal mottle virus. *Phytopathology* **86:**739–743.

Chod, J., Polak, J., Jokes, M., and Pivalova, J. (1992). Sensitivity of some cabbage hybrids (*Brassica oleracea* L. var. *capitata*) to *Turnip yellow mosaic virus*. *Zahradnictvi* **19:**249–255.

Clough, G. H., and Hamm, P. B. (1995). Coat protein transgenic resistance to watermelon mosaic and zucchini yellow mosaic virus in squash and cantaloupe. *Plant Dis.* **79:**1107–1109.

Coudriet, D. L., Kishaba, A. N., and Bohn, G. W. (1981). Inheritance of resistance to muskmelon necrotic spot virus in a melon aphid resistant breeding lines of muskmelon. *J. Am. Soc. Hort. Sci.* **106:**789–791.

Daryono, B. S., Somowiyarjo, S., and Natsuaki, K. T. (2005). Screening for resistance to Kyuri green mottle mosaic virus in various melons. *Plant Breed.* **124:**487–490.

Díaz, J. A., Nieto, C., Moriones, E., and Aranda, M. A. (2002). Spanish *Melon necrotic spot virus* isolate overcomes the resistance conferred by the recessive *nsv* gene of melon. *Plant Dis.* **86:**694.

Diaz-Pendon, J. A., Fernandez-Munoz, R., Gomez-Guillamon, M. L., and Moriones, E. (2005). Inheritance of resistance to *Watermelon mosaic virus* in *Cucumis melo* that impairs virus accumulation, symptom expression, and aphid transmission. *Phytopathology* **95:**840–846.

Dinant, S., Maisonneuve, B., Albouy, J., Chupeau, Y., Chupeau, M. C., Bellec, Y., Gaudefroy, F., Kusiak, C., Souche, S., Robaglia, C., and Lot, H. (1997). Coat protein gene-mediated protection in *Lactuca sativa* against lettuce mosaic potyvirus strains. *Mol. Breed.* **3:**75–86.

Dogimont, C., Slama, S., Martin, J., Lecoq, H., and Pitrat, M. (1996). Sources of resistance to *Cucurbit aphid borne yellows luteovirus* in a melon germ plasm collection. *Plant Dis.* **80:**1379–1382.

Drijfhout, E. (1978). Genetic Interaction Between *Phaseolus vulgaris* and *Bean Common Mosaic Virus* with Implications for Strain Identification and Breeding for Resistance. Center for Agricultural Publishing and Documentation, Wageningen, NL.

Eid, S., Abou-Jawdah, Y., El-Mohtar, C., Sobh, H., and Havey, M. (2006). Tolerance in cucumber to *Cucurbit yellow stunting disorder virus*. *Plant Dis.* **90:**645–649.

Elmer, W. H., Johnson, D. A., and Mink, G. I. (1996). Epidemiology and management of the diseases causal to asparagus decline. *Plant Dis.* **80:**117–125.

Enzie, W. D. (1943). A source of muskmelon mosaic resistance found in the oriental pickling melon, *Cucumis melo* var. *conomon*. *Proc. Am. Soc. Hort. Sci.* **43:**195–198.

Fang, G., and Grumet, R. (1993). Genetic engineering of potyvirus resistance using constructs derived from the zucchini yellow mosaic virus coat protein gene. *Mol. Plant Microbe Interact.* **6:**358–367.

FAOSTAT. http://faostat.fao.org.

Fuchs, M., and Gonsalves, D. (1995). Resistance of transgenic hybrid squash ZW-20 expressing the coat protein genes of zucchini yellow mosaic virus and watermelon mosaic virus 2 to mixed infections by both potyviruses. *Nat. Biotechnol.* **13:**1466–1473.

Fuchs, M., McFerson, J. R., Tricoli, D. M., McMaster, J. R., Deng, R. Z., Boeshore, M. L., Reynolds, J. F., Russell, P. F., Quemada, H. D., and Gonzalves, D. (1997). Cantaloupe line CZW-30 containing coat protein genes of cucumber mosaic virus, zucchini yellow mosaic virus, and watermelon mosaic virus-2 is resistant to these three viruses in the field. *Mol. Breed.* **3:**279–290.

Gao, Z., Eyers, S., Thomas, C., Ellis, N., and Maule, A. (2004). Identification of markers tightly linked to *sbm* recessive genes for resistance to *Pea seed-borne mosaic virus*. *Theor. Appl. Genet.* **109:**488–494.

Gilbert, R. Z., Kyle, M. M., Munger, H. M., and Gray, S. M. (1994). Inheritance of resistance to watermelon mosaic virus in *Cucumis melo* L. *HortScience* **29:**107–110.

Gilbert-Albertini, F., Lecoq, H., Pitrat, M., and Nicolet, J. L. (1993). Resistance of *Cucurbita moschata* to *Watermelon mosaic virus* type 2 and its genetic relation to resistance to *Zucchini yellow mosaic virus*. *Euphytica* **69:**231–237.

Gilbert-Albertini, F., Pitrat, M., and Lecoq, H. (1995). Inheritance of resistance to *Zucchini yellow fleck virus* in *Cucumis sativus* L. *HortScience* **30:**336–337.

Gillaspie, A. G., and Wright, J. M. (1993). Evaluation of *Citrullus* sp. germplasm for resistance to watermelon mosaic virus 2. *Plant Dis.* **77:**352–354.

Grube, R. C., and Ryder, E. J. (2003). Romaine lettuce breeding lines with resistance to lettuce dieback caused by tombusviruses. *HortScience* **38:**627–628.

Grube, R. C., Zhang, Y. P., Murphy, J. F., Loaiza-Figueroa, F., Lackney, V. K., Provvidenti, R., and Jahn, M. K. (2000). New source of resistance to *Cucumber mosaic virus* in *Capsicum frutescens*. *Plant Dis.* **84:**885–891.

Gubba, A., Gonsalves, C., Stevens, M. R., Tricoli, D. M., and Gonsalves, D. (2002). Combining transgenic and natural resistance to obtain broad resistance to tospovirus infection in tomato (*Lycopersicon esculentum* Mill). *Mol. Breed.* **9:**13–23.

Guner, N., and Wehner, T. C. (2008). Overview of Potyvirus resistance in watermelon. *In* "Cucurbitaceae 2008" (M. Pitrat, ed.),. Proceedings of the IXth EUCARPIA Meeting on Genetics and Breeding of Cucurbitaceae, Avignon, France, 21-24 May 2008, pp. 445–451.

Harrison, B. D., Mayo, M. A., and Baulcombe, D. C. (1987). Virus-resistance in transgenic plants that express Cucumber mosaic-virus satellite RNA. *Nature* **328:**799–802.

Hayes, R. J., Ryder, E., and Robinson, B. (2004). Introgression of Big vein tolerance from *Lactuca virosa* L. into cultivated lettuce (*Lactuca sativa* L.). *HortScience* **39:**881.

Herrington, M. E., Prytz, S., Wright, R. M., Walker, I. O., Brown, P., Persley, D. M., and Greber, R. S. (2001). 'Dulong QHI' and 'Redlands Trailblazer', PRSV-W-, ZYMV-, and WMV-resistant winter squash cultivars. *HortScience* **36:**811–812.

Ibn al–'Awwâm (XII century). "Le livre de l'agriculture" (translator Clément–Mullet, J. J.) Actes Sud, Arles, France

Janzac, B., Montarry, J., Palloix, A., Navaud, O., and Moury, B. (2010). A point mutation in the polymerase of *Potato virus Y* confers virulence toward the *Pvr4* resistance of pepper and a high competitiveness cost in susceptible cultivar. *Mol. Plant Microbe Interact.* **23:**823–830.

Jones, R. A. C., Smith, L. J., Gajda, B. E., Smith, T. N., and Latham, L. J. (2005). Further studies on *Carrot virus Y*: Hosts, symptomatology, search for resistance, and tests for seed transmissibility. *Aust. J. Agric. Res.* **56:**859–868.

Kaan, J. F. (1973). Recherches sur la résistance du melon aux maladies, notamment à la mosaïque de la pastèque et au *Pseudoperonospora*, appliquées au type variétal "Cantaloup Charentais" EUCARPIA Meeting on Melon, Avignon (France), pp. 41–49.

Kabelka, E., and Grumet, R. (1997). Inheritance of resistance to the Moroccan watermelon mosaic virus in the cucumber line TMG-1 and cosegregation with zucchini yellow mosaic virus resistance. *Euphytica* **95:**237–242.

Kaniewski, W., Lawson, C., Sammons, B., Haley, L., Hart, J., Delannay, X., and Tumer, N. E. (1990). Field-resistance of transgenic Russet Burbank potato to effects of infection by Potato virus-X and Potato virus-Y. *Biotechnology* **8:**750–754.

Karchi, Z., Cohen, S., and Govers, A. (1975). Inheritance of resistance to Cucumber Mosaic Virus in melons. *Phytopathology* **65:**479–481.

Kasimor, K., Baggett, J. R., and Hampton, R. O. (1997). Pea cultivar susceptibility and inheritance of resistance to the lentil strain (Pathotype P2) of *Pea seedborne mosaic virus*. *J. Am. Soc. Hort. Sci.* **122:**325–328.

Kawaide, T. (1975). Breeding for disease resistance of vegetable crops in Japan. I. Cucurbits. *Jap. Agric. Res. Q.* **9:**212–216.

Kishaba, A. N., Bohn, G. W., and Toba, H. H. (1971). Resistance to *Aphis gossypii* in muskmelon. *J. Econ. Entomol.* **64:**935–937.

Kooistra, E. (1969). The inheritance of resistance to *Cucumis* virus 1 in cucumber. *Euphytica* **18:**326–332.

Kousik, C. S., Adkins, S., Turechek, W. W., and Roberts, P. D. (2009). Sources of resistance in US plant introductions to Watermelon vine decline caused by *Squash Vein Yellowing Virus*. *HortScience* **44:**256–262.

Kyle, M. M., and Palloix, A. (1997). Proposed revision of nomenclature for potyvirus resistance genes in Capsicum. *Euphytica* **97:**183–188.

Lapidot, M., and Polston, J. E. (2006). Resistance to *Tomato yellow leaf curl virus* in Tomato. *In* "Natural Resistance Mechanisms of Plants to Viruses" (G. Loebenstein and J. P. Carr, eds.), pp. 503–520. Springer, Dordrecht, NL.

Larsen, R. C., and Porter, L. D. (2010). Identification of novel sources of resistance to *Pea enation mosaic virus* in chickpea germplasm. *Plant Pathol.* **59:**42–47.

Larsen, R., Porter, L., and McPhee, K. (2009). Evidence that a QTL may be involved in a partial resistance response to *Pea enation mosaic virus* in pea (*Pisum sativum* L.). *Phytopathology* **99**(Suppl. S):S69.

Laterrot, H., and Pecaut, P. (1969). Tm-2: New source. *Tomato Gen. Coop. Rep.* **19:**13–14.

Latham, L. J., and Jones, R. A. C. (2001). *Alfalfa mosaic* and *pea seed-borne mosaic viruses* in cool season crop, annual pasture, and forage legumes: Susceptibility, sensitivity, and seed transmission. *Aust. J. Agric. Res.* **52:**771–790.

Lebeda, A., Jurik, M., Matisova, J., and Mieslerova, B. (1999). Susceptibility of *Pisum* spp. germplasm to legume viruses and virus transmission by seeds. *Plant Var Seeds* **12:**43–51.

Lecoq, H., Cohen, S., Pitrat, M., and Labonne, G. (1979). Resistance to *Cucumber mosaic virus* transmission by aphids in *Cucumis melo*. *Phytopathology* **69:**1223–1225.

Lecoq, H., Desbiez, C., Wipf-Scheibel, C., and Girard, M. (2003). Potential involvement of melon fruit in the long distance dissemination of cucurbit potyviruses. *Plant Dis.* **87:**955–959.

Legnani, R., Gebre Selassie, K., Womdim, R. N., Gognalons, P., Moretti, A., Laterrot, H., and Marchoux, G. (1995). Evaluation and inheritance of the *Lycopersicon hirsutum* resistance against potato virus Y. *Euphytica* **86:**219–226.

Legnani, R., Gognalons, P., Gebre Selassie, K., Marchoux, G., Moretti, A., and Laterrot, H. (1996). Identification and characterization of resistance to Tobacco Etch Virus in *Lycopersicon* species. *Plant Dis.* **80:**306–309.

López-Sesé, A. I., and Gómez-Guillamón, M. L. (2000). Resistance to *Cucurbit Yellowing Stunting Disorder Virus* (CYSDV) in *Cucumis melo* L. *HortScience* **35:**110–113.

Lot, H., Chovelon, V., Souche, S., Delecolle, B., Messiaen, C. M., and Etoh, T. (2001). Resistance to onion yellow dwarf virus and leek yellow stripe virus found in a fertile garlic clone. *Acta Hort.* **555:**243–246.

Maisonneuve, B., Chovelon, V., and Lot, H. (1991). Inheritance of resistance to Beet Western Yellows virus in *Lactuca virosa* L. *HortScience* **26:**1543–1545.

Maisonneuve, B., Bellec, Y., Souche, S., and Lot, H. (1999). New resistance against downy mildew and Lettuce mosaic potyvirus in wild *Lactuca* spp. *In* "EUCARPIA Leafy Vegetables '99" (A. Lebeda and E. Kristkova, eds.),. Proceedings of the EUCARPIA Meeting on Leafy Vegetables Genetics and Breeding, Olomouc, Czech Republic, 8-11 June 1999, pp. 191–197.

Manzanares, M. J., Corre, J., and Hervé, Y. (1995). Evaluation of globe artichoke and related germplasm for resistance to *Artichoke latent virus*. *Euphytica* **84:**219–228.

McCreight, J. D. (1987). Resistance in wild lettuce to *Lettuce infectious yellows virus*. *HortScience* **22:**640–642.

McCreight, J. D., and Kishaba, A. N. (1991). Reaction of cucurbit species to squash leaf curl virus and sweetpotato whitefly. *J. Am. Soc. Hort. Sci.* **116:**137–141.

McCreight, J. D., and Wintermantel, W. M. (2008). Potential new sources of genetic resistance in melon to Cucurbit yellow stunting disorder virus. *In* "Cucurbitaceae 2008" (M. Pitrat, ed.),. Proceedings of the IXth EUCARPIA Meeting on Genetics and Breeding of Cucurbitaceae, Avignon, France, 21-24 May 2008, pp. 173–179.

McCreight, J. D., Kishaba, A. N., and Mayberry, K. S. (1986). Lettuce infectious yellows tolerance in lettuce. *J. Am. Soc. Hort. Sci.* **111:**788–792.

Monma, S., and Sakata, Y. (1997). Screening of *Capsicum* accessions for resistance to cucumber mosaic virus. *J. Jap. Soc. Hort. Sci.* **65:**769–776.

Mwanga, R. O. M., Kriegner, A., Cervantes-Flores, J. C., Zhang, D. P., Moyer, J. W., and Yencho, G. C. (2002). Resistance to *Sweetpotato chlorotic stunt virus* and *Sweetpotato feathery mottle virus* is mediated by two separate recessive genes in sweetpotato. *J. Am. Soc. Hort. Sci.* **127:**798–806.

Nakahara, K. S., Shimada, R., Choi, S. H., Yamamoto, H., Shao, J., and Uyeda, I. (2010). Involvement of the P1 cistron in overcoming eIF4E-mediated recessive resistance against *Clover yellow vein virus* in pea. *Mol. Plant Microbe Interact.* **23:**1460–1469.

O'Malley, P. J., and Hartmann, R. W. (1989). Resistance to tomato spotted wilt virus in lettuce. *HortScience* **24:**360–362.

Osorno, J. M., Munoz, C. G., Beaver, J. S., Ferwerda, F. H., Bassett, M. J., Miklas, P. N., Olezyk, T., and Bussey, B. (2007). Two genes from *Phaseolus coccineus* confer resistance to *Bean golden yellow mosaic virus* in common bean. *J. Am. Soc. Hort. Sci.* **132:**530–533.

Pang, S. Z., Jan, F. J., Carney, K., Stout, J., Tricoli, D. M., Quemada, H. D., and Gonsalves, D. (1996). Post-transcriptional transgene silencing and consequent tospovirus resistance in transgenic lettuce are affected by transgene dosage and plant development. *Plant J.* **9:**899–909.

Paris, H. S., and Cohen, S. (2000). Oligogenic inheritance for resistance to *Zucchini yellow mosaic virus* in *Cucurbita pepo*. *Ann. Appl. Biol.* **136:**209–214.

Paris, H. S., Cohen, S., Burger, Y., and Yoseph, R. (1988). Single gene resistance to zucchini yellow mosaic virus in *Cucurbita moschata*. *Euphytica* **37**:27–37.
Park, S. J., and Tu, J. C. (1991). Inheritance and allelism of resistance to a severe strain of Bean yellow mosaic virus in common bean. *Can. J. Plant Pathol.* **13**:7–10.
Parrella, G., Laterrot, H., Gebre Selassie, K., and Marchoux, G. (1998). Inheritance of resistance to alfalfa mosaic virus in *Lycopersicon hirsutum* f. *glabratum* PI 134417. *J. Plant Pathol.* **80**:241–243.
Pelham, J. (1966). Resistance in tomato to tomato mosaic virus. *Euphytica* **15**:258–267.
Picó, B., Sifres, A., Martinez-Perez, E., Leiva-Brondo, M., and Nuez, F. (2008). Genetics of the resistance to CVYV in cucumber. *In* "Modern Variety Breeding for Present and Future Needs" (J. Prohens and M. L. Badenes, eds.), pp. 452–456. Proceedings of the 18th EUCARPIA General Congress, Valencia, Spain, 9–12 September 2008.
Pink, D. A. C. (1987). Genetic control of resistance to cucumber mosaic virus in *Cucurbita pepo*. *Ann. Appl. Biol.* **111**:425–432.
Pink, D. A. C., and Walkey, D. G. A. (1988). The reaction of summer- and autumn-maturing cauliflowers to infection by *Cauliflower mosaic* and *Turnip mosaic viruses*. *J. Hort. Sci.* **63**:95–102.
Pink, D. A. C., Sutherland, R. A., and Walkey, D. G. A. (1986). Genetic analysis of resistance in Brussels sprout to *Cauliflower mosaic* and *Turnip mosaic viruses*. *Ann. Appl. Biol.* **109**:199–208.
Pink, D. A. C., Walkey, D. G. A., and McClement, S. J. (1991). Genetics of resistance to *Beet western yellows virus* in lettuce. *Plant Pathol.* **40**:542–545.
Pitrat, M. and Foury, C. (eds.) (2003). Histoires de légumes, des origines à l'orée du XXIème siècle. INRA, Paris, France.
Pitrat, M., and Lecoq, H. (1984). Inheritance of *Zucchini yellow mosaic virus* resistance in *Cucumis melo* L. *Euphytica* **33**:57–61.
Pitrat, M., Maestro, C., Ferrière, C., Ricard, M., and Alvarez, J. (1988). Resistance to *Aphis gossypii* in Spanish melon (*Cucumis melo*). *Cucurbit Gen. Coop. Rep.* **11**:50–51.
Porter, R. H. (1928). Further evidence of resistance to cucumber mosaic in the Chinese cucumber. *Phytopathology* **18**:143.
Provvidenti, R. (1985). Sources of resistance to viruses in two accessions of *Cucumis sativus*. *Cucurbit Gen. Coop. Rep.* **8**:12.
Provvidenti, R., and Hampton, R. O. (1993). Inheritance of resistance to *White lupin mosaic virus* in common pea. *HortScience* **28**:836–837.
Provvidenti, R., Robinson, R. W., and Shail, J. W. (1980). A source of resistance to a strain of *Cucumber mosaic virus* in *Lactuca saligna* L. *HortScience* **15**:528–529.
Rajamony, L., More, T. A., Seshadri, V. S., and Varma, A. (1987). Resistance to cucumber green mottle mosaic virus (CGMMV) in muskmelon. *Cucurbit Gen. Coop. Rep.* **10**:58–59.
Rast, A. T. B. (1988). Pepper tobamoviruses and pathotypes used in resistance breeding. *Capsicum Newslett.* **7**:20–23.
Raybould, A. F., Edwards, M. L., Clarke, R. T., Pallett, D., and Cooper, I. (2000). Heritable variation for the control of Turnip mosaic virus and Cauliflower mosaic virus replication in wild cabbage. *In* "New Aspects of Resistance Research on Cultivated Plants: Virus diseases", pp. 4–8. Proceedings of the 7th Aschersleben Symposium Aschersleben, Germany, 17-18 November 1999.
Risser, G., Pitrat, M., and Rode, J. C. (1977). Etude de la résistance du melon (*Cucumis melo* L.) au virus de la mosaïque du concombre. *Ann. Amélior. Plant* **27**:509–522.
Ryder, E. J. (1970). Inheritance of resistance to common lettuce mosaic. *J. Am. Soc. Hort. Sci.* **95**:378–379.
Ryder, E. J., and Robinson, B. J. (1995). Big-vein resistance in lettuce: Identifying, selecting, and testing resistant cultivars and breeding lines. *J. Am. Soc. Hort. Sci.* **120**:741–746.
Sato, M., Masuta, C., and Uyeda, I. (2003). Natural resistance to *Clover yellow vein virus* in beans controlled by a single recessive locus. *Mol. Plant Microbe Interact.* **16**:994–1002.

Sawada, H., Takeuchi, S., Matsumoto, K., Hamada, H., Kiba, A., Matsumoto, M., Watanabe, Y., Suzuki, K., and Hikichi, Y. (2005). A new Tobamovirus-resistance gene, Hk, in *Capsicum annuum*. *J. Jap. Soc. Hort. Sci.* **74:**289–294.
Schroeder, W. T., and Barton, D. W. (1958). The nature and inheritance of resistance to the pea enation mosaic virus in garden pea *Pisum sativum* L. *Phytopathology* **48:**628.
Shifriss, O. C., Myers, C. H., and Chupp, C. (1942). Resistance to the mosaic virus in the cucumber. *Phytopathology* **32:**773–784.
Smykal, P., Safarova, D., Navratil, M., and Dostalova, R. (2010). Marker assisted pea breeding: eIF4E allele specific markers to *Pea seed-borne mosaic virus* (PSbMV) resistance. *Mol. Breed.* **26:**425–438.
Soria, C., Moriones, E., Fereres, A., Garzo, E., and Gómez-Guillamón, M. L. (2003). New source of resistance to mosaic virus transmission by *Aphis gossypii* in melon. *Euphytica* **133:**313–318.
Stamova, B. S., and Chetelat, R. T. (2000). Inheritance and genetic mapping of *Cucumber mosaic virus* resistance introgressed from *Lycopersicon chilense* into tomato. *Theor. Appl. Genet.* **101:**527–537.
Stevens, M. R., Scott, S. J., and Gergerich, R. C. (1991). Inheritance of a gene for resistance to *Tomato spotted wilt virus* (TSWV) from *Lycopersicon peruvianum* Mill. *Euphytica* **59:**9–17.
Strange, E. B., Guner, N., Pesic-VanEsbroeck, Z., and Wehner, T. C. (2002). Screening the watermelon germplasm collection for resistance to *Papaya Ringspot Virus* type-W. *Crop Sci.* **42:**1324–1330.
Sugiyama, M., Ohara, T., and Sakata, Y. (2006). A new source of resistance to *Cucumber green mottle mosaic virus* in melon. *J. Jap. Soc. Hort. Sci.* **75:**469–475.
Suzuki, K., Kuroda, T., Miura, Y., and Murai, J. (2003). Screening and field trials of virus resistant sources in *Capsicum* spp. *Plant Dis.* **87:**779–783.
Teran, H., Lema, M., Webster, D., and Singh, S. P. (2009). 75 years of breeding pinto bean for resistance to diseases in the United States. *Euphytica* **167:**341–351.
Thomas, P. E., Evans, D. W., Fox, L., and Biever, K. D. (1990). Resistance to *Beet western yellows virus* among forage brassicas. *Plant Dis.* **74:**327–330.
Vandijk, P. (1993). Survey and characterization of potyviruses and their strains of *Allium* species. *Netherlands J. Plant Pathol.* **99**(Suppl. 2)**:**1–48.
Velez, J. J., Bassett, M. J., Beaver, J. S., and Molina, A. (1998). Inheritance of resistance to *Bean golden mosaic virus* in common bean. *J. Am. Soc. Hort. Sci.* **123:**628–631.
Wai, T., and Grumet, R. (1991). Genetic characterization of multiple potyvirus resistance in the cucumber line TMG-1. *Phytopathology* **81:**1208.
Wai, T., and Grumet, R. (1995). Inheritance of resistance to watermelon mosaic virus in the cucumber line TMG-1: Tissue specific expression and relationship to zucchini yellow mosaic virus resistance. *Theor. Appl. Genet.* **91:**699–706.
Walkey, D. G. A., and Pink, D. A. C. (1990). Studies on resistance to *Beet western yellows virus* in lettuce (*Lactuca sativa*) and the occurrence of field sources of the virus. *Plant Pathol.* **39:**141–155.
Wang, M., and Cho, J. J. (1992). Identification of resistance to *Tomato spotted wilt virus* in lettuce. *Plant Dis.* **76:**642.
Wasuwat, S. C., and Walker, J. C. (1961). Inheritance of resistance in cucumber to cucumber mosaic virus. *Phytopathology* **51:**423–424.
Watson, M. T., and Falk, B. W. (1994). Ecological and epidemiological factors affecting carrot motley dwarf development in carrots grown in the Salinas Valley of California. *Plant Dis.* **78:**477–481.
Webb, R. E. (1979). Inheritance of resistance to watermelon mosaic virus in *Cucumis melo* L. *HortScience* **14:**265–266.

Yoshioka, K., Hanada, K., Minobe, Y., Yakuwa, T., and Oosawa, K. (1991). Coat protein gene mediated resistance to Cucumber Mosaic Virus in transgenic melon and its progeny. *J. Jap. Soc. Hort. Sci.* **60**(Suppl. 2):206–207.

Yu, J., Gu, W. K., Provvidenti, R., and Weeden, N. F. (1995). Identifying and mapping 2 DNA markers linked to the gene conferring resistance to *Pea enation mosaic virus*. *J. Am. Soc. Hort. Sci.* **120:**730–733.

Zeven, A. C., and de Wet, J. M. J. (1982). Dictionary of Cultivated Plants and Their Regions of Diversity: Excluding most Ornementals, Forest Trees and Lower Plants. PUDOC, Wageningen, Netherlands.

Zink, F. W., and Duffus, J. E. (1973). Inheritance and linkage of *Turnip mosaic virus* and downy mildew (*Bremia lactucae*) reaction in *Lactuca serriola*. *J. Am. Soc. Hort. Sci.* **98:**49–51.

Zitter, T. A., and Guzman, V. L. (1977). Evaluation of Cos lettuce crosses, endive cultivars, and *Cichorium* cultivars for resistance to *Bidens mottle virus*. *Plant Dis. Rep.* **61:**767–770.

Zohary, D., and Hopf, M. (1993). Domestication of Plants in the Old World. Clarendon Press, Oxford, United Kingdom.

CHAPTER 2

Major Tomato Viruses in the Mediterranean Basin

Inge M. Hanssen* and **Moshe Lapidot†**

Contents

* Scientia Terrae Research Institute, Sint-Katelijne-Waver, Belgium
† Department of Vegetable Research, Institute of Plant Sciences, Agricultural Research Organization, Volcani Center, Bet Dagan, Israel

Advances in Virus Research, Volume 84
ISSN 0065-3527, DOI: 10.1016/B978-0-12-394314-9.00002-6

Abstract

Tomato (*Solanum lycopersicum* L.) originated in South America and was brought to Europe by the Spaniards in the sixteenth century following their colonization of Mexico. From Europe, tomato was introduced to North America in the eighteenth century. Tomato plants show a wide climatic tolerance and are grown in both tropical and temperate regions around the world. The climatic conditions in the Mediterranean basin favor tomato cultivation, where it is traditionally produced as an open-field plant. However, viral diseases are responsible for heavy yield losses and are one of the reasons that tomato production has shifted to greenhouses. The major tomato viruses endemic to the Mediterranean basin are described in this chapter. These viruses include *Tomato yellow leaf curl virus*, *Tomato torrado virus*, *Tomato spotted wilt virus*, *Tomato infectious chlorosis virus*, *Tomato chlorosis virus*, *Pepino mosaic virus*, and a few minor viruses as well.

I. INTRODUCTION

The cultivated tomato (*Solanum lycopersicum* L.) constitutes a major agricultural industry: it is grown worldwide and, in terms of vegetable crop production, is second only to potato. Tomato belongs to the nightshade family Solanaceae, and although botanically tomato is a fruit, it is generally regarded as a vegetable (Foolad, 2007). Tomato originated from South America, mainly Peru and Chile, and was probably domesticated in Mexico. The Spaniards brought it to Europe from Latin America in the sixteenth century following their colonization of Mexico. It was in the eighteenth century that tomato was introduced from Europe to North America. At first, the tomato was treated in Europe as an "ornamental" or a curiosity and was known as "golden apple," "love apple," or "Peruvian apple" (Jones, 2007). With time, however, it became an indispensable part of the diet. Tomato plants show wide climatic tolerance and are grown in both tropical and temperate regions around the world, as an open-field crop, under plastic covers or in greenhouses. The climatic conditions in the Mediterranean basin are highly suited to tomato production, and today, it is a major vegetable crop in the region. According to the Food and Agricultural Organization (FAOSTAT, 2009), of the top 10 major tomato-producing countries, 4 are Mediterranean—Turkey, Egypt, Italy, and Spain. The Mediterranean basin is responsible for approximately 26% of the world's tomato production, with a total harvested area in 2009 of 897,000 ha producing 39 million metric tons at an estimated gross production value of 13,373 million USD. Tomatoes are consumed fresh, alone or as part of a vegetable salad, as well as cooked or variously processed as paste, peeled or diced, juice, sauces, ketchup, soups, and even pickled, to name only a few products. It is hard to imagine a fresh Mediterranean salad, or Italy's cuisine, without tomatoes.

Tomato production is highly affected by viral diseases that are responsible for millions of dollars in production losses. The most severe viral disease is the notorious tomato yellow leaf curl disease (TYLCD) which is induced by a number of viruses that are all members of the *Geminiviridae*, genus *Begomovirus*. TYLCD, as well as criniviruses and torradoviruses, is transmitted by the whitefly *Bemisia tabaci*. Tomatoes grow well in the Mediterranean climate, usually as an open-field crop. However, whitefly-transmitted viruses are one of the major reasons for the shift in tomato production from open field to greenhouse or net-house production, in an attempt to protect the crop from whiteflies and the viruses they transmit.

This chapter describes the major tomato viruses that threaten tomato production in the Mediterranean basin.

II. *TOMATO YELLOW LEAF CURL VIRUS*—A LIMITING FACTOR IN TOMATO PRODUCTION

A. Symptoms and strains

Today, *Tomato yellow leaf curl virus* (TYLCV) is the most devastating virus in tomatoes in the Mediterranean basin as well as in many tropical and subtropical regions the world over (Lapidot and Friedmann, 2002;

Moriones and Navas-Castillo, 2000; Navas-Castillo *et al.*, 2011; Pico *et al.*, 1996). Two to three weeks after inoculation, the infected tomato plant displays pronounced disease symptoms that include upward cupping of the leaves, chlorosis of the leaf margins, erect shoots, smaller, misshaped leaflets, and severe stunting of the entire plant (Fig. 1). The virus induces severe yield losses which, depending on the age of the plant at the time of infection, can reach 100% (Levy and Lapidot, 2008). In many tomato-growing areas, particularly the Mediterranean coastal areas, TYLCV has become the limiting factor for production of both open-field and protected cultivation systems (Lapidot and Friedmann, 2002). Although TYLCV is mainly a pathogen of tomato, it has been shown to induce epidemics in other crops, such as beans (Lapidot, 2002; Navas-Castillo *et al.*, 1999).

TYLCV was first detected and identified in the northern part of Israel, following the outbreak of a new disease in tomatoes in 1959 (Cohen and Harpaz, 1964; Cohen and Lapidot, 2007). However, similar disease symptoms associated with high populations of whiteflies were observed on

FIGURE 1 Infected tomato plants exhibiting disease symptoms induced by viral diseases. (A) Open-field tomato plants severely stunted due to infection with *Tomato yellow leaf curl virus*; (B) fruit marbling and leaf symptoms induced by *Pepino mosaic virus*; (C) leaf chlorosis induced by *Tomato chlorosis virus*; (D) fruit necrosis induced by *Tomato torrado virus*; (E) mosaic and line patterns induced by *Pelargonium zonate spot virus*; (F) fruit displaying chlorotic and necrotic spots induced by *Tomato spotted wilt virus*. Pictures (B), (D), and (F) were kindly provided by David Levy (Hazera Genetics, Israel). (See Page 1 in Color Section at the back of the book.)

tomatoes grown in the Jordan Valley in the late 1930s (Avidov, 1946). The outbreaks of TYLCD, which were sporadic in the 1960s, became a serious economic problem in the early 1970s (Cohen and Lapidot, 2007). By the end of the 1970s, all tomato-growing regions in the eastern Mediterranean basin were affected by TYLCD (Makkouk, 1978; Makkouk *et al.*, 1979; Mazyad *et al.*, 1979). In the late 1980s, TYLCV particles were isolated and the virus was cloned and sequenced (GenBank accession no. X15656) and found to be a monopartite begomovirus (Navot *et al.*, 1991). Shortly thereafter, another Mediterranean viral strain inducing TYLCD was cloned and sequenced—*Tomato yellow leaf curl Sardinia virus* (TYLCSV; GenBank accession no. X61153) (Kheyr-Pour *et al.*, 1991). A few years later, the original TYLCV isolated in Israel in the early 1960s was found to be composed of two different viral strains, TYLCV and TYLCV-Mld (GenBank accession no. X76319) (Antignus and Cohen, 1994). Over the years, especially with the advent of sequencing as a routine procedure, it became apparent that the name TYLCV had been given to a heterogeneous group of more than 10 virus species and their strains, all of which induce very similar disease symptoms in tomato (Moriones and Navas-Castillo, 2000). The most prevalent strains in the Mediterranean basin are TYLCV, TYLCV-Mld, and TYLCSV. It was also found that following mixed infections, TYLCV strains have a tendency to recombine, thus creating new viral strains (Davino *et al.*, 2009; Garcia-Andres *et al.*, 2007; Monci *et al.*, 2002). The TYLCV "type-strain" itself is reported to be a recombinant virus between TYLCV-Mld and an ancestor of the begomovirus *Tomato leaf curl virus* (ToLCV; Navas-Castillo *et al.*, 2000b).

B. Virus properties

Like all members of the *Geminiviridae*, TYLCV has a twinned (geminate) particle that is 18–20 nm in diameter and 30 nm long, apparently consisting of two incomplete icosahedra fused together in a structure with 22 pentameric capsomers and 110 identical protein subunits (Gafni, 2003). The viral genome is comprised of a closed circular single-stranded (ss) DNA of nearly 2.8 kb in size (Gafni, 2003; Gronenborn, 2007; Lapidot and Polston, 2006; Moriones and Navas-Castillo, 2000).

The viral circular ssDNA genome contains six open-reading frames (ORFs) that are organized bidirectionally: two in the sense orientation and four in the complementary orientation. The bidirectional ORFs are separated by a ~230-bp intergenic region that contains elements for replication and bidirectional transcription (Gronenborn, 2007; Gutierrez, 1999; Hanley-Bowdoin *et al.*, 1999; Petty *et al.*, 1988).

On the complementary strand, the *C1* gene encodes Rep (replication-associated protein) which is a multifunctional protein involved in viral replication and transcriptional regulation. This is the only viral protein absolutely required for viral replication (Gronenborn, 2007). The *C2* gene encodes TrAP (transcriptional activator protein), which has been shown in other begomoviruses to enhance expression of the coat protein, and plays a role in the suppression of host defense responses as well as in viral systemic infection (Bisaro, 2006; Brough *et al.*, 1992; Etessami *et al.*, 1991). The *C3* gene encodes the REn (replication enhancer protein) which, in other begomoviruses, acts by enhancing viral DNA accumulation in infected plants and interacts with Rep (Sunter *et al.*, 1990). The *C4* gene is embedded within the *C1* gene, but in a different ORF. Mutagenesis experiments with TYLCSV and ToLCV have implicated C4 in viral pathogenicity and movement (Jupin *et al.*, 1994; Rigden *et al.*, 1994).

On the sense strand, the capsid protein (CP) encoded by *V1* is the most abundant protein produced by TYLCV. The CP is required for whitefly transmission, binds to viral ssDNA, may play a role in systemic movement, and acts as a nuclear shuttle protein that mediates movement of viral nucleic acid into the host-cell nucleus (Azzam *et al.*, 1994; Briddon *et al.*, 1990; Kunik *et al.*, 1998; Palanichelvam *et al.*, 1998; Rojas *et al.*, 2001). The product of the *V2* ORF is involved in viral movement (Rojas *et al.*, 2001; Wartig *et al.*, 1997), and recently, V2 has been shown to act as a suppressor of RNA silencing (Zrachya *et al.*, 2007).

C. Transmission and incidence

TYLCV is transmitted by the whitefly *B. tabaci* in a circulative and persistent manner (Cohen and Harpaz, 1964). The virus is retained throughout the life of the adult insect after acquisition and moves through the insect body to the salivary glands where it can leave the whitefly body via its saliva. However, it has been shown for TYLCV that transmission efficiency declines with time. The minimal time required for the virus to travel through the insect body and be available for inoculation, known as the latent period, was found to be 8 h (Czosnek, 2007). But for efficient transmission, the "practical" latent period was found to be between 21 and 24 h (Cohen and Nitzany, 1966). Female *B. tabaci* are more efficient virus transmitters than males: following 48 h of acquisition access feeding on infected tomato plants, only 5% of the male *B. tabaci* transmitted the virus as opposed to 32% of the females (Cohen and Nitzany, 1966).

Whether TYLCV can replicate in its insect vector is controversial. TYLCV transcripts accumulate in *B. tabaci* following feeding on infected tomato plants, but there is no direct evidence of replication (Czosnek, 2007; Sinisterra *et al.*, 2005).

Transovarial transmission of the virus from the viruliferous females to their progeny is also controversial (Accotto and Sardo, 2010; Czosnek, 2007; Diaz-Pendon *et al.*, 2010). Cohen and Nitzany (1966) allowed viruliferous insects to lay eggs on cotton plants, which are immune to TYLCV.

Following emergence, none of the adult offspring transferred to TYLCV-susceptible plants transmitted the virus. Thus it was concluded that TYLCV is not transmitted to *B. tabaci* progeny (Cohen and Lapidot, 2007; Cohen and Nitzany, 1966). However, using molecular tools which were unavailable in the 1960s, Ghanim *et al.* (1998) demonstrated that TYLCV DNA is transmitted transovarially to the progeny of viruliferous *B. tabaci*. In another study, Bosco *et al.* (2004) found that DNA of TYLCSV is transmitted to *B. tabaci* progeny, whereas DNA of TYLCV is not. However, whereas, according to one study (Ghanim *et al.*, 1998), the TYLCV-carrying *B. tabaci* progeny are able to transmit the virus to test plants, in other studies (Bosco *et al.*, 2004; Cohen and Nitzany, 1966; Polston *et al.*, 2001), the insect progeny are unable to transmit the virus.

TYLCV most probably emerged from the eastern Mediterranean, where its incidence, which was sporadic in the 1960s, became a serious economic problem in the early 1970s (Cohen and Lapidot, 2007; Hanssen *et al.*, 2010c; Lefeuvre *et al.*, 2010; Navas-Castillo *et al.*, 2011). Until nearly 1990, TYLCV was recognized as a tomato pathogen mainly in the fields of Cyprus, Egypt, Israel, Lebanon, Syria, and Turkey (Czosnek and Laterrot, 1997). The disease continued to spread westward, and today, it is present in all Mediterranean countries, as well as in most tomato-growing areas worldwide, with its most devastating economic effects in the Mediterranean coastal areas.

D. Detection and control methods

Early diagnosis of TYLCV was based on symptom production by indicator plants following transmission by whiteflies or grafting. However, TYLCD is caused by a number of begomoviruses that induce similar symptoms in infected tomato plants. Although it may differ in symptoms produced on other indicator plants, detection of TYLCV based on symptom production should be used with care and should be supported by other detection methods (for a review on TYLCV diagnosis, see Accotto and Noris, 2007; Polston and Lapidot, 2008).

Although the most common detection method for plant viruses is enzyme-linked immunosorbent assay (ELISA), serological methods have had little success in detecting TYLCV (as well as other begomoviruses). This may be due in part to the low antigenicity of the viral CP (Al-Bitar and Luisoni, 1995). Another limitation in detecting TYLCV by ELISA is the cross-reaction with other begomoviruses, due to the high level of homology between the CPs of different begomoviruses.

Recently, detection kits based on lateral flow (Danks and Barker, 2000) were developed for on-site detection and identification of TYLCV and are now available commercially. These systems were developed to answer the need for rapid pathogen detection and identification in the field or greenhouse. Although quite effective, they cannot discriminate between the different TYLCVs.

Nucleic acid hybridization is based on the application of nucleic acids to a membrane, followed by hybridization with a labeled viral nucleic acid probe. In most studies, the viral probe is radioactively labeled. Labeling with radioactivity is easy, simple, and highly sensitive, and results can be readily quantified. Depending on the probe used, hybridization can be highly sensitive as well as highly specific, enabling distinction between the different viruses inducing TYLCD (Navot *et al.*, 1989). As radioactive materials are sometimes undesirable, nonradioactive labels have been successfully used for TYLCV detection as well (Crespi *et al.*, 1991; Pico *et al.*, 1999).

Amplification by polymerase chain reaction (PCR) provides a very sensitive assay for the detection and amplification of nucleic acids. The rapid improvement in and availability of different DNA polymerases for PCR, combined with the small, closed circular ssDNA and published sequences, make it relatively easy to design primers for amplification of part or all of the TYLCV genome. Sequence differences between the viruses inducing TYLCD have allowed for the design of specific primers that can distinguish among them. Moreover, distinguishing primers have been designed to produce amplification products of different sizes so that two or even three different viruses can be identified in a single reaction (Anfoka *et al.*, 2005; Martinez-Culebras *et al.*, 2001).

Management of TYLCV is difficult and expensive. Control measures usually rely on vector control, mainly through the use of insecticides or physical barriers (Hilje *et al.*, 2001; Palumbo *et al.*, 2001; Polston and Lapidot, 2007). Chemical control is only partially effective, as the whitefly population can reach very high numbers and it only takes one whitefly to transmit the virus. There are concerns that the vector may develop insecticide resistance and that intense application of insecticides may have deleterious effects on the environment (Palumbo *et al.*, 2001). Physical barriers such as fine-mesh screens have been used in the Mediterranean basin to protect crops (Lapidot and Friedmann, 2002; Polston and Anderson, 1997). UV-absorbing plastic sheets and screens have been shown to inhibit penetration of whiteflies into greenhouses (Antignus *et al.*, 1996, 2001). Filtration of UV light has been shown to hinder the whiteflies' dispersal activity and, consequently, reduces viral spread (Antignus *et al.*, 2001). However, adoption of physical barriers creates other problems, such as shading, overheating, and high humidity, and it adds to production costs. The best solution for any virus problem, and especially for whitefly-transmitted viruses such as TYLCV, is the development of genetic resistance in the tomato host. Genetic resistance requires no chemical input and/or plant seclusion and may be stable and long lasting. Thus, the best way to reduce yield losses due to

TYLCV is by breeding tomatoes resistant or tolerant to the virus (Lapidot, 2007; Lapidot and Friedmann, 2002; Morales, 2001; Pico *et al.*, 1996). As no resistance to TYLCV has been found in the domesticated tomato (*S. lycopersicum*), wild *Solanum* species have been screened for their response to the virus and a number of TYLCV-resistant accessions have been identified (Lapidot, 2007; Lapidot and Friedmann, 2002; Morales, 2001; Nakhla and Maxwell, 1999; Pico *et al.*, 1996). TYLCV resistance was introgressed from wild into domesticated tomato, and today, commercial TYLCV-resistant tomato hybrids are available and used throughout the Mediterranean basin.

III. *TORRADOVIRUS*, A RECENTLY DESCRIBED VIRAL GENUS INFECTING TOMATO CROPS

A. Symptoms and strains

In the past decade, outbreaks of a tomato disease causing necrotic spots on leaves and/or fruits have been observed in tomato-production areas in Spain, Mexico, and Guatemala (Alfaro-Fernández *et al.*, 2007a; Batuman *et al.*, 2010; Verbeek *et al.*, 2007, 2008, 2010). The symptoms resemble those caused by *Tomato spotted wilt virus* (TSWV) and *Tobacco streak virus* (TSV), but samples from symptomatic plants tested negative for these and other known tomato viruses. Initial disease symptoms consist of necrotic spots at the base of the leaflets. At a later stage, leaves, stems, and fruits can display severe necrosis (Fig. 1) and plants suffer an overall growth reduction resulting in serious economic damage. In Spain, the disease was locally referred to as "torrado" (roasted) disease because of the burnt-like appearance of affected leaves (Verbeek *et al.*, 2007). The causal agent of the outbreak was characterized as a new viral species for which the name *Tomato torrado virus* (ToTV) was proposed (Verbeek *et al.*, 2007). A similar outbreak in Mexico, locally known as "marchitez" (wilted) disease and characterized by leaf, stem, and fruit necrosis, was shown to be caused by a related but clearly distinct viral species, for which the name *Tomato marchitez virus* (ToMarV) was proposed (Turina *et al.*, 2007; Verbeek *et al.*, 2008).

More recently, two new species from the *Torradovirus* genus have been isolated from tomato crops in Guatemala, with the proposed names *Tomato chocolate spot virus* (ToCSV; Batuman *et al.*, 2010) and *Tomato chocolàte virus* (ToChV; Verbeek *et al.*, 2010), referring to the local names "chocolate spot" and "chocolàte" disease, respectively. Disease symptoms attributed to ToCSV have been reported in tomato-production areas in Guatemala since 2007 and consist of leaf epinasty, necrotic spots on the basal part of newly emerging leaves, sometimes spreading to stems and petioles and resulting in dieback of entire shoots, and reduced fruit size (Batuman *et al.*, 2010). However, necrotic spots or other obvious symptoms were not seen on the fruits. In contrast, the more recent outbreak of ToChV (in 2007) was characterized by chocolate-brown spots on the fruits, in addition to leaf symptoms similar to those described for other torradoviruses (Verbeek *et al.*, 2010). The host range of the torradoviruses is relatively narrow as it is mainly limited to members of the Solanaceae (Batuman *et al.*, 2010; Turina *et al.*, 2007). ToTV has been shown to infect pepper (*Capsicum annuum*) and eggplant (*Solanum melongena*) in addition to tomato (Amari *et al.*, 2008).

B. Virus properties

Torradoviruses share virion characteristics and nucleotide sequence similarities with viruses from the genera *Sequivirus, Waikavirus, Sadwavirus,* and *Cheravirus,* but phylogenetic analyses of two different genome regions revealed a separate taxonomic position, for which the novel genus name *Torradovirus* was proposed, with ToTV as the type species (Sançafon *et al.*, 2009; Verbeek *et al.*, 2007). Viral particles are isometric, 28–30 nm in size. The genome consists of two ssRNA molecules containing three ORFs. RNA1 has a size of 7.2–7.8 kb and contains one ORF encoding a predicted polyprotein with a helicase and protease motif, and an RNA-dependent RNA polymerase (RdRp). RNA2 is 4.9–5.7 kb in size and contains two ORFs encoding predicted polyproteins resulting in three coat proteins (Vp35, Vp26, and Vp24) and a putative movement protein (MP). Both RNAs have a poly-A tail. All currently known torradoviruses share the unique, typical ORF1 of RNA1, with a similar length but with a high degree of sequence variability for the different species, ranging from 60.4% to 73.9% (Verbeek *et al.*, 2010). Complete genome sequences of the following torradoviruses are available in GenBank (accession numbers in parentheses): ToTV (RNA1: DQ388879; RNA2: DQ388880), ToMarV (RNA1: EF681764; RNA2: EF681765), ToCSV (RNA1: GQ305131; RNA2: GQ305132), and ToChV (RNA1: FJ560489; RNA2: FJ560490).

C. Transmission and incidence

ToTV is efficiently vectored by the greenhouse whitefly *Trialeurodes vaporariorum* (Pospieszny *et al.*, 2007) and by the whitefly *B. tabaci* (Amari *et al.*, 2008). The tentative species ToCSV was shown to be transmitted by the greenhouse whitefly *T. vaporariorum*, but the virus is also sap and graft transmissible. For ToMarV and the tentative species ToChV, information on mode of transmission is not yet available. Soon after the initial characterization of ToTV, the virus was reported in greenhouse tomato crops in the Canary Islands (Alfaro-Fernández *et al.*, 2007b), where the typical symptoms had already been observed in 2003. The virus was also found in 87 greenhouse tomato samples collected in peninsular Spain between 2003 and 2006, often with *Pepino mosaic virus* (PepMV) coinfection (Alfaro-Fernández *et al.*, 2007a). Natural infection of weed hosts, possibly serving as alternative hosts in close proximity to solanaceous crop-production systems, was shown to occur in Spain (Alfaro-Fernández *et al.*, 2008a). Recently, outbreaks of ToTV have been reported in tomato crops in Hungary, Southern France, Poland, Panama, and Australia (Alfaro-Fernández *et al.*, 2008a; EPPO, 2009; Herrera-Vasquez *et al.*, 2009; Pospieszny *et al.*, 2007; Verdin *et al.*, 2009). In contrast, ToMarV has thus far only been reported in Mexico, and the tentative species ToCSV and ToChV have only been reported in Guatemala.

D. Detection and control methods

Several indicator plants, including *Physalis floridana, Nicotiana glutinosa, Nicotiana benthamiana, Nicotiana occidentalis, Nicotiana hesperis* "67A," *Chenopodium quinoa*, and *Datura stramomium*, have been reported (Batuman *et al.*, 2010; Turina *et al.*, 2007). In Spain and France, where ToTV often coincides with PepMV (Alfaro-Fernández *et al.*, 2008a; Verdin *et al.*, 2009), *P. floridana* can be used to purify ToTV of PepMV contamination. *C. quinoa* is a good local lesion host for ToMarV, but not for ToTV (Turina *et al.*, 2007; Verbeek *et al.*, 2008). *N. hesperis* "67A," *N. occidentalis* "P1," and *P. floridana* are local lesion hosts for both ToMarV and ToTV (Verbeek *et al.*, 2008). General and species-specific *Torradovirus* primers for reverse-transcription PCR (RT-PCR) detection and identification of the different species have been designed by Batuman *et al.* (2010). As the virus is efficiently vectored in the field by whiteflies but not by mechanical transmission, controlling the whitefly population can reduce virus incidence. In addition, tomato varieties with resistance to ToTV and ToMarV are commercially available, although information on resistance sources has not yet been published.

IV. *TOMATO INFECTIOUS CHLOROSIS* AND *TOMATO CHLOROSIS CRINIVIRUSES*—FROM THE NEW WORLD TO THE MEDITERRANEAN BASIN

A. Symptoms and strains

The genus *Crinivirus*, family *Closteroviridae*, represents a group of viruses which (like begomoviruses) have emerged over the past few decades in association with the worldwide emergence of whiteflies (Hanssen *et al.*,

2010c; Navas-Castillo *et al.*, 2011; Wintermantel, 2004; Wisler *et al.*, 1998a). Two criniviruses are a problem for tomato production: *Tomato infectious chlorosis virus* (TICV) and *Tomato chlorosis virus* (ToCV). TICV was first identified in field-grown tomato crops in 1993 in California, with an estimated yield loss of 2 million USD in that year alone (Duffus *et al.*, 1994; Wisler *et al.*, 1996). Tomato plants showing symptoms resembling TICV infection were first reported in greenhouse tomato crops in Florida in 1989 (Wisler *et al.*, 1998b). The disease was referred to as "yellow leaf disorder" and was attributed to nutritional disorders or pesticide phytotoxicity, as initial analyses could not detect the presence of viruses. However, transmission experiments revealed that this "yellow leaf disorder" was efficiently transmitted by whiteflies. ToCV was isolated and found to be distinct from TICV in terms of RNA sequence, vector specificity, and host range (Wisler *et al.*, 1998b).

Both TICV and ToCV induce practically indistinguishable "yellowing disease" in tomato, which includes interveinal yellowing and thickening of mature leaves, while the new growth at the plant apex appears normal. Disease symptoms usually appear 3–4 weeks after inoculation and are readily mistaken for nutritional disorders or pesticide phytotoxicity. Although TICV and ToCV do not induce any symptoms on tomato fruit, fruits of infected plants are smaller, decreased in number, the ripening process is impeded, and the plants appear to go through early senescence, all of which result in yield and economic losses (Dalmon *et al.*, 2009; Wintermantel, 2004).

B. Virus properties

The *Closteroviridae* includes viruses with a linear, positive-sense ssRNA genome up to ~20 kb in length that is encapsidated in a long flexuous particle. The genome of both TICV and ToCV is composed of two molecules, termed RNA1 and RNA2, which are encapsidated independently. RNA1 of ToCV contains four ORFs that encode proteins involved in replication, and RNA2 of ToCV contains nine ORFs (RNA2 of TICV contains eight ORFs) that encode proteins involved in viral encapsidation and movement, and vector transmission (Dolja *et al.*, 2006; Karasev, 2000).

Three isolates of ToCV have been fully sequenced (GenBank accession numbers in parentheses): an American isolate (RNA1: 8595 nt, AY903447; RNA2: 8247 nt, AY903448) (Wintermantel *et al.*, 2005), a Spanish isolate (RNA1: 8594 nt, DQ983480; RNA2: 8244 nt, DQ136146) (Lozano *et al.*, 2006; 2007), and a Greek isolate (RNA1: 8594 nt, EU284745; RNA2: 8242 nt, EU284744) (Kataya *et al.*, 2008). All three isolates are quite similar, sharing high sequence identity of 97% for RNA1 and 99% for RNA2. For TICV, only RNA2 of an American isolate (GenBank accession no. FJ542306) and a Spanish isolate (GenBank accession no. FJ542305) have been sequenced. Both consist of 7914 nt with a nearly identical sequence (Orillio and Navas-Castillo, 2009).

C. Transmission and incidence

Criniviruses are transmitted by several species of *Bemisia* and *Trialeurodes* whiteflies in a semipersistent manner (Wintermantel, 2004). TICV is transmitted solely by the greenhouse whitefly *T. vaporariorum*, while ToCV is transmitted by a number of whitefly species which include *T. vaporariorum*; the banded-wing whitefly *Trialeurodes abutilonea*; and *B. tabaci* biotypes A, B, and Q (Navas-Castillo *et al.*, 2000a; Wintermantel and Wisler, 2006). *T. vaporariorum* is present in all temperate areas worldwide, while *T. abutilonea* has only been described in Cuba and the United States. *B. tabaci* was originally described in tropical and subtropical regions, but it has since spread to temperate regions as well. The B biotype is considered highly invasive and, with its worldwide spread, has been shown to be the most efficient vector for ToCV transmission (Wintermantel and Wisler, 2006).

In the Mediterranean basin, TICV was first observed in Italy in 1991, followed by Greece, France, Spain, and Jordan (Anfoka and Abhary, 2007; Dalmon *et al.*, 2005; Dovas *et al.*, 2002; Varia *et al.*, 2002). Although initially high TICV incidences were reported, it appears that with time, its incidence has decreased. This may be due to its being transmitted solely by *T. vaporariorum*, which is being displaced in certain areas by the more invasive *B. tabaci* (Navas-Castillo *et al.*, 2011).

Tomato plants infected with ToCV were first observed in the Mediterranean basin in 1997 in Spain, followed by Portugal, Italy, Morocco, Greece, France, Israel, Cyprus, Lebanon, and Turkey (Abou-Jawdah *et al.*, 2006; Accotto *et al.*, 2001; Dovas *et al.*, 2002; Louro *et al.*, 2000; Navas-Castillo *et al.*, 2000a; Segev *et al.*, 2004). It is tempting to speculate that, in the Mediterranean basin, ToCV was first introduced from the United States to Spain and/or Portugal and then spread from one country to the next by whiteflies. As symptoms of both viruses can be confused with nutritional disorders or poor growing conditions, it has been speculated that the virus might have been present in tomato crops long before its identification (Rojas and Gilbertson, 2008).

Interestingly, ToCV was identified in Israel in 2004, while in 2007, TICV was identified in tomato fields in the Jordan Valley in Jordan (Anfoka and Abhary, 2007; Segev *et al.*, 2004). Although the distance between the tomato fields in Jordan and Israel is only a few kilometers, ToCV has not yet been detected in Jordan and TICV has not yet been detected in Israel. This difference could be linked to different agricultural practices in the two countries: in Jordan, most tomatoes are grown in open fields, while in Israel, most tomatoes are grown in greenhouses.

D. Detection and control methods

Although both TICV and ToCV have a wide host range, only TICV infects lettuce. Both TICV and ToCV induce very similar disease symptoms in tomato plants, but they can be discriminated using the indicator plants *N. benthamiana* and *Nicotiana clevelandii* (Wisler *et al.*, 1998b). Moreover, both viruses are phloem limited, and infected plants have low viral titers, which complicates viral diagnostics.

As with begomoviruses, serological methods have had little success in detecting TICV and ToCV (as well as other criniviruses). Polyclonal antiserum raised against purified TICV virions failed to distinguish between TICV and ToCV in indirect ELISA (Wisler *et al.*, 1998a). Hence, the preferred detection method for TICV and ToCV (and criniviruses in general) is RT-PCR. Multiplex and degenerate primer RT-PCR methods have been developed, taking advantage of the finding that each *Crinivirus* RNA contains genomic regions that are highly conserved among all members of the genus. The most used regions are the RdRp encoded by ORF1 in RNA1 and the heat-shock protein 70 homologue (HSP70h) encoded by ORF2 in RNA2. Dovas *et al.* (2002) reported a multiplex RT-PCR in which degenerate primers are first used to amplify part of the HSP70h gene, followed by nested PCR using specific primers. Recently, Wintermantel and Hladky (2010) developed multiplex primers that can amplify both TICV and ToCV and distinguish between the two.

Like other whitefly-transmitted viruses, management of TICV and ToCV relies on vector control mainly through the use of insecticides, physical barriers, and cultural practices (Velasco *et al.*, 2008). There are no commercial tomato hybrids with resistance to either TICV or ToCV. However, García-Cano *et al.* (2010) recently screened wild tomato species for resistance to ToCV, and two sources of ToCV resistance were identified. The resistance was expressed by impaired virus accumulation and disease symptom expression. The hope is that these sources of ToCV resistance can be introgressed into the domesticated tomato for commercial use in the future.

V. *PEPINO MOSAIC VIRUS*, A TOMATO VIRUS WITH A RAPIDLY INCREASING INCIDENCE IN THE MEDITERRANEAN BASIN

A. Symptoms and strains

In 1999, the *Potexvirus* PepMV, which was initially isolated from pepino (*Solanum muricatum*) in 1974 in Peru (Jones *et al.*, 1980), was observed for the first time in tomato crops in the Netherlands (van der Vlugt *et al.*, 2000).

In only a few years' time, the virus became a major disease of greenhouse tomato crops worldwide, with outbreaks reported from the United Kingdom, France, Spain, United States, Poland, and Belgium (Aguilar *et al.*, 2002; Cotillon *et al.*, 2002; French *et al.*, 2001; Hanssen *et al.*, 2008; Ling, 2007; Maroon-Lango *et al.*, 2005; Mumford and Metcalfe, 2001; Pagán *et al.*, 2006).

The virus induces a wide range of symptoms, depending on environmental conditions, viral isolate, tomato variety, and other unknown factors (Hanssen and Thomma, 2010; Jordá *et al.*, 2001; Spence *et al.*, 2006). The most important PepMV symptom is the typical fruit marbling, which significantly impacts the economic value of the crop (Hanssen *et al.*, 2008; Mumford and Metcalfe, 2001; Spence *et al.*, 2006). Other fruit symptoms include blotchy ripening and the occurrence of open fruits (Hanssen *et al.*, 2009; Roggero *et al.*, 2001; Spence *et al.*, 2006). Leaf symptoms consist of yellow angular spots, mosaic patterns, scorching, deformation, and so-called nettlehead symptoms (Hanssen and Thomma, 2010).

The impact of environmental conditions and tomato genotype on PepMV symptom development is not yet fully understood. However, it is believed that PepMV symptoms are masked under high light conditions and warm temperatures (Jordá *et al.*, 2001). In a Mediterranean climate, the virus can therefore easily remain unnoticed and its incidence underestimated.

Between 2000 and 2003, several PepMV isolates originating from tomato crops in different European countries were completely sequenced. The isolates shared over 99% nucleotide sequence identity but were clearly distinct (96% nucleotide sequence identity) from the original pepino isolate, which was asymptomatic in tomato (Aguilar *et al.*, 2002; Cotillon *et al.*, 2002; French *et al.*, 2001; Mumford and Metcalfe, 2001; van der Vlugt *et al.*, 2000; Verhoeven *et al.*, 2003). The European isolates were thus considered distinct from the original pepino-infecting strain and designated "European tomato strain" of PepMV (Mumford and Metcalfe, 2001; Pagán *et al.*, 2006; Verhoeven *et al.*, 2003). Since 2005, new genotypes sharing around 80% nucleotide sequence identity with the European tomato strain have been identified (US1 and US2: Maroon-Lango *et al.*, 2005; CH1 and CH2: Ling, 2007).

Based on complete RNA sequence comparisons, four distinct PepMV genotypes have been recently proposed (Hanssen *et al.*, 2010a) (GenBank accession nos. in parentheses): the original Peruvian genotype (LP: AJ606361; López *et al.*, 2005), the European (tomato) genotype (EU: AJ438767; Cotillon *et al.*, 2002), the American genotype (US1: AY509926; Maroon-Lango *et al.*, 2005), and the Chilean genotype (CH2: DQ000985; Ling, 2007).

Initially, the EU genotype was dominant in European tomato production. The closely related LP genotype was also present, but less prominent. In recent years, a population shift toward the CH2 genotype has been reported in several European countries (Gómez *et al.*, 2009; Hanssen *et al.*, 2008). The US1 genotype was recently isolated in the Canary Islands from greenhouse tomato crops displaying leaf blistering and mosaic symptoms (Alfaro-Fernández *et al.*, 2008b).

Recent studies have shown that viral genotypes and isolates play important roles in differential symptomatology. Although no clear correlation between symptom severity and PepMV genotype could be found in viral population studies in Spain or Belgium (Hanssen *et al.*, 2008; Pagán *et al.*, 2006), simultaneous infection with the CH2 and EU genotypes was shown to enhance symptom severity when compared to crops infected with a single isolate (Hanssen *et al.*, 2008, 2010b). Recombinants of both genotypes have been reported in mixed infections (Hanssen *et al.*, 2008; Pagán *et al.*, 2006), and a study with an EU and CH2 isolate from Spain suggests that those mixed infections contribute to the shaping of the population structure (Gómez *et al.*, 2009). Studies with Polish and Belgian PepMV CH2 isolates sharing over 99% nucleotide sequence identity revealed clear differences in symptomatology and host range (Hanssen *et al.*, 2009; Hasiów-Jaroszewska *et al.*, 2009; Pospieszny *et al.*, 2008).

B. Virus properties

PepMV belongs to the genus *Potexvirus* within the family *Flexiviridae*. PepMV virions are nonenveloped, flexuous, rod-shaped particles with a length of 508 nm (Jones *et al.*, 1980). They contain a monopartite, positive-sense ssRNA genome of 6.4 kb with a 3′-poly-A tail. The genome contains five ORFs encoding a 164-kDa RdRp; three triple gene block proteins of 26, 14, and 9 kDa that are believed to play a role in viral movement; and a 25-kDa coat protein.

C. Transmission and incidence

PepMV is mechanically transmitted, with sap transmission during crop handling being the most important route of transmission within a crop. Spread between greenhouses and crops is thought to occur mainly through visitors, although strict hygiene and entrance control often fail to prevent infection (Hanssen and Thomma, 2010). Recently, bumblebees, the root-infecting fungus *Olpidium virulentus*, and recirculating drainage water have also been shown to transmit PepMV (Alfaro-Fernández *et al.*, 2010; Schwarz *et al.*, 2009; Shipp *et al.*, 2008). In addition, a low rate of transmission (0.026%) was found through seeds that had been cleaned to industry standards without disinfection (Hanssen *et al.*, 2010a). Given the high infectivity and persistence of the virus, seed transmission may have impacted PepMV epidemiology, despite the very low transmission rate.

For a more detailed overview on PepMV emergence and population dynamics, the reader is referred to a recent review (Hanssen and Thomma, 2010) and to Chapter 14 of this volume.

PepMV's presence in the Mediterranean basin has been evidenced since the beginning of the epidemics in the early 2000s in Spain, France, Italy, and Morocco, but the virus is currently spreading eastward, with recent outbreaks in Cyprus, Syria, and Greece (Efthimiou *et al.*, 2011; Fakhro *et al.*, 2010; Papayiannis, 2010).

D. Detection and control methods

Several commercial antisera are available for immunological detection of PepMV. In addition, conventional and quantitative RT-PCR methods have been described for PepMV detection, with the quantitative RT-PCR TaqMan assay developed by Ling *et al.* (2007) being the most commonly used, as this highly sensitive assay detects all currently known genotypes. For identification and discrimination of the different genotypes, quantitative TaqMan RT-PCR and RT-PCR–restriction fragment length polymorphism (RFLP) methods have been developed (Alfaro-Fernández *et al.*, 2009; Gutiérrez-Aguirre *et al.*, 2009; Hanssen *et al.*, 2008).

Although natural resistance sources have been identified in certain wild *Solanum* accessions (Ling and Scott, 2007; Soler-Aleixandre *et al.*, 2007), commercial tomato varieties with resistance to PepMV are not yet available, and transgenic PepMV resistance has not yet been reported. Control depends largely on hygiene measures, more specifically on strict entrance control with protective clothing for visitors, and on thorough cleaning and disinfection between crops. In some production areas with high infection pressure and a homogeneous PepMV population, cross-protection can be used as an efficient control strategy (Hanssen *et al.*, 2010b).

VI. *TOMATO SPOTTED WILT VIRUS* AND OTHER TOSPOVIRUSES: IS TOMATO PRODUCTION IN THE MEDITERRANEAN BASIN THREATENED BY NEWLY EMERGING TOSPOVIRUSES?

A. Symptoms and strains

TSWV is an important viral disease of tomato and pepper crops, among others, which causes significant losses worldwide. The first description of "spotted wilt of tomato" dates back to the beginning of the twentieth century, when the disease was observed in Australian tomato crops (Brittlebank, 1919). Samuel *et al.* (1930) identified the causal agent and gave the virus its current name. Symptoms of TSWV in tomato can range from mild to severe, depending on the tomato genotype, the viral isolate, the developmental stage of the plant, and the environmental conditions (Roselló *et al.*, 1996). Leaf symptoms consist of purpling on the lower leaf surface and yellowing combined with small necrotic spots that develop into the typical bronzing on the upper leaves. Tomato fruit sometimes displays concentric rings that vary in color depending on the ripening stage, and occasionally, circular necrotic spots or even general fruit necrosis can occur on the fruit surface (Roselló *et al.*, 1996).

Tospoviruses were formerly subdivided into serogroups I through V, but this system has been replaced by a classification referring to the type species of the three major serogroups: TSWV (formerly serogroup I), *Watermelon silver mottle virus* (WSMoV), and *Iris yellow spot virus* (IYSV; Knierim *et al.*, 2006). Alongside isolates of the TSWV serogroup, members of the WSMoV serogroup have also been reported to cause damage in tomato production since the mid-1980s, mainly in Asia (Dong *et al.*, 2008; Whitfield *et al.*, 2005). Recently, several new *Tospovirus* species, infecting tomato crops in Australia, Asia, and the Middle East, and (tentatively) named *Tomato yellow ring virus* (Hassani-Mehraban *et al.*, 2005; Iran; GenBank accession nos. for NS and N genes DQ462163 (strain s) and AY686718 (strain t)), *Capsicum chlorosis virus* (CaCV; McMicheal *et al.*, 2002; Dong *et al.*, 2008; Australia and Asia; GenBank accession nos. DQ256123 (S RNA), DQ256125 (M RNA), and DQ256124 (L RNA)); and *Tomato zonate spot virus* (Dong *et al.*, 2008; China; GenBank accession nos. EF552433 (S RNA), EF552434 (M RNA), and EF552435 (L RNA)), have been described.

B. Virus properties

TSWV is the type species of the genus *Tospovirus*, the only plant virus genus within the *Bunyaviridae*, a large viral family containing mainly vertebrate and invertebrate viruses. Tospoviruses are spherical or pleomorphic enveloped viruses with a tripartite ssRNA genome. They have a very broad host range, including many plant species from diverse families, and they also replicate in their invertebrate vector thrips. The largest RNA molecule, L RNA (8.9 kb), is negative sense, while the other two (S RNA, 2.9 kb and M RNA, 5.4 kb) have their two coding regions in an ambisense arrangement (Hull, 2002). In total, the viral genome encodes five proteins, the largest being the 330-kDa RNA polymerase on the L RNA. The M RNA encodes the nonstructural protein NSm and a precursor to the Gn and Gc glycoproteins. The S RNA encodes the nonstructural protein NSs and the nucleocapsid protein N (Hull, 2002). TSWV consists of spherical particles ranging from 80 to 110 nm in diameter (Roselló *et al.*, 1996).

C. Transmission and incidence

Although efficient transmission can be achieved by mechanical inoculation under laboratory conditions, transmission of TSWV and other tospoviruses under natural conditions occurs through thrips vectoring (genera *Thrips* and *Frankliniella* in the family Thysanoptera) in a persistent, circulative, and propagative manner (Pappu *et al.*, 2009; Roselló *et al.*, 1996). Thrips populations, mainly of the western flower thrips *Frankliniella occidentalis*, have notably increased over the past decades, thus facilitating the emergence of the thrips-transmitted tospoviruses (Prins and Goldbach, 1998). Under laboratory conditions, mechanical inoculation with crude plant extracts is efficiently performed. Tospoviruses are not seed transmitted but can be transmitted through vegetative propagation, a common practice in ornamental crop-production systems (Pappu *et al.*, 2009).

The first report of TSWV in Europe dates back to 1929 (United Kingdom), but at that time, the virus could efficiently be controlled through chemical control of the vector *Thrips tabaci* (Pappu *et al.*, 2009). In the 1980s, following the introduction of western flower thrips *F. occidentalis* into Europe, revival of the TSWV epidemics occurred (Pappu *et al.*, 2009; Roselló *et al.*, 1996). Currently, TSWV is well established in greenhouse production in Northern and Western Europe and in both open field and greenhouse production in the Mediterranean area (Italy, Spain, Portugal, and Greece) (Pappu *et al.*, 2009). In contrast, most of the newly described *Tospovirus* species are presently restricted to Asia and/or Australia and have not yet been reported in the Mediterranean basin. The high incidence of new *Tospovirus* species in tropical Asian regions suggests a "hot spot" of viral genetic diversity in reservoir host variants, from where they are transmitted to commercial crops through increasing vector populations (Rojas and Gilbertson, 2008).

D. Detection and control methods

TSWV inoculation of the indicator plant *Petunia hybrida* Vilm., displaying characteristic small, brown or black lesions 2 days after inoculation, used to be the most common diagnostic method (Francki and Hatta, 1981; Roselló *et al.*, 1996). Today, immunological and RT-PCR-based methods are widely available and are therefore more commonly used.

Control of *Tospoviruses* consists mainly of vector control and genetic resistance. However, *Tospovirus* resistance strategies are compromised by the continuous emergence of resistance-breaking strains and new species (Pappu *et al.*, 2009). The *Tospovirus* TSWV-resistance gene *Sw5* (Stevens, 1964) is widely used in commercial varieties, but resistance-breaking strains of TSWV have been isolated from tomato varieties carrying *Sw5* in Spain and Italy (Aramburu and Martí, 2003; Ciuffo *et al.*, 2005). The *Sw5* gene is specific for TSWV and does not provide resistance to the newly described *Tospovirus* species. Control of tospoviruses through vector control is hampered by the rapid development of insecticide resistance within thrips populations. Therefore, biological alternatives to chemical control, such as the use of thrips predators or thrips-proof insect nets, are gaining importance (Jones, 2004), although the suboptimal climatic conditions resulting from reduced ventilation in greenhouses with thrips-proof netting are an important downside of this approach.

VII. *CUCUMBER MOSAIC VIRUS*, AN OLD VIRUS THAT STILL CAUSES OUTBREAKS IN TOMATO CROPS WORLDWIDE

A. Symptoms and strains

Cucumber mosaic virus (CMV) was first identified in 1916 as the causal agent of a cucumber and muskmelon disease in the United States (Doolittle, 1916). The virus has an extremely broad host range of over 1300 different plant species in 500 genera of more than 100 families (García-Arenal and Palukaitis, 2008) and caused disease epidemics in the 1980s and 1990s in tomato, banana, cucumber, sweetpotato, melon, legumes, and many other crops worldwide. A huge variability in CMV strains, differing in host range and symptomatology, has been reported (Palukaitis *et al.*, 1992). These different strains and isolates are classified into subgroups I and II according to serological relationships, nucleic acid hybridization assays, peptide mapping of the coat protein, and sequence similarity of the genomic RNAs (García-Arenal *et al.*, 2000; Palukaitis *et al.*, 1992). The percentage identity in RNA sequence between isolates belonging to the different subgroups ranges from 69% to 77%, depending on the genomic region. For isolates within the same subgroup, sequence identity is above 88% for subgroup I and above 96% for subgroup II (García-Arenal and Palukaitis, 2008). Subgroup I can be further divided into subgroups IA and IB, based on sequence homology of RNA3. For more details on CMV taxonomy, see Palukaitis *et al.* (1992), Gallitelli (2000), García-Arenal and Palukaitis (2008), and Chapter 13 of this volume.

Next to the very common viral induced leaf mottling, the most typical symptom of CMV in tomato is the occurrence of "fern leaf," a filiform deformation of the leaves also referred to as "shoestring" syndrome. Since the 1970s, severe CMV outbreaks causing tomato necrosis have been reported in the Mediterranean region, with infected tomato plants displaying stunting, leaf curling, chlorosis, fruit discoloration, fruit necrosis, and even plant death (Gallitelli *et al.*, 1991; Jordá *et al.*, 1992).

CMV is a helper virus for a single-stranded satellite RNA (satRNA) of 333–405 nt. More than 100 satRNA variants have been characterized, associated with 65 CMV isolates from both subgroups (García-Arenal and Palukaitis, 2008). In general, the presence of satRNA will result in decreased accumulation of CMV particles and an attenuation of CMV symptoms. However, some satRNA variants enhance CMV symptoms (García-Arenal and Palukaitis, 1999). An outbreak of tomato necrosis in tomato produced in the Basilicata area of Italy was related to CMV strain PG carrying a 334-nt satRNA, PG-CARNA 5, which was shown to be directly responsible for the necrotic symptoms (Kaper *et al.*, 1990). Similarly, severe epidemics of tomato necrosis caused by CMV and a particular necrotic satRNA occurred in tomato, pepper, and cucurbit crops in Spain between 1986 and 1992, resulting in the disappearance of these crops in some of their traditional cultivation areas (Jordá *et al.*, 1992). A population study of 279 isolates collected from different vegetable crops in Spain revealed that most Spanish isolates (88%) belonged to subgroup I, with isolates from subgroup II only occurring in autumn in the northern half of Spain (García-Arenal *et al.*, 2000). In the same study, satRNAs were found in 27% of the analyzed CMV isolates, with the highest incidence in the eastern coastal area of Spain where tomato necrosis epidemics occurred in the late 1980s, early 1990s. The occurrence of satRNAs in Spain decreased after 1994 (García-Arenal *et al.*, 2000).

B. Virus properties

CMV is the type member of the genus *Cucumovirus* in the family *Bromoviridae*. Virion particles are isometric with a diameter of 29 nm and have a tripartite genome of positive-sense ssRNA (Palukaitis *et al.*, 1992). In total, the viral genome encodes five ORFs: RNA1 is about 3.3 kb in length and encodes the 111-kDa protein 1a; RNA2 (3 kb) encodes the 98-kDa protein 2a, as well as the multifunctional 13- to 15-kDa protein 2b which is expressed from a subgenomic RNA named RNA 4A; RNA3 is 2 kb in length and encodes the 30-kDa MP 3a and the 25-kDa coat protein 3b, the latter being expressed from the subgenomic RNA4 (García-Arenal and Palukaitis, 2008; Hull, 2002); CMV also produces a subgenomic RNA5 of unknown function. Subgenomic RNAs 5 and 4A are only encapsidated by subgroup II isolates (García-Arenal and Palukaitis, 2008).

C. Transmission and incidence

CMV is transmitted by more than 80 different aphid species in a non-persistent manner (Gallitelli, 2000). CMV isolates show worldwide distribution, in both temperate and tropical climate zones, in open-field crop production and in greenhouses. Isolates from subgroup I are more common than those from subgroup II, the latter being mainly limited to cooler areas or seasons in temperate climate zones (García-Arenal and

Palukaitis, 2008). In the Mediterranean area, mainly subgroup IB isolates occur, which might have been introduced from East Asia, the presumed region of origin of this subgroup (García-Arenal and Palukaitis, 2008). A severe, satellite-free CMV isolate, designated CMV-G and belonging to the subgroup IB, was isolated from tomato plants in Greece. Two different variants, one causing only mild symptoms (green mosaics) and the other causing more severe yellow mosaics in tobacco, were separated through subsequent inoculation on a local lesion host (Sclavounos *et al.*, 2006). CMV infects many weed species next to horticultural crops, resulting in major viral reservoirs in native weed populations (Laviña *et al.*, 1996).

D. Detection and control methods

Most isolates of CMV can be propagated in squash (*Curcubita pepo*), *N. clevelandii*, and *N. glutinosa*. Cowpea (*Vigna unguiculata*) and *C. quinoa* are local lesion hosts (García-Arenal and Palukaitis, 2008). Symptoms in tobacco (*Nicotiana tabacum* cv. *Xanthi* nc) consist mainly of stunting, leaf deformation, and mild mosaic (Sclavounos *et al.*, 2006). Immunological detection methods, such as ELISA kits, ELISA reagent sets, and lateral flow devices, are commercially available, for both general CMV detection and differentiation of subgroups I and II isolates. Nucleotide-based detection and identification methods include dot-blot hybridization with specific probes (Gallitelli *et al.*, 1991) and RT-PCR methods using specific primer pairs. An RT-PCR–RFLP method to differentiate CMV strains in subgroups IA, IB, and II has been described by Finetti-Sialer *et al.* (1999). Distinct satRNAs can be identified by RNase protection assay (Aranda *et al.*, 1995).

Control methods may include classical or transgenic genetic resistance (Abad *et al.*, 2000; Gallitelli, 2000), vector control (Gallitelli, 2000), or cross-protection with attenuated strains (Sclavounos *et al.*, 2006). Attenuated satRNAs can be used efficiently for cross-protection against the necrotic satRNAs (Gallitelli *et al.*, 1991). Chemical control of aphid vectors can reduce vector populations but is generally not effective enough to prevent CMV outbreaks (Gallitelli, 2000). Preventive strategies, including the use of healthy seeds or plantlets, insect netting of greenhouses, (biodegradable) mulching in open fields, and weed control, can help reduce disease incidence (Gallitelli, 2000). However, as these strategies are generally not sufficient to prevent losses, genetic resistance is warranted. Several resistance sources have been identified within the genus *Solanum*, but all were strain specific and/or quantitative (Abad *et al.*, 2000). Resistant tomato varieties are thus not yet commercially available. Transgenic resistance of tomato varieties expressing the CMV coat protein has been shown to be partially efficient in field trials (Murphy *et al.*, 1998; Tomassoli *et al.*, 1999).

VIII. IS *PELARGONIUM ZONATE SPOT VIRUS* EMERGING IN THE MEDITERRANEAN BASIN?

A. Symptoms and strains

Pelargonium zonate spot virus (PZSV) belongs to the family *Bromoviridae* and was recently assigned to the new genus *Anulavirus* (Gallitelli *et al.*, 2005). PZSV induces severe disease symptoms in tomato plants, characterized by chlorotic and necrotic ring and line patterns on the leaves and fruit, together with plant stunting, leaf malformation, and reduced fruit set. Moreover, infected tomato plants exhibit stem necrosis, which may result in plant death (Gallitelli, 1982; Lapidot *et al.*, 2010). PZSV symptoms are highly dependent on environmental conditions such as light intensity, temperature, and host species and are often weak under greenhouse conditions. Although tomato plants initially produce severe symptoms following inoculation with PZSV, 3–4 weeks later, some plants appear to undergo partial recovery as symptoms become milder (Lapidot *et al.*, 2010). PZSV was first isolated from *Pelargonium zonale* (Quacquarelli and Gallitelli, 1979), but in 1980, it was found to be the causal agent of a disease in tomato in southern Italy (Gallitelli, 1982).

B. Virus properties

The virus is a 25- to 35-nm diameter quasispherical particle with a genome composed of three ssRNA segments that encode four proteins. The PZSV isolate from Italy was cloned and fully sequenced, and RNA1 (3383 nt, GenBank accession no. NC003649) and RNA2 (2435 nt, GenBank accession no. NC003650) were found to be monocistronic and to encode nonstructural proteins involved in viral replication. RNA3 (2659 nt, GenBank accession no. NC003651) is bicistronic and encodes the viral MP (ORF 3a) and CP (ORF 3b) (Finetti-Sialer and Gallitelli, 2003).

Sequence variation among the different isolates is unknown because only the genomes of the Italian and Israeli isolates have been fully sequenced (Finetti-Sialer and Gallitelli, 2003; Lapidot *et al.*, 2010). The Spanish and French isolates of the virus were identified based on serology, whereas the American isolate has been partially (ORF 3b) sequenced (Gebre-Selassie *et al.*, 2002; Liu and Sears, 2007; Luis-Arteaga and Cambra, 2000).

Comparing the genomes of the Israeli and Italian isolates of PZSV revealed 92% nucleotide identity of RNA1, 98% nucleotide identity of RNA2, and 95% nucleotide identity of RNA3. Amino acid sequence comparison of the two strains revealed 93% identity of ORF 1a, 97% identity of ORF 2a, 98% identity of ORF 3a, and 96% identity of ORF 3b (Lapidot *et al.*, 2010). These results clearly show that both viruses are indeed isolates of PZSV.

C. Transmission and incidence

PZSV has been the causal agent of multiple disease outbreaks in commercial tomato crops in various geographic regions. Nearly 20 years after the report from Italy, the virus was detected in the late 1990s in greenhouse tomato crops and nearby weeds in Spain (Luis-Arteaga and Cambra, 2000) and, in 2000, in greenhouse-grown tomato plants in southeastern France (Gebre-Selassie *et al.*, 2002). More recently, the virus has been isolated from open-field tomato crops in California (Liu and Sears, 2007) and Israel (Lapidot *et al.*, 2010).

Recently, a high incidence of PZSV has been reported in greenhouse tomato plants in northeastern Spain (Escriu *et al.*, 2009). The virus also infected pepper plants that were in the tomato greenhouses. A partial 697-nt fragment (ORF 3a) of RNA3 was cloned and sequenced from the PZSV-infected pepper and tomato plants. Sequences representing viral ORF 3a from both infected plants were found to be identical (Escriu *et al.*, 2009).

The virus is readily transmitted mechanically from the sap of infected plants and has been shown to be seed transmitted in *N. glutinosa*, with an efficiency of ~5% (Gallitelli, 1982). Pollen from systemically infected *N. glutinosa* plants has also been shown to infect plants when mechanically inoculated onto their leaves (Gallitelli, 1982). Although Vovlas *et al.* (1989) found pollen collected from naturally infected tomato plants to be contaminated with PZSV, the virus was not found to be transmitted via tomato seed. While looking for alternative hosts, these authors found that *Diplotaxis erucoides* (L.) DC., an endemic weed in southern Italy, serves as a symptomless host for the virus. They also found that PZSV is seed transmitted in *D. erucoides* with an efficiency of ~5%. When pollen from infected *D. erucoides* was rubbed gently onto leaves of tomato plants, 8 of 10 plants became infected within 10 days (Vovlas *et al.*, 1989). These authors also presented preliminary results showing that a thrips species (*Melanothrips fuscus*) feeding on flowers of *D. erucoides* carries pollen grains on its body. When these thrips were transferred to tomato seedlings, 1 of 10 plants showed PZSV symptoms 25 days later (Vovlas *et al.*, 1989). Hence, it was suggested that PZSV might be transmitted by thrips in an unusual manner. Nevertheless, the possibility of accidental mechanical inoculation due to insect and plant handling during transfer from infected to test plants could not be ruled out.

Recently, the Israeli strain of PZSV was also tested for its transmission via seed and pollen of tomato plants. Seeds collected from PZSV-infected tomato plants were sown and gave rise to infected seedlings with a seed-transmission rate of PZSV of 11–29%. Attempts to disinfect seeds using hydrochloric acid and trisodium phosphate failed to eliminate this seed transmission. Seeds from infected tomato plants were also tested for the presence of PZSV by RT-PCR: the virus was detected in total RNA extracted from as few as five seeds (Lapidot *et al.*, 2010).

Pollen was also tested for the presence of PZSV: total RNA was extracted from pollen grains collected from flowers of infected plants and subjected to RT-PCR with PZSV-specific primers. Indeed, the pollen was found to contain PZSV. To determine whether the virus was pollen transmitted, pollen grains were collected from flowers of infected plants and used to hand-pollinate healthy mother tomato plants. Although none of the pollinated mother plants became infected with PZSV, 29% of the seedlings produced from seeds harvested from these plants were infected. This was the first demonstration of vertical transmission of PZSV via both pollen and seed in tomato plants (Lapidot *et al.*, 2010). This new mode of transmission has clear epidemiological consequences and may explain the recent spread of PZSV in Israel, the United States, and Spain.

D. Detection and control methods

As stated above, the Spanish and French isolates of PZSV were identified based on serology. Positive reactions in ELISA using a commercial antiserum developed against the Italian isolate of PZSV were obtained with extracts from leaves, stems, and fruits of infected tomato plants, as well as from infected indicator plants (Gebre-Selassie *et al.*, 2002; Luis-Arteaga and Cambra, 2000). Moreover, the CP of the American isolate of PZSV was expressed *in vitro* as a fusion protein, which was used for the production of antiserum. The antiserum was able to detect both the recombinant protein and PZSV CP in infected plants, showing very high specificity and sensitivity in both Western blotting and indirect ELISA (Gulati-Sakhuja *et al.*, 2009).

The availability of the full sequence of PZSV made it easy to design specific primers for RT-PCR detection and identification of the virus in infected plants (Escriu *et al.*, 2009; Lapidot *et al.*, 2010; Liu and Sears, 2007). Indeed, RT-PCR was specific and sensitive enough to allow the detection of PZSV in tomato seed and pollen (Lapidot *et al.*, 2010).

For the highly contagious, mechanically transmitted PZSV, control depends largely on hygienic measures because resistant cultivars are not yet available. Clearly, pollen and seed transmission of PZSV is of major concern for the tomato industry. Most commercial hybrid tomato seeds are grown and harvested in one geographical location and then marketed to others. For instance, most hybrid tomato seeds sold in Europe are produced elsewhere, usually in countries where production costs are low, such as China, Thailand, India, Chile, and so on. Thus, even low rates of viral seed transmission can facilitate the introduction of viruses into new geographical and previously noninfected tomato-production areas. The concern is even greater when the virus is also pollen transmitted. To protect precious hybrids, many seed companies grow their hybrid parents at different sites, sometimes even in different countries. Pollen is collected in the field from the male parent and transported to the field of the mother plants, which are hand pollinated. This may increase the risk of spread of a pollen-transmitted virus such as PZSV, particularly because it appears that tomato plants show some recovery from PZSV-induced symptoms, and these much milder symptoms may be overlooked under field conditions. Hence, pollen collectors may fail to notice that pollen is being collected from an infected plant.

The commercial use of insect pollinators, such as bumblebees, is widespread in many tomato-production greenhouses worldwide. This practice further increases the danger of PZSV spread to healthy plants through the pollen of infected plants. Moreover, pollinating insects can provide a means for dispersing the virus over long distances (hundreds of miles), as has been reported for other pollen-transmitted viruses (Mink, 1993).

IX. CONCLUDING REMARKS

Tomato is the most important vegetable crop in the Mediterranean basin. The plants have excellent yields in the Mediterranean climate and are grown both on large farms and in small family plots, although production can be significantly reduced by viral diseases. These diseases, especially those transmitted by whiteflies, are one of the major reasons for the shift from traditional open-field production to protected cultivation. Growing tomatoes in protected structures such as greenhouses and net houses creates problems of shading, overheating, and ventilation, and also increases production costs. Unfortunately, there is no ''cure'' for a viral disease. Disease prevention is usually achieved by reduction of viral reservoirs and vector populations, should they exist. Genetic resistance is the best way to cope with viral diseases, as it is stable, persistent, and requires no chemical input in the field. Virus-resistant tomato cultivars, especially TYLCV-resistant cultivars, are increasingly being grown in the Mediterranean region. However, new viruses and new strains of known viruses keep on emerging. This is largely a result of changes in cultivation habits, climate changes, and the increase in world trade of plant material, which can transmit diseases more rapidly than natural means. Another serious challenge resulting from the emergence of new viral diseases is that often two or more viruses coinfect a crop. These new challenges necessitate the development of plant varieties that express multiviral resistance. As more resistant cultivars become available, management of viral diseases will be less of a burden on the crop and on the farmer.

REFERENCES

Abad, J., Anastasio, G., Fraile, A., and García-Arenal, F. (2000). A search for resistance to *Cucumber mosaic virus* in the genus *Lycopersicon*. *J. Plant Pathol.* **82**:39–48.

Abou-Jawdah, Y., El Mohtar, C., Atamian, H., and Sobh, H. (2006). First report of Tomato chlorosis virus in Lebanon. *Plant Dis.* **90**:378.

Accotto, G. P., and Noris, E. (2007). Detection methods for TYLCV and TYLCSV. *In* "Tomato Yellow Leaf Curl Virus Disease" (H. Czosnek, ed.), pp. 241–249. Springer, The Netherlands.

Accotto, G. P., and Sardo, L. (2010). Transovarial transmission of begomoviruses in *Bemisia tabaci*. *In* "*Bemisia*: Bionomics and Management of a Global Pest" (P. A. Stansly and S. E. Naranjo, eds.), pp. 339–345. Springer, New York.

Accotto, G. P., Varia, A. M., Vecchiati, M., Finetti-Sialer, M. M., Gallitelli, D., and Davino, M. (2001). First report of *Tomato chlorosis virus* in Italy. *Plant Dis.* **85:**1208.

Aguilar, J. M., Hernandez-Gallardo, M. D., Cenis, J. L., Lacasa, A., and Aranda, M. A. (2002). Complete sequence of the *Pepino mosaic virus* RNA genome. *Arch. Virol.* **147:**2009–2015.

Al-Bitar, L., and Luisoni, E. (1995). Tomato yellow leaf curl geminivirus: Serological evaluation of an improved purification method. *EPPO Bull.* **25:**269–276.

Alfaro-Fernández, A., Córdoba-Sellés, C., Cebrián-Micó, M. C., Font, M., Juárez, V., Median, A., Lacasa, A., Sánchez-Navarro, J. A., Pallas, V., and Jordá-Gutiérrez, C. (2007a). Advances in the study of Tomato "Torrao" or "Cribado" syndrome *Bol. San. Veg. Plagas* **33:**99–109.

Alfaro-Fernández, A., Córdoba-Sellés, C., Cebrián, M. C., Sánchez-Navarro, J. A., Espino, A., Martín, R., and Jordá, C. (2007b). First report of tomato torrado virus in tomato in the Canary Islands, Spain. *Plant Dis.* **91:**1060.

Alfaro-Fernández, A., Córdoba-Sellés, C., Cebrián, M. C., Herrera-Vásquez, J. A., Sánchez-Navarro, J. A., Juárez, M., Espino, A., Martín, R., and Jordá, C. (2008a). First report of tomato torrado virus on weed hosts in Spain. *Plant Dis.* **92:**831.

Alfaro-Fernández, A., Córdoba-Sellés, C., Cebrián, M. C., Herrera-Vásquez, J. A., and Jordá, C. (2008b). First report of the US1 strain of Pepino mosaic virus in tomato in the Canary Islands, Spain. *Plant Dis.* **92:**11.

Alfaro-Fernández, A., Sanchez-Navarro, J. A., Cebrian, M. C., Córdoba-Sellés, M. C., Pallás, V., and Jordá, C. (2009). Simultaneous detection and identification of Pepino mosaic virus (PepMV) isolates by multiplex one-step RT-PCR. *Eur. J. Plant Pathol.* **125:**143–158.

Alfaro-Fernández, A., Córdoba-Sellés, M. C., Herrera-Vásquez, J. A., Cebrián, M. C., and Jordá, C. (2010). Transmission of *Pepino mosaic virus* by the fungal vector *Olpidium virulentus*. *J. Phytopathol.* **158:**217–226.

Amari, K., Gonzalez-Ibeas, D., Gómez, P., Sempere, R. N., Sanchez-Pina, M. A., Aranda, M. A., Diaz-Pendon, J. A., Navas-Castillo, J., Moriones, E., Blanca, J., Hernandez-Gallardo, M. D., and Anastasio, G. (2008). Tomato torrado virus is transmitted by *Bemisia tabaci* and infects pepper and eggplant in addition to tomato. *Plant Dis.* **92:**1139.

Anfoka, G. H., and Abhary, M. K. (2007). Occurrence of *Tomato infectious chlorosis virus* (TICV) in Jordan. *EPPO Bull.* **37:**186–190.

Anfoka, G. H., Abhary, M., and Nakhla, M. K. (2005). Molecular identification of species of the Tomato yellow leaf curl virus complex in Jordan. *J. Plant Pathol.* **87:**61–66.

Antignus, Y., and Cohen, S. (1994). Complete nucleotide sequence of an infectious clone of a mild isolate of tomato yellow leaf curl virus (TYLCV). *Phytopathology* **84:**707–712.

Antignus, Y., Mor, N., Ben-Joseph, R., Lapidot, M., and Cohen, S. (1996). Ultraviolet-absorbing plastic sheets protect crops from insect pests and from virus diseases vectored by insects. *Environ. Entomol.* **25:**919–924.

Antignus, Y., Nestel, D., Cohen, S., and Lapidot, M. (2001). Ultraviolet-deficient greenhouse environment affects whitefly attraction and flight-behavior. *Environ. Entomol.* **30:**394–399.

Aramburu, J., and Martí, M. (2003). The occurrence in north-east Spain of a variant of *Tomato spotted wilt virus* (TSWV) that breaks resistance in tomato (*Lycopersicon esculentum*) containing the *Sw5* gene. *Plant Pathol.* **52:**407.

Aranda, M. A., Fraile, A., García-Arenal, F., and Malpica, J. M. (1995). Experimental evaluation of the ribonuclease protection assay method for the assessment of genetic heterogeneity in populations of RNA viruses. *Arch. Virol.* **140:**1373–1383.

Avidov, H. Z. (1946). Tobacco whitefly in Israel. *Hassadeh (Hebrew)*1–33.

Azzam, O., Frazer, J., De La Rosa, D., Beaver, J. S., Ahlquist, P., and Maxwell, D. P. (1994). Whitefly transmission and efficient ssDNA accumulation of bean golden mosaic geminivirus require functional coat protein. *Virology* **204:**289–296.

Batuman, O., Kuo, Y.-W., Palmieri, M., Rojas, M. R., and Gilbertson, R. L. (2010). Tomato chocolate spot virus, a member of a new torradovirus species that causes a necrosis-associated disease of tomato in Guatemala. *Arch. Virol.* **155:**857–869.

Bisaro, D. M. (2006). Silencing suppression by geminivirus proteins. *Virology* **344:**158–168.

Bosco, D., Mason, G., and Accotto, G. P. (2004). TYLCSV DNA, but not infectivity, can be transovarially inherited by the progeny of the whitefly vector *Bemisia tabaci* (Gennadius). *Virology* **323:**276–283.

Briddon, R. W., Pinner, M. S., Stanley, J., and Markham, P. G. (1990). Geminivirus coat protein gene replacement alters insect specificity. *Virology* **177:**85–94.

Brittlebank, C. C. (1919). Tomato diseases. *J. Agric. Victoria* **17:**213–235.

Brough, C. L., Sunter, G., Gardiner, W. E., and Bisaro, D. M. (1992). Kinetics of tomato golden mosaic virus DNA replication and coat protein promoter activity in *Nicotiana tabacum* protoplasts. *Virology* **187:**1–9.

Ciuffo, M., Finetti-Sialer, M. M., Gallitelli, D., and Turina, M. (2005). First report in Italy of a resistance-breaking strain of Tomato spotted wilt virus infecting tomato cultivars carrying the Sw5 resistance gene. *Plant Pathol.* **54:**564.

Cohen, S., and Harpaz, I. (1964). Periodic, rather than continual acquisition of a new tomato virus by its vector, the tobacco whitefly (*Bemisia tabaci* Gennadius). *Entomol. Exp. Appl.* **7:**155–166.

Cohen, S., and Lapidot, M. (2007). Appearance and expansion of TYLCV: A historical point of view. *In* "Tomato Yellow Leaf Curl Virus Disease" (H. Czosnek, ed.), pp. 3–12. Springer, The Netherlands.

Cohen, S., and Nitzany, F. E. (1966). Transmission and host range of the tomato yellow leaf curl virus. *Phytopathology* **56:**1127–1131.

Cotillon, A. C., Girard, M., and Ducouret, S. (2002). Complete nucleotide sequence of the genomic RNA of a French isolate of *Pepino mosaic virus* (PepMV). *Arch. Virol.* **147:**2231–2238.

Crespi, S., Accotto, G. P., Caciagli, P., and Gronenborn, B. (1991). Use of digoxigenin-labeled probes for detection and host-range studies of tomato yellow leaf curl geminivirus. *Res. Virol.* **142:**283–288.

Czosnek, H. (2007). Interactions of tomato yellow leaf curl virus with its whitefly vector. *In* "Tomato Yellow Leaf Curl Virus Disease" (H. Czosnek, ed.), pp. 157–170. Springer, The Netherlands.

Czosnek, H., and Laterrot, H. (1997). A worldwide survey of tomato yellow leaf curl viruses. *Arch. Virol.* **142:**1391–1406.

Dalmon, A., Bouyer, S., Cailly, M., Girard, M., Lecoq, H., Desbiez, C., and Jacquemond, M. (2005). First report of Tomato chlorosis virus and Tomato infectious chlorosis virus in tomato crops in France. *Plant Dis.* **89:**1243.

Dalmon, A., Fabre, F., Guilbaud, L., Lecoq, H., and Jacquemond, M. (2009). Comparative whitefly transmission of *Tomato chlorosis virus* and *Tomato infectious chlorosis virus* from single or mixed infections. *Plant Pathol.* **28:**221–227.

Danks, C., and Barker, I. (2000). On-site detection of plant pathogens using lateral-flow devices. *EPPO Bull.* **30:**421–426.

Davino, S., Napoli, C., Dellacroce, C., Miozzi, L., Noris, E., Davino, M., and Accotto, G. P. (2009). Two natural begomovirus recombinants associated with tomato yellow leaf curl disease co-exist with parental viruses in tomato epidemics in Italy. *Virus Res.* **143:**15–23.

Diaz-Pendon, J. A., Canizares, M. C., Moriones, E., Bejarano, E. R., Czosnek, H., and Navas-Castillo, J. (2010). Tomato yellow leaf curl virus: Ménage a trios between the virus complex, the plant and the whitefly vector. *Mol. Plant Pathol.* **11:**441–450.

Dolja, V. V., Kreuze, J. F., and Valkonen, J. P. (2006). Comparative and functional genomics of closteroviruses. *Virus Res.* **117:**38–51.

Dong, J.-H., Cheng, X.-F., Yin, Y.-Y., Fang, Q., Ding, M., Li, T.-T., Zhang, L.-Z., Su, X.-X., McBeath, J. H., and Zhang, Z.-K. (2008). Characterization of tomato zonate spot virus, a new Tospovirus in China. *Arch. Virol.* **153:**855–864.

Doolittle, S. P. (1916). A new infectious mosaic disease of cucumber. *Phytopathology* **6:**145–147.

Dovas, C. I., Katis, N. I., and Avgelis, A. D. (2002). Multiplex detection of criniviruses associated with epidemics of a yellowing disease of tomato in Greece. *Plant Dis.* **86:**1345–1349.

Duffus, J. E., Liu, H.-Y., and Wisler, G. C. (1994). Tomato infectious chlorosis virus—A new clostero-like virus transmitted by *Trialeurodes vaporariorum. Eur. J. Plant Pathol.* **102:**219–226.

Efthimiou, K. E., Gatsios, A. P., Aretakis, K. C., Papayiannis, L. C., and Katis, N. I. (2011). First report of *Pepino mosaic virus* infecting greenhouse cherry tomatoes in Greece. *Plant Dis.* **95:**78.

EPPO (2009). *First record of tomato torrado virus in Australia. EPPO Reporting Service No. 2, Paris, February 01, 2009.* http://archives.eppo.org/EPPOReporting/2009/Rse-0902.pdf.

Escriu, F., Cambra, M. A., and Luis-Artega, M. (2009). First report of pepper as a natural host for *Pelargonium zonate spot virus* in Spain. *Plant Dis.* **93:**1346.

Etessami, P., Saunders, K., Watts, J., and Stanley, J. (1991). Mutational analysis of complementary-sense genes of African cassava mosaic virus DNA A. *J. Gen. Virol.* **72:**1005–1012.

Fakhro, A., Von Bargen, S., Bandte, M., and Büttner, C. (2010). *Pepino mosaic virus*, a first report of a virus infecting tomato in Syria. *Phytopathol. Mediterr.* **49:**99–101.

FAOSTAT (2009). http://faostat.fao.org/default.aspx/.

Finetti-Sialer, M., and Gallitelli, D. (2003). Complete nucleotide sequence of *Pelargonium zonate spot virus* and its relationship with the family *Bromoviridae. J. Gen. Virol.* **84:**3143–3151.

Finetti-Sialer, M. M., Cillo, F., Barbarossa, L., and Gallitelli, D. (1999). Differentiation of cucumber mosaic virus subgroups by RT-PCR RFLP. *J. Plant Pathol.* **81:**145–148.

Foolad, M. R. (2007). Genome mapping and molecular breeding of tomato. *Int. J. Plant Genomics.* Article ID 64358, volume noumber 2007, 52 pages, doi:10.1155/2007/64358.

Francki, R. I. B., and Hatta, T. (1981). Tomato spotted wilt virus. *In* "Handbook of Plant Virus Infections and Comparative Diagnosis" (E. Kurstak, ed.), pp. 491–512. North Holland Biomedical Press, Amsterdam.

French, C. J., Bouthillier, M., Bernardy, M., Ferguson, G., Sabourin, M., Johnson, R. C., Masters, C., Godkin, S., and Mumford, R. (2001). First report of *Pepino mosaic virus* in Canada and the United States. *Plant Dis.* **85:**1121.

Gafni, Y. (2003). *Tomato yellow leaf curl virus*, the intracellular dynamics of a plant DNA virus. *Mol. Plant Pathol.* **4:**9–15.

Gallitelli, D. (1982). Properties of the tomato isolate of *Pelargonium zonate spot virus. Ann. Appl. Biol.* **100:**457–466.

Gallitelli, D. (2000). The ecology of Cucumber mosaic virus and sustainable agriculture. *Virus Res.* **71:**9–21.

Gallitelli, D., Vovlas, C., Martelli, G., Montasser, M. S., Tousignant, M. E., and Kaper, J. M. (1991). Satellite-mediated protection of tomato against Cucumber mosaic virus: II. Field test under natural epidemic conditions in Southern Italy. *Plant Dis.* **75:**93–95.

Gallitelli, D., Finetti-Sialer, M., and Martelli, G. P. (2005). *Anulavirus*, a proposed new genus of plant viruses in the family *Bromoviridae. Arch. Virol.* **150:**407–411.

Garcia-Andres, S., Tomas, D. M., Sanchez-Campos, S., Navas-Castillo, J., and Moriones, E. (2007). Frequent occurrence of recombinants in mixed infections of tomato yellow leaf curl disease associated begomoviruses. *Virology* **365:**210–219.

García-Arenal, F., and Palukaitis, P. (1999). Structure and functional relationships of satellite RNAs of *Cucumber mosaic virus*. *In* "Satellites and Defective Viral RNAs" (P. K. Vogt and A. O. Jackson, eds.), pp. 37–63. Springer-Verlag, Berlin.

García-Arenal, F., and Palukaitis, P. (2008). Cucumber mosaic virus. *In* "Desk Encyclopedia of Plant and Fungal Virology" (B. W. J. Mahy and M. H. V. van Regenmortel, eds.), pp. 171–176. Elsevier, Amsterdam.

García-Arenal, F., Escriu, F., Aranda, M. A., Alonso-Prados, J. L., Malpica, J. M., and Fraile, A. (2000). Molecular epidemiology of *Cucumber mosaic virus* and its satellite RNA. *Virus Res.* **71:**1–8.

García-Cano, E., Navas-Castillo, J., Moriones, E., and Fernández-Muñoz, R. (2010). Resistance to Tomato chlorosis virus in wild tomato species that impair virus accumulation and disease symptom expression. *Phytopathology* **100:**582–592.

Gebre-Selassie, K., Dellecolle, B., Gognalons, P., Dufour, O., Gros, C., Cotillon, A. C., Parrella, G., and Marchoux, G. (2002). First report of an isolate of *Pelargonium zonate spot virus* in commercial glasshouse tomato crop in Southeastern France. *Plant Dis.* **86:**1052.

Ghanim, M., Morin, S., Zeidan, M., and Czosnek, H. (1998). Evidence for transovarial transmission of *Tomato yellow leaf curl virus* by its vector, the whitefly *Bemisia tabaci*. *Virology* **240:**295–303.

Gómez, P., Sempere, R. N., Elena, S. F., and Aranda, M. A. (2009). Mixed infections of *Pepino mosaic virus* strains modulate the evolutionary dynamics of this emergent virus. *J. Virol.* **83:**12378–12387.

Gronenborn, B. (2007). The tomato yellow leaf curl virus genome and function of its proteins. *In* "Tomato Yellow Leaf Curl Virus Disease" (H. Czosnek, ed.), pp. 67–84. Springer, The Netherlands.

Gulati-Sakhuja, A., Sears, J. L., Nunez, A., and Liu, H.-Y. (2009). Production of polyclonal antibodies against *Pelargonium zonate spot virus* coat protein expressed in *Escherichia coli* and application for immunodiagnosis. *J. Virol. Methods* **160:**29–37.

Gutierrez, C. (1999). Geminivirus DNA replication. *Cell. Mol. Life Sci.* **56:**313–329.

Gutiérrez-Aguirre, I., Mehle, N., Delíc, D., Gruden, K., Mumford, R., and Ravnikar, M. (2009). Real time quantitative PCR based sensitive detection and strain discrimination of *Pepino mosaic virus*. *J. Virol. Methods* **162:**46–55.

Hanley-Bowdoin, L., Settlage, S. B., Orozco, B. M., Nagar, S., and Robertson, D. (1999). Geminiviruses: Models for plant DNA replication, transcription, and cell cycle regulation. *Crit. Rev. Plant Sci.* **18:**71–106.

Hanssen, I. M., and Thomma, B. P. H. J. (2010). *Pepino mosaic virus*: A successful pathogen that rapidly emerged from emerging to endemic in tomato crops. *Mol. Plant Pathol.* **11:**179–189.

Hanssen, I. M., Paeleman, A., Wittemans, L., Goen, K., Lievens, B., Bragard, C., Vanachter, A. C. R. C., and Thomma, B. P. H. J. (2008). Genetic characterization of *Pepino mosaic virus* isolates from Belgian greenhouse tomatoes reveals genetic recombination. *Eur. J. Plant Pathol.* **121:**131–146.

Hanssen, I. M., Paeleman, A., Vandewoestijne, E., Van Bergen, L., Bragard, C., Lievens, B., Vanachter, A. C. R. C., and Thomma, B. P. H. J. (2009). *Pepino mosaic virus* isolates and differential symptomatology in tomato. *Plant Pathol.* **58:**450–460.

Hanssen, I. M., Mumford, R., Blystad, D.-R., Cortez, I., Hasiów-jareszewska, B., Hristova, D., Pagán, I., Pepeira, A.-M., Peters, J., Pospieszny, H., Ravnikar, M., Stijger, I., *et al.* (2010a). Seed transmission of *Pepino mosaic virus* in tomato. *Eur. J. Plant Pathol.* **126:**145–152.

Hanssen, I. M., Gutiérrez-Aguirre, I., Paeleman, A., Goen, K., Wittemans, L., Lievens, B., Vanachter, A. C. R. C., Ravnikar, M., and Thomma, B. P. H. J. (2010b). Cross-protection or enhanced symptom display in greenhouse tomato co-infected with different *Pepino mosaic virus* isolates. *Plant Pathol.* **59:**13–21.

Hanssen, I. M., Lapidot, M., and Thomma, P. H. J. B. (2010c). Emerging viral diseases of tomato crops. *Mol. Plant Microbe Interact.* **23:**539–548.

Hasiów-Jaroszewska, B., Pospieszny, H., and Borodynko, N. (2009). New necrotic isolates of *Pepino mosaic virus* representing the CH2 genotype. *J. Phytopathol.* **157:**494–496.

Hassani-Mehraban, A., Saaijer, J., Peters, D., Goldbach, R., and Kormelink, R. (2005). A new tomato-infecting *Tospovirus* from Iran. *Phytopathology* **95:**852–858.

Herrera-Vasquez, J. A., Alfaro-Fernández, A., Córdoba-Selles, M. C., Cebrian, M. C., Font, M. I., and Jorda, C. (2009). First report of tomato torrado virus infecting tomato in single and mixed infections with *Cucumber mosaic virus* in Panama. *Plant Dis.* **93:**198.

Hilje, L., Costa, H. S., and Stansly, P. A. (2001). Cultural practices for managing *Bemisia tabaci* and associated viral diseases. *Crop Prot.* **20:**801–812.

Hull, R. (2002). Matthews' Plant Virology. Academic Press, New York.

Jones, R. A. C. (2004). Using epidemiological information to develop effective integrated virus disease management strategies. *Virus Res.* **100:**5–30.

Jones, J. B. (2007). Tomato Plant Culture: In the Field, Greenhouse, and Home Garden. 2nd edn. CRC Press, Florida, USA.

Jones, R. A. C., Koenig, R., and Lesemann, D. E. (1980). *Pepino mosaic virus*, a new *Potexvirus* from pepino (*Solanum muricatum*). *Ann. Appl. Biol.* **94:**61–68.

Jordá, C., Alfaro, A., Aranda, M., Moriones, E., and García-Arenal, F. (1992). Epidemic of *Cucumber mosaic virus* plus satellite RNA in tomatoes in eastern Spain. *Plant Dis.* **76:**363–366.

Jordá, C., Lázaro-Perez, A., and Martinez-Culebras, P. V. (2001). First report of *Pepino mosaic virus* on tomato in Spain. *Plant Dis.* **85:**1292.

Jupin, I., De Kouchkovsky, F., Jouanneau, F., and Gronenborn, B. (1994). Movement of tomato yellow leaf curl geminivirus (TYLCV): Involvement of the protein encoded by ORF C4. *Virology* **204:**82–90.

Kaper, J. M., Gallitelli, D., and Tousignant, M. E. (1990). Identification of a 334 rubonucleotide viral satellite as principal aetiological agent in a tomato necrosis epidemic. *Res. Virol.* **141:**81–95.

Karasev, A. V. (2000). Genetic diversity and evolution of Closteroviruses. *Annu. Rev. Phytopathol.* **38:**293–324.

Kataya, A. R. A., Stavridou, E., Farhan, K., and Livieratos, I. C. (2008). Nucleotide sequence analysis and detection of a Greek isolate of Tomato chlorosis virus. *Plant Pathol.* **57:**819–824.

Kheyr-Pour, A., Bendahmane, M., Matzeit, M., Accotto, G. P., Crespi, S., and Gronenborn, B. (1991). Tomato yellow leaf curl virus from Sardinia is a whitefly-transmitted monopartite geminivirus. *Nucleic Acids Res.* **19:**6763–6769.

Knierim, D., Blawid, R., and Maiss, E. (2006). The complete nucleotide sequence of a capsicum chlorosis virus isolate from *Lycopersicon esculentum* in Thailand. *Arch. Virol.* **151:**1761–1782.

Kunik, T., Palanichelvam, K., Czosnek, H., Citovsky, V., and Gafni, Y. (1998). Nuclear import of the capsid protein of *Tomato yellow leaf curl virus* (TYLCV) in plant and insect cells. *Plant J.* **13:**393–399.

Lapidot, M. (2002). Screening common bean (*Phaseolus vulgaris*) for resistance to *Tomato yellow leaf curl virus*. *Plant Dis.* **86:**429–432.

Lapidot, M. (2007). Screening for TYLCV-resistant plants using whitefly-mediated inoculation. *In* "Tomato Yellow Leaf Curl Virus Disease" (H. Czosnek, ed.), pp. 329–342. Springer, The Netherlands.

Lapidot, M., and Friedmann, M. (2002). Breeding for resistance to whitefly-transmitted geminiviruses. *Ann. Appl. Biol.* **140:**109–127.

Lapidot, M., and Polston, J. E. (2006). Resistance to *Tomato yellow leaf curl virus* in tomato. *In* "Natural Resistance Mechanisms of Plants to Viruses" (G. Lobenstein and J. P. Carr, eds.), pp. 503–520. Springer, The Netherlands.

Lapidot, M., Guenoune-Gelbart, D., Leibman, D., Holdengreber, V., Davidovitz, M., Machbash, Z., Klieman-Shoval, S., Cohen, S., and Gal-On, A. (2010). *Pelargonium zonate spot virus* is transmitted vertically via seed and pollen in tomato. *Phytopathology* **100:**798–804.

Laviña, A., Aramburu, J., and Moriones, E. (1996). Occurrence of tomato spotted wilt and cucumber mosaic virus in field-grown tomato crops and associated weeds in northeastern Spain. *Plant Pathol.* **45:**837–842.

Lefeuvre, P., Martin, D. P., Harkins, G., Lemey, P., Gray, A. J. A., Meredith, S., Lakay, F., Monjane, A., Lett, J.-M., Varsani, A., and Heydarnejad, J. (2010). The spread of tomato yellow leaf curl virus from the Middle East to the world. *PLoS Pathog.* **6:**e1001164.

Levy, D., and Lapidot, M. (2008). Effect of plant age at inoculation on expression of genetic resistance to tomato yellow leaf curl virus. *Arch. Virol.* **153:**171–179.

Ling, K.-S. (2007). Molecular characterization of two *Pepino mosaic virus* variants from imported tomato seed reveals high levels of sequence identity between Chilean and US isolates. *Virus Genes* **34:**1–8.

Ling, K.-S., and Scott, J. W. (2007). Sources of resistance to *Pepino mosaic virus* in tomato accessions. *Plant Dis.* **91:**749–753.

Ling, K.-S., Wechter, W. P., and Jordan, R. (2007). Development of a one-step immunocapture real-time TaqMan RT-PCR assay for the broad spectrum detection of Pepino mosaic virus. *J. Virol. Methods* **144:**65–72.

Liu, H.-Y., and Sears, J. L. (2007). First report of *Pelargonium zonate spot virus* from tomato in the United States. *Plant Dis.* **91:**633.

López, C., Soler, S., and Nuez, F. (2005). Comparison of the complete sequences of three different isolates of *Pepino mosaic virus*: Size variability of the TGBp3 protein between tomato and *L. peruvianum* isolates. *Arch. Virol.* **150:**619–627.

Louro, D., Accotto, G. P., and Vaira, A. M. (2000). Occurrence and diagnosis of Tomato chlorosis virus in Portugal. *Eur. J. Plant Pathol.* **106:**589–592.

Lozano, G., Moriones, E., and Navas-Castillo, J. (2006). Complete nucleotide sequence of the RNA2 of the crinivirus tomato chlorosis virus. *Arch. Virol.* **151:**581–587.

Lozano, G., Moriones, E., and Navas-Castillo, J. (2007). Complete sequence of the RNA1 of a European isolate of tomato chlorosis virus. *Arch. Virol.* **152:**839–841.

Luis-Arteaga, M., and Cambra, M. (2000). First report of natural infection of greenhouse-grown tomato and weed species by *Pelargonium zonate spot virus* in Spain. *Plant Dis.* **91:**633.

Makkouk, K. M. (1978). A study on tomato viruses in the Jordan Valley with special emphasis on tomato yellow leaf curl. *Plant Dis. Rep.* **62:**259–262.

Makkouk, K. M., Shehab, S., and Madjalani, S. E. (1979). Tomato yellow leaf curl: Incidence, yield losses and transmission in Lebanon. *Phytopathol. Z.* **96:**263–267.

Maroon-Lango, C. J., Guaragna, M. A., Jordan, R. L., Hammond, J., Bandla, M., and Marquardt, S. K. (2005). Two unique US isolates of *Pepino mosaic virus* from a limited source of pooled tomato tissue are distinct from a third (EU like) US isolate. *Arch. Virol.* **150:**1187–1201.

Martinez-Culebras, P. V., Font, I., and Jorda, C. (2001). A rapid PCR method to discriminate between *Tomato yellow leaf curl virus* isolates. *Ann. Appl. Biol.* **139:**251–257.

Mazyad, H. M., Omar, F., Al-Taher, K., and Salha, M. (1979). Observations on the epidemiology of tomato yellow leaf curl disease on tomato plants. *Plant Dis. Rep.* **63:**695–698.

McMicheal, L. A., Persley, D. M., and Thomas, J. E. (2002). A new tospovirus serogroup IV species infecting capsicum and tomato in Queensland, Australia. *Aust. Plant Pathol.* **31:**231–239.

Mink, G. I. (1993). Pollen- and seed-transmitted viruses and viroids. *Annu. Rev. Phytopathol.* **31:**357–402.

Monci, F., Sanchez-Campos, S., Navas-Castillo, J., and Moriones, E. (2002). A natural recombinant between the geminiviruses Tomato yellow leaf curl Sardinia virus and Tomato yellow leaf curl virus exhibits a novel pathogenic phenotype and is becoming prevalent in Spanish populations. *Virology* **303:**317–326.

Morales, F. J. (2001). Conventional breeding for resistance to *Bemisia tabaci*-transmitted geminiviruses. *Crop Prot.* **20:**825–834.

Moriones, E., and Navas-Castillo, J. (2000). Tomato yellow leaf curl virus, an emerging virus complex causing epidemics worldwide. *Virus Res.* **71:**123–134.

Mumford, R. A., and Metcalfe, E. J. (2001). The partial sequencing of the genomic RNA of a UK isolate of *Pepino mosaic virus* and the comparison of the coat protein sequence with other isolates from Europe and Peru. *Arch. Virol.* **146:**2455–2460.

Murphy, J. F., Sikora, E. J., Sammons, B., and Kaniewski, W. K. (1998). Performance of transgenic tomatoes expressing cucumber mosaic virus CP gene under epidemic conditions. *HortScience* **33:**1032–1035.

Nakhla, M. K., and Maxwell, D. P. (1999). Epidemiology and management of tomato yellow leaf curl virus. *In* "Plant Virus Disease Control" (A. Hadidi, R. K. Khetarpal, and H. Koganezawa, eds.), pp. 565–583. APS Press, St. Paul, MN.

Navas-Castillo, J., Sanchez-Campos, S., and Diaz, J. A. (1999). Tomato yellow leaf virus-Is causes a novel disease of common bean and severe epidemics in tomato in Spain. *Plant Dis.* **83:**29–32.

Navas-Castillo, J., Camero, R., Bueno, M., and Moriones, E. (2000a). Severe yellowing outbreaks in tomato in Spain associated with infections of Tomato chlorosis virus. *Plant Dis.* **84:**835–837.

Navas-Castillo, J., Sanchez-Campos, S., Noris, E., Louro, D., Accotto, G. P., and Moriones, E. (2000b). Natural recombination between *Tomato yellow leaf curl virus*-Is and *Tomato leaf curl virus*. *J. Gen. Virol.* **81:**2797–2801.

Navas-Castillo, J., Fiallo-Olivé, E., and Sánchez-Campos, S. (2011). Emerging virus diseases transmitted by whiteflies. *Annu. Rev. Phytopathol.* **49:**219–248.

Navot, N., Ber, R., and Czosnek, H. (1989). Rapid detection of tomato yellow leaf curl virus in squashes of plants and insect vectors. *Phytopathology* **79:**562–568.

Navot, N., Pichersky, E., Zeidan, M., Zamir, D., and Czosnek, H. (1991). Tomato yellow leaf curl virus: A whitefly-transmitted geminivirus with a single genomic component. *Virology* **185:**151–161.

Orillio, A. F., and Navas-Castillo, J. (2009). The complete nucleotide sequence of the RNA2 of the crinivirus tomato infectious chlorosis virus: Isolates from North America and Europe are essentially identical. *Arch. Virol.* **154:**683–687.

Pagán, I., Córdoba-Sellés, M. C., Martinez-Priego, L., Fraile, A., Malpica, J. M., Jordá, C., and García-Arenal, F. (2006). Genetic structure of the population of *Pepino mosaic virus* infecting tomato crops in Spain. *Phytopathology* **96:**274–279.

Palanichelvam, K., Kunik, T., Citovsky, V., and Gafni, Y. (1998). The capsid protein of tomato yellow leaf curl virus binds cooperatively to single-stranded DNA. *J. Gen. Virol.* **79:**2829–2833.

Palukaitis, P., Roossinck, M. J., Dietzgen, R. G., and Francki, R. I. B. (1992). Cucumber mosaic virus. *Adv. Virus Res.* **41:**281–348.

Palumbo, J. C., Horowitz, A. R., and Prabhaker, N. (2001). Insecticidal control and resistance management for *Bemisia tabaci*. *Crop Prot.* **20:**739–766.

Papayiannis, L. C. (2010). *Pepino mosaic virus* (PepMV): A new virus disease infecting tomato crops in Cyprus. Agricultural Report 1. Agricultural Research Institute. ISSN: 1986-1370, 18pp (in Greek).

Pappu, H. R., Jones, R. A. C., and Jain, R. K. (2009). Global status of *Tospovirus* epidemics in diverse cropping systems: Successes achieved and challenges ahead. *Virus Res.* **141:**219–236.

Petty, I. T. D., Coutts, R. H. A., and Buck, K. W. (1988). Transcriptional mapping of the coat protein gene of *Tomato golden mosaic virus*. *J. Gen. Virol.* **69:**1359–1365.

Pico, B., Diez, M. J., and Nuez, F. (1996). Viral diseases causing the greatest economic losses to the tomato crop. II. The tomato yellow leaf curl virus—A review. *Sci. Hortic.* **67:**151–196.

Pico, B., Diez, M. J., and Nuez, F. (1999). Improved diagnostic techniques for tomato yellow leaf curl virus in tomato breeding programs. *Plant Dis.* **83:**1006–1012.

Polston, J. E., and Anderson, P. K. (1997). The emergence of whitefly-transmitted geminiviruses in tomato in the Western Hemisphere. *Plant Dis.* **81:**1358–1369.

Polston, J. E., and Lapidot, M. (2007). Management of Tomato yellow leaf curl virus: US and Israel perspectives. *In* "Tomato Yellow Leaf Curl Virus Disease" (H. Czosnek, ed.), pp. 251–262. Springer, The Netherlands.

Polston, J. E., and Lapidot, M. (2008). Tomato yellow leaf curl virus. *In* "Characterization, Diagnosis & Management of Plant Viruses" (G. P. Rao, L. Kumar, and R. J. Holguin-Pena, eds.), Vol. 3, pp. 41–161. Studium Press LLC, Texas, USA.

Polston, J. E., Sherwood, T. A., Rosell, R., and Nava, A. (2001). Detection of *Tomato yellow leaf curl* and *Tomato mottle viruses* in developmental stages of the whitefly vector, Bemisia tabaci. Third International Geminivirus Symposium, 24–28 July 2001, Norwich, England, p. 81.

Pospieszny, H., Borodynko, N., Obrepalska-Steplowska, A., and Hasiow, B. (2007). The first report of tomato torrado virus in Poland. *Plant Dis.* **91:**1364.

Pospieszny, H., Hasiów, B., and Borodyndo, N. (2008). Characterization of two distinct Polish isolates of *Pepino mosaic virus*. *Eur. J. Plant Pathol.* **122:**443–445.

Prins, M., and Goldbach, R. (1998). The emerging problem of Tospovirus infection and nonconventional methods of control. *Trends Microbiol.* **6:**31–35.

Quacquarelli, A., and Gallitelli, D. (1979). Tre virosi del geranio in Puglia. *Phytopathol. Mediterr.* **19:**61–70.

Rigden, J. E., Krake, L. R., Rezaian, M. A., and Dry, I. B. (1994). ORF C4 of tomato leaf curl geminivirus is a determinant of symptom severity. *Virology* **204:**847–850.

Roggero, P., Masenga, V., Lenzi, R., Coghe, F., Ena, S., and Winter, S. (2001). First report of *Pepino mosaic virus* in tomato in Italy. *Plant Pathol.* **50:**798.

Rojas, M. R., and Gilbertson, R. L. (2008). Emerging plant viruses: A diversity of mechanisms and opportunities. *In* "Plant Virus Evolution" (M. J. Roossicnk, ed.), pp. 27–51. Springer-Verlag, Berlin/Heidelberg.

Rojas, M. R., Jiang, H., Salati, R., Xoconostle-Cázares, B., Sudarshana, M. R., Lucas, W. J., and Gilbertson, R. L. (2001). Functional analysis of proteins involved in movement of the monopartite begomovirus, *Tomato yellow leaf curl virus*. *Virology* **291:**110–125.

Roselló, S., Díez, M. J., and Nuez, F. (1996). Viral disease causing the greatest economic losses to the tomato crop. I. The Tomato spotted wilt virus—A review. *Sci. Hortic.* **67:**117–150.

Samuel, G., Bald, J. G., and Pitman, H. A. (1930). Spotted wilt of tomatoes. *J. Aust. Counc. Sci. Ind. Res.* **44:**8–11.

Sançafon, H., Wellink, J., Le Gall, O., Karasev, A., van der Vlugt, R., and Wetzel, T. (2009). Secoviridae: A proposed family of plant viruses within the order Picornavirales that combines the families Sequiviridae and Comoviridae, the unassigned genera Cheravirus and Sadwavirus, and the proposed genus Torradovirus. *Arch. Virol.* **154:**899–907.

Schwarz, D., Paschek, U., Bandte, M., Büttner, C., and Obermeier, C. (2009). Detection, spread, and interactions of *Pepino mosaic virus* and *Pythium aphanidermatum* in the root environment of tomato in hydroponics. *Acta Hort.* **808:**163–170.

Sclavounos, A. P., Voloudakis, A. E., Arabatzis, C. H., and Kyriakopoulou, P. E. (2006). A severe Hellenic CMV tomato isolate: Symptom variability in tobacco, characterization and discrimination of variants. *Eur. J. Plant Pathol.* **115:**163–172.

Segev, L., Wintermantel, W. M., Polston, J. E., and Lapidot, M. (2004). First report of tomato chlorosis virus in Israel. *Plant Dis.* **88:**1160.

Shipp, J. L., Buitenhuis, R., Stobbs, L., Wang, K., Kim, W. S., and Ferguson, G. (2008). Vectoring of *Pepino mosaic virus* by bumble-bees in tomato greenhouses. *Ann. Appl. Biol.* **53:**149–155.

Sinisterra, X. H., McKenzie, C. L., Hunter, W. B., Powel, C. A., and Shatters, R. G. (2005). Differential transcriptional activity of plant-pathogenic begomoviruses in their whitefly vector (*Bemisia tabaci*, Gennadius; *Hemiptera*: *Aleyrodidae*). *J. Gen. Virol.* **86:**1525–1532.

Soler-Aleixandre, S., Lopez, C., Cebolla-Cornejo, J., and Nuez, F. (2007). Sources of resistance to *Pepino mosaic virus* (PepMV) in tomato. *HortScience* **42:**40–45.

Spence, N. J., Basham, J., Mumford, R. A., Hayman, G., Edmondson, R., and Jones, D. R. (2006). Effect of *Pepino mosaic virus* on the yield and quality of glasshouse-grown tomatoes in the UK. *Plant Pathol.* **55:**595–606.

Stevens, J. M. (1964). Tomato breeding.. Project report W-Vv1Department of Agricultural Technical Services. Republic of South Africa..

Sunter, G., Hartitz, M. D., Hormuzdi, S. G., Brough, C. L., and Bisaro, D. M. (1990). Genetic analysis of tomato golden mosaic virus: ORF AL2 is required for coat protein accumulation while ORF AL3 is necessary for efficient DNA replication. *Virology* **179:**69–77.

Tomassoli, L., Ilarda, V., Barba, M., and Kaniewski, W. (1999). Resistance of transgenic tomato to Cucumber mosaic cucumovirus under field conditions. *Mol. Breed.* **5:**121–130.

Turina, M., Ricker, M. D., Lenzi, R., Masenga, V., and Ciuffo, M. (2007). A severe disease of tomato in the Culiacan area (Sinaloa, Mexico) is caused by a new picorna-like viral species. *Plant Dis.* **91:**932–941.

van der Vlugt, R. A. A., Stijger, C. C. M. M., Verhoeven, J. T. J., and Lesemann, D. E. (2000). First report of *Pepino mosaic virus* on tomato. *Plant Dis.* **84:**103.

Varia, A. M., Accotto, G. P., Vecchiati, M., and Bragaloni, M. (2002). Tomato infectious chlorosis virus causes leaf yellowing and reddening of tomato in Italy. *Phytoparasitica* **30:**290–294.

Velasco, L., Simon, B., Janssen, D., and Cenis, J. L. (2008). Incidences and progression of tomato chlorosis virus disease and tomato yellow leaf curl virus disease in tomato under different greenhouse covers in southeast Spain. *Ann. Appl. Biol.* **153:**335–344.

Verbeek, M., Dullemans, A. M., van den Heuvel, J. F. J. M., Maris, P. C., and van der Vlugt, R. A. A. (2007). Identification and characterisation of tomato torrado virus, a new plant picorna-like virus from tomato. *Arch. Virol.* **152:**881–890.

Verbeek, M., Dullemans, A. M., van den Heuvel, J. F. J. M., Maris, P. C., and van der Vlugt, R. A. A. (2008). Tomato marchitez virus, a new plant picorna-like virus from tomato related to tomato torrado virus. *Arch. Virol.* **153:**127–134.

Verbeek, M., Dullemans, A., van den Heuvel, H., Maris, P., and van der Vlugt, R. (2010). Tomato chocolàte virus: A new plant virus infecting tomato and a proposed member of the genus *Torradovirus*. *Arch. Virol.* **155:**751–755.

Verdin, E., Gognalons, P., Wipf-Scheibel, C., Bornard, I., Ridray, G., Schoen, L., and Lecoq, H. (2009). First report of Tomato torrado virus in tomato crops in France. *Plant Dis.* **39:**1352.

Verhoeven, J. T. J., van der Vlugt, R., and Roenhorst, J. W. (2003). High similarity between tomato isolates of *Pepino mosaic virus* suggests a common origin. *Eur. J. Plant Pathol.* **109:**419–425.

Vovlas, C., Gallitelli, D., and Conti, M. (1989). Preliminary evidence for an unusual mode of transmission in the ecology of *Pelargonium zonate spot virus* (PZSV). 4th Plant Virus Epidemiology Workshop, Montpellier, France, pp. 302–305.

Wartig, L., Kheyr-Pour, A., Noris, E., De Kouchkovsky, F., Jouanneau, F., Gronenborn, B., and Jupin, I. (1997). Genetic analysis of the monopartite tomato yellow leaf curl geminivirus: Roles of V1, V2, and C2 ORFs in viral pathogenesis. *Virology* **228:**132–140.

Whitfield, A. E., Ullman, D. E., and German, T. L. (2005). Tospovirus-thrips interactions. *Annu. Rev. Phytopathol.* **43:**459–489.

Wintermantel, W. M. (2004). Emergence of greenhouse whitefly (*Trialeurodes vaporariorum*) transmitted criniviruses as threats to vegetable and fruit production in North America. *APSnet Features.* Online. doi:10.1094/APSnetFeature-2004-0604.

Wintermantel, W. M., and Hladky, L. (2010). Methods for detection and differentiation of existing and new crinivirus species through multiplex and degenerate primer RT-PCR. *J. Virol. Methods* **170:**106–114.

Wintermantel, W. M., and Wisler, G. C. (2006). Vector specificity, host range and genetic diversity of *Tomato chlorosis virus*. *Plant Dis.* **90:**814–819.

Wintermantel, W. M., Wisler, G. C., Anchieta, A. G., Liu, H.-Y., Karasev, A. V., and Tzanetakis, I. E. (2005). The complete nucleotide sequence and genome organization of tomato chlorosis virus. *Arch. Virol.* **150:**2287–2298.

Wisler, G. C., Liu, H.-Y., Klaassen, V. A., Duffus, J. E., and Falk, B. W. (1996). Tomato infectious chlorosis virus has a bipartite genome and induces phloem-limited inclusions characteristic of the closteroviruses. *Phytopathology* **86:**622–626.

Wisler, G. C., Dufus, J. E., Liu, H.-Y., and Li, R. H. (1998a). Ecology and epidemiology of whitefly-transmitted closteroviruses. *Plant Dis.* **82:**270–280.

Wisler, G. C., Li, R. H., Liu, H.-Y., Lowry, D. S., and Duffus, J. E. (1998b). Tomato chlorosis virus: A new whitefly-transmitted, phloem-limited bipartite closterovirus of tomato. *Phytopathology* **88:**402–409.

Zrachya, A., Glick, E., Levy, Y., Arazi, T., Citovsky, V., and Gafni, Y. (2007). Suppressor of RNA silencing encoded by Tomato yellow leaf curl virus-Israel. *Virology* **358:**159–165.

CHAPTER 3

Viruses of Cucurbit Crops in the Mediterranean Region: An Ever-Changing Picture

Hervé Lecoq and **Cécile Desbiez**

Contents

Abstract

Cucurbit crops may be affected by at least 28 different viruses in the Mediterranean basin. Some of these viruses are widely

INRA, UR407 Pathologie Végétale, Domaine Saint Maurice, Montfavet, France

Advances in Virus Research, Volume 84
ISSN 0065-3527, DOI: 10.1016/B978-0-12-394314-9.00003-8

distributed and cause severe yield losses while others are restricted to limited areas or specific crops, and have only a negligible economic impact. A striking feature of cucurbit viruses in the Mediterranean basin is their always increasing diversity. Indeed, new viruses are regularly isolated and over the past 35 years one "new" cucurbit virus has been reported on average every 2 years. Among these "new" viruses some were already reported in other parts of the world, but others such as *Zucchini yellow mosaic virus* (ZYMV), one of the most severe cucurbit viruses and *Cucurbit aphid-borne yellows virus* (CABYV), one of the most prevalent cucurbit viruses, were first described in the Mediterranean area. Why this region may be a potential "hot-spot" for cucurbit virus diversity is not fully known. This could be related to the diversity of cropping practices, of cultivar types but also to the important commercial exchanges that always prevailed in this part of the world. This chapter describes the major cucurbit viruses occurring in the Mediterranean basin, discusses factors involved in their emergence and presents options for developing sustainable control strategies.

I. INTRODUCTION

Four major cucurbit species are cultivated in the Mediterranean region, cucumber (*Cucumis sativus*), melon (*Cucumis melo*), watermelon (*Citrullus lanatus*), and different squash species (*Cucurbita pepo*, *C. moschata*, and *C. maxima*). These crops are cultivated as a great diversity of types and cultivars adapted to local consumers' demands. The botanical centers of origin of melon and watermelon are thought to be in the Sudano-Sahelian region of Africa, cucumber was probably domesticated in the foothills of the Himalayas and squashes in Central and South America (Robinson and Deckers-Walters, 1997). Melon and watermelon were cultivated by the ancient Egyptians several centuries BC, while cucumber was probably unknown in the Mediterranean region until a few centuries AD. Squashes were not known in the Mediterranean basin before Columbus and are therefore recent introductions in the region (see Chapter 1).

Cucurbits are eaten as mature or immature fruits, fresh or cooked and are an important part of many typical Mediterranean dishes. Seeds may be used as snacks or for oil production. Some cucurbits are grown for utilitarian uses (vegetable sponge) or as ornamentals (gourds and mini-squashes) (Robinson and Deckers-Walters, 1997).

Cucurbits are among the major vegetables produced in the Mediterranean basin (more that 23 million tons in 2009; FAOstat, 2009), ranking third just after tomatoes and potatoes. Turkey, Egypt, Spain, Morocco, and Italy are among the top 10 producers of cucumber, melon, watermelon, or squashes in the world (FAOstat, 2009).

Cucurbits worldwide are affected by at least 59-well characterized viruses belonging to the major plant virus groups (Brunt *et al.*, 1996; Fauquet *et al.*, 2005; Lecoq, 2003; Lovisolo, 1980; Provvidenti, 1986, 1996). In the Mediterranean region, 28 different viruses have been reported in cultivated cucurbits. This virus diversity probably results from the genetic and ecological diversity of their hosts. Indeed, cucurbits are grown in a variety of agroecosystems ranging from highly sophisticated soil-less cucumber production in heated glasshouses to more traditional watermelon rainfed cultivation. These various environments may provide more or less favorable conditions for specific viruses or for their vectors. However, these 28 different viruses do not have similar prevalence or economical impact. Only four viruses have been reported in more than half of the 26 Mediterranean countries (namely *Watermelon mosaic virus* (WMV), *Zucchini yellow mosaic virus* (ZYMV), *Cucumber mosaic virus* (CMV) and *Cucurbit aphid-borne yellows virus* (CABYV)). About 10 additional viruses are associated with severe losses locally or at a regional level. Some of these viruses infect only one or few cucurbit species, but most of them are found in all cucurbit crops.

In addition to plant growth reduction, typical viral symptoms in cucurbits fall within three major categories (Blancard *et al.*, 1994):

- Mosaics on leaves sometimes associated with leaf size reduction, blisters, lacinations, or enations; fruits may develop a range of discolorations and deformations altering their quality.
- Yellowings of older and mature leaves; fruit production may be reduced but fruit quality is generally not affected.
- Necrosis either as necrotic spots on leaves or as generalized necrosis or wilt. Fruits may also develop necrotic symptoms.

Fruit alteration caused by virus diseases may have a severe economic impact particularly when the production is for wholesalers or for export markets which generally require only top-quality products.

Cucurbit viruses in the Mediterranean basin constitute complex and changing pathosystems. In a comprehensive review on cucurbit viruses, Lovisolo listed in 1977, 11 viruses infecting naturally cultivated cucurbits in the Mediterranean basin, 3 being of economic importance: CMV, WMV, and *Squash mosaic virus* (SqMV; Lovisolo, 1977). Since then, at least 17 "new" viruses were described infecting cucurbits in this region. Some of them are already widespread and may cause major yield reductions: they constitute new challenges for plant virologists and breeders for developing reliable diagnostic tools and durable control strategies. Other "new" viruses remain restricted to limited geographical regions or to specific cropping systems and may remain—at least for a while—of minor importance.

Among these "new" viruses, some have been probably present in cultivated cucurbits for a long time. They may have remained unnoticed because of the lack of appropriate diagnostic method or because their symptoms were previously associated to other causes. This applies probably to CABYV, a widespread polerovirus causing yellowing symptoms on older leaves that were previously associated to nutritional or physiological disorders. CABYV was first identified in 1988 in France, but it was later found to be already very common in samples that had been collected in 1982 (Lecoq *et al.*, 1992). Now, CABYV is widely spread in the whole Mediterranean region.

Other "new" viruses may represent typical "emerging" viruses. Some became suddenly widely spread because they followed the rapid increase and the dissemination of their natural vector. Typical of this situation would be the crinivirus *Cucurbit yellow stunting disorder virus* (CYSDV) transmitted by the whitefly *Bemisia tabaci* (Wisler *et al.*, 1998). The situation of ZYMV is probably unique in cucurbits (Desbiez and Lecoq, 1997). This potyvirus, transmitted by many aphid species in a nonpersistent manner, was described in the mid-1970s as occurring locally in Northern Italy. Within a decade ZYMV spread to all the major cucurbit growing areas in the world. The rapid spread of ZYMV occurred by means which are not yet fully established but this could be through seed transmission (Simmons *et al.*, 2011). Whatsoever, it cannot be related to an extension of the distribution area of its aphid vectors which were already prevalent in the Mediterranean basin.

Often, in the fields, more than one virus infects plants, leading to a combination of symptoms and occasionally to synergism, making virus identification very difficult based on symptoms only. The complex viral pathosystems affecting cucurbit crops in the Mediterranean basin has been better characterized in recent years thanks to the improvement of diagnostic methods. The development of specific serological tests, including the double-antibody sandwich enzyme-linked immunosorbent assay (DAS-ELISA; Clark and Adams, 1977) has greatly simplified virus diagnosis. The availability of commercial diagnostic kits has allowed extensive virus surveys in cucurbits in many of the Mediterranean countries allowing a better appreciation of the prevalence and impact of the different virus diseases. More recently, molecular methods (including reverse transcription polymerase chain reaction (RT-PCR) followed by sequencing) have provided more insight in taxonomy, virus variability, and virus population structure.

II. APHID-BORNE VIRUSES

A. Potyviruses

Several potyvirus species cause severe economical losses in cucurbit crops in the Mediterranean region (Table I). Mosaic diseases of cucurbit crops caused by potyviruses were first reported in the 1920s in the United

TABLE I Geographical distribution of potyviruses infecting cultivated cucurbits

Virus	Distribution
Algerian watermelon mosaic virus, AWMV	Algeria
Clover yellow vein virus, ClYVV	Worldwide (occasional)
Melon vein-banding mosaic virus, MVBMV	Taiwan, China
Papaya ringspot virus, PRSV	Worldwide (Tropical, Mediterranean)
Telfairia mosaic virus, TeMV	Nigeria
Turnip mosaic virus, TuMV	Worldwide (occasional)
Moroccan watermelon mosaic virus, MWMV	Africa, Mediterranean Basin
Watermelon leaf mottle virus, WLMV	Florida
Watermelon mosaic virus, WMV	Worldwide (Temperate, Mediterranean)
Zucchini yellow fleck virus, ZYFV	Mediterranean Basin
Zucchini yellow mosaic virus, ZYMV	Worldwide

States, but the early literature contains a diversity of names for viruses or virus diseases that were only partially characterized. Webb and Scott (1965) differentiated two groups of WMV isolates, WMV1 and WMV2, based on crossprotection experiments, serological relationships, and host range reactions. In 1979, Purcifull and Hiebert clarified the situation by demonstrating that WMV1 and WMV2 were serologically distinct and that WMV1 was closely related to *Papaya ringspot virus* (PRSV; Purcifull and Hiebert, 1979). Now, WMV1 is considered as the W strain of PRSV, while WMV2 is referred to as *Watermelon mosaic virus* (WMV). The same authors showed that a WMV isolate from Morocco was a third serological entity which is now considered as a distinct species, *Moroccan watermelon mosaic virus* (MWMV; Yakoubi *et al.*, 2008b).

So, from the initial WMV complex emerged three different potyvirus species: WMV, PRSV type W, and MWMV. At least eight additional potyvirus species may infect cucurbit crops (Table I), but among them only ZYMV has been reported so far to be widely distributed in the Mediterranean basin.

Potyviruses are single-stranded, positive-sense RNA viruses. Their genome of about 9–11 kb is encapsidated in a flexuous particle of circa 750 nm long and 12 nm wide. They have a polyadenylated (polyA) tail at their 3′ extremity and a viral protein (VPg) linked at the 5′ extremity of the genome. The genome is translated as a single polyprotein of about 3000 amino acids, cleaved by autocatalytic activity into 10 functional proteins. The two proteins in the N-terminal part of the polyprotein are cleaved by *cis*-acting protease activity, whereas the other proteins are released by the *cis*- and *trans*-acting protease NIa-Pro. Protein P1 is the most variable in size and sequence among potyviruses. Besides its protease activity, its functions are but partially known. The helper component-protease (HC-Pro) is a multifunctional protein, required for aphid transmission, but also involved in virus movement in the plant and in the inhibition of the plant antiviral defenses mediated by posttranscriptional gene silencing (PTGS; Kasschau and Carrington, 2001). The functions of protein P3 are poorly known. It is involved in replication and plays a role in symptomatology. Protein CI is an RNA helicase required for virus replication. It forms pinwheel-like cytoplasmic inclusions in infected plant cells that constitute a hallmark of the *Potyviridae* family. Protein NIa is self-cleaved by the protease encoded in its C-terminal part to release VPg. The VPg binds covalently the 5′ extremity of the genome and is supposed to act like the cap of messenger RNAs. Direct interaction between the VPg and eukaryotic initiation factors eIF4E or eIF(iso)4E (Léonard *et al.*, 2000) is required for potyvirus infection, although it is not known whether the replication or translation processes are involved. Protein NIb is the viral RNA-dependent RNA-polymerase (RdRp). The coat protein (CP) of approximately 34–36 kDa is encoded at the C-terminal extremity of the polyprotein and is the structural protein required for encapsidation of the virions. It is necessary for virus movement (Dolja *et al.*, 1995) and for aphid transmission (Atreya *et al.*, 1991). Recently, an unsuspected 11th protein (PIPO) resulting from a translational frameshift in the region coding for the P3 protein, has been shown to be present in potyviruses (Chung *et al.*, 2008). It is involved at least in movement (Wei *et al.*, 2010; Wen and Hajimorad, 2010).

Potyviruses are efficiently transmitted in a nonpersistent manner by many different aphid species. Acquisition and transmission may occur during very short probes (less than a minute) and aphids retain infectivity for a few hours. Transmission probably relies on specific interactions between HC-Pro and the distal part of the aphid stylet on one side and HC-Pro and CP on the other side (Blanc *et al.*, 1997). HC-Pro preparations from WMV, PRSV, or ZYMV infected plants can mediate transmission of the three viruses although with different efficiencies, suggesting some level of specificity in this interaction (Lecoq and Pitrat, 1985). In laboratory conditions, cucurbit potyviruses are mechanically transmitted, and can be easily purified according to the method of Lecoq and Pitrat (1985).

Diagnosis for a specific potyvirus is generally based on serological methods (ELISA, Lateral flow) now commercially available. A potyvirus generic test using a monoclonal antibody was developed but some PRSV isolates escape detection (Jordan, 1992). Generic potyvirid RT-PCR primers permit the amplification of the 3′-nontranslated region, the CP gene and the 3′-part of the polymerase gene of all potyviruses (Gibbs and Mackenzie, 1997; Table II). Observation of viral inclusions in light microscopy is a reliable and cheap generic diagnostic method since all potyviruses induce typical fibrous inclusions with the Orange-Green staining technique (Christie and Edwardson, 1986).

1. Watermelon mosaic virus

The first complete genomic sequence of a WMV isolate was established less than 10 years ago (Desbiez and Lecoq, 2004). It revealed that although WMV is closely related molecularly to *Soybean mosaic virus* (SMV), a legume-infecting potyvirus, the 5′ half of its P1-coding region is more similar to *Bean common mosaic virus* (BCMV) and WMV probably arose through interspecific recombination between a SMV-like and a BCMV-like virus. By now, 31 complete sequences of WMV are available, including 24 from the Mediterranean basin, and they all present the same characteristic in the 5′ part of the genome, indicating that the recombination event is ancestral in the evolution of WMV. However, an isolate of SMV from China was recently found to have the same BCMV–SMV recombination in the P1 as WMV, without infecting cucurbits (Yang *et al.*, 2011), indicating that the recombination alone was not sufficient to explain the change in biological properties between SMV and WMV.

i. Symptoms, host range, and transmission WMV induces a diversity of symptoms according to the isolate and the host cultivar. On leaves, symptoms are mosaic, vein banding, more or less severe leaf deformations including blisters, filiformism, and important leaf size reduction. On fruits, severe discoloration and slight deformation are observed in some cultivar and with some isolates (Fig. 1A). Isolates causing necrosis on grafted watermelons were reported from Italy (Crescenzi *et al.*, 2001).

WMV has a relatively wide experimental and natural host range for a potyvirus. It infects over 170 species in 26 mono- or dicotyledonous families. Besides cucurbits, WMV can infect in natural conditions pea, carrot, and orchids (vanilla, *Habenaria radiata*) (Lecoq *et al.*, 2011; Parry and Persley, 2005). WMV also infects many weeds that can serve as alternative hosts, but generally, naturally infected weeds do not present evident symptoms of viral infection.

WMV is transmitted on the nonpersistent mode by at least 35 aphid species in 19 genera. *Aphis craccivora*, *Aphis gossypii*, and *Myzus persicae* are efficient WMV vectors (Lecoq and Desbiez, 2008).

WMV has not been reported to be seed transmitted.

ii. Distribution in the Mediterranean region In the Mediterranean region, WMV was first reported in Israel in 1963 by Cohen and Nitzany, under the name of melon mosaic virus (Cohen and Nitzany, 1963), and was successively identified in Yugoslavia (1967), Egypt (1969), Spain and Italy (1973), Tunisia (1975), and France (1976) (Lovisolo, 1977). Presently, WMV

FIGURE 1 Symptoms caused by some cucurbit viruses. (A) mosaic on a leaf and fruit of a melon plant infected by *Watermelon mosaic virus* (WMV); (B) Severe mosaic and deformation on leaves and fruits of a zucchini squash plant infected by *Zucchini yellow mosaic virus* (ZYMV); (C) Mosaic and deformation on leaves and young fruit of a zucchini squash plant infected by *Cucumber mosaic virus* (CMV); (D) Yellowing of older leaves of a melon plant infected by *Cucurbit aphid-borne yellows virus* (CABYV); (E) Vein clearing on a leaf of a cucumber plant infected by *Cucumber vein yellowing virus* (CVYV); (F) Severe yellow mosaic on a watermelon plant infected by WmCSV; (G) Necrotic spots on a leaf of a cucumber plant infected by *Melon necrotic spot virus* (MNSV); and (H) Vein banding on a leaf of a melon plant infected by *Squash mosaic virus* (SqMV). (See Page 2 in Color Section at the back of the book.)

has been reported in most Mediterranean countries and it is one of the most prevalent cucurbit viruses in the region (CABI/EPPO, 2003a).

*iii. **Biological and genetic variability*** Some biological variability has been reported for WMV. This concerns mainly differences in symptom intensity, some isolates causing very severe symptoms while others are mild. Also, slight differences in host range or aphid transmission efficiencies have been reported (Lecoq *et al.*, 2011).

Three molecular groups, namely G1 to G3, have been defined in the world, when comparing amino acid sequences of the CP of WMV isolates (Desbiez *et al.*, 2007a; Fig. 2). In the Mediterranean basin, until 2000 only G1 and G2 were observed, in France (Desbiez *et al.*, 2007a, 2009), Spain (Moreno *et al.*, 2004), Algeria, Israel, Italy, Morocco, Tunisia, and Turkey (Desbiez *et al.*, 2007a). When the information was available, most isolates were actually intraspecific recombinants between groups G1 and G2 (Desbiez and Lecoq, 2008; Moreno *et al.*, 2004). The third group of isolates (G3 or "EM" for emerging), probably originating from Asia, was first observed in South-eastern France in 2000, in relation with more severe symptoms on zucchini squash. Four subgroups of EM isolates were characterized, probably corresponding to different introduction events (Desbiez *et al.*, 2009). Recombinants between G1 and EM isolates, or between different subgroups of EM isolates, were detected in France in the few years following the introduction of EM isolates, but they did not seem to persist in the environment (Desbiez *et al.*, 2011). Preliminary data indicate that EM strains are now also present in Italy (Donato Gallitelli, unpublished), Spain (Miguel Aranda, unpublished) and Turkey (Muharrem Kamberoglu, unpublished), but their population dynamics is not known.

*iv. **Epidemiology*** Temporal pattern of WMV spread in melon or zucchini squash has been followed every year since 1981 at Montfavet (France). WMV spread rapidly every year in both crops reaching total infection 4–6 weeks after planting. Disease progress curves were S shaped and fit generally better the logistic model (Lecoq, 1992b; Lecoq *et al.*, 2005). Similar studies conducted in Spain showed that the WMV disease progress curves best fitted the Gompertz model (Alonso-Prados *et al.*, 2003). WMV infects many weeds (*Capsella bursa-pastoris*, *Senecio vulgaris*, *Lamium amplexicaule*) that can persist in winter, as well as a few winter crops (spinach) which can be virus reservoirs between two crop cycles (Lecoq, 1992a). A negative effect of minimum winter temperatures was noticed on epidemics, suggesting that WMV reservoirs or vectors may be cold sensitive (Alonso-Prados *et al.*, 2003).

In France, WMV is currently undergoing a change in population revealing a rapid replacement of local G1 (or "classic," CL) strains by recently introduced G3 (EM) ones. Surveys and population genetic

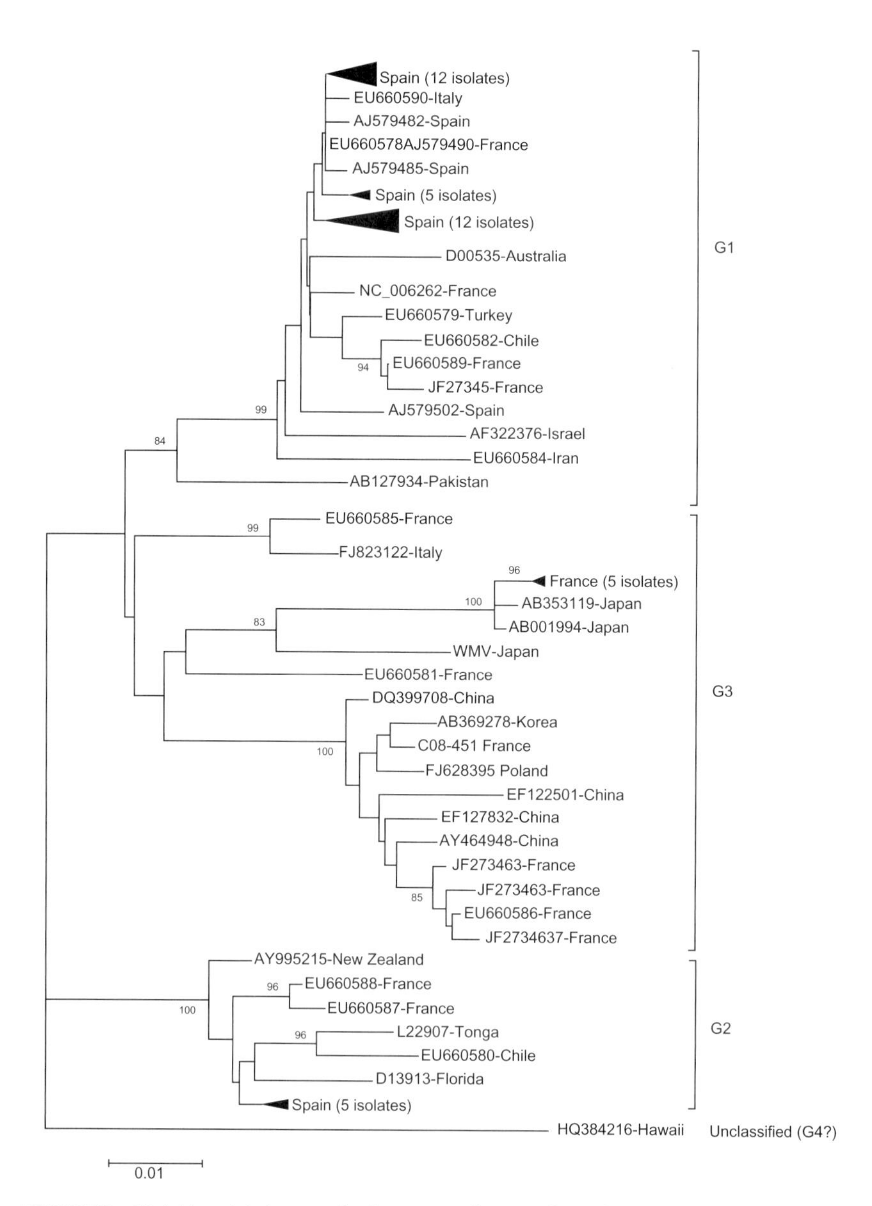

FIGURE 2 Neighbor-joining tree built on complete nucleotide sequence of *Watermelon mosaic virus* coat protein and indicating the main molecular clusters of isolates G1 to G3. Bootstrap support (500 bootstraps) above 80% are indicated for each node. The scale bar represents a genetic distance of 0.01.

approaches were used to study the structure of WMV populations in South-eastern France. These studies showed that WMV EM strains did not spread over long distances, but rapidly replaced preexisting CL strains in sites were both occurred (Desbiez *et al.*, 2009; Joannon *et al.*, 2010). Spatial spread of CL and EM strains was studied in zucchini squash fields. Analyzing and modeling these data identified an asymmetrical crossprotection (i.e., EM strains crossprotect better against CL than CL against EM) as a potential factor explaining the shift in WMV populations (Fabre *et al.*, 2010). A more classical approach investigated whether differences in host ranges or aphid transmissibility could be involved in this change. EM strains were significantly better transmitted from mixed CL–EM infection than CL strains, providing another potential epidemiological advantage to EM (Lecoq *et al.*, 2011).

v. Diagnostic The confusion that was prevalent in the early descriptions of watermelon mosaic diseases was mainly due to the convergence of symptoms caused by WMV, PRSV, and MWMV in cucurbits and to the lack of proper diagnostic tools. Symptomatology and virus particle morphology were clearly insufficient to differentiate these viruses. The production of specific polyclonal antisera, and the development of simple serological tests, such as the sodium dodecyl sulfate immunodiffusion test (SDS-ID; Purcifull and Hiebert, 1979) and DAS-ELISA brought a major contribution to the proper diagnosis of WMV. DAS-ELISA commercial kits are now available for WMV from several suppliers.

Monoclonal antibodies (MAbs) proved to be very useful to differentiate WMV subgroups and to study cross protection between WMV isolates (Desbiez *et al.*, 2007a; Lecoq *et al.*, 2011).

Many complete and partial nucleotide sequences are now available for WMV, particularly in the CP-coding region and specific primers have been designed (Table II). However, for routine testing and surveys, DAS-ELISA seems to be more reliable than RT-PCR (Lecoq and Desbiez, unpublished).

vi. Control

a. Prophylactic measures Prophylactic measures are intended to prevent or limit the contact of viruliferous aphids with cultivated plants. They are not specific for a particular virus and are generally efficient for all aphid-borne viruses. These include careful weeding near plantings and avoiding overlapping crops in the same area to reduce virus and aphid sources near new plantings. Plastic mulches have a repelling action on aphids and significantly delay WMV spread. However, they confer only a temporary protection that is limited to the early stages of the crop, because their efficiency decreases when the plant growth covers their surface. Row covers of different types (unwoven, perforated plastics, . . .) can also be used; they physically prevent winged aphids from reaching the plants, but they must be removed to allow insect pollination necessary for cucurbit fruit production (Lecoq, 1992b; Walters, 2003). Both methods have a

TABLE II Generic or specific primers for PCR detection of major cucurbit-infecting viruses present in the Mediterranean Basin

	Virus	Name	Sequence	Genome region	Approximate size (nt)	Reference
Generic primers	Potyvirids	Poty 5′	CCACGGATCCGGBAAYAAYAGYGGDCARCC	Nib-CP-3′UTR	1600–2100	Gibbs and Mackenzie (1997)
		Poty(dT)3′	CACGGATCCTTTTTTTTTTTTTTTTTV			
		HPFor	TGYGAYAAYCARYTIGAYIIIAAYG	HC-Pro	700	Ha *et al.* (2008)
		HPRev	GAICCRWAIGARTCIAIIACRTG			
	Poleroviruses	Pol-G-F	GAYTGCTCYGGYTTYGACTGGAG	MP-CP	1200	Knierim *et al.* (2010)
		Pol-G-R	GATYTTATAYTCATGGTAGGCCTTGAG			
	Tobamoviruses	Tobamo-1	TGATHAARMGDAAYWTBAAYDCDCC	RdRp	900	Gibbs *et al.* (1998)
		Tobamo-2	TTBGCYTCRAARTTCCA			
	begomoviruses	GemCP-V-5′	GCCYATRTAYAGRAAGCCMAG	CP (DNA-A)	600	Derived from Wyatt and Brown (1996)
		GemCP-C-3′	GGRTTNGANGCRTGHGTACAYG			
		PLB1v2040	GCCTCTGCAGCARTGRTCKATCTTCATACA	DNA-B	600	Rojas *et al.* (1993)
		PCRc154	GGTAATATTATAHCGGATGG			
	rhabdoviruses	FinDeg1	ACACCIRTHGGICCWTGGYTRCC	Glycoprotein	400	Katis *et al.* (2011)
		FinDeg2	CSGCCATCKGRGGGAACCA			
Specific primers	ZYMV	ZYMV-CP-5′	GGTTCATGTCCCACCAAGC	Nib-CP	600	Yakoubi *et al.* (2008a)
		ZYMV-CP-3′	ATGTCGAGTATCACATTTCC			
	WMV	WMV-5′	GGCTTCTGAGCAAAGATG	Nib-CP	400	Desbiez *et al.* (2009)
		WMV-3′	CCCAYCAACTGTYGGAAG			
	PRSV	PRSV-milCP-5′	TCTAACACTCGTGCCACTCA	CP	600	Cécile Désbiez and Hervé Lecoq *et al.* (unpublished)
		PRSV-finCP-3′	YARTTGCGCATACCCAGGAG			
	MWMV	MWMV-5′	AGCAAGCGCCATACTCTGA	Nib-CP	600	Yakoubi *et al.* (2008b)

	MWMV-3′	CAAACTCCATTAACATTCGG			
CMV	93–309	CATCGACCATGGACAAATCTGAATCAAC	CP	750	Hu *et al*. (1995)
	93–359	CTCTCCATGGCGTTTAGTGACTTCAGCAG			
CVYV	CV(+)	AGCTAGCGCGTATGGGGTGAC	CP	450	Cuadrado *et al*. (2001)
	CV(-)	GCGCCGCAAGTGCAAATAAAT			
CYSDV	CYSDV-CP-5′	ATGGCGAGTTCGAGTGAGAA	CP	700	Desbiez *et al*. (2003b)
	CYSDV-CP-3′	TCAATTACCACAGCCACCTG			
BPYV	BPYV-CP-5′	CTGACATATGGGAGATAATGATGATGG	CP	700	Yakoubi *et al*. (2007)
	BPYV-CP-5′	CTGACTCGAGTCAGTTTCCATAAGAAGC			
MNSV	MNSV-CP-5′	GTGAAGCTTGCTAARCAGGC	CP	700	Yakoubi *et al*. (2008d)
	MNSV-CP-3′	ACRTARAGATCACCRTGGGC			
SqMV	SqMV-CP-F2	AAGGCAACCATyGCTTTC	CP	420	Ling *et al*. (2011)
	SqMV-R1	GGGCTGTACTTTCTAAGGG			
CGMMV	CGMMV-F	GCTTACAATCCGATCACACCT	CP	500	Antignus *et al*. (2001)
	CGMMV-R	CTTTCGAGTGGTAGCCTCTGA			
EMDV	EMDVGup	GATCTGAGGGAACCATTTTGAGC	Glycoprotein	300	Katis *et al*. (2011)
	EMDVGdo	TCCCTTTATCTTACTGTGCGAAC			
CABYV	CABYV-CE9	GAATACGGTCGCGGCTAGAAATC	CP	600	Kassem *et al*. (2007)
	CABYV-CE10	CTATTTCGGGTTCTGGACCTGGC			
WCSV	WCSV-5′	GCCCATGTATCGAAAGCCCAG	CP	600	Desbiez and Lecoq (unpublished)
	WCSV-3′	GGATTCGATGCGTGTGTACACG			
SLCV	pEBV2359	ACACAATGGTCTATCTTCAT	DNA-B	450	Al-Musa *et al*. (2008)
	pEBC113	CCTAGAGGCTTTATGTGTCTTGAC			

Y=C or T; R=A or G; M=A or C; K=G or T; S=C or G; W=A or T; B=C, G, or T; D=A, G, or T; H=A, C, or T; V=A, C, or G; N=A, C, G, or T; I, deoxyinosine.

major drawback: they require a lot of plastic material that farmers must dispose in an ecologically sound way after the crop cycle. Insecticide applications have been generally found inefficient in limiting the spread of WMV. This can be related to the large number of winged aphids that land on the plants and to the rapidity of the transmission process. Oil applications can delay virus spread when inoculum pressure is moderate.

b. Resistant cultivars The use of virus-resistant cultivars is probably the easiest and cheapest way to control plant viral diseases at the farmer's level. Considerable efforts have been made to look for resistance to WMV in genetic resources but unfortunately with only limited success. Some WMV-resistant or -tolerant commercial cultivars are now available in cucumber and zucchini squash. In melon, a resistance to WMV transmission by *A. gossypii* was found to be governed by the single dominant gene *Vat* (Pitrat and Lecoq, 1980). This gene is now present in many commercial cultivars, but confers only limited protection in the fields, probably because WMV is also transmitted by many other aphid species in natural conditions. Partial resistances have been observed in melon accessions, but they do not provide a complete protection and have not yet been introduced into commercial cultivars (Diaz-Pendon *et al.*, 2005). One difficulty in breeding partially resistant cultivars is that there is no correlation between early tests (i.e., symptom intensity on young plants or virus multiplication rates estimated by DAS-ELISA) and the symptom intensity on mature fruit.

In the last two decades, attempts were made to obtain WMV transgenic-resistant plants using the pathogen-derived resistance approach. Different constructs were tested to obtain resistant WMV plants and the best results were obtained with the full-length *CP* gene and with ribozymes (Clough and Hamm, 1995; Fuchs and Gonsalves, 1995; Huttner *et al.*, 2001; Tricoli *et al.*, 1995). Freedom II, a transgenic squash hybrid containing the WMV and ZYMV *CP* genes was released in the United States in 1995, as the first virus-resistant transgenic crop to be commercially cultivated in the world. It proved to have a very efficient resistance to WMV in field conditions. Such hybrids, some also including transgenic resistance to CMV, are presently grown mainly in South Eastern United States but they cannot be cultivated in the Mediterranean basin due to the legal and social restrictions that ban the use of GMO crops in the region.

2. Zucchini yellow mosaic virus

i. Symptoms, host range, and transmission Since its first description, ZYMV was recognized as a virus causing extremely severe symptoms in cucurbits leading to complete yield losses when infection occurs early. In melon, leaf symptoms include vein clearing, yellow mosaic, leaf deformation with blisters and enations. Plants are severely stunted. Some ZYMV isolates induce a rapid wilting reaction in cultivars possessing the *Fn* gene (Lecoq and Pitrat, 1984). On fruits, a diversity of symptoms is observed: external mosaic or necrotic cracks, internal marbling and hardening of the flesh. Seeds are occasionally severely deformed and have poor germination rates. In zucchini squash, symptoms are very severe on leaves with vein clearing, mosaic, yellowing, leaf distortion and severe filiformism (Fig. 1B). Fruits are generally severely misshaped with prominent knobs. In cucumber and watermelon, severe mosaic and deformations are observed on leaves and fruits.

ZYMV has a relatively narrow host range. In natural conditions, it infects mostly cultivated or wild cucurbits but also a few ornamental species (*Althea*, *Begonia*, and *Delphinium*) or weeds (Lecoq and Desbiez, 2008).

ZYMV is transmitted on the nonpersistent mode with varying efficiencies by at least 26 aphid species (Katis *et al.*, 2006). *A. craccivora*, *A. gossypii*, *Macrosiphum euphorbiae*, and *M. persicae* are efficient ZYMV vectors (Lisa and Lecoq, 1984; Yuan and Ullman, 1996). Some aphid species (*Lipaphis erysimi*, *Myzus ascalonicus*) were not able to transmit ZYMV what suggests some level of specificity in the virus–vector interaction (Dombrovsky *et al.*, 2005; Katis *et al.*, 2006).

An interesting interaction has been observed between ZYMV and *A. gossypii*, an aphid vector colonizing cucurbit crops. *A. gossypii* lives longer and produces more offspring on ZYMV infected than on noninfected plants. In addition, more alatae are produced on infected plants, what may stimulate the spread of ZYMV (Blua and Perring, 1992).

ii. Distribution in the Mediterranean region ZYMV was first isolated in 1973 from a zucchini squash plant in a family garden in Northern Italy (Lisa *et al.*, 1981). In 1979, many melon crops were devastated in Southwestern France by an apparently new virus disease and the causal agent was tentatively named muskmelon yellow stunt virus (Lecoq *et al.*, 1981). Very rapidly it appeared that this tentatively new virus was a strain of ZYMV (Lecoq *et al.*, 1983). ZYMV was subsequently isolated in Lebanon (1979), Israel and Spain (1982), and Egypt and Turkey (1983) (Desbiez and Lecoq, 1997). Presently, ZYMV has been reported as a major cucurbit virus in 18 of the Mediterranean countries (CABI/EPPO, 2003b; Dukic *et al.*, 2002; Papayiannis *et al.*, 2005; Vucurovic *et al.*, 2009).

iii. Biological and genetic variability From its first description, ZYMV appeared to have a large biological diversity. When collections of field isolates were compared for their biological properties, an important variability was revealed in host range and symptomatology on susceptible hosts, with isolates producing mild or atypical mosaic symptoms, necrosis or wilting reactions. When these isolates were inoculated to melon or squash varieties possessing resistance genes, different pathotypes could be differentiated (Desbiez *et al.*, 2003a; Lecoq and Pitrat, 1984).

Important variability was also observed in aphid transmissibility. Several ZYMV isolates that have lost aphid transmissibility have been characterized, and a unique feature for this virus is that single amino acid mutants have been identified in the three major domains involved in transmission. ZYMV-NAT has an A to T substitution in the DAG motif in the CP, ZYMV-R1A a K to E substitution in the KLSC motif and ZYMV-PAT a T to A substitution in the PTK motif, both in the HC-Pro. This latter mutant, unique among potyviruses, led to the identification of a functional interaction between HC-Pro and CP through their PTK and DAG domains (Blanc *et al.*, 1997).

The nontransmissible isolate ZYMV-NAT (having the DTG motif in the CP) could be transmitted by aphids from plants infected concomitantly by a transmissible isolate of PRSV. This occurred through heteroencapsidation, a phenomenon by which the RNA of ZYMV is completely or partially encapsidated by the functional PRSV CP (Bourdin and Lecoq, 1991). An aphid nontransmissible isolate deficient for the HC-Pro could also be transmitted by aphids when in mixed infection with an isolate that has a functional HC-Pro (Lecoq *et al.*, 1991a). The transmissible isolate provides its functional HC-Pro to mediate the transmission of the deficient isolate. These two mechanisms, heteroencapsidation and heteroassistance, can contribute to the maintenance, in natural conditions, of variants which have lost their vector transmissibility. However, heteroassistance was observed to be more efficient than heteroencapsidation (Desbiez *et al.*, 1999).

More than 500 ZYMV partial sequences are available in GenBank, mostly from parts of the CP-coding region. At the world level, three major groups of ZYMV populations, namely Group A, Group B, and more recently Group C containing isolates from Viet-Nam and China, were defined (Fig. 3). Only Group A was detected so far in the Mediterranean basin. However, an isolate from Group C was found in Poland (Pospieszny *et al.*, 2007), indicating that this group is already present in Europe. Within Group A, three clusters (clusters 1–3) were originally defined (Desbiez *et al.*, 2002), and three others, namely clusters 4–6, containing mostly isolates from China and Korea, were distinguished later when the number of sequences available increased (Fig. 3; Coutts *et al.*, 2011). In the Mediterranean basin, cluster 1 was observed in France since 1979, Spain, Portugal, Tunisia, Turkey, Israel, Syria, and Jordan (Desbiez *et al.*, 2002); cluster 2 was observed in Italy since 1973 (Desbiez *et al.*, 2002), Tunisia (Yakoubi *et al.*, 2008a), Jordan and France since 1979; cluster 3 was detected in Spain and South-eastern France (Fig. 3). In addition to these three clusters, surveys have revealed the presence of clusters 4 and 5 in

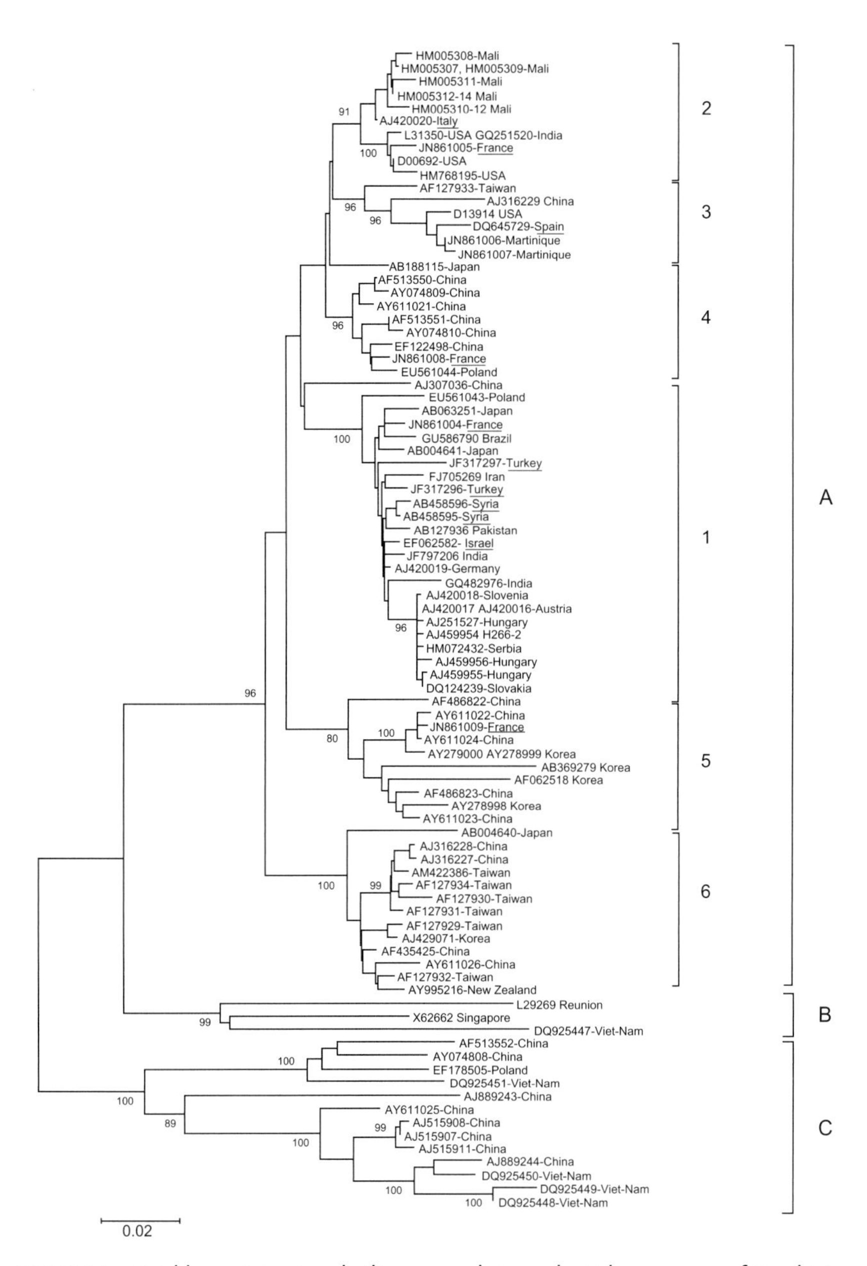

FIGURE 3 Neighbor-joining tree built on complete nucleotide sequence of *Zucchini yellow mosaic virus* coat protein showing the three molecular groups (A, B, and C) and the six clusters (1–6) within group A. Isolates from the Mediterranean Basin are underlined. Bootstrap support (500 bootstraps) above 80% are indicated for each node. The scale bar represents a genetic distance of 0.02.

South-eastern France since 2004 and 2005, respectively (Fig. 3; Lecoq *et al.*, 2009). Observations conducted in an experimental plot over a 30-year period indicated that ZYMV populations in one field are very homogenous, but population switches take place every few years (Lecoq *et al.*, 2009). The presence of the "Asian" clusters 4–6 in other countries from the Mediterranean basin has not been reported so far.

The development of MAbs against ZYMV allowed the characterization of serotypes closely correlated to the CP molecular variability (Desbiez and Lecoq, 1997).

iv. Epidemiology Seed transmission of ZYMV has remained an open question for a long time. In the 1990s, there were conflicting reports about possible ZYMV seed transmission. More recently, ZYMV seed transmission has been clearly confirmed in *C. pepo* var *styriaca* (hull-less seeded oil pumpkin) (Tobias and Palkovics, 2003), in *C. pepo* subsp. *texana* (Simmons *et al.*, 2011), and in zucchini squash (Coutts *et al.*, 2011). An important finding was that ZYMV infection through seeds could result in virtually symptomless infection, not detectable by standard serological methods but only by RT-PCR (Simmons *et al.*, 2011). This peculiar "masked" ZYMV seed transmission could explain the earlier conflicting reports, and highlights the need for specific detection methods for seed transmission. This is an important issue, since ZYMV seed transmission would be the simplest explanation for the rapid dissemination of ZYMV throughout the Mediterranean basin and the whole world in the 1980s.

Another possible way for long distance dissemination of ZYMV is through the globalization of vegetable production and trade. It has been shown that ZYMV infected fruits imported from Central America into Europe could be very efficient virus sources for aphids (Lecoq *et al.*, 2003).

Virus sources could also be infected overwintering weeds or crops. In tropical and subtropical regions, cucurbit crops are grown all year round, and viruses could easily move from an old infected crop or weed to a young planting. In the Mediterranean basin, only few noncucurbit weeds were found to be potential ZYMV reservoirs, but winter protected crops may efficiently contribute to ZYMV overwintering in the region.

v. Diagnosis ZYMV diagnosis is readily done by standard serological methods. DAS-ELISA commercial kits are available from several suppliers. Recently, serological tests based on the lateral flow technique have been developed that allow an easy and rapid diagnosis of ZYMV in the fields.

Many complete and partial nucleotide sequences are now available for ZYMV, particularly in the CP-coding region, which enabled the development of specific primers (Table II). However, in routine testing and surveys, as for WMV, DAS-ELISA seems to be more reliable than RT-PCR for ZYMV detection (Hervé Lecoq and Cécile Desbiez, unpublished).

vi. Control

a. Prophylactic measures The same as those proposed for WMV can be applied for ZYMV (see Section II.A.1).

b. Crossprotection Although the mechanism of crossprotection has not been fully elucidated, this method has been developed at a commercial level to protect cucurbit crops against ZYMV. The principle is simple: when a mild virus isolate (i.e., that has no significant impact on commercial yield) is inoculated to young seedlings, it protects the plant from subsequent contaminations by severe isolates of the same virus. The mild strain ZYMV-WK is a natural variant of a severe aphid nontransmissible isolate (Lecoq *et al.*, 1991b). Although efficient against most ZYMV isolates, ZYMV-WK does not protect against very divergent isolates such as those from Réunion Island indicating some specificity in the protection. A single amino acid change (R to I) in the FRNK conserved domain of the HC-Pro is responsible for symptom attenuation of ZYMV-WK (Gal-On and Raccah, 2000). A complete technological package (mild strain production, quality control protocols, inoculation machines) has been developed to implement commercially ZYMV crossprotection (Yarden *et al.*, 2000).

c. Resistant cultivars Since the first ZYMV description, breeding resistant cultivars has been a priority for cucurbit breeders. Resistance genes have been identified and they confer either a complete and durable resistance (such as the *zym* gene in cucumber) or partial and easily overcome resistances (such as the *Zym* gene in melon). An interesting situation was observed for ZYMV resistance in squash. Although the resistance level was high in the original accession of *C. moschata* in which the resistance was identified, when transferred through interspecific crosses to zucchini (*C. pepo*), a "tolerance" phenotype was observed: ZYMV can multiply but the plants display only mild symptoms (Desbiez *et al.*, 2003a). Tolerance appeared not to be stable since aggressive variants of the virus (i.e., causing severe symptoms in tolerant plants) were observed in the fields. A single amino acid change in the P3 coding region is sufficient to confer this aggressive phenotype (Desbiez *et al.*, 2003a). However, the aggressive variants are counter selected when in competition with common ZYMV isolates in susceptible cucurbits. This genetic load associated to aggressiveness could be a factor that will make the tolerance more durable (Desbiez *et al.*, 2003a). In melon, the single dominant gene *Vat* also confers resistance to ZYMV transmission by *A. gossypii*. This gene is present in many commercial cultivars, but confers only limited protection in the fields, probably because ZYMV is transmitted by many other aphid species in natural conditions.

Transgenic ZYMV-resistant squash cultivars have been commercially released in the United States (see Section II.A.1).

3. Papaya ringspot virus and related species

i. Symptoms, host range, and transmission PRSV causes severe mosaic, blistering, and malformations on leaves of squash, cucumber, melon, or watermelon. Fruits may also show a range of discoloration and deformation. In some melon cultivars, some PRSV isolates cause local and systemic necrosis leading to plant death. PRSV isolates infecting cucurbits (PRSV-W) have a host range mostly limited to cucurbits and to a few noncucurbit species including some *Chenopodium* sp. Other PRSV isolates infect papaya (PRSV-P) but cause only mild symptoms in cucurbits.

PRSV is transmitted by several aphid species on a nonpersistent mode, but not through seeds.

ii. Distribution in the Mediterranean region PRSV has been reported in cucurbits in many Mediterranean countries, including Bulgaria, Cyprus, France, Israel, Italy, Lebanon, Spain, Syria, Tunisia, and Turkey (CABI/EPPO, 2003c; Dikova, 1995; Koklu and Yilmaz, 2006; Mnari-Hattab *et al.*, 2008; Papayiannis *et al.*, 2005). PRSV has also been reported in papaya in a few countries of the Southern Mediterranean region.

iii. Biological and genetic variability A single mutation in NIa-Pro gene account for the host range differences between PRSV-W which does not infect papaya but cause severe symptoms in cucurbits and PRSV-P which causes severe symptoms in papaya and only mild infection in cucurbits (Chen *et al.*, 2008). A phylogenetic study of PRSV isolates suggested that PRSV-P has evolved on several occasions and in different places from PRSV-W populations (Bateson *et al.*, 2002). Slight differences in PRSV-W host range or reaction of specific hosts have also been reported (Russo *et al.*, 1979). Serologically very divergent PRSV isolates (PRSV-T) have been reported in Guadeloupe and other subtropical regions (Lecoq and Desbiez, unpublished; Quiot-Douine *et al.*, 1986).

Few PRSV sequences from the Mediterranean basin are available. At the world level, three clusters of PRSV were defined: a basal one from India (the probable center of origin of the virus), one from Asia, and one from America and Australia (Bateson *et al.*, 2002). Isolates from France, Spain, and Egypt all belonged to the "American–Australian" cluster.

iv. Epidemiology Probably because of its host range limited to cucurbits, PRSV is not a prevalent virus in cucurbit crops in the Mediterranean basin. It probably depends on protected cucurbit crops for its survival in winter.

v. Diagnosis PRSV diagnosis is readily done by standard serological methods. DAS-ELISA commercial kits are available from several suppliers. However, special care should be taken in analyzing the data because strong serological cross reactions are observed with PRSV related species (see below) (Quiot-Douine *et al.*, 1990)

Several complete and partial nucleotide sequences are now available for PRSV, particularly in the CP-coding region, what allowed the development of specific primers (Table II).

vi. Control

a. Prophylactic measures The same as those proposed for WMV can be applied for PRSV (see Section II.A.1). Since PRSV-W host range is mostly limited to cucurbits, a special care should be taken in avoiding overlapping cucurbit crops. Crossprotection has proved to be very effective in protecting papayas from PRSV-P. Unfortunately the PRSV-P mild strain is not efficiently protecting cucurbits against PRSV-W (You *et al.*, 2005).

b. Resistance PRSV resistances have been identified in cucumber, melon, and squash genetic resources (Lecoq *et al.*, 1998). Resistance is already present in some commercial cucumber cultivars.

vii. Potyvirus species related to PRSV Several potyviruses biologically or serologically related to PRSV have been described in the Mediterranean basin. They have similar host ranges mostly limited to cucurbits, may be controlled by the same resistance genes and share common epitopes (Quiot-Douine *et al.*, 1990). Molecular analyses have greatly contributed to determine whether they correspond to distinct species or to distant strains of PRSV-W. A commonly accepted criterion to define the species/isolate threshold in potyviruses is a nucleotide sequence identity in the CP of 76–77% or, more accurately, 76% identity in the complete genome nucleotide sequence (82% in amino acid sequence) (Adams *et al.*, 2005). Applying one or the other of these criteria, four species could be distinguished by molecular analysis among the "PRSV-like" isolates present the Mediterranean basin (Fig. 4): PRSV-W *sensu stricto*, MWMV (65% identity with PRSV in the complete sequence of a Tunisian isolate) (Lecoq *et al.*, 2001; Yakoubi *et al.*, 2008b), *Zucchini yellow fleck virus* (ZYFV) (69–71% identity with PRSV and 65–67% with MWMV in the CP-coding region) (Desbiez *et al.*, 2007b), and a new species closely related to MWMV and found only in Algeria so far, *Algerian watermelon mosaic virus* (AWMV) (65% identity with PRSV and 70% with MWMV in the complete sequence) (Yakoubi *et al.*, 2008c). Besides, a partially characterized virus common on the wild cucurbit *Bryonia dioica* in France and displaying 76–79% identity in the CP-coding region with ZYFV, is at the threshold between a very divergent strain of ZYFV and a new species (Cécile Desbiez and Hervé Lecoq, unpublished; Fig. 4).

MWMV causes very severe mosaic and deformation on leaves and fruits of squash, cucumber, and watermelon. In many melon cultivars,

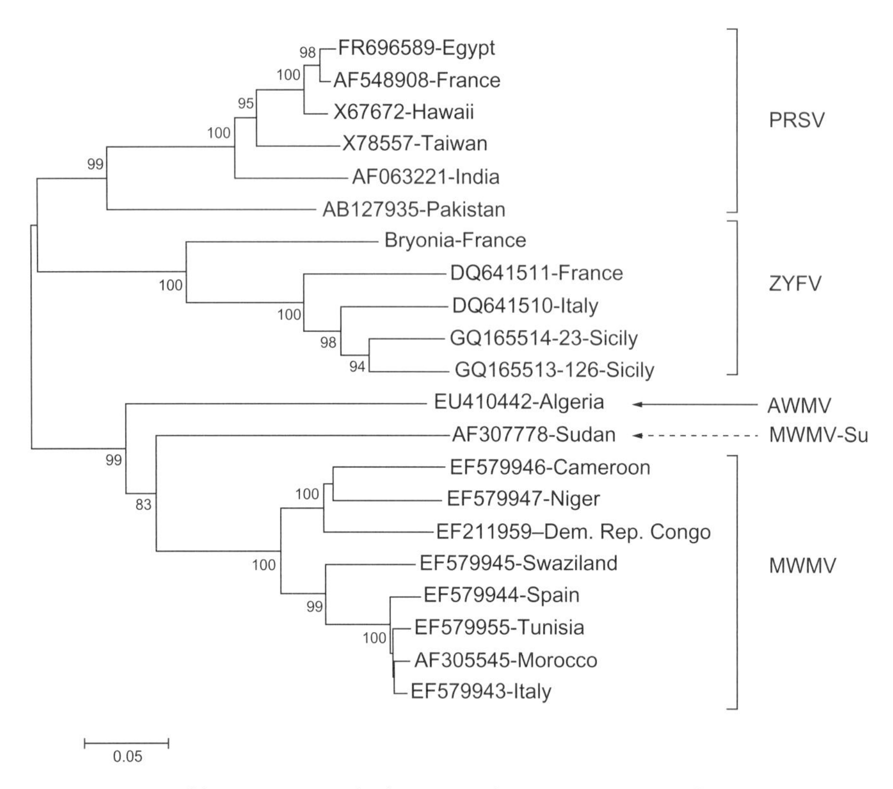

FIGURE 4 Neighbor-joining tree built on complete coat protein-coding sequence in nucleotides of potyviruses in the "PRSV" cluster. Bootstrap supports (500 bootstraps) above 80% are indicated for each node. The scale bar represents a genetic distance of 0.05. PRSV, *Papaya ringspot virus*; ZYFV, *Zucchini yellow fleck virus*; AWMV, *Algerian watermelon mosaic virus*; MWMV, *Moroccan watermelon mosaic virus*.

symptoms are systemic necrotic spots that are often followed by a complete collapse and necrosis of the plant. MWMV has already been reported in several western Mediterranean countries. It has been repeatedly reported in Central Italy and Southern France and could be emerging in more countries in a near future. MWMV partial sequences are available from Morocco, Italy, France, Spain, Tunisia, and Algeria. All these isolates are very closely molecularly related to each other (Yakoubi *et al.*, 2008b). This suggests that, contrary to isolates from sub-Saharan Africa that are much more divergent (Lecoq *et al.*, 2001; Yakoubi *et al.*, 2008b), MWMV isolates from the Mediterranean basin emerged recently from a common source and/or through rapid dissemination.

ZYFV has been reported so far only from the Mediterranean basin. It causes systemic flecks in zucchini squash, severe mosaic in cucumber and watermelon and systemic mosaic or necrotic spots and complete necrosis in melon. Four partial sequences of isolates from the wild cucurbit *Ecballium elaterium* in France and from zucchini squash and melon in Italy (Desbiez *et al.*, 2007b; Tomassoli *et al.*, 2010), are available. The isolates presented a particularly high molecular divergence: they share only 84–92% nucleotide sequence identity. However, the number of sequences available is too limited to determine if the molecular variability of ZYFV is structured by the host or by geography (Fig. 4). Regarding the virus isolate from white bryony, it has been reported only rarely in zucchini squash in France. It is therefore only of minor importance. All ZYFV isolates sequenced so far, as well as the highly divergent isolate/different virus from white bryony, have a "DAA" sequence instead of the highly conserved "DAG" motif involved in aphid transmission of potyviruses. Nevertheless, all the isolates tested were aphid transmissible.

AWMV causes symptoms similar to MWMV and has been reported so far only from Algeria.

B. Cucumber mosaic virus

CMV is the type species of genus *Cucumovirus*, family *Bromoviridae*. It was the causal agent of the first mosaic disease reported in cucurbits in the beginning of the twentieth century (Doolittle, 1916). Particles are isometric, circa 29 nm in diameter and are made up of 180 capsid subunits. The genome is divided in three positive-sense single-stranded RNAs of circa 3.4, 3, and 2.2 kb. (For more details on CMV see Chapter 13).

1. Symptoms, host range, and transmission

CMV causes typical mosaic symptoms on melon and cucumber leaves, plant stunting and fruit yield reduction. Mottle or mosaic is also often observed on fruits. In some cucumber cultivars, a rapid and complete wilt is observed on adult plants, a few days after CMV infection. In zucchini squash, CMV symptoms are very severe, including mosaic, yellow spots and leaf distortions (Fig. 1C). Infected plants remain stunted and generally fruit setting is drastically reduced and may even be stopped. Fruits are deformed with pinpoint depressions. CMV infection is rarely observed in watermelon, which reacts with dark necrotic lesions.

CMV infects a wide range of plants (over 1200 different species) and is transmitted by more than 60 aphid species in the nonpersistent mode. There are a few conflicting reports of CMV seed transmission in cucurbits. If confirmed, this could be significant for long distance dissemination of strains, but due to the abundance of CMV reservoirs, this should have probably no or only limited impact on local epidemics in crops.

2. Distribution in the Mediterranean region

CMV is present in all Mediterranean countries. However, its prevalence in cucurbits may vary according to the climatic conditions. Very common in cucurbits in temperate and Mediterranean regions, CMV is less prevalent in regions with warmer climatic conditions (subtropical or Sahelian regions)

3. Epidemiology

Temporal pattern of CMV spread in melon or zucchini squash has been followed every year since 1981 in Montfavet (France) (Fabre *et al.*, 2010; Lecoq, 1992b). CMV spreads rapidly every year in melon crops, often reaching total infection by 4–6 weeks after planting. In zucchini squash, CMV epidemics are often slow to develop and the final infection rate rarely exceeds 10%. This suggests limited secondary spread possibly in relation to a decreased plant susceptibility at the adult stage (Lecoq, 1992b). CMV infects a number of weeds that can persist in winter, as well as many winter crops which can be virus reservoirs between two cucurbit crop cycles (Quiot, 1980; Quiot *et al.*, 1983; Sacristan *et al.*, 2004).

4. Diagnosis

CMV diagnosis is readily done by standard serological methods. DAS-ELISA commercial kits for CMV are available from several suppliers. Due to its very high concentration in infected cucurbit plants, a special care should be taken to avoid contaminations between wells when testing CMV in DAS-ELISA. Serological tests based on the lateral flow technique have been developed allowing an easy and rapid diagnosis in the fields.

Many complete and partial nucleotide sequences are now available for CMV and specific primers have been developed (Table II).

5. Control

i. Prophylactic measures The same as those proposed for WMV can be applied for CMV (see Section II.A.1).

ii. Resistance Various levels of resistance to CMV are present in melon, cucumber, and squash. Oligogenic recessive resistances are available in melon and cucumber and partially dominant resistances have been described in squash (Lecoq *et al.*, 1998). In melon, some CMV strains belonging to the "Song" pathotype overcome the oligogenic resistance. They are already widely spread and represented about 30% of 1000 isolates collected from 1974 to 1978 in weeds or melons (Leroux *et al.*, 1979). A resistance to CMV transmission by one of its vectors, *A. gossypii*, is conferred by the dominant *Vat* gene in melon. When combined, *Vat* and the oligogenic resistance provided a very efficient and durable protection to melon against CMV in field conditions (Lecoq *et al.*, 2004).

Transgenic CMV-resistant squash cultivars have been commercially released in the United States (see Section II.A.1).

C. Cucurbit aphid-borne yellows virus

1. Virus properties

CABYV is a member of genus *Polerovirus*, family *Luteoviridae*. Viral particles are icosahedrical approximately 25 nm in diameter. Like other poleroviruses, CABYV is limited to phloem tissues in infected plants. CABYV genome consists of a single-stranded positive-sense RNA molecule of 5.7 kb (Guilley *et al.*, 1994). A VPg is covalently linked at the 5′ extremity of the RNA. The genome encodes six ORF. P0, at the 5′ extremity of the genome, is a suppressor of PTGS (Pfeffer *et al.*, 2002). P1 contains the VPg and a serine proteinase domain. P2 is expressed as a fusion protein with P1 through a translational frameshift, and encodes the RdRp. P3 and P5 are structural proteins: P3 (21 kDa) corresponds to the major CP, whereas a minor component of 75 kDa is a fusion of the CP and a readthrough domain encoded by ORF5 (Guilley *et al.*, 1994). The readthrough domain is exposed on the surface of the particle and is involved in virus movement in infected plants as well as aphid transmission specificity (Brault *et al.*, 2005). P4 protein, completely embedded in the P3-coding region but in a different frame, is a movement protein. ORF0, 1, and 1–2 are translated from the genomic RNA and ORF 3, 4, and 3–5 from a subgenomic RNA, in both cases using different frames (Fauquet *et al.*, 2005; Guilley *et al.*, 1994; Prüfer *et al.*, 1995).

2. Symptoms, host range, and transmission

CABYV infects cucumber, melon, squash, and watermelon. Following natural or experimental inoculations, commercial cucurbit cultivars develop typical symptoms of yellowing of the older leaves (Fig. 1D). Leaves with symptoms are thickened and brittle. There is a wide range of symptom intensity according to the cultivars, varying from a yellowing limited to a few older leaves to a complete discoloration of the plants (Lecoq *et al.*, 1992). CABYV incidence on yield seems to vary greatly according to the plant growing conditions and cultivars. Yield reduction can reach up to 50% of the marketable production in cucumber, but in melon, losses are more generally in the range of 10–15% (Lecoq, 1999). In contrast to mosaic inducing viruses, CABYV does not affect fruit quality, but rather induces flower abortions and reduces the number of fruit per plant (Lecoq *et al.*, 1992). Therefore fruits collected on infected plants can be readily marketed without any price depreciation.

CABYV has a moderately wide host range, infecting many cucurbits, but also fodder beet and lettuce. CABYV has also been detected in natural conditions in many weed species, which may be efficient alternative hosts (Lecoq *et al.*, 1992).

CABYV is transmitted in a persistent, circulative nonpropagative manner by a few aphid species (*A. gossypii*, *M. persicae*, and *M. euphorbiae*). Using transmission electron microscopy, it was shown that CABYV is internalized in its aphid vectors, both at the posterior midgut and hingut levels (Reinbold *et al.*, 2003), in contrast with other luteovirids that reach the hemocoel either through the posterior midgut epithelium or through the hindgut epithelium, but not through both. The consequences of such a peculiar behavior on transmission efficiency are not known. This gut tropism, as well as vector specificity were found to be associated to the readthrough domain of the minor CP (Brault *et al.*, 2005).

3. Distribution in the Mediterranean region

Although first described in France in 1992, CABYV was detected in extracts of field samples collected 10 years earlier (Lecoq *et al.*, 1992). Therefore, CABYV has likely been present but overlooked for several years, because its symptoms were attributed to nutrient deficiencies, aging, or infection by other pathogens. The development of a reliable DAS-ELISA diagnostic facilitated extensive surveys for CABYV in different parts of the Mediterranean region. CABYV was first detected in most cucurbit growing regions in France (Lecoq *et al.*, 1992). It was later found to be frequent in Algeria, Greece, and Turkey (Lecoq *et al.*, 1994), Lebanon (Abou-Jawdah *et al.*, 1997), Spain (Juarez *et al.*, 2004), Italy (Tomassoli and Meneghini, 2007), and Tunisia (Mnari-Hattab *et al.*, 2009). CABYV has now been detected in 15 Mediterranean countries and this number will probably increase in a near future. Whenever surveys were conducted, CABYV appeared to be one of the most common cucurbit viruses. In Tunisia, CABYV was detected in circa 70% of 330 cucurbit samples collected from 2000 to 2004 (Mnari-Hattab *et al.*, 2009). In Spain, CABYV was detected in 83% of melon samples and 66% of squash samples tested out of a total of 924 samples collected in 2003 and 2004. In France, in an extensive survey carried on from 2004 to 2008, CABYV was detected in 41% out of a total of 2660 cucurbit samples. CABYV was present in 20% and 57% of the samples tested in 2004 and 2008, respectively, indicative of a progressive increase in frequency in France (Hervé Lecoq and Catherine Wipf-Scheibel, unpublished).

4. Biological and genetic variability

Only a limited biological variability has been observed so far for CABYV. Differences were noticed between isolates for symptom intensity, capacity to infect differential hosts (such as *Physalis floridana*) or resistant melon lines. In Spain, where CABYV is prevalent, RFLP analysis revealed that CABYV isolates can be grouped into two genetic types. Both of them were present in 2003, but only one of the types was found in 2004. This grouping did not correlate with the original host or with the place of origin (Kassem *et al.*, 2007). In Tunisia, two CABYV subgroups were also observed, but the Tunisian isolates were more closely related to the French and Italian isolates than to the Spanish ones (Mnari-Hattab *et al.*, 2009).

It was suggested that recombinations in the readthrough domain have occurred with pea enation mosaic virus-1 (PEMV-1) during CABYV evolution (Gibbs and Cooper, 1995). More recently, a recombinant between CABYV and *Melon aphid-borne yellows virus* (MABYV), another polerovirus infecting cucurbits, has been identified in Taiwan (Knierim *et al.*, 2010). Such recombination events could contribute to increase CABYV variability.

5. Epidemiology

CABYV spread has been monitored in France in spring and summer during several years in melon or squash crops. It is characterized by early and massive primary infections what makes often secondary spread hardly detectable. In this respect, CABYV epidemics may be more rapid than those of viruses which are transmitted nonpersistently by aphids such as WMV. The rapid CABYV spread in cucurbit crops may be the result of the conjunction of several factors: the abundance of CABYV reservoirs around cultivated fields, the high population of *A. gossypii*, the major aphid species colonizing cucurbits, and the persistent mode of transmission of CABYV (Lecoq, 1999).

Several weed species were found infected by CABYV in natural conditions. Some belong to Cucurbitaceae (*B. dioica*, *E. elaterium*) and others belong to diverse botanical families (*C. bursa-pastoris*, *L. amplexicaule*, *S. vulgaris*) (Lecoq *et al.*, 1992). Interestingly, these three species are able to overwinter in Mediterranean climatic conditions and are alternative hosts for two other major cucurbit viruses, CMV and WMV.

6. Diagnosis

A purification protocol has been described for CABYV yielding circa 200 μg of purified virus per kg of *Montia perfoliata* infected tissue (Lecoq *et al.*, 1992). The purified virus proved to be highly immunogenic and an antiserum that could be used in DAS-ELISA was obtained. Now, DAS-ELISA kits for CABYV are commercially available.

Several complete and partial nucleotide sequences have been established for CABYV, and specific primers have been developed (Table II). CABYV detection by RT-PCR or immunocapture (IC)-RT-PCR was found to be more sensitive than by DAS-ELISA for routine testing in Tunisia (Mnari-Hattab *et al.*, 2009). Tissue-print hybridization also proved to be a very reliable method to detect CABYV in field samples (Kassem *et al.*, 2007).

7. Control

i. Prophylactic Several cultural practices have been tested in order to prevent or delay CABYV spread. Weekly insecticide applications had no significant effect on CABYV dissemination while plastic mulching delayed CABYV spread for circa 2 weeks (Lecoq, 1999).

ii. Resistance Resistances have been identified in cucumber, squash, and melon (Dogimont *et al.*, 1996; Lecoq *et al.*, 1998). In melon, there are two types of resistance: a resistance to the virus itself, governed by two recessive genes (*cab*1 and *cab*2) (Dogimont *et al.*, 1997) and a resistance to the transmission by one of its main vectors, *A. gossypii*, governed by the single dominant gene *Vat*. In the fields, the resistance to CABYV is very efficient, while the resistance to *A. gossypii* provides only a slight protection by delaying virus spread for circa 1 week (Lecoq, 1999). Melon cultivars with *Vat* are now commercially available. One zucchini squash cultivar is highly resistant to CABYV.

III. WHITEFLY-BORNE VIRUSES

A. Cucumber vein yellowing virus

1. Virus properties

Cucumber vein yellowing virus (CVYV) was first described in Israel in the late 1950s, causing a severe cucumber disease in the Jordan Valley (Cohen and Nitzany, 1960). Virus particles were shown to be rigid filaments circa 750 nm long (Sela *et al.*, 1980). CVYV genome and proper taxonomic position were assessed only 40 years after its discovery (Lecoq *et al.*, 2000). A partial sequence revealed that CVYV is a member of genus *Ipomovirus*, family *Potyviridae*. It is distinct from potyviruses by its whitefly (rather than aphid) transmission and by limited sequence similarities with potyviruses in the NIb-CP-coding region (Lecoq *et al.*, 2000). The complete sequence of CVYV revealed that this virus also differs from potyviruses in the organization of the 5′ extremity of its genome (Janssen *et al.*, 2005): whereas potyviruses have a P1 and HC-Pro in the N-terminal part of the polyprotein, CVYV lacks a HC-Pro but has an unusually long P1. This P1 was later shown to be cleaved autocatalytically in two functional proteins, P1a and P1b, none of them having any sequence similarity with HC-Pro or any of its conserved domains. P1b has been shown to be an inhibitor of the plant antiviral defense mediated by PTGS (Valli *et al.*, 2006). Further studies have shown that P1b is a serine protease that self-cleaves at its C-terminus, binds itself to form oligomeric structures and binds small interfering RNAs (siRNA) involved in PTGS (Valli *et al.*, 2008). Thus, despite its lack of structural similarity, P1b of

CVYV seems to have the same functions as the HC-Pro of potyviruses. However, unlike HC-Pro that is required for aphid transmission of potyviruses, the effect of P1b on whitefly transmission of CVYV is unknown. The lack of HC-Pro and presence of a P1b was reported for another ipomovirus (*Squash vein yellowing virus*), recently described in Florida, but not for the type member of genus *Ipomovirus*, *Sweet potato mild mottle virus* that has the same genetic organization as potyviruses (Colinet *et al.*, 1998).

2. Symptoms, host range, and transmission

In cucumber, CVYV causes typical vein clearing sometimes followed by a general chlorosis and necrosis (Fig. 1E). In melon, first symptoms are mild vein clearing followed by mosaic and vein banding. Symptom intensity both in cucumber and melon greatly varies according to the cultivar, some cultivars being almost symptomless carriers. In watermelon and zucchini squash symptoms are generally inconspicuous, although occasional splittings of watermelon fruits have been observed. Economic impact of CVYV varies according to the crop and the year; considerable losses were reported in protected cucumber crops in Almeria region, during the first outbreaks in Spain (Cuadrado *et al.*, 2001).

Natural host range of CVYV was first thought to be limited to cucurbits. Recent studies using a sensitive RT-PCR assay showed that CVYV could also infect in natural conditions several noncucurbit weeds including the perennial *Convolvulus arvensis* and the annual *Malva parviflora* (Janssen *et al.*, 2002)

CVYV is transmitted by the whitefly *B. tabaci* in a semipersistent manner (Cohen and Nitzany, 1960; Harpaz and Cohen, 1965). Higher transmission rates were obtained after acquisition and inoculation periods of 4 h, but acquisition and transmission could already occur after access periods of 15 min. CVYV was retained by its vector no longer than 4–6 h (Harpaz and Cohen, 1965).

3. Distribution in the Mediterranean region

CVYV was first reported in the Oriental part of the Mediterranean basin (Israel, Jordan, Turkey, and Cyprus). In the early 2000s, CVYV has been reported in the Western part of the Mediterranean basin, first in Spain and then in Portugal, France, and Tunisia (CABI/EPPO, 2005; Lecoq *et al.*, 2007; Papayiannis *et al.*, 2005; Yakoubi *et al.*, 2007)

4. Biological and genetic variability

Little biological variability has been reported for CVYV except unusually severe symptoms in cucumber caused by an isolate from Jordan (Lecoq *et al.*, 2000). A very limited molecular diversity was observed in the Northwestern part of the Mediterranean Basin (Spain, Portugal, France), indicating a recent introduction of the virus in these areas, probably from a common origin. Isolates from Israel and from Tunisia were more divergent from the ''North-western'' ones, and between each other, suggesting different origins and/or longer-term local evolution (Yakoubi *et al.*, 2007). The population structure and genetic diversity of CVYV was studied in Spain during a 5-year period following its first introduction. The isolates showed a low genetic diversity in three different regions of the genome, suggesting a genetic stability of CVYV populations and a probable single origin of introduction (Janssen *et al.*, 2007).

5. Epidemiology

CVYV spread has been monitored in 2001 and 2002 in cucumber protected crops in Almeria (Spain) (Ruiz *et al.*, 2006). The progress of CVYV was very rapid so that by the end of the season, 100% of the plants were infected. The logistic regression model best described CVYV disease progress curves. Whitefly numbers and indoor temperatures were shown to have a significant impact on CVYV epidemics (Ruiz *et al.*, 2006). In the Oriental Mediterranean basin, CVYV epidemics were also observed mostly during the hottest crop seasons, when whitefly populations were high (Cohen and Nitzany, 1960).

6. Diagnosis

Several methods were used to detect CVYV. First, the access to partial sequences allowed the development of specific RT-PCR primers and RNA probes (Cuadrado *et al.*, 2001; Lecoq *et al.*, 2000; Martinez-Garcia *et al.*, 2004). Highly purified CVYV preparations are difficult to obtain what precludes the production of polyclonal antiserum using virus particles as immunogen. Instead, recombinant CVYV CP expressed in *Escherichia coli* was used as immunogen. The antisera thus obtained proved to be useful to detect CVYV in western blot (Martinez-Garcia *et al.*, 2004) or DAS-ELISA (Desbiez and Lecoq, unpublished).

7. Control

i. Prophylactic In protected crops in Spain, disease prevention relies on the use of *Bemisia*- and CVYV-free seedlings, greenhouse insect-proof window screens, double doors, disinfection of crop residues and when applicable, a 1-month cucurbit-free period between plantings (Janssen *et al.*, 2003).

B. tabaci populations are difficult to control by classical pesticide treatments, because of the frequent resistances to insecticides. Biological control can efficiently limit *B. tabaci* populations but its impact on CVYV epidemics has not yet been investigated. UV-absorbing films that interfere with whitefly behavior could also be used for plastic tunnels (Antignus *et al.*, 1998).

ii. Resistance Resistance to CVYV is already available in some cucumber types. No resistance has been reported so far in other cucurbits.

B. Criniviruses

Criniviruses (family *Closteroviridae*) have flexuous filamentous particles of circa 800–850 nm long and 12 nm wide. They have a bipartite single-stranded RNA genome. The two genomic RNAs of 7.6–8 kb, both required for infectivity, are encapsidated separately in two particles. RNA 1 encodes two replication-related proteins. ORF 1a contains papain-like protease (P-Pro), methyltransferase (Met), and helicase (Hel) motifs. ORF 1b, expressed via a putative (-1) ribosomal frameshift, corresponds to the RdRp (Fauquet *et al.*, 2005). CYSDV has in addition three extra proteins of 5.2, 25, and 22 kDa, respectively, at the 3′ extremity of the RNA 1. P5.2 contains a putative signal peptide targeting the protein to the endoplasmic reticulum, P25 is a suppressor of PTGS and P22 function is not yet known (Kataya *et al.*, 2009). RNA 2 encodes seven ORFs, including a five-gene module typical of the family *Closteroviridae*: a 5-kDa small hydrophobic protein, a heat shock protein homologue (HSP70h), a 59-kDa product, and two structural proteins, a major CP and a minor CP (CPm). The major CP constitutes most of the virion protein, whereas CPm is present at one end of the particle and seems to be required for the assembly of virion tails. The two other ORFs correspond to proteins of unknown functions: a 9-kDa protein between the 59-kDa product and CP, and a 26-kDa protein at the 3′ end of RNA 2.

Criniviruses are all transmitted semipersistently by whiteflies (Wisler *et al.*, 1998). The CPm present at one end of the particles is involved in the whitefly transmission process (Tian *et al.*, 1999). Indeed, a single nucleotide deletion in *Lettuce infectious yellows virus* CPm prevented whitefly transmission but not systemic movement in the plant (Stewart *et al.*, 2010).

1. Beet pseudoyellows virus

i. Symptoms, host range, and transmission *Beet pseudoyellows virus* (BPYV) was first described in California in 1965. It was shown to be transmitted semipersistently by the greenhouse whitefly *Trialeurodes vaporariorum*, one of the most severe pests of greenhouse crops in the world (Duffus, 1965). In the early 1980s, severe yellowing diseases were described in cucumber and melon protected crops in France, Spain, and Bulgaria, associated with high *T. vaporariorum* populations. These diseases were named differently (muskmelon yellows, melon yellows, cucumber infectious chlorosis, cucumber chlorotic spot) but symptoms in cucurbits were similar to those caused by BPYV. Further molecular characterization of these different viruses showed that they had nearly identical sequences to

BPYV, indicating that they were indeed isolates of BPYV (Coffin and Coutts, 1995).

Symptoms start as chlorotic spots on older leaves sometimes near the petiole. The leaves become completely yellow except for the veins which remain green. Older leaves are thickened and brittle when crushed. Generally young leaves remain normal, but in some cultivars a complete plant chlorosis is observed. Yield reduction may reach 30–40%.

BPYV has a broad host range including other crops (lettuce, spinach, endive), ornamentals (marigold, zinnia), and weeds (dandelion, cheeseweed) (Duffus, 1965; Wisler *et al.*, 1998).

ii. Distribution in the Mediterranean region BPYV has been reported in several Mediterranean countries, mainly in protected crops where vector populations are often high, but also in open fields. Reports are from Bulgaria (Hristova *et al.*, 1983), Cyprus (Papayiannis *et al.*, 2005), France (Lot *et al.*, 1983), Greece (Boubourakas *et al.*, 2006), Italy (Tomassoli *et al.*, 2003), Spain (Berdiales *et al.*, 1999), and Turkey (Coffin and Coutts, 1995).

iii. Biological and genetic variability No biological variability has been reported for BPYV. Even if few sequences are available, the molecular variability of BPYV appears very limited (Rubio *et al.*, 1999), as for other criniviruses (Rubio *et al.*, 2001).

iv. Epidemiology BPYV has a wide host range including weeds that can be efficient virus reservoirs (Duffus, 1965). In addition, it infects winter crops (lettuce, spinach) which can act as "bridge" hosts between two cucurbit crops. BPYV epidemics were severe in the Mediterranean basin in the 1990s probably in relation with the high greenhouse whitefly populations that arose then through insecticide resistances. However in Southern Spain, as in other regions, *T. vaporarium* populations are progressively replaced by *B. tabaci* which does not transmit BPYV. Thus BPYV incidence tends to decrease (Berdiales *et al.*, 1999).

v. Diagnosis Initially, BPYV was identified through biological means (host range and greenhouse whitefly transmission). The convergence of symptoms caused by BPYV, CABYV, and CYSDV in cucurbits precludes any visual diagnosis. The progress in BPYV genome characterization, made possible the development of specific RT-PCR primers (Coffin and Coutts, 1995). An antiserum has been produced against recombinant BPYV CP expressed in *E. coli* that can be used in DAS-ELISA (Desbiez and Lecoq, unpublished).

vi. Control Methods that can limit whitefly populations (chemical, biological, or physical) may reduce BPYV epidemics. Careful weeding will limit virus sources near the crops. An important point is to dispose rapidly of old plants at the end of the crop cycle. Such abandoned plants can be efficient sources of virus and vectors. No commercial cucurbit cultivar is resistant to BPYV but partial resistances have been identified in melon genetic resources.

2. Cucurbit yellow stunting disorder virus

i. Symptoms, host range, and transmission CYSDV was first detected in cucurbit crops in the United Arab Emirates in 1982 (Hassan and Duffus, 1991) It was found in the Arabic peninsula both in outdoor melon and watermelon crops and in indoor cucumber crops. Symptoms were similar to those caused by BPYV including severe yellowing of the older leaves. Its host range initially thought to be limited to cucurbits has been recently extended to several noncucurbit hosts including cultivated species (lettuce, bean) or weeds (*Sida hederacea* and *Amaranthus retroflexus*) (Wintermantel *et al.*, 2009).

CYSDV is efficiently transmitted on a semipersistent mode by *B. tabaci* biotype B and Q prevalent in the Mediterranean basin, but only poorly by biotype A, that occurs in America. CYSDV can persist up to 9 days in the vector, the longest retention time reported for a whitefly-transmitted crinivirus (Wisler *et al.*, 1998). CYSDV is not transmitted by the greenhouse whitefly.

ii. Distribution in the Mediterranean region Since its first description, CYSDV has rapidly spread to the Eastern and Western Mediterranean areas. It has already been reported in 11 Mediterranean countries (Cyprus, Egypt, France, Greece, Israel, Lebanon, Morocco, Portugal, Spain, Tunisia, and Turkey) in protected and open field cucurbit crops (CABI/EPPO, 2004; Papayiannis *et al.*, 2005; Yakoubi *et al.*, 2007). CYSDV is one of the most damaging recent emerging viruses in cucurbits. It has rapidly replaced BPYV in Southern Spain and was recently introduced in the United States, where it causes severe epidemics.

iii. Biological and genetic variability No biological variability has been reported so far for CYSDV. Since the first genetic studies CYSDV genetic variability also appeared particularly low (Aguilar *et al.*, 2003; Rubio *et al.*, 2001). The average genetic distance in the CP or HSP70 regions between worldwide isolates was about 0.0025 (Rubio *et al.*, 2001). However, single strand chain polymorphism analyses on the CP gene suggested the presence of three groups, two of them being present in the Mediterranean basin (Rubio *et al.*, 1999, 2001). Sequencing of the CP or HSP70 genes revealed that there were indeed two main groups, a "Western" subpopulation in the Mediterranean basin (Spain, Jordan, Turkey, France, Tunisia, and Lebanon) and America, and an "Eastern" subpopulation present in UAE, Saudi Arabia, and Iran (Rubio *et al.*, 2001). All CYSDV isolates from the "Western" group shared 99–100% nucleotide sequence identity, suggesting a common and recent origin (Yakoubi *et al.*, 2007). A study of the local evolution of CYSDV in Spain over 8 years revealed a particularly low molecular diversity compared to the populations of other plant viruses; no evidence for selection related to host adaptation was found (Marco and Aranda, 2005). The year-round maintenance of vector populations in Southern Spain and the overlapping of susceptible crops may limit seasonal bottlenecks and explain this high genetic stability (Marco and Aranda, 2005).

iv. Epidemiology CYSDV epidemics depend on whitefly populations. The number of whiteflies infesting a crop at an early stage is a good predictor of the outcome of the disease (Ruiz *et al.*, 2006). In Spain, CYSDV was first reported in 1992, and since then it has become the most prevalent virus in protected cucurbit crops often reaching 100% infected plants at the end of the crop. Epidemic curves are generally best fitted to either the Gompertz or logistic regression models (Ruiz *et al.*, 2006).

v. Diagnosis Several methods have been used to detect CYSDV. The access to partial sequences allowed the development of specific RT-PCR primers (Célix *et al.*, 1996). A real-time RT-PCR assay was developed to detect the virus in individual whiteflies (Gil-Salas *et al.*, 2007). Since CYSDV is difficult to purify, recombinant CP expressed in *E. coli* has been used as immunogen to produce polyclonal antisera that can be used in DAS-ELISA (Cotillon *et al.*, 2005; Hourani and Abou-Jawdah, 2003). Commercial DAS-ELISA kits are now available.

vi. Control In protected crops, methods for controlling CYSDV are similar to those applied for CVYV (see Section III.A). In open fields, plastic mulches and floating covers may reduce virus spread. Resistances to CYSDV have been identified in genetic resources of cucumber and melon, and partially resistant commercial cucumber cultivars are available.

C. Begomoviruses

Begomoviruses (family *Geminiviridae*) are characterized by a monopartite or bipartite single-stranded DNA genome encapsidated in geminate particles of circa 30 × 18 nm consisting of two incomplete icosahedra. Begomoviruses can be distinguished molecularly as "Old World" and "New World" viruses. "Old World" begomoviruses can be monopartite or bipartite, whereas all "New World" begomoviruses known so far are bipartite. *Watermelon chlorotic stunt virus* (WmCSV) is a typical bipartite "Old World" virus. Its two single-stranded circular DNA molecules of 2.6 kb each, named DNA-A and DNA-B, share no sequence similarity except for a common intergenic region involved in replication initiation, and encompassing a stem loop that contains the TAATATTAC motif present in all begomoviruses. DNA-A encodes six ORFs, two in the viral sense and four in the complementary sense. ORFs AV1 and AV2 correspond to the CP and a movement protein respectively. ORF AC1 is a DNA helicase required for replication; AC2 and AC3 act as transcriptional activator and replication enhancer respectively. AC4, embedded in AC1, may act as a silencing inhibitor and symptom determinant, although its functions are more obvious in monopartite than in bipartite begomoviruses. DNA-B encodes two ORFs, BV1 and BC1, that function as nuclear shuttle protein and movement protein respectively (Nawaz-ul-Rehman and Fauquet, 2009). The genome structure and organization of *Squash leaf curl virus* (SLCV) is similar to that of WmCSV except that, like all "New World" begomoviruses, SLCV lacks ORF AV2 (Fauquet *et al.*, 2005; Nawaz-ul-Rehman and Fauquet, 2009).

Begomoviruses are transmitted by the whitefly *B. tabaci* on a persistent circulative and nonpropagative mode. It has been also shown that both SLCV and WmCSV can be transmitted horizontally among whiteflies during mating in a gender-dependant manner (Ghanim *et al.*, 2007).

1. Squash leaf curl virus

i. Symptoms, host range, and transmission SLCV causes severe leaf curling, yellow mottling, blisters, and fruit deformation in squash plants. Similar severe symptoms can be observed in melon and watermelon. Besides cucurbits SLCV infects and causes severe symptoms in a few noncucurbit species including *Nicotiana benthamiana*, *Datura stramonium*, and bean. SLCV has been recorded as occurring naturally in *Malva* sp. and in the wild cucurbit *E. elaterium* (Abudy *et al.*, 2010; Antignus *et al.*, 2003). SLCV is transmitted persistently by *B. tabaci*, and the first virus outbreaks were associated to very high whitefly populations.

ii. Distribution in the Mediterranean region SLCV had only been reported in Central and Northern America until it was recently identified in the Middle-East. First detected in Israel in 2002, SLCV has been identified in Egypt and Jordan in 2005, and in the Northern West Bank of the Palestinian Authority in 2008 (Ali-Shtayeh *et al.*, 2010; Al-Musa *et al.*, 2008; Antignus *et al.*, 2003; Idris *et al.*, 2006).

iii. Biological and genetic variability Some molecular and host range variability have been reported for SLCV isolates from America. Two groups were defined, differing in their host range and severity: SLCV-E (extended) and SLCV-R (restricted). Later on, molecular data indicated that they were two different but closely related viruses, and SLCV-R was renamed squash mild leaf curl virus (SMLCV) while SLCV-E retained the name SLCV. In Israel, Lebanon, Egypt, Syria, and Jordan, sequence data revealed the presence of SLCV only, even though SMLCV was diagnosed in Jordan by specific PCR (Al-Musa *et al.*, 2008). SLCV has been shown to be prone to produce reassortants with related viruses, what may modify significantly its biological properties (pathogenicity and host range) (Brown *et al.*, 2002).

iv. Epidemiology SLCV is a typical "New World" bipartite begomovirus that has been introduced recently by a still unknown route in the Mediterranean basin. It constitutes one of the major threats for cucurbit crops in the region wherever *B. tabaci* occurs.

v. Diagnosis Several methods were used to detect SLCV. Polyclonal antisera were used successfully in DAS-ELISA (Cohen *et al.*, 1983), and then specific PCR primers were developed (Table II). A commercial monoclonal antibody detecting TYLCV in triple antibody sandwich (TAS)-ELISA also detects SLCV.

vi. Control In protected crops, methods for controlling SLCV are similar to those applied for CVYV (see Section III.A). In open fields, plastic mulches and floating covers may reduce virus spread.

Resistance to SLCV has been described in squash genetic resources, and partially resistant squash commercial cultivars are available.

2. Watermelon chlorotic stunt virus

i. Symptoms, host range, and transmission WmCSV causes a severe disease of watermelon characterized by vein clearing, chlorotic mottling, stunting, and complete yellowing of the plant apex and drastic yield reduction (Fig. 1F; Kheyr-Pour *et al.*, 2000). Severe golden mosaic on leaves and mosaic and deformation on fruits are observed in some melon cultivars. Infected zucchini squash display leaf curling, apical stunt and fruit deformation. Besides cucurbits WmCSV infects and causes severe symptoms in *N. benthamiana* and bean. WmCSV is transmitted by *B. tabaci* in the persistent circulative mode.

ii. Distribution in the Mediterranean region WmCSV was first identified in Yemen and then in Sudan (Kheyr-Pour *et al.*, 2000). Outside the Red Sea region WmCSV was only reported from Iran. The first report in the Mediterranean Basin is from Israel (Abudy *et al.*, 2010). The virus was isolated in 2002 in Eilat on the Red Sea shore, and although the infected watermelon field was rapidly eradicated, the virus spread to the rest of the country (Abudy *et al.*, 2010). So far, WmCSV has not been reported in other Mediterranean countries.

iii. Biological and genetic variability Only limited genetic variability (below 2% and 5% divergence for DNA-A and DNA-B, respectively) has been reported. No biological variability has been reported except a cloned isolate that had lost its whitefly transmissibility. A single amino acid change in the CP was responsible for this property (Kheyr-Pour *et al.*, 2000).

iv. Epidemiology As for SLCV, WmCSV epidemics are generally associated with high whitefly populations.

v. Diagnosis DAS-ELISA with a polyclonal antiserum produced against purified virus preparations and molecular hybridization were used successfully for WmCSV surveys in Sudan (Kheyr-Pour *et al.*, 2000). Specific PCR primers are now commonly used. A commercial monoclonal antibody detecting TYLCV in TAS-ELISA also detects WmCSV.

vi. Control Plastic mulches, floating covers, and appropriate whitefly control may reduce virus spread.

High levels of resistance to WmCSV have been described in genetic resources of melon and watermelon, but no resistant commercial cultivar is available yet.

IV. OTHER VIRUSES

A. Melon necrotic spot virus

1. Virus properties

Melon necrotic spot virus (MNSV) belongs to genus *Carmovirus*, family *Tombusviridae*. It has a single-stranded positive-sense RNA genome of about 4300 bases, encapsidated in icosahedral particles of 32–35 nm diameter with a $T=3$ symmetry. The genomic RNA encodes four ORFs. ORF1 corresponds to a 29-kDa protein, and to a 89-kDa protein produced by the readthrough of an amber stop codon. P29 and P89 are required for replication and P89 is the actual RdRp, whereas P29 interacts with mitochondria membranes and appears involved in symptom expression (Mochizuki *et al.*, 2009). The other ORFs are expressed from subgenomic RNAs. ORFs 2 and 3 referred to as double gene block correspond to two small proteins (p7A and p7B) acting in *trans* and involved in virus cell-to-cell movement. p7A is a RNA-binding protein and p7B a membrane protein that is incorporated into the endoplasmic reticulum membrane. The 3′ proximal ORF 4 encodes a 42-kDa CP, related to those of tombusviruses (Genoves *et al.*, 2006). The CP is also a symptom determinant and is required for systemic movement of the virus. p7B and the CP display silencing inhibitor activity (Genoves *et al.*, 2006).

2. Symptoms, host range, and transmission

MNSV causes systemic necrotic spots on leaves and streaks on stem of melon, cucumber, and watermelon and occasionally a complete plant collapse ("sudden death") (Fig. 1G). Fruit quality can be affected (lower sugar content, necrotic spots on fruit). A range of susceptibility to MNSV has been reported among melon cultivars and symptom expression is highly season dependant. MNSV symptoms are more frequently observed from fall to spring in protected crops (glasshouses and plastic tunnels), probably due to a restriction of symptom expression at high temperatures (Kido *et al.*, 2008). MNSV host range is limited to a few cucurbit species and the virus does not infect squashes (Gonzalez-Garza *et al.*, 1979).

MNSV is transmitted by motile zoospores of the fungus *Olpidium bornovanus* (ex *O. radicale*). Zoospores may acquire the virus *in vitro*, that is, when mixed with a virus suspension or with a soil containing virus particles. MNSV is adsorbed to zoospore's head and flagellum and is transmitted when zoospores penetrate the roots of healthy plants (Campbell *et al.*, 1995). A single amino acid substitution in MNSV CP resulted in the loss of both binding to the surface of the zoospores and fungal transmission (Mochizuki *et al.*, 2008). More recently a protruding domain of the CP has been involved in the specificity of the virus–vector interaction (Ohki *et al.*, 2010).

MNSV is seed-borne through an unusual mode: the vector-assisted seed transmission (VAST; Campbell *et al.*, 1996). When contaminated seeds produced on infected plants are sown in a sterile soil, no or very limited seed transmission is observed. But when contaminated seeds are sown in a soil containing aviruliferous *O. bornovanus*, seed transmission occurs. VAST probably operates through the release of MNSV particles from the contaminated seed coats and their *in vitro* acquisition by zoospores prior to inoculation by the zoospores when penetrating seedling roots (Campbell *et al.*, 1996).

MNSV is easily mechanically transmitted in laboratory conditions. It is also transmitted through pruning operations and probably by leaf contact as well.

3. Distribution in the Mediterranean region

MNSV has been reported in many Mediterranean countries including France, Greece, Israel, Italy, Lebanon, Spain, Syria, Tunisia, and Turkey (CABI/EPPO, 2010). It is probably occurring in other countries since it is seed-borne but may remain overlooked because of its atypical symptoms that can be easily confused with those of fungal or bacterial diseases.

4. Biological and genetic variability

Limited biological diversity has been reported for MNSV with only differences in symptom intensity on some hosts. An atypical isolate has been reported in Southern Spain: MNSV-264 not only overcomes *nsv* resistance gene in melon but also has an extended host range including several noncucurbit hosts (Diaz *et al.*, 2004). At the world level, MNSV isolates cluster in three molecular groups, two from Japan associated with the cucurbit species (melon vs. watermelon), and a third one containing two geographic subgroups of isolates: "Latin America" and "Europe and Mediterranean Basin" (Herrera-Vasquez *et al.*, 2010)

5. Epidemiology

MNSV is a very stable virus. It does not appear to rely on reservoir plants for its survival. It is more likely to remain infectious in plant debris in the soil or on contaminated seeds. Resting spores of the fungus are also very resistant and can survive several years in dried root debris or in the soil. Soil-less cultivation may be particularly favourable to zoospores movement in nutrient solutions.

6. Diagnosis

MNSV diagnosis can be readily done by standard DAS-ELISA. Commercial kits for MNSV detection are available from several suppliers. Several complete and partial nucleotide sequences are now available for MNSV what allowed the development of specific primers (Table II). RT-PCR proved to be more sensitive than DAS-ELISA for MNSV detection (Herrera-Vasquez *et al.*, 2009).

A test based on virus precipitation and RT-PCR allowed detection of MNSV in irrigation water and nutrient solution (Gosalvez *et al.*, 2003).

7. Control

i. Prophylactic Some cultural practices can reduce *O. bornovanus* populations in soil or nutrient solutions. Soil disinfection is now difficult with the recent ban of methyl bromide, but can still be done by soil steaming or solarization. The surfactant Agral 20® was shown to efficiently eliminate zoospores in nutrient solutions thus preventing virus spread in soil-less cultivation (Tomlinson and Thomas, 1986). Crop rotations can limit virus and vector increase in the soil. Grafting cucumber or melon on resistant *Cucurbita* sp. rootstock is now a common practice to control MNSV as well as other soil-borne diseases. Disinfecting pruning tools with trisodium phosphate may limit mechanical spread through cultural operations.

Composting for 30 or 60 days infected melon plants after a crop cycle proved to be efficient to destroy MNSV and *O. bornovanus* infectivity (Aguilar *et al.*, 2010). Seed disinfection has been recommended, although a reliable test to evaluate both seed-transmission rates and seed-disinfection efficiency is still lacking (Herrera-Vasquez *et al.*, 2009).

*ii. **Resistance*** Two resistance mechanisms to MNSV have been observed in melon germplasm. One is governed by two dominant genes (*Mnr1* and *Mnr2*) and prevents systemic symptoms expression (Mallor Gimenez *et al.*, 2003). The other resistance conferred by the single recessive gene *nsv* is of the immunity type (Diaz *et al.*, 2004). The *nsv* gene is effective in controlling all MNSV isolates tested so far, except the Spanish isolate MNSV-264. The genetic determinant allowing MNSV-264 to overcome *nsv* resistance was mapped in the 3′-untranslated region (UTR) of the genome (Diaz *et al.*, 2004). The same determinant was found to be responsible for the ability of MNSV-264 to infect noncucurbit hosts including *N. benthamiana* (Diaz *et al.*, 2004). In melon, it was shown that the *nsv* gene encodes the initiation factor eIF4E, and that a single amino acid change in this protein led to the resistance to MNSV (Nieto *et al.*, 2006). Melon cultivars resistant to MNSV are commercially available.

8. Other fungus-borne viruses infecting cucurbits

Two other viruses transmitted by *O. bornovanus* and causing similar necrotic symptoms in cucumber have been reported in the Mediterranean region but only in limited geographical areas. *Cucumber leaf spot virus* (CLSV, *Aureusvirus, Tombusviridae*) is seed-borne and has been reported from Bulgaria, Greece, Israel, Spain, and recently from France. *Cucumber soil-borne virus* (CuSBV, *Carmovirus, Tombusviridae*) has been reported so far only from Lebanon. Twelve *O. bornovanus* isolates from different hosts (cucumber, melon, and squash) and different geographic origins were tested as vectors of MNSV, CLSV, and CuSBV. All *O. bornovanus* isolates transmitted MNSV, all cucumber and some melon isolates transmitted CLSV and only two squash isolates transmitted CuSBV, indicating a high specificity in virus–vector interactions (Campbell *et al.*, 1995).

Finally, a strain of *Tobacco necrosis virus* (TNV, *Necrovirus, Tombusviridae*) a virus transmitted by *Olpidium brassicae* has been reported as causing necrotic spots in cucumber and squash in protected crops in France and Italy, respectively (Campbell *et al.*, 1990; Roggero and Lisa, 1995).

B. Squash mosaic virus

1. Virus properties

SqMV belongs to genus *Comovirus*, family *Secoviridae*, within the recently defined order *Picornavirales*. It has a bipartite single-stranded positive-sense RNA genome, with a 5′ bound VPg and a 3′ polyadenylated tail. Virions are small isometric particles circa 30 nm in diameter. RNA 1 (5.9 kb) is translated in a single polyprotein that is processed into five domains: a 32-kDa protein (P1A) that regulates the processing of the rest of the polyprotein, a 58-kDa RNA helicase, a VPg, a 24-kDa protease and a 87-kDa RdRp. RNA 2 (3.3 kb) is translated into two largely overlapping polyproteins using different initiation codons in the same frame, and cleaved by RNA 1 protease, enhanced by P1A, into three domains: a 50-kDa/40 kDa movement protein and two CPs, CPL (42 kDa) and CPS (22 kDa).

2. Symptoms, host range, and transmission

Symptoms induced by SqMV in melon and squash include mosaic, pronounced vein banding and leaf deformation (Fig. 1H). After a few weeks, symptoms may become inconspicuous and the plant recovers. Severe mosaic and deformation can be observed on fruit.

SqMV host range in natural conditions is mainly limited to cucurbits, but the virus can also infect several *Chenopodium* species.

Seed transmission is probably the major way for SqMV long distance dissemination. Transmission rates ranging from 0.1% to 10% are commonly observed in infected commercial seed lots. SqMV can be detected in both the seed coat and the embryo, but seed transmission in melon occurs only when the embryo is contaminated (Alvarez and Campbell, 1978). Seed transmission has also been reported in *Chenopodium quinoa* and *C. murale* (Lockhart *et al.*, 1985).

SqMV is transmitted by several beetle species including *Acalymma vittatum* and *Diabrotica undecimpunctata* in North America and *Epilachna chrysomelina* in the Mediterranean basin. Vectors can acquire the virus within 5 min, and remain viruliferous during 20 days. SqMV is a very stable virus that can be mechanically transmitted in laboratory conditions. It can be transmitted in the fields through pruning operations (Lecoq *et al.*, 1988).

3. Distribution in the Mediterranean region

Probably in relation to its seed transmission, SqMV has been reported in many Mediterranean countries including Bulgaria, Egypt, France, Greece, Israel, Italy, Lebanon, Morocco, Syria, Tunisia, and Turkey (CABI/EPPO, 1997; Dikova, 1998; Kassem *et al.*, 2005; Lecoq *et al.*, 1988; Lockhart *et al.*, 1982; Mnari-Hattab *et al.*, 2008).

4. Biological and genetic variability

Two SqMV groups have been distinguished using immunodiffusion tests and molecular data (Haudendshield and Palukaitis, 1998): Group I causes severe symptoms in melon and mild symptoms in squash, while Group II induces severe symptoms in squash and milder symptoms in melon.

5. Epidemiology

Most often SqMV epidemics are related to the use of virus-infected seed lots. The secondary virus spread is achieved by the moderately efficient beetle vectors wherever they prevail (mainly in the Southern part of the Mediterranean region). In France where vectors are not present, secondary spread of the virus occurs through pruning operations (Lecoq *et al.*, 1988).

6. Diagnosis

SqMV diagnosis is readily done by standard serological methods. DAS-ELISA commercial kits are available from several suppliers. Recently, serological tests based on the lateral flow technique have been developed that allow an easy and rapid diagnosis of SqMV in the fields.

Accurate estimation of seed-transmission rates is better obtained by grow-out tests rather than by DAS-ELISA. Indeed, DAS-ELISA not only detects SqMV in the embryo but also in seed coats, greatly overestimating the number of infected plantlets (Lecoq *et al.*, 1988).

Several complete and partial nucleotide sequences are now available for SqMV what allowed the development of specific primers (Table II).

7. Control

Certification of virus-free seed lots is a major component of SqMV control strategy. Application of insecticides can reduce vector populations where they prevail, and limit virus spread. Disinfection of pruning tools with trisodium phosphate reduces the risk of transmission from one plant to another. No high level of resistance has been described either in melon or in squash.

C. Cucumber green mottle mosaic virus

1. Virus properties

Cucumber green mottle mosaic virus (CGMMV) belongs to genus *Tobamovirus*, family *Virgaviridae*. The monopartite single-stranded positive-sense genomic RNA of 6.4 kb is encapsidated in rod-shaped particles of circa 300×18 nm. A cap is covalently bound at the 5′ extremity of the genome. The RNA encodes four proteins. Two proteins of 129 and 186 kDa are translated directly from the genomic RNA. P129 contains methyltransferase and helicase domains, whereas P186, expressed as a readthrough of P129, also contains a RdRp domain. Both proteins are required for virus replication. P129 is also an inhibitor of PTGS. ORFs 3 and 4 translated from subgenomic RNAs encode a 29-kDa movement protein and a 17-kDa CP, respectively (Fauquet *et al.*, 2005; Ugaki *et al.*, 1991).

2. Symptoms, host range, and transmission

CGMMV causes a severe mosaic on cucumber leaves and fruits. CGMMV has also been reported in grafted watermelon, inducing chlorotic mottling on leaves, necrosis on fruit pedicels and fruit flesh deterioration (Boubourakas *et al.*, 2004). Natural host range includes cucurbits but also a few weed species (Boubourakas *et al.*, 2004).

CGMMV is seed-borne in cucumber, with transmission rates reaching 8% which may drop rapidly after a few months of seed storage. CGMMV, like all tobamoviruses, is very stable and can be mechanically transmitted during pruning or harvesting, or through leaf contacts. Infection may also occur through the roots, when soil or recycled substrates contain infected plant debris. Transmission through contaminated irrigation water or nutrient solution has also been reported (Dorst, 1988; Paludan, 1985).

3. Distribution in the Mediterranean region

CGMMV has been reported from few Mediterranean countries: France, Greece, Israel, Spain, and Syria (Antignus *et al.*, 1990; Avgelis and Vovlas, 1986; Blancard *et al.*, 1994; Celix *et al.*, 1996; Kassem *et al.*, 2005; Varveri *et al.*, 2002).

4. Biological and genetic variability

Differences were observed among CGMMV isolates in their response to differential hosts (Boubourakas *et al.*, 2004). Serological and molecular variability has also been reported between European and Asian isolates.

5. Epidemiology

In cucumber, CGMMV mainly occurs in protected crops. Primary contaminations are either through infected seeds or through root contamination by infested soil, substrate, or irrigation water. As for other tobamoviruses, secondary spread may be very rapid and occurs through mechanical or contact transmission.

6. Diagnosis

CGMMV diagnosis is readily done by standard serological methods. DAS-ELISA commercial kits are available. Recently, serological tests based on the lateral flow technique have been developed. Several complete and partial nucleotide sequences are now available what allowed the development of specific primers (Table II).

7. Control

Use of virus-free seeds (including for rootstock) is essential. Contamination of the seeds is mainly external, and the virus can be eliminated using dry heat treatment (i.e., 70 °C for 3 days) without adversely affecting seeds' germination. The major part of commercial seed lots are now submitted to dry heat or to chemical disinfection treatments prior to marketing. Disinfection of pruning and harvesting tools with trisodium phosphate will reduce mechanical transmission of the virus, but will not prevent transmission through leaf contact.

Resistance to CGMMV has been reported in melon and cucumber, and some commercial cucumber cultivars are partially resistant.

8. Other tobamoviruses infecting cucurbits

Cucumber fruit mottle mosaic virus (CFMMV) has been reported only from Israel as causing bright mottling and mosaic on fruit and severe mosaic, vein banding and yellow mottling on leaves. CFMMV symptoms in experimental host ranges were generally more severe than those caused by CGMMV (Antignus *et al.*, 2001).

Two other tobamovirus species have been reported in Asia: *Zucchini green mottle mosaic virus* and *Kyuri green mottle mosaic virus* infecting squash and cucumber, respectively (Lee *et al.*, 2003). Both are seed-borne but have not yet been reported in the Mediterranean region. However, they could be easily introduced through infected seeds.

D. Eggplant mottled dwarf virus

1. Virus properties

Eggplant mottled dwarf virus (EMDV) belongs to genus *Nucleorhabdovirus*, family *Rhabdoviridae*, order *Mononegavirales*. It has a monopartite negative-sense single-stranded RNA genome, encapsidated in bacilliform particles circa 100–400 × 40–100 nm with a double structure: the nucleocapsid, formed mostly by a viral (N) protein associated with viral RNA, and a bilayer lipidic envelope originating from the nuclear membrane but containing two viral proteins, an externally protruding glycoprotein (G) and an internal matrix protein (M). A viral phosphoprotein (P), cofactor of the polymerase and one copy of the viral RdRp (L) are associated with the nucleocapsid. Although the complete sequence of EMDV is not yet available, its genetic organization is supposed to be similar to that of other rhabdoviruses: the 14-kb genome with inverted complementary termini has five structural genes in the order 3′-N-P-M-G-L-5′. In nucleorhabdoviruses, a gene coding for a putative movement protein (Sc4) is present between the P and M genes.

2. Symptoms, host range, and transmission

In the early 1980s a severe disease of cucumber caused by a rhabdovirus was reported in South-eastern France (Lecoq, 1982). The symptoms were a vein clearing on older leaves, a severe inward curling of younger leaves with the main vein showing a sinuous course giving a typical "toad skin"-like appearance. Recently, the causal virus tentatively named cucumber toad skin virus was shown, based on DAS-ELISA and RT-PCR tests, to be a strain of EMDV (Lecoq and Desbiez, unpublished). EMDV has also been reported in Italy infecting melons with unusual symptoms of stunting, short internodes and vein yellowing (Ciuffo *et al.*, 1999). Fruit deformations were observed both in cucumber and melon. EMDV generally infects only a low percentage of scattered plants in a field. Therefore the overall crop damage is slight, but those plants which are affected cease to produce fruits.

Both natural and experimental host ranges of EMDV are rather wide including several solanaceous species (eggplant, pepper, tobacco, and tomato), perennial ornamentals (including *Pittosporum tobira*) and weeds. In laboratory conditions, EMDV is mechanically transmissible to noncucurbit indicator plants (i.e., *Nicotiana glutinosa*) but mechanical transmissions to cucumber or melon are more erratic.

The cucumber toad skin virus (and thus EMDV) was shown to be transmitted in France by two leafhopper species *Anaceratagallia laevis* and *A. ribauti* (Della Giustina *et al.*, 2000).

3. Distribution in the Mediterranean region

EMDV has been reported in Mediterranean countries and mainly in eggplant. EMDV was reported in cucumber in France (Lecoq, 1982), Italy (Roggero *et al.*, 1995), Greece (Katis *et al.*, 2000), Bulgaria (Kostova *et al.*, 2001), Spain (Aramburu *et al.*, 2006), and in melon in Italy only (Ciuffo *et al.*, 1999).

4. Biological and genetic variability

Slight differences in experimental host ranges have been observed among EMDV isolates.

5. Epidemiology

EMDV is often observed in protected crops on a few scattered plants sometimes near the openings. This spatial distribution suggests that the virus might be introduced by the vector, but that secondary spread either does not occur or is very poor. EMDV is also occasionally detected in cucumber and gherkin outdoor crops.

6. Diagnosis

EMDV diagnosis can be readily done by DAS-ELISA. Commercial kits for EMDV detection are available. A few partial nucleotide sequences are now in GenBank and specific RT-PCR primers have been developed (Katis *et al.*, 2011; Table II).

7. Control

No specific control method has been proposed. It could be advisable to remove infected plants which will anyway not produce marketable fruits.

V. CONCLUDING REMARKS

Viruses represent a major and shifting threat to cucurbit crops in many countries of the Mediterranean basin. To face present and future challenges, research efforts should continue to be deployed to improve simple and rapid high-throughput diagnostic methods, to identify factors involved in virus emergence and to develop durable and environmentally friendly control methods.

A. The need for improved diagnostic methods

The accurate and rapid identification of viruses infecting cucurbits is essential for effective control. Much progress has been achieved during the past 20 years with the generalization of DAS-ELISA tests to many cucurbit viruses and the development of (RT-) PCR assays giving rapidly access to partial viral sequences. These techniques can be used for screening several viruses in parallel. They have been extensively used in the last years to carry out surveys in many Mediterranean countries, providing important information on virus prevalence and distribution. Microarrays have been recently developed for the simultaneous detection of different viruses or virus strains (Lee *et al.*, 2003; Wei *et al.*, 2009). This new technique may represent the next generation diagnostic method and has a real potential for epidemiological or surveillance studies. However, microarrays as well as PCR still require sophisticated laboratory equipments and are rather time consuming.

Rapid, simple and rather inexpensive tests using the lateral flow technique can be used by farmers directly in the fields. Some of these kits are now commercially available but remain limited to only few cucurbit viruses (CMV, ZYMV, SqMV, CGMMV). Their use should be extended to more viruses, with devices for simultaneously testing for several viruses.

Rapid identification of a newly introduced virus is probably one of the key factors to prevent successful emergences. Several cucurbit viruses not yet present in the Mediterranean region are causing major crop losses in other parts of the world. These include several tospoviruses, begomoviruses, potyviruses, and a number of seed-borne viruses (Lecoq, 2003). Generic diagnostic methods are probably the best way to detect rapidly such potential new incomers. Both serological and molecular generic tests are already available (e.g., for potyviruses or begomoviruses) but more should be developed for an optimal coverage of cucurbit virus diversity.

B. Better evaluating factors involved in virus emergence

New viruses are regularly reported in cucurbit crops of the Mediterranean region. Therefore, assessing the risks of a successful viral emergence is an important issue for the future. This includes identifying introduction routes that can be multiple: seeds (for seed-borne viruses), imports of seedlings from contaminated areas, globalized trade of cucurbit vegetables for human consumption, ornamentals, or cut flowers.

The aetiology of potential emerging viruses should be thoroughly recorded including their host range, vector specificity as well as vector dynamics. Sharing experiences among Mediterranean countries could be very useful for preventing the spread of emerging viruses. For instance, it would be very useful to compare the ecological situations in Southern Spain and Southern France where CVYV was introduced at approximately the same time, but succeeded or failed, respectively, to establish durably (Cuadrado *et al.*, 2001; Lecoq *et al.*, 2007).

New incomers can also be more severe strains of a virus already present in the region. This has been the case for WMV in South-eastern France in the early 2000s (Desbiez *et al.*, 2009). Identifying parameters involved in population dynamics may help to predict whether there will be a shift in virus population. Modeling approaches can help to identify complex parameters difficult to identify experimentally. This can usefully complete studies of more classical biological parameters (host range, vector transmission efficiencies) (Fabre *et al.*, 2010; Lecoq *et al.*, 2011).

C. Developing durable and environmentally friendly control methods

Virus control strategies mainly include prophylactic measures, crossprotection, or breeding for resistant cultivars.

Prophylactic measures against cucurbit viruses are limited and mainly rely on the intensive use of plastic material (mulches, floating covers) that must be properly disposed after use by farmers. Efforts should be done to develop biodegradable and more environmentally friendly material to protect cucurbits from viruses.

Durability of crossprotection greatly depends on virus genetic diversity. Crossprotection against severe ZYMV strains using ZYMV-WK proved to be very efficient in France, Israel, and other Mediterranean countries in the 1990s (Yarden *et al.*, 2000). Since then, new divergent ZYMV strains have been introduced probably from Asia: they overcome the protection conferred by ZYMV-WK and jeopardize its future use.

Several virus resistances are already present in commercial cucurbit cultivars or have been identified in genetic resources. Their durability depends on the virus genetic diversity and plasticity, but also upon the effect on virus fitness of the genetic changes leading to virulence. The diversity of situations that may be encountered is well illustrated by ZYMV resistance (Lecoq *et al.*, 2002). ZYMV resistance must be durable in cucumber since no isolate overcoming the *zym* gene was found in nature or obtained in laboratory conditions. ZYMV resistance in melon is expected not to be durable since virulent isolates occurred in the fields before the release of resistant cultivars and are easily obtained in laboratory conditions. It was even suggested to breeders not to use this resistance gene until other genes that could enhance its efficiency are identified. In zucchini squash, the situation is more contrasted since the tolerance is overcome: aggressive variants emerge easily from wild type isolates in the fields or in laboratory conditions, but the gain in aggressiveness has a negative impact on virus fitness (Desbiez *et al.*, 2003a). Consequently, the aggressive variants are counter selected in the absence of tolerant genotypes, what should improve the tolerance durability, just by alternating cultivation of susceptible and tolerant cultivars.

Transgenosis provides a new avenue for creating virus-resistant cultivars. Different strategies have been proved promising to control viruses in cucurbits: CP-mediated resistance and ribozymes (Fuchs and Gonsalves, 1995; Huttner *et al.*, 2001). However, such resistances should first become acceptable to farmers, consumers, and citizens in Mediterranean countries and their durability should be evaluated before considering their use in the region.

Finally, there is also a need for better identifying the viral pathogenicity determinants and understanding the mechanisms of virus evolution. This will enable to elucidate the modes of action of virus resistances at a molecular level. It will also contribute to understanding the processes leading to the occurrence of new virus strains that are able to cause diseases in resistant/tolerant plants. Altogether, this will help to evaluate better resistance genes durability. In the future, combining classical breeding and biotechnology could provide composite resistances with a broader spectrum of action and an increased durability.

REFERENCES

Abou-Jawdah, Y., Sobh, H., Fayyad, A., and Lecoq, H. (1997). First report of cucurbit aphid-borne yellows luteovirus in Lebanon. *Plant Dis.* **81:**1331.

Abudy, A., Sufrin-Ringwald, T., Dayan-Glick, C., Guenoune-Gelbart, D., Livneh, O., Zaccai, M., and Lapidot, M. (2010). Watermelon chlorotic stunt and Squash leaf curl begomoviruses-New threats to cucurbit crops in the Middle East. *Israel J. Plant Sci.* **58:**33–42.

Adams, M. J., Antoniw, J. F., and Fauquet, C. (2005). Molecular criteria for genus and species discrimination within the family *Potyviridae*. *Arch. Virol.* **150:**459–479.
Aguilar, J. M., Franco, M., Marco, C. F., Berdiales, B., Rodriguez-Cerezo, E., Truniger, V., and Aranda, M. A. (2003). Further variability within the genus *Crinivirus*, as revealed by determination of the complete RNA sequence of *Cucurbit yellow stunting disorder virus*. *J. Gen. Virol.* **84:**2555–2564.
Aguilar, M. I., Guirado, M. L., Melero-Vara, J. M., and Gomez, J. (2010). Efficacy of composting infected plant residues in reducing the viability of Pepper mild mottle virus, Melon necrotic spot virus and its vector, the soil-borne fungus Olpidium bornovanus. *Crop Prot.* **29:**342–348.
Ali-Shtayeh, M. S., Jamous, R. M., Husein, E. Y., and Alkhader, M. Y. (2010). First Report of Squash leaf curl virus in Squash (*Cucurbita pepo*), Melon (*Cucumis melo*), and Cucumber (*Cucumis sativus*) in the Northern West Bank of the Palestinian Authority. *Plant Dis.* **94:**640.
Al-Musa, A., Anfoka, G., Misbeh, S., Abhary, M., and Ahmad, F. H. (2008). Detection and molecular characterization of squash leaf curl virus (SLCV) in Jordan. *J. Phytopathol.* **156:**311–316.
Alonso-Prados, J. L., Luis-Arteaga, M., Alvarez, J. M., Moriones, E., Batlle, A., Lavina, A., Garcia-Arenal, F., and Fraile, A. (2003). Epidemics of aphid-transmitted viruses in melon crops in Spain. *Eur. J. Plant Pathol.* **109:**129–138.
Alvarez, M., and Campbell, R. N. (1978). Transmission and distribution of squash mosaic virus in seeds of cantaloupe. *Phytopathology* **68:**257–263.
Antignus, Y., Pearlsman, M., Benyoseph, R., and Cohen, S. (1990). Occurrence of a variant of cucumber green mottle mosaic virus in Israel. *Phytoparasitica* **18:**50–56.
Antignus, Y., Lapidot, M., Hadar, D., Messika, Y., and Cohen, S. (1998). Ultraviolet-absorbing screens serve as optical barriers to protect crops from virus and insect pests. *J. Econ. Entomol.* **91:**1401–1405.
Antignus, Y., Wang, Y., Pearlsman, M., Lachman, O., Lavi, N., and Gal-On, A. (2001). Biological and molecular characterization of a new cucurbit-infecting tobamovirus. *Phytopathology* **91:**565–571.
Antignus, Y., Lachman, O., Pearlsman, M., Omar, S., Yunis, H., Messika, Y., Uko, O., and Koren, A. (2003). Squash leaf curl virus—A new illegal immigrant from the Western Hemisphere and a threat to cucurbit crops in Israel. *Phytoparasitica* **31:**415.
Aramburu, J., Galipienso, L., Tornos, T., and Matas, M. (2006). First report of Eggplant mottled dwarf virus in mainland Spain. *Plant Pathol.* **55:**565.
Atreya, P. L., Atreya, C. D., and Pirone, T. P. (1991). Amino acid substitutions in the coat protein result in loss of insect transmissibility of a plant virus. *Proc. Natl. Acad. Sci. USA* **88:**7887–7891.
Avgelis, A. D., and Vovlas, C. (1986). Occurrence of cucumber green mottle mosaic virus in the island of Crete (Greece). *Phytopathol. Mediterr.* **25:**166–168.
Bateson, M. F., Lines, R. E., Revill, P., Chaleeprom, W., Ha, C. V., Gibbs, A. J., and Dale, J. L. (2002). On the evolution and molecular epidemiology of the potyvirus Papaya ringspot virus. *J. Gen. Virol.* **83:**2575–2585.
Berdiales, B., Bernal, J. J., Saez, E., Woudt, B., Beitia, F., and Rodriguez-Cerezo, E. (1999). Occurrence of cucurbit yellow stunting disorder virus (CYSDV) and beet pseudo-yellows virus in cucurbit crops in Spain and transmission of CYSDV by two biotypes of *Bemisia tabaci*. *Eur. J. Plant Pathol.* **105:**211–215.
Blanc, S., Lopez-Moya, J.-J., Wang, R., Garcia-Lampasona, S., Thornbury, D. W., and Pirone, T. P. (1997). A specific interaction between coat protein and helper component correlates with aphid transmission of a potyvirus. *Virology* **231:**141–147.
Blancard, D., Lecoq, H., and Pitrat, M. (1994). A Colour Atlas of Cucurbit Diseases. Manson Publishing, London, UK.

Blua, M. J., and Perring, T. M. (1992). Alatae production and population increase of aphid vectors on virus-infected host plants. *Oecologia* **92:**65–70.

Boubourakas, I. N., Hatziloukas, E., Antignus, Y., and Katis, N. I. (2004). Etiology of leaf chlorosis and deterioration of the fruit interior of watermelon plants. *J. Phytopathol.* **152:**580–588.

Boubourakas, I. N., Avgelis, A. D., Kyriakopoulou, P. E., and Katis, N. I. (2006). Occurrence of yellowing viruses (Beet pseudo-yellows virus, Cucurbit yellow stunting disorder virus and Cucurbit aphid-borne yellows virus) affecting cucurbits in Greece. *Plant Pathol.* **55:**276–283.

Bourdin, D., and Lecoq, H. (1991). Evidence that heteroencapsidation between two potyviruses is involved in aphid transmission of a non-aphid-transmissible isolate from mixed infections. *Phytopathology* **81:**1459–1463.

Brault, V., Périgon, S., Reinbold, C., Erdinger, M., Scheidecker, D., Herrbach, E., Richards, K., and Ziegler-Graff, V. (2005). The polerovirus minor capsid protein determines vector specificity and intestinal tropism in the aphid. *J. Virol.* **79:**9685–9693.

Brown, J. K., Idris, A. M., Alteri, C., and Stenger, D. C. (2002). Emergence of a new cucurbit-infecting begomovirus species capable of forming viable reassortants with related viruses in the squash leaf curl virus cluster. *Phytopathology* **92:**734–742.

Brunt, A. A., Crabtree, K., Dallwitz, M. J., Gibbs, A. J., and Watson, L. (1996). Viruses of Plants. CAB International, Wallingford, UK.

CABI/EPPO (1997). *Squash Mosaic Virus*. Distribution Maps of Plant Dis.s No 749. CAB International, Wallingford, UK.

CABI/EPPO (2003a). *Watermelon Mosaic Virus*. Distribution Maps of Plant Dis.s No 906.CAB International, Wallingford, UK.

CABI/EPPO (2003b). *Zucchini Yellow Mosaic Virus*. Distribution Maps of Plant Dis.s No 893. CAB International, Wallingford, UK.

CABI/EPPO (2003c). *Papaya Ringspot Virus*. Distribution Maps of Plant Dis.s No 902. CAB International, Wallingford, UK.

CABI/EPPO (2004). *Cucurbit Yellow Stunting Disorder Virus*. Distribution Maps of Plant Dis. s No 910. CAB International, Wallingford, UK.

CABI/EPPO (2005). *Cucumber Vein Yellowing Virus*. Distribution Maps of Plant Dis.s No 949. CAB International, Wallingford, UK.

CABI/EPPO (2010). *Melon Necrotic Spot Virus*. Distribution Maps of Plant Dis.s No 1089. CAB International, Wallingford, UK.

Campbell, R. N., Lecoq, H., Wipf-Schiebel, C., and Sim, S. T. (1990). Soil-borne viruses of melons and cucumbers in France. Proceedings of the First Symposium of the International Working Group on Plant Viruses with Fungal Vectors, Braunschweig, Germany, 21–24 August 1990, pp. 153–154.

Campbell, R. N., Sim, S. T., and Lecoq, H. (1995). Virus transmission by host-specific strains of *Olpidium bornovanus* and *Olpidium brassicae*. *Eur. J. Plant Pathol.* **101:**273–282.

Campbell, R. N., Wipf-Scheibel, C., and Lecoq, H. (1996). Vector-assisted seed transmission of *Melon necrotic spot virus* in melon. *Phytopathology* **86:**1294–1298.

Célix, A., Luis-Arteaga, M., and Rodriguez-Cerezo, E. (1996). First report of cucumber green mottle mosaic tobamovirus infecting greenhouse-grown cucumber in Spain. *Plant Dis.* **80:**1303.

Célix, A., Lopez-Sesé, A., Almarza, N., Gomez-Guillamon, M. L., and Rodriguez-Cerezo, E. (1996). Characterization of cucurbit yellow stunting disorder virus, a *Bemisia tabaci*-transmitted closterovirus. *Phytopathology* **86:**1370–1376.

Chen, K. C., Chiang, C. H., Raja, J. A., Liu, F. L., Tai, C. H., and Yeh, S.-D. (2008). A single amino acid of NIaPro of Papaya ringspot virus determines host specificity for infection of papaya. *Mol. Plant Microbe Interact.* **21:**1046–1057.

Christie, R. G., and Edwardson, J. R. (1986). Light microscopic techniques for detection of plant-virus inclusions. *Plant Dis.* **70:**273–279.

Chung, B. Y.-W., Miller, W. A., Atkins, J. F., and Firth, A. E. (2008). An overlapping essential gene in the Potyviridae. *Proc. Natl. Acad. Sci. USA* **105:**5897–5902.

Ciuffo, M., Roggero, P., Masenga, V., and Stravato, V. M. (1999). Natural infection of muskmelon by eggplant mottled dwarf rhabdovirus in Italy. *Plant Dis.* **83:**78.

Clark, M. F., and Adams, A. N. (1977). Characteristics of the microplate method of enzyme-linked immunosorbent assay for the detection of plant viruses. *J. Gen. Virol.* **34:**475–483.

Clough, G. H., and Hamm, P. B. (1995). Coat protein transgenic resistance to watermelon mosaic and zucchini yellow mosaic virus in squash and cantaloupe. *Plant Dis.* **79:**1107–1109.

Coffin, R. S., and Coutts, R. H. A. (1995). Relationships among *Trialeurodes vaporariorum*-transmitted yellowing viruses from Europe and North America. *J. Phytopathol.* **143:**375–380.

Cohen, S., and Nitzany, F. E. (1960). A whitefly transmitted virus of cucurbits in Israel. *Phytopathol. Mediterr.* **1:**44–46.

Cohen, S., and Nitzany, F. E. (1963). Identity of viruses affecting cucurbits in Israel. *Phytopathology* **53:**193–196.

Cohen, S., Duffus, J. E., Larsen, R. C., Liu, H. Y., and Flock, R. A. (1983). Purification, serology, and vector relationships of squash leaf curl virus, a whitefly-transmitted geminivirus. *Phytopathology* **73:**1669–1673.

Colinet, D., Kummert, J., and Lepoivre, P. (1998). The nucleotide sequence and genome organization of the whitefly transmitted sweetpotato mild mottle virus: A close relationship with members of the family *Potyviridae*. *Virus Res.* **53:**187–196.

Cotillon, A.-C., Desbiez, C., Bouyer, S., Wipf-Scheibel, C., Gros, C., Delecolle, B., and Lecoq, H. (2005). Production of a polyclonal antiserum against the coat protein of *Cucurbit yellow stunting disorder crinivirus* expressed in *Escherichia coli*. *EPPO Bull.* **35:**99–103.

Coutts, B. A., Kehoe, M. A., Webster, C. G., Wylie, S. J., and Jones, A. L. (2011). Zucchini yellow mosaic virus: Biological properties, detection procedures and comparison of coat protein gene sequences. *Arch. Virol.* **156:**2119–2131.

Crescenzi, A., Fanigliulo, A., Comes, S., Masenga, V., Pacella, R., and Piazzolla, P. (2001). Necrosis of watermelon caused by *Watermelon mosaic virus*. *J. Plant Pathol.* **83:**227.

Cuadrado, I. M., Janssen, D., Velasco, L., Ruiz, L., and Segundo, E. (2001). First report of Cucumber vein yellowing virus in Spain. *Plant Dis.* **85:**336.

Della Giustina, W., Javoy, M., Bansept, P., Morel, E., Balasse, H., Goussard, N., and Passard, C. (2000). The leafhoppers of the genus *Anaceratagallia* vectors of the virus responsable for the cucumber toad skin disease. *PHM Revue Horticole* **420:**40–43.

Desbiez, C., and Lecoq, H. (1997). Zucchini yellow mosaic virus. *Plant Pathol.* **46:**809–829.

Desbiez, C., and Lecoq, H. (2004). The nucleotide sequence of *Watermelon mosaic virus* (WMV, Potyvirus) reveals interspecific recombination between two related potyviruses in the 5′ part of the genome. *Arch. Virol.* **149:**1619–1632.

Desbiez, C., and Lecoq, H. (2008). Evidence for multiple intraspecific recombinants in natural populations of Watermelon mosaic virus (WMV). *Arch. Virol.* **153:**1749–1754.

Desbiez, C., Wipf-Scheibel, C., and Lecoq, H. (1999). Reciprocal assistance for aphid transmission between non-transmissible strains of zucchini yellow mosaic potyvirus in mixed infections. *Arch. Virol.* **144:**2213–2218.

Desbiez, C., Wipf-Scheibel, C., and Lecoq, H. (2002). Biological and serological variability, evolution and molecular epidemiology of *Zucchini yellow mosaic virus* (ZYMV, *Potyvirus*) with special reference to Caribbean islands. *Virus Res.* **85:**5–16.

Desbiez, C., Gal-On, A., Girard, M., Wipf-Scheibel, C., and Lecoq, H. (2003a). Increase in *Zucchini yellow mosaic virus* symptom severity in tolerant zucchini cultivars is related to a point mutation in P3 protein and is associated with a loss of relative fitness on susceptible plants. *Phytopathology* **93:**1478–1484.

Desbiez, C., Lecoq, H., and Girard, M. (2003b). First report of *Cucurbit yellow stunting disorder virus* in commercial cucumber greenhouses in France. *Plant Dis.* **87:**600.
Desbiez, C., Costa, C., Wipf-Scheibel, C., Girard, M., and Lecoq, H. (2007a). Serological and molecular variability of watermelon mosaic virus (genus Potyvirus). *Arch. Virol.* **152:**775–781.
Desbiez, C., Justafre, I., and Lecoq, H. (2007b). Molecular evidence that zucchini yellow fleck virus is a distinct and variable potyvirus related to papaya ringspot virus and Moroccan watermelon mosaic virus. *Arch. Virol.* **152:**449–455.
Desbiez, C., Joannon, B., Wipf-Scheibel, C., Chandeysson, C., and Lecoq, H. (2009). Emergence of new strains of *Watermelon mosaic virus* in South-Eastern France: Evidence for limited spread but rapid local population shift. *Virus Res.* **141:**201–208.
Desbiez, C., Joannon, B., Wipf-Scheibel, C., Chandeysson, C., and Lecoq, H. (2011). Recombination in natural populations of watermelon mosaic virus: New agronomic threat or damp squib? *J. Gen. Virol.* **92:**1939–1948.
Diaz, J. A., Nieto, C., Moriones, E., Truniger, V., and Aranda, M. A. (2004). Molecular characterization of a Melon necrotic spot virus strain that overcomes the resistance in melon and nonhost plants. *Mol. Plant Microbe Interact.* **17:**668–675.
Diaz-Pendon, J. A., Fernandez-Munoz, R., Gomez-Guillamon, M. L., and Moriones, E. (2005). Inheritance of resistance to *Watermelon mosaic virus* in *Cucumis melo* that impairs virus accumulation, symptom expression, and aphid transmission. *Phytopathology* **95:**840–846.
Dikova, B. (1995). Establishment of viruses on cucurbit crops in Bulgaria. *Rasteniev'dni Nauki* **32:**99–100.
Dikova, B. (1998). Identification of the squash mosaic virus in Bulgaria. *Rasteniev'dni Nauki* **35:**474–479.
Dogimont, C., Slama, S., Martin, J., and Pitrat, M. (1996). Sources of resistance to cucurbit aphid-borne yellows luteovirus in a melon germplasm collection. *Plant Dis.* **80:**1379–1382.
Dogimont, C., Bussemakers, A., Martin, J., Slama, S., Lecoq, H., and Pitrat, M. (1997). Two complementary recessive genes conferring resistance to cucurbit aphid borne yellows luteovirus in an Indian melon line (*Cucumis melo* L.). *Euphytica* **96:**391–395.
Dolja, V. V., Haldeman-Cahill, R., Montgomery, A. E., Vandenbosch, K. A., and Carrington, J. C. (1995). Capsid protein determinants involved in cell-to-cell and long distance movement of tobacco etch potyvirus. *Virology* **206:**1007–1016.
Dombrovsky, A., Huet, H., Chejanovsky, N., and Raccah, B. (2005). Aphid transmission of a potyvirus depends on suitability of the helper component and the N terminus of the coat protein. *Arch. Virol.* **150:**287–298.
Doolittle, S. P. (1916). A new infectious mosaic disease of cucumber. *Phytopathology* **6:**145–147.
Dorst, H. J. M. V. (1988). Surface water as source in the spread of cucumber green mottle mosaic virus. *Neth. J. Agric. Sci.* **36:**291–299.
Duffus, J. E. (1965). Beet pseudo-yellows virus, transmitted by the greenhouse whitefly (*Trialeurodes vaporariorum*). *Phytopathology* **55:**450–453.
Dukic, N., Krstic, B., Vico, I., Katis, N. I., Papavassiliou, C., and Berenji, J. (2002). Biological and serological characterization of viruses of summer squash crops in Yugoslavia. *J. Agric. Sci. (Belgrade)* **47:**149–160.
Fabre, F., Chadœuf, J., Costa, C., Lecoq, H., and Desbiez, C. (2010). Asymmetrical over-infection as a process of plant virus emergence. *J. Theor. Biol.* **265:**377–388.
FAOSTAT (2009). http://faostat.fao.org/site/339/default.aspx.
Fauquet, C., Mayo, M. A., Maniloff, J., Desselberger, U., and Ball, L. A. (2005). Virus Taxonomy. Eigth Report of the International Committee on Taxonomy of Viruses. Elsevier, London.
Fuchs, M., and Gonsalves, D. (1995). Resistance of transgenic hybrid squash ZW-20 expressing the coat protein genes of ZYMV and WMV2 to mixed infection by both potyviruses. *Biotechnology* **13:**1466–1473.

Gal-On, A., and Raccah, B. (2000). A point mutation in the FRNK motif of the potyvirus helper component-protease gene alters symptom expression in cucurbits and elicits protection against the severe homologous virus. *Phytopathology* **90:**467–473.

Genoves, A., Navarro, J. A., and Pallas, V. (2006). Functional analysis of the five melon necrotic spot virus genome-encoded proteins. *J. Gen. Virol.* **87:**2371–2380.

Ghanim, M., Sobol, I., Ghanim, M., and Czosnek, H. (2007). Horizontal transmission of begomoviruses between *Bemisia tabaci* biotypes. *Arthropod Plant Inter.* **1:**195–204.

Gibbs, M. J., and Cooper, J. I. (1995). A recombinational event in the history of luteoviruses probably induced by base-pairing between the genomes of two distinct viruses. *Virology* **206:**1129–1132.

Gibbs, A., and Mackenzie, A. (1997). A primer pair for amplifying part of the genome of all potyvirids by RT-PCR. *J. Virol. Methods* **63:**9–16.

Gibbs, A., Armstrong, J., Mackenzie, A. M., and Weiller, G. F. (1998). The GPRIME package: Computer programs for identifying the best regions of aligned genes to target in nucleic acid hybridisation-based diagnostic tests, and their use with plant viruses. *J. Virol. Methods* **74:**67–76.

Gil-Salas, F. M., Morris, J., Colyer, A., Budge, G., Boonham, N., Cuadrado, I. M., and Janssen, D. (2007). Development of real-time RT-PCR assays for the detection of Cucumber vein yellowing virus (CVYV) and Cucurbit yellow stunting disorder virus (CYSDV) in the whitefly vector *Bemisia tabaci*. *J. Virol. Methods* **146:**45–51.

Gonzalez-Garza, R., Gumpf, D. J., Kishaba, A. N., and Bohn, G. W. (1979). Identification, seed transmission, and host range pathogenicity of a California isolate of melon necrotic spot virus. *Phytopathology* **69:**340–345.

Gosalvez, B., Navarro, J. A., Lorca, A., Botella, F., Sanchez-Pina, M. A., and Pallas, V. (2003). Detection of Melon necrotic spot virus in water samples and melon plants by molecular methods. *J. Virol. Methods* **113:**87–93.

Guilley, H., Wipf-Scheibel, C., Richards, K., Lecoq, H., and Jonard, G. (1994). Nucleotide sequence of cucurbit aphid-borne yellows luteovirus. *Virology* **202:**1012–1017.

Ha, C., Coombs, S., Revill, P. A., Harding, R. M., Vu, M., and Dale, J. L. (2008). Design and application of two novel degenerate primer pairs for the detection and complete genomic characterization of potyviruses. *Arch. Virol.* **153:**25–36.

Harpaz, I., and Cohen, S. (1965). Semipersistent relationship between cucumber vein yellowing virus (CVYV) and its vector, the tobacco whitefly (*Bemisia tabaci* Gennadius). *Phytopathol. Z.* **54:**240–248.

Hassan, A. A., and Duffus, J. E. (1991). A review of a yellowing and stunting disorder of cucurbits in the United Arab Emirates. *Emir. J. Agric. Sci.* **2:**1–16.

Haudendshield, J. S., and Palukaitis, P. (1998). Diversity among isolates of squash mosaic virus. *J. Gen. Virol.* **79:**2331–2341.

Herrera-Vasquez, J. A., Cordoba-Selles, M. C., Cebrian, M. C., Alfaro-Fernandez, A., and Jorda, C. (2009). Seed transmission of Melon necrotic spot virus and efficacy of seed-disinfection treatments. *Plant Pathol.* **58:**436–442.

Herrera-Vasquez, J. A., Cordoba-Selles, M. C., Cebrian, M. C., Rossello, J. A., Alfaro-Fernandez, A., and Jorda, C. (2010). Genetic diversity of Melon necrotic spot virus and *Olpidium* isolates from different origins. *Plant Pathol.* **59:**240–251.

Hourani, H., and Abou-Jawdah, Y. (2003). Immunodiagnosis of Cucurbit yellow stunting disorder virus using polyclonal antibodies developed against recombinant coat protein. *J. Plant Pathol.* **85:**197–204.

Hristova, D. P., Jankulova, M. D., and Natskova, V. S. (1983). 'Chlorosis' in cucumbers—A new viral disease in Bulgaria. *C. R. Acad. Bulg. Sci.* **36:**1093–1096.

Hu, J. S., Li, H. P., Barry, K., and Wang, M. (1995). Comparison of dot blot, ELISA and RT-PCR assays for detection of two cucumber mosaic virus isolates infecting banana in Hawaii. *Plant Dis.* **79:**902–906.

Huttner, E., Tucker, W., Vermeulen, A., Ignart, F., Sawyer, B., and Birch, R. (2001). Ribozyme genes protecting transgenic melon plants against potyviruses. *Curr. Issues Mol. Biol.* **3:**27–34.

Idris, A. M., Abdel-Salam, A., and Brown, J. K. (2006). Introduction of the New World Squash leaf curl virus to squash (*Cucurbita pepo*) in Egypt: A potential threat to important food crops. *Plant Dis.* **90:**1262.

Janssen, D., Ruiz, L., Velasco, L., Segundo, E., and Cuadrado, I. M. (2002). Non-cucurbitaceous weed species shown to be natural hosts of Cucumber vein yellowing virus in south-eastern Spain. *Plant Pathol.* **51:**797.

Janssen, D., Ruiz, L., Cano, M., Belmonte, A., Martin, G., Segundo, E., and Cuadrado, I. M. (2003). Physical and genetic control of *Bemisia tabaci*-transmitted cucurbit yellow stunting disorder virus and cucumber vein yellowing virus in cucumber. *Bull. OILB/SROP* **26:**101–106.

Janssen, D., Martin, G., Velasco, L., Gomez, P., Segundo, E., Ruiz, L., and Cuadrado, I. M. (2005). Absence of coding region for the helper component-proteinase in the genome of cucumber vein yellowing virus, a whitefly-transmitted member of the *Potyviridae*. *Arch. Virol.* **150:**1439–1447.

Janssen, D., Velasco, L., Martin, G., Segundo, E., and Cuadrado, I. M. (2007). Low genetic diversity among Cucumber vein yellowing virus isolates from Spain. *Virus Genes* **34:**367–371.

Joannon, B., Lavigne, C., Lecoq, H., and Desbiez, C. (2010). Barriers to gene flow between emerging populations of Watermelon mosaic virus in south-eastern France. *Phytopathology* **100:**1373–1379.

Jordan, R. (1992). Potyviruses, monoclonal antibodies, and antigenic sites. *Arch. Virol. Suppl.* **5:**81–95.

Juarez, M., Truniger, V., and Aranda, M. A. (2004). First report of Cucurbit aphid-borne yellows virus in Spain. *Plant Dis.* **88:**907.

Kasschau, K. D., and Carrington, J. C. (2001). Long-distance movement and replication maintenance functions correlate with silencing suppression activity of potyviral HC-Pro. *Virology* **285:**71–81.

Kassem, A. A. H., Halim, K. A., Rifai, O. E. G., and Warrak, W. (2005). The most important viruses affecting cucurbits in Syria. *Arab J. Plant Prot.* **23:**1–6.

Kassem, M. A., Sempere, R. N., Juarez, M., Aranda, M. A., and Truniger, V. (2007). Cucurbit aphid borne yellows virus is prevalent in field-grown cucurbit crops in southeastern Spain. *Plant Dis.* **91:**232–238.

Kataya, A. R. A., Suliman, M. N. S., Kalantidis, K., and Livieratos, I. C. (2009). Cucurbit yellow stunting disorder virus p25 is a suppressor of post-transcriptional gene silencing. *Virus Res.* **145:**48–53.

Katis, N. I., Chatzivassiliou, E. K., Avgelis, A. D., Manoussopoulos, I., and Lecoq, H. (2000). Occurrence of Eggplant mottled dwarf nucleorhabdovirus (EMDV) in tobacco and cucumber in Greece. *Phytopathol. Mediterr.* **39:**318.

Katis, N. I., Tsitsipis, J. A., Lykouressis, D. P., Papapanayotou, A., Margaritopoulos, J. T., Kokinis, G. M., Perdikis, D. C., and Manoussopoulos, I. N. (2006). Transmission of Zucchini yellow mosaic virus by colonizing and non-colonizing aphids in Greece and new aphid species vectors of the virus. *J. Phytopathol.* **154:**293–302.

Katis, N. I., Chatzivassiliou, E. K., Clay, C. M., Maliogka, V., Pappi, P., Efthimiou, K., Dovas, C. I., and Avgelis, A. D. (2011). Development of an IC-RT-PCR assay for the detection of eggplant mottled dwarf virus and partial characterization of isolates from various hosts in Greece. *J. Plant Pathol.* **93:**353–362.

Kheyr-Pour, A., Bananej, K., Dafalla, G., Caciagli, P., Noris, E., Ahoonmanesh, A., Lecoq, H., and Gronenborn, B. (2000). Watermelon chlorotic stunt virus from the Sudan and Iran:

Sequence Comparisons and Identification of a Whitefly-Transmission Determinant. *Phytopathology* **90:**629–635.

Kido, K., Tanaka, C., Mochizuki, T., Kubota, K., Ohki, T., Ohnishi, J., Knight, L. M., and Tsuda, S. (2008). High temperatures activate local viral multiplication and cell-to-cell movement of Melon necrotic spot virus but restrict expression of systemic symptoms. *Phytopathology* **98:**181–186.

Knierim, D., Deng, T. C., Tsai, W. S., Green, S. K., and Kenyon, L. (2010). Molecular identification of three distinct Polerovirus species and a recombinant Cucurbit aphid-borne yellows virus strain infecting cucurbit crops in Taiwan. *Plant Pathol.* **59:**991–1002.

Koklu, G., and Yilmaz, O. (2006). Occurrence of cucurbit viruses on field-grown melon and watermelon in the Thrace region of Turkey. *Phytoprotection* **87:**123–130.

Kostova, D., Masenga, V., Milne, R. G., and Lisa, V. (2001). First report of Eggplant mottled dwarf virus in cucumber and pepper in Bulgaria. *Plant Pathol.* **50:**804.

Lecoq, H. (1982). A new cucrumber disease caused by a rhabdovirus; toad's skin disease. *PHM Revue Horticole* **223:**15–17.

Lecoq, H. (1992a). Les virus des cultures de melon et de courgette de plein champ (I). *PHM Revue Horticole* **323:**23–27.

Lecoq, H. (1992b). Les virus des cultures de melon et de courgette de plein champ (II). *PHM Revue Horticole* **324:**15–25.

Lecoq, H. (1999). Epidemiology of Cucurbit aphid-borne yellows virus. *In* "The Luteoviridae" (H. G. Smith and H. Barker, eds.), pp. 243–248. CAB International, Wallingford, UK.

Lecoq, H. (2003). Cucurbits. *In* "Virus and Virus-Like Diseases of Major Crops in Developing Countries" (G. Loebenstein and G. Thottapilly, eds.), pp. 665–688. Kluwer Academic Publishers, Dordrecht, Netherlands.

Lecoq, H., and Desbiez, C. (2008). Watermelon mosaic virus and Zucchini yellow mosaic virus. *In* "Encyclopedia of Virology, Third Edition" (B. W. J. Mahy and M. H. V. Van Regenmortel, eds.), pp. 433–440. Elsevier, Oxford, UK, 5 Vols.

Lecoq, H., and Pitrat, M. (1984). Strains of zucchini yellow mosaic virus in muskmelon (*Cucumis melo* L.). *J. Phytopathol.* **111:**165–173.

Lecoq, H., and Pitrat, M. (1985). Specificity of the helper-component-mediated aphid transmission of three potyviruses infecting muskmelon. *Phytopathology* **75:**890–893.

Lecoq, H., Pitrat, M., and Clément, M. (1981). Identification et caractérisation d'un potyvirus provoquant la maladie du rabougrissement jaune du melon. *Agronomie* **1:**827–834.

Lecoq, H., Lisa, V., and Dellavalle, G. (1983). Serological identity of muskmelon yellow stunt and zucchini yellow mosaic viruses. *Plant Dis.* **67:**824–825.

Lecoq, H., Piquemal, J.-P., Michel, M. J., and Blancard, D. (1988). Virus de la mosaïque de la courge: une nouvelle menace pour les cultures de melon en France? *PHM Revue Horticole* **289:**25–30.

Lecoq, H., Bourdin, D., Raccah, B., Hiebert, E., and Purcifull, D. E. (1991a). Characterization of a Zucchini Yellow Mosaic Virus isolate with a deficient helper component. *Phytopathology* **81:**1087–1091.

Lecoq, H., Lemaire, J.-M., and Wipf-Scheibel, C. (1991b). Control of ZYMV in squash by cross protection. *Plant Dis.* **75:**208–211.

Lecoq, H., Bourdin, D., Wipf-Scheibel, C., Bon, M., Lot, H., Lemaire, O., and Herrbach, E. (1992). A new yellowing disease of cucurbits caused by a luteovirus, cucurbit aphid-borne yellows virus. *Plant Pathol.* **41:**749–761.

Lecoq, H., Gilbert-Albertini, F., Wipf-Scheibel, C., Pitrat, M., Bourdin, D., Belkhala, H., Katis, N., and Yilmaz, M. (1994). Occurence of a new yellowing disease of cucurbits in the mediterranean basin caused by a luteovirus, cucurbit aphid-borne yellows virus and prospects for control. 9th Congress of the Mediterranean Phytopathological Union, Kusadasi-Aydin, Turkey, pp. 461–463.

Lecoq, H., Wisler, G., and Pitrat, M. (1998). Cucurbit viruses: The classics and the emerging. *In* "Cucurbitaceae '98 Evaluation and Enhancement of Cucurbit Germplasm" (J. D. McCreight, ed.), pp. 126–142. ASHS Press, Alexandria, VA.

Lecoq, H., Desbiez, C., Delecolle, B., Cohen, S., and Mansour, A. (2000). Cytological and molecular evidence that the whitefly-transmitted *Cucumber vein yellowing virus* is a tentative member of the family *Potyviridae*. *J. Gen. Virol.* **81:**2289–2293.

Lecoq, H., Dafalla, G., Desbiez, C., Wipf-Scheibel, C., Delecolle, B., Lanina, T., Ullah, Z., and Grumet, R. (2001). Biological and molecular characterization of *Moroccan watermelon mosaic virus* and a potyvirus isolate from Eastern Sudan. *Plant Dis.* **85:**547–552.

Lecoq, H., Desbiez, C., Wipf-Scheibel, C., Girard, M., and Pitrat, M. (2002). Durability of Zucchini yellow mosaic virus resistances in cucurbits. *In* "Cucurbitaceae 2002" (D. N. Maynard, ed.), pp. 294–300. ASHS Press, Alexandria, VA.

Lecoq, H., Desbiez, C., Wipf-Scheibel, C., and Girard, M. (2003). Potential involvement of melon fruit in the long distance dissemination of cucurbit potyviruses. *Plant Dis.* **87:**955–959.

Lecoq, H., Moury, B., Desbiez, C., Palloix, A., and Pitrat, M. (2004). Durable virus resistance in plants through conventional approaches: A challenge. *Virus Res.* **100:**31–39.

Lecoq, H., Desbiez, C., Wipf-Scheibel, C., and Costa, C. (2005). Molecular epidemiology of Watermelon mosaic virus (WMV, Potyvirus) in cucurbits: From simple to complex patterns. Proceedings IX International Plant Virus Epidemiology Symposium, Lima, Peru, p. 53.

Lecoq, H., Dufour, O., Wipf-Scheibel, C., Girard, M., Cotillon, A.-C., and Desbiez, C. (2007). First report of Cucumber vein yellowing virus in melon in France. *Plant Dis.* **91:**909.

Lecoq, H., Wipf-Scheibel, C., Chandeysson, C., Le Van, A., Fabre, F., and Desbiez, C. (2009). Molecular epidemiology of *Zucchini yellow mosaic virus* in France: An historical overview. *Virus Res.* **141:**190–200.

Lecoq, H., Fabre, F., Joannon, B., Wipf-Scheibel, C., Chandeysson, C., Schoeny, A., and Desbiez, C. (2011). Search for factors involved in the rapid shift in watermelon mosaic virus (WMV) populations in south-eastern France. *Virus Res.* **159:**115–123.

Lee, G. P., Min, B. E., Kim, C. S., Choi, S. H., Harn, C. H., Kim, S. U., and Ryu, K. H. (2003). Plant virus cDNA chip hybridization for detection and differentiation of four cucurbit-infecting Tobamoviruses *J. Virol. Methods* **110:**19–24.

Léonard, S., Plante, D., Wittmann, S., Daigneault, N., Fortin, M. G., and Laliberté, J.-F. (2000). Complex formation between potyirus VPg and translation eukaryotic initiation factor 4E correlates with virus infectivity. *J. Virol.* **74:**7730–7737.

Leroux, J. P., Quiot, J.-B., Lecoq, H., and Pitrat, M. (1979). Mise en évidence et répartition dans le Sud-Est de la France d'un pathotype particulier du virus de la mosaïque du concombre. *Ann. Phytopathol.* **11:**431–438.

Ling, K.-S., Wechter, P., Walcott, R. R., and Keinath, A. P. (2011). Development of a Real-time RT-PCR Assay for Squash Mosaic Virus Useful for Broad Spectrum Detection of Various Serotypes and its Incorporation into a Multiplex Seed Health Assay. *J. Phytopathol.* **159:**649–656.

Lisa, V., and Lecoq, H. (1984). Zucchini Yellow Mosaic Virus. CMI/AAB Description of plant viruses, no 282.

Lisa, V., Boccardo, G., D'Agostino, G., Dellavalle, G., and D'Aquilio, M. (1981). Characterization of a potyvirus that causes zucchini yellow mosaic. *Phytopathology* **71:**667–672.

Lockhart, B. E. L., Ferji, Z., and Hafidi, B. (1982). Squash mosaic virus in Morocco. *Plant Dis.* **66:**1191–1193.

Lockhart, B. E. L., Jebbour, F., and Lennon, A. M. (1985). Seed transmission of squash mosaic virus in *Chenopodium* spp. *Plant Dis.* **69:**946–947.

Lot, H., Delécolle, B., and Lecoq, H. (1983). A whitefly transmitted virus causing muskmelon yellows in France. *Acta Hortic.* **127:**175–182.

Lovisolo, O. (1977). Virus diseases of Cucurbitaceae in the Mediterranean region. Comptes Rendus des 5ème Journées de Phytiatrie et de Phytopharmacie Circumméditerranéennes, Rabat, Morocco, pp. 1–13.

Lovisolo, O. (1980). Virus and viroid diseases of cucurbits. *Acta Hortic.* **88:**33–82.

Mallor Gimenez, C., Alvarez Alvarez, J. M., and Arteaga, M. L. (2003). Inheritance of resistance to systemic symptom expression of *Melon necrotic spot virus* (MNSV) in *Cucumis melo* L. 'Doublon'. *Euphytica* **134:**319–324.

Marco, C. F., and Aranda, M. A. (2005). Genetic diversity of a natural population of *Cucurbit yellow stunting disorder virus*. *J. Gen. Virol.* **86:**815–822.

Martinez-Garcia, B., Marco, C. F., Goytia, E., Lopez-Abella, D., Serra, M. T., Aranda, M. A., and Lopez-Moya, J.-J. (2004). Development and use of detection methods specific for *Cucumber vein yellowing virus* (CVYV). *Eur. J. Plant Pathol.* **110:**811–821.

Mnari-Hattab, M., Jebari, H., and Zouba, A. (2008). Identification and distribution of viruses responsible for mosaic diseases affecting cucurbits in Tunisia. *Bull. OEPP/EPPO Bull.* **38:**497–506.

Mnari-Hattab, M., Gauthier, N., and Zouba, A. (2009). Biological and molecular characterization of the cucurbit aphid-borne yellows virus affecting cucurbits in Tunisia. *Plant Dis.* **93:**1065–1072.

Mochizuki, T., Ohnishi, J., Ohki, T., Kanda, A., and Tsuda, S. (2008). Amino acid substitution in the coat protein of Melon necrotic spot virus causes loss of binding to the surface of *Olpidium bornovanus* zoospores. *J. Gen. Plant Pathol.* **74:**176–181.

Mochizuki, T., Hirai, K., Kanda, A., Ohnishi, J., Ohki, T., and Tsuda, S. (2009). Induction of necrosis via mitochondrial targeting of *Melon necrotic spot virus* replication protein p29 by its second transmembrane domain. *Virology* **390:**239–249.

Moreno, I. M., Malpica, J. M., Diaz-Pendon, J. A., Moriones, E., Fraile, A., and Garcia-Arenal, F. (2004). Variability and genetic structure of the population of watermelon mosaic virus infecting melon in Spain. *Virology* **318:**451–460.

Nawaz-ul-Rehman, M. S., and Fauquet, C. M. (2009). Evolution of geminiviruses and their satellites. *FEBS Lett.* **583:**1825–1832.

Nieto, C., Morales, M., Orjeda, G., Clepet, C., Monfort, A., Sturbois, B., Puigdomènech, P., Pitrat, M., Caboche, M., Dogimont, C., Garcia-Mas, J., Aranda, M. A., *et al.* (2006). An eIF4E allele confers resistance to an uncapped and non-polyadenylated RNA virus in melon. *Plant J.* **48:**452–462.

Ohki, T., Akita, F., Mochizuki, T., Kanda, A., Sasaya, T., and Tsuda, S. (2010). The protruding domain of the coat protein of Melon necrotic spot virus is involved in compatibility with and transmission by the fungal vector *Olpidium bornovanus*. *Virology* **402:**129–134.

Paludan, N. (1985). Spread of viruses by recirculated nutrient solutions in soilless cultures. *Tidsskr. Planteavl* **89:**467–474.

Papayiannis, L. C., Ioannou, N., Boubourakas, I. N., Dovas, C. I., Katis, N. I., and Falk, B. W. (2005). Incidence of viruses infecting cucurbits in Cyprus. *J. Phytopathol.* **153:**530–535.

Parry, J. N., and Persley, D. M. (2005). Carrot as a natural host of watermelon mosaic virus. *Australas. Plant Pathol.* **34:**283–284.

Pfeffer, S., Dunoyer, P., Heim, F., Richards, K. E., Jonard, G., and Ziegler-Graff, V. (2002). P0 of beet western yellows virus is a suppressor of posttranscriptional gene silencing. *J. Virol.* **76:**6815–6824.

Pitrat, M., and Lecoq, H. (1980). Inheritance of resistance to cucumber mosaic virus transmission by *Aphis gossypii* in *Cucumis melo*. *Phytopathology* **70:**958–961.

Pospieszny, H., Hasiow, B., and Borodynko, N. (2007). A Polish Isolate of Zucchini yellow mosaic virus from Zucchini is Distinct from Other European Isolates. *Plant Dis.* **91:**639.

Provvidenti, R. (1986). Viral diseases of cucurbits and sources of resistance. Food & Fertilizer Technology Center, Taiwan. Technical Bulletin N° 93.

Provvidenti, R. (1996). Diseases caused by viruses. *In* "Compendium of Cucurbit Diseases" (T. A. Zitter, D. L. Hopkins, and C. E. Thomas, eds.), pp. 37–45. APS, St Paul, MN.

Prüfer, D., Wipf-Scheibel, C., Richards, K., Guilley, H., Lecoq, H., and Jonard, G. (1995). Synthesis of a full-length infectious cDNA clone of Cucurbit aphid-borne yellows virus and its use in gene exchange experiments with structural proteins from other luteoviruses. *Virology* **214:**150–158.

Purcifull, D. E., and Hiebert, E. (1979). Serological distinction of watermelon mosaic virus isolates. *Phytopathology* **69:**112–116.

Quiot, J. B. (1980). Ecology of cucumber mosaic virus in the Rhône Valley of France. *Acta Hortic.* **88:**9–21.

Quiot, J. B., Labonne, G., and Quiot-Douine, L. (1983). The comparative ecology of cucumber mosaic virus in Mediterranean and tropical regions. *In* "Plant Virus Epidemiology" (R. T. Plumb and J. M. Tresh, eds.), pp. 177–183. Blackwell Scientific Publications, Oxford, UK.

Quiot-Douine, L., Purcifull, D. E., Hiebert, E., and De Meija, M. V. G. (1986). Serological relationships and *in vitro* translation of an antigenically distinct strain of papaya ringspot virus. *Phytopathology* **76:**346–351.

Quiot-Douine, L., Lecoq, H., Quiot, J.-B., Pitrat, M., and Labonne, G. (1990). Serological and biological variability of virus isolates related to strains of papaya ringspot virus. *Phytopathology* **80:**256–263.

Reinbold, C., Herrbach, E., and Brault, V. (2003). Posterior midgut and hindgut are both sites of acquisition of Cucurbit aphid-bome yellows virus in *Myzus persicae* and *Aphis gossypii*. *J. Gen. Virol.* **84:**3473–3484.

Robinson, R. W., and Deckers-Walters, D. S. (1997). Cucurbits. CAB International, Wallingford, UK.

Roggero, P., and Lisa, V. (1995). Characterization of an isolate of tobacco necrosis virus from zucchini. *J. Phytopathol.* **143:**485–489.

Roggero, P., Milne, R. G., Masenga, V., Ogliara, P., and Stravato, V. M. (1995). First reports of eggplant mottled dwarf rhabdovirus in cucumber and in pepper. *Plant Dis.* **79:**321.

Rojas, M. R., Gilbertson, R. L., Russell, D. R., and Maxwell, D. P. (1993). Use of degenerate primers in the polymerase chain reaction to detect whitefly-transmitted geminiviruses. *Plant Dis.* **77:**340–347.

Rubio, L., Soong, J., Kao, J., and Falk, B. W. (1999). Geographic distribution and molecular variation of isolates of three whitefly-borne closteroviruses of cucurbits: Lettuce infectious yellows virus, cucurbit yellow stunting disorder virus, and beet pseudo-yellows virus. *Phytopathology* **89:**707–711.

Rubio, L., Abou-Jawdah, Y., Lin, H.-X., and Falk, B. W. (2001). Geographically distant isolates of the crinivirus *Cucurbit yellow stunting disorder virus* show very low genetic diversity in the coat protein gene. *J. Gen. Virol.* **82:**929–933.

Ruiz, L., Janssen, D., Martin, G., Velasco, L., Segundo, E., and Cuadrado, I. M. (2006). Analysis of the temporal and spatial disease progress of *Bemisia tabaci*-transmitted *Cucurbit yellow stunting disorder virus* and *Cucumber vein yellowing virus* in cucumber. *Plant Pathol.* **55:**264–275.

Russo, M., Martelli, G. P., Vovlas, C., and Ragozzino, A. (1979). Comparative studies on Mediterranean isolates of watermelon mosaic virus. *Phytopathol. Mediter.* **18:**94–101.

Sacristan, S., Fraile, A., and Garcia-Arenal, F. (2004). Population dynamics of *Cucumber mosaic virus* in melon crops and in weeds in Central Spain. *Phytopathology* **94:**992–998.

Sela, I., Assouline, I., Tanne, E., Cohen, S., and Marco, S. (1980). Isolation and characterization of a rod-shaped, whitefly-transmissible, DNA-containing plant virus. *Phytopathology* **70:**226–228.

Simmons, H. E., Holmes, E. C., Gildow, F. E., Bothe-Goralczyk, M. A., and Stephenson, A. G. (2011). Experimental verification of seed transmission of zucchini yellow mosaic virus. *Plant Dis.* **95:**751–754.

Stewart, L. R., Medina, V., Tian, T., Turina, M., Falk, B. W., and Ng, J. C. K. (2010). A mutation in the lettuce infectious yellows virus minor coat protein disrupts whitefly transmission but not in planta systemic movement. *J. Virol.* **84**:12165–12173.

Tian, T. Y., Rubio, L., Yeh, H. H., Crawford, B., and Falk, B. W. (1999). Lettuce infectious yellows virus: *In vitro* acquisition analysis using partially purified virions and the whitefly *Bemisia tabaci*. *J. Gen. Virol.* **80**:1111–1117.

Tobias, I., and Palkovics, L. (2003). Characterization of Hungarian isolates of zucchini yellow mosaic virus (ZYMV, potyvirus) transmitted by seeds of *Cucurbita pepo* var *styriaca*. *Pest Manag. Sci.* **59**:493–497.

Tomassoli, L., and Meneghini, M. (2007). First report of Cucurbit aphid-borne yellows virus in Italy. *Plant Pathol.* **56**:537.

Tomassoli, L., Lumia, V., Siddu, G. F., and Barba, M. (2003). Yellowing disease of melon in Sardinia (Italy) caused by beet pseudoyellows virus. *J. Plant Pathol.* **85**:59–61.

Tomassoli, L., Tiberini, A., and Meneghini, M. (2010). Zucchini yellow fleck virus is an emergent virus on melon in Sicily (Italy). *J. Phytopathol.* **158**:314–316.

Tomlinson, J. A., and Thomas, B. J. (1986). Studies on melon necrotic spot virus disease of cucumber and on the control of the fungus vector (*Olpidium radicale*). *Ann. Appl. Biol.* **108**:71–80.

Tricoli, D. M., Carney, K. J., Russell, P. F., Russell McMaster, J., Groff, D. W., Hadden, K. C., Himmel, P. T., Hubbard, J. P., Boeshore, M. L., and Quemada, H. D. (1995). Field evaluation of transgenic squash containing single or multiple virus coat protein gene for resistance to CMV, WMV2, and ZYMV. *Biotechnology* **13**:1458–1465.

Ugaki, M., Tomiyama, M., Kakutami, T., Hidaka, S., Kiguchi, T., Nagata, R., Sato, T., Motoyoshi, F., and Nishiguchi, M. (1991). The complete nucleotide sequence of cucumber green mottle mosaic virus (SH strain) genomic RNA. *J. Gen. Virol.* **72**:1487–1495.

Valli, A., Martin-Hernandez, A. M., Lopez-Moya, J.-J., and Garcia, J. A. (2006). RNA silencing suppression by a second copy of the P1 serine protease of *Cucumber vein yellowing ipomovirus*, a member of the family *Potyviridae* that lacks the cysteine protease HCPro. *J. Virol.* **80**:10055–10063.

Valli, A., Dujovny, G., and Garcia, J. A. (2008). Protease activity, self interaction, and small interfering RNA binding of the silencing suppressor P1b from Cucumber vein yellowing ipomovirus. *J. Virol.* **82**:974–986.

Varveri, C., Vassilakos, N., and Bem, F. (2002). Characterization and detection of Cucumber green mottle mosaic virus in Greece. *Phytoparasitica* **30**:493–501.

Vucurovic, A., Bulajic, A., Ekic, I., Ristic, D., Berenji, J., and Krstic, B. (2009). Presence and distribution of oilseed pumpkin viruses and molecular detection of Zucchini yellow mosaic virus. *Pesticidi i Fitomedicina* **24**:85–94.

Walters, S. A. (2003). Suppression of watermelon mosaic virus in summer squash with plastic mulches and rowcovers. *HortTechnology* **13**:352–357.

Webb, R. E., and Scott, H. A. (1965). Isolation and identification of watermelon mosaic virus 1 and 2. *Phytopathology* **55**:895–900.

Wei, T., Pearson, M. N., Blohm, D., Nölte, M., and Armstrong, K. (2009). Development of a short oligonucleotide microarray for the detection and identification of multiple potyviruses. *J. Virol. Methods* **162**:109–118.

Wei, T., Zhang, C., Hong, J., Xiong, R., Kasschau, K. D., Zhou, X., Carrington, J. C., and Wang, A. (2010). Formation of complexes at plasmodesmata for potyvirus intercellular movement is mediated by the viral protein P3N-PIPO. *PLoS Pathog.* **6**:e1000962.

Wen, R.-H., and Hajimorad, M. R. (2010). Mutational analysis of the putative PIPO of soybean mosaic virus suggests disruption of PIPO protein impedes movement. *Virology* **400**:1–7.

Wintermantel, W. M., Hladky, L. L., Cortez, A. A., and Natwick, E. T. (2009). A new expanded host range of cucurbit yellow stunting disorder virus includes three agricultural crops. *Plant Dis.* **93**:685–690.

Wisler, G. C., Duffus, J. E., Liu, H.-Y., and Li, R. H. (1998). Ecology and epidemiology of whitefly-transmitted closteroviruses. *Plant Dis.* **82:**270–280.

Wyatt, S. D., and Brown, J. K. (1996). Detection of subgroup III geminivirus isolates in leaf extracts by degenerate primers and polymerase chain reaction. *Phytopathology* **86:**1288–1293.

Yakoubi, S., Desbiez, C., Fakhfakh, H., Wipf-Scheibel, C., Marrakchi, M., and Lecoq, H. (2007). Occurrence of *Cucurbit yellow stunting disorder virus* and *Cucumber vein yellowing virus* in Tunisia. *J. Plant Pathol.* **89:**417–420.

Yakoubi, S., Desbiez, C., Fakhfakh, H., Wipf-Scheibel, C., Fabre, F., Pitrat, M., Marrakchi, M., and Lecoq, H. (2008a). Molecular, biological and serological variability of Zucchini yellow mosaic virus in Tunisia. *Plant Pathol.* **57:**1146–1154.

Yakoubi, S., Desbiez, C., Fakhfakh, H., Wipf-Scheibel, C., Marrakchi, M., and Lecoq, H. (2008b). Biological characterization and complete nucleotide sequence of a Tunisian isolate of Moroccan watermelon mosaic virus. *Arch. Virol.* **153:**775–781.

Yakoubi, S., Lecoq, H., and Desbiez, C. (2008c). Algerian watermelon mosaic virus (AWMV): A new potyvirus species in the PRSV cluster. *Virus Genes* **37:**103–109.

Yakoubi, S., Desbiez, C., Fakhfakh, H., Wipf-Scheibel, C., Marrakchi, M., and Lecoq, H. (2008d). First report of Melon necrotic spot virus on melon in Tunisia. *Plant Pathol.* **57:**386.

Yang, Y., Gong, J., Li, H., Li, C., Wang, D., Li, K., and Zhi, H. (2011). Identification of a novel *Soybean mosaic virus* isolate in China that contains a unique 5′ terminus sharing high sequence homology with *Bean common mosaic virus*. *Virus Res.* **15:**13–18.

Yarden, G., Hemo, R., Livne, H., Maoz, F., Lev, E., Lecoq, H., and Raccah, B. (2000). Cross-protection of Cucurbitaceae from zucchini yellow mosaic potyvirus. *Acta Hortic.* **510:**349–356.

You, B. J., Chiang, C. H., Chen, L. F., Su, W. C., and Yeh, S. D. (2005). Engineered mild strains of Papaya ringspot virus for broader cross protection in cucurbits. *Phytopathology* **95:**533–540.

Yuan, C., and Ullman, D. E. (1996). Comparison of efficiency and propensity as measures of vector importance in zucchini yellow mosaic potyvirus transmission by *Aphis gossypii* and *A. craccivora*. *Phytopathology* **86:**698–703.

CHAPTER 4

Viruses of Pepper Crops in the Mediterranean Basin: A Remarkable Stasis

Benoît Moury and **Eric Verdin**

Contents

Abstract

Compared to other vegetable crops, the major viral constraints affecting pepper crops in the Mediterranean basin have been remarkably stable for the past 20 years. Among these viruses, the most prevalent ones are the seed-transmitted tobamoviruses; the aphid-transmitted *Potato virus Y* and *Tobacco etch virus* of the genus *Potyvirus*, and *Cucumber mosaic virus* member of the genus *Cucumovirus*; and thrips-transmitted tospoviruses. The last major viral emergence concerns the tospovirus *Tomato spotted wilt virus* (TSWV), which has undergone major outbreaks since the end of the 1980s and the worldwide dispersal of the thrips vector *Frankliniella occidentalis* from the western part of the USA. TSWV outbreaks in

INRA, UR407 Pathologie Végétale, Domaine Saint Maurice, Montfavet, France

Advances in Virus Research, Volume 84
ISSN 0065-3527, DOI: 10.1016/B978-0-12-394314-9.00004-X

the Mediterranean area might have been the result of both viral introductions from Northern America and local reemergence of indigenous TSWV isolates. In addition to introductions of new viruses, resistance breakdowns constitute the second case of viral emergences. Notably, the pepper resistance gene *Tsw* toward TSWV has broken down a few years after its deployment in several Mediterranean countries while there has been an expansion of L^3-resistance breaking pepper mild mottle tobamovirus isolates. Beyond the agronomical and economical concerns induced by the breakdowns of virus resistance genes in pepper, they also constitute original models to understand plant–virus interactions and (co) evolution.

I. INTRODUCTION

Viral diseases constitute the major limiting factor in pepper cultivation throughout the world (Florini and Zitter, 1987; Green and Kim, 1991; Martelli and Quacquarelli, 1983; Yoon *et al.*, 1989). Forty-nine virus species have been shown to infect pepper (Hanssen *et al.*, 2010) among which about 20, belonging to 15 different taxonomic groups, have been reported to cause damages in pepper crops. Mechanically transmitted viruses like tobamoviruses are predominant in protected crops, whereas insect-transmitted viruses like potyviruses, cucumoviruses, and tospoviruses are more frequent and severe in open fields.

Generally speaking, viral emergences, that is, outbreaks of novel virus entities or increase of the prevalence of, or damages induced by known viruses are favored by several factors frequently linked to the intensification of agricultural practices or of international trade (Gómez *et al.*, 2012). Given the wide geographical distribution of pepper crops, which exposes them to a large diversity of parasites, trade of infected plant material allows the introduction of viruses and their vectors in new countries or areas. Intensive and monovarietal pepper cultivations in Mediterranean regions favor virus adaptation, spread, and persistence from one crop cycle to the following. Finally, climate changes affect the distribution of virus vectors and/or the susceptibility of plants to virus infections and consequently increase the exposure of pepper crops to new viral diseases. In spite of these threats, few viral emergences have been recorded in pepper crops in the Mediterranean basin during the past 20 years which contrasts with the situation of other vegetable crops such as tomato (Hanssen *et al.*, 2010) or cucurbits (see Chapter 3 of this volume).

In this review, we describe the biological properties of the most important viruses which affect pepper crops in the Mediterranean surrounds and the methods used to control them, with an emphasis on varietal resistances.

II. APHID-TRANSMITTED VIRUSES

A. Potyviruses

The genus *Potyvirus* constitutes one of the largest groups of plant viruses as a whole. Potyviruses are responsible for particularly important diseases in a wide range of plant species all over the world. All of them involve aphid vectors for their transmission, while some of them are additionally seed transmitted. Potyviruses are single-stranded RNA viruses with flexuous particles 680–900 nm long and 11–13 nm wide possessing a helical symmetry with a pitch of about 3.4 nm. Virions contain a linear, positive-sense RNA of about 9.7 kb in size, with a polyadenylated tract at the 3′end, and a viral protein, the VPg, covalently linked at the 5′end. The RNA encodes a single polyprotein which is subsequently cleaved into 10 proteins by three viral proteinases. A shift during translation allows the synthesis of an 11th protein (Chung *et al.*, 2008). Diversification of potyviruses was estimated to be quite recent and contemporaneous to the development of agriculture, around 10,000 years ago (Gibbs *et al.*, 2008).

Species that infect solanaceous plants belong to three separate clades within the genus *Potyvirus* (Fig. 1). The largest clade includes *Potato virus Y* (PVY), the type member of the genus, and a number of virus species infecting or not solanaceous plants (Figs. 1 and 2). The fact that potyvirus species infecting or not solanaceous plants are interspersed in this part of the phylogenetic tree is indicative of several host jumps during evolution of this group (Figs. 1 and 2), a feature that is observed in other virus genera (Fig. 4). The two other clades include a lower number of virus species all infecting solanaceous plants: the *Tobacco etch virus* (TEV) group which contains *Potato virus A* and *Tobacco vein mottling virus* and a group including *Pepper veinal mottle virus*, *Chilli veinal mottle virus*, *Wild tomato mosaic virus*, and *Tobacco vein banding mosaic virus* (Fig. 1). Only two potyvirus species are prevalent in pepper crops in the Mediterranean basin: PVY, which is widespread in all this area, and TEV, which is prevalent only in Turkey. There are four major clades among PVY isolates named N, O, C, and Chile (Moury, 2010). Among them, only members of clades C and Chile can infect pepper crops efficiently and only clade C isolates are prevalent in the Mediterranean basin. Members of the other PVY clades are mostly prevalent on potato or other solanaceous species and poorly infectious in pepper after inoculation in laboratory conditions (Gebre Selassie *et al.*, 1985; Moury, 2010). Recombinant isolates possessing genome regions that belong to different clades are also frequent in PVY (Glais *et al.*, 2002; Hu *et al.*, 2009a,b; Moury *et al.*, 2002; Ogawa *et al.*, 2008; Revers *et al.*, 1996; Schubert *et al.*, 2007). However, they are rather rare in pepper crops, except an Italian isolate that induces veinal necrosis in pepper leaves (Fanigliulo *et al.*, 2005). The largest part

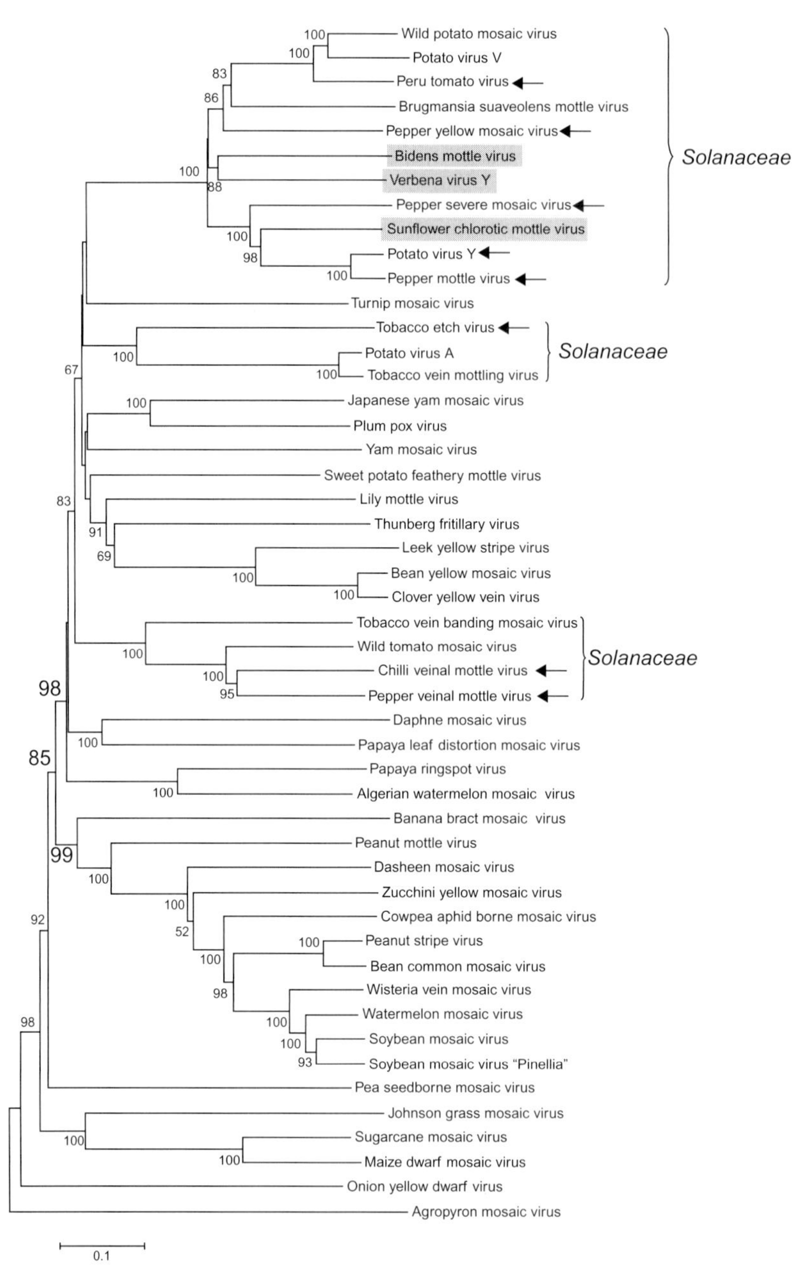

FIGURE 1 Unrooted neighbor joining phylogenetic tree of full-length genomes of potyviruses. The arrows indicate pepper-infecting potyviruses and potyviruses that do not infect solanaceous plants but cluster with *Solanaceae* groups are shaded in gray. Bootstrap percentages above 50% are shown. The scale bar indicates branch lengths in substitutions per nucleotide.

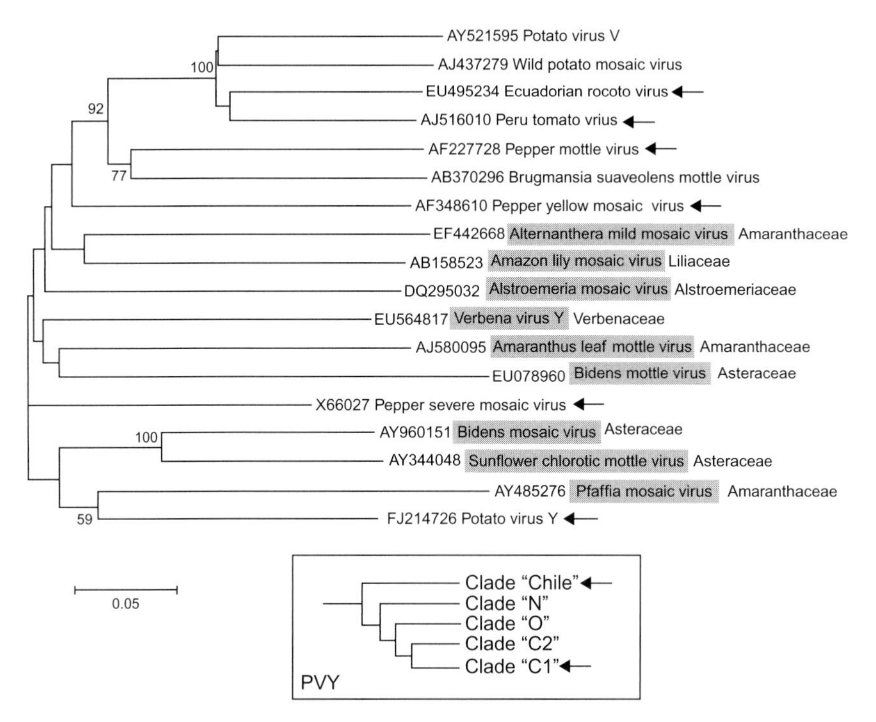

FIGURE 2 Unrooted neighbor joining phylogenetic tree of the coat protein (CP) coding region of potyviruses in the *Potato virus Y* (PVY) cluster. The arrows indicate pepper-infecting potyviruses and potyviruses that do not infect solanaceous plants but cluster with *Solanaceae* groups are shaded in gray. The tree topology of the PVY clades is boxed (see Moury, 2010). Bootstrap percentages above 50% are shown. The scale bar indicates branch lengths in substitutions per nucleotide.

of the genome of that isolate belongs to clade C while the 5′untranslated region (UTR) and part of the P1-coding region cluster with clade O and the 3′UTR clusters with clade N (Schubert *et al.*, 2007). This isolate is therefore the result of at least two independent recombination events involving three different PVY isolates.

PVY is common in open field or plastic tunnel pepper cultivation in warm climates. Its prevalence is quite high all around the Mediterranean basin, especially where traditional cultivars that are devoid of the most durable resistance genes (*pvr2*^{2} or *Pvr4*) are grown or where growers produce their own seeds which can therefore be the result of cross-pollination with susceptible cultivars (Ben Khalifa *et al.*, 2009; Buzkan *et al.*, 2006).

PVY host range includes mostly solanaceous plants but also plants in the families Amaranthaceae, Asteraceae, Chenopodiaceae, and Fabaceae. The most common symptom induced by PVY in pepper is systemic vein clearing progressing into a mosaic or mottle and generally dark green vein banding in leaves (Pernezny *et al.*, 2003). Vein and petiole necroses often develop, depending on the pepper genotype and, possibly, on the PVY isolate (Dogimont *et al.*, 1996). In some pepper genotypes, systemic necrosis upon PVY infection was shown to depend on the presence of one major dominant gene (Dogimont *et al.*, 1996). In some extreme cases, stem and apical bud necrosis can lead to plant death. Necrotic spots, mosaic patterns, and distortions may develop on fruits of some cultivars. However, fruit symptoms do not always occur in PVY-infected plants. Yield losses greatly depend on the earliness of infection and can reach 100% (Avilla *et al.*, 1997a). *Myzus persicae* is one of the most efficient PVY vectors. In Spain, the most frequent alate aphids landing on open-field pepper crops were *Aphis* spp., *Aphis fabae*, *M. persicae*, and *Diuraphis noxia* (Pérez *et al.*, 2003), all being potential PVY vectors (Pérez *et al.*, 1995). Usually, *M. persicae* and *Aphis* spp. colonize pepper plants only in warm climate areas (Rahman *et al.*, 2010; Satar *et al.*, 2008) or in greenhouse or plastic tunnel crops.

TEV induces severe mosaic and mottle in leaves in addition to leaf distortion and general stunting of pepper plants. It can also provoke abortion of floral buds and distortion and mosaic on fruits, especially in plants infected at a young stage. On some cultivars (*Capsicum frutescens* cv. Tabasco), root necrosis and severe wilting symptoms can be followed by death of plants. These specific symptomatology traits have been shown to be determined by two regions in the TEV genome: the 3′ one-third of the P3 coding region and a region spanning the 3′end of the CI coding region, the region coding the 6K2 protein and the 5′end of the VPg coding region (Chu *et al.*, 1997). In the Mediterranean basin, TEV was reported only in Turkey. Its prevalence reaches 23% of plants in pepper crops in southeast Anatolia and the eastern Mediterranean region of Turkey (Buzkan *et al.*, 2006). Yield reduction caused by TEV infection can reach 70% (Pernezny *et al.*, 2003).

Two main control methods have been used against potyviruses in pepper: prophylactic methods aiming to reduce the inoculation of viruses to plants and genetic control where plants are resistant to inoculated potyviruses. Among prophylactic measures, insecticides or biological control of the aphid vectors have often failed to prevent the virus spread because potyviruses are nonpersistently transmitted viruses that can be acquired and inoculated by aphids during superficial and brief probes in the plant epidermal cells (Collar *et al.*, 1997; Raccah, 1986). However, the use of these methods at a regional scale could reduce the aphid population size and partially decrease secondary infections. Application of mineral oil in pepper fields was shown to reduce the incidence of PVY and other nonpersistent viruses by about 40% (Marco, 1993) and is used in pepper integrated pest management programs in eastern Spain

(Martín-López *et al.*, 2006). Mineral oil sprays cause high mortality (>80%) of colonizing aphids, probably by asphyxia, and are also efficient to control aphid transmission of nonpersistent viruses (Martín-López *et al.*, 2006).

Weeds, especially in the family Solanaceae like black nightshade (*Solanum nigrum*), can also increase the PVY inoculum pressure in pepper fields and even be more efficient virus sources than infected pepper plants (Fereres *et al.*, 1996). Consequently, weeding the pepper fields and their edges is of prime importance to reduce epidemics. Polypropylene floating row covers were also shown to reduce access of vectors to pepper plants and to control virus infections in the field and are economically profitable in areas with a high virus pressure (Avilla *et al.*, 1997b). Barrier crops (sunflower, maize, vetch, or sorghum) used within pepper fields or around them were also shown to reduce PVY and *Cucumber mosaic virus* (CMV) (see below) spread in pepper (Anandam and Doraiswamy, 2002; Avilla *et al.*, 1996; Fereres, 2000; Simons, 1957). The effect of the barrier crop on viruses was to act as a sink for the virus, thereby reducing the virus secondary spread and inoculum pressure, to prevent the aphid colonization on pepper plants or to reduce the landing of alate aphids (Hooks and Fereres, 2006). Insecticide sprays on the barrier plants could further reduce the disease spread.

In addition to these approaches, two major resistance systems have been used to protect pepper against potyviruses. Recessive resistance genes at the *pvr2* locus, mapping on chromosome P4, have been used successfully for more than 50 years. The *pvr2* gene was the first cloned natural recessive gene conferring resistance to viruses in plants (Ruffel *et al.*, 2002). It was shown to encode a translation initiation factor (eIF4E; eukaryotic initiation factor 4E) essential to the translation of cellular mRNAs (Ruffel *et al.*, 2002). Plant eIF4Es interact directly with the cap of mRNAs as an initial step toward the building of a scaffold of cellular proteins allowing the recruitment of ribosomes. Together with other virus groups, potyviruses evolved to highjack this system to enhance their multiplication. The key step allowing PVY or TEV multiplication is the direct physical interaction between the virus VPg, which replaces the cap of mRNAs at the 5′end of the viral RNA, and plant eIF4E (Charron *et al.*, 2008; Kang *et al.*, 2005). Nucleotide substitutions in the *pvr2* gene can create amino acid substitutions in the encoded eIF4E that can disrupt the interaction with the virus VPg and thus create a recessive resistance gene. At least nine such *pvr2* alleles have been described in the pepper germplasm. They differ from one another and from the wild-type susceptibility *pvr2*$^{+}$ allele by a small number of nucleotide substitutions, almost all of them being nonsynonymous (Charron *et al.*, 2008). In turn, nonsynonymous substitutions in PVY or TEV VPg coding regions can restore the physical interaction between the mutated eIF4E and virus VPg (Charron *et al.*, 2008) and be the cause of resistance breakdowns. Two *pvr2* alleles have been used extensively to breed potyvirus resistant pepper cultivars for more than 50 years, $pvr2^1$ and $pvr2^2$. Both alleles confer efficient resistances toward PVY, while only $pvr2^2$ is effective against TEV. The resistance of $pvr2^2$ proved extremely durable against PVY. After over 50 years of usage, only two $pvr2^2$-breaking PVY isolates have been described. The first one was selected in laboratory conditions from a $pvr2^1$-breaking isolate by serial passages in a $pvr2^2$-carrying pepper cultivar (Gebre Selassie *et al.*, 1985). The other one was found in Málaga, Spain, in 1988 (Luis-Arteaga *et al.*, 1993) but was not the cause of epidemics since. $pvr2^1$-breaking PVY isolates are much more frequent but usually less prevalent than avirulent isolates (Luis-Arteaga and Gil-Ortega, 1986). Consequently, cultivars carrying the $pvr2^1$ resistance continue to be used and are economically satisfactory in many growing regions. Other *pvr2* alleles are almost not used to breed elite pepper F_1 hybrids but can be found in traditional pepper populations. Though possessing extremely similar sequences, the different *pvr2* alleles show extremely contrasted durability. For example, plants carrying the $pvr2^3$ allele, which has not been used to breed cultivars, can be infected by PVY isolates presently prevalent in pepper crops (Ben Khalifa *et al.*, in press). One reason of these durability differences seems to be the number of mutational pathways toward resistance breakdown and the number of mutations involved in these pathways. At least five nucleotide substitutions in the VPg coding region have been shown to confer independently $pvr2^3$-breaking properties (Ayme *et al.*, 2006; Montarry *et al.*, 2011; Fig. 3). In contrast, only one pathway involving a single-nucleotide substitution allows breakdown of $pvr2^1$ and breakdown of $pvr2^2$ involves two or more consecutive substitutions, depending on the wild-type avirulent PVY isolate (Ayme *et al.*, 2007; Ben Khalifa *et al.*, in press Fig. 3). In accordance, the fixation of two given nucleotide substitutions in a virus seems to be a threshold that determines plant resistance durability in general (Harrison, 2002; Lecoq *et al.*, 2004). When two or more substitutions are needed for resistance breakdown, the resistance is usually quite durable while it is not if a single substitution suffices. Breakdown of the $pvr2^2$ resistance by TEV seems to be much more frequent than by PVY (Depestre *et al.*, 1993; Muhyi *et al.*, 1993). However, no Turkish isolate breaking down $pvr2^2$ has been described yet (Palloix *et al.*, 1994).

The other resistance that is widely used in pepper cultivars is that based on the *Pvr4* dominant gene that maps on chromosome P10. *Pvr4* controls an extreme resistance phenotype to all natural PVY isolates but not to TEV. It has a very broad range of action since it also confers resistance to five additional potyvirus species that are prevalent in the Americas. The mechanism of the *Pvr4* resistance is related to hypersensitivity, but hypersensitive reactions (HR) appear only on inoculated cotyledons or under very high inoculum pressure such as through graft

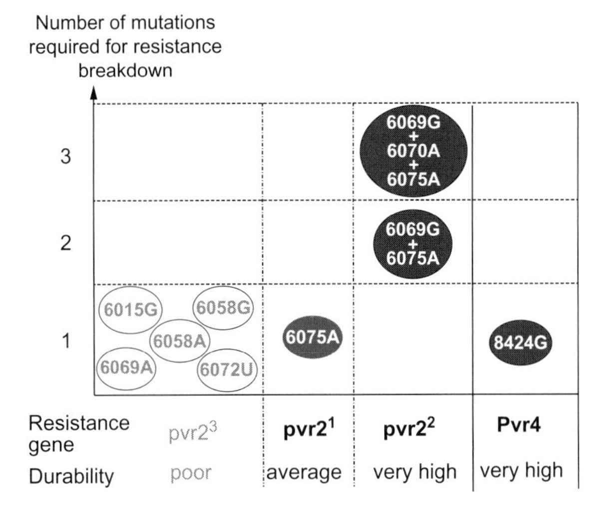

FIGURE 3 Nucleotide substitutions in the *Potato virus Y* (PVY) genome (numbered according to isolate SON41p; accession number AJ439544) involved in the breakdown of pepper resistances at the *pvr2* or *Pvr4* loci and durability level of these resistances. For breakdown of $pvr2^2$, the two combinations of mutations correspond to French (two nucleotide substitutions) or Tunisian (three nucleotide substitutions) isolates with VPg sequences closest to SON41p, the only $pvr2^2$-breaking PVY isolate characterized. (See Page 3 in Color Section at the back of the book.)

inoculation (Janzac *et al.*, 2009). Though being used in pepper crops for about 20 years, the *Pvr4* resistance is highly durable since no field PVY isolate can infect any *Pvr4* cultivars so far. Graft inoculations have consequently been undertaken to favor the selection of *Pvr4*-breaking PVY isolates (Janzac *et al.*, 2009) and, further, to identify causal mutations (Janzac *et al.*, 2010). That mutation maps to the NIb coding region which encodes the RNA-dependent RNA polymerase (RdRp) (Fig. 3). The fact that only one nucleotide substitution suffices for the breakdown of the *Pvr4* resistance seems to contradict the high durability of this resistance. However, that mutation induces a very high competitiveness cost to PVY in *Pvr4*-devoid pepper cultivars, and this cost cannot be easily compensated by the virus (Janzac *et al.*, 2010).

In addition to these monogenic resistances, polygenic resistances to PVY have also been investigated. A number of resistance quantitative trait loci (QTLs) have been mapped in a cross involving the Indian line "Perennial" as a resistant parent (Caranta *et al.*, 1997a). By themselves, these QTLs reduced PVY symptom intensity and progress but do not impair the invasion of the plant by the virus. However, these minor-effect QTLs were shown to improve greatly the durability of the major-effect gene $pvr2^3$, which alone is rapidly broken down (Palloix *et al.*, 2009). This was one of the first demonstrations of the higher durability of a polygenic over a monogenic resistance. However, a two-step adaptation of PVY, first to *pvr2*^{3} and second to the gene "pyramid" consisting of *pvr2*^{3} and QTLs, was still possible. This showed that the components of a polygenic resistance should better be used together than separately in order not to jeopardize simultaneously all of these resistance factors.

B. Cucumoviruses

CMV, the type member and most widespread species of the genus *Cucumovirus*, is one of the most prevalent viruses worldwide, infecting 85 distinct plant families and more than 1000 species experimentally, including a large number of weeds or wild plant species that can act as reservoirs between successive growing periods.

The genome of CMV consists of three linear, positive-sense single-stranded RNAs with 5′terminal cap structures. The 3′ends are not polyadenylated but show aminoacylated tRNA-like structures. Virions are icosahedral, 29 nm in diameter, uniform in size, and have electron dense centers. They encapsidate the three genomic RNAs separately and occasionally subgenomic or satellite RNAs. CMV is the only cucumovirus that affects significantly pepper crops. It is distributed all over the world and can induce severe economic losses by affecting the growth of the vegetative parts of the plants and also by inducing symptoms in fruits. Symptoms include punctiform mosaic and dull leaves, filiformism of young leaves associated to a curling of the nerves, necrotic "oak-leaf" symptoms in older leaves, misshaped fruits with annular discolorations, and sterility when infection occurred at plantlet stage. In fruits, CMV can induce distortions, reduction of size, and irregular maturation. Early infections can induce stem necrosis and death of plants. Yield losses greatly depend on the earliness of infection and can reach 80% (Avilla *et al.*, 1997b). Moreover, synergism between CMV and other viruses coinfecting the same plant can increase CMV accumulation or symptom intensity (Mascia *et al.*, 2010). Reassortment between two CMV isolates induced more severe symptoms on pepper plants as compared to single inoculations of the parental viruses (Hellwald *et al.*, 2000). Coinfection by CMV can also enhance infection by other viruses, as in *Pepper mottle virus* (PepMoV)-resistant pepper plants by stimulating systemic movement of PepMoV in the internal phloem (Guerini and Murphy, 1999).

CMV isolates are classified into two subgroups named I and II which are distantly related according to molecular analyses of the genome and can be distinguished serologically (Owen and Palukaitis, 1988). Members of these two subgroups share from 69% to 77% nucleotide identity, depending on the RNA and isolates compared (Palukaitis and García-Arenal, 2003). Within subgroup I, clade IA containing closely related isolates was separated from the rest of subgroup I (themselves included in the polyphyletic subgroup IB), on the basis of phylogenetic evidence (Roossinck *et al.*, 1999). Isolates from subgroups IA and II are distributed worldwide while almost all isolates of subgroup IB are from East Asia. A few subgroup IB isolates have probably been introduced in Italy, Spain, and California (Bonnet *et al.*, 2005; Gallitelli, 2000; Lin *et al.*, 2003). The prevalence of isolates from subgroups I and II varies also according to climatic conditions, with isolates from group I being mostly found under warm climate and isolates from group II under colder conditions (Marchoux *et al.*, 1976; Quiot *et al.*, 1979). Given that pepper crops are more widely distributed in geographic regions with a hot climate, they are therefore probably more frequently affected by subgroup I isolates.

Like potyviruses, CMV is transmitted by aphids in a nonpersistent manner. At least 86 aphid species have been described as CMV vectors, two of the most efficient and prevalent being *M. persicae* and *Aphis gossypii*. In contrast with potyviruses, a single viral protein, the coat protein (CP), is directly involved in aphid transmissibility (Chen and Francki, 1990; Gera *et al.*, 1979). As a consequence, mutations in the CP of CMV were shown to affect the specificity of transmission by different aphid species (Perry *et al.*, 1998) and, consistently, aphid species and populations seem to exert strong diversifying selection and population differentiation on CMV (Martínez-Torres *et al.*, 1998; Moury, 2004).

CMV is seed transmitted in a number of host species including possibly pepper. Using RT-PCR for detection, Ali and Kobayashi (2010) showed that both the coat (53–83%) and embryo (10–46%) of seeds were CMV positive after inoculation of pepper plants with isolate Fny, which belongs to subgroup IA. The observed transmission rate to the progeny was 10–14%. Since no sanitary problems linked to seed infection by CMV in pepper had been mentioned before, it will be important to estimate this risk on a larger scale.

Since they share many similarities at the epidemiological level, prophylactic methods used to control CMV are the same as those used against potyviruses. Both kinds of viruses are transmitted in a nonpersistent manner by numerous aphid species, and many of these species can transmit both CMV and potyviruses. Consequently, methods that reduce aphid population sizes, virus transmission, the ability of aphid vectors to move in fields, and the inoculum sources as weed reservoirs can also be used to decrease the impact of CMV infections in pepper crops.

Genetic resistance has been largely exploited to control CMV infections in pepper. Only one major-effect resistance gene has been described in the pepper germplasm so far (Kang *et al.*, 2010). This dominant resistance gene, named *Cmr1*, is located on the pepper chromosome P2 and was identified in a Korean cultivar. *Cmr1* was shown to restrict the systemic movement of CMV isolates belonging to subgroup IA, but plants carrying *Cmr1* can be infected by an isolate (named P1) belonging to subgroup IB (Kang *et al.*, 2010). Similarly, following the deployment of pepper cultivars resistant to subgroup IA CMV isolates in South Korea during the 1990s, resistance breakdowns involving subgroup IB CMV isolates have been observed in the mid-2000 (Lee *et al.*, 2006). It remains to be determined if these CMV infections observed in resistant cultivars correspond to "true" resistance breakdowns or to a narrow spectrum of action of the resistance. Almost all other CMV resistance sources in pepper are partial and polygenic. They act by inhibiting virus establishment in inoculated tissues (Caranta *et al.*, 1997b), or virus movement (Caranta *et al.*, 2002), or by decreasing symptoms (Ben Chaim *et al.*, 2001). Several of these resistances are ontogenic, that is, they depend on a particular developmental stage of plants (Dufour *et al.*, 1989; García-Ruiz and Murphy, 2001). Seven QTLs with additive or epistatic effects are involved in resistance to CMV systemic movement, and three in CMV establishment in inoculated tissues (Caranta *et al.*, 1997b, 2002). The QTLs controlling these two resistance traits are independent of each other. These resistances restrict only partially the virus translocation within plants, but proved to confer a good level of resistance in the field, particularly when different resistance sources were combined into a cultivar (Nono Womdim *et al.*, 1993; Lapidot *et al.*, 1997; Palloix *et al.*, 1997). Though polygenic resistances are thought more durable than monogenic and qualitative ones, there has been no investigation of their durability concerning CMV. Since these resistances are largely based on mechanisms controlling the systemic movement of CMV within plants and since resistant cultivars carrying similar mechanisms of resistance have been infected by CMV isolates from subgroup IB in Korea, the recent introduction of subgroup IB isolates in the Mediterranean basin (Italy and Spain) possibly from Asia, questions their durability.

C. Other aphid-transmitted viruses

In addition to the previous and economically most important viruses, *Alfalfa mosaic virus* (AMV) and *Broad bean wilt virus* (BBWV) are two nonpersistently transmitted viruses that are widely distributed in pepper crops in the Mediterranean basin although at a usually low prevalence. The persistently transmitted poleroviruses are confined to a few areas but could represent a future threat to pepper crops.

AMV is the type species of the genus *Alfamovirus* and belongs to the family *Bromoviridae.* Its natural host range includes over 250 plant species belonging to 48 families, mostly herbaceous plants. More than 20 aphid species are known to transmit AMV in a nonpersistent manner. AMV occurs mainly in temperate climates, causing significant diseases in alfalfa and sweet clover and also occasionally in neighboring crops like soybean, tobacco, tomato, or pepper. AMV has been described in pepper plants since 1939 in Italy and 1965 in France (Marchoux *et al.*, 2003). Infected plants develop local chlorotic and necrotic rings on the leaves followed by a systemic foliar mosaic, often brilliant yellow or white. Some AMV isolates cause severe necrosis in pepper cultivars like "Yolo Wonder," "Yolo Y," "Tabasco," or "Sucette." Bleaching can also occur on pepper fruits.

BBWV, which is the type species of the genus *Fabavirus*, has been described for the first time in *Vicia faba* in Australia in 1947. BBWV comprises two distinct viral species recognizable by the divergence of their genome: BBWV-1 and BBWV-2 (van Regenmortel *et al.*, 2000). The natural host range of BBWV includes more than 200 plant species in 41 families, and its dissemination is by aphids in a nonpersistent manner. In Mediterranean countries, BBWV is frequently found in tomato and pepper plants, especially in France, Italy, and Spain. Sequence analyses have demonstrated the presence of BBWV-1 isolates in Spain (Rubio *et al.*, 2002). Infected pepper plants develop mosaic and concentric rings on leaves and fruits. BBWV can also cause partial to general bleaching of pepper fruits reducing their commercial value. However, in pepper crops, BBWV shows a rather low prevalence and is thus not an economical concern.

Viruses belonging to the genus *Polerovirus* are restricted to the phloem and are transmitted by aphids in a persistent, circulative, and nonpropagative mode. The high specificity of aphid transmission is mediated by the minor capsid protein of poleroviruses which has specific intestinal tropism into the aphid (Brault *et al.*, 2005). Among poleroviruses, one species, pepper yellow leaf curl virus (PYLCV), caused devastating diseases in pepper crops in Israel since 1998 (Dombrovsky *et al.*, 2010). PYLCV symptoms in pepper include shortening of the internodes, interveinal yellowing, and upward rolling of leaf margins, accompanied by fruit discoloration and size reduction. PYLCV is closely related to *Tobacco vein distorting virus* (TVDV) and is transmitted by *M. persicae* and *A. gossypii*. The virus was not mentioned yet in pepper crops in other countries. However, it is genetically close and probably belongs to the same species as a Japanese isolate tentatively named pepper vein yellows virus (Murakami *et al.*, 2011) and as a Turkish isolate of pepper yellows virus collected in Antalya (L. Lotos *et al.*, unpublished data; GenBank accession number FN600344) since it shares more than 90% identity with these viruses at the nucleotide level (Murakami *et al.*, 2011).

III. THRIPS-TRANSMITTED TOSPOVIRUSES

The genus *Tospovirus* in the family *Bunyaviridae* includes important species of plant viruses with pleomorphic particles (80–120 nm) enveloped by a double-membrane layer and which contain tripartite single-stranded

RNAs, designated L, M, and S (De Haan *et al.*, 1990, 1991; German *et al.*, 1992; Law *et al.*, 1992). The negative L RNA (8.9 kb) consists of a single open reading frame (ORF) in the viral complementary sense that encodes a 331-kDa protein containing RdRp motifs required for virus replication (Adkins *et al.*, 1995; De Haan *et al.*, 1991; Van Poelwijk *et al.*, 1997). The ambisense M RNA (4.8 kb) (Kormelink *et al.*, 1992; Law *et al.*, 1992) consists of two ORFs that encode a 36-kDa nonstructural protein, NSm, in the viral sense, and a 127.4-kDa precursor for G1 and G2 glycoproteins in the viral complementary sense. NSm protein may be involved in cell-to-cell movement (Storms *et al.*, 1995), whereas G1 and G2 structural glycoproteins are included in the outer membrane of the virion generating spikes (Adkins *et al.*, 1996; Law *et al.*, 1992) and are essential for thrips transmission (Sin *et al.*, 2005) probably through their interaction with thrips receptor proteins. The ambisense S RNA (2.9 kb) (De Haan *et al.*, 1990) encodes two proteins: the 52.4-kDa nonstructural NSs and the 29-kDa structural nucleocapsid protein (NP) in the viral and the complementary senses, respectively. The NSs protein is associated to fibrous inclusions in infected plant cells (Kormelink *et al.*, 1991) and has been shown to be a suppressor of posttranscriptional gene silencing (Takeda *et al.*, 2002).

Tospoviruses cause great losses in many economically important crops, including pepper, worldwide. Two virus species belonging to this genus infect pepper in the Mediterranean surroundings: *Tomato spotted wilt virus* (TSWV) and *Impatiens necrotic spot virus* (INSV) (Fig. 4). In pepper, TSWV is more prevalent than INSV, which infects mainly ornamentals (Daughtrey *et al.*, 1997). TSWV is also the most widespread, occurring in all countries of the Mediterranean region, even if it has not been confirmed in a few ones (Morocco and Tunisia) and has one of the largest host ranges among plant viruses (Parrella *et al.*, 2003). INSV seems to be restricted to France, Spain, Italy, and Israel and has not been described in North Africa. Several studies have shown that high temperatures (>30 °C) promote TSWV infections (Llamas-Llamas *et al.*, 1998; Roggero *et al.*, 1999) and the resistance of some pepper cultivars can be impaired under continuous high temperatures, resulting in systemic infection of the plants (Black *et al.*, 1991; Moury *et al.*, 1998). By contrast, high temperatures decrease the systemic movement of INSV in *Capsicum chinense* and *Capsicum annuum* (Roggero *et al.*, 1999), which may explain the rarity of natural INSV infections in pepper crops in the Mediterranean area. Most of the time, symptoms caused by TSWV are similar to those due to INSV. Symptoms in *C. annuum* include stunting and yellowing of the whole plant, mosaic or necrotic spots, and curling of the leaves. Infected fruits often show deformations, necrotic ring patterns, and arabesque-like discolorations.

In nature, tospoviruses are transmitted from plant to plant almost exclusively by thrips (order Thysanoptera; family Thripidae) in a

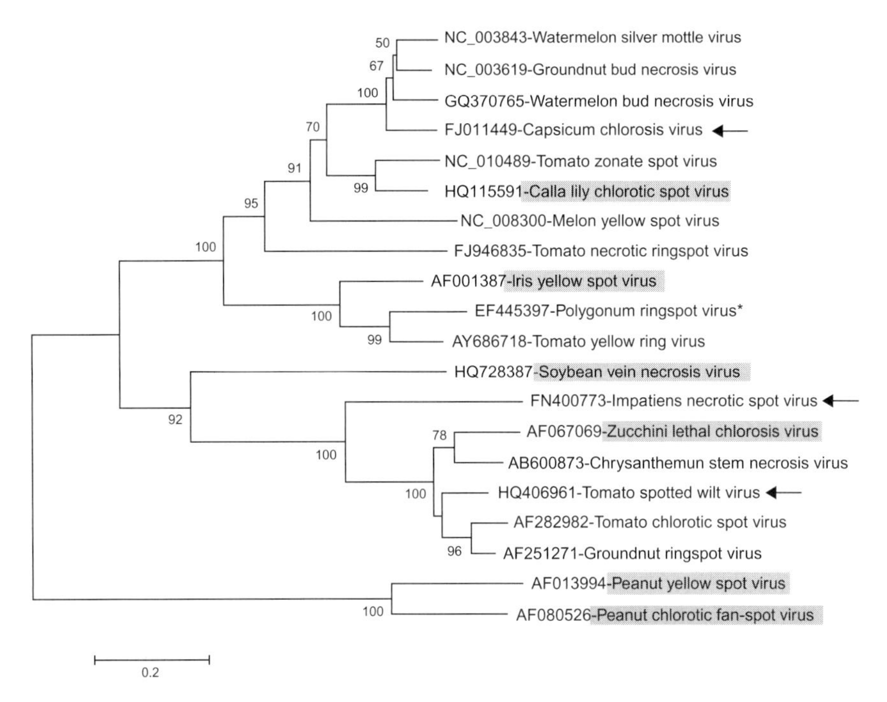

FIGURE 4 Unrooted neighbor joining phylogenetic tree of the coat protein gene of tospoviruses. The arrows indicate pepper-infecting tospoviruses and tospoviruses that do not infect solanaceous plants are shaded in gray. Bootstrap percentages above 50% are shown. The scale bar indicates branch lengths in substitutions per nucleotide.

persistent and multiplicative manner. TSWV, like INSV, is transmitted mainly by the western flower thrips (*Frankliniella occidentalis*) (De Angelis *et al.*, 1994), but other Thripidae, like *Thrips tabaci* and *Frankliniella intonsa*, can also participate to the spread of TSWV. Outbreaks of TSWV in Europe have been associated to the introduction of *F. occidentalis* from western USA in the early 1980s. An estimation of the speed of spread across Europe and northern Africa was 229 ± 20 km/year (Kirk and Terry, 2003). The western flower thrips not only is established in glasshouses but also outdoors in areas with mild winters like the Mediterranean basin. Vectors can acquire tospoviruses only during larval stages while both larval (late stage) and adult thrips can transmit the virus. For TSWV, it has been shown that the ability to acquire and transmit tospoviruses was lost during the development to adults, probably because of the formation of a midgut barrier (Ullman *et al.*, 1992). No transovarial transmission has been reported for tospoviruses.

Various management procedures have been undertaken during the past decades to reduce the spread of tospoviruses. Control of vectors is complicated by the high fecundity of thrips and their capacity to develop insecticide resistances. The wide host range of tospoviruses, including weeds that constitute virus reservoirs, increases the difficulties to control the disease. The application of sanitation measures must be intensified in glasshouses, particularly the eradication of weeds inside and outside the cultivated area, the use of blue, more attractive, or yellow sticky cards to monitor the presence of winged adults thrips (Matteson and Terry, 1992; Roditakis *et al.*, 2001), the regular examination of the crops, and the eradication of infected plants. Biological control of thrips on pepper crops relies on the use of predatory mites like *Neoseiulus cucumeris* or predatory bugs (*Orius* spp.) (Hatala Zseller and Kiss, 1999; Maisonneuve and Marrec, 1999) and can decrease the virus inoculum pressure. Genetic resistance has been developed on several plants species, especially tomato and pepper, to control the dissemination of viral diseases associated with tospoviruses. Concerning pepper, several *C. chinense* lines possess monogenic resistances conferred by the *Tsw* gene (Boiteux, 1995; Moury *et al.*, 1997) and have been used to breed *C. annuum* cultivars resistant to TSWV. *Tsw* controls HR against most TSWV isolates and prevents the virus movement from cell to cell (Soler *et al.*, 1999). It is not efficient against other tospovirus species like INSV.

Appearance of TSWV isolates adapted to the *Tsw* resistance has been observed first in laboratory conditions (Black *et al.*, 1991; Moury *et al.*, 1997). In southern Europe, breakdowns of the resistance were observed very rapidly after the release of *Tsw*-carrying cultivars and have been described in 1999 in Italy and Spain (García-Arenal and McDonald, 2003). The TSWV genetic factors involved in the breakdown of the *Tsw* resistance are still under investigation. Some authors designated the NSs nonstructural protein (Margaria *et al*, 2007; Tentchev *et al.*, 2011), whereas others demonstrated the role of the NP encoded by the N gene (Lovato *et al.*, 2008). A way to reconcile these findings would be that two separate TSWV genes interact with the *Tsw* resistance in pepper: a gene which induces the resistance process (potentially the NP gene which was shown to be a specific elicitor of hypersensitive response in *Tsw* pepper plants) and a second gene which is targeted by the defence reactions and where resistance-breaking mutations can occur (presumably the NSs gene).

Resistance to the thrips vectors is known in several pepper (*C. annuum*) accessions and affects the level of feeding damage, host preference, and host suitability for reproduction. Some authors have shown that TSWV transmission was little affected by vector resistance under experimental conditions (Maris *et al.*, 2003). Owing to the lower reproduction rate and the lower attraction of thrips for resistant pepper plants, these authors supposed that beneficial effects might be expected from resistant cultivars under field conditions. Relationships between TSWV and its vectors are complex since TSWV-infected pepper plants increase attraction for female thrips compared to noninfected plants, thus improving TSWV dissemination (Maris *et al.*, 2004). Also, male thrips infected with TSWV fed more than uninfected males, with a threefold increase of noningestion probes during which they salivate, thus increasing the probability of virus inoculation (Stafford *et al.*, 2011).

Finally, *Polygonum ringspot virus*, a new tospovirus species, was recently discovered in northern and central Italy in wild buckwheat (*Polygonum convolvulus*) and in *P. dumetorum* (Ciuffo *et al.*, 2008). *Polygonum ringspot virus* is closely related to *Tomato yellow ring virus* (Fig. 4), a tospovirus infecting ornamental and vegetable crops in Iran (Hassani-Mehraban *et al.*, 2007; Rasoulpour and Izadpanah 2007). Although this virus was found only in wild plants and was not detected in neighboring crops, it was shown to infect a large number of solanaceous plants, including pepper, after mechanical inoculation in laboratory conditions and could be a future threat for Italian and Mediterranean horticulture.

IV. WHITEFLY-TRANSMITTED VIRUSES

Two groups of whitefly-transmitted viruses, belonging to the genera *Begomovirus* and *Crinivirus*, are found in pepper in the Mediterranean area but have presently low economical impacts on this crop.

Begomoviruses are single-stranded DNA viruses transmitted by the whitefly *Bemisia tabaci* in a persistent and circulative manner that infect important crops, including *Cucurbitaceae* (watermelon, melon, and squash), *Euphorbiaceae* (cassava), *Fabaceae* (common bean), *Malvaceae* (cotton), and *Solanaceae* (tomato, tobacco, and pepper) and are the cause of devastating plant diseases, particularly in many tropical and subtropical regions of the world. Among them, *Tomato yellow leaf curl virus* (TYLCV) is one of the most destructive viruses affecting tomato crops throughout the Mediterranean region since its first description in the 1930s in Israel. In 1999, TYLCV has been reported for the first time on *C. annuum* plants in southeastern Spain (Reina *et al.*, 1999) and later in Tunisia (Gharsallah Chouchane *et al.*, 2007; Gorsane *et al.*, 2004). TYLCV affecting pepper plants is also strongly suspected in Egypt and Morocco. The prevalence was estimated from 2% to 6% in southern Spain, much lower than that estimated in Florida. Morilla *et al.* (2005) detected also the related species *Tomato yellow leaf curl Sardinia virus* in some pepper plants in Spain. TYLCV-infected pepper plants are frequently symptomless (Morilla *et al.*, 2005; Polston *et al.*, 2006). Although many pepper cultivars appear susceptible to TYLCV, large differences in infection rates have been observed, both in field conditions and after inoculation under controlled conditions. In transmission experiments with the Q biotype of *B. tabaci* and infected pepper plants as virus sources, Morilla *et al.* (2005) did not succeed to transmit TYLCV-Mld to tomato or pepper plants, which could be related to the low virus titer in the source plants compared to that in infected tomato plants. An uneven distribution of TYLCV in infected pepper plants was also observed (Morilla *et al.*, 2005; Polston *et al.*, 2006). Consequently, Morilla *et al.* (2005) suggested that pepper could be a dead-end host for TYLCV. However, using the B biotype of *B. tabaci* and a larger set of pepper cultivars, Polston *et al.* (2006) observed high rates of TYLCV transmission from infected pepper plants to tomato plants. However, they did not observe any TYLCV transmission when acquisition was from infected pepper fruits. Pepper and tomato crops are often close to each other, and *B. tabaci* populations are able to feed and reproduce on pepper as well as on tomato plants. These data suggest a potential role of pepper plants in the epidemiology of TYLCV, as pepper plants could act as reservoirs for TYLCV dissemination, but the economical incidence of TYLCV on pepper production is probably very low.

Tomato infectious virus (ToCV) is a phloem-restricted bipartite *Closteroviridae* and belongs to the genus *Crinivirus*. It is transmitted by a number of whiteflies including *B. tabaci* and *Trialeurodes vaporariorum*. ToCV was first described on tomato plants in Florida in 1989 (Wisler *et al.*, 1998). In the Mediterranean region, ToCV has been reported in Spain, Italy, Greece, France, Turkey, Israel, Lebanon, and Morocco. Although tomato is the main crop affected by ToCV, the virus has been reported on sweet pepper plants in greenhouses of southern Spain in 1999 (Lozano *et al.*, 2004). Infected pepper plants developed interveinal yellowing, leaf curling, and stunting. The HSP70h gene of the pepper isolate of ToCV was 100% identical to ToCV isolates collected from tomato plants in the same region of Spain.

V. TOBAMOVIRUSES

Among pepper viruses that are not transmitted by biological vectors, tobamoviruses are, by far, the most important ones even though the potexvirus *Potato virus X* can locally reach high prevalence (up to 70% of plants in some areas in Turkey; Buzkan *et al.*, 2006). There are very few epidemiological data on this latter virus in other countries.

Tobamovirus is a genus of single-stranded RNA viruses whose particles are particularly stable. Tobamovirus particles are elongated rigid cylinders approximately 18 nm in diameter and 300 nm long, with a central cavity and a helical symmetry (2.3-nm pitch), containing the genomic RNA. Shorter virions, constituting a minor component of the virus population, encapsidate the subgenomic RNAs. The genome of tobamoviruses consists of one linear positive-sense single-stranded RNA, 6.3–6.6 kb in size. A cap structure and a tRNA-like structure are found at the 5′ and 3′ends, respectively, of the genomic RNA. The CP is the only structural protein. Two nonstructural proteins are produced from the genomic

RNA: an ORF allows the production of a 124- to 132-kDa protein and a 181- to 189-kDa protein is produced by occasional readthrough of the stop codon of this ORF. These two nonstructural proteins are required for virus replication and contain methyltransferase or guanylyl transferase, helicase, and RdRp motifs. The CP and a third nonstructural movement protein of 28–31 kDa required for cell-to-cell and systemic movements of the virus are expressed from subgenomic RNAs. Molecular clock analyses of tobamovirus genomes sampled over the past century have indicated that the genus is probably not older than 100,000 years (Pagán *et al.*, 2010).

Species infecting solanaceous crops build a distinct clade among tobamoviruses, and several of these have been isolated from pepper all over the world and induce important economic losses in pepper crops (Alonso *et al.*, 1989; Avgelis, 1986; Beczner *et al.*, 1997; Nagai *et al.*, 1981; Pares, 1985; Rast, 1988; Wetter *et al.*, 1984; Fig. 5). The presently most prevalent pepper tobamoviruses in the Mediterranean basin are *Tobacco mild green mosaic virus* (TMGMV) (Font *et al.*, 2009; Fraile *et al.*, 2011), *Pepper mild*

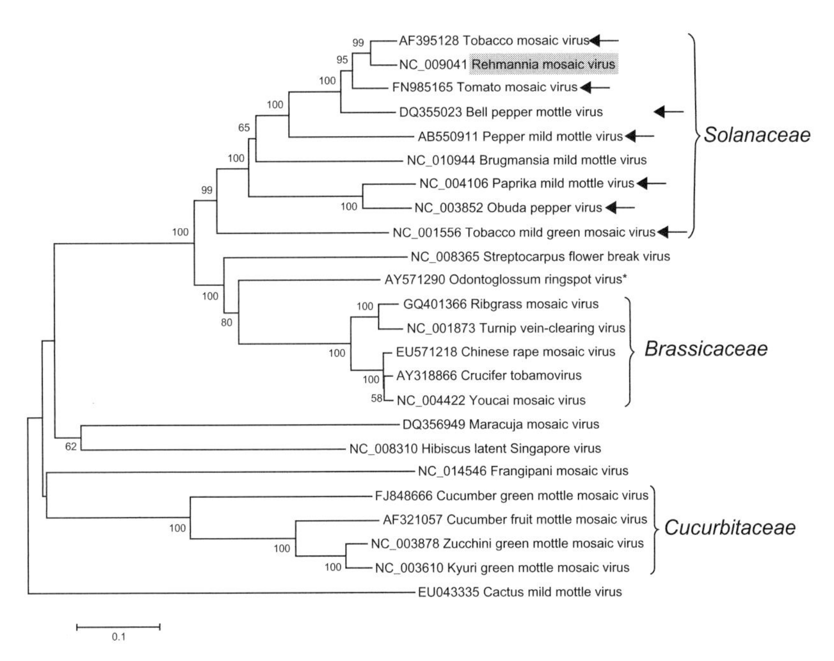

FIGURE 5 Unrooted neighbor joining phylogenetic tree of the replicase coding region of tobamoviruses. The arrows indicate pepper-infecting tobamoviruses and a tobamovirus that does not infect solanaceous plants but clusters with the *Solanaceae* group is shaded in gray. The asterisk indicates that *Odontoglossum ringspot virus* is an interspecific recombinant tobamovirus that clusters with different groups depending on the genome region examined. Bootstrap percentages above 50% are shown. The scale bar indicates branch lengths in substitutions per nucleotide.

mottle virus (PMMoV) (Buzkan *et al.*, 2006; Fraile *et al.*, 2011; Güldür and Çağlar, 2006), *Tobacco mosaic virus* (TMV), and *Tomato mosaic virus* (ToMV) (Arli-Sokmen *et al.*, 2005; Fig. 5). Both the yield and quality of pepper production can be severely reduced upon tobamovirus epidemics. Tobamoviruses induce leaf chlorotic mosaic or mottling, leaf distortion and surface reduction, and irregular shapes and colors associated with a reduction of size of fruits which, consequently, cannot be commercialized. Necroses can also be observed on leaves and fruits.

Due to their high stability, tobamoviruses remain infectious in contaminated plant debris, compost, soil, and irrigation water. Tobamoviruses cannot infect the seed embryo or albumen but are found in maternal tissues such as seed coat or residual perisperm, or as contaminant on the seed surface allowing infection of seedlings during germination. PMMoV is more efficiently seed transmitted than TMV or ToMV, probably due to a more internal contamination of seeds. Seed transmission has facilitated the introduction of tobamoviruses into different parts of the world through the international trade of pepper or tomato seeds. Given the absence of a biological vector of tobamoviruses, the primary measure of control involves prophylaxis, including disinfection and/or control of seed lots, eradication of infected plants, and care during handling of plants since the viruses can be transmitted by the physical contact between plants. Several seed disinfection methods can reduce significantly tobamovirus transmission by seeds, such as a 2-h treatment of seeds in 10% (w/v) trisodium phosphate (Na_3PO_4) solution (Rast and Stijger, 1987) or a 15-min treatment in trisodium phosphate followed by a 30-min treatment in a 0.525% (w/v) sodium hypochlorite (NaOCl) solution (Gooding, 1975). A heat treatment (3 days at 76 °C) can also avoid seed transmission of tobamoviruses but adversely affects germination (Rast and Stijger, 1987).

TMV or PMMoV variants with attenuated virulence have also been used to develop cross protection control methods in pepper (Goto *et al.*, 1984; Hagiwara *et al.*, 2002; Ichiki *et al.*, 2009; Yoon *et al.*, 2006) but have not yet been used in commercial conditions in the Mediterranean region. Mutations responsible for attenuated symptoms have been mapped in the 126-kDa protein or in the 3′UTR (Hagiwara *et al.*, 2002; Ichiki *et al.*, 2009; Yoon *et al.*, 2006).

Another important control method against tobamoviruses in pepper is growing resistant cultivars. Different dominant alleles at the *L* locus, located on chromosome P11, have been identified in different *Capsicum* species (Table I) and shown recently to encode a coiled-coil, nucleotide-binding, leucine-rich repeat type of resistance protein (Tomita *et al.*, 2011). These alleles differ by their specificity toward tobamovirus species and isolates (Table I) and by their efficiency under temperature stresses. Several of these alleles at the *L* locus lack efficiency above 30 °C (Boukema, 1982) and show a higher efficiency at the homozygous state (Boukema,

TABLE I Reaction of different tobamovirus species (see Fig. 5) and pathotypes toward *Capsicum* spp. genotypes carrying different alleles at the *L* locus

		Pepper species and genotype				
		C. annuum	*C. annuum*	*C. frutescens*	*C. chinense*	*C. chacoense*
Virus	Pathotype	L^+/L^+	$L^1/-$	$L^2/-$	$L^3/-$	$L^4/-$
TMV, ToMV, TMGMV, BPeMV	P_0	S	R	R	R	R
PaMMV, ObPV	P_1	S	S	R	R	R
PMMoV	$P_{1,2}$	S	S	S	R	R
PMMoV	$P_{1,2,3}$	S	S	S	S	R
PMMoV	$P_{1,2,3,4}$	S	S	S	S	S

S: susceptibility, that is, systemic infection; R: resistance, that is, necrotic local lesions without systemic infection.
TMV: *Tobacco mosaic virus;* ToMV: *Tomato mosaic virus;* TMGMV: *Tobacco mild green mosaic virus;* BPeMV: *Bell pepper mottle virus;* PaMMV: *Paprika mild mottle virus;* ObPV: *Obuda pepper virus;* PMMoV: *Pepper mild mottle virus.*

1980, 1984). Other alleles show an enhanced efficiency at high temperature (Sawada *et al.*, 2005). The CP gene of tobamoviruses was shown to be responsible for the specificity and breakdown of the resistances conferred by the *L* gene (Berzal-Herranz *et al.*, 1995; Cruz *et al.*, 1997; Gilardi *et al.*, 1998, 2004). The *Hk* dominant gene was also shown to confer a temperature-dependent resistance to *Paprika mild mottle virus* (PaMMV) (Sawada *et al.*, 2005) but, in contrast, is not efficient at lower temperatures (24 °C). It is also incompletely dominant. The methyltransferase domain of the replicase of PaMMV was shown to be involved in breakdown of *Hk* (Matsumoto *et al.*, 2009).

Infections of peppers carrying the L^1 or L^2 resistance genes seem to be the result of the emergence of new tobamovirus species rather than by "classical" resistance breakdowns (i.e., breakdown resulting from accumulation of a small number of mutations in the genome of nonadapted, avirulent, virus isolates). Indeed, no TMV, ToMV, or TMGMV isolate breaking down the L^1 or L^2 resistances and no PaMMV or ObPV isolate breaking down the L^2 resistance have been described so far (Table I). Consequently, the wide use of L^1- or L^2-carrying pepper cultivars together with the efficiency and durability of these genes against ToMV, TMV, and TMGMV could have created free "host niches" for other tobamoviruses and might have contributed to the emergence of PMMoV.

In contrast, breakdown of the L^3 and L^4 resistances by PMMoV variants has been observed and mutations involved in these events identified

(Antignus *et al.*, 2008; Genda *et al.*, 2007; Hamada *et al.*, 2002, 2007; Tsuda *et al.*, 1998). PMMoV is composed of three major clades (Fig. 6), and evolution of resistance breakdown capacities occurred independently in all of them. One PMMoV clade (clade 3; Fig. 6) is composed essentially of Mediterranean isolates of pathotype $P_{1,2,3}$ that all carry the $M_{139}N$ substitution in their CP, sufficient for breaking down the L^3 resistance (Berzal-Herranz *et al.*, 1995). Until now, only resistance-breaking isolates from this group have spread in different countries.

One Israeli isolate from this group was also shown to break down the L^4 resistance. A few Asiatic PMMoV isolates belonging to other clades were also shown to break the L^3 or L^4 resistances (Genda *et al.*, 2007) but have not been observed in Mediterranean countries. These resistance breakdown events show several remarkable common points: (i) they were frequently observed first in pepper varieties heterozygous at the *L* locus, (ii) they result from independent events (except clade 3) involving different mutations, and (iii) they were the results of particular mutational pathways usually involving several nucleotide and amino acid substitutions. The latter two points suggest that these events are rather rare and could explain why the L^4 resistance was rather durable. Breakdown of the L^4 resistance was caused apparently by a (former) pathotype $P_{1,2}$ isolate in Japan and by a (former) pathotype $P_{1,2,3}$ isolate in Israel (Fig. 6).

The capacity to infect pepper cultivars with the L^1, L^2, or L^3 resistance genes seems to confer a high fitness cost to tobamoviruses. In susceptible L^+/L^+ pepper plants, TMGMV is more competitive and accumulates to higher rates than PMMoV. Similarly, pathotype $P_{1,2}$ PMMoV isolates are more competitive and accumulate to higher rates than pathotype $P_{1,2,3}$ PMMoV isolates (Fraile *et al.*, 2011). As a consequence, the relative acreage grown in peppers with different alleles at the *L* locus during the past 15 years has had a strong influence on the composition in, and prevalence of the different tobamovirus species and pathotypes (Fraile *et al.*, 2011). This suggests that management of tobamovirus epidemics at a regional scale by growing pepper cultivars carrying different alleles at the *L* locus, including the susceptible L^+/L^+ genotypes, should be feasible, at least in theory.

VI. OTHER VIRUSES

In 2004, *Parietaria mottle virus* (PMoV), a member of the genus *Ilarvirus*, has been identified in bell pepper plants grown in greenhouses in southeast Spain (Janssen *et al.*, 2005). Infected pepper plants develop stem necrosis as well as brown patches and corky rings on the fruit surface. More recently, similar symptoms have been observed in pepper plants collected in southeastern France and were associated with PMoV using serological and molecular diagnostic tools (E. Verdin, unpublished

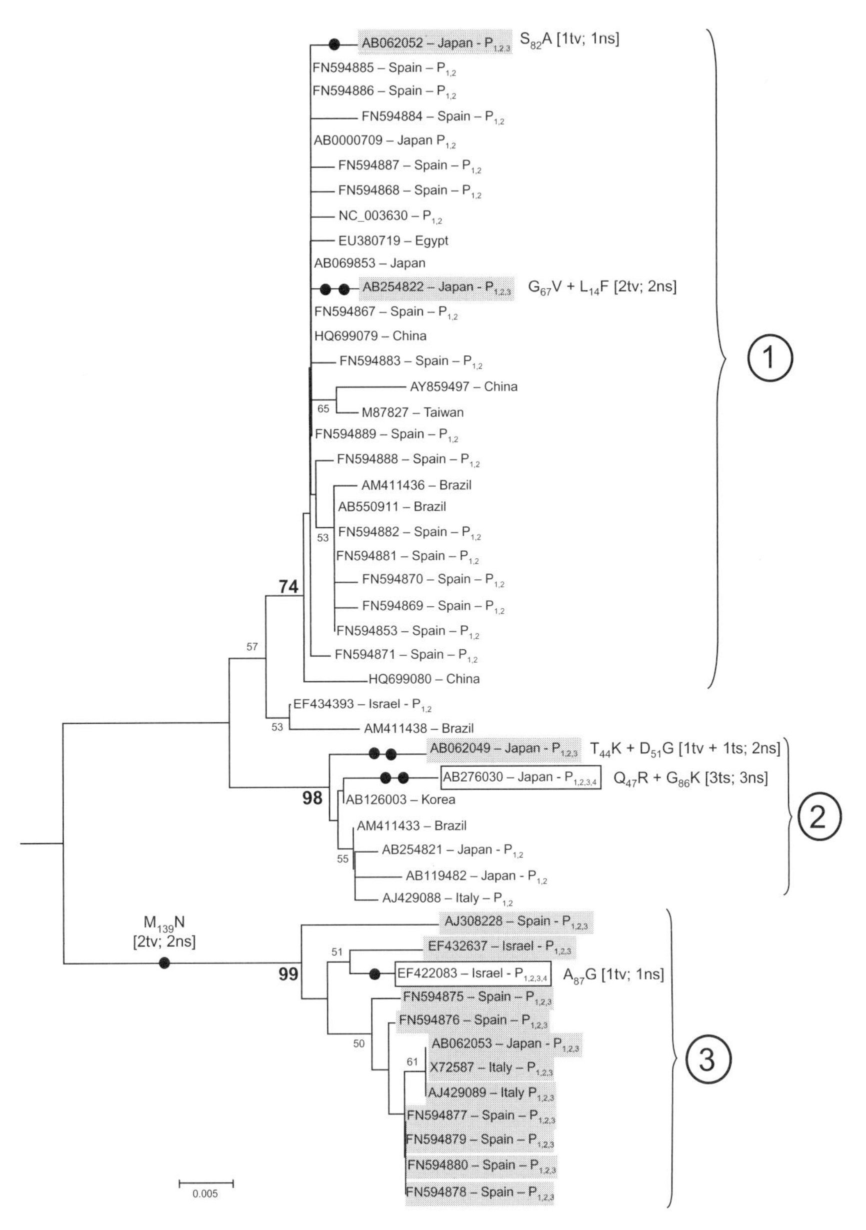

FIGURE 6 Rooted neighbor joining phylogenetic tree of the coat protein (CP) coding region of *Pepper mild mottle virus* (PMMoV). PMMoV pathotypes defined according to the behavior against pepper resistance genes at the *L* locus are indicated when known (see Table I). Pathotype $P_{1,2,3}$ PMMoV isolates are shaded in gray and pathotype $P_{1,2,3,4}$ PMMoV isolates are boxed. The number of amino acid changes in PMMoV CP involved in the breakdown of the L^3 and L^4 resistance genes are indicated (black circles), together with the corresponding number of transitions (ts), transversions (tv), and nonsynonymous (ns) nucleotide substitutions. Bootstrap percentages above 50% are shown. The scale bar indicates branch lengths in substitutions per nucleotide.

results). Initially, PMoV was described on the wild plant species *Parietaria officinalis* (pellitory-of-the-wall) showing yellow mosaic and mottling in Italy (Caciagli *et al.*, 1989) and further in tomato crops in Italy, Southern France, Greece, and the Mediterranean coast of Spain (Aramburu, 2001; Lisa *et al.*, 1998; Marchoux *et al.*, 1999; Roggero *et al.*, 2000) as well as in *Mirabilis jalapa* plants in Italy (Parrella, 2002). PMoV, as the other ilarviruses, is a single-stranded RNA virus with a tripartite genome. Tomato and pepper PMoV isolates have distinct genome sequences (92% nucleotide identity in the P1 gene) and do not share the same biological host range (Caciagli *et al.*, 1989; Janssen *et al.*, 2005). Several studies described the transmission of ilarviruses by seed or pollen from infected plants (Mink, 1993), but no specific natural vectors are known even if transmission by thrips, mites, or nematodes have been reported for some ilarviruses (Fulton, 1981). Mechanical transmission of PMoV using pollen collected on PMoV-infected tomato plants has also been described (Verdin *et al.*, 2005). Transmission of PMoV by several insects (including the thrips species *F. occidentalis*) has been reported to pepper and tomato plants using PMoV-infected *P. officinalis* plants at the flowering stage as sources for virus acquisition (Aramburu *et al.*, 2010). These results suggest that the high incidence of PMoV observed in some Mediterranean countries, especially Italy and Spain, could be reduced by removing *P. officinalis* plants around pepper and tomato crops.

VII. CONCLUSION

Compared to other vegetable crops, very few viral emergences or novel threats have been mentioned in pepper crops in the Mediterranean basin during the past 20 years (Hanssen *et al.*, 2010). Two major explanations can be invoked.

Pepper, as many other crop species from the family Solanaceae, originates from the Americas and was introduced to Europe and other areas of the Old World approximately 500 years ago. As a consequence, viral diseases that affect pepper in the Mediterranean basin are only a small subset of those that are prevalent in the rest of the world. Pepper viruses that are prevalent in the Mediterranean area belong to two main categories. The first one consists of viruses that are particularly stable and/or seed-borne, like tobamoviruses, since they are easily introduced and spread in new areas by human activities and trade. The second category consists of viruses that have a broad host range, particularly if this host range includes vegetatively propagated crops. These viruses could have been introduced into the Mediterranean area through other crops, like PVY that could have been introduced with potato tubers and could have adapted secondarily to pepper (Moury, 2010). Others like CMV or TSWV may have been present in the Old World before the introduction of pepper and may have jumped and adapted to this crop afterward. Viruses that have narrower host ranges and that are not transmitted vertically by seeds in pepper were less likely to be introduced or to emerge in the Mediterranean area. For example, only 2 *Potyvirus* species of 10 described worldwide in pepper are present in the Mediterranean basin.

The second explanation to the apparent stasis of viral diseases in pepper in the Mediterranean region compared to other crops is the lower susceptibility of this plant species to whiteflies and notably to *B. tabaci*, which was recently responsible for the emergence of a large number of virus species from the genus *Begomovirus* and to a lower extent from the genera *Crinivirus*, *Ipomovirus*, and *Torradovirus*, in other vegetable crops such as tomato or cucurbits (Hanssen *et al.*, 2010).

The most recent viral emergences in pepper crops in the Mediterranean area correspond to tospoviruses and/or to isolates breaking down specific resistances of pepper cultivars. Phylogenetic arguments suggest that some TSWV variants could have been introduced in Spain, Italy, France, and Algeria from Northern America, whereas other TSWV variants have probably reemerged in the Old World following the worldwide spread of the thrips species *F. occidentalis* (Tentchev *et al.*, 2011). Recent resistance breakdowns include the tobamovirus resistance gene L^3, estimated to have occurred between 24 and 56 years ago (Fraile *et al.*, 2011), and the TSWV resistance gene *Tsw* which was broken down only a few years after being deployed (García-Arenal and McDonald, 2003; Margaria *et al.*, 2004; Roggero *et al.*, 2002; Sharman and Persley, 2006; Thomas-Carroll and Jones, 2003). No alternative resistance gene is available yet against TSWV, making of this virus the primary constraint on pepper production in some areas.

Future viral threats to pepper production in the Mediterranean area could include CMV isolates from subgroup IB, which have been introduced probably from Asia and have been observed in pepper crops in Spain (Bonnet *et al.*, 2005); poleroviruses that have so far been isolated from pepper crops in the eastern part of the Mediterranean area only, but can induce severe outbreaks; and, potentially, the tospovirus *Polygonum ringspot virus* that is confined to wild plants in several areas in Italy but has the potential to infect many pepper cultivars. In addition, the risk that begomoviruses that affect severely pepper crops in other areas of the world (Ala-Poikela *et al.*, 2005; De Barro *et al.*, 2008; Tiendrebeogo *et al.*, 2011; Torres-Pacheco *et al.*, 1996) could be introduced, for example, with whiteflies, and spread in the Mediterranean area should not be underestimated.

Finally, pepper has also been a model plant for the study of plant resistances to viruses. Natural eIF4E-mediated recessive resistances against viruses were first characterized in pepper (Ruffel *et al.*, 2002). Pepper is also a model plant for the study of polygenic resistances against viruses (Palloix *et al.*, 2009; St Clair, 2010). Finally, two series of virus resistance alleles in pepper correspond to the two main plant–parasite coevolutionary models: resistance to tobamoviruses controlled by the *L* locus fits the ''gene-for-gene'' model of interaction and the resistance to potyviruses controlled by the *pvr2* locus rather corresponds to the ''matching-allele'' model of interaction (Sacristán and García-Arenal, 2008).

ACKNOWLEDGMENTS

We thank Valérie Fontaine (Gautier Semences) and Cécile Marchal (Clause) for reading the chapter.

REFERENCES

Adkins, S., Quadt, R., Choi, T. J., Ahlquist, P., and German, T. L. (1995). An RNA-dependent RNA polymerase activity associated with virions of tomato spotted wilt virus, a plant- and insect-infecting *Bunyavirus*. *Virology* **207**:308–311.

Adkins, S., Choi, T. J., Israel, B. A., Bandla, M. D., Richmond, K. E., Schulz, K. T., Sherwood, J. L., and German, T. L. (1996). Baculovirus expression and processing of tomato spotted wilt tospovirus glycoproteins. *Phytopathology* **86**:849–855.

Ala-Poikela, M., Svensson, E., Rojas, A., Horko, T., Paulin, L., Valkonen, J. P. T., and Kvarnheden, A. (2005). Genetic diversity and mixed infections of begomoviruses infecting tomato, pepper and cucurbit crops in Nicaragua. *Plant Pathol.* **54**:448–459.

Ali, A., and Kobayashi, M. (2010). Seed transmission of *Cucumber mosaic virus* in pepper. *J. Virol. Methods* **163**:234–237.

Alonso, E., García-Luque, I., Avila-Rincón, M. J., Wicke, B., Serra, M. T., and Díaz-Ruíz, J. R. (1989). A tobamovirus causing heavy losses in protected pepper crops in Spain. *J. Phytopathol.* **125**:67–76.

Anandam, R. J., and Doraiswamy, S. (2002). Role of barrier crops in reducing the incidence of mosaic disease in chilli. *J. Plant Dis. Prot.* **109**:109–112.

Antignus, Y., Lachman, O., Pearlsman, M., Maslenin, L., and Rosner, A. (2008). A new pathotype of *Pepper mild mottle virus* (PMMoV) overcomes the L^4 resistance genotype of pepper cultivars. *Plant Dis.* **92**:1033–1037.

Aramburu, J. (2001). First report of *Parietaria mottle virus* on tomato in Spain. *Plant Dis.* **85**:1210.

Aramburu, J., Galipienso, L., Aparicio, F., Soler, S., and López, C. (2010). Mode of transmission of *Parietaria mottle virus*. *Plant Pathol. J.* **92**:679–684.

Arli-Sokmen, M., Mennan, H., Sevik, M. A., and Ecevit, O. (2005). Occurrence of viruses in field-grown pepper crops and some of their reservoir weed hosts in Samsun, Turkey. *Phytoparasitica* **33**:347–358.

Avgelis, A. D. (1986). A pepper strain of TMV who is new in Crete (Greece). *Phytopath. Medit.* **25**:33–38.

Avilla, C., Collar, J. L., Duque, M., Hernáiz, P., Martín, B., and Fereres, A. (1996). Cultivos barrera como método de control de virus no persistentes en pimiento. *Bol. San. Veg. Plagas* **22**:301–307.

Avilla, C., Collar, J. L., Duque, M., and Fereres, A. (1997a). Yield of bell pepper (Capsicum annuum) inoculated with CMV and/or PVY at different time intervals. *J. Plant Dis. Prot.* **104**:1–8.

Avilla, C., Collar, J. L., Duque, M., Pérez, P., and Fereres, A. (1997b). Impact of floating rowcovers on bell pepper yield and virus incidence. *Hortscience* **32:**882–883.

Ayme, V., Souche, S., Caranta, C., Jacquemond, M., Chadoeuf, J., Palloix, A., and Moury, B. (2006). Different mutations in the genome-linked protein VPg of *Potato virus Y* confer virulence on the *pvr2*^{3} resistance in pepper. *Mol. Plant Microbe Interact.* **19:**557–563.

Ayme, V., Petit-Pierre, J., Souche, S., Palloix, A., and Moury, B. (2007). Molecular dissection of the *Potato virus Y* VPg virulence factor reveals complex adaptations to the *pvr2* resistance allelic series in pepper. *J. Gen. Virol.* **88:**1594–1601.

Beczner, L., Rochon, D. M., and Hamilton, R. I. (1997). Characterization of an isolate of pepper mild mottle tobamovirus occurring in Canada. *Can. J. Plant Pathol.* **19:**83–88.

Ben Chaim, A., Grube, R. C., Lapidot, M., Jahn, M., and Paran, I. (2001). Identification of quantitative trait loci associated with resistance to cucumber mosaic virus in *Capsicum annuum. Theor. Appl. Genet.* **102:**1213–1220.

Ben Khalifa, M., Simon, V., Marrakchi, M., Fakhfakh, H., and Moury, B. (2009). Contribution of host plant resistance and geographic distance to the structure of *Potato virus Y* (PVY) populations in pepper in Northern Tunisia. *Plant Pathol.* **58:**763–772.

Ben Khalifa, M., Simon, V., Fakhfakh, H., and Moury, B. (in press). Tunisian *Potato virus Y* isolates with unnecessary pathogenicity towards pepper: Support for the matching allele model in eIF4E resistance—Potyvirus interactions. *Plant Pathol.* **61**(3):441–447.

Berzal-Herranz, A., de la Cruz, A., Tenllado, F., Díaz-Ruíz, J. R., López, L., Sanz, A. I., Vaquero, C., Serra, M. T., and García-Luque, I. (1995). The *Capsicum* L^3 gene-mediated resistance against the tobamoviruses is elicited by the coat protein. *Virology* **209:**498–505.

Black, L. L., Hobbs, H. A., and Gatti, J. M., Jr. (1991). *Tomato spotted wilt virus* resistance in *Capsicum chinense* PI 152225 and 159236. *Plant Dis.* **75:**863.

Boiteux, L. S. (1995). Allelic relationships between genes for resistance to tomato spotted wilt tospovirus in *Capsicum chinense. Theor. Appl. Genet.* **90:**146–149.

Bonnet, J., Fraile, A., Sacristan, S., Malpica, J. M., and Garcia-Arenal, F. (2005). Role of recombination in the evolution of natural populations of *Cucumber mosaic virus*, a tripartite RNA plant virus. *Virology* **332:**359–368.

Boukema, I. W. (1980). Allelism controlling resistance to TMV in *Capsicum* L. *Euphytica* **29:**433–439.

Boukema, I. W. (1982). Resistance to a new strain of TMV in *Capsicum chacoense* Hunz. *Capsicum Newsl.* **1:**49–51.

Boukema, I. W. (1984). Resistance to TMV in *Capsicum chacoense* Hunz. is governed by an allele of the *L* locus. *Capsicum Newsl.* **3:**47–48.

Brault, V., Perigon, S., Reinbold, C., Erdinger, M., Scheidecker, D., Herrbach, E., Richards, K., and Ziegler-Graff, V. (2005). The polerovirus minor capsid protein determines vector specificity and intestinal tropism in the aphid. *J. Virol.* **79:**9685–9693.

Buzkan, N., Demir, M., Oztekin, V., Mart, C., Çağlar, B. K., and Yilmaz, M. A. (2006). Evaluation of the status of capsicum viruses in the main growing regions of Turkey. *OEPP/EPPO Bull.* **36:**15–19.

Caciagli, P., Boccardo, G., and Lovisolo, O. (1989). *Parietaria mottle virus*, a possible new ilarvirus from *Parietaria officinalis* (Urticaceae). *Plant Pathol.* **38:**577–584.

Caranta, C., Lefebvre, V., and Palloix, A. (1997a). Polygenic resistance of pepper to potyviruses consists of a combination of isolate-specific and broad-spectrum quantitative trait loci. *Mol. Plant Microbe Interact.* **10:**872–878.

Caranta, C., Palloix, A., Lefebvre, V., and Daubèze, A. M. (1997b). QTLs for a component of partial resistance to *Cucumber mosaic virus* in pepper: Restriction of virus installation in host-cells. *Theor. Appl. Genet.* **94:**431–438.

Caranta, C., Pflieger, S., Lefebvre, V., Daubèze, A. M., Thabuis, A., and Palloix, A. (2002). QTLs involved in the restriction of *Cucumber mosaic virus* (CMV) long-distance movement in pepper. *Theor. Appl. Genet.* **104:**586–591.

Charron, C., Nicolaï, M., Gallois, J. L., Robaglia, C., Moury, B., Palloix, A., and Caranta, C. (2008). Natural variation and functional analyses provide evidence for co-evolution between plant eIF4E and potyviral VPg. *Plant J.* **54:**56–68.

Chen, B., and Francki, R. I. B. (1990). Cucumovirus transmission by the aphid *Myzus persicae* is determined solely by the viral coat protein. *J. Gen. Virol.* **71:**939–944.

Chu, M. H., López-Moya, J. J., Llave-Correas, C., and Pirone, T. P. (1997). Two separate regions in the genome of the *Tobacco etch virus* contain determinants of the wilting response of Tabasco pepper. *Mol. Plant Microbe Interact.* **10:**472–480.

Chung, B. Y. W., Miller, W. A., Atkins, J. F., and Firth, A. E. (2008). An overlapping essential gene in the Potyviridae. *Proc. Natl. Acad. Sci. U.S.A.* **105:**5897–5902.

Ciuffo, M., Tavella, L., Pacifico, D., Masenga, V., and Turina, M. (2008). A member of a new *Tospovirus* species isolated in Italy from wild buckwheat (*Polygonum convolvulus*). *Arch. Virol.* **153:**2059–2068.

Collar, J. L., Avilla, C., and Fereres, A. (1997). New correlations between aphid stylet paths and non-persistent virus transmission. *Environ. Entomol.* **26:**537–544.

Cruz, A., López, L., Tenllado, F., Díaz-Ruíz, J. R., Sanz, A. I., Vaquero, C., Serra, M. T., and García-Luque, I. (1997). The coat protein is required for the elicitation of the *Capsicum* L^2 gene-mediated resistance against the tobamoviruses. *Mol. Plant Microbe Interact.* **10:**107–113.

Daughtrey, M. L., Jones, R. K., Moyer, J. W., Daub, M. E., and Baker, J. R. (1997). Tospoviruses strike the greenhouse industry: INSV has become a major pathogen on flower crops. *Plant Dis.* **81:**1220–1230.

De Angelis, J. D., Sether, D. M., and Rossignol, P. A. (1994). Transmission of impatiens necrotic spot virus in peppermint by western flower thrips (*Thysanoptera: Thripidae*). *J. Econ. Entomol.* **87:**197–201.

De Barro, P. J., Hidayat, S. H., Frohlich, D., Subandiyah, S., and Ueda, S. (2008). A virus and its vector, pepper yellow leaf curl virus and *Bemisia tabaci*, two new invaders of Indonesia. *Biol. Invasions* **10:**411–433.

De Haan, P., Wagemakers, L., Peters, D., and Goldbach, R. (1990). The S RNA segment of *Tomato spotted wilt virus* has an ambisense character. *J. Gen. Virol.* **71:**1001–1007.

De Haan, P., Kormelink, R., de Oliveira Resende, R., van Poelwijk, F., Peters, D., and Goldbach, R. (1991). *Tomato spotted wilt virus* L RNA encodes a putative RNA polymerase. *J. Gen. Virol.* **72:**2207–2216.

Depestre, T., Palloix, A., Camino, V., and Gebre Selassie, K. (1993). Identification of virus isolates and of *Tobacco etch virus* (TEV) pathotypes infecting green pepper in Caujeri Valley (Guantanamo, Cuba). *Capsicum Eggplant Newsl.* **12:**73–74.

Dogimont, C., Palloix, A., Daubèze, A. M., Marchoux, G., Gebre Selassie, K., and Pochard, E. (1996). Genetic analysis of broad spectrum resistance to potyviruses using doubled haploid lines of pepper (*Capsicum annuum* L.). *Euphytica* **88:**231–239.

Dombrovsky, A., Glanz, E., Pearlsman, M., Lachman, O., and Antignus, Y. (2010). Characterization of *Pepper yellow leaf curl virus*, a tentative new Polerovirus species causing a yellowing disease of pepper. *Phytoparasitica* **38:**477–486.

Dufour, O., Palloix, A., Gebre Selassie, K., Pochard, E., and Marchoux, G. (1989). The distribution of *Cucumber mosaic virus* in resistant and susceptible plants of pepper. *Can. J. Bot.* **67:**655–660.

Fanigliulo, A., Comes, S., Pacella, R., Harrach, B., Martin, D. P., and Crescenzi, A. (2005). Characterisation of *Potato virus Y* nnp strain inducing veinal necrosis in pepper: A naturally occurring recombinant strain of PVY. *Arch. Virol.* **150:**709–720.

Fereres, A. (2000). Barrier crops as a cultural control measure of non-persistently transmitted aphid-borne viruses. *Virus Res.* **71:**221–231.

Fereres, A., Avilla, C., Collar, J. L., Duque, M., and Fernandez-Quintanilla, C. (1996). Impact of various yield-reducing agents on open-field sweet peppers. *Environ. Entomol.* **25:**983–986.

Florini, D. A., and Zitter, T. A. (1987). *Cucumber mosaic virus* (CMV) in peppers (*C. annuum* L.) in New York and associated yield losses. *Phytopathology* **77:**652.

Font, M. I., Cordoba-Selles, M. C., Cebrian, M. C., Herrera-Vasquez, J. A., Alfaro-Fernandez, A., Boubaker, A., Soltani, I., and Jorda, C. (2009). First report of *Tobacco mild green mosaic virus* infecting *Capsicum annuum* in Tunisia. *Plant Dis.* **93:**761.

Fraile, A., Pagán, I., Anastasio, G., Sáez, E., and García-Arenal, F. (2011). Rapid genetic diversification and high fitness penalties associated with pathogenicity evolution in a plant virus. *Mol. Biol. Evol.* **28:**1425–1437.

Fulton, R. W. (1981). Ilarviruses. *In* ''Handbook of Plant Virus Infections and Comparative Diagnosis'' (E. Kurstak, ed.), pp. 377–421. Elsevier/North Holland Medical Press, Amsterdam.

Gallitelli, D. (2000). The ecology of *Cucumber mosaic virus* and sustainable agriculture. *Virus Res.* **71:**9–21.

García-Arenal, F., and McDonald, B. A. (2003). An analysis of the durability of resistance to plant viruses. *Phytopathology* **93:**941–952.

García-Ruiz, H., and Murphy, J. F. (2001). Age-related resistance in bell pepper to *Cucumber mosaic virus*. *Ann. Appl. Biol.* **139:**307–317.

Gebre Selassie, K., Marchoux, G., Delecolle, B., and Pochard, E. (1985). Variabilité naturelle des souches du virus Y de la pomme de terre dans les cultures de piment du sud-est de la France. Caractérisation et classification en pathotypes. *Agronomie* **5:**621–630.

Genda, Y., Kanda, A., Hamada, H., Sato, K., Ohnishi, J., and Tsuda, S. (2007). Two amino acid substitutions in the coat protein of *Pepper mild mottle virus* are responsible for overcoming the L^4 gene-mediated resistance in *Capsicum* spp. *Phytopathology* **97:**787–793.

Gera, A., Loebenstein, G., and Raccah, B. (1979). Protein coats of 2 strains of *Cucumber mosaic virus* affect transmission by *Aphis gossypii*. *Phytopathology* **69:**396–399.

German, T. L., Ullman, D. E., and Moyer, J. W. (1992). Tospoviruses: Diagnosis, molecular biology, phylogeny, and vector relationships. *Annu. Rev. Phytopathol.* **30:**315–348.

Gharsallah Chouchane, S., Gorsane, F., Nakhla, M. K., Maxwell, D. P., Marrakchi, M., and Fakhfakh, H. (2007). First report of *Tomato yellow leaf curl virus*-Israel species infecting tomato, pepper and bean in Tunisia. *J. Phytopathol.* **155:**236–240.

Gibbs, A. J., Ohshima, K., Phillips, M. J., and Gibbs, M. J. (2008). The prehistory of potyviruses: Their initial radiation was during the dawn of agriculture. *PLoS One* **3:**e2523.

Gilardi, P., García-Luque, I., and Serra, M. T. (1998). *Pepper mild mottle virus* coat protein alone can elicit the *Capsicum* spp. *L-3* gene-mediated resistance. *Mol. Plant Microbe Interact.* **11:**1253–1257.

Gilardi, P., García-Luque, I., and Serra, M. T. (2004). The coat protein of tobamovirus acts as elicitor of both L^2 and L^4 gene-mediated resistance in *Capsicum*. *J. Gen. Virol.* **85:**2077–2085.

Glais, L., Tribodet, M., and Kerlan, C. (2002). Genomic variability in Potato potyvirus Y (PVY): Evidence that (PVYW)-W-N and PVYNTN variants are single to multiple recombinants between PVYO and PVYN isolates. *Arch. Virol.* **147:**363–378.

Gooding, G. V. (1975). Inactivation of tobacco mosaic virus on tomato seed with trisodium orthophosphate and sodium hypochlorite. *Plant Dis. Rep.* **59:**770–772.

Gorsane, F., Fekih-Hassen, I., Gharsallah-Chouchene, S., Nakhla, M. K., Maxwell, D. P., Marrakchi, M., and Fakhfakh, H. (2004). Typing of *Tomato yellow leaf curl virus* spreading on pepper in Tunisia. Proceedings of the XIIth EUCARPIA Meeting on Genetics and Breeding of Capsicum and Eggplant, Noordwijkerhout, Netherlands, 17–19 May, 2004, p. 182.

Goto, T., Iizuka, N., and Komochi, S. (1984). Selection and utilization of an attenuated isolate of pepper strain of tobacco mosaic virus. *Annu. Phytopathol. Soc. Jpn.* **50:**221–228.

Green, S. K., and Kim, J. S. (1991). Characteristics and control of viruses infecting peppers, a literature review. *Asian Vegetable Research and Development Centre, Technical Bulletin* **18:** p. 60.

Guerini, M. N., and Murphy, J. F. (1999). Resistance of *Capsicum annuum* 'Avelar' to pepper mottle potyvirus and alleviation of this resistance by co-infection with cucumber mosaic cucumovirus are associated with virus movement. *J. Gen. Virol.* **80:**2785–2792.

Güldür, M. E., and Çağlar, B. K. (2006). Outbreaks of *Pepper mild mottle virus* in greenhouses in Sanliurfa, Turkey. *J. Plant Pathol.* **88:**341.

Hagiwara, K., Ichiki, T. U., Ogawa, Y., Omura, T., and Tsuda, S. (2002). A single amino acid substitution in 126-kDa protein of *Pepper mild mottle virus* associates with symptom attenuation in pepper; the complete nucleotide sequence of an attenuated strain, C-1421. *Arch. Virol.* **147:**833–840.

Hamada, H., Takeuchi, S., Kiba, A., Tsuda, S., Hikichi, Y., and Okuno, T. (2002). Amino acid changes in *Pepper mild mottle virus* coat protein that affect L3 gene-mediated resistance in pepper. *J. Gen. Plant Pathol.* **68:**155–162.

Hamada, H., Tomita, R., Iwadate, Y., Kobayashi, K., Munemura, I., Takeuchi, S., Hikichi, Y., and Suzuki, K. (2007). Cooperative effect of two amino acid mutations in the coat protein of *Pepper mild mottle virus* overcomes *L3*-mediated resistance in *Capsicum* plants. *Virus Genes* **34:**205–214.

Hanssen, I. M., Lapidot, M., and Thomma, B. P. H. J. (2010). Emerging viral diseases of tomato crops. *Mol. Plant Microbe Interact.* **23:**539–548.

Harrison, B. D. (2002). Virus variation in relation to resistance-breaking in plants. *Euphytica* **124:**181–192.

Hassani-Mehraban, A., Saaijer, J., Peters, D., Goldbach, R., and Kormelink, R. (2007). Molecular and biological comparison of two *Tomato yellow ring virus* (TYRV) isolates: Challenging the *Tospovirus* species concept. *Arch. Virol.* **152:**85–96.

Hatala Zseller, I., and Kiss, E. F. (1999). Control of *Frankliniella occidentalis* and TSWV in capsicum crops in Hungary. *EPPO Bull.* **29:**63–67.

Hellwald, K. H., Glenewinkel, D., and Hauber, S. (2000). Increased symptom severity in pepper plants after co-inoculation with two *Cucumber mosaic virus* subgroup I strains in comparison to single inoculations of the corresponding wildtype strains. *J. Plant Dis. Prot.* **107:**368–375.

Hooks, C. R. R., and Fereres, A. (2006). Protecting crops from non-persistently aphid-transmitted viruses: A review on the use of barrier plants as a management tool. *Virus Res.* **120:**1–16.

Hu, X. X., He, C. Z., Xiao, Y., Xiong, X. Y., and Nie, X. Z. (2009a). Molecular characterization and detection of recombinant isolates of *Potato virus Y* from China. *Arch. Virol.* **154:**1303–1312.

Hu, X. J., Karasev, A. V., Brown, C. J., and Lorenzen, J. H. (2009b). Sequence characteristics of *Potato virus Y* recombinants. *J. Gen. Virol.* **90:**3033–3041.

Ichiki, T. U., Nagaoka, E. N., Hagiwara, K., Sasaya, T., and Omura, T. (2009). A single residue in the 126-kDa protein of *Pepper mild mottle virus* controls the severity of symptoms on infected green bell pepper plants. *Arch. Virol.* **154:**489–493.

Janzac, B., Fabre, M.-F., Palloix, A., and Moury, B. (2009). Phenotype and spectrum of action of the *Pvr4* resistance in pepper against potyviruses, and selection for virulent variants. *Plant Pathol.* **58:**443–449.

Janzac, B., Montarry, J., Palloix, A., Navaud, O., and Moury, B. (2010). A point mutation in the polymerase of *Potato virus Y* confers virulence toward the *Pvr4* resistance of pepper and a high competitiveness cost in susceptible cultivar. *Mol. Plant Microbe Interact.* **23:**823–830.

Janssen, D., Saez, E., Segundo, E., Martín, G., Gil, F., and Cuadrado, I. M. (2005). Capsicum annuum - a new host of Parietaria mottle virus in Spain. *Plant Pathol* **54:**567.

Kang, B. C., Yeam, I., Frantz, J. D., Murphy, J. F., and Jahn, M. M. (2005). The *pvr1* locus in *Capsicum* encodes a translation initiation factor eIF4E that interacts with *Tobacco etch virus* VPg. *Plant J.* **42:**392–405.

Kang, W. H., Hoang, N. H., Yang, H. B., Kwon, J. K., Jo, S. H., Seo, J. K., Kim, K. H., Choi, D., and Kang, B. C. (2010). Molecular mapping and characterization of a single dominant gene controlling CMV resistance in peppers (*Capsicum annuum* L.). *Theor. Appl. Genet.* **120:**1587–1596.

Kirk, W. D. J., and Terry, L. I. (2003). The spread of the western flower thrips *Frankliniella occidentalis* (Pergande). *Agric. For. Entomol.* **5:**301–310.

Kormelink, R., Kitajima, E. W., de Haan, P., Zuidema, D., Peters, D., and Goldbach, R. (1991). The nonstructural protein (NSs) encoded by the ambisense S RNA segment of *Tomato spotted wilt virus* is associated with fibrous structures in infected plant cells. *Virology* **181:**459–468.

Kormelink, R., de Haan, P., Meurs, C., Peters, D., and Goldbach, R. (1992). The nucleotide sequence of the M RNA segment of *Tomato spotted wilt virus*, a bunyavirus with two ambisense RNA segments. *J. Gen. Virol.* **73:**2795–2804.

Lapidot, M., Paran, I., Ben-Joseph, R., Ben-Harush, S., Pilowsky, M., Cohen, S., and Shifriss, C. (1997). Tolerance to *Cucumber mosaic virus* in pepper: Development of advanced breeding lines and evaluation of virus level. *Plant Dis.* **81:**185–188.

Law, M. D., Speck, J., and Moyer, J. W. (1992). The M RNA of impatiens necrotic spot tospovirus (*Bunyaviridae*) has an ambisense genomic organization. *Virology* **188:**732–741.

Lecoq, H., Moury, B., Desbiez, C., Palloix, A., and Pitrat, M. (2004). Durable virus resistance in plants through conventional approaches: A challenge. *Virus Res.* **100:**31–39.

Lee, M. Y., Lee, J. H., Ahn, H. I., Kim, M. J., Her, N. H., Choi, J. K., Harn, C. H., and Ryu, K. H. (2006). Identification and sequence analysis of RNA3 of a resistance-breaking isolate of *Cucumber mosaic virus* from *Capsicum annuum*. *Plant Pathol. J.* **22:**265–270.

Lin, H. X., Rubio, L., Smythe, A., Jiminez, M., and Falk, B. W. (2003). Genetic diversity and biological variation among California isolates of *Cucumber mosaic virus*. *J. Gen. Virol.* **84:**249–258.

Lisa, V., Ramasso, E., Ciuffo, M., and Roggero, P. (1998). Tomato apical necrosis caused by a strain of parietaria mottle ilarvirus. Proceedings 9th Conference of the International Society for Horticultural Science. VVWG, Recent Advances in Vegetable Virus Research, Turin, 1998, pp. 290–291.

Llamas-Llamas, M. E., Zavaleta-Mejia, E., Gonzalez-Hernandez, V. A., Cervantes-Diaz, L., Santizo-Rincon, J. A., and Ochoa-Martinez, D. L. (1998). Effect of temperature on symptom expression and accumulation of tomato spotted wilt virus in different host species. *Plant Pathol.* **47:**341–347.

Lovato, F. A., Inoue-Nagata, A. K., Nagata, T., de Avila, A. C., Pereira, L. A., and Resende, R. O. (2008). The N protein of *Tomato spotted wilt virus* (TSWV) is associated with the induction of programmed cell death (PCD) in *Capsicum chinense* plants, a hypersensitive host to TSWV infection. *Virus Res.* **137:**245–252.

Lozano, G., Moriones, E., and Navas-Castillo, J. (2004). First report of sweet pepper (*Capsicum annuum*) as a natural host plant for *Tomato chlorosis virus*. *Plant Dis.* **88:**224.

Luis-Arteaga, M., and Gil-Ortega, R. (1986). Biological characterization of PVY as isolated from pepper in Spain. VI Meeting on Capsicum and eggplant, Zaragoza, Spain, October 21–24, 183–188.

Luis-Arteaga, M., Gil-Ortega, R., and Pasko, P. (1993). Presence of PVY1-2 pathotype in pepper crops in Spain. *Capsicum and Eggplant Newsl.* **12:**67–68.

Maisonneuve, J. C., and Marrec, C. (1999). The potential of Chrysoperla lucasina for IPM programmes in greenhouses. *In* "OILB/SROP Bulletin" (J. C. van Lenteren, ed.), IOBC/WPRS Working Group 'Integrated Control in Glasshouses'. Proceedings of the Meeting at Brest, France, 25–29 May, 1999, Vol. 22, pp. 165–168.

Marchoux, G., Douine, L., and Quiot, J.-B. (1976). Comportement thermique différentiel de certaines souches de CMV. Hypothèse d'un mécanisme pléiotropique reliant plusieurs propriétés. *C.R. Acad. Sci. Paris Sér. D* **283:**1601–1604.

Marchoux, G., Parrella, G., Gebre-Selassie, K., and Gognalons, P. (1999). Identification de deux ilarvirus sur tomate dans le sud de la France. *Phytoma* **522:**53–55.

Marchoux, G., Parrella, G., and Gognalons, P. (2003). An alfalfa virus is causing damage to vegetable and aromatic crops in France and Italy. *Phytoma* **559:**41–45.

Marco, S. (1993). Incidence of nonpersistently transmitted viruses in pepper sprayed with whitewash, oil and insecticide, alone or combined. *Plant Dis.* **77:**1119–1122.

Margaria, P., Ciuffo, M., and Turina, M. (2004). Resistance breaking strain of *Tomato spotted wilt virus* (*Tospovirus*; Bunyaviridae) on resistant pepper cultivars in Almería, Spain. *Plant Pathol.* **53:**795.

Margaria, P., Ciuffo, M., Pacifico, D., and Turina, M. (2007). Evidence that the nonstructural protein of *Tomato spotted wilt virus* is the avirulence determinant in the interaction with resistant pepper carrying the Tsw gene. *Mol. Plant Microbe Interact.* **20:**547–558.

Maris, P. C., Joosten, N. N., Peters, D., and Goldbach, R. W. (2003). Thrips resistance in pepper and its consequences for the acquisition and inoculation of *Tomato spotted wilt virus* by the western flower thrips. *Phytopathology* **93:**96–101.

Maris, P. C., Joosten, N. N., Goldbach, R. W., and Peters, D. (2004). *Tomato spotted wilt virus* infection improves host suitability for its vector *Frankliniella occidentalis*. *Phytopathology* **94:**706–711.

Martelli, G. P., and Quacquarelli, A. (1983). The present status of tomato and pepper viruses. *Acta Hort.* **127:**39–64.

Martínez-Torres, D., Carrio, R., Latorre, A., Simón, J. C., Hermoso, A., and Moya, A. (1998). Assessing the nucleotide diversity of three aphid species by RAPD. *J. Mol. Evol.* **10:**459–477.

Martín-López, B., Varela, I., Marnotes, S., and Cabaleiro, C. (2006). Use of oils combined with low doses of insecticide for the control of *Myzus persicae* and PVY epidemics. *Pest Manag. Sci.* **62:**372–378.

Mascia, T., Cillo, F., Fanelli, V., Finetti-Sialer, M. M., De Stradis, A., Palukaitis, P., and Gallitelli, D. (2010). Characterization of the interactions between *Cucumber mosaic virus* and *Potato virus Y* in mixed infections in tomato. *Mol. Plant Microbe Interact.* **23:**1514–1524.

Matsumoto, K., Johnishi, K., Hamada, H., Sawada, H., Takeuchi, S., Kobayashi, K., Suzuki, K., Kiba, A., and Hikichi, Y. (2009). Single amino acid substitution in the methyltransferase domain of *Paprika mild mottle virus* replicase proteins confers the ability to overcome the high temperature-dependent *Hk* gene-mediated resistance in *Capsicum* plants. *Virus Res.* **140:**98–102.

Matteson, N. A., and Terry, L. I. (1992). Response to color by male and female *Frankliniella occidentalis* during swarming and non-swarming behavior. *Entomol. Exp. Appl.* **63:**187–201.

Mink, G. I. (1993). Pollen and seed-transmitted viruses and viroids. *Annu. Rev. Phytopathol.* **31:**375–402.

Montarry, J., Doumayrou, J., Simon, V., and Moury, B. (2011). Genetic background matters: A plant-virus gene-for-gene interaction is strongly influenced by genetic contexts. *Mol. Plant Pathol.* **12:**911–920.

Morilla, G., Janssen, D., Garcia-Andres, S., Moriones, E., Cuadrado, I. M., and Bejarano, E. R. (2005). Pepper (*Capsicum annuum*) is a dead-end host for *Tomato yellow leaf curl virus*. *Phytopathology* **95:**1089–1097.

Moury, B. (2004). Differential selection of genes of *Cucumber mosaic virus* subgroups. *Mol. Biol. Evol.* **21:**1602–1611.

Moury, B. (2010). A new lineage sheds light on the evolutionary history of *Potato virus Y*. *Mol. Plant Pathol.* **11:**161–168.

Moury, B., Palloix, A., Gebre-Selassie, K., and Marchoux, G. (1997). Hypersensitive resistance to *Tomato spotted wilt virus* in three *Capsicum chinense* accessions is controlled by a single gene and is overcome by virulent strains. *Euphytica* **94:**45–52.

Moury, B., Gebre Selassie, K., Marchoux, G., Daubèze, A. M., and Palloix, A. (1998). High temperature effects on hypersensitive resistance to tomato spotted wilt tospovirus (TSWV) in pepper (*Capsicum chinense* Jacq.). *Eur. J. Plant Pathol.* **104**:489–498.
Moury, B., Morel, C., Johansen, E., and Jacquemond, M. (2002). Evidence for diversifying selection in *Potato virus Y* and in the coat protein of other potyviruses. *J. Gen. Virol.* **83**:2563–2573.
Muhyi, R. I., Bosland, P. W., and Pochard, E. (1993). Difference between USA and French isolates of *Tobacco etch virus* and *Pepper mottle virus* displayed by doubled haploid line analysis. *Euphytica* **72**:23–29.
Murakami, R., Nakashima, N., Hinomoto, N., Kawano, S., and Toyosato, T. (2011). The genome sequence of *Pepper vein yellows virus* (family Luteoviridae, genus *Polerovirus*). *Arch. Virol.* **156**:921–923.
Nagai, Y., Takeuchi, T., and Tochihara, H. (1981). A new mosaic disease of sweet pepper caused by pepper strain of *Tobacco mosaic virus*. *Ann. Phytopathol. Soc. Jpn.* **47**:541–546.
Nono Womdim, R., Palloix, A., Gebre Selassie, K., and Marchoux, G. (1993). Partial resistance of bell pepper to *Cucumber mosaic virus* movement within plants: Field evaluation of its efficiency in Southern France. *J. Phytopathol.* **137**:125–132.
Ogawa, T., Tomitaka, Y., Nakagawa, A., and Ohshima, K. (2008). Genetic structure of a population of *Potato virus Y* inducing potato tuber necrotic ringspot disease in Japan; comparison with North American and European populations. *Virus Res.* **131**:199–212.
Owen, J., and Palukaitis, P. (1988). Characterization of *Cucumber mosaic virus*. 1. Molecular heterogeneity mapping of RNA-3 in eight CMV strains. *Virology* **166**:495–502.
Pagán, I., Firth, C., and Holmes, E. C. (2010). Phylogenetic analysis reveals rapid evolutionary dynamics in the plant RNA virus genus tobamovirus. *J. Mol. Evol.* **71**:298–307.
Palloix, A., Abak, K., Daubèze, A. M., Güldür, M., and Gebre Selassie, K. (1994). Survey of pepper diseases affecting the main production regions of Turkey with special interest in viruses and potyvirus pathotypes. *Capsicum and Eggplant Newsl.* **13**:78–81.
Palloix, A., Daubèze, A. M., Lefebvre, V., Caranta, C., Moury, B., Pflieger, S., Gebre Selassie, K., and Marchoux, G. (1997). Construction de systèmes de résistance aux maladies adaptés aux conditions de cultures chez le piment. *C.R. Acad. Agric. Fr.* **83**:87–98.
Palloix, A., Ayme, V., and Moury, B. (2009). Durability of plant major resistance genes to pathogens depends on the genetic background, experimental evidence and consequences for breeding strategies. *New Phytol.* **183**:190–199.
Palukaitis, P., and García-Arenal, F. (2003). Cucumoviruses. *Adv. Virus Res.* **62**:241–323.
Pares, R. D. (1985). A tobamovirus infecting *Capsicum* in Australia. *Ann. Appl. Biol.* **106**:469–474.
Parrella, G. (2002). First report of *Parietaria mottle virus* in *Mirabilis jalapa*. *Plant Pathol.* **51**:401.
Parrella, G., Gognalons, P., Gebre-Selassie, K., Vovlas, C., and Marchoux, G. (2003). An update of the host range of *Tomato spotted wilt virus*. *J. Plant Pathol.* **85**:227–264.
Pérez, P., Collar, J. L., Avilla, C., Duque, M., and Fereres, A. (1995). Estimation of vector propensity of *Potato virus Y* in open-field pepper crops of central Spain. *J. Econ. Entomol.* **88**:986–991.
Pérez, P., Duque, M., Collar, J. L., Avilla, C., and Fereres, A. (2003). Activity of alatae aphids landing on open-field pepper crops in Spain. *J. Plant Dis. Prot.* **110**:195–202.
Pernezny, K. L., Roberts, P. D., Murphy, J. F., and Goldberg, N. P. (2003). Compendium of Pepper Diseases. American Phytopathology Society Press, St Paul, MN.
Perry, K. L., Zhang, L., and Palukaitis, P. (1998). Amino acid changes in the coat protein of *Cucumber mosaic virus* differentially affect transmission by the aphids *Myzus persicae* and *Aphis gossypii*. *Virology* **242**:204–210.
Polston, J. E., Cohen, L., Sherwood, T. A., Ben-Joseph, R., and Lapidot, M. (2006). *Capsicum* species: Symptomless hosts and reservoirs of *Tomato yellow leaf curl virus*. *Phytopathology* **96**:447–452.

Quiot, J.-B., Devergne, J.-C., Marchoux, G., Cardin, L., and Douine, L. (1979). Ecologie et épidémiologie du virus de la mosaïque du concombre dans le sud-est de la France. VI. Conservation de deux types de populations sauvages dans les plantes sauvages. *Annu. Phytopathol.* **11:**349–357.

Raccah, B. (1986). Non-persistent viruses: Epidemiology and control. *Adv. Virus Res.* **31:**387–429.

Rahman, T., Roff, M. N. M., and Bin Abd Ghani, I. (2010). Within-field distribution of *Aphis gossypii* and aphidophagous lady beetles in chili, *Capsicum annuum*. *Entomol. Exp. Appl.* **137:**211–219.

Rasoulpour, R., and Izadpanah, K. (2007). Characterisation of cineraria strain of *Tomato yellow ring virus* from Iran. *Australas. Plant Pathol.* **36:**286–294.

Rast, A. T. B. (1988). Pepper tobamovirus and pathotypes used in resistance breeding. *Capsicum Newsl.* **7:**20–23.

Rast, A. T. B., and Stijger, C. C. M. M. (1987). Disinfection of pepper seed infected with different strains of capsicum mosaic virus by trisodium phosphate and heat treatment. *Plant Pathol.* **36:**583–588.

Reina, J., Morilla, G., Bejarano, E. R., Rodriguez, M. D., Janssen, D., and Cuadrado, I. M. (1999). First report of *Capsicum annuum* plants infected by *Tomato yellow leaf curl virus*. *Plant Dis.* **83:**1176.

Revers, F., Le Gall, O., Candresse, T., Le Romancer, M., and Dunez, J. (1996). Frequent occurrence of recombinant potyvirus isolates. *J. Gen. Virol.* **77:**1953–1965.

Roditakis, N. E., Lykouressis, D. P., and Golfinopoulou, N. G. (2001). Color preference, sticky trap catches and distribution of western flower thrips in greenhouse cucumber, sweet pepper and eggplant crops. *Southwest. Entomol.* **26:**227–238.

Roggero, P., Dellavalle, G., Ciuffo, M., and Pennazio, S. (1999). Effects of temperature on infection in *Capsicum sp* and *Nicotiana benthamiana* by impatiens necrotic spot tospovirus. *Eur. J. Plant Pathol.* **105:**509–512.

Roggero, P., Ciuffo, M., Katis, N., Alioto, D., Crescenzi, A., Parrella, G., and Gallitelli, D. (2000). Necrotic disease in tomatoes in Greece and south Italy caused by tomato strain of *Parietaria mottle virus*. *J. Plant Pathol.* **82:**159.

Roggero, P., Masenga, V., and Tavella, L. (2002). Field isolates of *Tomato spotted wilt virus* overcoming resistance in pepper and their spread to other hosts in Italy. *Plant Dis.* **86:**950–954.

Roossinck, M. J., Zhang, L., and Hellwald, K.-H. (1999). Rearrangements in the 59 nontranslated region and phylogenetic analyzes of *Cucumber mosaic virus* RNA 3 indicate radial evolution of three subgroups. *J. Virol.* **73:**6752–6758.

Rubio, L., Luis-Arteaga, M., Cambra, M., Serra, J., Moreno, P., and Guerri, J. (2002). First report of *Broad bean wilt virus* 1 in Spain. *Plant Dis.* **86:**698.

Ruffel, S., Dussault, M.-H., Palloix, A., Moury, B., Bendahmane, A., Robaglia, C., and Caranta, C. (2002). A natural recessive resistance gene against *Potato virus Y* in pepper corresponds to the eukaryotic initiation factor 4E (eIF4E). *Plant J.* **32:**1067–1075.

Sacristán, S., and García-Arenal, F. (2008). The evolution of virulence and pathogenicity in plant pathogen populations. *Mol. Plant Pathol.* **9:**369–384.

Satar, S., Kersting, U., and Uygun, N. (2008). Effect of temperature on population parameters of *Aphis gossypii* Glover and *Myzus persicae* (Sulzer) (Homoptera:Aphididae) on pepper. *J. Plant Dis. Prot.* **115:**69–74.

Sawada, H., Takeuchi, S., Matsumoto, K., Hamada, H., Kiba, A., Matsumoto, M., Watanabe, Y., Suzuki, K., and Hikichi, Y. (2005). A new *Tobamovirus*-resistance gene, *Hk*, in *Capsicum annuum*. *J. Jpn. Soc. Hort. Sci.* **74:**289–294.

Schubert, J., Fomitcheva, V., and Sztangret-Wisniewska, J. (2007). Differentiation of *Potato virus Y* strains using improved sets of diagnostic PCR-primers. *J. Virol. Methods* **140:**66–74.

Sharman, M., and Persley, D. M. (2006). Field isolates of *Tomato spotted wilt virus* overcoming resistance in *Capsicum* in Australia. *Australas. Plant Pathol.* **35:**123–128.

Simons, J. (1957). Effects of insecticides and physical barriers on field spread of *Pepper vein banding mosaic virus*. *Phytopathology* **47:**139–145.

Sin, S. H., McNulty, B. C., Kennedy, G. G., and Moyer, J. W. (2005). Viral genetic determinants for thrips transmission of *Tomato spotted wilt virus*. *Proc. Natl. Acad. Sci. U.S.A.* **102:**5168–5173.

Soler, S., Diez, M. J., Rosello, S., and Nuez, F. (1999). Movement and distribution of *Tomato spotted wilt virus* in resistant and susceptible accessions of *Capsicum* spp. *Can. J. Plant Pathol.* **21:**317–325.

St Clair, D. A. (2010). Quantitative disease resistance and quantitative resistance loci in breeding. *Annu. Rev. Phytopathol.* **48:**247–268.

Stafford, C. A., Walker, G. P., and Ullman, D. E. (2011). Infection with a plant virus modifies vector feeding behavior. *Proc. Natl. Acad. Sci. U.S.A.* **108:**9350–9355.

Storms, M. M. H., Kormelink, R., Peters, D., Van Lent, J. W. M., and Goldbach, R. (1995). The nonstructural NSm protein of *Tomato spotted wilt virus* induces tubular structures in plant and insect cells. *Virology* **214:**485–493.

Takeda, A., Sugiyama, K., Nagano, H., Mori, M., Kaido, M., Mise, K., Tsuda, S., and Okuno, T. (2002). Identification of a novel RNA silencing suppressor, NSs protein of *Tomato spotted wilt virus*. *FEBS Lett.* **532:**75–79.

Tentchev, D., Verdin, E., Marchal, C., Jacquet, M., Aguilar, J. M., and Moury, B. (2011). Evolution and structure of *Tomato spotted wilt virus* populations: Evidence of extensive reassortment and insights into emergence processes. *J. Gen. Virol.* **92:**961–973.

Thomas-Carroll, M. L., and Jones, R. A. C. (2003). Selection, biological properties and fitness of resistance-breaking strains of *Tomato spotted wilt virus* in pepper. *Ann. Appl. Biol.* **142:**235–243.

Tiendrebeogo, F., Lefeuvre, P., Hoareau, M., Traore, V. S. E., Barro, N., Perefarres, F., Reynaud, B., Traore, A. S., Konate, G., Lett, J.-M., and Traore, O. (2011). Molecular and biological characterization of *Pepper yellow vein Mali virus* (PepYVMV) isolates associated with pepper yellow vein disease in Burkina Faso. *Arch. Virol.* **156:**483–487.

Tomita, R., Ken-Taro, S., Hiroyuki, M., Sakamoto, M., Murai, J., Kiba, A., Hikichi, Y., Suzuki, K., and Kobayashi, K. (2011). Genetic basis for the hierarchical interaction between *Tobamovirus* spp. and *L* resistance gene alleles from different pepper species. *Mol. Plant Microbe Interact.* **24:**108–117.

Torres-Pacheco, I., Garzon-Tiznado, J. A., Brown, J. K., Becerra-Flora, A., and Rivera-Bustamante, R. F. (1996). Detection and distribution of geminiviruses in Mexico and the southern United States. *Phytopathology* **86:**1186–1192.

Tsuda, S., Kirita, M., and Watanabe, Y. (1998). Characterization of a pepper mild mottle tobamovirus strain capable of overcoming the L^3 gene-mediated resistance, distinct from the resistance-breaking Italian isolate. *Mol. Plant Microbe Interact.* **1:**327–331.

Ullman, D. E., Cho, J. J., Mau, R. F. L., Westcot, D. M., and Custer, D. M. (1992). A midgut barrier to *Tomato spotted wilt virus* acquisition by adult western flower thrips. *Phytopathology* **82:**1333–1342.

van Poelwijk, F., Prins, M., and Goldbach, R. (1997). Completion of the *Impatiens necrotic spot virus* genome sequence and genetic comparison of the L proteins within the family *Bunyaviridae*. *J. Gen. Virol.* **78:**543–546.

van Regenmortel, M. H. V., Fauquet, C. M., Bishop, D. H. L., Carstens, E. B., Estes, M. K., Lemon, S. M., Maniloff, J., Mayo, M. A., McGeoch, D. J., Pringle, C. R., and Wickner, R. B. (2000). Virus taxonomy: Classification and nomenclature of viruses. *In* "International Committee on Taxonomy of Viruses" (M. H. V. van Regenmortel, C. M. Fauquet, D. H. L. Bishop, E. B. Carstens, M. K. Estes, S. M. Lemon, J. Maniloff, M. A. Mayo,

D. J. McGeoch, C. R. Pringle, and R. B. Wickner, eds.), Seventh Report of the International Committee on Taxonomy of Viruses, p. 1162. Academic Press, San Diego.

Verdin, E., Gognalons, P., Cardin, L., Moretti, A., Parrella, G., Cotillon, A. C., Jacquemond, M., and Marchoux, G. (2005). Production d'un sérum polyclonal spécifique du *Parietaria mottle ilarvirus* (PMoV) et étude de sa transmission par le pollen. 10ème Rencontres de Virologie Végétale, 6–10 mars 2005, Aussois, France.

Wetter, C., Conti, M., Altschuh, D., Tabillion, R., and van Regenmortel, M. H. V. (1984). *Pepper mild mottle virus*, a *Tobamovirus* infecting pepper cultivars in Sicily. *Phytopathology* **74:**405–410.

Wisler, G. C., Li, R. H., Liu, H. Y., Lowry, D. S., and Duffus, J. E. (1998). *Tomato chlorosis virus*: A new whitefly-transmitted, phloem-limited, bipartite *Closterovirus* of tomato. *Phytopathology* **88:**402–409.

Yoon, J. Y., Green, S. K., Tschanz, A. T., Tsou, S. C. S., and Chang, L. C. (1989). Pepper Improvement for the Tropics, Problems and the AVRDC Approach. Asian Vegetable Research and Development Center, Tomato and Pepper Production in the Tropics, AVRDC, Shanhua, Tainan, 86–98.

Yoon, J. Y., Il Ahn, H., Minjea, K., Tsuda, S., and Ryu, K. H. (2006). *Pepper mild mottle virus* pathogenicity determinants and cross protection effect of attenuated mutants in pepper. *Virus Res.* **118:**23–30.

CHAPTER 5

Viruses of the Genus *Allium* in the Mediterranean Region

Nikolaos I. Katis,* Varvara I. Maliogka,* and **Chrysostomos I. Dovas**[†]

Contents

Abstract

Allium species are economically important crops in the Mediterranean basin. Viruses are among the most important pathogens affecting their yield and especially those belonging to the genera *Potyvirus*, *Carlavirus*, and *Allexivirus*. Members of the genus *Potyvirus* are usually the most abundant and cause most of the damage induced. Nevertheless, coinfections with different viruses are not scarce, especially in garlic, and can have synergistic effects that lead

* Plant Pathology Laboratory, School of Agriculture, Aristotle University of Thessaloniki, Thessaloniki, Greece
† Laboratory of Microbiology and Infectious Diseases, Faculty of Veterinary Medicine, Aristotle University of Thessaloniki, Thessaloniki, Greece

Advances in Virus Research, Volume 84
ISSN 0065-3527, DOI: 10.1016/B978-0-12-394314-9.00005-1

to even greater crop losses. Vegetative propagation of alliums and the transmission of most of their viruses by arthropod vectors have significantly contributed to their wide dissemination in the Mediterranean region and elsewhere in the world. Here, we review the general biological and molecular features, the epidemiology, incidence, and methods of diagnosis of the most widespread allium viruses in the basin. Control measures are proposed depending on the mode of propagation of the various alliums, the epidemiology of their viruses and the cultivation procedures adapted by the Mediterranean farmers. The importance of the production and use of virus-free propagative material in order to combat viral diseases of allium crops is especially highlighted. A final discussion focuses on the main shortages identified in the research area of allium viruses, and proposals are made for putative future developments.

I. IMPORTANCE OF ALLIUM CROPS AND THEIR HISTORY

The genus *Allium* is one of the largest plant genera, which includes more than 600 species. The genus name comes from the Greek αλεω which means to avoid because of its offensive smell (Boswell, 1883). Most of the plant species in the genus are bulb forming, and several of them are important edible crops all over the world (Block, 2010). Among the most economically important cultivated species, are the onion (*Allium cepa*), particularly the varieties grown for bulbs, chive (*Allium schoenoprasum*), garlic (*Allium sativum* var. *sativum*), shallot (*Allium cepa* var. ascalonicum), and leek (*Allium ampeloprasum* var. *porrum*) (Brewster, 1994). Onion is considered as the principal *Allium* species and occupies the third place in world production of vegetables, with a volume of 57.9 million tons (FAO, 2005). Garlic is the second most widely consumed allium produced at an annual level of 14 million tons (FAO, 2005).

The origin of *Allium* species remains speculative. Evidence suggests that garlic and onion were first domesticated in the central Asian mountainous regions (Block, 2010) and most likely brought to the Middle East by Marco Polo and other travelers. In the Mediterranean basin, garlic was transported through trading routes (Block, 2010). The Middle East and Mediterranean are considered a secondary center for onion in which its big-headed types were selected.

Onion, garlic, and their relatives have always been popular foods in Egypt and elsewhere in the Mediterranean basin (Block, 2010). Although primarily grown for food, they are also used in traditional medicine as the sulfur compounds they produce have antimicrobial action. Recent research has also shown that onion and garlic extracts may prevent cardiovascular and other diseases (Schwartz *et al.*, 2006).

II. PRODUCTION OF ALLIUMS IN THE MEDITERRANEAN COUNTRIES

Even though China is the world's largest producer of garlic and dry onions (FAO, 2007), the countries of the Mediterranean region are also among the biggest allium producers. Turkey and Egypt are fifth (2,007,120 tons) and seventh (1,728,417 tons) producers of dry onion worldwide, whereas Tunisia, Turkey, and, to a less extent, Spain and France are important producers of green onions including shallots (FAO, 2008). In garlic production, Egypt and Spain are fourth (258,608 tons) and ninth (142,500 tons), respectively (FAO, 2008). Turkey, France, and Spain are in the top 10 producers of leeks and other allium vegetables.

III. THE MAIN VIRUSES INFECTING *ALLIUM* SPP.

The most economically important and worldwide distributed viruses of *Allium* species are those affecting onion and garlic, whereas other species such as leek and shallot are less affected and usually their diseases are of local interest.

For many years, mosaic symptoms have been observed on the leaves of garlic, onion, and shallot crops and were attributed to some diseases of unknown etiology named "garlic mosaic," "onion mosaic," and "shallot mosaic," respectively (Van Dijk, 1993a; Walkey, 1990). In addition, a yellow stripe disease has been observed on leek crops in different countries (Bos *et al.*, 1978a). The identification of the causal agents of these diseases was difficult and for a long time incomplete. Problems related with the isolation and purification of the implicated viruses were mainly due to their limited host range (it includes few wild *Allium* spp.), the absence of specific indicator plants for their differentiation, and the simultaneous infections with many viruses (especially in garlic). This resulted to difficulties in their identification and confusion in bibliography. Nowadays, intensive studies conducted by a great number of researchers have shown that mainly members of the genera *Potyvirus*, *Carlavirus*, and the recently ratified genus *Allexivirus* (Table I) often found forming viral complexes, prevail in allium crops, and are implicated in the etiology of the abovementioned diseases.

Several of these viruses are often found in high incidence in almost all allium-cultivating regions of the world and are responsible for severe yield losses, while others may be localized in a few geographical regions. Members of the genus *Potyvirus* are usually the most abundant and cause most of the damage induced, while carla- and allexiviruses are mainly latent. However, coinfection of the latter viruses with the potyviruses can

TABLE I Families and genera of viruses infecting *Allium* species in the Mediterranean basin

Family/genera	Virus species	Transmission manner
Potyviridae/Potyvirus	*Leek yellow stripe virus* (LYSV)	Aphid-borne (nonpersistently)
	Onion yellow dwarf virus (OYDV)	
	Turnip mosaic virus (*TuMV*)	
Betaflexiviridae/Carlavirus	*Garlic common latent virus* (GarCLV)	Aphid-borne (nonpersistently)
	Shallot latent virus (SLV) [synonym for *Garlic latent virus* (GLV)]	
Alphaflexiviridae/Allexivirus	*Garlic mite-borne filamentous virus* (GarMbFV)[a]	Mites (*Aceria tulipae*)
	Shallot virus X (ShVX)	
	Garlic virus-A, -*B*, -*C*, -*D*, and -*E* (GarV-A, GarV-B, GarV-C, GarV-D, and GarV-E)	
	Garlic virus-X (GarV-X)	
Reoviridae/Fijivirus	*Garlic dwarf virus* (GDV)	Vector not known
Comoviridae/Nepovirus	*Artichoke yellow ringspot virus* (AYRV)	Seed (in onion and other hosts), Pollen, vector not known[b]
Bunyaviridae/Tospovirus	*Iris yellow spot virus* (IYRV)	Thrips
	Tomato spotted wilt virus (TSWV)	
Bromoviridae/Cucumovirus	*Cucumber mosaic virus* (CMV)	Aphid-borne (nonpersistently), seed transmission in some species
Tombusviridae/Necrovirus	*Leek white stripe virus* (LWSV)	Unknown

[a] From the known allexiviruses so far to our knowledge, GarV-B, -C, -D, -X, and ShVX have been identified in the Mediterranean region.
[b] Although belongs to the *Nepovirus* genus, no nematode vector was identified.

have synergistic effects. Especially in garlic, it was estimated that viruses can reduce the yield up to 50% during successive cultivation (Conci *et al.*, 2003; Lot *et al.*, 1998) and garlic mosaic disease has been reported to cause losses in bulb weight up to 88% (Canavelli *et al.*, 1998; Lot *et al.*, 1998; Walkey and Antill, 1989). In addition, the bulb weight of virus-free garlic plants has been shown to be 32–216% higher than that from infected plants for most cultivars tested (Conci, 1997; Conci *et al.*, 2003; Melo Filho *et al.*, 2006; Walkey and Antill, 1989).

Arthropods play an important role in the epidemiology of allium viruses as poty- and carlaviruses are transmitted by aphids in a nonpersistent manner and allexiviruses by mites. It has been found that virus-free garlic is quickly reinfected once planted in the field (Conci *et al.*, 2003; Lot *et al.*, 1998; Lunello *et al.*, 2007; Melo Filho *et al.*, 2006), indicating that the viruses are efficiently transmitted by arthropod vectors from adjacent infected plots. In addition, vegetative propagation of alliums is also a major mode for their transmission. The viruses can accumulate in the bulbs and perpetuate the infection from one season to the next. Therefore, species such as garlic and shallot which are exclusively vegetatively propagated are infected with a complex of viruses (Walkey, 1990). This transmission manner of allium viruses resulted in their worldwide distribution. All traditional commercial clones of garlic are infected with more than one virus (Walkey, 1990), while, in most cases, garlic crops are highly or almost totally infected (Dovas *et al.*, 2001a; Klukakova *et al.*, 2007). Most allium viruses are not seed transmitted (Bos *et al.*, 1978a); thus plant species such as leek and onion, propagated by seed, start their life cycle free of viruses. However, an important source of infection for this type of plants comes from overwintering crops, volunteer plants, or bulbs replanted for seed production.

It is interesting to note that probably due to long-term vegetative propagation of allium crops and their distribution to areas with very different climatic conditions, host specific and genetically diverse strains of different viruses have been selected and adapted to many local allium cultivars of diverse geographical and climatic regions (Barg *et al.*, 1995). This finding came from a great bulk of intensive studies on the biological characterization of the allium viruses and provided evidence for striking host specificities, especially in the case of potyviruses, which can be particularly important in their epidemiology and spread.

Reliable diagnosis and characterization of the allium viruses have been largely hampered by the lack of appropriate specific methods and have often resulted in misidentifications. Thus similar or even identical viruses from different countries were described as different species (Salomon *et al.*, 1996; Yamashita *et al.*, 1995). Even though serology has been the main diagnostic method for routine testing of allium viruses, unreliable results were obtained in some cases due to serological cross-reactions (Koch and Salomon, 1994). The production of highly specific antisera against garlic viruses is troublesome, as the plants are usually coinfected with a mixture of viruses and their isolation due to lack of differential hosts is rather difficult (Van Dijk, 1993a,b). Especially in the case of allexiviruses, isolation and purification is particularly difficult (Van Dijk and Van der Vlugt, 1994; Van Dijk *et al.*, 1991). Monoclonal antibodies have been used to overcome this problem and can also differentiate among various virus strains. Serological variability between strains of different poty- and carlaviruses, reflecting variability in the N-terminal region of the viral coat protein (CP) (Tsuneyoshi *et al.*, 1998b), which carries the major virus-specific epitopes (Shukla *et al.*, 1988) has been reported (Barg, 1996; Barg *et al.*, 1994) leading to diagnostic deficiencies in ELISA and decoration experiments. The recent development of species-specific and generic molecular assays contributed significantly to the reliable and sensitive diagnosis of allium viruses and also to the species differentiation and characterization.

Different viruses of the three genera and the related diseases have been reported from various Mediterranean countries such as France, Egypt, Morocco, Turkey, Syria, Tunisia, Israel, Spain, Slovenia, Italy, and Greece. Even though there are several reports on the presence of different viruses, information related to their incidence in each country is rather limited due to the lack of large-scale epidemiological surveys. The greatest bulk of information concerns viruses of garlic which is an economically important crop for several countries in the Mediterranean basin. The fact that traditionally growers produce their own propagative material accounts for the observed heavy viral infections in garlic (Dovas *et al.*, 2001a; Mahmoud *et al.*, 2007). Furthermore, more data concern incidence of poty- and carlaviruses and less allexiviruses, as the first cause severe symptoms and yield losses and detection of the latter is still problematic. Overall, observations and surveys indicate that garlic is almost totally infected by mixtures of viruses in which potyviruses (mainly *Onion yellow dwarf virus* (OYDV), followed by *Leek yellow stripe virus* (LYSV), are prevalent. Carlaviruses (mainly *Garlic common latent virus*, GarCLV) and allexiviruses are also very frequent depending on the region and the cultivar. Onion and leek are mainly affected by potyviruses (OYDV and LYSV, respectively), but their importance seems to be local and especially in areas where certain cultivation procedures are favoring virus propagation.

A description of the main viruses infecting *Allium* in the Mediterranean basin is presented in Table I.

A. Potyviruses

Potyviruses such as OYDV and LYSV were the first described and characterized viruses in *Allium* plant species (Bos, 1976, 1981; Melhus *et al.*, 1929). OYDV was first described in onion crops in the USA (Melhus *et al.*,

1929), and later it was reported in Europe, New Zealand, North Africa, South and North America, but its diagnosis was based on symptomatology (Bos, 1983; Walkey, 1990). Polyclonal antibodies against OYDV were first produced in Holland and then in France (Delecolle *et al.*, 1985), and serological testing revealed that symptoms in leek attributed to OYDV were actually due to LYSV (Bos *et al.*, 1978a). Leek yellow stripe symptoms were also previously reported in other countries, but in all cases, they were considered to be caused by OYDV. Epidemics reported in Holland, Belgium, Denmark, and Germany contributed to the study and identification of LYSV (Bos *et al.*, 1978a).

Identification, isolation, and purification of these viruses from garlic were very difficult; thus, the causal agent of "garlic mosaic disease" stayed for a long time unknown. Single infections of OYDV in onion and of LYSV in leek facilitated their isolation and allowed further characterization of garlic isolates. Using serological methods in 1981 OYDV was recognized as a pathogenic component of the Garlic mosaic disease complex (Delecolle and Lot, 1981). LYSV was first reported in garlic in 1987 (Walkey *et al.*, 1987), but it is very well known that this virus was previously reported in New Zealand, as part of the garlic virus complex (Mohamed and Young, 1981).

Two additional potyviruses namely *Shallot yellow stripe virus* (SYSV) and *Turnip mosaic virus* (TuMV) were described infecting *Allium* species (Stefanac and Plese, 1980; Van Dijk, 1993a). Several other virus isolates closely related to SYSV were previously erroneously identified as separate species (Tsuneyoshi *et al.*, 1998b; Van Dijk, 1993a), but finally, they were all shown to be different strains of SYSV (Van der Vlugt *et al.*, 1999). In different *Allium* species, the viruses *Alstroemeria mosaic virus*, *Hippeastrum mosaic virus* (Walkey, 1990), *Potato virus A*, and *Potato virus Y* (Rongchang *et al.*, 1992) have also been reported, but their incidence is low and therefore of minor importance.

LYSV and OYDV are considered to be the most important viruses worldwide due to their wide distribution in *Allium* species (Barg *et al.*, 1997; Chen *et al.*, 2001; Conci *et al.*, 1992; Klukakova *et al.*, 2007; Tsuneyoshi *et al.*, 1998b; Van Dijk, 1993a) and the yield losses they cause. In garlic propagative material collected from all over the world, LYSV and OYDV incidence was 45% and 73%, respectively, (Van Dijk, 1993a). High incidence in onion of more than 90% has been reported for OYDV from different countries (Conci *et al.*, 1992; Hoa *et al.*, 2003; Melhus *et al.*, 1929). Even though studies on the effect of viruses on yield loss (mainly garlic) have been hampered by the difficulties in isolating them to obtain single virus strains, it is now known that OYDV is causing the greatest economic crop losses in garlic crops followed by LYSV. Garlic infection by OYDV and LYSV may cause bulb weight reduction up to 69% and 54%, respectively (Canavelli *et al.*, 1998; Lot *et al.*, 1998; Lunello *et al.*, 2007).

In mixed infections, viruses act synergistically reducing even more the yield (Lot *et al.*, 1998). In garlic, mixed virus infection causes losses of about 30–50% in the plants weight and size (Walkey, 1990). Yield losses may also depend on the virus strain as reported for OYDV strains infecting the garlic variety Fructidor (Walkey and Antill, 1989). OYDV causes also great reduction of bulb yield and seed production in onion (Bos, 1981; Kumar *et al.*, 2010). SYSV and TuMV incidence in allium crops is lower, and they have been reported only in certain geographic regions (Dovas *et al.*, 2001a; Gera *et al.*, 1997; Van der Vlugt *et al.*, 1999; Van Dijk, 1993a). SYSV has not been identified in the Mediterranean region.

OYDV, LYSV, TuMV, and SYSV are transmitted nonpersistently by aphids (Bos *et al.*, 1978a; Diekmann, 1997). OYDV is transmitted by more than 50 aphid species, with most of them noncolonizing the virus host (Drake *et al.*, 1933). In Holland, garlic, onion, and leek crops were never found infested with aphids, and thus spread of aphid-borne viruses is attributed to migratory aphid species which visit the crops only occasionally. In the USA and Europe, spread of OYDV usually takes place in spring and summer due to the high populations of immigrant aphids (Drake *et al.*, 1933; Marrou *et al.*, 1972; Van Dijk, 1994).

Host-specific strains of OYDV and LYSV which differ biologically and serologically and infect different *Allium* species have been described (Barg *et al.*, 1995). This does not allow infection of a different *Allium* species which is cultivated nearby (Van Dijk, 1993a), and virus sources of potyviruses come from infected propagative material of the same species or from crops of the same species in the surroundings. Wild *Allium* species do not act as virus sources of OYDV and LYSV for further spread to cultivated allium, at least in Holland (Van Dijk, 1993a).

The potyviruses identified to infect alliums in the Mediterranean region are described below.

1. *Leek yellow stripe virus*

Synonyms: Garlic mosaic virus, Garlic virus 2, Garlic yellow streak virus.

i. Host range, strains, and symptomatology LYSV infects *Allium* species such as leek, garlic, *A. ampeloprasum* L. var. *holmense*, and different wild and ornamental *Allium* species (Diekmann, 1997; Dovas *et al.*, 2001a; Salomon *et al.*, 1996). *Allium vineale* is very susceptible; onion is rarely infected, whereas shallot is immune (Diekmann, 1997; Lunello *et al.*, 2002; Van Dijk, 1993a).

Except from the leek infecting isolates of LYSV (LYSV-L), there are host-specific ones that infect garlic (LYSV-G) and *A. ampeloprasum* var. *holmense* (LYSV-GhG) (Van Dijk 1993a). LYSV-G and LYSV-GhG are highly host specific and do not infect onion or leek. Likewise LYSV, isolates from leek hardly infect garlic (Diekmann, 1997). Recently, a putatively new strain was reported from leek in Argentina (LYSV-L-Arg) which is capable to infect garlic and also onion, though with a low efficiency (Lunello *et al.*, 2002). Two potyviruses, namely GV-2 and GV-7, that have been isolated from garlic were proved to be isolates of LYSV (LYSV-G) (Yamashita *et al.*, 1995).

Symptoms depend on virus strain and host genetic background (Van Dijk, 1994). In leek, it causes chlorotic stripes in leaves, which initiate from their base, whereas yellowing of the whole leaf has been reported. Affected plants are stunted and less juicy. The stem is lusterless, and the storing quality of the harvested product is impaired. Infected leek plants in autumn and winter crops are more susceptible to low temperatures and may be killed (Walkey, 1990). Symptoms are more severe when leeks are coinfected with *Shallot latent virus* (SLV). In garlic, LYSV-G causes mild streaking up to severe yellow streaking and reduction of bulb size.

ii. Transmission LYSV is transmitted experimentally by *Myzus persicae, Aphis fabae, Aphis gossypii, Aphis nerii, Hyperomyzus carduellinus, Rhopalosiphum maidis, Rhopalosiphum padi, Schizaphis graminum*, and *Uroleucon sonchi* (Bos *et al.*, 1978a; Lunello *et al.*, 2002; Verhoyen and Horvat, 1973) in the nonpersistent manner. The virus is not seed borne (Bos *et al.*, 1978a). Under laboratory conditions, it is transmitted mechanically to a number of susceptible plant species.

iii. Presence in Mediterranean countries LYSV is one of the most important and widespread viruses of garlic and leek worldwide (Bos, 1981; Diekmann, 1997). It is endemic in different countries of the Mediterranean basin. LYSV is one of the most common garlic viruses in Israel (Salomon *et al.*, 1996; Shiboleth *et al.*, 2001), Greece (83.7% of the tested samples) (Dovas *et al.*, 2001a), Italy (83%) (Dovas and Vovlas, 2003), Syria (Mohammad *et al.*, 2007), Turkey (21.2% of the tested samples) (Fidan and Baloglu, 2009a) and is also present in garlic crops in France, Egypt, Slovenia, and Morocco (Delecolle and Lot, 1981; Lot *et al.*, 1998; Mavrič and Ravnikar, 2005; Van Dijk, 1993a). Infection of leek has been reported from Italy (Grancini, 1951), Israel (Shiboleth *et al.*, 1997), in most vegetable growing areas in France (Cornuet, 1959), and in most of the fields surveyed in Turkey (incidence 30–40%–sometimes 90% based on visual symptoms) (Korkmaz and Cevik, 2009). In Greece, leek was found to be solely infected by LYSV with its incidence ranging from 10% to 90%, and it was higher in fields located near overwintering leek crops (Dovas *et al.*, 2001a). LYSV was also isolated from *A. ampeloprasum* var. *holmense* in Tunisia and Israel (Salomon *et al.*, 1996; Van Dijk, 1993a) and was detected with a low incidence in the wild *Allium* species *A. ampeloprasum* ssp. *ampeloprasum* and *A. flavum* in Greece (Dovas *et al.*, 2001a).

iv. Particle morphology and properties LYSV virions are flexuous filamentous with a length of 820 nm. The thermal inactivation point (TIP) is 50–60 °C, dilution end point (DEP) 10^{-2}–10^{-3}, and longevity *in vitro* 3 to 4 days.

v. Nucleic acid component The genome is monopartite linear, positive-sense, single-stranded RNA of 10,142 nucleotides excluding the poly (A) tail (Chen *et al.*, 2002). It codes a large polyprotein that is further proteolytically cleaved into 10 mature proteins, typical of the genus potyvirus.

Several partial nucleotide sequences (i.e., GenBank accession nos. X89711, AJ307057, and FJ358732[1]) and complete genome sequences (GenBank accession no. AJ307057) are available.

vi. Propagation hosts and purification The most commonly used maintenance and propagation host species are *A. ampeloprasum* var. *porrum* (leek) and *Chenopodium quinoa* (Bos *et al.*, 1978a). The virus can be purified using different methods (Bos *et al.*, 1978a; Helguera *et al.*, 1997b; Huttinga, 1975; Lunello *et al.*, 2002).

vii. Detection methods

a. Diagnostic species Some strains cause necrotic local lesions on *Chenopodium album*, *Chenopodium murale*, *Chenopodium amaranticolor*, and *C. quinoa*. Salomon *et al.* (1996) used *C. amaranticolor* as indicator for LYSV-G. According to Van Dijk (1993a), only few LYSV-G isolates infect leek and one or two *Chenopodium* spp. One strain of LYSV-G was isolated from garlic mixed infected with SLV-G and OYDV-G by using *C. amaranticolor*. Therefore, the use of indicator plants as diagnostic species is not considered reliable.

b. Serological tests Due to the virus sequence variability in the N-terminal region of the CP (Tsuneyoshi *et al.*, 1998b) serologically different strains have been detected which cannot be recognized using a single antiserum (Barg, 1996). This should be taken under consideration when serology is used as a diagnostic assay.

ISEM decoration (ISEM-D) tests are particularly useful to detect LYSV in plants coinfected with other viruses (Diekmann, 1997). Moreover, various other serological assays have been developed based on different antibodies produced by different researchers (e.g., E. Barg, K. Yamashita, D. Maat, H. Lot, B. Delecolle, *et al.*). Tsuneyoshi and Sumi (1996) were able to determine the nucleotide sequence of the LYSV (GV-7) *CP* and produced virus-specific antiserum against the bacterially expressed recombinant CP which was further used in ELISA and direct tissue blotting

[1] Only few of the available in the GenBank partial sequences are reported.

immunoassays (DTBIA). Polyclonal antiserum was also prepared against an LYSV isolate (LYSV-L-Arg) from leek, which was applied in ISEM-D, PTA/DAS-ELISA, and Western blotting (Lunello *et al.*, 2002). A quantitative DAS-ELISA was reported by Dovas *et al.* (2002) for the detection of LYSV and OYDV in garlic. Since these viruses are unevenly distributed in garlic plants, this study highlighted the importance of sampling conditions for their reliable detection by ELISA. The optimum period for sampling is during rapid plant growth, when plants have more than three fully developed leaves and before plant growth is retarded (under environmental conditions of the Mediterranean region the best period is during March). The best tissue is the tip of the first or second fully expanded leaves from the top of the plant. Samples should be tested during the first 12 days of storage at 4 °C, as a rapid reduction of virus antigen was observed afterwards.

A broad range of DAS-ELISA kits is commercially available for routine tests of bulbs and leaves.

c. Molecular tests PCR methodology has been used for the specific detection of LYSV in infected allium plants by designing several primer sets (Fajardo *et al.*, 2001; Shiboleth *et al.*, 2001; Takaichi *et al.*, 1998; Tsuneyoshi and Sumi, 1996; Tsuneyoshi *et al.*, 1998b). Dovas *et al.* (2001b) have optimized and developed highly sensitive PCR-based detection methods for the large-scale testing of leaves and bulbs for the presence of LYSV and OYDV. They showed that RT-PCR and IC-RT-PCR techniques are 10^2–10^4 times more sensitive than DAS-ELISA for detecting these viruses in *Allium* spp. The polyvalence of the LYSV-specific primer pair allowed consistent detection of LYSV both in garlic and in leek samples, originating from different regions. A pair of degenerate primers was also designed for improved simultaneous PCR detection of various potyviruses, including LYSV, in *Allium* species (Lunello *et al.*, 2005). The application of further immunocapture and nested PCR steps in this assay increased the detection sensitivity in leaf extracts 10^4 times compared to DAS-ELISA. A multiplex RT-PCR that simultaneously detects six garlic viruses, among which are the potyviruses LYSV and OYDV, and thus reduces consumable costs, labor, and time has been reported in Korea (Park *et al.*, 2005).

More recently, TaqMan and SYBR Green real-time RT-PCR assays have been developed to achieve even higher detection sensitivity levels (10^6 times over DAS-ELISA, five gene copies, respectively) both in *in vitro* and in commercially grown plants (Leisova-Svobodova and Karlova-Smekalova, 2011; Lunello *et al.*, 2004). Finally, a short oligonucleotide microarray was developed for the simultaneous detection and identification of four distinct potyviruses (Wei *et al.*, 2009). This method showed high specificity and detection range when tested with geographically diverse virus isolates.

A list of various DNA primers used for the detection of LYSV in different countries and laboratories is presented in Table II.

2. *Onion yellow dwarf virus*

Synonyms: Allium virus 1, garlic potyvirus, Marmor cepae virus.

i. Host range, strains, and symptomatology In the field, the host range of OYDV is restricted to the genus *Allium*, mainly onion and garlic. It also infects *Allium scorodoprasum* L., but not leek. The virus was not detected in wild *Allium* species such as *A. vineale*, *A. oleraceum*, *A. ursinum*, and *A. scorodoprasum* (Van Dijk, 1993a).

Except from the presence of OYDV in onion, there are also other host-specific strains that infect garlic (OYDV-G), and *A. ampeloprasum* var. *holmense* (OYDV-GhG) (Van Dijk, 1993a). These strains are strictly host specific, and they do not infect onion and leek. Furthermore, OYDV isolates from onion do not infect garlic. Strains of OYDV ($OYDV^{vir}$, $OYDV^{com}$) differing in virulence were also reported in onion and shallot (Van Dijk, 1993a).

In *Allium* species, the virus causes stunting of the plants, with the leaves showing yellow striping, crinkling, and flaccidity, whereas more specifically in onion, it causes distortion of flower stems, reduction in the number of flowers and seeds, and impairment of seed quality. Affected onion bulbs deteriorate during storage and show premature sprouting, while the affected plants are more susceptible to low temperatures (Paludan, 1980). In garlic, symptoms depend mainly on the virus strain and plant genotype. OYDV causes mainly yellow and chlorotic striping, whereas mild strains cause milder symptoms (Van Dijk, 1993a).

ii. Transmission OYDV is transmitted by over 50 aphid species in a nonpersistent manner (Bos, 1981) with *Myzus ascalonicus*, *M. persicae*, *R. maidis*, and *Acyrthosiphon pisum* being the most important virus vectors (Drake *et al.*, 1933; Van Dijk, 1993a). The virus is mechanically transmitted under lab conditions. Seed transmission is not reported in onion (Diekmann, 1997).

iii. Presence in Mediterranean countries OYDV occurs in all countries where onion and garlic are cultivated (Bos, 1981; Van Dijk 1993a). In garlic samples collected worldwide, OYDV-G incidence was very high (73% of the tested samples), ranging from 52% to 86% in samples originating from Europe and Asia, respectively (Van Dijk, 1993a). The virus is endemic in the Mediterranean basin. In France, it has been a major problem for garlic plantations for many years causing high infestation rates (almost 100% of French varieties infected) (Delecolle and Lot, 1981; Delecolle *et al.*, 1985; Messiaen and Marrou, 1965; Messiaen *et al.*, 1994). Spanish garlic cultivars

TABLE II Primers used for the molecular detection of LYSV

Primers	Sequence	Amplicon size (bp)	Genomic region	Assay	References
LYSV81-410F	5′-AAGAACACCAGTTAGAGCGCG-3′	126	*CP*	SYBR Green real-time RT-PCR	Leisova-Svobodova and Karlova-Smekalova (2011)
LYSV81-535R	5′-TGCCTCTCCGTGTCCTCATC-3′				
F-LYSV	5′-TCTCGCACGGTATGCATTTG-3′	66	*CP*	TaqMan real-time RT-PCR	Lunello *et al.* (2004)
R-LYSV	5′-GCCTCGCGCGCTCTAA-3′				
VIC probe	5′-AGTCACATCAAGAACACCA-3′				
P1	5′-CGGGGCCCTWGGNMAAGCRCCATWYAT-3′	566	*NIb/CP*	RT-PCR/IC-RT-PCR	Lunello *et al.* (2005)
P2	5′-CGGAGCTCTNCCRTTYTCRATRCACCA-3′				
LYSV 1	5′-CACATCAAGAACACCAGTTAG AGC-3′	304	*CP*-3′UTR	RT-PCR/IC-RT-PCR	Dovas *et al.* (2001b)
LYSV 2	5′-GTAGAAACTGCCTTGAACGAG TG-3′				
(+)	5′-GAACTAGATGCAGGGACAC-3′	768	*CP*	Single/multiplex RT-PCR	Park *et al.* (2005)
(−)	5′-GCCGTCTAATCCAAACAGC-3′				

were found to be chronically infected by a potyvirus which was probably OYDV (Pena-Iglesias and Ayuso, 1982), while it is also prevalent in field grown garlic in Israel (Koch and Salomon, 1994; Shiboleth *et al.*, 2001). During surveys conducted in Italy and Greece, OYDV was found to be the most abundant garlic virus in all surveyed regions with an incidence ranging around 98% (Dovas and Vovlas, 2003; Dovas *et al.*, 2001a). The virus was also the most frequently detected (52–72.7% of the samples tested) in field surveys of garlic crops in Syria (Mohammad *et al.*, 2007) and Turkey (28.2%) (Fidan and Baloglu, 2009a). OYDV-G has been detected and isolated from garlic in Egypt, Morocco, and Slovenia (Mahmoud *et al.*, 2007; Mavrič and Ravnikar, 2005; Van Dijk, 1993a).

The virus was detected also in onion in France, Morocco, Spain, Egypt, Greece, and Italy (Dovas and Vovlas, 2003; Dovas *et al.*, 2001a; El-Kewey and Sidaros, 1996; Van Dijk, 1993a). It was found in 92% of the onion samples tested in Italy but only rarely in Greece (1 of 30 inspected fields). According to Van Dijk (1993a), $OYDV^{vir}$ strain may be prevalent in Spain because of resistance of sweet Spanish onion cultivars to $OYDV^{com}$, and it also occurs in its neighbor Morocco. Infection rates of OYDV in Spanish and French shallots reached 100% (Van Dijk, 1993a). Finally, OYDV-GhG strain was found infecting *A. ampeloprasum* var. *holmense* in Israel (Salomon *et al.*, 1996; Van Dijk, 1993a).

iv. Particle morphology and properties OYDV virions are filamentous, flexuous with a length of 772–823 nm and width 12 nm. The virus TIP is at 60–65 °C, DIP 10^{-3}–10^{-4} and longevity *in vitro* is 2 to 3 days.

v. Propagation hosts and purification Onion and garlic have been used for propagation of OYDV (Bos *et al.*, 1978a; Mahmoud *et al.*, 2007), while different approaches can be followed for its purification (Bos *et al.*, 1978a; Helguera *et al.*, 1997b; Huttinga, 1975). *C. amaranticolor* was used for virus isolation from mixed infections in garlic (Mahmoud *et al.*, 2007).

vi. Nucleic acid component The monopartite, linear, positive-sense, single-stranded RNA genome is 10,538 nts long and is predicted to encode a 3403 aa polyprotein (Chen *et al.*, 2003). The polyprotein is further proteolytically cleaved into 10 mature proteins, typical of the genus potyvirus.

Several partial nucleotide sequences (i.e., GenBank accession nos. L28079, AB219833, AJ292223, AY170321, HM473189[1]) and complete genome sequences (GenBank accession no. AJ510223) are available.

vii. Detection methods

a. Diagnostic species OYDV-G can hardly infect other plant species than garlic, and this makes difficult its purification from this host by using classical methods, as garlic is usually coinfected with other viruses

(Salomon *et al.*, 1996; Van Dijk, 1993a). Mechanical transmission of OYDV from garlic onto onion was unsuccessful or very difficult and resulted to mild local symptoms (Mahmoud *et al.*, 2007; Van Dijk, 1993a). OYDV-G and OYDV which infects onion hardly infect other *Allium* species apart from *A. vineale*.

According to Van Dijk (1993a), the virus cannot infect or very rarely infects *Chenopodium* species. Thus, local symptoms on *C. amaranticolor* or *C. quinoa* which were initially attributed to garlic mosaic or OYDV (garlic isolate) were possibly due to GarCLV, SLV, or LYSV-G (Van Dijk, 1993a). However, a garlic virus isolate reported from Egypt was able to induce local lesions onto *Chenopodium* species (Mahmoud *et al.*, 2007). Furthermore, $OYDV^{vir}$ strain from onion may cause small pinpoint necrotic lesions on *C. murale* (Van Dijk, 1993a). Therefore, *Chenopodium* species are not reliable diagnostic species of OYDV.

b. Serological tests ELISA is the main diagnostic method for large-scale routine detection of OYDV. However, as reported earlier for LYSV, serologically different strains of OYDV may occur (Barg, 1996).

A range of different polyclonal and monoclonal antibodies have been produced against OYDV (e.g., E. Barg, R. Salomon, H. Lot, B. Delecolle, *et al.*). DAS-ELISA (Delecolle *et al.*, 1985; Lot *et al.*, 1998) and antigen-coated plate-ELISA (ACP-ELISA) (Koch and Salomon, 1994) have been used for the reliable detection of the virus in garlic and onion. As reported earlier, a quantitative DAS-ELISA was developed for the detection of LYSV and OYDV in garlic (Dovas *et al.*, 2002). Due to fluctuations in the concentration of these viruses during the growing period, the sampling conditions are crucial for their reliable serological detection (see also Section III.A.1) (Dovas *et al.*, 2002).

A broad range of DAS-ELISA kits is commercially available for routine tests of bulbs and leaves.

c. Molecular tests RT-PCR and IC-RT-PCR assays were developed for OYDV detection in *Allium* spp. (Dovas *et al.*, 2001b; Fajardo *et al.*, 2001; Mahmoud *et al.*, 2007; Shiboleth *et al.*, 2001; Takaichi *et al.*, 1998). These methods are more sensitive (see Section III.A.1) and may show a wider detection range than ELISA (Dovas *et al.*, 2001b). The primers designed by Dovas *et al.* (2001b) can efficiently amplify OYDV in leaf and bulb tissue using double-tube RT-PCR even when crude plant extract is used, thus alleviating the need for nucleic acid extraction. More recently, an RT-PCR-based method was standardized for detection of the virus in garlic and onion leaves and also garlic bulbs (Arya *et al.*, 2006). The same primer pair was successfully used in a duplex RT-PCR for the simultaneous detection of OYDV and SLV both in infected leaves and in garlic cloves (Majumder *et al.*, 2008). A duplex RT-PCR was also developed to detect mixed infections of OYDV with an allexivirus in onion leaves and bulbs

(Kumar *et al.*, 2010), and a multiplex RT-PCR simultaneously amplifies parts of the genomes of OYDV and LYSV (Park *et al.*, 2005). The pair of degenerate primers designed by Lunello *et al.* (2005) for improved simultaneous molecular detection of various potyviruses in *Allium* species is amplifying OYDV both in RT-PCR and in IC-RT-PCR.

Ultra-sensitive detection of this potyvirus can be achieved by real-time RT-PCR assays. Simultaneous detection of OYDV and LYSV has been reported combining IC-RT-PCR/RT-PCR with the use of TaqMan probes (Lunello *et al.*, 2004). The developed method is 10^6-fold more sensitive than ELISA. More recently, an SYBR Green real-time RT-PCR assay has been developed for the detection of OYDV along with other four garlic viruses, which shows a detection limit for OYDV as low as five gene copies (Leisova-Svobodova and Karlova-Smekalova, 2011).

A list of various DNA primers used for the detection of OYDV in different countries and laboratories is presented in Table III.

3. *Turnip mosaic virus*

TuMV has occasionally been reported to infect wild *Allium* species. It was identified in ex-Yugoslavia infecting *A. ampeloprasum* and *A. roseum* (Stefanac and Plese, 1980), and in Israel infecting *A. ampeloprasum* (Gera *et al.*, 1997). TuMV was also detected with a low incidence in wild *Allium* spp. (*A. sphaerocephalon*, *A. guttatum*, *A. subhirsutum*, and *A. neapolitanum*) but not in any of the garlic and leek crops surveyed in Greece (Dovas *et al.*, 2001a).

Although it is known that TuMV causes reduction of plant growth and yellow striping in the leaves of *A. ampeloprasum* (Salomon *et al.*, 1996), generally, it is not considered to be an important virus pathogen of *Allium* species as its incidence is usually very low (Dovas *et al.*, 2001a; Gera *et al.*, 1997).

B. Carlaviruses

In *Allium* species, the first reports of carlaviruses are those of SLV, isolated from "*A. cepa* var. *ascalonicum*" in Holland (Bos *et al.*, 1978b), and *Garlic latent virus* (GLV), isolated from garlic in Japan (Lee *et al.*, 1979). More specifically, SLV has been reported to affect garlic in the United Kingdom (Walkey *et al.*, 1987) together with another unknown carlavirus. In Germany, a carlavirus identified as GLV (Graichen and Leistner, 1987) was different from GLV reported from Japan (Van Dijk, 1993b). For this reason, Van Dijk renamed this new virus as Garlic common latent virus (GarCLV).

GarCLV and SLV have been reported to infect garlic crops in various countries (Barg *et al.*, 1994; Nieto *et al.*, 2004; Tsuneyoshi *et al.*, 1998a; Van Dijk, 1993b). Van Dijk (1993b) described another allium carlavirus,

TABLE III Primers used for the molecular detection of OYDV

Primers	Sequence	Amplicon size (bp)	Genomic region	Assay	References
OYDV81F OYDV81R	5′-TTTAGCACGTTACGCATTCGA-3′ 5′-TTACCATCCAGGCCAAACAA-3′	132	*CP*	SYBR Green real-time RT-PCR	Leisova-Svobodova and Karlova-Smekalova (2011)
F-OYDV R-OYDV FAM probe	5′-AGTGATGCAGCTGAAGCATACATT-3′ 5′-ACGTTACCATCCAGGCCAAA-3′ 5′-TGCAAATGAAGGCGGC -3′	227	*CP*	TaqMan real-time RT-PCR	Lunello *et al.* (2004)
P1 P2	5′-CGGGGCCCTWGGNMAAGCRCCATWYAT-3′ 5′-CGGAGCTCTNCCRTTYTCRATRCACCA-3′	489	*NIb/CP*	RT-PCR/IC-RT-PCR	Lunello *et al.* (2005)
OYDV 1 OYDV 2	5′-GAAGCACAYATGCAAATGAAGG-3′ 5′-GCCACAACTAGTGGTACACCAC-3′	283	*CP*-3′UTR	RT-PCR/IC-RT-PCR	Dovas *et al.* (2001b)
OYDVVKBF OYDVVKBR	5′-ATAGCAGAAACAGCTCTTA-3′ 5′-GTCTCYGTAATTCACGC-3′	1111	*NIb*-3′UTR	RT-PCR	Arya *et al.* (2006), Majumder *et al.* (2008)
(+) (−)	5′-AAGGATAAAGACGTTGATG-3′ 5′-CGTGTGTTGGTTCTTGTTTA-3′	693	*CP*	Single/multiplex RT-PCR	Park *et al.* (2005)

namely *Sint-Jan's onion latent virus* (SjoLV) which reacts with the antisera prepared against SLV and GarCLV in electron microscopy decoration tests. Tsuneyoshi *et al.* (1998a) suggested that SjoLV might be a strain of SLV or GarCLV, but sequence data of virus isolates previously identified as SjoLV based on biological characteristics (host range, symptomatology) are required to confirm this.

In garlic, *Narcissus latent virus* another carlavirus was also detected in the United Kingdom (Walkey, 1990) even though this isolate needs further comparison with SLV (Van Dijk, 1993b). Furthermore, a carlavirus from garlic serologically related to *Carnation latent virus* (CaLV) was reported in Argentina (Conci *et al.*, 1992), and another (CG) with a similar reaction and closely related to GarCLV was reported from garlic crops in Slovenia (Mavrič and Ravnikar, 2005). It is not clear whether this is due to a cross-reaction of CaLV antiserum with GarCLV.

Carlaviruses have a wide host range within members of the genus *Allium*. They cause latent infections and, although the effect on crop yields has not been determined (Perotto *et al.*, 2010), it seems that they cause rather limited crops losses. However, they can cause significant yield losses when the plants are coinfected with potyviruses due to synergistic effects (Paludan, 1980; Sako, 1989).

Carlaviruses are transmitted nonpersistently by aphids but probably less efficiently than potyviruses (Van Dijk, 1993b). The main source of carlaviruses in the field is probably the infected propagative planting material as weeds do not contribute to their epidemiology (Van Dijk, 1993b). Natural transmission of SLV and GarCLV from shallot and garlic crops to leek shows that carlaviruses have lower degree of host specificity than potyviruses. Nevertheless, failures of SLV transmission from shallot to garlic (Van Dijk, 1993b) and of SLV-G (garlic strain) from garlic in onion (Lee *et al.*, 1979) show some degrees of specificity.

Carlaviruses found in Mediterranean countries are described below.

1. *Shallot latent virus* (synonym for *Garlic latent virus*)

Other synonyms: Garlic mosaic virus, Garlic virus 1.

i. Host range, strains, and symptomatology SLV was the first carlavirus detected in *Allium* species (Bos *et al.*, 1978b). Another carlavirus, namely, GLV was isolated from garlic in Japan (Lee *et al.*, 1979), but it came out that SLV and GLV are serologically closely related and differ in host reactions (Lee *et al.*, 1979; van Dijk, 1993b). Based on serology, host range, and symptomatology, Van Dijk (1993b) concluded that GLV was a garlic strain of SLV (SLV-G). Apart from SLV-G, an Asian shallot and a type strain were recognized by Van Dijk (1993b) and were differing mainly on their biological characteristics. SLV-G infects onion and *C. murale* where it causes local lesions, whereas it causes systemic infection in broadbean (*Vicia faba*).

Partial sequencing and phylogenetic analysis of different isolates of SLV and GLV showed that they are strains of the same virus (Tsuneyoshi *et al.*, 1998a). Furthermore, Tsuneyoshi *et al.* (1998a) confirmed that previously reported carlaviruses (GV-H, GV-1) isolated from garlic in Japan (Nagakubo *et al.*, 1994; Tsuneyoshi and Sumi, 1996) show high sequence similarity and are both strains of GLV and a virus reported from Israel similar to GV-1 (Salomon *et al.*, 1996). The high sequence variation observed between SLV strains depends mainly on their geographic origin and not on their natural host (Tsuneyoshi *et al.*, 1998a).

In natural conditions, SLV host range is restricted to members of the family Alliaceae. This includes species of the genus *Allium*, such as onion, leek, garlic, shallot, *A. scorodoprasum*, *A. cepa* var. *perutile*, *A. cepa* var. *agregatum*, *A. ampeloprasum* var. *sectivum*, *A. chinense*, and *A. fistulosum* (Van Dijk, 1993b).

The virus occurs apparently symptomlessly, in shallot, onion, and garlic, but it acts synergistically with potyviruses. In leek, mild chlorotic streaking appears in singly infected plants and severe chlorotic or white streaking and even plant death in plants coinfected with LYSV (Paludan, 1980).

ii. Transmission SLV is transmitted in the nonpersistent manner by *Aulacorthum solani*, *M. ascalonicus*, *M. persicae*, *Neutoxoptera formosana*, and *A. gossypii* (Bos *et al.*, 1978b; Sako *et al.*, 1990), whereas it is not transmitted by *A. fabae* (Bos *et al.*, 1978b). Transmission of SLV from plants coinfected with OYDV (*Potyvirus* genus) is lower compared to OYDV (Van Dijk, 1993b). In the laboratory, it can be mechanically transmitted to a number of plant species.

iii. Presence in Mediterranean countries SLV occurs mainly in Asia, Europe, and also in Mexico (Barg *et al.*, 1994; Bos, 1982; Van Dijk 1993b). Van Dijk (1993b) reported that garlic from different parts of the world was infected by SLV by 25%, whereas in Japan its incidence was even higher. In the Mediterranean basin, SLV infects garlic in a low incidence in Greece, Syria, Slovenia, and Italy (Dovas and Vovlas, 2003; Dovas *et al.*, 2001b; Mavrič and Ravnikar, 2005; Mohammad *et al.*, 2007). Especially in Greece, the virus was mainly detected in fields which were cultivated with propagative material imported from Iran and China (Dovas *et al.*, 2001b). SLV was also isolated from garlic in Egypt and from shallot in France and Spain (Van Dijk, 1993b).

iv. Particle morphology and properties SLV virions are flexuous, filamentous with a length of 650 nm. The virus TIP is at 80 °C, DEP 10^{-4}–10^{-5} and longevity *in vitro* 8–11 days (Bos *et al.*, 1978b).

v. Nucleic acid component The monopartite, linear, positive-sense, single-stranded RNA genome is 8353–8363 nts long (Chen *et al.*, 2001; Song *et al.*, 2002) and is predicted to code six ORFs namely RdRp, TGB1, TGB2, TGB3, CP, and NABP.

Several partial nucleotide sequences (i.e., GenBank accession nos. D11161, AB004456, and GU355922[1]) and complete genome sequences (GenBank accession nos. AJ292226 and Z68502) are available.

vi. Propagation hosts and purification Leek is suitable for maintaining SLV and is a good source of virus for purification (Bos *et al.*, 1978b). Garlic has also been used for virus isolation (Song *et al.*, 2002), but it has the disadvantage to be infected with many viruses. *A. fistulosum* is a good filter host to free SLV from contamination with LYSV (Paludan, 1980). A method, including Sephadex chromatography, which was successful for the purification of OYDV and LYSV (Huttinga, 1975), also proved useful for SLV purification (Bos *et al.*, 1978b). For virus isolation in mixed infections, repeated transfers on *N. occidentalis* or *Celosia argentea* can be used (Van Dijk, 1993b) before the final passage on leek for propagation.

vii. Detection methods

a. Diagnostic species SLV causes local lesions on *Chenopodium* spp., *C. argentea* var. *plumosa* "Geisha," *V. faba*, *Nicotiana occidentalis*, and *N. hesperis*, whereas it causes latent systemic infections to a wide host range of species of the family *Alliaceae* such as leek, shallot, and *A. vineale* (Van Dijk, 1993b).

In Japan, *C. amaranticolor*, *C. quinoa*, *Tetragonia expansa*, and *V. faba* are mainly used as indicator plants (Lee *et al.*, 1979).

b. Serological tests High serological variability was revealed, by using monoclonal antibodies, between SLV isolates originating from different countries, and four serotypes were described (Barg *et al.*, 1994). ISEM-D and DAS/TAS-ELISA have been largely used for monitoring the virus (Conci *et al.*, 2003; Dovas *et al.*, 2001a; Shahraeen *et al.*, 2008). Polyclonal antibodies, produced by expressing the *CP* gene of SLV in bacteria, were used successfully in various methods (Immuno-electron Microscopy, Immunoblot, ELISA) (Song *et al.*, 2002; Tsuneyoshi and Sumi, 1996). Dot-blot immunoassay has been proposed for large-scale screening of the virus since it is convenient, simple, and as sensitive as ELISA (Tsuneyoshi and Sumi, 1996). DIBA (dot-immunobinding assay) was also used for its detection in *A. chinense* (Sako *et al.*, 1990). Commercial kits for DAS-ELISA and TAS-ELISA are available and have been used for virus identification in different countries (Mituti *et al.*, 2011; Torrico *et al.*, 2010).

c. Molecular tests The application of serological methods for SLV diagnosis is sometimes problematic, as they are not always reliable (Mituti *et al.*, 2011). For this reason, molecular assays, which combine higher sensitivity and broader detection range, have been developed. An RT-PCR method has been reported for the detection of SLV (isolates GV-1 and GV-H) (Takaichi *et al.*, 1998; Tsuneyoshi and Sumi, 1996). In addition, a duplex RT-PCR was developed and successfully used for the simultaneous detection of OYDV and SLV in both garlic leaves and cloves (Majumder *et al.*, 2008), and a multiplex RT-PCR enables the simultaneous detection of SLV and other five garlic viruses (Park *et al.*, 2005). Two pairs of primers have been designed for the molecular identification of SLV in garlic in Brazil (Mituti *et al.*, 2011). Finally, a SYBR Green real-time RT-PCR assay was developed and optimized for the sensitive detection of the virus (Leisova-Svobodova and Karlova-Smekalova, 2011) and can be used for screening during the production of virus-free plants.

A list of various DNA primers used for the detection of SLV in different countries and laboratories is presented in Table IV.

2. *Garlic common latent virus*
Synonyms: GLV—France; GLV—Germany.

i. Host range, strains, and symptomatology GarCLV has been first described in France (Delecolle and Lot, 1981) and later in Germany (Graichen and Leistner, 1987), and it was erroneously called GLV. The name GarCLV was proposed to be used for this pathogen in order to avoid confusion with the Japanese GLV (Van Dijk, 1993a). GarCLV is not serologically related to GLV and SLV-type strain.

It is restricted to *Allium* species where it infects more than 50 species with garlic being the main host. It also infects leek, onion, and *A. ampeloprasum* var. *holmense* (Van Dijk, 1993b). GarCLV is latent in garlic, leek, and onion, but it acts synergistically with potyviruses.

ii. Transmission Aphid transmission is suspected (Barg *et al.*, 1994, 1997; Van Dijk, 1993b), but no data are available. Seed transmission has not been reported. The virus is mechanically transmitted in the laboratory.

iii. Presence in Mediterranean countries GarCLV is widely distributed all over the world where *Allium* species are cultivated, and usually, its incidence is higher than SLV (Barg *et al.*, 1994, 1997; Brunt *et al.*, 1996). The virus is prevalent in garlic crops in France, Spain, Israel, and Slovenia (Delecolle and Lot, 1981; Mavrič and Ravnikar, 2005; Pena-Iglesias and Ayuso, 1982; Shiboleth *et al.*, 2001; Van Dijk, 1993b). It was also detected in Italy (Apulia, Emilia-Romagna), with its incidence ranging from 23% to 98% (Bellardi *et al.*, 1995; Dovas and Vovlas, 2003), Turkey (Fidan and Baloglu, 2009b),

TABLE IV Primers used for the molecular detection of SLV

Primers	Sequence	Amplicon size (bp)	Genomic region	Assay	References
SLV-1	5′-GTGGTNTGGAATTAC-3′	308	*CP*	RT-PCR	Majumder *et al*. (2008)
SLV-2	5′-CAACATCGATTYTCTC-3′				
SLV-7044	5′-CTTTTGGTTCACTTTAGG-3′	960	*CP*	One-tube RT-PCR	Mituti *et al*. (2011)
SLV-8004	5′-GCACGCAATAGTCTACGG-3′				
SLV-6737	5′-YCCSGCCARGAAYTTCCC-3′	340	*TGB2–TGB3*	RT-PCR	Mituti *et al*. (2011)
SLV-7060	5′-TTAGAGCGCTGTWAACC-3′				
(+)	5′-TATACAGCGCTCTAAATTGA-3′	921	*CP*	Single/multiplex RT-PCR	Park *et al*. (2005)
(−)	5′-TTCTCTGTTTGATCAACATC-3′				
SLV-81-569F	5′-AACAAAGCAGCGATTCAACC-3′	160	*CP*	SYBR Green real-time RT-PCR	Leisova-Svobodova and Karlova-Smekalova (2011)
SLV-81-730R	5′-ACATCCGAAGAAACCTCCAGT-3′				

Syria (37–39%) (Mohammad *et al.*, 2007), and central and southern Greece (incidence ranged from 18 to 97.6%) (Dovas *et al.*, 2001a). GarCLV was isolated from *A. ampeloprasum* var. *holmense* in Israel (Van Dijk, 1993b).

iv. Particle morphology and properties GarCLV virions are flexuous, filamentous with a length of 650nm.

v. Nucleic acid component The genome is monopartite, linear, positive-sense, single-stranded RNA. The full nucleotide sequence is not available yet.

Several partial nucleotide sequences are available (i.e., GenBank accession nos. X81138, JF320828, AF228416, FJ154841, and AB004566[1]).

vi. Propagation hosts and purification The most commonly used maintenance and propagation host species of GarCLV are leek and *N. occidentalis*. *N. occidentalis* is useful for isolation of the virus and leek for its propagation. For virus isolation single local lesion transfers on *C. amaranticolor* or *C. argentea* can also be done before transfer on leek or onion for propagation (Van Dijk, 1993b) and purification (Helguera *et al.*, 1997b).

vii. Detection methods

a. Diagnostic species In contrast to SLV, GarCLV does not cause local lesions on *V. faba* and infects systemically *N. occidentalis* leading to veinal necrosis. It causes chlorotic and necrotic local lesions onto *C. quinoa* and *C. murale*, whereas on *C. amaranticolor* it causes faint green local rings. *C. argentea* reacts with chlorotic and necrotic lesions, but some isolates cause very mild symptoms.

b. Serological methods GarCLV is serologically related to CaLV since polyclonal antibodies prepared against CaLV also react with GarCLV (Barg *et al.*, 1994).

DAS-ELISA has widely been used for the detection of GarCLV (Barg *et al.*, 1997; Dovas and Vovlas, 2003; Dovas *et al.*, 2001b; Shahraeen *et al.*, 2008). Commercial kits are available for DAS-ELISA and have been used for large-scale surveys of allium crops (Fidan and Baloglu, 2009b; Klukakova *et al.*, 2007; Pappu *et al.*, 2008). ISEM-D was also applied in programs for production of virus-free garlic plants (Conci *et al.*, 2005; Torres *et al.*, 2000).

c. Molecular methods A limited number of molecular detection assays have been developed. RT-PCR methods were reported from different countries for virus identification in garlic (Fidan and Baloglu, 2009b; Majumder and Baranwal, 2009; Pappu *et al.*, 2008; Shiboleth *et al.*, 2001) even though their polyvalence has not been evaluated. A very sensitive and polyvalent SYBR Green real-time RT-PCR assay (detection limit five gene copies) was developed for the virus detection in garlic (Leisova-Svobodova and Karlova-Smekalova, 2011). Finally, a degenerate primer was recently reported that allows amplification of part of the 3′-terminus of the *Carlavirus* genome (Gaspar *et al.*, 2008) and could be possibly used for the nonspecific amplification of GarCLV.

A list of various DNA primers used for the detection of GarCLV in different countries and laboratories is presented in Table V.

C. Allexiviruses

In the early 1990s, new types of highly flexuous filamentous virus particles were observed in tissues of different *Allium* species (Barg *et al.*, 1994; Kanyuka *et al.*, 1992; Sumi *et al.*, 1993; Van Dijk *et al.*, 1991; Vishnichenko *et al.*, 1993). Unlike the other groups of viruses infecting *Allium* species, these ones are transmitted by mites (Barg *et al.*, 1994; Van Dijk *et al.*, 1991), while they usually are latent or they cause very mild symptoms. Based on host range and serological reactions, Van Dijk *et al.* (1991) described three types of these viruses namely *Onion mite-borne latent virus* (OMbLV), *Onion mite-borne latent virus-Garlic strain* (in onion and garlic, respectively), and *Shallot mite-borne latent virus* (SMbLV). Initially, they were erroneously classified as rymoviruses of the family *Potyviridae* based on their transmission by mites, the morphology and the length of their virus particles (700–800 nm), and the presence of granular inclusion bodies in infected tissues.

Barg *et al.* (1994) also reported the occurrence of at least five serologically distinguishable types of the so-called mite-borne filamentous viruses (MBFV), which were morphologically similar to potyviruses and closteroviruses but differing by the absence of pinwheel inclusion bodies and vesicles in infected cells. In Russia, another virus named *Shallot virus X* (ShVX) was described based on its properties and partial nucleotide sequence (Kanyuka *et al.*, 1992; Vishnichenko *et al.*, 1993). The particle morphology of ShVX resembled that of OMbLV, SMbLV, and MBFV. All these viruses are closely serologically related. In addition, Van Dijk and Van der Vlugt (1994) have shown that ShVX could be reidentified as OMbLV, SMbLV, or a complex of the two using antiserum against ShVX. The determination of the genome organization of ShVX revealed that it combines characteristics of carlaviruses and potexviruses, except in possessing an unusual ORF4. The identification of genome organization and nucleotide sequence of four other similar viruses (GV-A, GV-B, GV-C, and GV-D) isolated from mosaic-diseased garlic, and their phylogenetic analysis showed that they are related to ShVX and probably constitute a new virus genus closely related to the carlaviruses (Sumi *et al.*, 1993).

The characterization of these viruses and their species differentiation has been largely hampered by the fact that they often occur in multiple

TABLE V Primers used for the molecular detection of GarCLV

Primers	Sequence	Amplicon size (bp)	Genomic region	Assay	References
GCLV2F	5′-CGACCACCTGCTGGTTGG-3′	106	*CP*	SYBR Green real-time RT-PCR	Leisova-Svobodova and Karlova-Smekalova (2011)
GCLV1R	5′-TCAAGTGGCTGCACACAAGC-3′				
–	5′-AAATGTTAATCGCTAAACGACC-3′	500	*CP-NABP*	RT-PCR	Majumder and Baranwal (2009)
–	5′-CTTTGTGGATTTTCGGTAAG-3′				
GCLV-F	5′-GCACCAGTGGTTTGGAATGA-3′	481	*CP*	RT-PCR	Fidan and Baloglu (2009b)
GCLV-R	5′-AGCACTCCTAGAACAACCATTA-3′				

infections and their separation for further studies is difficult. Van Dijk *et al.* (1991) isolated some of them through successive local lesion passages onto *Chenopodium* spp. or mite transmission. Nevertheless, the biological data concerning these viruses are limited. Therefore, sequencing data have been mainly used to justify the creation of several species namely *Garlic virus*-A, -*B*, -*C*, -*D*, -*E*, -*X* (GarV-A, GarV-B, GarV-C, GarV-D, GarV-E, GarV-X) (Chen *et al.*, 2001; Ryabov *et al.*, 1996; Song *et al.*, 1998; Sumi *et al.*, 1993, 1999), *Garlic mite-borne filamentous virus* (GarMbFV), ShVX (Kanyuka *et al.*, 1992; Vishnichenko *et al.*, 1993). Western blot analysis revealed that GarV-D and GarV-E are closely related serologically, while weaker relationships exist between GarV-E and GarV-A, GarV-X and GarV-A, GarV-X and GarV-B, and GarV-X and GarV-C (Lu *et al.*, 2008). The novel mite-borne *Allium* viruses are distinguished by their characteristic genomic organization and have been classified to the recently ratified genus *Allexivirus* (Pringle, 1999) in the family *Alphaflexiviridae*. Nevertheless, their characterization is still in progress.

The economic significance of allexiviruses is not well known so far. Since they mainly cause latent infections, it seems that their effect on crop losses is rather limited although some of them may have a significant economic impact (Van Dijk and Van der Vlugt, 1994). Recent studies in Argentina showed that probably not all of them affect yield similarly. More specifically, GarV-A causes significant crop losses to two garlic cultivars (Blanco IFFIVE and Morado INTA) ranging from 14% to 32%, while GarV-C, under the conditions studied and in the cultivars tested, caused mild or no damage on garlic yield (Cafrune *et al.*, 2006a).

Very often allexiviruses persist in the infected plants as multiple infections with carlaviruses and potyviruses. This coexistence may have a synergistic effect and lead to even higher yield losses. A recent study indicated that garlic yield decreases more rapidly in plants previously infected with at least one allexivirus and then reinfected with other naturally occurring viruses than in plants that initially are virus free (Perotto *et al.*, 2010).

1. Host range, virus species strains, and symptomatology

Allexiviruses as potyviruses have a limited host range restricted mostly to *Allium* species. They infect garlic, leek, onion, and different wild alliums. Generally, allexiviruses are latent or very rarely they are causing very mild chlorotic stripes and mild mosaic in leaves of *Allium* species (Van Dijk and Van der Vlugt, 1994; Van Dijk *et al.*, 1991; Yamashita *et al.*, 1996). It should be noted that often, stripes or other mild leaf deformations are caused by *Aceria tulipae*, the vector of allexiviruses.

Except from the eight virus species (GarV-A, GarV-B, GarV-C, GarV-D, GarV-E, GarV-X, GarMbFV, ShVX) that were molecularly characterized, other virus isolates that were described are probably to be classified among the already reported species. Thus the nucleotide sequence of another mite-borne virus, namely Japanese garlic virus deposited to the database (Ryabov *et al.*, 1996; GenBank accession no. L38892), shows nucleotide similarity higher than 96% with GarV-D, and based on phylogenetic analysis, it is classified as a GarV-D isolate (Chen *et al.*, 2001). In Japan, the sequence of another mosaic causing mite-borne virus from garlic was determined, and the virus was named *Garlic mite-borne mosaic virus* (GMbMV) (Yamashita *et al.*, 1996), which is finally an isolate of GarV-C (99% nucleotide homology) (see also Chen *et al.*, 2001). A GarMbFV described in Argentina was shown to be a strain of GarV-A (Chen *et al.*, 2001; Helguera *et al.*, 1997a). Finally, OMbLV and SMbLV viruses reported by Van Dijk *et al.* (2001) should be possibly considered as ShVX (Barg *et al.*, 1997; Van Dijk and van der Vlugt, 1994) even though sequence data are not available.

2. Transmission

Allexiviruses are thought to be transmitted by mites both in the field where the vectors are moved passively by the wind and in the bulbs storage rooms (Barg *et al.*, 1997; Van Dijk *et al.*, 1991). The only vector known is the dry bulb mite, A. *tulipae* which was proved to transmit GarV-C and GarV-D (Zavriev, 2008). However, information related to transmission characteristics is not available. It was speculated that certain ways of storing bulbs may favor population buildup of the bulb mites, which occur under the bulb scales, and then the concomitant spread of the viruses (Barg *et al.*, 1997). Allexiviruses are not transmitted by aphids (Van Dijk *et al.*, 1991; Yamashita *et al.*, 1996) but are manually transmissible by sap under laboratory conditions.

3. Presence in Mediterranean countries

Allexiviruses have a worldwide distribution (Barg *et al.*, 1994; Van Dijk *et al.*, 1991). However, information related to their incidence and distribution in the Mediterranean basin is limited. The so-called OMbLV (Van Dijk, 1993b) was isolated from shallot in France and from garlic in Spain. GarV-C, GarV-D, and GarV-B were also detected in garlic crops in Greece in all regions surveyed, and their incidence depending on the region and the virus species ranged from 62.5% to 70.5% (Dovas *et al.*, 2001a). In Italy, GarV-C and GarV-D were detected in all regions surveyed, and their incidence depending on the viral species and the geographic region ranged from 10% to 20% (Dovas and Vovlas, 2003). GarV-X seems to exist in Italy based on GenBank data (Afunian M.R. *et al.*, unpublished). Allexiviruses were also reported to be prevalent in field grown garlic in Israel and occasionally in Slovenia (Mavrič and Ravnikar, 2005; Shiboleth *et al.*, 2001).

4. Particle morphology and properties

Allexivirus is a new virus genus belonging to the family *Alphaflexiviridae* (Pringle, 1999). Virions are very flexuous, filamentous with a length of about 700–800 nm and width 12 nm.

5. Nucleic acid component

The genome is a linear, positive-sense, single-stranded polyadenylated RNA.

The complete nucleotide sequences of the genomes of GarV-A, GarV-C (Sumi *et al.*, 1999), GarV-E (Chen *et al.*, 2001), GarV-X (Chen *et al.*, 2001; Song *et al.*, 1998), and ShVX (Kanyuka *et al.*, 1992) have been determined (GenBank accession nos. AB010300, AB010302, AJ292230, U89243, AJ292229 and M97264). They are about 9 kb in length encoding a large alpha-like replicase and five smaller ORFs. Also, partial nucleotide sequences of the genomes of GarMbFV, GarV-B, and GarV-D have been determined (GenBank accession nos. AY390254, FJ643475, AB010301, AF519572, and FJ643476).

Allexiviruses have a genome organization similar to that of carlaviruses, but the third triple gene block ORF lacks a classical initiation codon and there is an ORF4 encoding a 40-kDa serine-rich protein with no known homology to other reported proteins.

6. Propagation hosts and purification

Garlic and *C. murale* have been mainly used as propagation hosts of most known allexiviruses (Cafrune *et al.*, 2006b; Helguera *et al.*, 1997a; Perotto *et al.*, 2010; Yamashita *et al.*, 1996), whereas garlic is their purification host except from ShVX for which shallot was used (Vishnichenko *et al.*, 1996). A list of virus purification methods applied is provided in Table VI. Since allexiviruses are usually occurring in mixed infections, virus isolation is often necessary for further characterization and can be achieved by successive local lesion passages through *C. quinoa* (Melo Filho *et al.*, 2004), *C. murale* (Cafrune *et al.*, 2006a), or by mite transfer (Van Dijk *et al.*, 1991).

TABLE VI Purification methods used for different allexiviruses

Virus species	Purification method
GarV-A	Mohamed and Young (1981), Helguera *et al.* (1997a)
GarV-B	Mohamed and Young (1981), E. Barg, BBA
GarV-C	Mohamed and Young (1981), Yamashita *et al.* (1996)
GarV-D	Mohamed and Young (1981), E. Barg, BBA
GarV-E	Cohen *et al.* (2000)
ShVX	Vishnichenko *et al.* (1993)

7. Detection methods

i. Diagnostic species Allexiviruses cause chlorotic local lesions onto *C. murale*, *C. amaranticolor*, *C. quinoa*, and *Atriplex hortensis*. However, not all allexiviruses infect all indicator plants (Van Dijk and Van der Vlugt, 1994, Van Dijk *et al.*, 1991). Hence, their value as diagnostic tools is rather limited.

ii. Serological tests Due to the already reported problems of virus isolation and purification from garlic, serology is not a very good tool for virus species differentiation among allexiviruses. However, polyclonal antibodies were raised against GMbMV (Yamashita *et al.*, 1996), which was actually proved to be GarV-C (Tsuneyoshi and Sumi, 1996; Yamashita *et al.*, 1996), and GarV-D (E. Barg, personal communication). Polyclonal and/or monoclonal antisera have also been produced against ShVX, GarV-A, GarV-B, and GarV-C (D. E. Lesemann and E. Barg, personal communication). In order to overcome the problem of virus purification, most of the antisera raised against allexiviruses were produced through the expression of the recombinant viral CP in heterologous systems. This type of antibodies has been used for the detection of GarV-A (Helguera *et al.*, 1997a; Lu *et al.*, 2008), GarV-B (Lu *et al.*, 2008), GarV-C (Alves *et al.*, 2008; Lu *et al.*, 2008; Tsuneyoshi and Sumi, 1996), GarV-D (Lu *et al.*, 2008), GarV-E (Lu *et al.*, 2008), and GarV-X (Lu *et al.*, 2008; Song *et al.*, 1997). It should be noted that the polyclonal antibody prepared by Tsuneyoshi and Sumi (1996) expressing the *CP* gene of GarV-C in *Escherichia coli* detected four viruses, namely GarV-A, GarV-B, GarV-C, and GarV-D.

Different serological tests have been applied for the detection of allexiviruses such as ISEM and/or decoration (Bertaccini *et al.*, 2004; Cafrune *et al.*, 2006a,b; Helguera *et al.*, 1997a; Kang *et al.*, 2007; Koo *et al.*, 2002; Shahraeen *et al.*, 2008; Song *et al.*, 1997; Tsuneyoshi and Sumi, 1996), Western blot analysis (Alves *et al.*, 2008; Helguera *et al.*, 1997a; Kang *et al.*, 2007; Lu *et al.*, 2008; Song *et al.*, 1997; Yamashita *et al.*, 1996), DAS-ELISA (Cafrune *et al.*, 2006a,b; Dovas *et al.*, 2001b; Helguera *et al.*, 1997a; Shahraeen *et al.*, 2008), Dot-ELISA (Alves *et al.*, 2008; Melo Filho *et al.*, 2004), and tissue printing/DTBIA (Helguera *et al.*, 1997a; Koo *et al.*, 2002; Tsuneyoshi and Sumi, 1996). Interestingly, a DTBIA kit for detecting viruses in garlic in the field has been developed (Sumi *et al.*, 2001). Cafrune *et al.* (2006b) used a semiquantitative DAS-ELISA for monitoring concentration fluctuation of GarV-A during the garlic crop cycle. Their results showed that the optimum sampling period for GarV-A is about 64–81 days post-planting or toward the end of the crop cycle using tip section of the youngest leaves. Concerning the bulbs, the best sampling time is in devernalized cloves.

Commercial kits of DAS-ELISA are available for serological detection of GarV-A, GarV-B, GarV-C, and GarV -D. However, it is reported that most of these antisera cross-react with GarV-A.

*iii. **Electron microscopy*** Electron microscopy can discriminate allexiviruses from potyviruses and carlaviruses due to different particle morphology (Diekmann, 1997) but cannot differentiate virus species of the genus.

*iv. **Molecular tests*** Due to the difficulties reported for a reliable serological diagnosis of allexiviruses, molecular assays were considered to be an alternative for both the characterization and the detection of the different species. An RT-PCR using nonspecific primers that detect all of the GarV-A, GarV-B, GarV-C, and GarV-D isolates was developed (Tsuneyoshi and Sumi, 1996). The application of restriction enzyme analysis makes possible the identification of individual viruses (Sumi *et al.*, 2001; Tsuneyoshi and Sumi, 1996). Such a broad-range RT-PCR was also reported by Shiboleth *et al.* (2001). Later on, another pair of polyvalent degenerate primers was used for the generic detection of allexiviruses in different versions of RT-PCR (Dovas *et al.*, 2001b). This method allows the detection even directly in leaf extracts, thus avoiding the RNA extraction. Double-tube RT-PCR using leaf extract was shown to be at least 100 times more sensitive than ELISA for the detection of GarV-C (Dovas *et al.*, 2001b). The same primer pair was used in an optimized duplex RT-PCR for the simultaneous detection of an allexivirus and OYDV (Kumar *et al.*, 2010). Degenerate genus-specific primers were also designed (Chen *et al.*, 2004) and used for the detection and classification of allexiviruses. A molecular assay allowing the detection of all allexiviruses would be useful for testing propagation plant material.

On the other hand, species-specific primers have also been reported. Melo Filho *et al.* (2004) developed RT-PCR assays for the specific detection of GarV-A, GarV-B, GarV-C, and GarV-D. Furthermore, primers specific to the *CP* gene of GarV-D and GarV-B were designed and used in RT-PCR testing of garlic cloves (Gieck *et al.*, 2009). A differential RT-PCR analysis was also applied using a combination of a universal upstream primer with species-specific downstream ones (Koo *et al.*, 2002). Specific primers were designed for the amplification of part of the ShVX genome and used in both RT-PCR and IC-RT-PCR assays (Perez-Egusquiza *et al.*, 2008). A multiplex RT-PCR was reported in Korea that simultaneously detects and differentiates six garlic viruses including GarV-X and GarMbFV (Park *et al.*, 2005). However, it should be noted that the sequence used to design the primers for amplification of GarMbFV is probably corresponding to GarV-A (Chen *et al.*, 2001; Helguera *et al.*, 1997a). Finally, the degenerate primers reported from Dovas *et al.* (2001b) were further optimized for the detection of GMbFV in garlic using SYBR Green real-time RT-PCR (Leisova-Svobodova and Karlova-Smekalova, 2011).

A list of various DNA primers used for the detection of allexiviruses in different countries and laboratories is presented in Table VII.

TABLE VII Generic and species-specific primers used for the molecular detection of allexiviruses

Virus species	Primers	Sequence	Amplicon size (bp)	Genomic region	Assay	References
Allexivirus spp.	ALLEX1[a]	5′-CYGC TAA GCT ATA TGC TGA ARG G-3′	183–192	*ORF6*-3′UTR	RT-PCR	Dovas *et al.* (2001b)
	ALLEX2[a]	5′-TGTT RCA ARG TAA GTT TAG YAA TAT CAA CA-3′				
	Allex-CP(+)	5′-TGG RCX TGC TAC CACAAY GG-3′	750	*CP-ORF6*	RT-PCR	Chen *et al.* (2004)
	Allex-NABP(−)	5′-CCY TTC AGC ATA TAG CTT AGC-3′				
GarV-A	GarV-A1	5′-CCCAAGCTTACTGGAAGGGTGAATTAGAT-3′	800	*CP*	RT-PCR	Melo Filho *et al.* (2004)
	GarV-A2	5′-CCCAAGCTTAGGATATTAAAGTCTTGAGG-3′				
GarV-B	GV-B-F1	5′-GAGGAGAACTAACGCCACAC-3′	–	*CP*	RT-PCR	Gieck *et al.* (2009)
	GV-B-R2	5′-ACGACCTAGCTTCCTACTTG-3′				
	GarV-B1	5′-CCAAGCTT TTAATTTACACTGGCTTAGA-3′	800	*CP*	RT-PCR	Melo Filho *et al.* (2004)
	GarV-B2	5′-CCAAGCTT TATGCATTTCTGGGTCAAGA-3′				
GarV-C	GarV-C1	5′-CCCAAGCTTCATCTACAACAACAAAGGCG-3′	800	*CP*	RT-PCR	Melo Filho *et al.* (2004)
	GarV-C2	5′-CCCAAGCTTATAAGGGTGCATGATTGTGG-3′				
GarV-D	JGV-F2	5′-GCTCACTCRGATGTGTTAGC-3′	–	*CP*	RT-PCR	Gieck *et al.* (2009)
	JGV-R2	5′-CGCGTGGACATAAGTTGTTG-3′				
	GarV-D1	5′-CCAAGCTTAAGCAAGTGAAGAGTGTAAG-3′	800	*CP*	RT-PCR	Melo Filho *et al.* (2004)
	GarV-D2	5′-CCAAGCTTTTTGGAAGAGGAGGTTGAGA-3′				
GarV-X	(+)	5′-GATCGGAACCAAGGAATAA-3′	661	*CP*	Single/multiplex RT-PCR	Park *et al.* (2005)
	(−)	5′-GAGTGGAAACCATATTCGAG-3′				
ShVX	ShVX-CPF	5′-ATTTAGGGGTGAAGGTCTGT-3′	912	*CP*	RT-PCR	Perez-Egusquiza *et al.* (2008)
	ShVX-CPR	5′-GAGTTTTGAGGTCGTTGG-3′				

X = A/T/C/G; Y = T/C; R = A/G.
[a] These primers have also been used for the detection of GMbFV in SYBR Green real-time RT-PCR (Leisova-Svobodova and Karlova-Smekalova, 2011).

D. Other viruses infecting *Allium* species in the Mediterranean region

These viruses have rather limited economic importance due to their limited incidence, and for this reason, only a short description is given.

1. Tospoviruses

i. **Iris yellow spot virus** *Iris yellow spot virus* (IYSV), a tospovirus infecting a number of cultivated crops, has also been found in different *Allium* species (onion, shallot, leek, garlic, chive) worldwide such as Germany, Holland, Austria, USA, Brazil, India, Australia, Japan, South Africa, Peru, Mexico, and Chile (Cortês *et al.*, 1998; Coutts *et al.*, 2003; Creamer *et al.*, 2004; du Toit *et al.*, 2007; Gent *et al.*, 2006; Kumar and Rawal, 1999; Leinhos *et al.*, 2007; Mullis *et al.*, 2004, 2006; Plenk and Grausgruber-Gröger, 2011; Pozzer *et al.*, 1999; Ravi *et al.*, 2006; Rosales *et al.*, 2005; Schwartz *et al.*, 2002).

In the Mediterranean region, IYSV was first reported in Israel in onions (Gera *et al.*, 1998a). Subsequently, it has been reported infecting onion and shallot in France (Huchette *et al.*, 2008) and Serbia (Bulajic *et al.*, 2008); onion and leek in Spain (Cordoba-Selles *et al.*, 2005, 2007), Slovenia (Mavrič and Ravnikar, 2000), and northern Greece (Chatzivassiliou *et al.*, 2009); onion, leek, Egyptian leek, and garlic in Egypt (Elnagar *et al.*, 2005); and onion crops in Italy (Tomassoli *et al.*, 2009) and Tunisia (Ben Moussa *et al.*, 2005).

IYSV is exclusively transmitted by the onion thrips (*Thrips tabaci*) (Kritzman *et al.*, 2001; Mumford *et al.*, 1996; Nagata *et al.*, 1999). The virus is not transmitted through the seed and also was not located in bulbs of infected plants (Kritzman *et al.*, 2001).

The disease that IYSV causes to onions is severe. In Israel, it is known as "onion straw bleaching" and is characterized by straw-colored ring spot on leaves and flower stalks that sometimes coalesce and cause premature plant death (Gera *et al.*, 1998b; Kritzman *et al.*, 2001). IYSV causes reduction of bulb size (Gent *et al.*, 2004a). In Israel and Brazil, the incidence of the disease often reached 50–60% and 100%, respectively, resulting in heavy losses of onion yield production (Gera *et al.*, 1998a; Pozzer *et al.*, 1999). In Colorado, USA, IYSV has become a serious threat to onion production (Gent *et al.*, 2004a) and has been characterized an emerging threat to onion bulb and seed production (Gent *et al.*, 2006). The biology, epidemiology, and impact of IYSV were recently reviewed by Gent *et al.* (2006).

DAS-ELISA allows the reliable detection of the virus by using the available polyclonal antibodies prepared against the N protein (Kritzman *et al.*, 2001; Pozzer *et al.*, 1999). However, recently, the presence of serologically distinct isolates of IYSV has been reported, which may hamper the virus diagnosis through different antisera (Tomassoli *et al.*, 2009). IYSV has also been detected by Dot-ELISA in onion leaf discs (Nagata *et al.*, 1999). RT-PCR assays were applied using primers that amplify the *N* gene of the virus (Cortês *et al.*, 1998; Kritzman *et al.*, 2001; Pozzer *et al.*, 1999; Tomassoli *et al.*, 2009).

Tomato spotted wilt virus (TSWV), another tospovirus infecting allium crops, has also been found occasionally to infect onion in Italy (Tomassoli *et al.*, 2009), Brazil (de Avila *et al.*, 1981), and Georgia (USA) (Mullis *et al.*, 2004).

ii. **Garlic dwarf virus** *Garlic dwarf virus* (GDV), the only fijivirus infecting *Allium* species, was only reported to occur in southeastern France where occasionally it reaches epidemics since 1988 (Lot *et al.*, 1994). Although worldwide it is considered to be of low economic importance because of its limited distribution, it causes high yield losses to the affected areas (Lot *et al.*, 1994). Garlic is the only host of GDV. The virus causes a symptom complex described as garlic dwarf syndrome. Diseased plants are severely dwarfed, and the leaves are thickened. The majority of the leaves are showing darker green color, and some of them show vein swellings or enations. The plants resemble tulips as the leaves appear to insert at the same point, and no false stem is observed as on unaffected plants. The young leaves often have purple tips. The bulb is soft, containing cloves of reduced size, surrounded by thickened, and wrinkled tunics. Later in the season, plants with apparently normal older leaves sometimes develop shortened internodes, a thin appearance, and spongy pear-shaped bulbs containing few cloves. GDV belongs to genus *Fijivirus* of the family *Reoviridae* with isometric virus particles with diameter 65–70 nm (Lot *et al.*, 1994). So far no vector has been found. The virus is not transmitted mechanically or by aphids. However, the spread in the field and its low incidence do not support the idea of an efficient insect vector (Lot *et al.*, 1994). DAS-ELISA and ISEM-D have been used successfully for GDV detection (Diekmann, 1997).

iii. **Artichoke yellow ringspot virus** An isometric virus ca. 25 nm in diameter with angular contour was isolated from onion plants, showing yellow leaf striping and necrotic tips in Greece (Maliogka *et al.*, 2006). *Artichoke yellow ringspot virus* (AYRSV), a member of the family *Comoviridae*, was identified as the causal agent of the disease. Identification was based on symptomatology onto indicator plants, cytopathological studies in infected *Nicotiana benthamiana* tissues, and RT-PCR using degenerate primers specific for the family *Comoviridae*. Sequence comparisons of the amplified region with those from other AYRSV isolates provided evidence that the nepovirus is naturally infecting onion in the field. AYRSV was found to be seed transmitted in onion (Maliogka *et al.*,

2006); thus, it is the only seed-transmitted virus in this host. The molecular detection of the virus is feasible using AYRSV-specific RT-PCR (Maliogka *et al.*, 2006). So far, the virus has been only detected in onion crops in Greece (Maliogka *et al.*, 2006) although it has been reported to occur in other crops in Italy (Rana *et al.*, 1978, 1983). Therefore, it seems to be of local economic importance.

iv. **Cucumber mosaic virus** It was isolated in ex-Yugoslavia from garlic plants showing stunting (Štefanac, 1980), but its economic significance is limited and of local importance.

v. **Leek white stripe virus** White stripe is a relatively new viral disease affecting leek in France with which an isometric virus ca. 30 nm in diameter is associated (Lot *et al.*, 1996). The disease was observed in 1987 in leek winter crops of the Loire valley (France) and has spread epidemically since then. The most evident symptom is the presence of white stripes on the leaves extending to the stem. Occasionally, some plants turn completely white and rotten. Even though the distribution of infected plants in the field was suggestive of a possible soil-borne nature of the disease, no such a mode of transmission was demonstrated. The virus isolated was a new distinct species in the genus *Necrovirus*, for which the name *Leek white stripe virus* (LWSV) was proposed. It can be mechanically transmitted with difficulties under laboratory conditions and has a restricted host range.

IV. CONTROL MEASURES OF ALLIUM VIRUSES

The main sources of the viruses infecting *Allium* species are the crops themselves, such as garlic, onion, and leek, and they are endemic in areas where these crops are cultivated throughout the whole year. Depending on the mode of propagation of the various alliums, different measures are necessary for the efficient control of their viruses.

Control of tospoviruses (IYSV, TSWV) affecting alliums is not included in this chapter, as it is included in Chapter 12. Viruses belonging to the genera *Carlavirus*, *Potyvirus*, and *Allexivirus* are not seed borne. Therefore, in *Allium* species propagated by true seeds, such as leek and onion, the control of viruses is based on the following:

1. Nurseries and fields with virus-free propagative material should be isolated, more than few hundred meters away from susceptible allium crops.

2. Use of virus-free seed. As mentioned, AYRSV is the only virus transmitted through seed. Therefore, in countries such as Greece where it is endemic, it is recommended to use virus-free onion seed.
3. Control of arthropod vectors. As already mentioned, aphids and mites are important vectors of allium viruses, and it seems that reducing their population might result to lower virus incidence. However, it is very well established that nonpersistently transmitted viruses such as poty- and carlaviruses cannot be controlled by aphid chemical control (Perring *et al.*, 1999). Control of mites during bulb storage might reduce spread of allexiviruses.
4. Introduction of a host-free period. Continued year-round cultivation of the same crop is a potential cause of virus epidemics, particularly for viruses with a very narrow host range, such as LYSV. Overlapping of onion and shallot crops in New Zealand resulted in outbreaks of OYDV (Chamberlain and Bayliss, 1948). Similarly, continual cropping of leek in the Netherlands has also led to outbreaks of LYSV (Bos, 1983). Therefore, OYDV and LYSV can be controlled efficiently by a break period during which the susceptible crop is not grown. This strategy is mainly effective for controlling LYSV in leek crops and OYDV in onion crops.
5. Roguing of affected plants. It is recommended to monitor the growing plants and remove all those infected. This is of great importance when it is applied early in the growing season when virus incidence is low and when the propagative material used is of high quality (low virus incidence). This is also important for vegetatively propagated *Allium* species, as it reduces the overall percentage of infected bulbs.
6. Use of resistant or tolerant cultivars. Leek cultivars differ in their susceptibility to LYSV, and the cultivars Lancelot, Bandit, Laura, Winta, and Ligina are tolerant to the virus (Matthieu *et al.*, 1984). Also, differences in susceptibility of onion cultivars to IYSV have been reported (du Toit and Pelter, 2005; Gent *et al.*, 2004b). Although genetic engineering has been used for many years for the production of virus-resistant cultivars, this technology has not been applied so far for the control of allium viruses. This is mainly attributed to the fact that alliums such as onion, leek, and garlic are recalcitrant to genetic transformation (Eady *et al.*, 1996). However, transformation of *Allium* species has been improved in the past few years (Kondo *et al.*, 2000; Sawahel, 2002), and in the near future, transfer of virus genes in *Allium* might result to virus resistance.

For vegetatively propagated *Allium* species, such as garlic, the main source of viruses is the propagative material (propagules). For this reason, the above measures are rather inadequate for controlling viruses. The first step toward this direction is the selection and use of large bulbs which are possibly infected with fewer viruses or with mild virus strains (Van Dijk, 1994).

Furthermore, the global trade of alliums, mainly garlic and onion, between countries is a potential means of virus dissemination worldwide. Thus, imported alliums should always be inspected for viral diseases.

A. Virus elimination techniques

The production and use of virus-free propagative planting material is the most effective method for controlling viruses infecting vegetatively propagated *Allium* species (Salomon, 2002). Messiaen *et al.* (1994) were pioneers in developing the first certification scheme, including sanitary and clonal selection, of a French garlic cultivar back in the early 1960s. A few years later symptomless plants were also obtained from cultivars in which OYDV infestation was 100% by application of virus-elimination techniques and they were thus incorporated in the national certification scheme to attain virus-free mother bulbs.

When one cultivar is totally infected, attempts must be made to free a plant from the viruses. To achieve this, meristem/shoot/scape tip culture and thermotherapy either alone or in combination have been used successfully (Bhojwani *et al.*, 1982; Conci and Nome, 1991; Ma *et al.*, 1994; Ucman *et al.*, 1998). Alternative methods such as root meristem culture, cryopreservation, or even chemotherapy have been evaluated for the production of virus-free alliums (Haque *et al.*, 2007; Kim *et al.*, 2007; Shiboleth, 1998). The efficiency of virus elimination depends mainly on the virus involved, the plant species (genotype), and the adopted treatment. For example, potyviruses are eliminated from garlic propagative material more easily compared to allexiviruses which eradication is more difficult (Conci *et al.*, 2004; Luciani *et al.*, 1998; Perotto *et al.*, 2003). Furthermore, garlic genotypes vary in their response to tissue culture conditions (Salomon, 2002); thus, specific procedures should be applied to the local cultivars of each region.

After the application of the elimination strategy, the obtained explants should be tested by the most sensitive detection method in order to verify and separate the healthy from the infected ones (Lunello *et al.*, 2005). Having obtained a virus-free stock a foundation stock should be propagated in isolated areas where no commercial garlic crops are cultivated in order to avoid reinfection. Also, special measures should be undertaken to protect virus-free mother plants used for vegetative propagation. For example, mother plants are grown in gauze houses to exclude the aerial vectors (aphids and mites). The use of garlic virus-free propagative material is highly recommended and results in an increase of bulb weight ranging from 32% to 216% for most cultivars tested (Conci, 1997;

Conci *et al.*, 2003; Melo Filho *et al.*, 2006; Walkey and Antill, 1989). Production of virus-free propagative material has been successfully used for the allium viruses' control worldwide (Messiaen *et al.*, 1994; Peiwen *et al.*, 1994; Verbeek *et al.*, 1995).

B. Cross protection

Cross protection, a method used for combating viral diseases by preimmunizing the plants with mild virus strains, has been suggested for the effective control of OYDV in garlic (Messiaen *et al.*, 1981; Van Dijk, 1994). However, the method has not been adopted in practice.

V. FUTURE DEVELOPMENTS AND PERSPECTIVES

Future developments on allium crop viruses should focus on their further characterization. Especially for allexiviruses, more studies are needed on their both molecular and biological features. The molecular basis of the virus–host–vector interactions needs clarification in order to better understand and face the induced diseases. Furthermore, more surveys for the presence of viral infections, including allexiviruses, are necessary especially in the Mediterranean region, using appropriate up-to-date detection techniques.

Since the available information on the effect of each virus on yield losses is scarce, future research will have to determine the extent of damage induced by each individual virus as well as from the different virus combinations.

It is obvious that the use of virus-free certified material, especially in the case of garlic, is an improved horticultural practice in order to combat viral diseases of allium crops. Research is required to modify and apply traditional or alternative virus-elimination methods in order to free the local clones of each country from viruses so as to avoid using imported material of foreign cultivars that may not be suitable. Sensitive polyvalent diagnostic assays should be developed to ensure the efficiency and reliability of the applied procedures. Finally, since the genetic transformation efficiency of allium plants has been improved, expression of viral transgenes could be applied to achieve plant resistance against the most damaging garlic and onion viruses.

REFERENCES

Alves, M., Marraccini, F. M., Melo, P. D., Dusi, A. N., Pio-Riberio, G., and Ribeiro, B. M. (2008). Recombinant expression of *Garlic virus C* (GarV-C) capsid protein in insect cells and its potential for the production of specific antibodies. *Microbiol. Res.* **163**:354–361.

Arya, M., Baranwal, V. K., Ahlawat, Y. S., and Singh, L. (2006). RT-PCR detection and molecular characterization of *Onion yellow dwarf virus* associated with garlic and onion. *Curr. Sci.* **91:**1230–1234, India.

Barg, E. 1996. Serologiche und moleculargenetische Untersuchungen zur Variabilität Allum-Arten infizierender, filamentöser Viren/vorgelegt von Erchard Barg.-Clausthal-Zellerfeld: Papierflieger, 1996. Zugl., Göttingen, Univ., Diss. ISBN: 3-931986-40-3.

Barg, E., Lesemann, D. E., Vetten, H. J., and Green, S. K. (1994). Identification, partial characterization and distribution of viruses infecting *Allium* crops in South and Southeast Asia. *Acta Hortic.* **358:**251–258.

Barg, E., Lesemann, D. E., Vetten, H. J., and Schonfelder, M. (1995). Differentiation of potyviruses infecting cultivated Allium species. Proceedings of the 8th Conference on Virus Diseases of Vegetables, Prague July 9–15, pp. 29–31.

Barg, E., Lesemann, D. E., Vetten, H. J., and Green, S. K. (1997). Viruses of alliums and their distribution in different Allium crops and geographical regions. *Acta Hortic.* **433:**607–616.

Bellardi, M. G., Marani, F., Betti, L., and Rabiti, A. L. (1995). Detection of *garlic common latent virus* (GCLV) in *Allium sativum* L. in Italy. *Phytopathol. Mediterr.* **34:**58–61.

Ben Moussa, A., Marrakchi, M., and Makni, M. (2005). Characterisation of *Tospovirus* in vegetable crops in Tunisia. *Infect. Genet. Evol.* **5:**312–322.

Bertaccini, A., Botti, S., Tabanelli, D., Dradi, G., Fogher, C., Previati, A., and Dare, F. (2004). Micropropagation and establishment of mite-borne virus-free garlic (*Allium sativum*). *Acta Hortic.* **631:**201–206.

Bhojwani, S. S., Cohen, D., and Fry, P. F. (1982). Production of virus free garlic and field performance of micropropagate plants. *Sci. Hortic.* **18:**39–43.

Block, E. (2010). Garlic and Other Alliums: The Lore and the Science. The Royal Society of Chemistry, Cambridge, UK, p. 454.

Bos, L. (1976). *Leek yellow stripe virus.* CMI/AAB Descriptions of Plant Viruses, p. 240.

Bos, L. (1981). *Onion yellow dwarf virus.* CMI/AAB Descriptions of Plant Viruses, p. 158.

Bos, L. (1982). *Shallot latent virus.* CMI/AAB Descriptions of Plant Viruses. No. 250.

Bos, L. (1983). Viruses and virus diseases of *Allium* species. *Acta Hortic.* **127:**11–29.

Bos, L., Huijberts, N., Huttinga, H., and Maat, D. Z. (1978a). *Leek yellow stripe virus* and its relationships to *onion yellow dwarf virus*; characterization, ecology and possible control. *Neth. J. Plant Pathol.* **84:**185–204.

Bos, L., Huttinga, H., and Maat, D. Z. (1978b). *Shallot latent virus*, a new *carlavirus*. *Neth. J. Plant Pathol.* **84:**227–237.

Boswell, J. J. (ed.) (1883). English Botany Vol. 9, George Bell and Sons, London.

Brewster, J. L. (1994). Onions and Other Vegetable Alliums. CAB International, Oxon, UK.

Brunt, A. A., Crabtree, K., Dallwitz, M. J., Gibbs, A. J., and Watson, L. (1996). Garlic Common Latent Carlavirus. Viruses of Plants Descriptions and Lists from the VIDE Database. CAB International, WallingFord, UK, pp. 601–603.

Bulajic, A., Jovic, J., Krnjajic, S., Petrov, M., Djekic, I., and Krstic, B. (2008). First report of *Iris yellow spot virus* on onion (*Allium cepa*) in Serbia. *Plant Dis.* **92:**1247.

Cafrune, E. E., Perotto, M. C., and Conci, V. C. (2006a). Effect of two Allexivirus isolates on garlic yield. *Plant Dis.* **90:**898–904.

Cafrune, E. E., Balzarini, M., and Conci, V. C. (2006b). Changes in the concentration of an *Allexivirus* during the crop cycle of two garlic cultivars. *Plant Dis.* **90:**1293–1296.

Canavelli, A., Nome, S. F., and Conci, V. C. (1998). Efecto de distintos virus en la produccion de ajo (*Allium sativum*) Rosado Paraguayo. *Fitopatol. Bras.* **23:**354–358.

Chamberlain, E. E., and Baylis, G. T. S. (1948). Onion yellow dwarf. Successful eradication. *N. Z. Ji. Sci. Technol.* **A29:**300–301.

Chatzivassiliou, E. K., Giavachtsia, V., Mehraban, A. H., Hoedjes, K., and Peters, D. (2009). Identification and Incidence of *Iris yellow spot virus*, a new pathogen in onion and leek in Greece. *Plant Dis.* **93:**761.

Chen, J., Chen, J., and Adams, M. J. (2001). Molecular characteristics of a complex mixture of viruses in garlic with mosaic symptoms in China. *Arch. Virol.* **146:**1841–1853.
Chen, J., Chen, J. P., and Adams, M. J. (2002). Characterisation of some carla- and potyviruses from bulb crops in China. Brief report. *Arch. Virol.* **147:**419–428.
Chen, J., Adams, M. J., Zheng, H. Y., and Chen, J. P. (2003). Sequence analysis demonstrates that *Onion yellow dwarf virus* isolates from China contain a P3 region much larger than other potyviruses. *Arch. Virol.* **148:**1165–1173.
Chen, J., Zheng, H. Y., Antoniw, J. F., Adams, M. J., Chen, J., and Lin, L. (2004). Detection and classification of allexiviruses from garlic in China. *Arch. Virol.* **149:**435–445.
Cohen, J., Zeidan, M., Rosner, A., and Gera, A. (2000). Biological and molecular characterization of a new carlavirus isolated from an *Aconitum* sp. *Phytopathology* **90:**340–344.
Conci, V. C. (1997). Virus y Fitoplasmas de ajo. *In* "50 Temas Sobre Produccion de Ajo" (J. L. Burba, ed.), pp. 267–293. EEA-INTA La Consulta, Mendoza.
Conci, V. C., and Nome, S. F. (1991). Virus-free garlic (*Allium sativum* L.) plants obtained by thermotherapy and meristem tip culture. *J. Phytopathol.* **132:**186–192.
Conci, V. C., Nome, S. F., and Milne, R. G. (1992). Filamentous viruses of garlic in Argentina. *Plant Dis.* **76:**594–596.
Conci, V. C., Canavelli, A., and Lunello, P. (2003). Yield losses associated with virus-infected garlic plants during five successive years. *Plant Dis.* **87:**1411–1415.
Conci, V. C., Cafrune, E. E., Lunello, P., Nome, S., and Perotto, C. (2004). Produccion de plantas de ajo libres de virus. *In* "Biotecnologia y Mejoramiento Vegetal" (V. Echenique, C. Rubinstein, and L. Mroginski, eds.), pp. 313–316. INTA, Buenos Aires.
Conci, V. C., Perotto, M. C., Cafrune, E., and Lunello, P. (2005). Program for intensive production of virus-free garlic plants. *Acta Hortic.* **688:**195–200.
Cordoba-Selles, C., Marinez-Priego, L., Munoz-Gomez, R., and Jorda-Gutierrez, C. (2005). *Iris yellow spot virus*: Onion disease in Spain. *Plant Dis.* **89:**1243.
Cordoba-Selles, C., Cebrian-Mico, C., and Alfaro-Fernadez, A. (2007). First report of *Iris yellow spot virus* in commercial leek (*Allium porrum*) in Spain. *Plant Dis.* **91:**1365.
Cornuet, P. (1959). Maladies a Virus des Plantes Cultivees et Methodes de Lutte. Inst. Natn. Rech. Agron, Paris, p. 440.
Cortês, I., Livieratos, I. C., Derks, A., Peters, D., and Kormelink, R. (1998). Molecular and serological characterization of *iris yellow spot virus*, a new and distinct tospovirus species. *Phytopathology* **88:**1276–1282.
Coutts, B. A., McMichael, L. A., Tesoriero, L., Rodoni, B. C., Wilson, C. R., Wilson, A. J., Persley, D. M., and Jones, R. A. C. (2003). *Iris yellow spot virus* found infecting onions in three Australian states. *Australas. Plant Pathol.* **32:**355–357.
Creamer, R., Sanogo, S., Moya, A., Romero, J., Molina-Bravo, R., and Cramer, C. (2004). Iris yellow spot virus on onion in New Mexico. *Plant Dis.* **88:**1049.
de Avila, A. C., Gama, M. I. C. S., Kitajima, E. W., and Pereira, W. (1981). Um virus do grupo vira-cabeca do tomateiro isolado de cebola (*Allium cepa* L.). *Fitopatol. Bras.* **6:**525.
Delecolle, B., and Lot, H. (1981). Viroses de l'ail: I. Mise en evidence et essais de caracterisation par immunoe1ectromicroscopie d'un complexe de trois virus chez differentes populations d'ail atteintes de mosaique. *Agronomie* **1:**763–769.
Delecolle, B., Lot, H., and Michel, M. J. (1985). Application of ELISA for detecting *onion yellow dwarf virus* in garlic and shallot seeds and plants. *Phytoparasitica* **13:**266–267.
Diekmann, M. (1997). *FAO/IPGRI Technical Guidelines for the Safe Movement of Germplasm. No. 18.* Allium *spp.* Food and Agriculture Organization of the United Nations/International Plant Genetic Resources Institute, Rome.
Dovas, C. I., and Vovlas, C. (2003). Viruses infecting Allium spp. in southern Italy. *J. Plant Pathol.* **85:**135.
Dovas, C. I., Hatziloukas, E., Salomon, R., Barg, E., Shiboleth, Y., and Katis, N. I. (2001a). Incidence of viruses infecting *Allium* spp. in Greece. *Eur. J. Plant Pathol.* **107:**677–684.

Dovas, C. I., Hatziloukas, E., Salomon, R., Barg, E., Shiboleth, Y., and Katis, N. I. (2001b). Comparisons of methods for virus detection in *Allium* spp. *J. Phytopathol.* **149:**731–737.

Dovas, C. I., Mamolos, A. P., and Katis, N. I. (2002). Fluctuation in concentration of two potyviruses in garlic during the growing period and sampling conditions for reliable detection by ELISA. *Ann. Appl. Biol.* **140:**21–28.

Drake, C. J., Tate, H. D., and Harris, H. M. (1933). The relationship of aphids to the transmission of yellow dwarf of onions. *J. Econ. Entomol.* **26:**841–846.

du Toit, L. J., and Pelter, G. Q. (2005). Susceptibility of storage onion cultivars to Iris yellow spot in the Columbia Basin of Washington. *Biol. Cult. Tests* **20:**V006.

du Toit, L. J., Burger, J. T., McLeod, A., Engelbrecht, M., and Viljoen, A. (2007). *Iris yellow spot virus* in onion seed crops in South Africa. *Plant Dis.* **91:**1203.

Eady, C. C., Lister, C. E., Suo, Y., and Schaper, D. (1996). Transient expression of uidA constructs in *in vitro* onion (*Allium cepa* L.) cultures following particle bombardment and Agrobacterium-mediated DNA delivery. *Plant Cell Rep.* **15:**958–962.

El-Kewey, S. A., and Sidaros, S. A. (1996). *Onion Yellow Dwarf Virus* (OYDV) affecting onion in middle delta Egypt. *In* Fourth Arabic Conference, pp. 291–302. Minia, Egypt.

Elnagar, S., El-Sheikh, M. A. K., and Abdel Wahab, A. S. (2005). *Iris yellow spot virus* (IYSV): A newly isolated thrips-borne tospovirus in Egypt. Proceedings of the 7th International Conference on Pests in Agriculture, October 26–27, Montpellier, FR, p. 8.

Fajardo, T. V. M., Nishijima, M., Buso, J. A., Torres, A. C., Avila, A. C., and Resende, R. O. (2001). Garlic viral complex: Identification of potyviruses and carlavirus in central Brazil. *Fitopatol. Bras.* **26:**619–626.

FAO (Food and Agricultural Organization of the United Nations), 2005 statistics. http://faostat.fao.org.

FAO (Food and Agricultural Organization of the United Nations), 2007 statistics. http://faostat.fao.org.

FAO (Food and Agricultural Organization of the United Nations), 2008 statistics. http://faostat.fao.org.

Fidan, H., and Baloglu, S. (2009a). First report of *Onion yellow dwarf virus* and *Leek yellow stripe virus* in Garlic in Turkey. *Plant Dis.* **93:**672.

Fidan, H., and Baloglu, S. (2009b). First report of *Garlic common latent virus* in garlic in turkey. *J. Plant Pathol.* **91**(4 Suppl.)**:**S4.99.

Gaspar, J. O., Belintani, P., Almeida, A. M. R., and Kitajima, E. W. (2008). A degenerate primer allows amplification of part of the 3′-terminus of three distinct carlavirus species. *J. Virol. Methods* **148:**283–285.

Gent, D. H., Schwartz, H. F., and Khosla, R. (2004a). Distribution and incidence of *Iris yellow spot virus* in Colorado and its relation to onion plant population and yield. *Plant Dis.* **88:**446–452.

Gent, D. H., Schwartz, H. F., and Khosla, R. (2004b). Managing *Iris yellow spot virus* of onion with cultural practices, host genotype, and novel chemical treatments. *Phytopathology* **94:** S34.

Gent, D. H., du Toit, L. J., Fichner, S. F., Mohan, S. K., Pappu, H. R., and Schwartz, H. F. (2006). *Iris yellow spot virus*: An emerging threat to onion bulb and seed production. *Plant Dis.* **90:**1468–1480.

Gera, A., Lesemann, D. E., Cohen, J., Franck, A., Levy, S., and Salomon, R. (1997). The natural occurrence of *turnip mosaic potyvirus* in *Allium ampeloprasum*. *J. Phytopathol.* **145:**289–293.

Gera, A., Cohen, J., Salomon, R., and Raccah, B. (1998a). *Iris yellow spot tospovirus* detected in onion (*Allium cepa*) in Israel. *Plant Dis.* **82:**127.

Gera, A., Kritzman, A., Cohen, J., and Raccah, B. (1998b). Tospoviruses infecting bulb crops in Israel. *In* "Recent Progress in Tospovirus and Thrips Research" (D. Peters and R. Goldbach, eds.), Fourth International Symposium on Tospoviruses and Thrips in Floral and Vegetable Crops, pp. 86–87. Wageningen, the Netherlands.

Gieck, S. L., Hamm, P. B., David, N. L., and Pappu, H. R. (2009). First report of *Garlic virus B* and *Garlic virus D* in Garlic in the Pacific Northwest. *Plant Dis.* **93**:431.

Graichen, K., and Leistner, H. U. (1987). Zwiebelgelbstreifen-Virus (*onion yellow dwarf virus*) verursacht Knoblauchmosaik. *Arch. Phytopathol. Pflanzensch.* **23**:165–168.

Grancini, P. (1951). Malattie da virus degli orteiggi il mosaic della cipolla. *Flora* **6**:19.

Haque, M. S., Hattori, K., Suzuki, A., and Tsuneyoshi, T. (2007). An efficient novel method of producing virus free plants from garlic root meristem. Proceedings of the 11th IAPTC&B Congress, August 13–18, 2006, Beijing, China, Biotechology and Sustainable Agriculture 2006 and Beyond, Vol. 2, pp. 107–110. doi: 10.1007/978-1-4020-6635-1_11.

Helguera, M., Bravo-Almonacid, F., Kobayashi, K., Rabinowicz, P. D., Conci, V. C., and Mentaberry, A. (1997a). Immunological detection of a GarV-type virus in Argentine garlic cultivars. *Plant Dis.* **81**:1005–1010.

Helguera, M., Lunello, P., Nome, C., and Conci, V. C. (1997b). Advances in the purification of filamentous viruses from garlic and in antisera production. *Acta Hortic.* **433**:623–630.

Hoa, N. V., Ahlawat, Y. S., and Pant, R. P. (2003). Partial characterization of *Onion yellow dwarf virus* from onion in India. *Indian Phytopathol.* **56**:276–282.

Huchette, O., Bellamy, C., Filomenko, R., Pouleau, B., Seddas, S., and Pappu, H. R. (2008). *Iris yellow spot virus* on shallot and onion in France. *Plant Health Prog.* doi: 10.1094/PHP 2008-0610-01-BR. (online).

Huttinga, H. (1975). Purification by molecular sieving of a leek virus related to onion yellow dwarf virus. *Neth. J. Plant Pathol.* **81**:81–83.

Kang, S. G., Koo, B. J., Lee, T., and Chang, M. U. (2007). Allexivirus transmitted by eriophyid mites in garlic plants. *J. Microbiol. Biotechnol.* **17**:1833–1840.

Kanyuka, K. V., Vishnichenko, V. K., Levay, K. E., Kondrikov, E. V., Ryabov, E. V., and Zavriev, S. K. (1992). Nucleotide sequence of *shallot virus X* RNA reveals a 5′-proximal cistron closely related to those of the potexviruses and a unique arrangement of the 3′-proximal cistrons. *J. Gen. Virol.* **73**:2553–2560.

Kim, H. H., Lee, J. K., Hwang, H. S., and Engelmann, F. (2007). Cryopreservation of garlic germplasm collections using the droplet-vitrification technique. *Cryo Lett.* **28**:471–482.

Klukakova, J., Navratil, M., and Duchoslav, M. (2007). Natural infection of garlic (*Allium sativum* L.) by viruses in the Czech Republic. *J. Plant Dis. Prot.* **114**:97–100.

Koch, M., and Salomon, R. (1994). Serological detection of *onion yellow dwarf virus* in garlic. *Plant Dis.* **78**:785–788.

Kondo, T., Hasegawa, H., and Suzuki, M. (2000). Transformation and regeneration of garlic (*Allium sativum* L.) by Agrobacterium-mediated gene transfer. *Plant Cell Rep.* **19**:989–993.

Koo, B. J., Kang, S.-G., and Chang, M. U. (2002). Survey of garlic virus disease and phylogenetic characterization of garlic viruses of the genus *Allexivirus* isolated in Korea. *Plant Pathol. J.* **18**:237–243.

Korkmaz, S., and Cevik, B. (2009). *Leek yellow stripe virus* newly reported in Turkey. *Plant Pathol.* **58**:787.

Kritzman, A., Lampel, M., Raccah, B., and Gera, A. (2001). Distribution and transmission of *Iris yellow spot virus*. *Plant Dis.* **85**:838–842.

Kumar, N. K. K., and Rawal, R. D. (1999). Onion thrips, *Thrips tabaci*, a vector of onion tospovirus. *Insect Environ.* **5**:52.

Kumar, S., Baranwal, V. K., Joshi, S., Arya, M., and Majunder, S. (2010). Simultaneous detection of mixed infection of *Onion yellow dwarf virus* and an allexivirus in RT-PCR for ensuring virus free onion bulbs. *Indian J. Virol.* **21**:64–68.

Lee, Y. W., Yamazaki, S., Osaki, T., and Inouye, T. (1979). Two elongated viruses in garlic, garlic latent virus and garlic mosaic virus. *Ann. Phytopathol. Soc. Jap.* **45**:727–734.

Leinhos, G., Muller, J., Heupel, M., and Krauthausen, H. J. (2007). *Iris yellow spot virus* an Bund- und Speisezwiebeln-erster Nachweis in Deutschland. *Nachrichtenbl. Deut. Pflanzenschuttzd* **59**:310–312.

Leisova-Svobodova, L., and Karlova-Smekalova, K. (2011). Detection of garlic viruses using SYBR Green real-time reverse transcription-polymerase chain reaction. *J. Phytopathol.* **159:**429–434.

Lot, H., Dellecolle, B., Boccardo, G., Marzachi, C., and Milne, R. G. (1994). Partial characterization of reovirus-like particles associated with garlic dwarf disease. *Plant Pathol.* **43:**537–546.

Lot, H., Rubino, L., Delecolle, B., Jacquemond, M., Turturo, C., and Russo, M. (1996). Characterization, nucleotide sequence and genome organization of *leek white stripe virus*, a putative new species of the genus *Necrovirus. Arch. Virol.* **141:**2375–2386.

Lot, H., Chevelon, V., Souche, S., and Dellecolle, B. (1998). Effects of *Onion yellow dwarf virus* and *Leek yellow dwarf virus* on symptomatology and yield loss of three French garlic cultivars. *Plant Dis.* **82:**1381–1385.

Lu, Y.-W., Chen, J., Zheng, H.-Y., Adams, M. J., and Chen, J.-P. (2008). Serological relationships among the overexpressed coat proteins of allexiviruses. *J. Phytopathol.* **156:**251–255.

Luciani, G. F., Cafrune, E. E., Curvetto, N. R., and Conci, V. C. (1998). Obtencion de plantas de ajo colorado (*Allium sativum* L.) libres de virus. *Fitopatologia* **33:**165–169.

Lunello, P., Ducasse, D. A., Helguera, M., Nome, S. F., and Conci, V. C. (2002). An Argentinean isolate of *leek yellow stripe virus* from leek can be transmitted to garlic. *J. Plant Pathol.* **84:**11–17.

Lunello, P., Mansilla, C., Conci, V., and Ponz, F. (2004). Ultra-sensitive detection of two garlic potyviruses using a real-time fluorescent (Taqman) RT-PCR assay. *J. Virol. Methods* **118:**15–21.

Lunello, P., Ducasse, D., and Conci, V. (2005). Improved PCR detection of potyviruses in *Allium* species. *Eur. J. Plant Pathol.* **112:**371–378.

Lunello, P., Di Rienzo, J., and Conci, V. (2007). Yield loss in garlic caused by *Leek yellow stripe virus* Argentinean isolate. *Plant Dis.* **91:**153–158.

Ma, Y., Wang, H. I., Zhang, C. J., and Kang, Y. Q. (1994). High rate of virus-free plantlet regeneration via garlic scape-tip culture. *Plant Cell Rep.* **14:**65–68.

Mahmoud, S. Y. M., Abo-El Maaty, S. A., El-Borollosy, A. M., and Abdel-Ghaffar, M. H. (2007). Identification of *Onion yellow dwarf potyvirus* as one of the major viruses infecting garlic in Egypt. *Am. Eurasian J. Agric. Environ. Sci.* **2:**746–755.

Majumder, S., and Baranwal, V. K. (2009). First report of *Garlic common latent virus* in garlic from India. *Plant Dis.* **93:**106.

Majumder, S., Baranwal, V. K., and Joshi, S. (2008). Simultaneous detection of *Onion yellow dwarf virus* and *Shallot latent virus* in infected leaves and cloves of garlic by duplex RT-PCR. *J. Plant Pathol.* **90:**371–374.

Maliogka, V. I., Dovas, C. I., Lesemann, D. E., Winter, S., and Katis, N. I. (2006). Molecular identification, reverse transcription-polymerase chain reaction detection, host reactions, and specific cytopathology of *Artichoke yellow ringspot virus* infecting onion crops. *Phytopathology* **96:**622–629.

Marrou, J., Leclant, F., and Leroux, J. P. (1972). Epidemiologie du virus de la mosa'ique de l'ail. *In* Actas do 1II Congresso da Uniao Fitopatologica Mediterranea, 22–28 Outubro, pp. 53–55. Oeiras, Portugal.

Matthieu, J. L., Meurens, M., Ceustermans, N., Benoit, M., and Verhoyen, M. (1984). Influence of the epicuticular wax layer on the susceptibility of different leek varieties against *leek chlorotic streak virus. Meded. Fac. Landbouw. Rijks.* **36:**433–441.

Mavrič, I., and Ravnikar, M. (2000). *Iris yellow spot tospovirus* in Slovenia. *In* Proceedings of the 5th Congress of the European Foundation for Plant Pathology: Biodiversity in Plant Pathology, pp. 223–225. Taormina-Giardini, Naxos.

Mavrič, I., and Ravnikar, M. (2005). A carlavirus serologically closely related to Carnation latent virus in Slovenian garlic. *Acta Agric. Slov.* **85**(2)**:**343–349.

Melhus, I. E., Reddy, C., Shenderson, W. J., and Vestal, E. (1929). A new virus disease epidemic on onions. *Phytopathology* **19:**73–77.

Melo Filho, P., Nagata, T., Dusi, A. N., Buso, J., Torres, A. C., Eiras, M., and Resende, R. O. (2004). Detection of three *Allexivirus* species infecting garlic in Brazil. *Pesqui. Agropecu. Bras.* **39:**735–740.

Melo Filho, P., Resende, R. O., Torres Cordeiro, C. M., Buso, J., Torres, A. C., and Dusi, A. N. (2006). Viral reinfection affecting bulb production in garlic after seven years of cultivation under field condtions. *Eur. J. Plant Pathol.* **116:**95–101.

Messiaen, C. M., and Marrou, J. (1965). Selection sanitaire de Fail: Deux solutions possibles au probleme de la mosaique de l'ail, plantes sensibles saines, ou plantes virosees tolerantes. C.R. 1eres Journees de Phytiatrie et de Phytopharmacie Circummediterraneenne, Marseille, pp. 204–207.

Messiaen, C. M., Youcef-Benkada, M., and Beyries, A. (1981). Potential yield and tolerance to virus disease in garlic (*Allium sativum* L.). *Agronomie* **1:**759–762.

Messiaen, C. M., Lot, H., and Delecolle, B. (1994). Thirty years of France' experience in the production of disease-free garlic and shallot mother bulbs. *Acta Hortic.* **358:**275–279.

Mituti, T., Marubayashi, J. M., Moura, M. F., Krause-Sakate, R., and Pavan, M. A. (2011). First report of Shallot *latent virus* in garlic in Brazil. *Plant Dis.* **95:**227.

Mohamed, N. A., and Young, B. R. (1981). *Garlic yellow streak virus*, a potyvirus infecting garlic in New Zealand. *Ann. Appl. Biol.* **97:**65–74.

Mohammad G., Kawas, H., and Al-Safadi, B. (2007). Survey of garlic viruses in southern Syria. http://www.damascusuniversity.sy/mag/farm/images/stories/2550.pdf.

Mullis, S. W., Langston, D. B., Gilaitis, R. D., Sherwood, J. L., Csinos, A. C., Sparks, A. N., Torrance, R. L., and Cook, M. J. (2004). First report of Vidalia onion (*Allium cepa*) naturally infected with *Tomato spotted wilt virus* and *Iris yellow spot virus* (Family Bunyaviridae, genus *Tospovirus*) in Georgia. *Plant Dis.* **88:**1285.

Mullis, S. W., Gitaitis, R. D., Nischwitz, C., Csinos, A. S., Rafael Mallaupoma, Z. C., and Inguil Rojas, E. H. (2006). First report of onion (*Allium cepa*) naturally infected with *Iris yellow spot virus* in Peru. *Plant Dis.* **90:**377.

Mumford, R. A., Barker, I., and Wood, K. R. (1996). The biology of the tospoviruses. *Ann. Appl. Biol.* **128:**159–183.

Nagakubo, T., Kubo, M., and Oeda, K. (1994). Nucleotide sequences of the 30 region of two major viruses from mosaic-diseased garlic: Molecular evidence of mixed infection by a potyvirus and a carlavirus. *Phytopathology* **84:**640–645.

Nagata, T., Almedia, A. L., de Resende, R., and de Avila, C. (1999). The identification of the vector species of *Iris yellow spot tospovirus* occurring in onion in Brazil. *Plant Dis.* **83:**399.

Nieto, A. M., Conci, V. C., and Conci, L. R. (2004). Nucleotide sequence of the 3′ region of a carlavirus from Argentinian garlic mosaic. 4th International ISHS Symposium on Edible Alliaceae (ISEA), Bejing, China, Abstracts, p. 103.

Paludan, N. (1980). Virus attack on leek: survey, diagnosis, tolerance of varieties and winterhardiness. *Tidsskr. Planteavl* **84:**371–385.

Pappu, H. R., Hellier, B. C., and Dugan, F. M. (2008). Evaluation of the national plant germplasm system's garlic collection for seven viruses. *Plant Health Prog.* doi: 10.1094/PHP-2008-0919-01-RS. (online).

Park, K.-S., Bae, Y.-J., Jung, E.-J., and Kang, S.-J. (2005). RT-PCR-based detection of six garlic viruses and their phylogenetic relationships. *J. Microbiol. Biotechnol.* **15:**1110–1114.

Peiwen, X., Huisheng, S., Ruijie, S., and Yuanjun, Y. (1994). Strategy for the use of virus-free seed garlic in field production. *Acta Hortic.* **358:**307–311.

Pena-Iglesias, A., and Ayuso, P. (1982). Characterization of Spanish garlic viruses and their elimination by *in vitro* shoot apex culture. *Acta Hortic.* **127:**183–193.

Perez-Egusquiza, Z., Ward, L. I., Clover, G. R. G., Fletcher, J. D., and van der Vlugt, R. A. A. (2008). First report of *Shallot virus X* in shallot in New Zealand. *New Dis. Rep.* **18:**29.

Perotto, M. C., Conci, V. C., Cafrune, E. E., Alochis, P., and Bracamonte, R. (2003). Differences in the response of garlic cultivars to the eradication of five viruses. *Phyton* **126:**489–495.

Perotto, M. C., Cafrune, E. E., and Conci, V. C. (2010). The effect of additional viral infections on garlic plants initially infected with *Allexiviruses*. *Eur. J. Plant Pathol.* **126:**489–495.
Perring, T. M., Gruenhagen, N. M., and Farrar, C. A. (1999). Management of plant virus diseases through chemical control of insect vectors. *Annu. Rev. Entomol.* **44:**457–481.
Plenk, A., and Grausgruber-Gröger, S. (2011). First report of *Iris yellow spot virus* in onions (*Allium cepa*) in Austria. *New Dis. Rep.* **23:**13. doi: 10.5197/j.2044-0588.2011.023.013.
Pozzer, L., Bezerra, I. C., Kormelink, R., Prins, M., Peters, D., de Resende, R. O., and de Avila, A. C. (1999). Characterization of a tospovirus isolate of *Iris yellow spot virus* associated with a disease in onion fields in Brazil. *Plant Dis* **83:**345–350.
Pringle, C. R. (1999). Virus taxonomy at the XIth International congress of Virology in Sydney, Australia. *Arch. Virol.* **144:**2065–2070.
Rana, G. L., Rosciglione, B., and Cannizzaro, G. (1978). La maculatura anulare gialla del cardo e del carciofo. *Phytopathol. Mediterr.* **17:**63–64.
Rana, G. L., Kyriakopoulou, P. E., and Martelli, G. P. (1983). *Artichoke yellow ringspot virus.* CMI/AAB Descriptions of Plant VirusesCommonw. Mycol. Inst./Assoc. Appl. Biol, Kew, England, No. 271.
Ravi, K. S., Kitkaru, A. S., and Winter, S. (2006). *Iris yellow spot virus* in onions: A new tospovirus record from India. *Plant Pathol.* **55:**288.
Rongchang, C., Xiaolong, L., and Xuezhan, L. (1992). Evaluation of mosaic virus and meristem tissue culture for garlic (In Chinese). *Chin. Vegetables* **52:**10–14.
Rosales, M., Pappu, H. R., Lopez, L., Mora, R., and Aljaro, A. (2005). *Iris yellow spot virus* in onion in Chile. *Plant Dis.* **89:**1245.
Ryabov, E. V., Generozov, E. V., Vetten, H. J., and Zavriev, S. K. (1996). Analysis of the 3′-region of the mite born filamentous virus genome testifies its relation to the shallot virus X group. *Mol. Biol. (Moscow)* **30:**103–110.
Sako, I. (1989). Occurrence of *garlic latent virus* in *Allium* species. *Plant Prot.* **43:**389–392.
Sako, I., Taniguchi, T., Osaki, T., and Inouye, T. (1990). Transmission and translocation of *garlic latent virus* in rakkyo (*Allium chinense* G. Don). *Proc. Kansai Plant Prot. Soc.* **32:**21–27.
Salomon, R. (2002). Virus Diseases in Garlic and the Propagation of virus-free plants. *In* "Allium Crop Science: Recent Advances" (H. D. Currah Rabinowitch and L. Currah, eds.), CABI Publishing, New York, USA.
Salomon, R., Koch, M., Levy, S., and Gal-On, A. (1996). Detection and identification of the viruses forming mixed infection in garlic. *In* Symposium Proceedings No. 65: Diagnostics in Crop production, pp. 193–198. British Crop Protection Council, Farnham, UK.
Sawahel, W. A. (2002). Stable genetic transformation of garlic plants using particle bombardment. *Cell. Mol. Biol. Lett.* **7:**49–59.
Schwartz, H. F., Brown, W. M., Jr., Blunt, T., and Gent, D. H. (2002). *Iris yellow spot virus* on onion in Colorado. *Plant Dis.* **86:**560.
Schwartz, H. F., Mohan, S. K., Havey, M. J., and Crowe, F. J. (2006). Introduction. *In* "Compendium of Onion and Garlic Diseases and Pests" (H. F. Schwartz and S. K. Mohan, eds.). 2nd edn. American Phytopathological Society, St. Paul, MN.
Shahraeen, N., Lesemann, D. E., and Ghotbi, T. (2008). Survey for viruses infecting onion, garlic and leek crops in Iran. *OEPP/EPPO Bull.* **38:**131–135.
Shiboleth, Y. M. (1998). Molecular diagnosis of garlic (*Allium sativum* L.) viruses in Israel and evaluation of tissue culture methods for their elimination. MSc thesis, The Hebrew University of Jerusalem, Faculty of Agricultural, Food and Environmental Sciences, Rehovot, Israel.
Shiboleth, Y. M., Gal-On, A., Levy, S., Koch, M., Rabinowitch, H. D., and Salomon, R. (1997). Identification of viruses in garlic (*Allium sativum* L.) and closely related *Allium* species grown in Israel. *In* Proceedings of the 10th congress of the Mediterranean Phytopathological union, 1–5 June, Montpellier, France, pp. 313–317.

Shiboleth, Y. M., Gal-On, A., Koch, M., Rabinowitch, H. D., and Salomon, R. (2001). Molecular characterization of *Onion yellow dwarf virus* (OYDV) infecting garlic (*Allium sativum* L.) in Israel: Thermotherapy inhibits virus elimination by meristem tip culture. *Ann. Appl. Biol.* **138:**187–195.

Shukla, D. D., Strike, P. M., Tracy, S. L., Gough, K. H., and Ward, C. W. (1988). The N and C termini of the coat proteins of potyviruses are surface-located and the N terminus contains the major virus specific epitopes. *J. Gen. Virol.* **69:**1497–1508.

Song, S., Song, J., Chang, M., Lee, J., and Choi, Y. (1997). Identification of one of the major viruses infecting garlic plants, garlic virus X. *Mol. Cell* **7:**705–709.

Song, S. I., Song, J. T., Kim, C. H., Lee, J. S., and Choi, Y. D. (1998). Molecular characterization of the *garlic virus X* genome. *J. Gen. Virol.* **79:**155–159.

Song, S. I., Choi, J. N., Song, J. T., Ahn, J. H., Lee, J. S., Kim, M., Cheong, J. J., and Choi, Y. D. (2002). Complete genome sequence of *garlic latent virus*, a member of the carlavirus family. *Mol. Cells* **14:**205–213.

Stefanac, Z. (1980). *Cucumber mosaic virus* in garlic. *Acta Bot. Croat.* **39:**21.

Stefanac, Z., and Plese, N. (1980). *Turnip mosaic virus* in two Mediterranean *Allium* species. *In* Proceedings of the Fifth Congress of the Mediterranean Phytopathological Union, 21–27 September, Patras, Greece, pp. 37–38.

Sumi, S., Tsuneyoshi, T., and Furutani, H. (1993). Novel rod-shaped viruses isolated from garlic, *Allium sativum*, possessing a unique genome organization. *J. Gen. Virol.* **74:**1879–1885.

Sumi, S., Matsumi, T., and Tsuneyoshi, T. (1999). Complete nucleotide sequences of garlic viruses A and C, members of the newly ratified genus *Allexivirus. Arch. Virol.* **144:**1819–1826.

Sumi, S., Tsuneyoshi, T., Suzuki, A., and Ayabe, M. (2001). Development and establishment of practical tissue culture methods for production of virus-free garlic seed bulbs, a novel field cultivation system and convenient methods for detecting garlic infecting viruses. *Plant Biotechnol.* **18:**179–190.

Takaichi, M., Yamamoto, M., Nagacubo, T., and Oeda, K. (1998). Four garlic viruses identified by the reverse transcription-polymerase chain reaction and their regional distribution in northern Japan. *Plant Dis.* **82:**694–698.

Tomassoli, L., Tiberini, A., Masenga, V., Vicchi, V., and Turina, M. (2009). Characterization of *Iris yellow spot virus* isolates from onion crops in Northern Italy. *J. Plant Pathol.* **91:**733–739.

Torres, A. C., Fajardo, T. V., Dusi, A. N., Resende, R. O., and Buso, J. A. (2000). Shoot tip culture and thermotherapy in recovering virus free plants of garlic. *Hortic. Bras.* **18:**192–195.

Torrico, A. K., Cafrune, E. E., and Conci, V. C. (2010). First report of *Shallot latent virus* in Garlic in Argentina. *Plant Dis.* **94:**915.

Tsuneyoshi, T., and Sumi, S. (1996). Differentiation among garlic viruses in mixed infections based on RT-PCR procedures and direct tissue bloting immunoassays. *Phytopathology* **86:**253–259.

Tsuneyoshi, T., Matsumi, T., Deng, T. C., Sako, I., and Sumi, S. (1998a). Differentiation of *Allium* carlaviruses isolated from different parts of the world based on the viral coat protein sequence. *Arch. Virol.* **143:**1093–1107.

Tsuneyoshi, T., Matsumi, T., Natsuaki, K. T., and Sumi, S. (1998b). Nucleotide sequence analysis of virus isolates indicates the presence of three potyvirus species in *Allium* plants. *Arch. Virol.* **143:**97–113.

Ucman, R., Zel, J., and Ravnikar, M. (1998). Thermotherapy in virus elimination from garlic: Influence on shoot multiplication from meristems and bulb formation *in vitro*. *Sci. Hortic.* **73:**193–202.

Van der Vlugt, R. A. A., Steffens, P., Cuperus, C., Barg, E., Lesemann, D. F., Bos, L., and Vetten, H. J. (1999). Further evidence that *shallot yellow stripe virus* (SYSV) is a distinct

potyvirus and reidentification of Welsh onion yellow stripe virus as SYSV strain. *Phytopathology* **89:**148–155.
Van Dijk, P. (1993a). Survey and characterization of potyviruses and their strains of *Allium* species. *Neth. J. Plant Pathol.* **99**(Suppl. 2)**:**1–48.
Van Dijk, P. (1993b). Carlavirus isolates from cultivated *Allium* species represent three viruses. *Neth. J. Plant Pathol.* **99:**233–257.
Van Dijk, P. (1994). Virus diseases of *Allium* species and prospects for their control. *Acta Hortic.* **358:**299–306.
Van Dijk, P., and Van der Vlugt, R. A. A. (1994). New mite-borne virus isolates from rakkyo, shallot and wild leek species. *Eur. J. Plant Pathol.* **100:**269–277.
Van Dijk, P., Verbeek, M., and Bos, L. (1991). Mite-borne virus isolates from cultivated *Allium* species, and their classification into two new rymoviruses in the family *Potyviridae*. *Neth. J. Plant Pathol.* **97:**381–399.
Verbeek, M., Van Dijk, P., and Well, P. M. A. (1995). Efficiency of eradication of four viruses from garlic (*Allium sativum*) by meristem –tip culture. *Eur. J. Plant Pathol.* **101:**231–239.
Verhoyen, M., and Horvat, F. (1973). La striure chlorotique du porreau. 1. Identification de l'agent causal. *Parasitica* **29:**16–28.
Vishnichenko, V. K., Konareva, T. N., and Zavriev, S. K. (1993). A new filamentous virus in shallot. *Plant Pathol.* **42:**121–126.
Vishnichenko, V. K., Stelmashchuk, V. Y., and Zavriev, S. K. (1996). Cloning of full-length cDNA of the Shallot virus X genome and infectivity of its transcripts in sugar beet protoplasts. *Mol. Biol. (Russia)* **30:**959.
Walkey, D. G. A. (1990). Virus diseases. *In* "Onion and allied crops" (H. D. Rabinowitch and J.-L. Brewster, eds.), Vol II, pp. 191–212. CRC Press, Inc, Boca Raton, FL.
Walkey, D. G. A., and Antill, D. N. (1989). Agronomic evaluation of virus-free and virus infected garlic (*Allium sativum* L.). *J. Hortic. Sci.* **64:**53–60.
Walkey, D. G. A., Webb, M. J. W., Bolland, C. J., and Miller, A. (1987). Production of virus-free garlic (*Allium sativum* L.) and shallot (*A. ascalonicum* L.) by meristem-tip culture. *J. Hortic. Sci.* **62:**211–220.
Wei, T., Pearson, M. N., Blohm, D., Nolte, M., and Armstrong, K. (2009). Development of a short oligonucleotide microarray for the detection and identification of multiple potyviruses. *J. Virol. Methods* **162:**109–118.
Yamashita, K., Sakai, J., and Hanada, K. (1995). *Leek yellow stripe virus* (LYSV) isolated from garlic and its relationship to garlic mosaic (GMV). *Ann. Phytopathol. Soc. Jap.* **61:**273–278.
Yamashita, K., Sakai, J., and Hanada, K. (1996). Characterization of a new virus from garlic (*Allium sativum* L.), garlic mite-borne mosaic virus. *Ann. Phytopathol. Soc. Jap.* **62:**483–489.
Zavriev, S. K. (2008). *Allexivirus*. *In* "Encyclopedia of Virology" (B. W. J. Mahy and M. H. V. Van Regenmortel, eds.). Elsevier, Amsterdam, The Netherlands, pp. 96–98.

CHAPTER 6

Viruses of Potato

Gad Loebenstein and **Victor Gaba**

Contents

Abstract

Potatoes are an important crop in Mediterranean countries both for local consumption and for export to other countries, mainly during the winter. Many Mediterranean countries import certified seed potato in addition to their own seed production. The local seeds are mainly used for planting in the autumn and winter, while the imported seed are used for early and late spring plantings. *Potato virus Y* is the most important virus in Mediterranean countries, present mainly in the autumn plantings. The second important

Department of Virology, Agricultural Research Organization, Bet Dagan, Israel

Advances in Virus Research, Volume 84
ISSN 0065-3527, DOI: 10.1016/B978-0-12-394314-9.00006-3

virus is *Potato leafroll virus*, though in recent years its importance seems to be decreasing. *Potato virus X*, *Potato virus A*, *Potato virus S*, *Potato virus M*, and the viroid, Potato spindle tuber viroid, were also recorded in several Mediterranean countries. For each virus the main strains, transmission, characterization of the virus particle, its genome organization, detection, and control methods including transgenic approaches will be discussed.

I. INTRODUCTION

Potato is the fourth most important food crop in the world, with an annual production of about 300 million tons. Potato (*Solanum tuberosum*) originated in the highlands of South America, where it has been consumed. Spanish explorers brought the plant to Europe in the late sixteenth century. Potatoes were sold in Seville, Spain, as early as 1573. Carmelite friars took the potato from Spain into Italy and from these Mediterranean countries it spreads to the rest of Europe. By the nineteenth century, it had spread throughout the continent, providing cheap and abundant food. The "degeneration" disease of potato crops over successive vegetative generations, known in Europe since the eighteenth century was recognized by Salaman (1921, 1955) to be the result of virus infection.

In Germany, Appel (1906) first described potato leafroll, without, however, identifying its cause. Quanjer *et al*. (1916) showed that leafroll was an infectious agent, and aphids were later identified as its vector (Schultz and Folsom, 1923). Pioneering research on the application of serological methods for detection of potato viruses was first conducted in the Netherlands (van Slogteren, 1955). Antisera to *Potato virus X* (PVX) and *Potato virus S* (PVS) were produced for testing about one million potato plants annually intended for production of virus-free seed potatoes. A major advance in potato virus detection occurred with the application of the robust enzyme-linked immunosorbent assay (ELISA), enabling detection of as little as 10–100 ng/ml virus. Meristem cultures to obtain virus-free potato plants became a standard procedure (Stace-Smith and Mellor, 1968), after the pioneering work of Morel and Martin (1952) with dahlia plants. During the past decades, transgenic plants resistant to viruses have been produced. These and additional data on historical perspectives of potato virology were summarized by Lawson and Stace-Smith (2001).

Many Mediterranean countries import certified seed potato in addition to their own seed production (Table I). The local seeds are mainly used for planting in the autumn and winter, while the imported seed are used for early and late spring plantings. In Turkey, however, imported seeds are either directly used for planting (which occurs rarely), or multiplied

TABLE I Production of ware, local, and imported seed potatoes (in Tons) in some Mediterranean countries in 2008

Country	Ware potatoes[a]	Local seed potatoes[a]	Imported seed potatoes: From UK[b]	Imported seed potatoes: From Netherlands[b]
Algeria	1,800,000	96,300	120	81,435
Cyprus	131,695	12,042	100	8254
Egypt	3,567,050	200,000	29,914	57,259
Greece	848,000	90,000	2585	14,750
Israel	557,917	6000	200	16,537
Italy	1,603,828	185,000	1509	57,786
Jordan	139,787	14,607	50	2333
Lebanon	514,600	20,100	177	16,112
Morocco	1,536,560	106,760	101	5255
Palestinian authority	62,840	4872	1210	81,436
Spain	2,365,400	14,000	13,460	43,294
Syria	570,128	37,298	29,915	57,259
Tunisia	370,000	24,800	–	13,352
Turkey	4,225,168	306,000	13,460	43,294

[a] From FAOstat (2010) for year 2009.
[b] From Eurostat (2010) for year 2009.

according to the national certification scheme. Spain also produced seed potato (37,300 tons in 2009), the majority in the Castilla y León region, mainly local bred cultivars. Some potato seed facilities used micropropagation and molecular testing for viruses (Isla *et al.*, 2008). In Greece, certified potato seed is produced in 11 delimited areas scattered in isolated areas with different climatic conditions, either in mountains or in an island (Panayotou and Katis, 1996). In Cyprus, the first stages for local potato tuber seed production from *in vitro* culture of apical stems were made. Minitubers were produced for three generations after which seed potatoes were given to growers for one generation multiplication (Minas *et al.*, 2007). Most Mediterranean countries do not produce certified seed, and attempts during the past decades to become more self-sufficient for producing seed potatoes were only limited. This is apparently due to climatic conditions, use of specific cultivars protected by breeder's rights, and the growing of potatoes in 2–3 seasons (spring, autumn, and winter) where seed tubers—out of dormancy and not too old, cannot be obtained locally. Therefore, certification schemes over 6–7 generations, as practiced in Northern European countries, starting from pre-nuclear and first year clones, with 4 additional generations to reach elite seed potato grade, followed by class A seed potato—altogether seven yearly multiplication generations—are not applicable in Mediterranean countries. In warmer climates where potatoes are grown for at least two seasons a year, a dormancy period of the potato tuber for at least 3–4 months does not enable the operation of the European certification scheme.

Nevertheless, viruses, especially *Potato virus Y* (PVY) and *Potato leafroll virus* (PLRV) can be a problem in the second crop after planting certified seed (generally the autumn or winter planting seasons) if infected plants remain in the field or in neighboring plots.

In the Basque Country and neighboring areas, extraordinarily high rates of PVY infection have been disrupting the seed potato industry. Tuber infection did increase from about 1% (basic seed grade) to well over 10% (ware grade), in one single cycle of field multiplication (Legorburu *et al.*, 1996).

II. THE MAIN VIRUSES

A. *Potato virus Y*: Genus *Potyvirus*

PVY, the type member of the *Potyvirus*, is the most important of the three potyviruses infecting potatoes. Potato isolates have been divided into three main strains according to the symptoms they induce on potato and tobacco (De Bokx and Huttinga, 1981). The PVY^{N} strain induces typical "tobacco veinal necrosis" on *Nicotiana tabacum*, whereas PVY^{O} induces a nonnecrotic mosaic on this host. Some PVY^{N} isolates cause the potato tuber necrotic disease (Le Romancer and Kerlan, 1991), and are referred to as PVY^{NTN}. PVY^{C} comprises isolates that induce "stipple streak" symptoms on potato cultivars bearing the *Nc* gene. In *N. tabacum*, these isolates induce symptoms similar to those of the PVY^{O} strain. In many cultivars, the primary symptoms of PVY^{O} and PVY^{C} are leaf mottling and necrosis, and secondary symptoms are leaf mottling and crinkling, and plant dwarfing. PVY^{N} causes mosaic primary symptoms, and, usually, leaf mottling and mosaic secondary symptoms. Infected plants are more severely affected if PVY occurs in complex with other viruses, especially PVX. Yield losses of 10–80% have been reported (de Bokx and Huttinga, 1981).

PVY^{NTN} causes a damaging disease, designated potato tuber necrotic ringspot disease, in which tubers develop superficial rings that initially are raised but later are sunken and necrotic and often become more conspicuous during storage; such symptoms can develop in as much as 90% of susceptible cultivars (Le Romancer and Nedelec, 1997). The cylindrical cytoplasmic inclusions (CIs) of PVY isolates are of Edwardson's Division IV.

PVY has a wide natural host range; it occurs naturally in at least 43 species of four families (Brunt, 2001). Various strains occur in potato worldwide and, in some Mediterranean, tropical, and subtropical countries, in pepper, tobacco, tomato, and several other solanaceous species.

1. Transmission

More than 40 aphid species are known to transmit PVY (Kennedy *et al.*, 1962; Sigvald, 1984), but they vary in efficiency. The more important vectors are probably those that also colonize potatoes such as *Macrosiphum euphorbiae*, *Aphis fabae*, *Myzus persicae*, and *Rhopalosiphoninus latysiphon*; however, others such as *Myzus certus*, *Phorodon humuli*, and *Rhopalosiphum insertum* are also important natural vectors (Kennedy *et al.*, 1962). Aphid transmissibility is dependent on virus-encoded helper component proteinase and a DAG triplet in the coat protein.

2. PVY in Mediterranean countries

PVY^{O} occurs worldwide. PVY^{N} occurs in Europe, Africa, New Zealand, and S. America. PVY^{C} occurs in Australia, Ecuador, Europe, North America, South Africa, and New Zealand. PVY^{NTN} has been reported from Europe. In the Mediterranean basin, PVY^{NTN} was reported from Italy (Tomassoli and Lumia, 1998), Israel, and Lebanon (Boonham *et al.*, 2002; Jeffries, 1998). PVY^{O} and PVY^{N} were found in Spain (Blanco-Urgoiti *et al.*, 1996). In Greece, in 2006 high infection rates by PVY were observed in both noncertified seeds and ware potatoes, at rates up to 100% of the suspected plants or tubers. Autumn potatoes were more infected than the spring crop (Chatzivassiliou *et al.*, 2008). PVY^{NTN}, causing potato tuber necrotic ringspot disease, has also been recorded from Greece (Bem *et al.*, 1999). Some isolates belonged to the Wilga type (Varveri, 2006). PVY^{N}, PVY^{O}, and PVY^{C} were also observed in Turkey (Bostan and Haliloglu, 2004). PVY isolates were also found in Syria. Recombination analysis grouped isolates of PVY^{SYR} into three recombination patterns, SYR-I, SYR-II, and SYR-III, which varied in the first 700 nucleotides of their genomes, with the second recombination pattern, SYR-II, the most frequent. PVY^{SYR} isolates shared highest genomic identity and close phylogenetic relationships with PVY^{NTN} and $PVY^{N}W$, suggesting a common origin and local emergence of these isolates in Syria (Chikh Ali *et al.*, 2010). PVY was detected also in Lebanon including an isolate causing superficial necrosis on potato tubers (Le Romancer *et al.*, 1994), probably a PVY^{NTN} isolate. A unique PVY^{O} strain was described from Egypt (Abdel-Halim *et al.*, 2000), as well as a PVY^{N} strain (El-Mohsen *et al.*, 2003), PVY^{NTN} (Amer *et al.*, 2004). Partial sequencings of the coat protein regions were used to identify isolates of PVY (Lorenzen *et al.*, 2006a).

3. The virus

PVY can be purified according to Hammond and Lawson (1988) or Huttinga (1973). This virus is characterized by flexuous particles mostly 730 × 11 nm, which sediment as a single component with a sedimentation coefficient of ca. 150S. They consist of a 10-kb positive-sense RNA, with a

5′-covalently linked protein (VPg) and a 3′ poly (A) tail (Hari *et al.*, 1979). This RNA is enclosed in a protein capsid (CP) composed of up to 2000 subunits of 30–37 kDa (Riechmann *et al.*, 1992). The sequences of about 30 PVY isolates have been reported with homologies of 93–99% (van der Vlugt *et al.*, 1993). RNA of potyviruses is translated into a large polyprotein that is cleaved co- and posttranslationally into mature proteins (Dougherty and Carrington, 1988; Fig. 1). The gene products of the potyviral genome from the 5′ to the 3′ are: the P1 protein that contains a protease domain at its C-terminal region, cleaving itself from the adjacent helper component protease (HC-Pro) protein, with another presently unknown function. The HC-Pro protein is multifunctional—it is involved in aphid transmission of the virus, movement of the virus in the plant and is a suppressor of the RNA-dependent gene silencing antivirus defense (Kasschau and Carrington, 1998).

The C-terminal part of the HC-protein also acts as a protease, cleaving itself from the precursor polyprotein, the P3 protein of unknown function, which is a "replication complex block" composed of the CI protein, VPg protease, and polymerase (Glais *et al.*, 2002). The CI protein forms pinwheels, it has an associated helicase activity, and is perhaps involved in viral cell-to-cell movement. CI is bordered by two small proteins 6K1 and 6K2 of unknown function. Protein NIa forms nuclear inclusions and has two domains: the VPg (apparently involved in replication and translation), and a protease domain that cleaves all proteins at the C-terminal half of the precursor. The NIb protein is the putative RNA-dependent RNA polymerase, which also forms nuclear inclusions. The coat protein is involved in cell-to-cell movement and vector transmission together with HC-Pro.

4. Detection methods

An indication of the strain can be obtained by the reactions induced in a range of test species that include *Solanum demissum* × *S. tuberosum* A6, *S. demissum* Y and *Solanum chacoense* "TEL." Serological techniques,

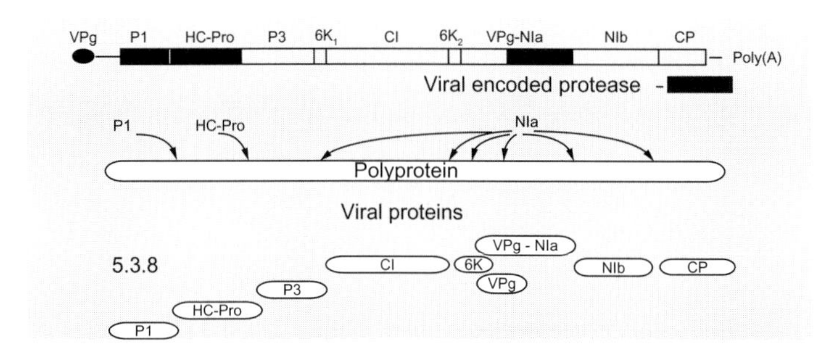

FIGURE 1 Genome organization of *Potato virus Y*. (For color version of this figure, the reader is referred to the online version of this chapter.)

however, have long been used for rapid detection and identification especially various formats of ELISA are used for bulk sampling (e.g., Barker *et al.*, 1993; Gugerli and Gehringer, 1980; Singh *et al.*, 1993; Vetten *et al.*, 1983; Weidemann, 1988). Dot immuno-binding assay has also great potential for mass screening (Singh *et al.*, 1993). Serological differentiation of strains has been improved by the use of monoclonal antibodies (Cerovska, 1998; Ellis *et al.*, 1996; Fernandez-Northcote and Gugerli, 1988; Singh *et al.*, 1993). The strains can also be detected and differentiated by nucleic acid spot hybridization using radioactive and nonradioactive probes (Baulcombe and Fernandez-Northcote, 1988; LeClerc *et al.*, 1992), RT- and IC-PCR (e.g., Barker *et al.*, 1993; Weidemann and Maiss, 1996). PVY^{NTN} can be distinguished from other PVY strains by differences in the electrophoretic mobility of its RNA transcripts (Rosner and Maslenin, 1999). Recently, a multiplex PCR assay was described that characterizes PVY isolates and identifies strain mixtures, including NTN- and NTN-like isolates (Lorenzen *et al.*, 2006b).

5. Control

Control of PVY in potato fields is difficult. Of course, the first approach is to use high grade seed potatoes, with a minimal infection by PVY (and other viruses). However, if infection pressure by aphids from outside sources is high, current season infection rates may reach 10% (Legorburu *et al.*, 1996), or even 30–70% (Loebenstein, unpublished observation; Gaba, unpublished observation). Oil sprays (Shands, 1977) alone or in combination with a pyrethroid (Gibson and Cayley, 1984) reduced virus infection. To increase the effectiveness of these sprays, aphid monitoring is advised to determine optimal schedules for spraying (Ragsdale *et al.*, 2001).

The best way to control viruses is by resistance breeding. Two main types of resistance to PVY in potato have been described: extreme resistance (ER) and hypersensitive resistance (HR). ER protects potato plants against all strains of PVY. It strongly suppresses virus accumulation in infected cells, and no visible symptoms or detectable amounts of PVY are observed in inoculated plants (Barker and Harrison, 1984; Cockerham, 1970). Thus, the Ry_{adg} gene from *S. tuberosum* ssp. *andigena* has been shown to provide ER to PVY, defined as resistance against all strains (Whitworth *et al.*, 2009). HR to PVY is strain specific and characterized by development of necrotic lesions (cell death) at the initial infection sites on inoculated leaves. HR may prevent the spread of virus within and from the inoculated leaf. Sometimes, HR fails to restrict virus movement resulting in larger necrotic lesions or vein necrosis.

Probably in the future once the phobia against GM food has diminished transgenic potatoes resistant to viruses will become available and widespread. Transforming cv. Russet Burbank with the coat proteins of PVY and PVX rendered them resistant to both viruses when inoculated mechanically. One line was also resistant when PVY was inoculated with viruliferous green peach aphids (Lawson *et al.*, 1990). These transgenic plants were resistant also in field experiments (Kaniewski *et al.*, 1990). Attempts to transform potato plants with the pokeweed antiviral protein (PAP), a ribosome-inhibiting protein found in the cell walls of *Phytolacca americana* (pokeweed), thought to give a broad-spectrum protection against several viruses, gave only limited protection against aphid-transmitted PVY. When PAP was applied exogenously it protected potato plants from mechanical inoculation with PVY. It was proposed that exogenously applied PAP enters the damaged cell together with the virus and inhibits translation of viral gene products during the first steps after inoculation. PAP applied to the leaf surface may not be sufficient to enter the cell during aphid feeding (Lodge *et al.*, 1993). Further, transgenic potato plants highly resistant to the major strains of PVY through RNA silencing have already been obtained and are awaiting field tests and evaluation (Missiou *et al.*, 2004). It was interesting to see that infection of the transgenic plants with PVX simultaneously or prior to the challenge with PVY did not interfere with PVY resistance. High levels of resistance to PVY were obtained by transforming potato with the P1 sequence of PVY^{O} in sense or antisense orientation (Mäki-Valkama *et al.*, 2000). The transgenic resistance fully protected the crop from infection with PVY^{O} transmitted by aphids. These plants were, however, not resistant to field isolates of the PVY^{N} strain group.

In an interesting research (though with tobacco as a model plant), Waterhouse *et al.* (1998) showed that when sense or antisense gene constructs derived from the Pro gene of PVY, transformed into plants, immunity is conferred to the virus from which the transgene was derived. Their experiments showed that coexpression of sense and antisense Pro mRNAs, from a single T-DNA construct or by introduction through crossing, was much more effective at inducing PVY immunity than by transforming plants with only Pro[s] or Pro[a/s] *genes.*

Also, replicase-mediated resistance to PVY was obtained in transgenic tobacco plants (Audy *et al.*, 1994).

B. *Potato leafroll virus*: Genus *Polerovirus*

PLRV is the second most important virus infecting potato. Symptoms of primary infection by PLRV consist of pallor and upward rolling of young leaves, especially at the base, with an upright habit. The edges of young leaflets of some cultivars may develop reddening. Secondary symptoms, in plants grown from infected tubers, are stunting of the shoots and upward rolling of leaflets, especially of basal leaves which become rigid and leathery. Necrosis may develop in the phloem tissue of stems and petioles, and excessive callose occurs in the sieve tubes of stems and tubers. Callose staining with resorcin blue, the Igel-Lange test (de Bokx, 1967), was often used for detecting PLRV in tubers before the introduction of ELISA. Carbohydrates accumulate in the leaves—sometimes two to three times more than in healthy leaves, with a corresponding reduction in the tubers—due to impaired phloem transport. This is probably due not only to necrosis in the phloem cells but perhaps also because of blockage of photoassimilate movement from the chloroplast into the cytosol by the triose-phosphate translocator (TPT). Thus, in transgenic plants in which cDNA for TPT was expressed in reverse orientation, leaves accumulated five times as much starch as leaves of control plants (Riesmeier *et al.*, 1993). This could also be due to blockage of sucrose loading into the phloem of PLRV-infected plants by the sucrose transporter protein. Thus, in sucrose transporter antisense potato plants starch content increased up to 10-fold (Riesmeier *et al.*, 1994). It was suggested that the p28 (p0, see later) protein of PLRV may be involved in leaf symptom expression (van der Wilk *et al.*, 1991) as healthy potato plants transformed with the PLRV p28 protein gene displayed an altered phenotype resembling virus-infected plants. In tobacco plants transformed with the PLRV MP17 movement protein (MP), sugars and starch increased markedly in the leaves. It was suggested that the phloem tissue is the primary site of the PLRV MP17 protein action in altering the carbohydrate metabolism of the infected plant (Herbers *et al.*, 1997).

Tuber symptoms are rare, but some cultivars such as Green Mountain and Russet Burbank have internal net necrosis, which can be seen when tubers are cut. Sometimes tubers from infected plants may develop thin sprouts—"spindling sprouts." This is an inconsistent symptom and depends on cultivar, virus strain, and environment.

PLRV is of great economic importance, although in recent years damages by PLRV seem to be decreasing. In plants grown from infected tubers (secondary infection), yields may be reduced by 33–50%. Tubers of infected plants are small to medium. Even greater losses are observed when PLRV occurs in complex with PVX or PVY.

1. Transmission

Several aphid species have been reported to transmit PLRV (Kennedy *et al.*, 1962). *M. persicae* seems to be the most efficient and important vector. *M. euphorbiae* transmits potato strains less efficiently. PLRV is transmitted in a persistent manner, characterized by minimum access times for acquisition and inoculation of ca. 1 h each. There is a latency period between acquisition and transmission of the virus, and the minimum time for transmission is ca. 12 h. Transmission frequency increases with an increase in the access-feeding period of up to 2 days or more. Both larvae and adults can transmit the virus. Aphids remain infective after molting and are viriferous for life. PLRV is circulative in the vector aphids and can be found in the hemolymph of *M. persicae*.

Endosymbiotic bacteria (*Buchnera* sp.) are involved in the persistent transmission of PLRV by *M. persicae* (Hogenhout *et al.*, 1996, 1998). PLRV displayed a strong affinity for symbionin, the major protein synthesized and released into the aphid's hemolymph by the bacterial endosymbiont. This protein has a high degree of homology with *Escherichia coli* heat shock protein GroEl. PLRV-binding sites of *Buchnera* GroEl are located in the equatorial domain of the protein. Mutants lacking this domain lost the ability to bind PLRV. Absence of *Buchnera* GroEl in the hemolymph of aphids treated with antibiotics led to the rapid degradation of the viral capsid protein and a concomitant loss of infectivity. Also *Azadirachta indica* metabolites, which interfered with the host–endosymbiont relationship inhibited transmission of PLRV by *M. persicae* (van den Heuvel *et al.*, 1998). Availability of virus for acquisition by aphids generally declines with increases in plant age and symptom severity (van den Heuvel and Peters, 1990).

PLRV is not transmitted by mechanical inoculation, by seed or pollen; it is transmitted experimentally by grafting.

2. Presence in Mediterranean countries

The virus is probably distributed worldwide in potato growing areas. In the Mediterranean basin, PLRV was reported from Spain (de San and Roman, 1963), Tunisia (Djilani-Khouadja *et al.*, 2005), France (Guyader and Giblot, 2002), Greece (Chatzivassiliou *et al.*, 2008), Egypt (El-Araby *et al.*, 2009), Turkey (Bostan *et al.*, 2006), Cyprus (Ioannou, 1989), Israel (Zimmerman-Gries and Harpaz, 1970), Lebanon (Abou-Jawdah *et al.*, 2001), Syria (Chikh and Natsuaki, 2008), and Morocco (Hanafi *et al.*, 1995).

3. Particle morphology and properties

Virions are isometric ca. 24 nm in diameter, and not enveloped. Purified preparations contain one sedimenting component of 115S (Takanami and Kubo, 1979).

A260/A280: 1.78; A260/A240: 1.43.

Specific absorbance at 260: 8.6.

Density in CsCl: 1.39 g/cm^3 (Rowhani and Stace-Smith, 1979).

Density in Cs_2SO_4: 1.34 g/cm^3 (Thomas, 1984).

The virus contains 30% nucleic acid and 70% protein.

Molecular weight of the coat protein 26.3 kDa (Rowhani and Stace-Smith, 1979).

4. Nucleic acid component

The genome consists of a single stranded, linear messenger sense RNA molecule of 5.88–5.99 kb (mol. wt. 2.0×10^6), covalently linked to a small protein (VPg) of 32 amino acids at its N-terminal (Rowhani and Stace-Smith, 1979). The RNA does not contain a poly A sequence at its 3′-terminus, and is infectious when used to inoculate tobacco mesophyll protoplasts (Mayo *et al.*, 1982). Sequence database accession codes include: D00530 Em(40)_vi:PLVGR Gb(84)_vi:PLVGR 5987 bp, D00733 Em(40)_vi:PLRVA (Australian isolate) 5882 bp, D00734 Em(40)_vi: PLVRC (Canadian isolate) 5883 bp, and X13906 Em(40)_vi:PLRVCOAT Gb(84) (PLRV coat protein gene) 5987 bp.

The genomic sequence of PLRV consists of eight open reading frames (ORFs) (Ashoub *et al.*, 1998; Martin *et al.*, 1990; Mayo and Ziegler-Graff, 1996; Fig. 2). The six major ORFs are separated by a small intergenic region into two gene clusters—ORFs 0, 1, and 2, and ORFs 3, 4, and 5. ORF 0 encodes a factor involved in symptom development. ORFs 1 and 2, with motifs characteristic of helicases (ORF 1) and polymerases (ORF 2), form part of the viral replicase. The other ORFs are located in the 3′ half of the genome. ORF 3 encodes the capsid protein. Initiation of an internally located AUG codon within the CP gene, but in a different reading frame, codes for the MP (ORF 4). Suppression of the CP amber stop codon results in formation of an ORF 3/ORF 5 read-through protein (Tacke *et al.*, 1990), which is supposedly the aphid transmission factor (Wang *et al.*, 1995).

ORFs 0, 1, and 2 are translated from genomic RNA. In addition, two subgenomic (sg) RNAs have been observed—sgRNA 1 (~2.3 kb) (Smith and Harris, 1990) and sgRNA 2 (~0.8 kb) (Ashoub *et al.*, 1998). sgRNA 1 serves as mRNA for ORF 3, ORF 3/5, and ORF 4. sgRNA 2 may code for two viral proteins of 7.1 kDa (ORF 6) and 14 kDa (ORF 7), respectively. The VPg sequence has been mapped to position 400–431 of ORF 1, downstream of the putative protease domain and in front of the RNA-dependent RNA polymerase (van der Wilk *et al.*, 1997). The structural organization of the PLRV genome in the coat protein region is similar to that of beet western yellows virus and barley yellow dwarf virus (Tacke *et al.*, 1989).

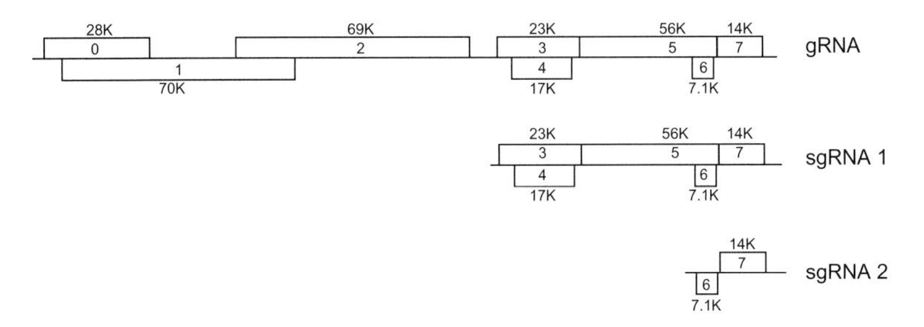

FIGURE 2 Genome organization of *Potato leafroll virus*.

A full-length cDNA copy of the virus has been expressed in protoplasts of *Chenopodium quinoa*, and in transgenic potato plants that accumulated the 17 kDa MP and had symptoms typical of PLRV (Prüfer *et al.*, 1997).

5. Propagation species and purification

Physalis pubescens (syn. *P. floridana*), *Datura stramonium*, and *S. tuberosum* (especially, for maintenance by storing dormant infected tubers at 4 °C).

PLRV can be purified from *P. pubescens* or *D. stramonium* as described by Takanami and Kubo (1979).

6. Diagnostic species

D. stramonium—systemic interveinal yellowing, *P. pubescens*—severe stunting when young seedlings become infected; systemically infected leaves develop mild interveinal necrosis, older leaves become slightly rolled. *Physalis* spp. close to *P. heterophylla*—stunting, interveinal necrosis, and epinasty become evident within 8–12 days of infection (de Souza-Dias *et al.*, 1993).

S. tuberosum ssp. *tuberosum* (potato)—stunting and leafroll.

7. Detection methods

PLRV can be detected in potato leaves by serology, using commercial ELISA kits. However, the concentration of PLRV varies and, in plants grown at temperatures of ca. 30 C or in older plants, ELISA may not always detect infection. It is also difficult to detect PLRV by ELISA in unsprouted tubers (Hill and Jackson, 1984).

Various methods based on PCR or nucleic acid probes are now being developed and evaluated. Thus, PLRV could be detected in tubers within 1 day by immunocapture and a fluorogenic 5′ nuclease RT-PCR assay (Russo *et al.*, 1999; Schoen *et al.*, 1996). Using a digoxigenin (DIG)-labeled cRNA probe, PLRV was easily detected in dormant tubers (Loebenstein *et al.*, 1997). The limit of PLRV detection with this probe was 1 pg/ml compared with 2 ng/ml by ELISA. Such methods, if adopted by testing laboratories, may become important in inspection schemes by eliminating the present necessity of sprouting dormant tubers.

8. Therapy of individual plants

PLRV can be eliminated by keeping tubers for ca. 25 days at 37.5 C in a humid atmosphere (Kassanis, 1950; Lizárraga *et al.*, 1991). Heat treatment causes tubers to deteriorate and is therefore not used for commercial stocks, but is mainly employed to obtain virus-free plants of small special potato lines which can then be propagated. In certain regions in India, an increase in PLRV was observed when cold storage replaced open-air storage; apparently, the high ambient temperatures eliminated PLRV from the tubers (Quak, 1987). Tissue culture, with or without thermotherapy, is the most common practice to retrieve virus-free potatoes in general and PLRV-free plants in particular (Lizárraga *et al.*, 1991). It should, however, be emphasized that following the therapy step reliable testing for viruses, preferably over two growing periods and evaluation of trueness-to-type of the cured lines are essential.

9. Control of PLRV

As PLRV is transmitted in a persistent manner, requiring minimum access times for acquisition and inoculation of ca. 1 h each, it is possible to decrease its spread by using insecticides. Thus in Scotland granular insecticides decreased PLRV spread to a quarter or less of that in control plots (Woodford *et al.*, 1983).

Resistance to PLRV has been reported in both cultivated and wild *Solanum* species, though at present only very few commercial varieties with good resistance are available. Plants with high field resistance to PLRV inoculated with the virus either by grafting or aphids attained only 1–10% of the virus concentration than genotypes with low resistance (Barker and Harrison, 1985). Further, infected plants of resistant genotypes were less potent as a source of PLRV for transmission by aphids than infected plants of a susceptible genotype (Barker and Harrison, 1986), and virus is less likely to be transported from infected foliage to tubers in resistant genotypes (Barker, 1987).

Resistance to PLRV has been reported in *Solanum brevidens*, a wild nontuber bearing species which is sexually incompatible with *S. tuberosum* (Jones, 1979). Somatic fusion hybrids were produced between a dihaploid potato *S. tuberosum* and the sexually incompatible wild *S. brevidens* using both chemical and electrical fusion techniques. Therefore this approach might be used to introduce potentially useful characters from such wild species into commercial potato cultivars (Gibson *et al.*, 1988).

High levels of resistance to PLRV were obtained in genetically modified potatoes. Thus, ER to PLRV in potato cv. Russet Burbank was obtained by expression of the unmodified, full-length PLRV replicase gene (Thomas *et al.*, 2000). Also transforming cv. Desiree plants with the coat protein of PLRV inhibited virus infection (van der Wilk *et al.*, 1991). Although, the mRNA for the CP was detected in the transgenic plants, no CP was detected. Kawchuk *et al.* (1991) showed that the accumulation of CP mRNA in transgenic Russet Burbank was able to confer protection against PLRV, and the aphid transmission of the virus from transgenic plants was reduced compared with that in the controls. Since the CP was not detected in any of the transgenic lines, it seems that the mRNA alone was sufficient to confer protection.

However, so far the European Commission for cultivation of GM potato has approved only cv. Amflora for production of amylopectin (GMO compass—http://www.gmo-compass.org/eng/search/).

Presently, potato growers in Mediterranean countries will continue to import seed potatoes for their main growing season and grow certified seeds only for their autumn or winter crops.

C. *Potato virus X*: Genus *Potexvirus*

Symptoms of PVX are variable and depend on cultivar, virus strain, environmental conditions, and synergism in mixed infections. Most isolates cause only mild leaf mosaic or symptomless infections, especially at higher temperatures. For example, a strain that in cv. Alpha causes interveinal mosaic when plants are grown below 22 °C, remained symptomless in those grown above 22 °C (Beemster and de Bokx, 1987). The yield reductions in such symptomless plants are generally less than 10%, even though they may contain high virus titers. These plants remain carriers of the virus and are important sources of infection. Other strains induce necrotic streaks, severe mosaic, crinkling, and rugosity of leaves and may cause significant yield losses. In potato cultivars with genes *Nx* and *Nb*, some strains of PVX induce a hypersensitive severe top necrosis (Cockerham, 1955; Jones, 1985). In mixed infections with PVY, yields are reduced by up to 50%.

PVX induces a unique type of cellular inclusion, described as an aggregate of laminated inclusions or beaded sheets (Kozar and Sheludko, 1969; Shalla and Shepard, 1972). The sheets are ca. 3 nm thick and may or may not be studded on both sides with beads ca. 11–14 nm in diameter. The sheets and beads are antigenically unrelated to the viral coat protein.

PVX is restricted mainly to solanaceous species including *S. tuberosum*, *D. stramonium*, *Nicotiana* spp., *Solanum esculentum*, *Petunia hybrida*, *Solanum nigrum*, and *Cyphomandra betacea*. PVX also infects *Brassica campestris* ssp. *rapa* plants causing stunting, mild mosaic, and leaf distortion.

Strains of PVX have been divided into three groups on the basis of their thermal inactivation point (Köhler, 1962), into four groups according to their serological properties (Matthews, 1949) and into four groups based on their ability to overcome resistance conferred by two dominant resistance genes *Nx* and *Nb* in *S. tuberosum* (Cockerham, 1955). Group 1 strains induce a hypersensitive (HR) reaction in potatoes carrying either the *Nx* or *Nb* gene; group 2 strains induce HR only in *Nb* potatoes, whereas group 3 strains induce HR only in *Nx* potatoes; group 4 strains overcome both *Nx*- and *Nb*-mediated resistance (Table II). The ER gene (*Rx*) can be overcome by strain PVXHB found in 7% of Bolivian clones of *S. tuberosum* ssp. *andigena* (Jones, 1985; Moreira *et al.*, 1980).

Salicylic acid is involved in local and systemic defense responses mediated by the *Nb* gene potato (Sánchez *et al.*, 2010).

A major class of disease-resistance (R) genes in plants encode nucleotide-binding site/leucine-rich repeat (LRR) proteins. The LRR

TABLE II Genes for hypersensitivity in potatoes and strain groups of PVX (Cockerham, 1955)

Cultivar	Genotype		Strain group 1	2	3	4
Arran Banner	*nx*	*nb*	s	s	s	s
Epicure	*nx*	*Nb*	R	s	s	s
Arran Victory	*Nx*	*nb*	R	R	s	s
Craigs Defiance	*Nx*	*Nb*	R	R	R	s

R, hypersensitive; s, susceptible.

domains mediate recognition of pathogen-derived elicitors. The original Rx protein confers resistance only against a subset of PVX strains, whereas selected mutants were effective against an additional strain of PVX (Farnham and Baulcombe, 2006).

1. Transmission

PVX is transmitted by mechanical inoculation. It is spread naturally by contact between infected and healthy plants, on contaminated farm equipment and workers or by animals that have been in contact with diseased plants. Healthy "seed" potato tubers may become infected during storage by contact of their sprouts with those of infected tubers. Contact between roots may result in infection in the field, but is probably unimportant.

Although the concentration of PVX in plants is very high, the virus is not transmitted by aphids. Plant hoppers and other chewing insects can transmit the virus due to mechanical contact. Transmission by zoospores of the fungus *Synchytrium endobioticum* has also been reported (Nienhaus and Stille, 1965), and infection of *Nicotiana debneyi* and *Nicotiana benthamiana* by PVX was frequently obtained when they were planted in soil from the Andes infested by *S. endobioticum* (Salazar, 1996). PVX can be transmitted experimentally by *Cuscuta campestris* (Ladeburg *et al.*, 1950).

PVX is not transmitted through true seed or pollen.

2. Presence in Mediterranean countries

PVX is distributed worldwide in potato growing areas.

The virus in potato has been reported from Spain (González-Jara *et al.*, 2004), Egypt (El-Araby *et al.*, 2009), Turkey (Bostan and Haliloglu, 2004), Italy (Faccioli, 2001), Tunisia (Djilani-Khouadja *et al.*, 2009), Lebanon (Choueiri *et al.*, 2004), Greece (Chatzivassiliou *et al.*, 2008), Syria (Haj Kassem *et al.*, 2006), Turkey (Tahtacioglu and Ozbek, 1997), Cyprus (Ioannou and Vakis, 1988), and Israel (Zimmerman-Gries and Harpaz, 1967).

3. The virus

PVX can be purified according to Francki and McLean (1968) and simplified by Sadvaksova *et al.* (1996).

Virions are not enveloped. Nucleocapsids filamentous, usually flexuous, 470–580 nm long with a modal length of 515 nm and 13 nm in diameter (Brandes, 1964). Helical symmetry. Axial canal 3.4 nm in diameter. Basic helix obvious. Pitch of helix 3.4 nm (Varma *et al.*, 1968), 8 7/8 subunits per turn (Tollin *et al.*, 1980). Purified preparations contain one sedimenting component.

Sedimentation coefficient (s20, W): 115–130S.
Isoelectric point: pH 4.4.
A260/A280: 1.20. Specific absorbance at 260: 2.97 (Paul, 1959).
Density in CsCl: 1.31 g/cm^3.
Mol. wt.: 3.5×10^6, particles stable.
The virus contains 6% RNA (Knight, 1963) and 94% protein.

Molecular weight of the coat protein: ca. 25 kDa, as deduced from the amino acid sequence obtained from the nt sequence of the coat protein gene. In SDS-PAGE intact, protein appears to have an apparent molecular weight of ca. 30 kDa, but for preparations which have been kept at room temperature faster moving bands are observed, with calculated molecular weight of ca. 22–24 kDa (Koenig *et al.*, 1970). The intact PVX coat protein shows an anomalous behavior in SDS-PAGE, depending on the gel concentration (Koenig, 1972). The coat protein is an *O*-glycoprotein, based on periodate oxidation, digestion with glycosidases, and trifluoromethanesulfonic acid treatment of several PVX strains (Tozzini *et al.*, 1994). The amino acid composition of the coat protein was determined by Shaw *et al.* (1962) and Miki and Knight (1968). A summary of some structural characteristics of PVX, including a model of PVX coat protein's (CP) tertiary structure in the virion was published (Atabekov *et al.*, 2007). The N-terminal segment of the PVX coat protein subunits is glycosylated and mediates formation of a bound water shell on the virion surface (Baratova *et al.*, 2004).

Infectivity is retained when PVX is deproteinized with proteases, phenol, or detergent.

4. Nucleic acid component

One molecule of linear positive-sense ssRNA (mol. wt. 2.1×10^6; 6435 nt). Base composition: 22% G, 32% A, 24% C, and 22% U. The 5′-terminus has a methylated nucleotide cap m7G5 pppGpA, and poly (A) at the 3′-terminus. Nucleotide sequence deposited at EMBL/GenBank under the following accession numbers M63141; M28049. Sequence database accession codes include: D00344 Em(40)_vi:PVXX3 Gb(84)_vi:PVXX3

PVX genomic RNA 6435 bp. M31541 Em(40)_vi:PVXKPA PVX 166K RNA, complete cds; 24K RNA, complete cds; 12K RNA, complete cds; 8K RNA. M38655 Em(40)_vi:PVXCT23 Gb(84)_vi:PVXCT23 PVX coat protein gene, complete cds, 861 bp. X72214 Em(40)_vi:PVXHBRNA Gb (84)_vi:PVXHBRNA PVX strain HB RNA 6432 bp. For additional accession codes see Brunt *et al.* (1996) or http://life.anu.edu.au/viruses/CTVdB/56010001.htm.

The genomic RNA of PVX contains five ORFs (Fig. 3). ORF 1 codes for a 166-kDa polypeptide that presumably possesses an RNA-dependent RNA polymerase activity (Mentaberry and Orman, 1995; Morozov *et al.*, 1990). Products of the overlapping ORFs 2, 3, and 4 (the "triple block") are involved in cell-to-cell transport of PVX (Lough *et al.*, 1998), together with a functional coat protein (Santa Cruz *et al.*, 1998). Cell-to-cell movement is dependent on suppression of RNA silencing (Bayne *et al.*, 2005). The polypeptide encoded by ORF 2 contains a second NTPase/helicase motif similar to that found in the product of ORF 1. Products of ORF 3- and 4-polypeptides of 12 and 8 kDa, respectively, contain sequences of uncharged amino acids, resembling membrane-spanning domains (Morozov *et al.*, 1987). ORF 5 encodes the coat protein. The gRNA serves as a messenger for the 5′ proximal ORF, while several 3′ coterminal sgRNAs direct the synthesis of the other proteins. Thus, the 0.9-kb sgRNA serves as a messenger for coat protein. It was suggested that the 1.4 kb sgRNA acts as a bicistronic mRNA for the products of ORFs 3 and 4. The 25K protein is expressed as a single translation product of the 2.1 kb sgRNA (Morozov *et al.*, 1991). It is generally assumed that the transcription of the sgRNAs is controlled by promoter regions located in the minus-strand RNA (Mentaberry and Orman, 1995). The complete

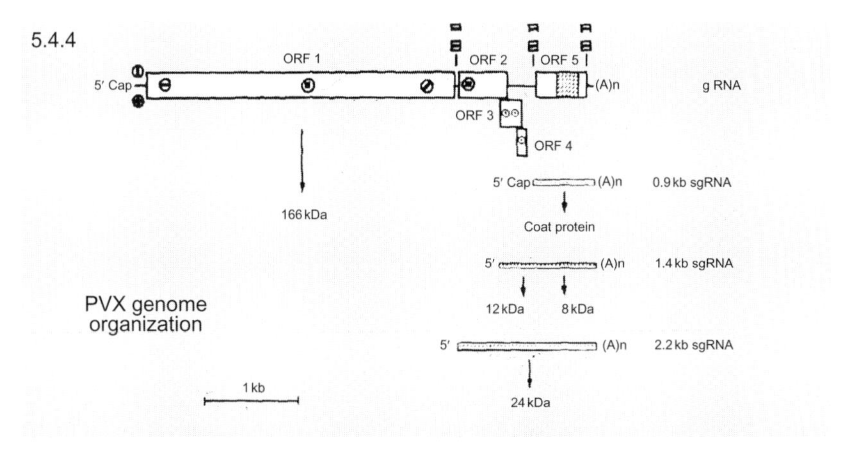

FIGURE 3 Genome organization of *Potato virus X*.

genomic sequence of 14 PVX isolates, including a Chinese one have been determined (Yu *et al.*, 2008). Phylogenetic analysis of the complete genomic sequence of the 14 isolates reveals two groups: the Eurasian group and the American group. Similar analyses of the coat protein genes of 37 PVX isolates also revealed two major groups. All PVX isolates from Asia are clustered to group I, whereas isolates from Europe and America are clustered to both groups.

Nucleotide sequence diversity analyses show that there is no geographical differentiation between PVX isolates and that constraint on the ORF encoding RNA-dependent RNA polymerase is much higher than those on the other four ORFs (Yu *et al.*, 2008).

5. Diagnostic species

D. stramonium—systemic chlorotic rings, mosaic, and mottling; symptoms differ according to strain. As this host is immune to PVY it may be used to separate PVX from the latter when the two viruses occur in complex.

Gomphrena globosa—all strains of PVX except PVXHB induce local lesions. Four to five days after inoculation small gray spots appear, which later increase in size and develop a reddish halo. The middle leaves of *G. globosa* plants with 8–10 leaves on the main shoot are the most susceptible.

N. tabacum cvs. Samsun, Samsun NN, White Burley, etc.—systemic vein clearing, mottling, and ringspots.

Other hosts include *S. esculentum* (systemic), *Nicotiana glutinosa* (systemic), and *Chenopodium amaranticolor* (local lesions).

6. Propagation species

N. tabacum, N. glutinosa (to avoid contamination with TMV), other *Nicotiana* species.

7. Detection methods

PVX can be detected easily in potato leaves and dormant and sprouted tubers by serology, using commercial ELISA kits, or by immunoelectron microscopy (Milne and Luisoni, 1977). Diffusion tests in agar can be used both for detection and determining strain relationships, but PVX virions have to be dissociated into soluble antigens, which migrate more readily into the agar medium. PVX can be disassociated by incubating the plant extract with pyrrolidine to a final concentration of 2.5% (Shepard, 1970).

PLRV, PVA, PVX, and PVY from dormant potato tubers can be detected by TaqMan® real-time RT-PCR (Agindotan *et al.*, 2007) thus eliminating the need to sprout the tubers for biological or serological assays. A cDNA macroarray was developed for simultaneous detection of 12 different potato viruses, including PVX (Maoka *et al.*, 2010).

The macroarray method developed was 5×10^2 to 4×10^6 times more sensitive than ELISA and 5 to 5×10^4 times more sensitive than RT-PCR.

8. Therapy of individual plants

PVX-free plants can be obtained by culturing 0.1 mm meristems, with or without a leaf primordium (Kassanis and Varma, 1967), although only a small percentage of plantlets was regenerated from the meristems. Combining heat treatment before the excision of the meristem tip enhanced the proportion of virus-free plants obtained (Stace-Smith and Mellor, 1968). Incorporation of Virazole (ribavirin) into the culture medium yielded a higher percentage of virus-free progeny than comparable cultures without Virazole (Faccioli and Marani, 1998). It is essential that following therapy, progeny should be thoroughly tested for viruses, preferably over two growing periods, and evaluated for trueness-to-type.

9. Control

Although durable resistance genes to PVX (e.g., the *Rx* gene) are known, commercial cultivars, even if virus resistant, have to be superior for many other agricultural traits to become accepted. The development of transgenic plants may provide an additional means to control viruses in the future, if allowed by official regulations and accepted by farmers and consumers.

Several approaches have been investigated.

Thus, Monsanto transformed Mexican potato lines with a triple gene construct to provide resistance to PVX, PVY, and PLRV, including the Alpha, Rosita, and Nortena varieties, which are now at the line selection stage (Kaniewski and Thomas, 2004).

Production of dysfunctional or partially active MPs in transgenic plants is assumed to confer resistance to the wild-type (wt) virus by competition between wt virus-coded MP and the preformed modified MP. Seppänen *et al.* (1997) evaluated two mutant PVX MP genes (m12K-Sal and m12K-Kpn) obtained by inserting specific linkers at the boundary between the N-terminal hydrophobic and putative transmembrane segment, and the central invariant hydrophilic region of the respective 12 kDa, 12K, triple gene block (TGB) protein. Several transgenic potato lines which expressed m12K-Sal or m12K-Kpn to different degrees were resistant to infection by PVX, potato aucuba mosaic potexvirus and the carlaviruses *Potato virus M* and *S* (Seppänen *et al.*, 1997).

ER to PVX infection in plants was obtained by expressing a modified component of the putative viral replicase (Longstaff *et al.*, 1993).

When tobacco plants were transformed with constructs in which the transgene was a cDNA of replicating PVX RNA, several transformed plants showed no symptoms and extreme strain-specific resistance against PVX (Angell and Baulcombe, 1997).

Artificial miRNAs (amiRNAs) that exploit the endogenous gene silencing mechanism can be designed to target the transcript of any gene of interest and provide a highly specific approach for effective posttranscriptional gene silencing in plants. Two types of amiRNA targeting sequences that encode the silencing suppressor HC-Pro of PVY and the TGBp1/p25 (p25) of PVX were tested. The detected amiRNAs efficiently inhibited HC-Pro and p25 gene expression and conferred highly specific resistance against PVY or PVX infection in transgenic *N. tabacum*; this resistance was also maintained under conditions of increased viral pressure (Ai *et al.*, 2010).

At present, the conventional approach of supplying certified potato seed tubers is still the main accepted approach. Schemes for producing ''seed'' potatoes by rapid propagation (Watad *et al.*, 2001) are beginning to be used. The rapid propagation technology to produce ''minitubers'' seems to be of advantage in regions where natural infection by viruses is not extremely low. Disease-indexed plantlets produced by meristem culture are grown in glass vessels in temperature-controlled chambers to produce microtubers. These are planted in an insect-proof greenhouse to produce minitubers. In this scheme, the potato plants are exposed to only two growing seasons outside, compared with four field generations in the conventional scheme. Thus, chances for reinfection by viruses are greatly reduced.

D. *Potato virus A*: Genus *Potyvirus*

Symptoms—mild mosaic, rough leaf surface with waxy margins. PVA together with PVX or PVY show crinkle symptoms.

PVA is restricted mainly to solanaceous species including *S. tuberosum*, *Nicotiana* spp., *Solanum pimpinellifolium*, and *Nicandra physalodes.*

Transmission—PVA is transmitted by aphids, as *Aphis frangulae*, *Aphis nasturtii*, and *M. persicae* in a nonpersistent manner. PVA is also transmitted by mechanical inoculation.

1. Presence in Mediterranean countries

Although PVA is present worldwide only one report only from Lebanon (Choueiri *et al.*, 2004) was found, though very probably it is present also in other Mediterranean countries.

2. Particle morphology and properties

Virions are filamentous, not enveloped, usually flexuous with a clear modal length of 730 nm and 11 nm wide (Brandes and Paul, 1957).

The genome consists of a single-stranded 9513 bp linear RNA.

3. Sequence database accession code

PVA isolate 143 complete genome sequence: *GU144321.1.*

4. Experimental hosts

N. tabacum cv. Samsun—systemic vein-clearing and diffuse mottling.
N. tabacum cv. White Burley—systemic vein-clearing and dark green vein-banding.
N. physalodes—systemic slight vein-clearing and mottle to severe necrosis, rugosity, and stunting.
S. pimpinellifolium—systemic necrosis and death.
S. demissum × *S. tuberosum* cv. Aquila (=A6 hybrid) and *S. demissum* SdA—many local lesions.

5. Detection

PVA can be detected by ELISA and commercial kits are available. *S. demissum* × *S. tuberosum* cv. Aquila can serve as a local lesion assay host.

Multiplex RT-PCR can be used to detect PVA simultaneously with other viruses (Nie and Singh, 2000).

6. Control

Use of resistant cultivars as Shepody (Singh *et al.*, 2000).

Use of virus-tested seed-potatoes.

E. *Potato virus S*: Genus *Carlavirus*

Two major (PVS^A and PVS^O) and several minor virus strains are recognized. Many isolates of PVS^O alone induce no conspicuous symptoms in several potato cultivars. However, in susceptible cultivars some isolates cause undulation of leaf margins and some rugosity of leaf surfaces. The importance of PVS is mainly dependent upon its local incidence, the virulence of the isolate, tolerance of the potato cultivar, and environmental conditions. The virus, however, often occurs in complex with other viruses and may then exacerbate the severity of infection.

Infected plant cells contain aggregated virus particles, seen as paracrystalline inclusions or banded bodies, and irregular inclusions consisting of virus particles, ribosomes, and proliferated endoplasmic reticulum (Edwardson and Christie, 1997).

1. Transmission

Some isolates are not aphid-transmissible but others are readily transmitted by several aphid species including *A. fabae, A. nasturtii, M. persicae,* and *Rhopalosiphum padi* (Weidemann, 1986). PVS is not seed transmitted but is sufficiently infectious to be transmitted mechanically from infected to healthy field-grown potato plants (Franc and Banttari, 1984).

2. Presence in Mediterranean countries

Although PVS occurs worldwide and especially in Eastern European countries, in the Mediterranean countries it has so far been reported from Syria (Chikh and Natsuaki, 2008) and Turkey (Bostan and Peker, 2009). Apparently, the imported seed seems to be free to a large extent from PVS.

3. Experimental hosts

Although the virus has a very restricted natural host range, it is experimentally transmissible by mechanical inoculation to at least 56 other solanaceous species and to 33 species in 12 other families (Edwardson and Christie, 1997). Useful indicator plant species are:

Chenopodium album, C. amaranticolor, C. quinoa—chlorotic local lesions, often with a green halo on older leaves. PVS^A, but not PVS^O, infect *C. quinoa* systemically and induce chlorotic spotting.
Cyamopsis tetragonoloba—small brown necrotic lesions in inoculated cotyledons, but no subsequent systemic infection.
S. esculentum—symptomless systemic infection by PVS^A but immune to PVS^O.
Nicotiana clevelandii—conspicuous chlorosis of systemically infected leaves. This is a very useful host for the propagation and maintenance of virus cultures.
N. debneyi—symptomless local infection but vein clearing, mottling, and necrosis of systemically infected leaves. This species is immune to PVM, and so allows PVS to be separated from complexes of the two viruses.

4. Particle morphology and properties

PVS has particles mostly measuring ca. 650 × 12 nm which contain ca. 5% single-stranded RNA and 95% protein. The protein subunits have a molecular weight of 33–34 kDa.

5. Nucleic acid component

The genomic RNA contains ca. 7.500 nucleotides, the sequence of which at the 3′-terminus have been sequenced (Foster and Mills, 1992); the RNA has six ORFs which encode proteins of 10, 33, 7, 11, 25, and 41 kDa (Fig. 4).

6. Detection methods

ELISA is used routinely for the detection of the virus in tubers and leaves (Banttari and Franc, 1982). Direct tissue blotting has been reported to be more sensitive, cheaper, and faster than conventional ELISA (Samson

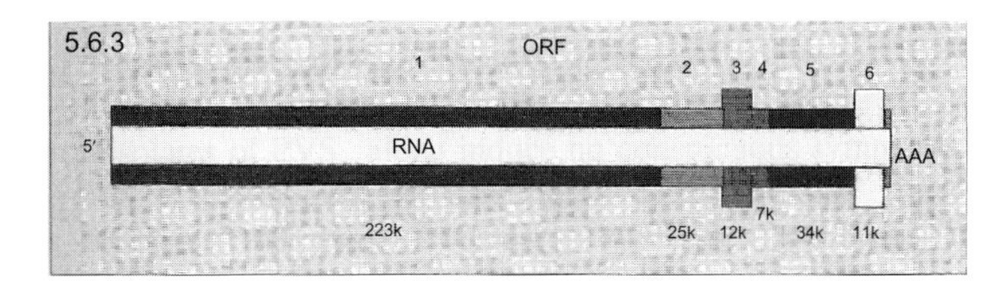

FIGURE 4 Genome organization of *Potato virus M*. (Modified from Zavriev *et al*., 1991, courtesy of Association of Applied Biologists.)

et al., 1993). Monoclonal antibodies have been produced for the specific identification of PVS^{A} (Cerovska and Filigarova, 1995).

Nucleic acid spot hybridization (NASH) is a sensitive and reliable procedure for the detection of PVS using radioactive (Foster and Mills, 1990) or nonradioactive probes (Audy *et al*., 1991). RT-PCR has been shown to be ca. 1000-fold more sensitive than ELISA as a diagnostic procedure (Badge *et al*., 1996).

7. Therapy

Virus-induced yield losses are best minimized by the production, large-scale propagation and distribution of virus-free stocks. These are obtainable by meristem tip culture (Faccioli and Colombarini, 1996), especially if preceded by thermotherapy (Stace-Smith and Mellor, 1968) or if antiviral agents such as Virazole or Ribavirin are included in the culture media (Faccioli and Colombarini, 1996). Although the use of elite virus-tested stocks in production schemes is well established in many countries, virus-resistant cultivars are also utilized.

F. Other viruses

No publications of other viruses in the Mediterranean countries have come to our attention. This is probably due to the relatively limited research on potato viruses in these countries compared to Northern European countries. In addition to this, the import of potato seed tubers of high quality also prevented introduction of potato virus diseases. Nevertheless, *Potato mop-top virus* was observed several times in potato fields in one place in Upper Galilee (Israel) in the middle 1960s (Loebenstein, unpublished observation). The disease was not observed in other areas. It may be that in this area soils have a little lower pH than the majority of the calcareous soils in the vicinity and in the majority of Mediterranean soils in general.

G. Conclusion

Virus diseases of potato in Mediterranean countries are generally a minor problem in the first generation after importing tuber seeds from a good source (especially from Ireland, United Kingdom, and the Netherlands—personal observation). However, heavy infections especially with PVY and PLRV (in recent years, PLRV seems to be less of a problem) are often observed in fields planted with second generation potatoes, if the first generation fields were planted near ''old'' potato fields with remnants of plants. Projects to grow local tuber seeds for several generations in Mediterranean countries are limited, as imported seeds, because of their right physiological age give much higher yields, than local grown seed potatoes which are either too young (not completely out of their dormancy) or too old.

III. *POTATO SPINDLE TUBER VIROID*: GENUS *POSPIVIROID*

Viroids are the smallest known agents of infectious disease and consist of a unique nucleic acid molecule without a protein capsid.

Different strains of potato spindle tuber viroid (PSTVd) exist and symptoms range from mild to severe. They can be evaluated by using Fernow's tomato crossprotection test (Singh *et al.*, 1970). Common symptoms of severe infections include color changes in the foliage, smaller leaves, and spindle-like elongation of the tubers. Sprouting also occurs at a slower rate than in unaffected potatoes.

A. Transmission

It can be achieved by mechanical transmission of PSTVd to a susceptible plant host. Under field conditions, a sick plant may infect a nearby plant by simple contact. The viroid can also be transmitted to other plants during handling (i.e., by using contaminated tools) or through true seed and pollen. Transmission via aphids (*M. persicae*) also occurs but only in the presence of PLRV, by encapsidation of PSTVd in particles of PLRV (Querci *et al.*, 1997).

B. PSTVd in Mediterranean countries

PSTVd has been found infecting potato in Africa (e.g., Nigeria), Asia (e.g., Afghanistan, China, and India), parts of Eastern Europe including the former USSR, North America, and Central America. Systematic indexing over the past 10–20 years has apparently eliminated or significantly reduced PSTVd from commercial seed and ware production schemes in North America and parts of Eastern Europe, but is still widespread in

Russia and poses problems for seed potato production. PSTVd has been reported from Turkey (in *Physalis peruviana*) (Bostan *et al.*, 2010; Verhoeven *et al.*, 2009), Egypt (Mahfouze *et al.*, 2010), Italy (Navarro *et al.*, 2009), and Greece (Malandraki *et al.*, 2010). In France, PSTVd has been eradicated (Promed-mail, 2003).

C. The viroid

Purification of PSTVd requires large amount of leaf tissue and subsequent polyacrylamide gel electrophoresis (Diener, 2003), as the viroid constitutes only a tiny fraction of the total cellular RNA, and has to be separated from other host nucleic acid components. PSTVd was also partially purified from infected Rutgers tomato leaves (Singh and Boucher, 1987) and was further purified by a 2 M LiCl extraction (Diener, 1979), and nondenaturing polyacrylamide gel electrophoresis (Lakshman and Tavantzis, 1993).

PSTVd has 359 nucleotides. Viroid nucleic acid is a single stranded, covalently closed, circular RNA molecule with a rod-like secondary structure, characterized by an alternate arrangement of double-helical regions and single-stranded loops (Salazar *et al.*, 2001). Present within the viroidal RNA is the Pospiviroid RY motif stem loop. The secondary structure of PSTVd is shown in Fig. 5.

D. Replication of viroids

PSTVd-infected cells accumulate partially double-stranded RNA structures that contain multimeric strands of both polarities. Viroids lack protein-coding ability and PSTV replication involves copying of the input circular plus strand by way of a rolling circle mechanism (Branch and Robertson, 1984). The plus circular RNA unit of PSTVd serves as a template to synthesize multimeric minus strands, which are cleaved by an RNase to obtain monomers. The monomers are circularized by an RNA ligase, resulting in copies of the original plus circular RNA PSTVd units.

E. Detection of PSTVd

1. Tomato test: Plant tissue to be tested is macerated in distilled water and then mechanically inoculated onto young tomato leaves (*S. esculentum* cv. Rutgers). Inoculated tomato plants are kept at

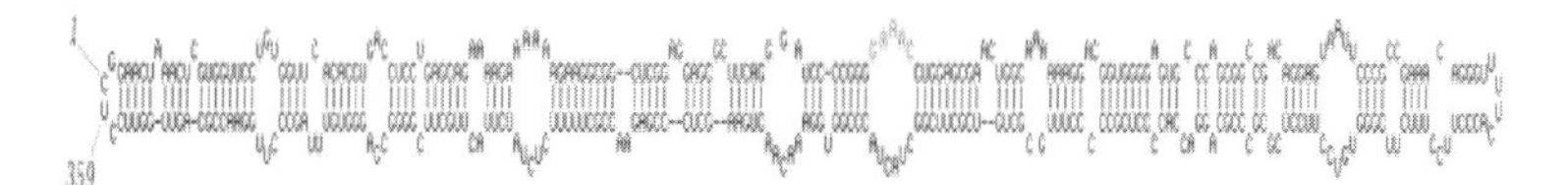

FIGURE 5 Secondary structure of *Potato spindle tuber viroid*. (For color version of this figure, the reader is referred to the online version of this chapter.)

30 °C. The infected tomatoes show initial epinasty, rugosity of the leaves (which then roll), dwarfism, and initial vein necrosis. Later, vein necrosis spreads to the whole leaf causing a brown color.

2. Return-polyacrylamide gel electrophoresis (r-PAGE) is based on the difference in conformation of the viroid molecule under native and denaturing conditions (Reisner, 1987). The first run is performed under native conditions (temperature below the main transition temperature), when the electrophoretic mobility of the viroid is similar to that of linear nucleic acid molecules of the plant. However, during the return run under denaturing conditions (temperature above the transition temperature), the viroid assumes its circular form, which results in decreased electrophoretic mobility in comparison with linear molecules. Therefore, a clear separation between the viroid band and plant nucleic acids is obtained. The viroid RNA is visualized by staining with silver nitrate. The test may be sensitive enough to allow bulking of 10 plants (Jeffries and James, 2010).
3. NASH. The International Potato Center (CIP) has used this technique for routine work due to its high sensitivity, and because it is possible to simultaneously analyze a large number of samples. NASH is based on the principle of duplex formation (hybridization) between a "reporter" nucleic acid (probe) and the nucleic acid of the viroid (target). The probe molecule carries a tag, which can be radioactive or nonradioactive. A typical NASH protocol involves the following steps: sample preparation and application to membranes, sample immobilization and prehybridization, hybridization with a probe, removal of the excess probe through washing, and finally detection of the hybridized probe by colorimetric or chemiluminescent agents (Singh, 1989). DIG is a system for nonradioactive labeling. Compared to radioactive labeling and detection it offers many advantages including:
   - The labeling and detection technology has none of the health and safety hazards of working with radioactive probes.
   - Labeled probes are stable and can be stored for at least a year.
   - Hybridization solutions can be reused several times.

   The method described is based on that of Podleckis *et al.* (1993). The standardized probe is available from Agdia Inc. For details see Jeffries and James (2010).
4. TaqMan® This method combines RT-PCR with real-time fluorescent detection (Mumford *et al.*, 2000). The primers/probe will probably allow detection of most PSTVd isolates listed on sequence databases and some other closely related Pospiviroids. For details see Jeffries and James (2010).

F. Therapy of individual plants

PSTVd can be eliminated by keeping the potato plants at low temperatures (5–8 °C) and subsequent meristem culture. Viroid-free plantlets obtained from meristems were grown at 5–6 C or 8 C for 6 and 4 months. The percentage of plants free of PSTV reached 53% in plantlets derived from nodal cuttings kept at 5–6 °C more than in those developed from infected tubers at 8 °C (30%). Keeping the meristems and plantlets at temperatures above 20 °C did not cure the plants from PSTVd. PSTVd can be successfully eliminated by growing infected mother plants under conditions of low temperature and low light intensity. Apparently, under low temperature the multiplication rate of PSTVd is lowered, thus permitting the development of viroid-free meristems (Lizarraga *et al.*, 1980).

G. Control

PSTVd can be controlled effectively by planting PSTVd-free potato tubers in fields free of diseased tubers. Equipment (knives, cutters, sprayers, cultivators) have to be washed with sanitizers before moving to another field or sowing another seed-line. Viroids are resistant to heat sterilization so sometimes freezing or chemical treatments are more effective. So equipment, surfaces, and tools can be disinfected with sodium or calcium hypochloride (bleach) and 2% sodium hydroxide.

Transgenic potato plants bearing a hammerhead ribozyme [R(−)] targeting the minus strand RNA of PSTVd and a mutated nonfunctional ribozyme [mR(−)] had a high levels of resistance to PSTVd, and were free of PSTVd accumulation after challenge inoculation with PSTVd (Yang *et al.*, 1997).

In Mediterranean basin where PSTVd is not endemic, it is of primary importance to import tuber seed potatoes only from countries without PSTVd and from sources of assured quality.

REFERENCES

Abdel-Halim, M. M., Sadik, A. S., Salama, M. L., and Salem, A. M. (2000). Isolation and characterization of an Egyptian strain of potato virus Y. *Arab J. Biotech.* **3:**115–132.

Abou-Jawdah, Y., Sobh, H., and Saad, A. (2001). Incidence of potato virus diseases and their significance for a seed certification program in Lebanon. *Phytopathol. Mediterr.* **40:**113–118.

Agindotan, B. O., Shiel, P. J., and Berger, P. H. (2007). Simultaneous detection of potato viruses, PLRV, PVA, PVX and PVY from dormant potato tubers by TaqMan® real-time RT-PCR. *J. Virol. Methods* **149:**1–9.

Ai, T., Zhang, L., Gao, Z., Zhu, C. X., and Guo, X. (2010). Highly efficient virus resistance mediated by artificial microRNAs that target the suppressor of PVX and PVY in plants. *Plant Biol.* doi: 10.1111/j.1438-8677.2010.00374.x.

Amer, M. A., El-Hammady, M. H., Mazyad, H. M., Shalaby, A. A., and Abo El-Abbas, F. M. (2004). Cloning, expression and nucleotide sequence of coat protein gene of an Egyptian isolate of potato virus Y strain NTN infecting potato plants. *Egypt. J. Virol.* **1**:39–50.

Angell, S. M., and Baulcombe, D. C. (1997). Consistent gene silencing in transgenic plants expressing a replicating potato virus X RNA. *EMBO J.* **16**:3675–3684.

Appel, O. (1906). Neuere Untersuchungen der Kartoffel-ubd Tomaten Erkrankungen. *Jahresber. Vereinig. Angewandten Bot.*122–136. 3 Jahrg. 1904/5.

Ashoub, A., Rohde, W., and Prüfer, D. (1998). *In planta* transcription of a second subgenomic RNA increases the complexity of the subgroup 2 luteovirus genome. *Nucleic Acids Res.* **26**:420–426.

Atabekov, J., Dobrov, E., Karpova, O., and Rodinova, N. (2007). *Potato virus X*: Structure, disassembly and reconstitution. *Mol. Plant Pathol.* **8**:667–675.

Audy, P., Parent, J. G., and Asselin, A. (1991). A note on four non-radioactive labelling systems for dot hybridization of potato viruses. *Phytoprotection* **72**:81–86.

Audy, P., Palukaitis, P., Slack, S. A., and Zaitlin, M. (1994). Replicase-mediated resistance to potato virus Y in transgenic tobacco plants. *Mol. Plant Microbe Interact.* **7**:15–22.

Badge, J., Brunt, A., Carson, R., Dagless, E., Karamagioli, M., Phillips, S., Seal, S., Turner, R., and Foster, G. D. (1996). A carlavirus-specific PCR primer and partial nucleotide sequence provides further evidence for the recognition of cowpea mild mottle virus as a whitefly-transmitted carlavirus. *Eur. J. Plant Pathol.* **102**:305–310.

Banttari, E. E., and Franc, G. D. (1982). ELISA with single or combined antisera for viruses S and X in potato tubers and plants. *Am. Potato J.* **59**:375–387.

Baratova, L. A., Fedorova, N. V., Dobrov, E. N., Lukashina, E. V., Kharlanov, A. N., Nasonov, V. V., Serebryakova, M. V., Kozlovsky, S. V., Zayakina, O. V., and Rodionova, N. P. (2004). N-terminal segment of potato virus X coat protein subunits is glycosylated and mediates formation of a bound water shell on the virion surface. *Eur. J. Biochem.* **271**:3136–3145.

Barker, H. (1987). Multiple components of the resistance of potatoes to potato leafroll virus. *Ann. Appl. Biol.* **111**:641–648.

Barker, H., and Harrison, B. D. (1984). Expression of genes for resistance to potato virus Y in potato plants and protoplasts. *Ann. Appl. Biol.* **105**:539–545.

Barker, H., and Harrison, B. D. (1985). Restricted multiplication of potato leafroll virus in resistant potato genotypes. *Ann. Appl. Biol.* **107**:205–212.

Barker, H., and Harrison, B. D. (1986). Restricted distribution of potato leafroll virus antigen in resistant potato genotypes and its effects on transmission of the virus by aphids. *Ann. Appl. Biol.* **109**:595–604.

Barker, H., Websyer, K. D., and Reavy, B. (1993). Detection of potato virus Y in potato tubers: A comparison of polymerase chain reaction and enzyme-linked immunosorbent assay. *Potato Res.* **36**:13–20.

Baulcombe, D. C., and Fernandez-Northcote, E. N. (1988). Detection of strains of potato virus X and of a broad spectrum of potato virus Y isolates by nucleic acid spot hybridization (NASH). *Plant Dis.* **72**:307–309.

Bayne, E. H., Rakitina, D. V., Morozov, S. Y., and Baulcombe, D. C. (2005). Cell-to-cell movement of potato Potexvirus X is dependent on suppression of RNA silencing. *Plant J.* **44**:471–482.

Beemster, A. B. R., and de Bokx, J. A. (1987). Survey of properties and symptoms. *In* "Viruses of Potato and Seed Potato Production" (J. A. de Bokx and J. P. H. van der Want, eds.), pp. 84–113. Pudoc, Wageningen, The Netherlands.

Bem, F., Varveri, C., Eleftheriadis, I., and Karafyllidis, D. (1999). First report of occurrence of potato tuber necrotic ringspot disease in Greece. *Plant Dis.* **83**:488.

Blanco-Urgoiti, B., Sánchez, F., Dopazo, J., and Ponz, F. (1996). A strain-type clustering of potato virus Y based on the genetic distance between isolates calculated by RFLP analysis of the amplified coat protein gene. *Arch. Virol.* **141**:2425–2442.

Boonham, N., Walsh, K., Hims, M., Preston, S., North, J., and Barker, I. (2002). Biological and sequence comparisons of *Potato virus Y* isolates associated with potato tuber necrotic ringspot disease. *Plant Pathol.* **51:**117–126.

Bostan, H., and Haliloglu, K. (2004). Distribution of PLRV, PVS, PVX and PVY (PVY^N, PVY^o and PVY^c) in the seed potato tubers in Turkey. *Pak. J. Biol. Sci.* **7:**1140–1143.

Bostan, H., and Peker, P. K. (2009). The feasibility of tetraplex RT-PCR in the determination of PVS, PLRV, PVX and PVY from dormant potato tubers. *Afr. J. Biotechnol.* **8:**4043–4047.

Bostan, H., Guclu, C., Ozturk, E., Ozdemir, I., and Ilbagi, H. (2006). Influence of aphids on the epidemiology of potato virus diseases (PVY, PVS and PLRV) in the high altitude areas of Turkey. *Pak. J. Biol. Sci.* **9:**759–765.

Bostan, H., Gazel, M., Elibuyuk, I. O., Atabeyoğlu, O., and Cağlayan, K. (2010). Occurrence of *Pospiviroid* in potato, tomato and some ornamental plants in Turkey. *Afr. J. Biotechnol.* **9:**2613–2617.

Branch, A. D., and Robertson, H. D. (1984). A replication cycle for viroids and other small infectious RNA's. *Science* **223:**450–455.

Brandes, R. (1964). Identifizierung von gestrekten pflanzenpathogenen Viren auf morphologischer Grundlage. *Mitt. Biol. Bundesanstalt Land-forstwirtsch. Berlin-Dahlem* **110:**1–130.

Brandes, J., and Paul, H. L. (1957). Das Elektronenmikroskop als Hilfsmittel bei der Diagnose pflanzlicher Virosen. In Betrachtungen zur Vermessung faden und stäbchenförmiger Virusteilchen. *Arch. Mikrobiol.* **26:**358–368.

Brunt, A. A. (2001). Potyviruses. *In* "Virus and Virus-like Diseases of Potatoes and Production of Seed-Potatoes" (G. Loebenstein, P. H. Berger, A. A. Brunt, and R. H. Lawson, eds.), pp. 77–86. Kluwer Academic Publishers, Dordrecht.

Brunt, A. A., Crabtree, K., Dallwitz, M. J., Gibbs, A. J., and Watson, L. (1996). Viruses of Plants. CAB International, Wallingford, UK.

Cerovska, N. (1998). Production of monoclonal antibodies to potato virus Y^{NTN} strain and their use for strain differentiation. *Plant Pathol.* **4:**505–509.

Cerovska, N., and Filigarova, M. (1995). Specific detection of the Andean strain of potato virus S by monoclonal antibodies. *Ann. Appl. Biol.* **127:**87–93.

Chatzivassiliou, E. K., Moschos, E., Gazi, S., Koutretsis, P., and Tsoukaki, M. (2008). Infection of potato crops and seed with Potato virus Y and Potato leafroll virus in Greece. *J. Plant Pathol.* **90:**253–261.

Chikh Ali, M., Maoka, T., Natsuaki, T., and Natsuaki, K. T. (2010). $PVY^{NTN\text{-}NW}$, a novel recombinant strain of *Potato virus Y* predominating in potato fields in Syria. *Plant Pathol.* **59:**31–41.

Chikh, M. T., and Natsuaki, K. (2008). The occurrence of potato viruses in Syria and the molecular detection and characterization of Syrian potato virus s isolates. *Potato Res.* **51:**151–161.

Choueiri, E., El-Zammar, S., Jreijiri, F., Mnayer, D., Massaad, R., Saad, A. T., Hanna, L., and Varveri, C. (2004). Phytosanitary status of potato in the Bekaa valley in Lebanon. *EPPO Bull.* **34:**117–121.

Cockerham, G. (1955). Strains of potato virus X. Proc. 2nd Conf. Potato Virus Diseases, Lisse-Wageningen 1954, pp. 89–92.

Cockerham, G. (1970). Genetical studies on resistance to potato viruses X and Y. *Heredity* **25:**309–348.

De Bokx, J. A. (1967). The callose test for the detection of potato leafroll virus in potato tubers. *Eur. Potato J.* **10:**221–234.

de Bokx, J. A., and Huttinga, H. (1981). Potato Virus Y. *CMI/AAB Descriptions of Plant Viruses. No. 242 (No. 37 Revised).* Commonwealth Mycol. Inst./Assoc. Applied Biol., Kew, Surray, England.

de San, Perez, and Roman, F. (1963). Vulnerabilidad de la patata al virus del enrollado. *Anales INIA* **11:**89–156.

de Souza-Dias, J. A. C., Costa, A. S., and Nardin, A. M. (1993). Potato leafroll virus in solanaceous weeds in Brazil explains severe outbreaks of the disease in absence of known potato donor sources. *Summa Phytopathol.* **19:**80–85.

Diener, T. O. (1979). Viroids and Viroid Diseases. Wiley, New York.

Diener, T. O. (2003). Discovering viroids—A personal perspective. *Nat. Rev. Microbiol.* **1:**75–80.

Djilani-Khouadja, F. D., Rouzé-Jouan, J., Guyader, S., Marrakchi, M., and Fakhfakh, H. (2005). Biological and molecular characterization of Tunisian isolates of *Potato leafroll virus*. *J. Plant Pathol.* **87:**91–99.

Djilani-Khouadja, F., Glais, L., Tribodet, M., Kerlan, C., and Fakhfakh, H. (2009). Incidence of potato viruses and characterisation of *Potato virus Y* variability in late season planted potato crops in Northern Tunisia. *Eur. J. Plant Pathol.* **126:**479–488.

Dougherty, W. G., and Carrington, J. C. (1988). Expression and function of potyviral gene products. *Annu. Rev. Phytopathol.* **26:**123–143.

Edwardson, J. R., and Christie, R. G. (1997). Viruses infecting peppers and other solanaceous crops. *Monogr. Agric. Exp. Stn. Univ. Florida 18* **1:**106–123.

El-Araby, W. S., Ibrahim, I. A., Hemeida, A. A., Mahmoud, A., Soliman, A. M., El-Attar, A. K., and Mazyad, H. M. (2009). Biological, serological and molecular diagnosis of three major potato viruses in Egypt. *Int. J. Virol.* **5:**77–88.

Ellis, P., Stace-Smith, R., Bowler, G., and MacKenzie, D. J. (1996). Production of monoclonal antibodies for detection and identification of strains of potato virus Y. *Can. J. Plant Pathol.* **18:**64–70.

El-Mohsen, A., Nashwa, M. A., El-Din, A. S. G., Sohair, I. E. A., Sadik, A. S., and Abdelmaksoud, H. M. (2003). Characterization o f potato virus Y strain "N-Egypt" *Ann. Agric. Sci. Cairo* **48:**485–504.

European Union statistics http://epp.eurostat.ec.europa.eu/portal/page/portal/statistics/search_database

Faccioli, G. (2001). Control of potato viruses using meristem and stem-cutting cultures, thermotherapy and chemotherapy. *In* "Virus and Virus-like Diseases of Potatoes and Production of Seed-Potatoes" (G. Loebenstein, P. H. Berger, A. A. Brunt, and R. H. Lawson, eds.), pp. 365–390. Kluwer Academic Publishers, Dordrecht.

Faccioli, G., and Colombarini, A. (1996). Correlation of potato virus S and potato virus M contents of potato meristem tips with the percentage of virus-free plantlets produced in vitro. *Potato Res.* **39:**129–140.

Faccioli, G., and Marani, F. (1998). Virus elimination by meristem tip culture and tip micrografting. *In* "Plant Virus Disease Control" (A. Hadidi, R. K. Khetarpal, and H. Koganezawa, eds.), pp. 346–380. APS Press, St. Paul, Minnesota.

Farnham, G., and Baulcombe, D. C. (2006). Artificial evolution extends the spectrum of viruses that are targeted by a disease-resistance gene from potato. *Proc. Natl. Acad. Sci. USA* **103:**18828–18833.

Fernandez-Northcote, E. N., and Gugerli, P. (1988). Reaction of a broad spectrum of potato virus Y isolates to monoclonal antibodies in ELISA. *Fitopatologia brasileira* **22:**33–36.

Food and Agriculture Organization of the United Nations statistics. www.fao.org/corp/statistics.

Foster, G. D., and Mills, P. R. (1990). Detection of strains of potato virus S by nucleic acid spot hybridisation. *Potato Res.* **33:**487–495.

Foster, G. D., and Mills, P. R. (1992). The 3′-nucleotide sequence of an ordinary strain of potato virus S. *Virus genes* **6:**213–220.

Franc, G. D., and Banttari, E. E. (1984). The transmission of potato virus S by the cutting knife and retention time of infectious PVS on common surfaces. *Am. Potato J.* **61:**253–260.

Francki, R. I. B., and McLean, G. D. (1968). Purification of potato virus X and preparation of infectious ribonucleic acid by degradation with lithium chloride. *Aust. J. Biol. Sci.* **21:**1311–1318.

Gibson, R. W., and Cayley, G. R. (1984). Improved control of potato virus Y by mineral oils plus the pyrethroid cypermethrin applied electrostatically. *Crop Prot.* **3:**469–478.

Gibson, R. W., Jones, M. G. K., and Fish, N. (1988). Resistance to potato leaf roll virus and potato virus Y in somatic hybrids between dihaploid *Solanum tuberosum* and *S. brevidens*. *Theor. Appl. Genet.* **76:**113–117.

Glais, L., Tribodet, M., and Kerlan, C. (2002). Genomic variability in potato potyvirus Y PVY: Evidence that $PVY^{N}W$ and PVY^{NTN} variants are single to multiple recombinants between PVY^{O} and PVY^{N} isolates. *Arch. Virol.* **147:**363–378.

González-Jara, P., Tenllado, F., Martínez-García, B., Atencio, F. A., Barajas, D., Vargas, M., Diaz-Ruiz, J., and Díaz-Ruíz, J. R. (2004). Host-dependent differences during synergistic infection by Potyviruses with potato virus X. *Mol. Plant Pathol.* **5:**29–35.

Gugerli, P., and Gehringer, W. (1980). Enzyme-linked immunosorbent assay (ELISA) for the detection of PLRV and PVY in potatoes after artificial break of dormancy. *Potato Res.* **23:**353–359.

Guyader, S., and Giblot Ducray, D. (2002). Sequence analysis of Potato leafroll virus isolates reveals genetic stability, major evolutionary events and differential selection pressure between overlapping reading frame products. *J. Gen. Virol.* **83:**1799–1807.

Haj Kassem, A. A., Abdul Halim, K., Rifai, O. E. G., and Kassem, M. (2006). First report on viruses which infect potato in Syria. 9th Arab Congress of Plant Protection, V25, Damascus, Syria.

Hammond, J., and Lawson, R. H. (1988). An improved purification procedure for preparing potyviruses and cytoplasmic inclusions from the same tissue. *J. Virol. Methods* **20:**203–217.

Hanafi, A., Radcliffe, E. B., and Ragsdale, D. W. (1995). Spread and control of potato leafroll virus in the Souss Valley of Morocco. *Crop Prot.* **14:**145–153.

Hari, V., Siegel, A., Rzek, C., and Timberlake, W. E. (1979). The RNA of tobacco etch virus contains poly (A). *Virology* **92:**568–571.

Herbers, K., Tacke, E., Hazirezaei, M., Krause, K., Melzer, M., Rohde, W., and Sonnewald, U. (1997). Expression of luteoviral movement protein in transgenic plants leads to carbohydrate accumulation and reduced photosynthetic capacity in source leaves. *Plant J.* **12:**1045–1056.

Hill, S. A., and Jackson, E. A. (1984). An investigation on the reliability of ELISA as a practical test for detecting potato leafroll virus and potato virus Y in tubers. *Plant Pathol.* **33:**21–26.

Hogenhout, S. A., Verbeek, M., Hans, F., Houterman, P. M., Fortass, M., van der Wilk, F., Huttinga, H., and van den Heuvel, J. F. J. M. (1996). Molecular bases of the interactions between luteoviruses and aphids. *Agronomie* **16:**167–173.

Hogenhout, S. A., van der Wilk, F., Verbeek, M., Goldbach, R. W., and van den Heuvel, J. F. M. (1998). Potato leafroll virus binds to the equatorial domain of the aphid endosymbiotic GroEl homolog. *J. Virol.* **72:**358–365.

Huttinga, H. (1973). Properties of viruses of the potyvirus group. 1. A simple method to purify bean yellow mosaic virus, pea mosaic virus, lettuce mosaic virus and potato virus Y^{N}. *Eur. J. Plant Pathol.* **29:**125–129.

Ioannou, N. (1989). The infection pressure of potato leafroll virus and potato virus Y in relation to aphid populations in Cyprus. *Potato Res.* **32:**33–47.

Ioannou, N., and Vakis, N. J. (1988). Production of seed potatoes in Cyprus: Incidence and economic importance of virus diseases. *Potato Res.* **31:**55–65.

Isla, S., Calderon, L. J., Ortega, F., and Carrasco, A. (2008). New planting systems for prebasic and basic potato seed production in Spain. *Ann. Natl. Inst. Res. Dev.*(Suppl):314–315. Potato Sugar Beet (INCDCSZ) Braşov.

Jeffries, C. (1998). *FAO/IPGRI Technical Guidelines for the Safe Movement of Germplasm. No. 19. Potato*. Food and Agriculture Organization of the United Nations, Rome/International Plant Genetic Resources Institute, Rome.

Jeffries, C., and James, T. (2010). Protocol for the diagnosis of quarantine organism- potato spindle tuber viroid (PSTVD). http://www.fera.defra.gov.uk/plants/plantHealth/pestsDiseases/documents/protocols/PSTVd.pdf.

Jones, R. A. C. (1979). Resistance to potato leaf roll virus in *Solanum brevidens*. *Potato Res* **22**:149–152.

Jones, R. A. C. (1985). Further studies on resistance-breaking strains of potato virus X. *Plant Pathol.* **34**:182–189.

Kaniewski, W. K., and Thomas, P. E. (2004). The potato story. *AgBioForum* **7**:41–46. http://www.agbioforum.org.

Kaniewski, W., Lawson, C., Sammons, B., Haley, L., Hart, J., Delannay, X., and Tumer, N. (1990). Field resistance of transgenic Russet Burbank potato to effects of infection by Potato Virus X and Potato Virus Y. *Biotechnology* **8**:750–754.

Kassanis, B. (1950). Heat inactivation of leaf-roll virus in potato tubers. *Ann. Appl. Biol.* **37**:339–341.

Kassanis, B., and Varma, A. (1967). The production of virus-free clones of some British potato varieties. *Ann. Appl. Biol.* **59**:447–450.

Kasschau, K. D., and Carrington, J. C. (1998). A counter defensive strategy of plant viruses: Suppression of posttranslational gene silencing. *Cell* **95**:461–470.

Kawchuk, L. M., Martin, R. R., and Macpherson, J. (1991). Sense and antisense RNA-mediated resistance to potato leafroll virus in Russet Burbank potato plants. *Mol. Plant Microbe Interact.* **4**:247–253.

Kennedy, J. S., Day, M. F., and Eastop, V. F. (1962). A Conspectus of Aphids as Vectors of Plant Viruses. Commonwealth Institute of Entomology, London, p. 114.

Knight, C. A. (1963). Chemistry of viruses. *Protoplasmatologia* **4**:1–177.

Koenig, R. (1972). Anomalous behaviour of the coat proteins of potato virus X and cactus virus X during electrophoresis in dodecylsulfate containing polyacrylamide gels. *Virology* **50**:263–266.

Koenig, R., Stegemann, H., Francksen, H., and Paul, H. (1970). Protein subunits in the potato virus X group. Determination of the molecular weights by polyacrylamide electrophoresis. *Biochim. Biophys. Acta* **207**:184–189.

Köhler, E. (1962). Über die unterschiedlichen Temperaturempfindlichkeit von Typen und Stämmen des Kartoffel-X-Virus. *Phytopathol. Z.* **44**:189–199.

Kozar, F., and Sheludko, Y. H. (1969). Ultrastructure of potato and *Datura stramonium* plant cells infected by potato virus X. *Virology* **38**:220–229.

Ladeburg, R. C., Larson, R. H., and Walker, J. C. (1950). Origin, interrelation and properties of ringspot strains of virus X in American potato varieties. *Res. Bull. Wisc. Agr. Exp. Stn.* **165**:47.

Lakshman, D. K., and Tavantzis, S. M. (1993). Primary and secondary structure of a 360-nucleotide isolate of potato spindle tuber viroid. *Arch. Virol.* **128**:319–331.

Lawson, R. H., and Stace-Smith, R. (2001). Historical perspectives of potato virus research. *In* "Virus and Virus-like Diseases of Potatoes and Production of Seed-Potatoes" (G. Loebenstein, P. H. Berger, A. A. Brunt, and R. H. Lawson, eds.), pp. 53–63. Kluwer Academic Publishers, Dordrecht.

Lawson, C., Kaniewski, W., Haley, L., Rozman, R., Newell, C., Sanders, P., and Tumer, N. E. (1990). Engineering resistance to mixed infection in a commercial potato cultivar: Resistance to Potato virus X and Potato virus Y in transgenic Russet Burbank. *Nat. Biotechnol.* **8**:127–134.

Le Romancer, M., and Kerlan, C. (1991). La maladie des nécroses annulaires superficielles des tubercules: Une affection de la pomme de terre, due au virus Y. *Agronomie* **11**:889–900.

Le Romancer, M., and Nedelec, M. (1997). Effect of plant genotype, virus isolate and temperature on the expression of the potato necrotic ringspot disease (PTNRD). *Plant Pathol.* **46:**104–111.

Le Romancer, M., Kerlan, C., and Nedelec, M. (1994). Biological characterisation of various geographical isolates of potato virus Y inducing superficial necrosis on potato tubers. *Plant Pathol.* **43:**138–144.

LeClerc, D., Eweida, M., Singh, R. P., and AbouHaidar, M. G. (1992). Biotinylated probes for detecting virus Y and aucuba mosaic virus in leaves and dormant tubers of potato. *Potato Res.* **35:**173–182.

Legorburu, F. J., Marquínez, R., and Ruiz de Gauna, J. A. (1996). Spatiotemporal gradients of PVY infection from an external source. Abstr 13th Triennial Conf. EAPR, Veldhoven, The Netherlands, pp. 120–121.

Lizarraga, R. E., Salazar, L. F., Roca, W. M., and Schilde-Rentschler, L. (1980). Elimination of potato spindle tuber viroid by low temperature and meristem culture. *Phytopathology* **70:**754–755.

Lizárraga, R., Panta, A., Jayasinghe, U., and Dodds, J. (1991). Tissue culture for elimination of pathogens. *CIP Research Guide 3.* International Potato Center (CIP), Lima, Peru, p. 21.

Lodge, J. K., Kaniewski, W. K., and Tumer, N. E. (1993). Broad-spectrum virus resistance in transgenic plants expressing pokeweed antiviral protein. *Proc. Natl. Acad. Sci. USA* **90:**7089–7093.

Loebenstein, G., Akad, F., Filatov, V., Sadvakasova, G., Manadilova, A., Bakelman, H., Teverovsky, E., Lachman, O., and David, A. (1997). Improved detection of potato leafroll luteovirus in leaves and tubers with a digoxigenin-labeled cRNA probe. *Plant Dis.* **81:**489–491.

Longstaff, M., Brigneti, G., Boccard, F., Chapman, S., and Baulcombe, D. (1993). Extreme resistance to potato virus X infection in plants expressing a modified component of the putative viral replicase. *EMBO J.* **12:**379–386.

Lorenzen, J. H., Piche, L. M., Gudmestad, N. C., Meacham, T., and Shiel, P. (2006a). A multiplex PCR assay to characterize *Potato virus Y* isolates and identify strain mixtures. *Plant Dis.* **90:**935–940.

Lorenzen, J. H., Meacham, T., Berger, P. H., Shiel, P. J., Crosslin, J. M., Hamm, P. B., and Kopp, H. (2006b). Whole genome characterization of Potato virus Y isolates collected in the western USA and their comparison to isolates from Europe and Canada. *Arch. Virol.* **151:**1055–1074.

Lough, T. J., Shash, K., Xoconostle-Cazares, B., Hofstra, K. R., Beck, D. L., Balmori, E., Forster, R. L. S., and Lucas, W. J. (1998). Molecular dissection of the mechanism by which potexvirus triple gene block proteins mediate cell-to-cell transport of infectious RNA. *Mol. Plant Microbe Interact.* **11:**801–814.

Mahfouze, S. A., El-Dougdoug, K. A., and Allam, E. K. (2010). Production of *Potato Spindle Tuber Viroid*—Free potato plant materials *in vitro*. *New York Sci. J.* **3:**60–67.

Mäki-Valkama, T., Valkonen, J. P. T., Kreuze, J. F., and Pehu, E. (2000). Transgenic resistance to PVY^{o} associated with post-translational silencing of P1 transgene is overcome by PVY^{N} strains that carry highly homologous P1 sequences and recover transgene expression at infection. *Mol. Plant Microbe Interact.* **13:**366–373.

Malandraki, I., Papachristopoulou, M., and Vassilakos, N. (2010). First report of *Potato spindle tuber viroid* (PSTVd) in ornamental plants in Greece. *New Dis. Rep.* **21:**9.

Maoka, T., Sugiyama, S., Maruta, Y., and Hataya, T. (2010). Application of cDNA macroarray for simultaneous detection of 12 potato viruses. *Plant Dis.* **94:**1248–1254.

Martin, R. R., Keese, P. K., Young, M. J., Waterhouse, P. M., and Gerlach, W. L. (1990). Evolution and molecular biology of luteoviruses. *Annu. Rev. Phytopathol.* **28:**341–363.

Matthews, R. E. F. (1949). Studies on potato virus X. *Ann. Appl. Biol.* **36:**449–474.

Mayo, M. A., and Ziegler-Graff, V. (1996). Molecular biology of luteoviruses. *Adv. Virus Res.* **46:**413–460.

Mayo, M. A., Barker, H., Robinson, D. J., Tamada, T., and Harrison, B. D. (1982). Evidence that potato leafroll virus RNA is positive-stranded, is linked to a small protein and does not contain polyadenylate. *J. Gen. Virol.* **59:**63–67.

Mentaberry, A., and Orman, B. (1995). Potexviruses. *In* ''Pathogenesis and Host Specificity in Plant Diseases. Vol. 3. Viruses & Viroids'' (R. P. Singh, U. S. Singh, and K. Kohmoto, eds.), pp. 19–33. Pergamon, Elsevier Science, UK.

Miki, T., and Knight, C. A. (1968). The protein subunit of potato virus X. *Virology* **36:**168–173.

Milne, R. G., and Luisoni, E. (1977). Rapid immune electron microscopy of virus preparations. *In* ''Methods in Virology'' (K. Maramorosch and H. Koprowsky, eds.), Vol. 6, pp. 264–281. Academic Press, New York.

Minas, G. J., Gregoriou, S., Kapari-Isaia, T. H., and Papayiannis, L. (2007). Seed potato production in Cyprus starting from in vitro culture of apical meristem. *Acta Hort.* **741:**283–288.

Missiou, A., Kalantidis, K., Boutla, A., Tzortzakaki, S., Tabler, M., and Tsagris, M. (2004). Generation of transgenic potato plants highly resistant to potato virus Y (PVY) through RNA silencing. *Mol. Breed.* **14:**185–197.

Moreira, A., Jones, R. A. C., and Fribourg, C. E. (1980). Properties of a resistance-breaking strain of potato virus X. *Ann. Appl. Biol.* **95:**93–101.

Morel, G., and Martin, C. (1952). Guérison de Dahlia attaints d'une maladie à virus. *C.R. Acad. Sci. Paris* **235:**1324–1325.

Morozov, S. Y., Lukasheva, L. I., Chernov, B. K., Skryabin, K. G., and Atabekov, J. G. (1987). Nucleotide sequence of the open reading frames adjacent to the coat protein cistron in potato virus X genome. *FEBS Lett.* **213:**438–442.

Morozov, S. Y., Kanyuka, K. V., Levay, K. E., and Zavriev, S. K. (1990). The putative RNA replicase of potato virus M: Obvious sequence similarities with potex- and tymoviruses. *Virology* **179:**911–914.

Morozov, S. Y., Miroshnichenko, N. A., Solovyev, A. G., Fedorkin, O. N., Zelenina, D. A., Lukasheva, L. I., Karasev, A. V., Dolja, V. V., and Atabekov, J. G. (1991). Expression strategy of the potato virus X triple gene block. *J. Gen. Virol.* **72:**2039–2042.

Mumford, R. A., Walsh, K., and Boonham, N. (2000). A comparison of molecular methods for the routine detection of viroids. *EPPO Bull.* **30:**431–435.

Navarro, B., Silletti, M. R., Trisciuzzi, V. N., and Di Serio, F. (2009). Identification and characterization of *Potato spindle tuber viroid* infecting tomato in Italy. *J. Plant Pathol.* **91:**723–726.

Nie, X., and Singh, R. P. (2000). Detection of multiple potato viruses using an oligo(dT) as a common cDNA primer in multiplex RT-PCR. *J. Virol. Methods* **86:**179–185.

Nienhaus, F., and Stille, B. (1965). Übertragung des Kartoffel-X-Virus durch Zoosporen von *Synchytrium endobioticum. Phytopahol. Z.* **54:**335–337.

Panayotou, F. C., and Katis, N. (1996). Contribution to the study of potato aphids in Greece. *Entom. Hellenica* **4:**11–14.

Paul, H. L. (1959). Spektralphotometrische Untersuchungen am Kartoffel-X-Virus. *Arch. Microbiol.* **32:**416–422.

Podleckis, E. V., Hammond, R. W., Hurtt, S. S., and Hadidi, A. (1993). Chemiluminescent detection of potato and pome fruit viroids by digoxigenin-labelled dot blot and tissue blot hybridization. *J. Virol. Methods* **43:**147–158.

Promed-Mail 2003. Results of the 2002/2003 survey on potato spindle tuber pospiviroid in France: Eradication has been achieved. NPPO of France, 2003-07. Archive No. 20030814.2021.

Prüfer, D., Schmitz, L., Tacke, E., Kull, B., and Rohde, W. (1997). In vivo expression of a full-length cDNA copy of potato leafroll virus (PLRV) in protoplasts and in transgenic plants. *Mol. Gen. Genet.* **253:**609–614.

Quak, F. (1987). Therapy of individual plants. *In* ''Viruses of Potatoes and Seed-Potato Production'' (J. A. de Bokx and J. P. H. van der Want, eds.), pp. 151–161. Pudoc, Wageningen, The Netherlands.

Quanjer, H. M., van der Lek, H. A. A., and Oortwijn Botjes, J. (1916). On the nature, mode of dissemination and control of phloem-necrosis (leaf-roll) and related diseases. *Meded. R. Hoog. Land-Tuin en Boschboruwsch. Wageningen* **10:**84–138.

Querci, M., Owens, R. A., Bartolini, I., Lazarte, V., and Salazar, L. F. (1997). Evidence for heterologous encapsidation of potato spindle tuber viroid in particles of potato leafroll virus. *J. Gen. Virol.* **78:**1207–1211.

Ragsdale, D. W., Radcliffe, E. B., and DiFono, C. D. (2001). Epidemiology and field control of PVY and PLRV. *In* "Virus and Virus-like Diseases of Potatoes and Production of Seed-Potatoes" (G. Loebenstein, P. H. Berger, A. A. Brunt, and R. H. Lawson, eds.), pp. 237–270. Kluwer Academic Publishers, Dordrecht.

Reisner, D. (1987). Structure formation (physical-chemical properties). *In* "The Viroids" (T. O. Diener, ed.), pp. 63–98. Plenum Press, New York (US).

Riechmann, J. L., Lain, S., and Garcia, J. A. (1992). Highlights and prospects of potyvirus molecular biology. *J. Gen. Virol.* **73:**1–16.

Riesmeier, J. W., Flügge, U. I., Schulz, B., Heineke, D., Heldt, H. W., Willmitzer, L., and Frommer, W. B. (1993). Antisense repression of the chloroplast triose phosphate translocator affects carbon partitioning in transgenic potato plants. *Proc. Natl. Acad. Sci. USA* **90:**6160–6164.

Riesmeier, J. W., Willmitzer, L., and Frommer, W. B. (1994). Evidence for an essential role of the sucrose transporter in phloem loading and assimilate partitioning. *EMBO J.* **13:**1–7.

Rosner, A., and Maslenin, L. (1999). Transcript conformation polymorphism: A novel approach for differentiating PVY^{NTN}. *J. Phytopathol.* **147:**661–664.

Rowhani, A., and Stace-Smith, R. (1979). Purification and characterization of potato leafroll virus. *Virology* **98:**45–54.

Russo, P., Miller, L., Singh, R. P., and Slack, S. A. (1999). Comparison of potato leafroll virus and potato virus Y detection in seed potato samples tested by Florida winter field inspection and RT-PCR. *Am. Potato J.* **76:**313–316.

Sadvaksova, G., Tam, Y., and Loebenstein, G. (1996). A simple method for purification of potato virus X. *Appl. Biochem. Microbiol.* **32:**393–396 (in Russian).

Salaman, R. N. (1921). Degeneration of potatoes. Report on the International Potato Conference, Royal Horticultural Society, London, pp. 79–91.

Salaman, R. N. (1955). Half century of potato research. Proc. Second Conf. Potato Virus Diseases, Lisse-Wageningen, pp. 14–25.

Salazar, L. F. (1996). Potato Viruses and Their Control. International Potato Center, Lima, Peru, p. 214.

Salazar, L. F., Bartolini, I., and Hurtada, A. (2001). Viroids. *In* "Virus and Virus-like Diseases of Potatoes and Production of Seed-Potatoes" (G. Loebenstein, P. H. Berger, A. A. Brunt, and R. H. Lawson, eds.), pp. 135–144. Kluwer Academic Publishers, Dordrecht, The Netherlands.

Samson, R. G., Allen, T. C., and Whitworth, J. L. (1993). Evaluation of direct tissue blotting to detect potato viruses. *Am. Potato J.* **70:**257–265.

Sánchez, G., Gerhardt, N., Siciliano, F., Vojnov, A., Malcuit, I., and Marano, M. R. (2010). Salicylic acid is involved in the *Nb*-mediated defense responses to *Potato virus X* in *Solanum tuberosum. Mol. Plant Microbe Interact.* **25:**394–405.

Santa Cruz, S., Roberts, A. G., Prior, D. A. M., Chapman, S., and Oparka, K. J. (1998). Cell-to-cell and phloem-mediated transport of potato virus X: The role of virions. *Plant Cell* **10:**495–510.

Schoen, C. D., Knorr, D., and Leone, G. (1996). Detection of potato leafroll virus in dormant potato tubers by immunocapture and a fluorogenic 5′ nuclease RT-PCR assay. *Phytopathology* **86:**993–999.

Schultz, E. S., and Folsom, D. (1923). Transmission, variation, and control of certain degeneration diseases of Irish potatoes. *J. Agric. Res.* **25:**43–147.

Seppänen, P., Puska, R., Honkanen, J., Tyulkina, L. G., Fedorkin, O., Morozov, S. Y., and Atabekov, J. G. (1997). Movement protein-derived resistance to triple gene block-containing plant viruses. *J. Gen. Virol.* **78:**1241–1246.

Shalla, T. A., and Shepard, J. F. (1972). The structure and antigenic analysis of amorphous inclusion bodies induced by potato virus X. *Virology* **49:**654–666.

Shands, W. A. (1977). Control of aphid-borne Potato virus Y in potatoes with oil emulsions. *Am. J. Potato Res.* **54:**179–187.

Shaw, J. G., Reichmann, M. E., and Hatt, D. L. (1962). Amino acid analysis of the proteins of two strains of potato virus X. *Virology* **18:**79–88.

Shepard, J. F. (1970). A radial-immunodiffusion test for the simultaneous diagnosis of potato viruses S and X. *Phytopathology* **60:**1669–1671.

Sigvald, R. (1984). The relative efficiency of some aphid species as vectors of potato virus Y^o (PVY^o). *Potato Res.* **27:**285–290.

Singh, R. P. (1989). Molecular hybridization with complementary DNA for plant viruses and viroid detection. *In* "Perspectives in Phytopathology" (V. P. Agnihotri, N. Singh, H. S. Chaube, U. S. Singh, and T. S. Dwivedi, eds.), pp. 51–60. Today & Tomorrow's Printers & Publishers, New Delhi, India.

Singh, R. P., and Boucher, A. (1987). Electrophoretic separation of a severe from mild strains of potato spindle tuber viroid. *Phytopathology* **77:**1588–1589.

Singh, R. P., Finnie, R. E., and Bagnal, R. H. (1970). Relative prevalence of mild and severe strains of potato spindle tuber virus in eastern Canada. *Am. J. Potato Res.* **47:**289–293.

Singh, R. P., Boucher, A., Somerville, T. H., and Dhar, A. K. (1993). Selection of a monoclonal antibody to detect PVY/N and its use in ELISA and DIBA. *Can. J. Plant Pathol.* **15:**293–300.

Singh, R. P., Nie, X., and Tai, G. C. C. (2000). A novel hypersensitive resistance response against potato virus A in cultivar 'Shepody'. *Theor. Appl. Genet.* **100:**401–408.

Smith, O. P., and Harris, K. F. (1990). Potato leafroll virus 3′ genome organization: Sequence of the coat protein gene and identification of a viral subgenomic RNA. *Phytopathology* **80:**609–614.

Stace-Smith, R., and Mellor, F. C. (1968). Eradication of potato viruses X and S by thermotherapy and axillary bud culture. *Phytopathology* **58:**199–203.

Tacke, E., Sarkar, S., and Rohde, W. (1989). Cloning of the gene for the capsid protein of potato leafroll virus. *Arch. Virol.* **105:**153–163.

Tacke, E., Prüfer, D., Salamini, F., and Rohde, W. J. (1990). Characterization of a potato leafroll luteovirus subgenomic RNA: Differential expression by internal translation initiation and UAG suppression. *J. Gen. Virol.* **71:**2265–2272.

Tahtacioglu, L., and Ozbek, H. (1997). Monitoring aphid (Homoptera: Aphidoidea) species and their population changes on potato crop in Erzurum (Turkey) province throughout the growing season. *Turk. Entomol. Derg.* **21:**9–25.

Takanami, Y., and Kubo, S. (1979). Enzyme-assisted purification of two phloem-limited plant viruses: Tobacco necrotic dwarf and potato leafroll. *J. Gen. Virol.* **44:**153–159.

Thomas, J. E. (1984). Characterisation of an Australian isolate of tomato yellow top virus. *Ann. appl. Biol.* **104:**79–86.

Thomas, P. E., Lawson, E. C., Zalewski, J. C., Reed, G. L., and Kaniewski, W. K. (2000). Extreme resistance to *Potato leafroll virus* in potato cv. Russet Burbank mediated by the viral replicase gene. *Virus Res.* **71:**49–62.

Tollin, P., Wilson, H. R., and Bancroft, J. B. (1980). Further observations on the structure of potato virus X. *J. Gen. Virol.* **49:**407–410.

Tomassoli, L., and Lumia, V. (1998). Occurrence of Potato Tuber Necrotic Ringspot Disease (PTNRD) in Italy. *Plant Dis.* **82:**350.

Tozzini, A. C., Ek, B., Palva, E. T., and Hopp, H. E. (1994). Potato virus X coat protein is a glycoprotein. *Virology* **202:**651–658.

van den Heuvel, J. F. M., and Peters, D. (1990). Transmission of potato leafroll virus in relation to the honey dew excretion of *Myzus persicae*. *Ann. Appl. Biol.* **116:**403–502.

van den Heuvel, J. F. J. M., Hogenhout, S. A., Verbeek, M., and van der Wilk, F. (1998). *Azadirachta indica* metabolites interfere with the host-endosymbiont relationship and inhibit the transmission of potato leafroll virus by *Myzus persicae*. *Entomol. Exp. Appl.* **86:**253–260.

Van der Vlugt, R. A. A., Leunissen, J., and Goldbach, R. (1993). Taxonomic relationships between distinct potato virus Y isolates based on detailed comparisons of the viral coat proteins and 3′-nontranslated regions. *Arch. Virol.* **131:**361–375.

van der Wilk, F., Willink, D. P. L., Huisman, M., Huttinga, H., and Goldbach, R. (1991). Expression of the potato leafroll luteovirus coat protein gene in transgenic potato plants inhibits viral infection. *Plant Mol. Biol.* **17:**431–439.

van der Wilk, F., Verbeek, M., Dullemans, A. M., and van den Heuvel, J. F. J. M. (1997). The genome-linked protein of potato leafroll virus is located downstream of the putative protease domain of the ORF1 product. *Virology* **234:**300–303.

van Slogteren, D. H. M. (1955). Preparation of antisera against potato virus S with special reference to the preparation of non-toxic virus-suspensions for the immunization of rabbits from extracts of infected potato plants. Proc. 2nd Conf. Potato Virus Diseases, Lisse-Wageningen 1954, , pp. 35–39.

Varma, A., Gibbs, A. J., Woods, R. D., and Finch, J. T. (1968). Some observations on the structure of the filamentous particles of several plant viruses. *J. Gen. Virol.* **2:**107–114.

Varveri, C. (2006). Biological, serological and molecular characterization of Potato virus Y isolates in Greece. *Ann. Benaki Phytopath. Inst.* **20:**67–81.

Verhoeven, J. Th.J., Botermans, M., Roenhorst, J. W., Westerhof, J., and Meekes, E. T. M. (2009). First Report of *Potato spindle tuber viroid* in Cape Gooseberry (*Physalis peruviana*) from Turkey and Germany. *Plant Dis.* **93:**316.

Vetten, H. J., Ehlers, U., and Paul, H. L. (1983). Detection of potato virus Y and A in tubers by enzyme-linked immunosorbent assay after natural and artificial break of dormancy. *Phytopathol. Z.* **108:**41–53.

Wang, J. Y., Chay, C., Gildow, F. E., and Gray, S. M. (1995). Readthrough protein associated with virions of barley yellow dwarf luteovirus and its potential role in regulating the efficiency of aphid transmission. *Virology* **206:**954–962.

Watad, A. A., Sluis, C., Nachmias, A., and Levin, R. (2001). Rapid propagation of virus-tested potatoes. *In* "Virus and Virus-like Diseases of Potatoes and Production of Seed-Potatoes" (G. Loebenstein, P. H. Berger, A. A. Brunt, and R. H. Lawson, eds.), pp. 391–406. Kluwer Academic Publishers, Dordrecht, The Netherlands.

Waterhouse, P. M., Graham, M. W., and Wang, M.-B. (1998). Virus resistance and gene silencing in plants can be induced by simultaneous expression of sense and antisense RNA. *Proc. Natl. Acad. Sci. USA* **95:**13959–13964.

Weidemann, H. L. (1986). Die Ausbreitung der Kartoffel Viren S und M unter Feldbedingungen. *Potato Res.* **29:**109–118.

Weidemann, H. L. (1988). Rapid detection of potato viruses by dot-ELISA. *Potato Res.* **31:**485–492.

Weidemann, H. L., and Maiss, E. (1996). Detection of the potato necrotic ringspot strain of potato virus Y (PVY^{NTN}) by reverse transcription and immunocapture polymerase chain reaction. *J. Plant. Dis. Prot.* **103:**337–345.

Whitworth, J. L., Novy, R. G., Hall, D. G., Crosslin, J. M., and Brown, C. R. (2009). Characterization of broad spectrum Potato Virus Y resistance in a Solanum tuberosum ssp. andigena-derived population and select breeding clones using molecular markers, grafting, and field inoculations. *Am. J. Potato Res.* **86:**286–296.

Woodford, A. T., Harrison, B. D., Aveyard, C. S., and Gordon, S. C. (1983). Insecticidal control of aphids and the spread of potato leafroll virus in potato crops in eastern Scotland. *Ann. Appl. Biol.* **103:**117–130.

Yang, X., Yie, Y., Zhu, F., Liu, Y., Kang, L., Wang, X., and Tien, P. (1997). Ribozyme-mediated high resistance against potato spindle tuber viroid in transgenic potatoes. *Proc. Natl. Acad. Sci. USA* **94:**4861–4865.

Yu, X. Q., Wang, H. Y., Lan, Y. F., Zhu, X. P., Li, X. D., Fan, Z. F., Li, H. F., and Wang, Y. (2008). Complete genome sequence of a Chinese isolate of *Potato virus X* and analysis of genetic diversity. *J. Phytopathol.* **156:**346–351.

Zavriev, S. K., Kanyuka, K. V., and Leavy, K. E. (1991). The genomic organisation of potato virus M RNA. *J. Gen. Virol.* **72:**9–14.

Zimmerman-Gries, S., and Harpaz, I. (1967). Spread of potato virus Y (PVY) in relation to aphid population trends in potato fields in Israel. *Potato Res.* **10:**108–115.

Zimmerman-Gries, S., and Harpaz, I. (1970). Incidence of virus diseases in trials to grow foundation-stock seed potatoes in Israel. *Potato Res.* **13:**91–100.

CHAPTER 7

Virus Diseases in Lettuce in the Mediterranean Basin

Aranzazu Moreno and **Alberto Fereres**

Contents

Department of Plant Protection, Instituto de Ciencias Agrarias, ICA-CSIC, Madrid, Spain

Advances in Virus Research, Volume 84
ISSN 0065-3527, DOI: 10.1016/B978-0-12-394314-9.00007-5

Abstract

Lettuce is frequently attacked by several viruses causing disease epidemics and considerable yield losses along the Mediterranean basin. Aphids are key pests and the major vectors of plant viruses in lettuce fields. *Lettuce mosaic virus* (LMV) is probably the most important because it is seed-transmitted in addition to be transmissible by many aphid species that alight on the crop. *Tomato spotted wilt virus* (TSWV) is another virus that causes severe damage since the introduction of its major vector, the thrips *Frankliniella occidentalis*. In regions with heavy and humid soils, *Lettuce Mirafiori big-vein virus* (LMBVV) can also produce major yield losses.

I. INTRODUCTION

Lettuce is an important dietary vegetable, which is primarily consumed fresh in salads. Consumption of lettuce has some health benefits attributed to the presence of vitamin C, phenolic compounds, and fiber content (Mulabagal *et al.*, 2010). Considering the global market, Spain and Italy are two of the top-ranking lettuce producers of the world, responsible for about 9% of global lettuce (*Lactuca sativa* L.) production (Table I). Spain, Italy, Turkey, and France are the major lettuce-producing countries in the Mediterranean basin reaching a yield of almost 3 million tons/year. Since 2000 until 2007, 77% of the total lettuce production in the European Union was harvested in Spain, Italy, Turkey, and France (Eurostat, 2012). This chapter describes the particle properties, the strains, symptoms, purification methods, different transmission strategies, geographical distribution, and the epidemiology and control options of the most important viruses infecting lettuce in the Mediterranean basin.

TABLE I Lettuce and chicory production in the world, 2009 (Faostat, 2012)

Rank	Area	Production (Int $1000)	Production (MT)
1	China	6,009,978	12,855,211
2	USA	1,918,879	3,827,390
3	India	433,547	927,349
4	*Spain*	*409,073*	*875,000*
5	*Italy*	*442,173*	*844,976*
6	Japan	250,119	549,800
7	*Turkey*	*204,788*	*438,038*
8	*France*	*194,811*	*421,264*
9	Iran	182,720	390,836
10	Germany	162,022	346,562
11	Mexico	148,566	317,781
12	Australia	76,925	164,543
13	Republic of Korea	78,074	146,061
14	United Kingdom	58,439	134,300
15	Canada	50,598	108,228
16	Portugal	47,826	102,300
17	Chile	47,078	100,700
18	Iraq	58,439	100,496
19	*Greece*	*42,076*	*90,000*
20	Netherlands	40,206	86,000

Countries of the Mediterranean basin indicated in italic.

II. *ALFALFA MOSAIC VIRUS* (*BROMOVIRIDAE, ALFAMOVIRUS*)

A. Particle morphology and properties

Alfalfa mosaic virus (AMV) particles have a bacilliform structure with a diameter of 19 nm and a length varying from 30 to 56 nm (Hull, 1969). The virion is stabilized primarily by protein–RNA interaction and thus is salt labile. The presence of coat protein (CP) or the corresponding subgenomic RNA is required for virus replication. Specific association of the CP with the RNA 3′-terminal sequences or with the subgenomic mRNA is required for infection (Hull, 2002). Virions contain 16% nucleic acid, 84% protein, and 0% lipid.

B. Nucleic acid components

AMV belongs to the genus *Alfamovirus*, one of the five genera in the family *Bromoviridae*, whose genome consists of three genomic RNAs (RNAs 1, 2, and 3) that are capped at the 5′ end and lack polyadenylation at the 3′

terminus (Bol, 2005). RNAs 1 (accession number NC_001495) and 2 (accession number NC_002024) encode the viral subunits P1 and P2 of the replicase, respectively. RNA3 (accession number NC_002025) encodes the movement protein and serves as a template for the synthesis of subgenomic RNA4, which encodes the CP that is required for infectivity (Chen and Olsthoorn, 2010; http://www.ncbi.nlm.nih.gov).

C. Symptoms and damages

Expression of symptoms in lettuce depends on the strain and environmental conditions. Virus infection causes bright yellow spots of different sizes located along the veins. Often, the damage affects one or two leaves only. Less often the plants will display mosaic symptoms with light yellow areas. Also, some rings and spots may appear on the lower leaves. The spots are often irregular and small in size. Bright yellow mosaic patterns called "calico" may appear in some lettuce cultivars.

The damage of AMV in lettuce is often restricted to plots close to alfalfa fields, which is the major host for the virus. Therefore, aphids flying from infected alfalfa to lettuce fields may spread the disease rapidly.

D. Variability/strains

Since the description of AMV in the early 1930s, numerous strains or variants have been identified. These strains can be differentiated by symptoms, host range, and differences in their physicochemical and biological properties. Most isolates of AMV differ in some way from each other, although strains differing in symptoms do not necessarily differ in amino acid composition and serological behavior (Hull, 1969; Tremaine and Stace-Smith, 1969). Although differences in the host range between the different AMV strains have been observed, there is no restriction in host range for a particular AMV strain. In this way, the same AMV strain can infect several plant species belonging to different families (Price, 1940; Schmelzer, 1962).

Phylogenetic analysis of *CP* genes has revealed that AMV isolates can be clustered in two subgroups (I and II) related to their geographic origin (Parrella *et al.*, 2000). Complete and partial nucleotide sequence data of different AMV strains are available in GenBank (i.e., strain 425/isolate Leiden, strain 425/isolate Madison, strain Strasbourg, and strain YSMV; http://www.ncbi.nlm.nih.gov).

E. Propagation host and purification

AMV has a wide host range, but the number of field crops seriously damaged by this virus in natural conditions is relatively low. There are many families susceptible to AMV infection containing about 150 plant species including commercially important crops as alfalfa (*Medicago sativa*), lettuce (*Lactuca sativa*), potato (*Solanum tuberosum*), and tomato (*Lycopersicon esculentum*). Experimentally, AMV infects about 430 species belonging to 51 dicotyledon families (Gibbs, 1987).

An often-used purification method was published by van Vloten-Doting and Jaspars (1972). According to this procedure, the virus components are separated by differential precipitation in 0.03 M $MgSO_4$ and by centrifugation in sucrose gradients. Alternatively, the virus can be precipitated directly with polyethylene glycol, in 0.2 M NaCl (Clark, 1968).

F. Transmission, epidemiology, and geographic distribution in the Mediterranean basin

The virus is transmitted in a nonpersistent manner by *Myzus persicae* and at least 13 other aphid species (Kennedy *et al.*, 1962). The virus is transmitted after superficial probes lasting less than a minute. The most efficient aphid species transmitting the virus to lettuce are *M. persicae* and *Aphis craccivora*. Seed transmission is frequent in pepper and alfalfa, but has never been reported in lettuce. Dissemination of the virus in the field occurs mainly by transient winged aphids landing and probing on lettuce plants and leaving the field (noncolonizer species). Consequently, a randomly distributed spatial pattern of virus-infected plants is frequently observed. The proximity of virus reservoirs, such as susceptible crops (alfalfa) or noncultivated susceptible plants, are the major factors driving virus epidemics.

The extent of damage in the Mediterranean basin is limited. In Spain and France, the virus is present in lettuce but its damage is insignificant (Blancard *et al.*, 2006; Moreno *et al.*, 2004). AMV causes more damage in other crops such as alfalfa and pepper. The virus has been found infecting tomatoes in Italy but not lettuce (Parrella and Crescenzi, 2005). In Greece, AMV is present in tobacco fields (Chatzivassiliou *et al.*, 2004).

G. Detection methods

The susceptible host species most frequently used for the evaluation of AMV symptoms and host range studies are *Chenopodium amaranticolor* and *Chenopodium quinoa* (where AMV produces chlorotic local lesions), *Phaseolus vulgaris* (which is susceptible to necrotic and chlorotic local lesions, systemic mottle, vein necrosis, and leaf malformation), *Pisum sativum* (local lesions and/or wilting of inoculated leaves, systemic stem necrosis, and plant death), and *Nicotiana tabacum* (systemic mottle, vein banding, and ringspots) (Gibbs, 1987; Jaspars and Bos, 1980).

AMV-specific antibodies are commercially available for serological detection by enzyme-linked immunosorbent assay (ELISA) and tissue blot immunoassay (TBIA). The polymerase chain reaction (preceded by reverse transcription; RT-PCR) is the most widely used nucleic acid-based assay for the detection of AMV (Parrella *et al.*, 2000; Untiveros *et al.*, 2010; Xu and Nie, 2006). Other techniques, such as the tandem mass spectrometry of the CP, are less commonly used but complemented with molecular or immunological procedures provide a rapid and convenient way to detect the virus (Luo *et al.*, 2010).

H. Control methods: Present and future

As for other nonpersistently transmitted plant viruses, the main way to prevent disease outbreaks is by using virus-free seedlings, removal of virus reservoirs, and the deployment of other cultural control tactics. Controlling the virus in the sources (nurseries, infected fields) will prevent the spread of the disease. For protected lettuce, it is important to use physical and optical barriers such as UV-absorbing plastic films, which reduce the entry and dispersal of aphid vectors in lettuce tunnels (Diaz *et al.*, 2006; Legarrea *et al.*, 2009). Other types of barriers such as nonwoven agrotextile covers are useful in limiting the spread of aphids and protecting lettuce fields from nonpersistently transmitted virus diseases (Nebreda *et al.*, 2005). The use of insecticides is ineffective because they are not fast enough to prevent virus inoculation by viruliferous aphids landing and probing on the crop. However, insecticides and biocontrol agents can be useful in limiting secondary spread of the virus by limiting the number of potential viruliferous aphids present on a given field. No resistance to AMV (either conventional or transgenic) in lettuce or other *Lactuca* species has been reported.

III. *BEET WESTERN YELLOWS VIRUS* (*LUTEOVIRIDAE*, *POLEROVIRUS*)

A. Particle morphology and properties

Virus with isometric particles 25–30 nm in diameter comprises subunits of a major capsid protein (around 24 kDa) and a minor capsid protein or readthrough protein (around 75 kDa) (Astier *et al.*, 2007; Hull, 2002). Virions contain 30% nucleic acid, 70% protein, and 0% lipid.

B. Nucleic acid components

The particles contain a single molecule of positive-sense RNA with a genome-linked protein (VPg) and no polyA tail in 3′ or tRNA-like structure (Veidt *et al.*, 1988). The *Beet western yellows virus* (BWYV) genome (accession number NC_004756) contains six ORFs. The gene products from ORF1 and 2 are involved in the viral replicase. An important feature of all luteoviruses is the expression of the putative viral replicase by a −1 shift reading frame in the region of overlap between ORF1 and ORF2 that produce an ORF1–ORF2 fusion protein. The product from ORF3 is the major CP; the termination codon of this ORF is read through to give a fusion protein together with ORF5. The product of the ORF5 is involved in the transmission process by aphids. ORF4 produces a protein involved in cell-to-cell movement of virus particles within the vascular bundle tissues. The product of ORF0 is only found in *Polerovirus* and it is needed for virion accumulation. Product from ORF6 is not present in *Polerovirus* (Hull, 2002; Ziegler-Graff *et al.*, 1996; http://www.ncbi.nlm.nih.gov).

C. Symptoms and damages

Characteristic symptoms of BWYV are interveinal yellowing on the lower leaves of the lettuce plant. The veins and a limited band of adjacent tissue remain green. The lower leaves affected in this way may become totally discolored and will turn white. The chlorotic interveinal areas sometimes turn reddish in winter time in some cultivars.

Early infections may cause significant damage and a reduction in weight of lettuce head. No differences in aggressiveness have been noticed between strains except in the United States where a particular isolate with a satellite RNA causes very severe symptoms. The yield losses due to BWYV infection in lettuce crops may reach up to 30% (Walkey and Payne, 1990). Mixed infections of BWYV with other lettuce viruses such as LMV, *Cucumber mosaic virus* (CMV), and TSWV (see Fig. 1) may appear causing severe symptoms including necrosis.

FIGURE 1 Lettuce plant mixed infected with BWYV, LMV, CMV, and TSWV in the region of Murcia, Spain in 2002. (See Page 3 in Color Section at the back of the book.)

A BWYV isolate not infecting sugar beet has been reclassified by the International Committee for the Taxonomy of Viruses (ICTV) as an independent virus species and renamed as *Turnip yellows virus* (TuYV) (Stevens *et al.*, 2005). This distinct species appears to be the one that occurs in Europe infecting lettuce crops as well as oilseed rape. The interveinal yellow chlorosis in the older leaves of lettuce plants is the most characteristic symptom (Fig. 2).

D. Variability/strains

The term "beet polerovirus" refers to sugar beet-pathogenic isolates of *Beet mild yellowing virus* (BMYV) as well as of BWYV. BMYV was first reported by Russell (1958) in Europe as a disease affecting sugar beets. BWYV was first reported in *Beta vulgaris*, *L. sativa*, *Spinacia oleracea*, and *Raphanus sativus* from California, United States (Duffus, 1961, 1977). Many variants have been distinguished on the basis of host range and virulence (Duffus, 1964). Host range studies conducted with BMYV isolates showed a characteristic reaction and were able to infect sugar beet, *Capsella bursa-pastoris*, and *Montia perfoliata*. However, BMYV isolates failed to infect radish, lettuce, and *Brassica napus*, which are known to be hosts of BWYV (Duffus, 1973).

At a later stage, several attempts were made to differentiate BMYV and BWYV based on their host range and biological, molecular, and genetic proprieties (mainly by comparison of the CP nucleotide and ORF0 sequences) (Hauser *et al.*, 2000). The ability of the different BWYV strains to infect sugar beet has led to discussion over the characterization and nomenclature of closely related virus isolates belonging to the "beet polerovirus-like" subgroup. To solve the debate, the ICTV has recently

FIGURE 2 Lettuce leaf showing typical interveinal yellow chlorosis symptoms caused by *Turnip yellows virus*. (See Page 4 in Color Section at the back of the book.)

reclassified and renamed as TuYV the BWYV isolates that do not infect sugar beet (Stevens *et al.*, 2005).

The complete biological, serological, and molecular characterization of an American isolate of BWYV (BWYV-USA) and its comparison with other European isolates prove that BWYV-USA is a distinct virus species of the *Polerovirus* genus and is related to BMYV and *Beet chlorosis virus* (BChV; Beuve *et al.*, 2008). According to this study, the three distinct *Polerovirus* spp. that do infect sugar beet and induce the mild yellowing disease alone or as a viral complex are BWYV, BMYV, and BChV.

Genome sequences of several BWYV isolates are available in the GenBank (i.e., isolates GB1 and FL1; http://www.ncbi.nlm.nih.gov).

E. Propagation host and purification

More than 150 species in 23 dicotyledonous families are susceptible, including economically important crops such as beet, cabbage, rape, soya, lettuce, pea, potato, turnip, and a few *Cucurbitaceae*. Some common weeds associated with lettuce fields such as wild lettuce, sow thistle, and mallow are susceptible to this virus although most of them are asymptomatic. Moreover, many species belonging to more than nine families are considered as experimental hosts showing yellowing, reddening, and stunting symptoms (Duffus, 1972; Duffus and Johnstone, 1983).

Several strains have been purified by chloroform clarification, differential centrifugation, density gradient centrifugation, and density gradient electrophoresis (D'Arcy *et al.*, 1983; Duffus, 1981; Gold and Duffus, 1967; Govier, 1985; Smith *et al.*, 1966).

F. Transmission, epidemiology, and geographic distribution in the Mediterranean basin

The virus is transmitted by several aphid species in a persistent circulative manner, which needs long feeding periods, access to the phloem, and a latent period of more than 2–3 days for the aphids to become viruliferous. The virus needs to cross the gut epithelium cells and circulate through the hemocoel to reach the aphid's salivary glands. Aphids will remain viruliferous for many days or their entire life cycle although the virus does not replicate in the vector and is not transmitted to the progeny (no transovarial transmission).

Among the 12 or more aphid species transmitting BWYV, *M. persicae* and *Macrosiphum euphorbiae* seem to be the most important. The virus is spread by aphid species that colonize lettuce plants and feed for long periods. Aphid species that form large colonies and are capable of transmitting the virus to lettuce include *Aulacorthum solani* that can be easily misidentified with *M. persicae*. As aphids remain viruliferous for their entire cycle, the virus can spread over long distances by aphids aided by the wind. Most of the aphids transmitting the virus can form large colonies and move from plant to plant spreading the disease to neighboring plants forming aggregated typical patterns. This virus is not seed-transmitted.

There are several wild plants that are susceptible and act as virus reservoirs. Most of them belong to the *Asteraceae* family including wild lettuce and to the *Brassicaceae* family. Overwintering aphids in noncultivated plants can migrate in spring and transmit the virus to lettuce fields.

This disease is frequent in the Mediterranean area, mainly in the open field. The virus has been reported in lettuce-growing areas in the Central region and in the Southeastern Spain (Moreno *et al.*, 2004). This virus is also frequent in France (Lot and Maury-Chovelon, 1985) and Israel (Marco, 1984).

G. Detection methods

Beet poleroviruses identification may require host range comparisons (Hauser *et al.*, 2000). Most frequent susceptible host species used for BWYV diagnosis are *C. bursa-pastoris* (where BWYV induces severe chlorosis and moderate leaf curl, thick and brittle leaves yellowing developing acropetally. On some biotypes of *Capsella*, purpling or reddening accompanies the chlorosis), *Senecio vulgaris* (where purple coloration of the margins of mature leaves is produced by BWYV infection), and *Claytonia* or *M. perfoliata* (where BWYV infection modifies the coloration of the edges of older leaves; Duffus, 1972; Duffus and Johnstone, 1983).

BWYV is detected by serological assays with polyclonal antisera and monoclonal antibodies (Ellis and Wieczorek, 1992; Oshima and Shikata, 1990; Smith *et al.*, 1996). Molecular detection by hybridization assays with RNA probes (Herrbach *et al.*, 1991; Lemaire *et al.*, 1995), RNA amplification by conventional and quantitative RNA-PCR (Chomič *et al.*, 2010; Xiang *et al.*, 2010), and sequence comparisons of their genomes are also used (Hauser *et al.*, 2000).

H. Control methods: present and future

The main way to prevent BWYV infection as for any other persistently transmitted plant virus is by reducing vector numbers at the source (virus reservoirs) and at the sink (virus-free receptor plants). In other words, there is a good chance to limit the spread of BWYV by reducing aphid population in the target crop and in nearby weeds and other potentially infected virus sources. Biocontrol agents and selective systemic aphicides are good options to reduce the aphid numbers. When possible, the use of agrotextiles, nets, and other mechanical or photoselective barriers are also an option to limit the aphid movement and reduce virus spread in the protected lettuce crops. No genetic resistance to BWYV is available yet in lettuce, although batavias and icebergs seem to be less susceptible than romaine or butterhead lettuce. A recessive gene named *bwy* has been identified in a batavia and a butterhead accession. In addition, a dominant gene (*Bw*) conferring almost complete resistance has been identified in an accession of *Lactuca virosa* and introgressed into *L. sativa* (Maisonneuve, 2003).

IV. *BROAD BEAN WILT VIRUS (COMOVIRIDAE, FABAVIRUS)*

A. Particle morphology and properties

Broad bean wilt virus 1 (BBWV-1) and BBWV-2 are the two most important virus species in the genus *Fabavirus*. They have icosahedral virions, about 25 nm in diameter, and bipartite, single-stranded positive-sense RNA genomes (Lisa and Boccardo, 1996). Both genomic RNAs are individually encapsidated in two species of icosahedral shells with two distinctive polypeptides of molecular mass 41–45 kDa (large coat protein, L-CP) and 23–27 kDa (small coat protein, S-CP), respectively (Lee *et al.*, 2000).

B. Nucleic acid component

The genome is composed of two single-stranded positive-sense RNA molecules designated as RNA1 and RNA2 of about 6 and 4 kb, respectively. Both RNAs are translated into single polyprotein precursors from which functional proteins are derived by proteolytic cleavage. RNA1 (accession number NC_005289 for BBWV1 and NC_003003 for BBWV2) encodes a single large protein containing a nucleoside triphosphate (NTP)-binding motif, a viral genome-linked protein (VPg), a protease, and a polymerase. RNA2 (accession number NC_005290 for BBWV1 and NC_003004 for BBWV2) encodes the cell-to-cell movement protein and the two coat proteins. Structural and nonstructural proteins are generated from the polyprotein precusor by protease cleavage (Ferrer *et al.*, 2011; Lee *et al.*, 2000; http://www.ncbi.nlm.nih.gov)

C. Symptoms and damages

The symptoms caused by this virus are mild and sometimes pass unnoticed. They are characterized by mottling of the young leaves of lettuce and slight discoloration. Plant growth is slowed down. When present together with LMV, the symptoms can be exacerbated. BBWV is quite rare in lettuce and its damage is not serious in the Mediterranean basin.

D. Variability/strains

Analysis of genetic variability of BBWV-1 and BBWV-2 has provided relevant information for understanding their evolution and epidemiology. To date, the complete genome of two BBWV-1 isolates (PV132 and

Ben) (Ferrer *et al.*, 2005; Kobayashi *et al.*, 2003) and six BBWV-2 isolates (ME, IA, MB7, P, K, and B935) (Ikegami *et al.*, 1998, 2001; Koh *et al.*, 2001; Kuroda *et al.*, 2000; Nakamura *et al.*, 1998; Qi *et al.*, 2000a,b) have been determined. Moreover, several partial sequences of both viruses have been published in GenBank (http://www.ncbi.nlm.nih.gov).

The complete sequence of the BBWV-1, Spanish isolate Ben differs considerably from that of the other BBWV-1 isolates (Ferrer *et al.*, 2005).

Phylogenetic and genetic structure analyses showed that BBWV-2 has a high genetic variation in comparison with most plant viruses. The phylogenetic analyses suggested no correlation between genetic clustering and host or geographical origin (Ferrer *et al.*, 2011).

E. Propagation hosts and purification

BBWV host range is wide, infecting more than 150 plant species of the *Compositae*, *Brassicaceae*, *Fabaceae*, and *Solanaceae* families (Bowyer, 1980; Zhou, 2002).

Several hosts, such as *Nicotiana clevelandii*, *C. quinoa*, and some broad bean cultivars are suitable for maintaining cultures of BBWV and as a source for virus purification. The virus can be readily purified and fractionated in sucrose density gradients according to different protocols (Makkouk *et al.*, 1990; Taylor *et al.*, 1968).

F. Transmission, epidemiology, and geographic distribution in the Mediterranean basin

The virus is transmitted in a nonpersistent manner by at least 20 aphid species, being *M. persicae* the most efficient. It is not seed-transmitted. The epidemiology of this virus is similar to other nonpersistently transmitted viruses as previously described for AMV. There are more than 100 host plant species, including many in the *Solanaceae* and *Fabaceae* families. The distribution of the virus includes Europe, although the virus is not widely spread in the Mediterranean region. It has been reported in Spain and France (Blancard *et al.*, 2006; Moreno *et al.*, 2004), although it does not produce disease epidemics in lettuce or cause economic damage.

G. Detection methods

The most frequent susceptible host species used for diagnosis of BBWV are: *C. quinoa*, where BBWV induces chlorotic local lesions in inoculated leaves followed by systemic chlorosis of the whole plant or of the apical leaves. Also, the leaves are malformed and mottled; *Vicia faba*, where the systemic vein-clearing induced by the virus is rapidly followed by necrosis of terminal leaves wilting and mosaic; and *Datura stramonium*, where

BBWV produces coalescent, necrotic local lesions, systemic ringspots, and line patterns (Taylor and Stubbs, 1972).

Electron microscopy (EM) and ELISA with polyclonal antibodies are suitable for BBWV detection. However, the nucleotide sequencing of BBWV has allowed the development of new and very reliable detection techniques based on nucleic acid detection such as RT-PCR or molecular hybridization (Ferrer *et al.*, 2008; Ferriol *et al.*, 2011; Miller and Martin, 1988).

H. Control methods: Present and future

Control tactics include the same ones as previously described for AMV, which is also transmitted in a nonpersistent manner. The use of physical and optical barriers is a good option together with the use of resistant cultivars. Controlling the virus at the source of infection, such as the removal of weeds and nursery inspection to limit the number of infected plants before reaching the field are good options. Some cultivars resistant to BBWV are presently available.

V. *CUCUMBER MOSAIC VIRUS* (*BROMOVIRIDAE*, *CUCUMOVIRUS*)

A. Particle morphology and properties

CMV, as other cucumoviruses, has isometric particles, about 30 nm in diameter, that are stabilized by protein–RNA interactions. The capsid comprises 180 copies of a single protein species of about 24 kDa in pentamer–hexamer clusters (Hull, 2002). Virions contain 18% nucleic acid, 82% protein, and 0% lipid.

B. Nucleic acid component

CMV has a three-partite, single-stranded linear RNA genome of messenger polarity with a total genome size of 8.621 kb. RNA1 (accession number NC_002034) is monocistronic, while RNAs 2 (accession number NC_002035) and 3 (accession number NC_001440) encode two different proteins each. Most of RNAs 1 and 2 encode proteins 1a and 2a, respectively, which are part of the virus replicase. Protein 2b, encoded by an ORF partially overlapping that of protein 2a, is a suppressor of posttranscriptional gene silencing. The two ORFs in RNA3 are separated by a noncoding intergenic region (IR) and encode, respectively, protein 3a, which is the movement protein required for cell-to-cell movement, and the CP (Bonnet *et al.*, 2005; Palukaitis *et al.*, 1992; http://www.ncbi.nlm.nih.gov).

C. Symptoms and damages

This virus is often present all the year round in lettuce, but symptoms become more apparent in autumn as the differences between night and day temperatures become greater. Symptoms of CMV are difficult to differentiate from those of LMV, and mixed infections are frequent. Both viruses cause mottling and mosaic symptoms and reduce plant growth. Chlorotic mottling and crinkling of the lamina can also be seen in CMV-infected plants. The damage can be serious in central and northern Europe, but in the Mediterranean regions crop losses are generally low. Losses are more frequent in autumn crops, especially when CMV is present in mixed infections with LMV.

D. Variability/strains

Since its discovery, numerous and extremely heterogeneous CMV strains have been characterized on the basis of their biological properties (symptomatology, thermosensitivity). Based on different criteria, CMV strains were classified into two subgroups named subgroup I and subgroup II (Palukaitis *et al.*, 1992). Subgroup I is more heterogeneous than subgroup II. Subsequent molecular analyses of the RNA3 ORF have distinguished two subgroups within subgroup I: a monophyletic cluster formed by closely related isolates named IA, and a nonmonophyletic subgroup, named IB (Palukaitis and García-Arenal, 2002; Roossinck, 2002; Roossinck *et al.*, 1999).

Different strains have complete nucleotide sequence data published or entered in GenBank and several protein sequences are available (http://www.ncbi.nlm.nih.gov). One of the best-known CMV strains, Fny-CMV, is frequently used for molecular and biological research and has been used to determine the virus structure (Smith *et al.*, 2000).

E. Propagation hosts and purification

CMV has the largest recorded range of hosts and can infect over 800 different species representing 85 botanical families from both the monocotyledons and the dicotyledons, particularly the *Cucurbitaceae* and some *Solanaceae* (Price, 1940). All plant tissues are susceptible to CMV infection except, possibly, meristematic regions. Virions are found in the inclusions present in the cytoplasm of the infected cells, which are often rhomboidal, hexagonal, or roughly spherical.

Most frequent propagation species hosts are *Nicotiana glutinosa* or *N. tabacum* cv. *Xanthi-nc* which are suitable plants for maintaining cultures; *N. tabacum*, *N. clevelandii*, and *Cucurbita pepo* are good sources of virus for purification. Several methods seem satisfactory to purify CMV (Clark *et al.*, 1974; Lot *et al.*, 1972; Peden and Symons, 1973; Scott, 1963).

F. Transmission, epidemiology, and geographic distribution in the Mediterranean basin

The virus is frequently present in most of the lettuce-growing regions of France, Italy, Spain, and Israel. In the central region of Spain, CMV was frequently detected in lettuce in the autumn of 2001 and 2002 (Moreno *et al.*, 2004).

CMV is transmitted by aphids in a nonpersistent manner, in a similar way as described for AMV. There are more than 80 aphid species transmitting CMV, *Aphis gossypii* and *M. persicae* being the most efficient (Edwardson and Christie, 1991).

The epidemiology of this virus has been studied in several crops, mainly in melons and other cucurbits where CMV causes severe disease outbreaks (Alonso-Prados *et al.*, 2003). As described before, aphids disseminate this virus from infected sources to nearby plants. There are a large number of virus reservoirs because of the wide host range of this virus. In some species, but not in lettuce, the virus is seed-transmissible.

G. Detection methods

Diagnostically susceptible host species for CMV detection are *Cucumis sativus* (where CMV induces green or yellow/green systemic mosaic), *L. esculentum* (where a systemic mosaic and much narrowed leaf laminae is produced as a consequence of CMV infection), and *C. amaranticolor* and *C. quinoa* (where CMV infection is expressed as chlorotic or necrotic local lesions but not as systemic lesions; Gibbs and Harrison, 1970).

ELISA and dot immunobinding assay (DIBA) among other serological methods are commonly used for detection of CMV (Yu *et al.*, 2005; Zein and Miyatake, 2009). However, in recent years, the use of RT-PCR has increased for detection of CMV due to its greater sensitivity (Bhat and Siju, 2007; Chen *et al.*, 2011; Niimi *et al.*, 2003). The absolute and relative copy numbers of CMV genomic RNAs contained in purified virions can be determined by real-time RT-PCR (Feng *et al.*, 2006).

Other methods used for CMV detection and identification are the oligonucleotide microarray-based technique (Tiberini *et al.*, 2010) and the immunogold–silver staining (IGSS) technique in combination with epifluorescence detection (Helliot *et al.*, 2007).

H. Control methods: Present and future

As for any other nonpersistently transmitted plant virus, the main way to prevent the disease is to deploy control tactics at the source of infection. Farmers should use virus-free seedlings, remove virus reservoirs, and use physical barriers to prevent aphids from entering the field in a similar way as described for AMV. For protected lettuce, it is important to use optical barriers such as UV-absorbing plastic films (Diaz *et al.*, 2006) and UV-absorbing nets (Legarrea *et al.*, 2009; Legarrea *et al.*, 2012), which can limit secondary spread of both CMV and LMV in lettuce tunnels. Other types of barriers such as nonwoven agrotextiles can be useful in limiting the spread of aphids and protecting lettuce fields from nonpersistently transmitted viruses (Nebreda *et al.*, 2005). The use of insecticides or biocontrol agents is generally ineffective in preventing primary infections but could limit secondary spread of the virus by reducing the number of viruliferous aphids within the field. However, recent experiments conducted in small arenas and simulation models of plant virus epidemics show that parasitoids may enhance the spread of nonpersistent viruses (Hodge *et al.*, 2011; Jeger *et al.*, 2011). No resistance to CMV (either conventional or transgenic) is available on lettuce although some type of moderate resistance or tolerance from wild *Lactuca* species has been reported.

VI. LETTUCE BIG-VEIN DISEASE: *LETTUCE BIG-VEIN ASSOCIATED VIRUS* (*VARICOSAVIRUS*) AND *MIRAFIORI LETTUCE BIG-VEIN VIRUS* (*OPHIOVIRUS*)

A. Particle morphology and properties

Characterization of the causal agent has been difficult. Two viruses, *Lettuce big-vein associated virus* (LBVaV) and *Mirafiori lettuce big-vein virus* (MLBVV) are associated with lettuce big-vein disease. They belong to the genera *Varicosavirus* and *Ophiovirus*, respectively (Fauquet *et al.*, 2005).

LBVaV, previously called *Lettuce big-vein virus*, was first described by Kuwata *et al.* (1983) and was thought to be the causal agent of big-vein disease for nearly two decades. However, there was no evidence that LBVaV induced big-vein symptoms in lettuce. The second virus associated with lettuce big-vein disease, MLBVV, was found by Roggero *et al.* (2000). Later, Lot *et al.* (2002) and Sasaya *et al.* (2008) reported that lettuce plants infected with LBVaV did not develop big-vein symptoms in the absence of MLBVV and that lettuce plants infected with MLBVV developed big-vein symptoms regardless of the presence or absence of LBVaV. Therefore, MLBVV, but not LBVaV, is now regarded as the causal agent of lettuce big-vein disease.

LBVaV has labile rod-shaped particles about 320–350 nm in length and 18 nm in width. This virus contains a single CP with an M_r of 48 kDa and does not have an enveloped structure (Kuwata *et al.*, 1983). Particles of MLBVV appear as thin filamentous either collapsed double-stranded forms of circular particles or as linear spiral filaments (van der Wilk *et al.*, 2002).

B. Nucleic acid component

LBVaV is a single-stranded negative-sense RNA virus with a bipartite genome RNA1 (accession number NC_011558) and RNA2 (accession number NC_01568), formerly designated *ss-1* and *ss-2*, respectively. Viral (negative sense) and virus-complementary (positive sense) RNAs are separately encapsidated in the virions (Sasaya *et al.*, 2001, 2002, 2005).

MLBVV consists of a multipartite genome composed of four single-stranded RNAs of approximately 7.8, 1.7, 1.5, and 1.4 kb (Roggero *et al.*, 2000; van der Wilk *et al.*, 2002). The virus genome contains seven ORFs. The gene encoding the viral polymerase is located on the RNA segment 1 (accession number NC_004779). The CP is encoded by the RNA3 (accession number NC_004782). RNA2 (accession number NC_004781) and RNA4 (accession number NC_004780) encode other proteins without significant sequence similarities with other viral proteins known (van der Wilk *et al.*, 2002; http://www.ncbi.nlm.nih.gov).

C. Symptoms and damages

Although symptomatic plants serologically negative for MLBVV have been reported in the past to be positive for LBVaV in Italy (Roggero *et al.*, 2003), nowadays it is assumed that MLBVV is the only viral agent associated with big-vein symptoms. MLBVV causes characteristic symptoms, located mainly on the veins, which become progressively clear. When the disease becomes severe, a ''big-vein'' clearing appears especially in the ''Batavia'' type cultivars. The symptoms become more apparent in lettuce grown in winter under protected environments where the fungus vector develops much faster. Expression of symptoms requires temperatures below 18 °C. Leaves commonly become twisted as shown in Fig. 3.

This virus was first recorded in California, but has expanded all over the world. It is present in the Mediterranean basin causing severe disease epidemics in certain regions where lettuce is grown year after year in open and protected environments. Damage up to 50–70% has been reported in certain fields.

D. Variability/strains

Two MLBVV isolate groups (A and B) were found after phylogenetic analysis of the complete CP nucleotide sequences of different MLBVV isolates (mostly from Europe) (Maccarone *et al.*, 2010; Navarro *et al.*, 2005b). Several complete and partial sequences of the virus have been published in GenBank (http://www.ncbi.nlm.nih.gov).

FIGURE 3 Symptoms of lettuce big-vein disease (Courtesy of Vicente Pallas, IBMCP, CSIC, Spain). (See Page 4 in Color Section at the back of the book.)

E. Propagation hosts and purification

Susceptible host species are found in the families *Chenopodiaceae*, *Compositae*, *Solanaceae*, and *Tetragoniaceae*. Its known natural host range includes lettuce, endive (*Cichorium endivia*), spiny sow thistle (*Sonchus asper*), and common sow thistle (*Sonchus oleraceus*) (Latham *et al.*, 2004; Navarro *et al.*, 2005a). Symptoms on these natural hosts vary seasonally (Campbell, 1965; Huijberts *et al.*, 1990).

The most commonly used maintenance and propagation host species are *L. sativa* and *Nicotiana occidentalis*. Virus can be purified following procedures by Kuwata *et al.* (1983), Sasaya *et al.* (2001), and van der Wilk *et al.* (2002).

F. Transmission, epidemiology, and geographic distribution in the Mediterranean basin

MLBVV and LBVaV are transmitted by viruliferous zoospores of the fungus *Olpidium brassicae*, a chytridiomycete fungus which is an obligate parasite living on roots. The fungus has resting spores that may survive for many years and thus perpetuate the presence of the virus. The fungus infects the roots of lettuce and many other cultivated and noncultivated plants. Primary infections occur via the zoospores, which are released by the resting spores. The mobile spores infect the epidermal cells of the young roots of lettuce. The fungus remains on the roots and forms numerous zoosporangia that produce a large number of zoospores, some of which become viruliferous. When soil humidity is high, secondary infections occur in nearby plants. *O. brassicae* grows best in cool clay soils, usually heavy and poorly drained.

This virus is widely present in the Mediterranean basin, causing severe epidemics especially in Italy (Roggero *et al*., 2003), Spain (Navarro *et al*., 2004), and France (Blancard *et al*., 2006). A survey conducted in several weeds present in lettuce-growing regions of Spain showed that some of them are important reservoirs of these viruses causing big-vein disease. *S. oleraceus* was the weed species identified as the main virus reservoir in the Almeria and Granada regions (Navarro *et al*., 2005a).

G. Detection methods

The most useful biological method for diagnosis is to test the soil for virus transmissibility. One of the most used diagnostically host species for lettuce big-vein disease is *L. sativa*, which is susceptible not only to the infection by the fungus *O. brassicae* but also to the two viruses associated with the disease. Other species used for the diagnosis of lettuce big-vein disease is *S. oleraceus*, which can be naturally infected by both viruses showing a slight vein banding. However, since symptomatology is highly dependent on environmental conditions, the infected plants may or may not show big-vein symptoms. Therefore, the diagnosis of lettuce big-vein disease should be confirmed by other alternative detection methods (Kuwata, 1991; Navarro *et al*., 2005b; Sasaya *et al*., 2008). Extracts from leaf or root tissue of the infected plants may be diagnosed by serological tests such as DAS-ELISA, immunosorbent electron microscopy (ISEM), and western blot analysis (Kawazu *et al*., 2006). Also, direct observation of the viral particles by electronmicroscopy is another diagnostic procedure for the diagnosis of LBVaV disease (Kuwata *et al*., 1983).

In the past years, the determination of the whole genome sequence of MLBVV and the partial genome sequence of LBVaV has allowed the development of the following molecular methods to improve the specific detection of both viruses in the infected tissue: nonisotopic molecular dot blot hybridization, Southern blot analysis, northern blot analysis, and multiplex RT-PCR (Kawazu *et al*., 2009; Navarro *et al*., 2004).

H. Control methods: Present and future

Preventive strategies based on cultural tactics can be effective to control the disease, particularly those related with sanitary soil disinfection and elimination of plant debris at the end of the growing season. The use of virus-free seedlings is also highly recommended. Eradicating the disease from the diseased plots is almost impossible because resting spores retain their capacity to transmit the virus for more than 15 years. Soil disinfection through solarization, or other cultural practices focused on avoiding excess humidity in the soil will help to control the fungus vector and reduce the disease incidence. Latham *et al.* (2004) have provided an effective way to control and supress LBVaV-associated disease by combining black plastic mulches with partially resistant cultivars. Total resistance to LBVaV has only been described in *L. virosa* lines, but it could not be successfully introgressed into cultivated *L. sativa* (Hayes and Ryder, 2007).

VII. *LETTUCE MOSAIC VIRUS* (*POTYVIRIDAE*, *POTYVIRUS*)

A. Particle morphology and properties

LMV has flexuous filamentous particles with a length of 750 nm and a width of 15 nm (German-Retana *et al.*, 2008; Hull, 2002).

B. Nucleic acid component

The genomic organization of LMV is typical of potyviruses, with a single positive-sense genomic RNA of 10,080 nucleotides (nt) (accession number NC_003605). The viral genomic RNA has a viral encoded protein (VPg) linked covalently at its 5′-end, a polyA tail at its 3′-end, and contains a single open reading frame (ORF) which encodes a large polyprotein processed by three virus-encoded proteases: P1-Pro, HC-Pro, and NIa. The HC-Pro protein is multifunctional acting as a protease, as a helper component for transmission by aphids, and as an inhibitor of gene silencing. The NIa protein (Nuclear Inclusion a) liberates the VPg protein by an autocatalytic cleavage. As result of the protease activity, several proteins are produced. The CI-Hel protein or cytoplasmic inclusion helicase forms the *Potyvirus* typical inclusion bodies in infected cells. The NIb (Nuclear Inclusion b) is involved in genome replication (RNA-dependent RNA polymerase). The CP (capsid protein) is involved in different roles such as to be part of the virion structure, to help in virus transmission by aphids, and in virus cell-to-cell movement. Finally, LMV has a P3 protein whose functions are unknown (Astier *et al.*, 2007; López-Moya and García, 1999; Revers *et al.*, 1997; http://www.ncbi.nlm.nih.gov).

C. Symptoms and damages

LMV causes the most damaging viral disease of lettuce. A mosaic and crinkling of the lamina and blistering of leaves are the typical symptoms associated with LMV. Vein yellowing and some necrotic damage (Fig. 4) can be sometimes observed in certain cultivars. Some strains can cause leaf deformation and yellow mosaic patterns. Plants have reduced growth and become stunted (Fig. 5).

FIGURE 4 Necrotic symptoms in a leaf of a lettuce plant infected with LMV. (See Page 5 in Color Section at the back of the book.)

FIGURE 5 Plant on the top showing symptoms of LMV in a Batavia-type lettuce. Reduced growth, mosaic symptoms and plants fail to form the head are observed; for comparison see the healthy plant at the bottom. (See Page 5 in Color Section at the back of the book.)

D. Variability/strains

There are numerous strains of LMV, which are differentiated from each other by their biological and serological properties. LMV isolates differ in the symptoms they induce, in their ability of being seedborne or in their ability to infect lettuce varieties carrying the *mo1* resistance gene (Krause-Sakate *et al.*, 2002, 2005).

By means of a molecular characterization of the variable regions of the genome of LMV, the different isolates could be clustered in three main groups: a single isolate from Yemen, a group from the Balkans (Greece and Croatia, named Gr), and a third group with very diverse geographical origins (including the Middle East and Greece, called "Rest of the World" or RoW) (German-Retana *et al.*, 2008).

Genome sequences of several LMV strains are available nowadays in the GenBank (i.e., strain 0 and strain E) (http://www.ncbi.nlm.nih.gov).

E. Propagation hosts and purification

LMV host range is wide. Susceptible species in 20 genera belonging to 10 different families including *Compositae*, *Chenopodiaceae*, and *Cruciferae* have been identified (Dinant and Lot, 1992).

Lettuce, safflower (*Carthamus tinctorius*), and *C. quinoa* are the best propagation species used for maintenance of cultures and LMV purification (Tomlinson, 1970a). The methods mainly used for purifying the filamentous particles of LMV have been described by Tomlinson (1964) and Moghal and Francki (1976).

F. Transmission, epidemiology, and geographic distribution in the Mediterranean basin

The transmission of LMV has been studied extensively (Broadbent *et al.*, 1951; Nebreda *et al.*, 2004). It is transmitted in a nonpersistent manner by many aphid species, being *M. persicae* and *M. euphorbiae* the most efficient vectors worldwide. The main vectors of LMV in the central region of Spain were *M. persicae*, *A. gossypii*, and *M. euphorbiae*, while *Aphis fabae* and *Hyperomyzus lactucae* transmitted with low efficiency and *Rhopalosiphum padi* and *Nasonovia ribisnigri* did not transmit the virus. This virus is seed-transmitted in lettuce and several weed species that are commonly present in lettuce fields (Nebreda *et al.*, 2004).

The most important factors causing LMV epidemics are the presence of virus-infected seeds, which are involved in the primary spread of the virus. The secondary spread of the virus occurs when noncolonizing aphid species (mainly *M. persicae* and *Aphis* spp.) land on infected plants and move the virus to neighboring plants. LMV epidemics were best described by a Gompertz model and followed a polycyclic disease progression curve in the central region of Spain (Moreno *et al.*, 2007b).

This virus is frequent in the Mediterranean basin and causes important economic damage especially in regions where growing virus-free seeds is not a general practice.

G. Detection methods

The presence of LMV in its host plants and in its aphid vectors has been successfully detected by different methods. For diagnosis, the best host species for LMV detection in plant are *L. sativa* that shows vein clearing, yellow mottling sometimes with venal necrosis, and bronzing. Plants fail to "heart" and inner leaves remain dwarfed and rosette. Also, *C. amaranticolor* and *C. quinoa* produce chlorotic or necrotic local lesions, systemic yellow veinal flecks, or yellow netting of the younger leaves. *C. quinoa* is more sensitive than *C. amaranticolor* (Tomlinson, 1970a, 1984).

ELISA and other serological detection methods have been used to detect LMV in plant tissues (Hull, 2002). Nucleic acid amplification and other molecular techniques are being used to detect viruses in plants and insect vectors, including heminested and nested RT-PCR combined with immunocapture or print/squash-capture (Moreno *et al.*, 2007a; Peypelut *et al.*, 2004; Revers *et al.*, 1997).

H. Control methods: Present and future

The control of LMV is mainly based on the use of certificated virus-free seeds. The threshold allowed for LMV-infected seeds may vary depending on the country and the production conditions. In Europe, this threshold is limited to 1 per 1000. The use of resistant varieties is another good alternative to control the spread of LMV. Other additional control strategies against LMV have been mentioned above under the sections of AMV and BBWV, which are also transmitted in a nonpersistent manner by aphids. In summary, the virus needs to be controlled at the source not at the sink, by using virus-free seeds and virus-free seedlings and by removing virus-infected reservoirs such as weeds in the *Compositae* family. Additionally, limiting the number of viruliferous aphids reaching the crop by using physical or photoselective barriers can also help to control the disease (Diaz *et al.*, 2006). Also, the use of crop barriers that can act as a virus sink can reduce the number of effective viruliferous aphids reaching the target crop (Hooks and Fereres, 2006). There are a series of resistant cultivars available, most of them carrying one or more dominant genes for resistance against the main virus isolates. A resistant gene identified in *L. virosa* has been introgressed into commercial lettuce. The only resistant genes currently used to protect lettuce crops worldwide are the recessive genes *mo1(1)* and *mo1(2)* corresponding to mutant alleles of the gene encoding the translation initiation factor eIF4E in lettuce (German-Retana *et al.*, 2008). However, LMV has evolved producing new pathotypes that can overcome the resistance genes. Recent studies have identified the viral determinants involved in resistance breaking (Decroocq *et al.*, 2009).

VIII. *TOMATO SPOTTED WILT VIRUS* (*BUNYAVIRIDAE*, *TOSPOVIRUS*)

A. Particle morphology and properties

Particles of TSWV are roughly spherical and membrane-bound with a diameter of approximately 85 nm. The virion contains at least four different proteins: an internal nucleocapsid protein (N; 29 kDa), two membrane glycoproteins (G1 and G2; 78 and 58 kDa, respectively), and a large protein (L). Virus particles are relatively unstable and during purification are stabilized by the addition of reducing agents. Particles contain about 65% protein, 20% lipid, 7% carbohydrate, and 5% RNA (Best, 1968).

B. Nucleic acid components

The virus possesses a tripartite ssRNA genome (with a total size of 16.6 kb) of which one segment is of negative polarity and the other two are ambisense (called L for large, M for medium, and S for small, respectively). Negative-sense L RNA segment (8897 nt) (accession number NC_002052) encodes for the protein L, which may represent the viral polymerase. The ambisense M RNA segment (4821 nt) (accession number NC_002050) encodes for the nonstructural protein NSm (that is supposed to be the movement protein) and for the glycoprotein precursor of G1 and G2. The ambisense S RNA segment (2918 nt) (accession number NC_002051) encodes for the nonstructural protein NSs and for the protein N or nucleocapsid protein (De Haan *et al.*, 1989, 1990, 1991; Kormelink *et al.*, 1992; http://www.ncbi.nlm.nih.gov).

C. Symptoms and damages

Lettuce plants show necrotic lesions, varying from light brown to black on the young leaves and petioles. Old leaves display chlorotic spots, which become necrotic at a later stage. The discoloration extends to the heart leaves and cessation of growth on one side of the plant produces characteristic symptoms. In some cases, plants grow slowly, get stunted, and finally die. If they survive, their heads fail to form (Fig. 6).

This virus causes severe damage to lettuce, especially when plants become infected at an early stage. Heavy losses almost reaching 100% were observed on specific lettuce fields in France and Spain.

D. Variability/strains

Heterogeneity and rapid adaptability are the two prominent phenotypic characteristics that distinguish TSWV from other plant viruses (Tsompana *et al.*, 2005). Several strains of TSWV have been characterized and studied worldwide, showing a geographical structuring of the isolated

FIGURE 6 Lettuce plant (right) showing severe symptoms of TSWV. (See Page 6 in Color Section at the back of the book.)

populations. Most of the data used for this virus description has been obtained from studies on the best-known strain, BR-01, originally isolated from tomato (De Ávila *et al.*, 1990).

Nucleotide analysis of a number of worldwide TSWV isolates showed a relatively high genetic variability which are grouped in two main genotypes (groups of isolates based on genetic similarity) named genotype A and genotype B. Nucleotide identities are greater than 98% between isolates of the same genotype and about 93% between isolates of different genotypes (Lopez *et al.*, 2011; Tsompana *et al.*, 2005). Different TSWV strains and isolates have complete or partial nucleotide sequence data published or entered in GenBank (http://www.ncbi.nlm.nih.gov).

E. Propagation host and purification

TSWV was first reported in tomato (Brittlebank, 1919). TSWV is polyphagous infecting more than 1000 different plant species from more than 80 botanical families, including both monocotyledonous and dicotyledonous plants. Pepper (*Capsicum annuum*), lettuce, tobacco, tomato, and various ornamental crops are the main hosts of TSWV (Cho *et al.*, 1987; Debreczeni *et al.*, 2011; EPPO/CABI, 1997).

Hosts used for virus maintenance and propagation are *Gomphrena globosa*, *N. glutinosa*, *Nicotiana rustica*, *Tropaeolum majus*, and *Emilia sonchifolia* (Inoue-Nagata *et al.*, 1997).

Several methods have been published for complete virus particles isolation from systemically infected leaves (Gonsalves and Trujillo, 1986; Tas *et al.*, 1977; Verkleij and Peters, 1983).

F. Transmission, epidemiology, and geographic distribution in the Mediterranean basin

TSWV is transmitted by several thrips species, but *F. occidentalis* is by large the most important vector. The virus is acquired by the first or second instar larvae, replicates in the insect's body, and is mainly inoculated by adults. First-instar larvae are unable to transmit the virus, and adults are unable to acquire the virus. The virus is transmitted in a circulative propagative manner by adult thrips when piercing the plant epidermis after injecting their saliva during brief superficial probes lasting few minutes. The adults, who live 30–45 days, remain viruliferous during their entire life. TSWV can be transmitted by seed in several crops but has never been reported to be seed-transmissible in lettuce.

There are many weeds acting as reservoirs of TSWV and of their main vector, *F. occidentalis*. The virus has a broad host range of more than 1000 different plant species including many weeds that are commonly present in lettuce fields (Jorda *et al.*, 1995). Plant species that act as hosts for both the virus and the thrips larvae are the ones that serve as a reservoir of the disease.

Epidemics of TSWV in protected and outdoor crops in the Mediterranean basin are closely associated with the establishment and rapid infestation of its main vector, *F. occidentalis*. It is extended in all of the Mediterranean regions were lettuce is grown including Spain, France, Italy, and Greece. It has also been recently cited in Turkey (Yardimci and Kiliç, 2009). TSWV together with LMV causes the most important virus disease of lettuce grown in the Mediterranean basin.

G. Detection methods

Although difficulties are sometimes encountered in transmitting the virus from old infected plants, TSWV identification is usually efficient if young test plants are used. Most frequent diagnostic species are *C. sativus* (where cotyledons develop local chlorotic spots with necrotic centers by TSWV infection) and several *Nicotiana* species including *N. clevelandii*, *N. glutinosa*, and *N. tabacum* (where local necrotic lesions, followed by systemic necrotic patterns and leaf deformation are induced by LMV infection) (Ie, 1970).

Antisera for TSWV detection are available. These antisera can be used for ELISA, Dot blot, and western blotting, as well as *in situ* localization studies, with generally specific reactions and low backgrounds, both in extract from infected plants and in viruliferous thrips (Bandla *et al.*, 1994; Cho *et al.*, 1988; Gonsalves and Trujillo, 1986; Hsu and Lawson, 1991). The double antibody sandwich (DAS) direct ELISA using polyclonal antibodies to the whole virion is recommended over the use of antibodies to structural proteins for detecting a specific selection of TSWV isolates (Huguenot *et al.*, 1990; Sherwood *et al.*, 1989; Wang and Gonsalves, 1990).

Different PCR-based assays have been developed for TSWV detection (Mumford *et al.*, 1994; Weekes *et al.*, 1996). Recently, a quantitative real-time RT-PCR procedure has been published for general and genotype-specific detection and quantitation of the genomic M segment of TSWV (Debreczeni *et al.*, 2011).

H. Control methods: Present and future

Reducing thrips population is the most effective way to control TSWV. Insecticides can be an effective way to control thrips population. The use of predators such as *Orius laevigatus* and *Amblyseius swirskii* have provided an effective way to control thrips and to reduce the damage caused by TSWV in the Mediterranean region under protected environments (Lacasa and Sánchez, 2002). The use of virus-free seedlings and removal of symptomatic plants are also essential to control the disease. Installation of fine mesh nets in greenhouses and UV-absorbing plastic films also reduce the incidence of TSWV in lettuce (Diaz *et al.*, 2006).

Sources of resistance to TSWV have been found in *Lycopersicon* spp. In lettuces, two cultivars (Tinto and Ancora) are reported to be resistant to TSWV in Hawaii (United States) (O'Malley and Hartmann, 1989). In addition, wild species of *Lactuca perennis*, incompatible with *L. sativa*, has been identified as being resistant to TSWV. Transgenic tomato lines expressing the nucleocapsid (N) protein gene of the lettuce isolate of tomato spotted wilt virus (TSWV-BL) were evaluated for their resistance to tospovirus infection in greenhouse inoculation tests (Gubba *et al.*, 2002). Their results show that it is possible to obtain broad resistance to tospoviruses by combining transgenic and natural resistance in a single plant.

IX. *TURNIP MOSAIC VIRUS* (*POTYVIRIDAE, POTYVIRUS*)

A. Particle morphology and properties

Turnip mosaic virus (TuMV) has filamentous and flexuous particles the same as the other potyviruses, with an average length of 720 nm (Tomlinson, 1970b; Urcuqui-Inchima *et al.*, 2001).

B. Nucleic acid component

TuMV genome, which is a single-stranded, is a positive-sense RNA molecule of approximately 10,000 nt (accession number NC_002509). TuMV has the same typical potyviruses genetic organization as previously described for LMV in Section VI.B (Urcuqui-Inchima *et al.*, 2001; http://www.ncbi.nlm.nih.gov/).

C. Symptoms and damages

The symptoms of TuMV are mottling in broad, yellow, circular, and irregular areas. The oldest leaves often become bright yellow all over. The lamina often becomes necrotic. Unlike BWYV, veins do not become green, and leaves do not become brittle. In Europe, this virus is frequently detected in iceberg lettuce. TuMV is mainly present in iceberg and Batavia types, but rarely observed in romaine lettuce.

D. Variability/strains

TuMV is a highly variable potyvirus. Many biological and serological classifications of TuMV isolates have been reported. Four TuMV isolates have been described according to their ability to infect different hosts: (B) host-type isolates that infect *Brassica* plants latently and occasionally and do not infect *Raphanus* plants; B host-type isolates that infect many *Brassica* plants systemically, producing mosaic symptoms on uninoculated leaves, but do not infect *Raphanus* plants; B(R) host-type isolates that infect many *Brassica* plants systemically, with mosaic symptoms on uninoculated leaves, but infect *Raphanus* plants latently and occasionally; and BR host-type isolates that infect both *Brassica* and *Raphanus* plants systemically, with mosaic symptoms on uninoculated leaves (Ohshima *et al.*, 2002; Tomimura *et al.*, 2003).

Most recently, four TuMV genogroups have been revealed by phylogenetic analysis of numerous TuMV isolates collected around the world: basal-B (*Brassica*), basal-BR (*Brassica/Raphanus*), Asian-BR, and world-B (Ohshima *et al.*, 2010).

TuMV displays a large degree of intraspecific variation comprising several genetic strains. The prevalence of the different strains seems to be associated with the involvement of different host plant species (Tomimura *et al.*, 2004). Genome sequences for UK1 isolate and Quebec strain of TuMV are available in GenBank (http://www.ncbi.nlm.nih.gov/).

E. Propagation hosts and purification

TuMV was first reported in *Brassica campestris* ssp. *chinensis*, *Brassica japonica* and *B. campestris* ssp. *rapa* in the United States (Schultz, 1921). It has a very wide host range infecting at least 318 species in 156 genera of 43 families including various cultivated *Cruciferae* (cabbage, cauliflower, radish, turnip) and numerous wild plants such as *C. bursa-pastoris* or *Stellaria media* (Blancard *et al.*, 2006; Tomlinson, 1970b).

TuMV isolates can be maintained in *Brassica perviridis*, *Petunia hybrida*, or *N. glutinosa*. *Brassica pekinensis*, *B. perviridis*, and *B. rapa* are good sources of virus for purification. Methods for TuMV purification have been described by Shepherd and Pound (1960) and Tomlinson (1964). Some modifications were later proposed by Thompson *et al.* (1988).

F. Transmission, epidemiology, and geographic distribution in the Mediterranean basin

The virus is transmitted in a nonpersistent manner by over 40 aphid species, particularly by *M. persicae*, *A. craccivora*, and *M. euphorbiae*. The same mode of transmission as described for LMV does apply for TuMV. The virus is not seed-transmissible in lettuce. It has a very broad host range particularly among the brassicas.

The virus is present all over the Mediterranean basin including Spain, Italy, and Greece (Moreno *et al.*, 2004; Sanchez *et al.*, 2007), although disease outbreaks are not common in lettuce.

G. Detection methods

Even though it is a laborious and time-consuming method, TuMV detection based on symptoms on diagnostic species is possible. Some susceptible hosts are *C. quinoa*, *C. amaranticolor* (chlorotic and necrotic local lesions and systemic veinal flecks and spots), and *N. tabacum* cv. White Burley (infection is not systemic but chlorotic local spots, becoming centrally necrotic) (Provvidenti, 1983; Tomlinson, 1970b).

Although it is not routinely used nowadays, the EM is another technique that allows virus detection by means of identification of the viral particles and the typical potyvirus inclusion bodies in the cytoplasm of the infected cells (Martin *et al.*, 2004; Ohi *et al.*, 2004).

Moreover, serological techniques including ELISA, western blot, and TBIA, using TuMV or *Potyvirus* specific antibodies, are reliable techniques to detect the virus in infected plants (Chang *et al.*, 2011; Ling *et al.*, 1995).

However, the most powerful tools for the detection and diagnosis of the virus are the nucleic acid-based approaches such as RT-PCR-based techniques or microarray systems using oligonucleotide probes (Chang *et al.*, 2011; Grover *et al.*, 2010; Lunello *et al.*, 2005).

H. Control methods: Present and future

The virus is controlled in a similar manner as previously described for other nonpersistently transmitted viruses (e.g., AMV). There are resistant lettuce cultivars (both butterhead and romaine) available. Resistance to TuMV is controlled by a single dominant gene designated *Tu*, which confers complete immunity and has been introduced in commercial cultivars to protect lettuce. Furthermore, it is interesting to indicate the TuMV susceptibility in *L. sativa*, and *Lactuca serriola* is associated with mildew-resistant progenies, which express the dominant gene *Dm* (Robbins *et al.*, 1994; Zink and Duffus, 1969, 1970).

X. *LETTUCE NECROTIC YELLOWS VIRUS (RHABDOVIRIDAE, CYTORHABDOVIRUS)*

Lettuce necrotic yellows virus (LNYV) particles are bacilliform, about 227 nm long and 66 nm wide (Harrison and Crowley, 1965). The LNYV genome consists of a monopartite, negative-sense, single-stranded RNA of 12,807 nt. The physical map of the LNYV genome is 3′ leader-*N-P-4b-M-G-L*-5′ trailer, where *N* is the nucleocasid gene, *P* is the phosphoprotein gene, *4b* encodes a putative movement protein, *M* is the matrix protein gene, *G* is the glycoprotein gene, and *L* is the polymerase gene (accession number NC_007642) (Dietzgen *et al.*, 2006; Wetzel *et al.*, 1994; http://www.ncbi.nlm.nih.gov/). It has a narrow host range, is transmitted by sap inoculation and naturally by the aphid *H. lactucae*, in a persistent propagative manner (Francki *et al.*, 1989). It is not transmitted by seeds. This virus was recorded for the first time in Australia (Stubbs and Grogan, 1963) and has been cited from time to time in New Zealand, Italy, Spain, and Great Britain. Its distribution appears to coincide with that of some hosts, such as *S. oleraceus*, which are potentially important sources of the disease. LNYV has a very low incidence on lettuce. Typical symptoms are chlorosis, mottling on lower leaves, and sometimes necrosis. When plants are infected before the head has formed, internal necrotic damage can be observed (Blancard *et al.*, 2006; Randles, 1987). *N. glutinosa* and naturally infected lettuce are good virus sources for purification (McLean and Francki, 1967). Control measures to reduce the incidence of LNYV are host and/or vector orientated. The high degree of association between the main vector *H. lactucae* and the primary host plant *S. oleraceus* should allow the deployment of control measures for disease control. For instance, the eradication of *S. oleraceus* from the vicinity of lettuce fields provides effective control of LNYV (Stubbs *et al.*, 1963). Removing of weeds and application of insecticides close to crops is also an option (Davis *et al.*, 2002). Detection of LNYV in host plants and in its aphid vector is possible by serological methods, such as ELISA or immunobloting tests (Chu and Francki, 1982; Dietzgen and Francki, 1990), as well as by nucleic acid-based techniques such as northern blot or RT-PCR (Callaghan and Dietzgen, 2005; Dietzgen *et al.*, 1989; Thomson and Dietzgen, 1995).

XI. CONCLUSIONS

Lettuce is frequently attacked by several virus diseases along the Mediterranean basin. The most damaging and serious virus diseases are caused by TSWV and LMV. In regions with heavy and humid soils, MLBVV can also produce major yield losses. There are a series of control strategies to limit the spread of the virus and to limit the yield losses depending on the type of virus–vector relationship and the epidemiological properties of the disease. Reducing thrips population by using insecticides or physical barriers can provide an effective way to reduce the spread of TSWV. Certification of virus-free seeds is the most effective way to reduce the incidence of LMV. Sanitary soil disinfection and elimination of plant debris at the end of the growing season is one of the most effective ways to remove the fungal vector and the incidence of MLBVV in lettuce fields.

REFERENCES

Alonso-Prados, J. L., Luis-Arteaga, M., Alvarez, J. M., Moriones, E., Batlle, A., Lavina, A., Garcia-Arenal, F., and Fraile, A. (2003). Epidemics of aphid-transmitted viruses in melon crops in Spain. *Eur. J. Plant Pathol.* **109:**129–138.

Astier, S., Albouy, J., Maury, Y., Robaglia, C., and Lecoq, H. (2007). Principles of Plant Virology: Genome, Pathogenicity, Virus Ecology. Science Publishers, New Hampshire, p. 472.

Bandla, M. D., Westcot, D. M., Chenault, K. D., Ullman, D. E., German, T. L., and Sherwood, J. L. (1994). Use of monoclonal antibody to the nonstructural protein encoded by the small RNA of *tomato spotted wilt tospovirus* to identify viruliferous thrips. *Phytopathology* **84:**1427–1431.

Best, R. J. (1968). Tomato spotted wilt virus. *Adv. Virus Res.* **13:**65–145.

Beuve, M., Stevens, M., Liu, H. Y., Wintermantel, W. M., Hauser, S., and Lemaire, O. (2008). Biological and molecular characterization of an American sugar beet-infecting *Beet western yellows virus* isolate. *Plant Dis.* **92:**51–60.

Bhat, A. I., and Siju, S. (2007). Development of a single-tube multiplex RT-PCR for the simultaneous detection of *Cucumber mosaic virus* and *Piper yellow mottle virus* associated with stunt disease of black pepper. *Curr. Sci.* **93:**973–976.

Blancard, D., Lot, H., and Maisonneuve, B. (2006). A Colour Atlas of Diseases of Lettuce and Related Salad Crops: Observation, Biology and Control. Manson Publishing, London, p. 375.

Bol, J. F. (2005). Replication of alfamo- and ilarvirus: Role of the coat protein. *Annu. Rev. Phytopathol.* **43:**39–62.

Bonnet, J., Fraile, A., Sacristán, S., Malpica, J. M., and García-Arenal, R. (2005). Role of recombination in the evolution of natural populations of *Cucumber mosaic virus*, a tripartite RNA plant virus. *Virology* **332:**359–368.

Bowyer, J. W. (1980). *Broad bean wilt fabavirus*. Plant Viruses Online. Descriptions and Lists from the VIDE Database. http://www.agls.uidaho.edu/ebi/vdie/refs.htm.

Brittlebank, C. C. (1919). Tomato diseases. *J. Agric. Victoria* **27:**231–235.

Broadbent, L., Tinsley, T. W., Buddin, W., and Roberts, E. T. (1951). The spread of lettuce mosaic in the field. *Ann. Appl. Biol.* **38:**689–706.

Callaghan, B., and Dietzgen, R. G. (2005). Nucleocapsid gene variability reveals two subgroups of *Lettuce necrotic yellows virus*. *Arch. Virol.* **150:**1661–1667.

Campbell, R. N. (1965). Weeds as reservoir hosts of the *Lettuce big-vein virus*. *Can. J. Bot.* **43:**1141–1149.

Chang, P. G. S., McLaughlin, W. A., and Tolin, S. A. (2011). Tissue blot immunoassay and direct RT-PCR of cucumoviruses and potyviruses from the same NitroPure nitrocellulose membrane. *J. Virol. Methods* **171:**345–351.

Chatzivassiliou, E. K., Efthimiou, K., Drossos, E., Papadopoulou, A., Poimenidis, G., and Katis, N. I. (2004). A survey of tobacco viruses in tobacco crops and native flora in Greece. *Eur. J. Plant Pathol.* **110:**1011–1023.

Chen, S. C., and Olsthoorn, R. C. L. (2010). In vitro an in vivo studies of the RNA conformational switch in *Alfalfa mosaic virus. J. Virol.* **84:**1423–1429.

Chen, S., Gu, H., Wang, X., Chen, J., and Zhu, W. (2011). Multiplex RT-PCR detection of *Cucumber mosaic virus* subgroups and Tobamoviruses infecting Tomato using 18S rRNA as an internal control. *Acta Biochim. Biophys. Sin.* **43:**465–471.

Cho, J. J., Mau, R. F. L., Mitchell, W. C., Gonsalves, D., and Yudin, L. S. (1987). Host List of Plants Susceptible to *Tomato Spotted Wilt Virus* (TSWV). University of Hawaii, Research Extension Series, RES-078, Honolulu, HI, p. 10.

Cho, J. J., Mau, R. F. L., Hamasaki, R. T., and Gonsalves, D. (1988). Detection of *tomato spotted wilt virus* in individual thrips by enzyme-linked immunosorbent assay. *Phytopathology* **78:**1348–1352.

Chomič, A., Pearson, M. N., Clover, G. R. G., Farreyrol, K., Saul, D., Hampton, J. G., and Armstrong, K. F. (2010). A generic RT-PCR assay for the detection of *Luteoviridae. Plant Pathol.* **59:**429–442.

Chu, P. W. G., and Francki, R. I. B. (1982). Detection *of lettuce necrotic yellows virus* by an enzyme-linked immunosorbent assay in plant hosts and the insect vector. *Ann. Appl. Biol.* **100:**149–156.

Clark, M. F. (1968). Purification and fractionation of *Afalfa mosaic virus* with polyethylene glycol. *J. Gen. Virol.* **3:**427–432.

Clark, G. L., Peden, K. W. C., and Symons, R. H. (1974). *Cucumber mosaic virus*-induced RNA polymerase: Partial purification and properties of the template-free enzyme. *Virology* **62:**434–443.

D'Arcy, C. J., Hewings, A. D., Burnet, P. A., and Jedlinski, H. (1983). Comparative purification of two luteoviruses. *Phytopathology* **73:**755–759.

Davis, R. M., Subbarao, K. V., Raid, R. N., and Kurtz, E. A. (2002). Lettuce Pests and Diseases. The American Phytopathological Society, Mundi-Prensa, Madrid, p. 79.

de Ávila, A. C., Huguenot, C., Resende, R. O., Kitajima, E. W., Goldbach, R. W., and Peters, D. (1990). Serological differentiation of 20 isolates of *Tomato spotted wilt virus. J. Gen. Virol.* **71:**2801–2807.

de Haan, P., Wagemakers, L., Peters, D., and Goldbach, R. (1989). Molecular cloning and terminal sequence determination of the S and M RNAs of *Tomato spotted wilt virus. J. Gen. Virol.* **70:**3469–3473.

de Haan, P., Wagemakers, L., Peters, D., and Goldbach, R. (1990). The S RNA segment of spotted wilt virus has an ambisense character. *J. Gen. Virol.* **71:**1001–1007.

de Haan, P., Kormelink, R., de Oliveira Resende, R., van Poelwijk, F., Peters, D., and Goldbach, R. (1991). *Tomato spotted wilt virus* L RNA encodes a putative RNA polymerase. *J. Gen. Virol.* **72:**2207–2216.

Debreczeni, D. E., Ruiz-Ruiz, S., Aramburu, J., López, C., Belliure, B., Galipienso, L., Soler, S., and Rubio, L. (2011). Detection, discrimination and absolute quantitation of *Tomato spotted wilt virus* isolates using real time RT-PCR with TaqMan®MGB probes. *J. Virol. Methods* **176:**32–37.

Decroocq, V., Salvador, B., Sicard, O., Glasa, M., Cosson, P., Svanella-Dumas, L., Revers, F., Garcia, J. A., and Candresse, T. (2009). The determinant of potyvirus ability to overcome the RTM resistance of *Arabidopsis thaliana* maps to the N-terminal region of the coat protein. *Mol. Plant Microbe Interact.* **22:**1302–1311.

Diaz, B. M., Biurrun, R., Moreno, A., Nebreda, M., and Fereres, A. (2006). Impact of ultraviolet-blocking plastic films on insect vectors of virus diseases infesting crisp lettuce. *HortScience* **41:**711–716.

Dietzgen, R. G., and Francki, R. I. (1990). Reducing agents interfere with the detection of *lettuce necrotic yellows virus* in infected plants by immunoblotting with monoclonal antibodies. *J. Virol. Methods* **28:**199–206.

Dietzgen, R. G., Hunter, B. G., Francki, R. I. B., and Jackson, A. O. (1989). Cloning of *lettuce necrotic yellows virus* RNA and identification of virus-specific polyadenylated RNAs in infected *Nicotiana glutinosa* leaves. *J. Gen. Virol.* **70:**2299–2307.

Dietzgen, R. G., Callaghan, B., Wetzel, T., and Dale, J. L. (2006). Completion of the genome sequence of *Lettuce necrotic yellows virus*, type species of the genus *Cytorhabdovirus*. *Virus Res.* **118:**16–22.

Dinant, S., and Lot, H. (1992). Lettuce mosaic virus. *Plant Pathol.* **41:**528–542.

Duffus, J. E. (1961). Economic significance of beet western yellows (Radish yellows) on sugar beet. *Phytopathology* **51:**605–607.

Duffus, J. E. (1964). Host relationships of *Beet western yellows virus* strains. *Phytopathology* **54:**736–738.

Duffus, J. E. (1972). *Beet western yellows virus*. Descriptions of Plant Viruses. http://www.dpvweb.net/index.php.

Duffus, J. E. (1973). The yellowing viruse diseases of beet. *Adv. Virus Res.* **18:**347–386.

Duffus, J. E. (1977). Serological relationships among Beet western yellows, Barley yellow dwarf, and Soybean dwarf viruses. *Phytopathology* **67:**1197–1201.

Duffus, J. E. (1981). *Beet western yellows virus*—A major component of some potato leaf roll-affected plants. *Phytopathology* **71:**193–196.

Duffus, J. E., and Johnstone, G. R. (1983). *Beet western yellows luteovirus*. Plant Viruses Online. Descriptions and List from the VIDE Database. http://www.agls.uidaho.edu/ebi/vdie/refs.htm.

Edwardson, J. R., and Christie, R. G. (1991). CRC Handbook of Viruses Infecting Legumes. CRC Press, Boca Raton, FL, p. 293.

Ellis, P. J., and Wieczorek, A. (1992). Production of monoclonal antibodies to *beet western yellows virus* and *potato leafroll virus* and their use in luteovirus detection. *Plant Dis.* **76:**75–78.

EPPO/CABI (1997). *Tomato spotted wilt tospovirus* . *In* Quarantine Pests for Europe, 2nd edn. pp. 1379–1387. CAB International, Wallingford, CT.

Eurostat. (2012). Fruit and vegetables—Yield—(Annual data). http://epp.eurostat.ec.europa.eu/portal/page/portal/product_details/dataset?p_product_code=APRO_CPP_FRUVEG (accessed on 2nd January 2012).

Faostat. (2012). Crop production data. http://faostat.fao.org (accessed on 2nd January 2012).

Fauquet, C. M., Mayo, M. A., Maniloff, J., Desselberger, U., and Ball, L. A. (2005). Virus taxonomy. VIIIth Report of the International Committee on the Taxonomy of Viruses. Elsevier Academic Press, San Diego, CA, p. 1259.

Feng, J. L., Chen, S. N., Tang, X. S., Ding, X. F., Du, Z. Y., and Chen, J. S. (2006). Quantitative determination of *Cucumber mosaic virus* genome RNAs in virions by real-time reverse transcription-polymerase chain reaction. *Acta Biochim. Biophys. Sin.* **38:**669–676.

Ferrer, R. M., Escriu, F., Luis-Arteaga, M., Guerri, J., Moreno, P., and Rubio, L. (2008). New molecular methods for identification of *Broad bean wilt virus* 1. *Mol. Cell. Probes* **22:**223–227.

Ferrer, R. M., Ferriol, I., Moreno, P., Guerra, J., and Rubio, L. (2011). Genetic variation and evolutionary analysis of *Broad bean wilt virus* 2. *Arch. Virol.* **156:**1445–1450.

Ferrer, R. M., Guerri, J., Luis-Arteaga, M. S., Moreno, P., and Rubio, L. (2005). The complete sequence of a Spanish isolate of *Broad bean wilt virus* 1 (BBWV-1) reveals a high variability and conserved motifs in the genus *Fabavirus*. *Arch. Virol.* **150:**2109–2116.

Ferriol, I., Ruiz-Ruiz, S., and Rubio, L. (2011). Detection and absolute quantitation of *Broad bean wilt virus* 1 (BBWV-1) and BBWV-2 by real time RT-PCR. *J. Virol. Methods* **177:**202–205.

Francki, R. I. B., Randles, J. W., and Dietzgen, R. G. (1989). *Lettuce necrotic yellows virus. CMI/AAB Descr. Pl. Viruses* No. 343, p. 5

German-Retana, S., Walter, J., and Le Gall, O. (2008). Pathogen profile *Lettuce mosaic virus*: From pathogen diversity to host interactors. *Mol. Plant Pathol.* **2:**127–136.

Gibbs, A. J. (1987). *Alfalfa mosaic alfamovirus*. Plant Viruses Online. Descriptions and List from the VIDE Database. http://www.agls.uidaho.edu/ebi/vdie/refs.htm.

Gibbs, A. J., and Harrison, B. D. (1970). *Cucumber mosaic virus*. Descriptions of Plant Viruses. http://www.dpvweb.net/index.php.

Gold, A. H., and Duffus, J. E. (1967). Infectivity neutralization—A serological method as applied to persistent viruses of beets. *Virology* **31:**308–313.

Gonsalves, D., and Trujillo, E. E. (1986). *Tomato spotted wilt virus* in papaya and detection of the virus by ELISA. *Plant Dis.* **70:**501–506.

Govier, D. A. (1985). Purification and partial characterisation of *beet mild yellowing virus* and its serological detection in plants and aphids. *Ann. Appl. Biol.* **107:**439–447.

Grover, V., Pierce, M. L., Hoyt, P., Zhang, F., and Melcher, U. (2010). Oligonucleotide-based microarray for detection of plant viruses employing sequence-independent amplification of targets. *J. Virol. Methods* **163:**57–67.

Gubba, A., Gonsalves, C., Stevens, M. R., Tricoli, D. M., and Gonsalves, D. (2002). Combining transgenic and natural resistance to obtain broad resistance to tospovirus infection in tomato (Lycopersicon esculentum mill). *Mol. Breed.* **9:**13–23.

Harrison, B. D., and Crowley, N. C. (1965). Properties and structure of *lettuce necrotic yellows virus*. *Virology* **26:**297–310.

Hauser, S., Stevens, M., Mougel, C., Smith, H. G., Fritsch, C., Herrbach, E., and Lemaire, O. (2000). Biological, serological, and molecular variability suggest three distinct polerovirus species infecting beet or rape. *Phytopathology* **90:**460–466.

Hayes, R. J., and Ryder, E. J. (2007). Introgression of novel alleles for partial resistance to big vein disease from *Lactuca virosa* into cultivated lettuce. *HortScience* **42:**35–39.

Helliot, B., Panis, B., Busogoro, J. P., Sobry, S., Poumay, Y., Raes, M., Swennen, R., and Lepoivre, P. (2007). Immunogold silver staining associated with epi-fluorescence for *cucumber mosaic virus* localisation on semi-thin sections of banana tissues. *Eur. J. Histochem.* **51:**153–158.

Herrbach, E., Lemaire, O., Ziegler-Graff, V., Lot, H., Rabenstein, F., and Bouchery, Y. (1991). Detection of BMYV and BWYV isolates using monoclonal antibodies and radioactive RNA probe, and relationships among luteoviruses. *Ann. Appl. Biol.* **118:**127–138.

Hodge, S., Hardie, J., and Powell, G. (2011). Parasitoids aid dispersal of a nonpersistently transmitted plant virus by disturbing the aphid vector. *Agric. For. Entomol.* **13:**83–88.

Hooks, C. R. R., and Fereres, A. (2006). Protecting crops from non-persistently aphid-transmitted viruses: A review on the use of barrier plants as a management tool. *Virus Res.* **120:**1–16.

Hsu, H. T., and Lawson, R. H. (1991). Direct tissue blotting for detection of *tomato spotted wilt virus* in *Impatiens*. *Plant Dis.* **75:**292–295.

Huguenot, C., Dobbelsteen, G., Van Den Haan, P., De Wagemakers, C. A. M., Drost, G. A., Osterhaus, A. D. M. E., and Peters, D. (1990). Detection of *tomato spotted wilt virus* using monoclonal antibodies and riboprobes. *Arch. Virol.* **110:**47–62.

Huijberts, N., Blystad, D. R., and Bos, L. (1990). Lettuce big-vein virus: Mechanical transmission and relationships to tobacco stunt virus. *Ann. Appl. Biol.* **116:**463–475.

Hull, R. (1969). Alfalfa mosaic virus. *Adv. Virus Res.* **25:**365–433.

Hull, R. (2002). Matthews' Plant Virology. 4th edn Academic Press, London. p. 1001.

Ie, T. S. (1970). *Tomato spotted wilt virus*. Descriptions of Plant Viruses. http://www.dpvweb.net/index.php.

Ikegami, M., Kawashima, H., Natsuaki, T., and Sugimura, N. (1998). Complete nucleotide sequence of the genome organization of RNA2 of *patchouli mild mosaic virus*, a new fabavirus. *Arch. Virol.* **143:**2431–2434.

Ikegami, M., Onobori, Y., Sugimura, N., and Natsuaki, T. (2001). Complete nucleotide sequence and the genome organization of *patchouli mild mosaic virus* RNA1. *Intervirology* **44:**355–358.

Inoue-Nagata, A. K., Kormelink, R., Nagata, T., Kitajima, E. W., Goldbach, R., and Peters, D. (1997). Effects of temperature and host on the generation of *tomato spotted wilt virus* defective interfering RNAs. *Phytopathology* **87:**1168–1173.

Jaspars, E. M. J., and Bos, L. (1980). *Alfalfa mosaic virus*. Descriptions of Plant Viruses. http://www.dpvweb.net/index.php.

Jeger, M. J., Chen, Z., Powell, G., Hodge, S., and van den Bosch, F. (2011). Interactions in a host plant-virus-vector-parasitoid system: Modelling the consequences for virus transmission and disease dynamics. *Virus Res.* **159:**183–193.

Jorda, C., Ortega, A., and Juarez, M. (1995). New hosts of *tomato spotted wilt virus*. *Plant Dis.* **79:**538.

Kawazu, Y., Fujiyama, R., Sugiyanta, K., and Sasaya, T. (2006). A transgenic lettuce line with resistance to both *lettuce big-vein associated virus* and *mirafiori lettuce virus*. *J. Am. Soc. Hortic. Sci.* **131:**760–763.

Kawazu, Y., Fujiyama, R., and Noguchi, Y. (2009). Transgenic resistance to *Mirafiori lettuce virus* in lettuce carrying inverted repeats of the viral coat protein gene. *Transgenic Res.* **18:**113–120.

Kennedy, J. S., Day, M. F., and Eastop, V. F. (1962). A Conspectus of Aphids as Vectors of Plant Viruses. Commonwealth Institute of Entomology, London.

Kobayashi, Y. O., Kobayashi, A., Nakano, M., Hagiwara, K., Honda, Y., and Omura, T. (2003). Analysis of genetic relations between *Broad bean wilt virus* 1 and *Broad bean wilt virus* 2. *J. Gen. Plant Pathol.* **69:**320–326.

Koh, L. H., Cooper, J. I., and Wong, S. M. (2001). Complete sequences and phylogenetic analyses of a Singapore isolate of *broad bean wilt fabavirus*. *Arch. Virol.* **146:**135–147.

Kormelink, R., de Haan, P., Meurs, C., Peters, D., and Goldbach, R. (1992). The nucleotide sequence of the M RNA segment of *tomato spotted wiltvirus*, a bunyavirus with two ambisense RNA segments. *J. Gen. Virol.* **73:**2795–2804.

Krause-Sakate, R., Le Gall, O., Fakhfakh, H., Peypelut, M., Marrakchi, M., Varveri, C., Pavan, M. A., Souche, S., Lot, H., Zerbini, F. M., and Candresse, T. (2002). Molecular and biological characterization of *Lettuce mosaic virus* (LMV) isolates reveals a distinct and widespread type of resistance-breaking isolate: LMV-Most. *Phytopathology* **92:**563–572.

Krause-Sakate, R., Redondo, E., Richard-Forget, F., Jadao, A. S., Houvenaghel, M. C., German-Retana, S., Pavan, M. A., Candresse, T., Zerbini, F. M., and Le Gall, O. (2005). Molecular mapping of the viral determinants of systemic wilting induced by a *Lettuce mosaic virus* (LMV) isolate in some lettuce cultivars. *Virus Res.* **109:**175–780.

Kuroda, T., Okumura, A., Takeda, I., Miura, Y., and Suzuki, K. (2000). Nucleotide sequence and synthesis of infectious RNA from cloned cDNA of *broad bean wilt virus* 2 RNA 2. *Arch. Virol.* **145:**787–793.

Kuwata, S. (1991). *Lettuce big-vein varicosavirus*. Plant viruses online. Descriptions and list from the VIDE Database. http://www.agls.uidaho.edu/ebi/vdie/refs.htm.

Kuwata, S., Kubo, S., Yamashita, S., and Doi, Y. (1983). Rod-shaped particles, a probable entity of *lettuce big-vein virus*. *Ann. Phytopathol. Soc. Japan* **49:**246–251.

Lacasa, A., and Sánchez, J. A. (2002). El estado actual del control integrado de los tisanópteros en cultivos de invernadero. La situación del pimiento. *Phytoma España* 101–105.

Latham, L. J., Jones, R. A. C., and McKirdy, S. J. (2004). Lettuce big-vein disease: Sources, patterns of spread, and losses. *Aust. J. Agric. Res.* **55:**125–130.

Lee, U., Hong, J. S., Choi, J. K., Kim, K. C., Kim, Y. S., Curtis, I. S., Nam, H. G., and Lim, P. O. (2000). *Broad bean wilt virus* causes necrotic and generates defective RNAs in *Capsicum annuum*. *Phytopathology* **90:**1390–1395.

Legarrea, S., Fraile, A., Garcia-Arenal, F., Viñuela, E., and Fereres, A. (2009). Spatio-temporal dynamics of aphid-transmitted viruses in lettuce crops under UV-absorbing nets. *IOBC/WPRS Bull.* **49:**253–258.

Legarrea, S., Betancourt, M., Plaza, M., Fraile, A., García-Arenal, F., and Fereres, A. (2012). Dynamics of nonpersistent aphid-borne viruses in lettuce crops covered with UV-absorbing nets. *Virus Res.* **165:**1–8.

Lemaire, O., Herrbach, E., Stevens, M., Bouchery, Y., and Smith, H. G. (1995). Detection of sugar beet-infecting beet mild yellowing luteovirus isolates with a specific RNA probe. *Phytopathology* **85:**1513–1518.

Ling, K. S., Provvidenti, R., and Gonsalves, D. (1995). Detection of *turnip mosaic virus* isolates using an antiserum to coat protein breakdown product. *Plant Dis.* **79:**809–812.

Lisa, V., and Boccardo, G. (1996). Fabaviruses: Broad bean wilt virus and allied viruses. *In* "Polyhedral Virions and Bipartite RNA Genomes, The Plant Viruses" (B. D. Harrison and A. F. Murant, eds.), Vol. 5, pp. 229–250. Plenum Press, New York.

Lopez, C., Aramburu, J., Galipienso, L., Soler, S., Nuez, F., and Rubio, L. (2011). Evolutionary analysis of tomato Sw-5 resistance-breaking isolates of *Tomato spotted wilt virus*. *J. Gen. Virol.* **92:**210–215.

López-Moya, J. J., and García, J. A. (1999). *Potyvirus* (*Potyviridae*). *In* Encyclopedia of Virology, pp. 1369–1375. Academic Press, San Diego, CA.

Lot, H., and Maury-Chovelon, V. (1985). New data on the two major virus diseases of lettuce in France: *Lettuce mosaic virus* and *beet western yellows virus*. *Phytoparasitica* **13:**277.

Lot, H., Marrou, J., Quiot, J. B., and Esvan, C. (1972). Contribution a l'etude du virus de la mosaique du concombre. I. Méthode de purification rapide du virus. *Ann. Phytopathol.* **4** (25)**:**38.

Lot, H., Campbell, R. N., Souche, S., Milne, R. G., and Roggero, P. (2002). Transmission by *Olpidium brassicae* of *Mirafiori lettuce virus* and *Lettuce big-vein virus*, and their roles in lettuce big-vein etiology. *Phytopathology* **92:**288–293.

Lunello, P., Ducasse, D., and Conci, V. (2005). Improved PCR detection of potyviruses in Allium species. *Eur. J. Plant Pathol.* **112:**371–378.

Luo, H., Wylie, S. J., and Jones, M. G. K. (2010). Identification of plant viruses using one-dimensional gel electrophoresis and peptide mass fingerprints. *J. Virol. Methods* **165:**297–301.

Maccarone, L. D., Barbetti, M. J., Sivasithamparam, K., and Jones, R. A. C. (2010). Comparison of the coat protein genes of *Mirafiori lettuce big-vein virus* isolates from Australia with those of isolates from other continents. *Arch. Virol.* **155:**1519–1522.

Maisonneuve, B. (2003). *Lactuca virosa*, a source of disease resistance genes for lettuce breeding: Results and difficulties for gene introgression. Eucarpia Leafy Vegetables Conf. Noordwijkerhout, the Netherlands, 19–21 March 2003, pp. 61–67.

Makkouk, K. M., Kumari, S. G., and Bos, L. (1990). *Broad bean wilt virus*: Host range, purification, serology, transmission characteristics, and occurrence in faba bean in West Asia and North Africa. *Neth. J. Plant Pathol.* **96:**291–300.

Marco, S. (1984). *Beet western yellows virus* in Israel. *Plant Dis.* **68:**162–163.

Martin, E. M., Cho, J. D., Kim, J. S., Goeke, S. C., Kim, K. S., and Gergerich, R. C. (2004). Novel cytopathological structures induced by mixed infection of unrelated plant viruses. *Phytopathology* **94:**111–119.

McLean, G. D., and Francki, R. I. B. (1967). Purification of *lettuce necrotic yellows virus* by column chromatography on calcium phosphate gel. *Virology* **31:**585–591.

Miller, S. A., and Martin, R. R. (1988). Molecular diagnosis of plant disease. *Annu. Rev. Phytopathol.* **26:**409–432.

Moghal, S. M., and Francki, R. I. B. (1976). Towards a system for the identification and classification of potyviruses: I. Serology and amino acid composition of six distinct viruses. *Virology* **73:**350–362.

Moreno, A., de Blas, C., Biurrun, R., Nebreda, M., Palacios, I., Duque, M., and Fereres, A. (2004). The incidence and distribution of viruses infecting lettuce, cultivated Brassica and associated natural vegetation in Spain. *Ann. Appl. Biol.* **144:**339–346.

Moreno, A., Bertolini, E., Olmos, A., Cambra, M., and Fereres, A. (2007a). Estimation of vector propensity for *Lettuce mosaic virus* based on viral detection in single aphids. *Span. J. Agric. Res* **5:**376–384.

Moreno, A., Nebreda, M., Diaz, B. M., Garcia, M., Salas, F., and Fereres, A. (2007b). Temporal and spatial spread of *Lettuce mosaic virus* in lettuce crops in central Spain: Factors involved in *Lettuce mosaic virus* epidemics. *Ann. Appl. Biol.* **150:**351–360.

Mulabagal, V., Ngouajio, M., Nair, A., Zhang, Y., Gottumukkala, A. L., and Nair, M. G. (2010). In vitro evaluation of red and green lettuce (*Lactuca sativa*) for functional food properties. *Food Chem.* **118:**300–306.

Mumford, R. A., Barker, I., and Wood, K. R. (1994). The detection of *tomato spotted wilt virus* using the polymerase chain reaction. *J. Virol. Methods* **46:**303–311.

Nakamura, S., Iwai, T., and Honkura, R. (1998). Complete nucleotide sequence and genome organization of *broad bean wilt virus* 2. *Ann. Phytopathol. Soc. Japan* **64:**565–568.

Navarro, J. A., Botella, F., Maruhenda, A., Sastre, P., Sanchez-Pina, M. A., and Pallas, V. (2004). Comparative infection progress analysis of *Lettuce big-vein* virus and *Mirafiori lettuce virus* in lettuce crops by developed molecular diagnosis tecnhiques. *Phytopathology* **94:**470–477.

Navarro, J. A., Botella, F., Maruhenda, A., Sastre, P., Sanchez-Pina, M. A., and Pallas, V. (2005a). Identification and partial characterisation of *Lettuce big-vein associated virus* and *Mirafiori lettuce big-vein virus* in common weeds found amongst Spanish lettuce crops and their role in lettuce big-vein disease transmission. *Eur. J. Plant Pathol.* **113:**25–34.

Navarro, J. A., Torok, V. A., Vetten, H. J., and Pallas, V. (2005b). Genetic variability in the coat protein genes of *Lettuce big-vein associated virus* and *Mirafiori lettuce big-vein virus*. *Arch. Virol.* **150:**681–694.

Nebreda, M., Moreno, A., Pérez, N., Palacios, I., Seco Fernández, M. V., and Fereres, A. (2004). Activity of aphids associated with lettuce and broccoli in Spain and their efficiency as vector of *Lettuce mosaic virus*. *Virus Res.* **100:**83–88.

Nebreda, M., Biurrun, R., Moreno, A., Díaz, B., and Fereres, A. (2005). Impacto de cubiertas agrotextiles en el control de pulgones, mosca blanca y virus en cultivos de lechuga y bróculi. *Phytoma España* **166:**16–26.

Niimi, Y., Han, D. S., Mori, S., and Kobayashi, H. (2003). Detection of *Cucumber mosaic virus*, *Lily symptomless virus* and *Lily mottle virus* in Lilium species by RT-PCR technique. *Sci. Hortic.* **97:**57–63.

Ohi, M., Li, Y., Cheng, Y. F., and Walz, T. (2004). Negative staining and image classification—Powerful tools in modern electron microscopy. *Biol. Proced. Online* **6:**23–34.

Ohshima, K., Yamaguchi, Y., Hirota, R., Hamamoto, T., Tomimura, K., Tan, Z., Sano, T., Azuhata, F., Walsh, J. A., Fletcher, J., Chen, J., Gera, A., *et al.* (2002). Molecular evolution of *Turnip mosaic virus*: Evidence of host adaptation, genetic recombination and geographical spread. *J. Gen. Virol.* **83:**1511–1521.

Ohshima, K., Akaishi, S., Kajiyama, H., Koga, R., and Gibbs, A. J. (2010). Evolutionary trajectory of *turnip mosaic virus* populations adapting to a new host. *J. Gen. Virol.* **91:** 788–801.

O'Malley, P. J., and Hartmann, R. W. (1989). Resistance to *tomato spotted wilt virus* in lettuce. *HortScience* **24:**360–362.

Oshima, K., and Shikata, E. (1990). On the screening procedures of ELISA or monoclonal antibodies against three luteoviruses. *Ann. Phytopathol. Soc. Japan* **56:**219–228.

Palukaitis, P., and García-Arenal, F. (2002). Cucumoviruses. *Adv. Virus Res.* **62:**241–323.

Palukaitis, P., Roossinck, M. J., Dietzgen, R. G., and Francki, R. I. B. (1992). Cucumber mosaic virus. *Adv. Virus Res.* **41:**281–348.

Parrella, G., and Crescenzi, A. (2005). The present status of tomato viruses in Italy. (M. T. Momol, P. Ji, and J. B. Jones, eds.),Proceedings of the 1st International Symposium on Tomato Diseases*Acta Horticulturae* **695:**37–42.

Parrella, G., Lanave, C., Marchoux, G., and Finetti, M. M. (2000). Evidence for two distinct subgroups of *Alfalfa mosaic virus* (AMV) from France and Italy and their relationships with other AMV Straits. *Arch. Virol.* **145:**2659–2667.

Peden, K. W. C., and Symons, R. H. (1973). *Cucumber mosaic virus* contains a functionally divided genome. *Virology* **53:**487–492.

Peypelut, M., Krause-Sakate, R., Guiraud, T., Pavan, M. A., Candresse, T., Zerbini, F. M., and Le Gall, O. (2004). Specific detection of *Lettuce mosaic virus* isolates belonging to the "Most" type *J. Virol. Methods* **121:**119–124.

Price, W. C. (1940). Comparative host ranges of six plant viruses. *Am. J. Bot.* **27:**530–541.

Provvidenti, P. (1983). *Turnip mosaic potyvirus*. Plant Viruses Online. Descriptions and Lists from the VIDE Database. http://www.agls.uidaho.edu/ebi/vdie/refs.htm.

Qi, Y., Zhou, X., and Li, D. (2000a). Complete nucleotide sequence and infectious cDNA clone of the RNA1 of a Chinese isolate of *broad bean wilt virus* 2. *Virus Genes* **20:**201–207.

Qi, Y., Zhou, X., Xue, C., and Li, D. (2000b). Nucleotide sequence of RNA2 and polyprotein processing sites of a Chinese isolate of *broad bean wilt virus* 2. *Prog. Nat. Sci.* **10:**680–686.

Randles, J. W. (1987). *Lettuce necrotic yellows cytorhabdovirus*. Plant Viruses Online. Descriptions and Lists from the VIDE Database. http://www.agls.uidaho.edu/ebi/vdie/refs.htm.

Revers, F., Lot, H., Souche, S., Le Gall, O., Candresse, T., and Dunez, J. (1997). Biological and molecular variability of *Lettuce mosaic virus* isolates. *Phytopathology* **87:**397–403.

Robbins, M. A., Witsenboer, H., Michelmore, R. W., Laliberte, J. F., and Fortin, M. G. (1994). Genetic mapping of *turnip mosaic virus* resistance in *Lactuca sativa*. *Theor. Appl. Genet.* **89:**583–589.

Roggero, P., Ciuffo, M., Vaira, A. M., Accotto, G. P., Masenga, V., and Milne, R. G. (2000). An Ophiovirus isolated from lettuce with big-vein symptoms. *Arch. Virol.* **145:** 2629–2642.

Roggero, P., Lot, H., Souche, S., Lenzi, R., and Milne, R. G. (2003). Occurrence of *Mirafiori lettuce virus* and *Lettuce big-vein virus* in relation to development of big-vein symptoms in lettuce crops. *Eur. J. Plant Pathol.* **109:**261–267.

Roossinck, M. J. (2002). Evolutionary history of *Cucumber mosaic virus* deduced by phylogenetic analyses. *J. Virol.* **76:**3382–3387.

Roossinck, M. J., Zhang, L., and Hellwald, K. H. (1999). Rearrangements in the 5_ nontranslated region and phylogenetic analyses of *cucumber mosaic virus* RNA 3 indicate radial evolution of three subgroups. *J. Virol.* **73:**6752–6758.

Russell, G. E. (1958). Sugar beet yellows: A preliminary study of the distribution and interrelationships of viruses and virus strains found in East Anglia, 1955–1957. *Ann. Appl. Biol.* **46:**393–398.

Sanchez, F., Rodriguez-Mateos, M., Tourino, A., Fresno, J., Gomez-Campo, C., Jenner, C. E., Walsh, J. A., and Ponz, F. (2007). Identification of new isolates of *Turnip mosaic virus* that cluster with less common viral strains. *Arch. Virol.* **152:**1061–1068.

Sasaya, T., Ishikawas, K., and Koganezawa, H. (2001). Nucleotide sequence of the coat protein gene of *Lettuce big-vein virus*. *J. Gen. Virol.* **82:**1509–1515.

Sasaya, T., Ishikawa, K., and Kogenezawa, H. (2002). The nucleotide sequence of RNA1 of *Lettuce big-vein virus*, genus *Varicosavirus* reveals its relation to nonsegmented negative-strand RNA viruses. *Virology* **297:**289–297.

Sasaya, T., Ishikawa, K., Kuwata, S., and Koganezawa, H. (2005). Molecular analysis of coat protein coding region of tobacco stunt virus shows that it is a strain of *Lettuce big-vein virus* in the genus *Varicosavirus*. *Arch. Virol.* **150:**1013–1021.

Sasaya, T., Fujii, H., Ishikawa, K., and Koganezawa, H. (2008). Further evidence of *Mirafiori lettuce big-vein virus* but not of *Lettuce big-vein associated virus* with big-vein disease in lettuce. *Phytopathology* **98:**464–468.

Schmelzer, K. (1962). Untersuchungen an Viren der Zier-und Wildgehölze: 1. Mitteilung Virosen an Viburum und Ribes. *Phytopath. Z.* **46:**17–52.

Schultz, E. S. (1921). A transmissible mosaic disease of Chinese cabbage, mustard and turnip. *J. Agric. Res.* **22:**173–177.

Scott, H. (1963). Purification of *Cucumber mosaic virus*. *Virology* **20:**103–106.

Shepherd, R. J., and Pound, G. S. (1960). Purification of *Turnip mosaic virus*. *Phytopathology* **50:**797–803.

Sherwood, J. L., Sanborn, M. R., Keyser, G. C., and Myers, L. D. (1989). Use of monoclonal antibodies in detection of *tomato spotted wilt virus*. *Phytopathology* **79:**61–64.

Smith, S. H., Duffus, J. E., and Gold, A. H. (1966). Butanol treatment and density-gradient electrophoresis as independent methods for purification of *beet western yellows virus*. *Phytopathology* **56:**902.

Smith, H. G., Barker, I., Brewer, G., Stevens, M., and Hallsworth, P. B. (1996). Production and evaluation of monoclonal antibodies for the detection of *beet mild yellowing luteovirus* and related strains. *Eur. J. Plant Pathol.* **102:**163–169.

Smith, T. J., Chase, E., Schmidt, T., and Perry, K. L. (2000). The structure of *cucumber mosaic virus* and comparison to *cowpea chlorotic mottle virus*. *J. Virol.* **74:**7578–7586.

Stevens, M., Freeman, B., Liu, H. Y., Herrbach, E., and Lemaire, O. (2005). Beet poleroviruses: Close friends or distant relatives? *Mol. Plant Pathol.* **6:**1–9.

Stubbs, L. L., and Grogan, R. G. (1963). Necrotic Yellows—A newly recgonized virus disease of lettuce. *Aust. J. Agr. Res.* **14:**439–459.

Stubbs, L. L., Guy, J. A. D., and Stubbs, K. J. (1963). Control of *lettuce necrotic yellows virus* disease by the destruction of common sowthistle (*Sonchus oleraceus*). *Aust. J. Exp. Agric.* **3:**215–218.

Tas, P. W. L., Boerjan, M. L., and Peters, D. (1977). Structural proteins of *Tomato spotted wilt virus*. *J. Gen. Virol.* **36:**267–279.

Taylor, R. H. and Stubbs, L. L. (1972). *Broad bean wilt virus*. Descriptions of Plant Viruses. http://www.dpvweb.net/index.php.

Taylor, R. H., Smith, P. R., Reinganu, C., and Gibbs, A. J. (1968). Purification and properties of *Broad bean wilt virus*. *Aus. J. Biol. Sci.* **21:**929–935.

Thompson, S., Fraser, R. S. S., and Barnden, K. L. (1988). A beneficial effect of trypsin on the purification of *turnip mosaic virus* (TuMV) and other potyviruses. *J. Virol. Methods* **20:**57–64.

Thomson, D., and Dietzgen, R. G. (1995). Detection of DNA and RNA plant viruses by PCR and RT-PCR using a rapid virus release protocol without tissue homogenization. *J. Virol. Methods* **54:**85–95.

Tiberini, A., Tomassoli, L., Barba, M., and Hadidi, A. (2010). Oligonucleotide microarray-based detection and identification of 10 major tomato viruses. *J. Virol. Methods* **168:**133–140.

Tomimura, K., Gibbs, A. J., Jenner, C. E., Walsh, J. A., and Ohshima, K. (2003). The phylogeny of *Turnip mosaic virus*; comparisons of 38 genomic sequences reveal a Eurasian origin and a recent 'emergence' in east Asia. *Mol. Ecol.* **12:**2099–2111.

Tomimura, K., Spak, J., Katis, N., Jenner, C. E., Walsh, J. A., Gibbs, A. J., and Ohshima, K. (2004). Comparisons of the genetic structure of populations of *Turnip mosaic virus* in West and East Eurasia. *Virology* **330:**408–423.

Tomlinson, J. A. (1964). Purification and properties of *lettuce mosaic virus*. *Ann. Appl. Biol.* **53:**95–102.

Tomlinson, J. A. (1970a). *Lettuce mosaic virus*. Descriptions of Plant Viruses. http://www.dpvweb.net/index.php.

Tomlinson, J. A. (1970b). *Turnip mosaic virus*. Description of Plant Viruses. http://www.dpvweb.net/index.php.

Tomlinson, J. A. (1984). *Lettuce mosaic potyvirus*. Plant Viruses Online. Descriptions and Lists from the VIDE Database. http://www.agls.uidaho.edu/ebi/vdie/refs.htm.

Tremaine, J. H., and Stace-Smith, R. (1969). Amino acid analyses of two strains of *alfalfa mosaic virus*. *Phytopathology* **59:**521–522.

Tsompana, M., Abad, J., Purugganan, M., and Moyer, J. W. (2005). The molecular population genetics of the *Tomato spotted wilt virus* (TSWV) genome. *Mol. Ecol.* **14:**53–66.

Untiveros, M., Perez-Egusquiza, Z., and Clover, G. (2010). PCR assays for the detection of members of the genus *Ilarvirus* and family *Bromoviridae*. *J. Virol. Methods* **165:**97–104.

Urcuqui-Inchima, S., Haenni, A. L., and Bernardi, F. (2001). Potyvirus proteins: A wealth of functions. *Virus Res.* **74:**157–175.

van der Wilk, F., Dullemans, A. M., Verbeek, M., and van den Heuvel, J. F. J. M. (2002). Nucleotide sequence and genomic organization of an *Ophiovirus* associated with lettuce big-vein disease. *J. Gen. Virol.* **83:**2869–2877.

Van Vloten-Doting, L., and Jaspars, E. M. J. (1972). The uncoating of *Alfalfa mosaic virus* by its own RNA. *Virology* **48:**699–708.

Veidt, I., Lot, H., Leiser, M., Scheidecker, D., Guilley, H., Richards, K., and Jonard, G. (1988). Nucleotide sequence of *Beet western yellows virus* RNA. *Nucleic Acids Res.* **16:**9917–9932.

Verkleij, F. N., and Peters, D. (1983). Characterization of a defective form of *tomato spotted wilt virus*. *J. Gen. Virol.* **64:**677–686.

Walkey, D. G. A., and Payne, C. J. (1990). The reaction of two lettuce cultivars to mixed infection by*beet western yellows virus, lettuce mosaic virus* and *cucumber mosaic virus*. *Plant Pathol.* **39:**156–160.

Wang, M., and Gonsalves, D. (1990). ELISA detection of various *tomato spotted wilt virus* isolates using specific antisera to structural proteins of the virus. *Plant Dis.* **75:**154–158.

Weekes, R., Barker, I., and Wood, K. R. (1996). An RT-PCR test for the detection of *tomato spotted wilt virus* incorporating immunocapture and colorimetic estimation. *J. Phytopathol.* **144:**575–580.

Wetzel, T., Dietzgen, R. G., and Dale, J. L. (1994). Genomic organization of lettuce necrotic yellows rhabdovirus. *Virology* **200:**401–412.

Xiang, H. J., Dong, S. W., Zhang, H. Z., Wang, W. L., Li, M. Q., Han, C. G., Li, D. W., and Yu, J. L. (2010). Molecular characterization of two Chinese isolates of *Beet western yellows virus* infecting sugar beet. *Virus Genes* **41:**105–110.

Xu, H., and Nie, J. (2006). Identification, characterization, and molecular detection of *Alfalfa mosaic virus* in potato. *Phytopathology* **96:**1237–1242.

Yardimci, N., and Kiliç, H.Ç. (2009). *Tomato spotted wilt virus* in vegetable growing areas in the west Mediterranean region of Turkey. *Afr. J. Biotechnol.* **8:**4539–4541.

Yu, C., Wu, J., and Zhou, X. (2005). Detection and subgrouping of *Cucumber mosaic virus* isolates by TAS-ELISA and immunocapture RT-PCR. *J. Virol. Methods* **123:**155–161.

Zein, H. S., and Miyatake, K. (2009). Development of rapid, specific and sensitive detection of *Cucumber mosaic virus*. *Afr. J. Biotechnol.* **8:**751–759.

Zhou, X. (2002). *Broad bean wilt virus 2*. Descriptions of Plant Viruses. http://www.dpvweb.net/index.php.

Ziegler-Graff, V., Brault, V., Mutterer, J. D., Simonis, M. T., Herrbach, E., Guilley, H., Richards, K. E., and Jonard, G. (1996). The coat protein of *Beet western yellows luteovirus* is essential for systemic infection but the viral gene products P29 and P19 are dispensable for systemic infection and aphid transmission. *Mol. Plant Microbe Interact.* **9:**501–510.

Zink, F. W., and Duffus, J. E. (1969). Relationship of *turnip mosaic virus* susceptibility and downy mildew (*Bremia lactucae*) resistance in lettuce. *J. Am. Soc. Hortic. Sci.* **94:**403–407.
Zink, F. W., and Duffus, J. E. (1970). Linkage of *turnip mosaic virus* susceptibility and downy mildew, *Bremia lactucae*, resistace in lettuce. *J Am. Soc. Hortic. Sci.* **95:**420–422.

CHAPTER 8

Viruses in Artichoke

Donato Gallitelli,*,† Tiziana Mascia,*,† and **Giovanni P. Martelli†**

Contents

* Dipartimento di Biologia e Chimica Agroforestale ed Ambientale, Università degli Studi di Bari "Aldo Moro", Bari, Italy
† Istituto di Virologia Vegetale del CNR, U.O. di Bari, Bari, Italy

Advances in Virus Research, Volume 84
ISSN 0065-3527, DOI: 10.1016/B978-0-12-394314-9.00008-7

Abstract

Most of the 25 viruses found in globe artichoke (*Cynara scolymus* L.) and cardoon (*Cynara cardunculus* L.) were recorded from Europe and the Mediterranean basin, where they decrease both the productivity and the quality of the crop. Although, sometimes, these viruses are agents of diseases of different severity, most often their infections are symptomless. These conditions have contributed to spread virus-infected material since farmers multiply traditional artichoke types vegetatively with no effective selection of virus-free plants. This review reports the main properties of these viruses and the techniques used for their detection and identification. ELISA kits are commercially available for most of the viruses addressed in this review but have seldom been used for their detection in artichoke. Conversely, nucleic acid-based diagnostic reagents, some of which are commercially available, have successfully been employed to identify some viruses in artichoke sap. Control measures mainly use virus-free stocks for new plantations. A combined procedure of meristem-tip culture and thermotherapy proved useful for producing virus-free regenerants of the reflowering southern Italian cultivar Brindisino, which kept earliness and typical heads shape.

I. INTRODUCTION

Globe artichoke (*Cynara scolymus* L.), a perennial species of the family *Compositeae* (*Asteraceae*), is grown for its large fleshy immature inflorescences (flower heads). Production is mainly concentrated in Mediterranean countries, among which Italy, Spain, and Egypt rank first. Extensive cropping also takes place in the Near East, South America (mainly Peru and Argentina), the United States, and China (FAO, 2010).

Two varietal groups are known. Early-flowering types produce heads during autumn and spring from offshoots collected in autumn–winter of the previous year or from "ovuli" collected in summer. Late-flowering types produce edible heads in spring–early summer. Traditional artichoke types are vegetatively multiplied keeping a significant genetic variation because of the limited selection adopted by farmers (Lanteri *et al.*, 2001; Portis *et al.*, 2005). By contrast, the recently introduced hybrids are reproduced by seeds and show high genetic uniformity. As with most vegetatively propagated crops, the lack of active selection for mother plants with a good sanitary status has led overtime several pathogens to accumulate, such as vascular fungi (mainly *Verticillium* spp.) and viruses. This has led to a progressive degeneration of the crop, which reflects on the poor quantity and quality of produce.

Studies on artichoke viruses began in the late 1950s, and continued till these days with a peak in the 1960s and 1970s, when most of the viruses currently known to infect this species where identified and characterized. At the 1st International Congress on Globe Artichoke (ICGA) (held in Bari, Italy in 1967), Ciccarone (1967) reported two filamentous viruses, that is, *Artichoke curly dwarf virus* (ACDV) (Leach and Oswald, 1950; Morton, 1961) and Artichoke latent virus S (ArLVS) (Costa *et al.*, 1959), both recorded from California, and an isometric virus—*Artichoke mottled crinkle virus* (AMCV), recorded from Italy (Martelli, 1965; Quacquarelli and Martelli, 1966). At that time, the viral nature of a bright yellow mosaic present in Sicily (Gigante, 1951), a generalized yellowing condition reported from California (Smith, 1940), and a chlorotic mottling with a degenerative condition observed Tunisia (Laudanki, 1958) was not supported by experimental evidence.

Five years later, at the 2nd ICGA (Bari, 1972) Martelli and Rana (1976) listed nine more viruses from artichoke, cardoon, and *Cynara syriaca*, that is, *Artichoke Italian latent virus* (AILV), *Artichoke yellow ringspot virus* (AYRSV), Artichoke degeneration virus (ADV), *Artichoke latent virus* (ArLV), *Cucumber mosaic virus* (CMV), *Tobacco streak virus* (TSV), *Tobacco mosaic virus* (TMV), *Tobacco rattle virus* (TRV), and *Cynara* virus (CraV). At the 3 rd ICGA (Bari, 1979), the number of viruses had grown to 16 (Martelli *et al.*, 1981), to reach 23 at the fourth ICGA (Bari, 2000) (Rana *et al.*, 2005).

So far 25 viruses belonging to 15 genera in 10 families (Gallitelli *et al.*, 2004; Martelli and Gallitelli, 2008) are known to infect globe artichoke. Most of them are endemic to the Mediterranean basin and affect the industry to different extents.

II. VIRUSES WITH ISOMETRIC PARTICLES

Four of the 10 isometric particle viruses that infect globe artichoke belong to the genus *Nepovirus*, family *Secoviridae* (Carstens, 2010; Le Gall *et al.*, 2005a; Mayo and Jones, 1999; Murant *et al.*, 1996). All have isometric particles with angular contours and diameters of 28–30 nm, containing two types of positive-sense single-stranded RNA (ssRNA) molecules (RNA-1 and RNA-2), encapsidated separately but both needed for infectivity. Virus preparations sediment as three types of components differing in their buoyant properties: T (top, empty particles), M (middle, particles usually containing one molecule of RNA-2), and B (bottom, particles containing a single molecule of RNA-1 or two molecules of RNA-2). Based on the size of RNA-2 and its presence only in M component or in both B and M components, nepoviruses cluster in subgroups A, B, and C (Le Gall *et al.*, 2005a). Two of the nepoviruses found in artichoke AILV and *Tomato black ring virus* (TBRV) have recognized nematode vectors.

A. *Artichoke Aegean ringspot virus*

Artichoke Aegean ringspot virus (AARSV) was originally found in symptomless plants from Turkey (Smyrna) (Kyriakopoulou and Bem, 1982) and Greece (Marathon and Nauplion) and in cvs Black and Argitiki, showing yellow blotches and mild mottling. It was identified as a serotype of *Raspberry ringspot virus* (RpRSV) (Rana *et al.*, 1985). The Greek and Turkish isolates, named AARSV-G and AARSV-T, respectively, have two RNA species with estimated M_r of 2.4×10^6 (RNA-1) and 1.4×10^6 (RNA-2) and coat protein (CP) subunits with M_r of 54 kDa. They are serologically closely related (SDI = 1) and are distantly related to RpRSV-S and RpRSV-E (SDI = 5) (Rana *et al.*, 1985). However, when analyzed molecularly, AARSV-G and AARSV-T proved to share 73% sequence homology with each other but had only 7% homology with RpRSV-E (Robinson and Clark, 1987). According to species demarcation criteria in the genus *Nepovirus* (Le Gall *et al.*, 2005a), this difference suggests that they represent different species. Murant *et al.* (1996) gave the virus its present name and included it in subgroup A of the genus *Nepovirus*. AARSV has a moderately wide experimental host range, infecting 22 species in seven dicotyledonous families (Rana *et al.*, 1985). It can be grown in and purified from *Chenopodium quinoa* and distinguished from RpRSV-S and RpRSV-E by reaction on cucumber. AARSV induces chlorotic and/or necrotic local lesions, while RpRSV isolates cause chlorotic pinpoints on newly formed leaves. Back inoculation of AARSV to artichoke cvs Romanesco (late-flowering type) and Violet de Provence (early-flowering type) always yielded symptomless infections. AARSV vector is unknown (Roca *et al.*, 1986).

B. *Artichoke Italian latent virus*

AILV was recovered from symptomless artichokes in Apulia (southern Italy) (Majorana and Rana, 1970a) and occasionally from stunted plants with mild yellowing of the leaves. Later, again in Apulia, AILV was found in chicory showing chlorotic mottling of the leaves often with bright yellow spots (Vovlas *et al.*, 1971), in grapevine with fanleaf-like symptoms in Bulgaria (Jankulova *et al.*, 1977), in pelargonium with severe leaf deformation (Martelli, 1977), and in artichoke in northeast Peloponnesus (Greece), where it causes a disease called artichoke patchy chlorotic stunting (Brown *et al.*, 1997; Kyriakopoulou, 1995; Rana and Kyriakopoulou, 1982). *Longidorus apulus* transmits AILV-type strain (Apulian) (Rana and Roca, 1976; Roca *et al.*, 1975), while *Longidorus fasciatus* transmits the Greek strain (Roca *et al.*, 1982). Results from a recent survey in Apulian artichoke crops suggest that AILV is now more widespread than in the past and often in mixed infection with the potyvirus ArLV. This condition is damaging for the reflowering types (Brindisino, Violet de Provence, Locale di Mola) where it is difficult to find AILV-free plants.

The physicochemical properties of AILV are: isometric particles ca. 30 nm in diameter, sedimenting as three components with coefficient of 55*S* (T), 95*S* (M), and 121*S* (B); two species of functional ssRNA making up 34% of M particle and 41% of B particle weight, with estimated M_r of 2.4×10^6 (RNA-1) and 1.5×10^6 (RNA-2), a single species of CP subunits with M_r of 54 kDa (Savino *et al.*, 1977). Virions are stable at 60 °C but are sensitive to organic solvents and CsCl. A partial nucleotide sequence of RNA-2 is available under the Acc. No. X87254. AILV is a member of subgroup B of the genus *Nepovirus* (Le Gall *et al.*, 2005a).

The natural and experimental host ranges are wide, including 63 species in 12 dicotyledonous families (Camele *et al.*, 1991; Quacquarelli *et al.*, 1976; Savino *et al.*, 1977). In horticultural areas of southern Italy, AILV survives in susceptible weeds in which it is seed-transmitted, and in artichoke and chicory often cropped in succession in the same plots. AILV is transmitted in ca. 5% of globe artichoke seeds where it was detected in both seed coats and fully expanded cotyledons (Bottalico *et al.*, 2002). *Crocus sativus* and *Phaseolus vulgaris* (French bean) are useful diagnostic species, while the virus can be maintained in *Nicotiana tabacum* "Samsun." In this host, AILV infection induces chlorotic spots with necrotic ring-like patterns of inoculated leaves and systemic symptoms consisting of necrotic and/or chlorotic ringspots, line pattern, and leaf blade distortion. Recent and current studies (Mascia *et al.*, 2009; Santovito *et al.*, 2010) show that after 21 days postinoculation tobacco plants recover and symptoms do not appear on the new vegetation. This was not noted in infected plants grown at 15 °C. These results suggest that AILV is sensitive to RNA silencing, as it may not have an RNA-silencing suppressor (see Section VI) (E. Santovito, T. Mascia, and D. Gallitelli, unpublished results).

There are no commercially available antisera for the diagnosis and/or identification of AILV. The virus can be identified in artichoke by reverse transcription-polymerase chain reaction (RT-PCR) using either the following primer pair AILV fwd 5′-ATT CAC TAG TCC CTA TTT AG-3′ and AILV rev 5′-GGT CTG GGG TGC CCG TGG CG-3′, which produce an amplicon of 760 bp, or by Dot blot hybridization with a digoxigenin (DIG)-labeled DNA probe.

C. *Artichoke yellow ringspot virus*

AYRSV was first reported from Greece in plants showing bright yellow blotches, ringspots, and line patterns on the leaves, occasionally followed by extensive necrosis (Kyriakopoulou and Bem, 1973). It was later isolated in Sicily (southern Italy) from cardoon (*C. cardunculus* L.) cv. Bianco avorio with striking yellow ringspots and line patterns (Rana *et al.*, 1978). Isolates of Italian and Greek origin differ in host range but not serologically. AYRSV contains two species of functional ssRNA with estimated M_r of 2.3×10^6 (RNA-1) and 1.9×10^6 (RNA-2) and CP subunits with M_r of 53 kDa. Cytopathological changes (Russo *et al.*, 1978) consist of cytoplasmic inclusion bodies made up of tangled membranes and vesicles containing finely stranded material resembling nucleic acid. Virus particles occur in the cytoplasm at the margin of the inclusions in stacked layers, or between membranes, or within tubular structures.

AYRSV is a member of subgroup C of the genus *Nepovirus* (Le Gall *et al.*, 2005a) and is not related serologically to 28 isometric plant viruses including 14 nepoviruses. In nature, it infects 37 plant species in 15 families, seven of which are agricultural crops (artichoke, broad bean, cardoon, cucumber, French bean, onion, and tobacco) and 30 are wild plants (Avgelis and Vovlas, 1989; Avgelis *et al.*, 1992; Kyriakopoulou, 1981; Kyriakopoulou *et al.*, 1985; Maliogka *et al.*, 2006). The experimental host range is also wide, including 60 species in 13 dicotyledonous families (Maliogka *et al.*, 2006; Rana *et al.*, 1980). The virus can be maintained in and purified from *C. quinoa*, *P. vulgaris*, and *C. sativus*. It is not is not sensitive to organic solvents but particles aggregate during purification. AYRSV has no known vector but was shown to be seed-transmitted in tobacco and onion and to infect *Nicotiana clevelandii* by pollen.

Three isolates of the virus have been studied in detail, AYRSV-AtG from artichoke, AYRSV-F from broad bean, and AYRSV-On from onion. Artichoke and broad bean isolates are serologically related, while there is no obvious similarity with the onion isolate. This lack of reactivity in ELISA and immunosorbent electron microscopy (ISEM) tests reflects the poor quality of available antisera (Maliogka *et al.*, 2006). *C. sativus*, which responds with enations following inoculation, is an indicator for all isolates (Maliogka *et al.*, 2006; Rana *et al.*, 1983). *Nicotiana rustica* is immune to AYRSV-On but susceptible to the other isolates, and *Physalis floridana* reacts with necrotic local lesions to AYRSV-On and with different symptoms to two of the other isolates tested, suggesting that a biological divergence exists within AYRSV isolates (Maliogka *et al.*, 2006). Partial sequence data for the fragment of RNA-1 coding for RNA-dependent RNA polymerase (RdRp) of AYRSV-F and AYRSV-AtG are available under the Acc. No. AM087672 and AM087673, respectively. The sequence of a 5783nt fragment of AYRSV-On RNA-1 has been deposited under the

Acc. No. AM087671. Comparison between partial RdRp sequences of the three AYRSV isolates revealed 100% amino acid (aa) identity and 93% nucleotide identity.

Specific detection of AYRSV in symptomatic onion plants and in *C. scolymus*, *Vicia faba*, *Gomphrena globosa*, *Nicotiana benthamiana*, and *Silybum marianum* was obtained by RT-PCR using two specific primers designed on the RdRp sequence, that is, AYRSVup11 (5′-GAACGCAACATCGGAGA-3′) and AYRSVd12 (5′-GCTCCACAAAGTGACTTG-3′), that produce a 530 bp amplicon (Maliogka *et al.*, 2006). Amino acid sequence analysis revealed significant homologies with similar RdRp regions of other nepoviruses, with 60% similarity with *Blackcurrant reversion associated virus* (BRAV) (Swiss Protein Translated European Molecular Biology Laboratory, SPTREMBL, database Acc. No. Q8V5E0) and 52% with *Grapevine fanleaf virus* (GFLV, SPTREMBL Acc. No. P29149). Phylogenetic analysis based on available sequence data showed that AYRSV groups with BRAV, a virus classified as a mite-transmitted member of nepovirus subgroup C. Although the patchy distribution of diseased plants in artichoke fields suggests that AYRSV has a nematode vector, this has not been ascertained. Interestingly, the phylogenetic relationship with BRAV and the epidemic behavior in onion, which resembles that of an airborne pathogen, suggest that AYRSV might be transmitted by mites (Maliogka *et al.*, 2006).

D. *Tomato black ring virus*

TBRV infects in nature several plant species, including globe artichoke (Murant *et al.*, 1996). It was first isolated in France in an experimental plot of cv. Camus de Bretagne that showed mild mottling of the leaves (Migliori *et al.*, 1984b). TBRV infection in this artichoke variety caused 19% decrease of the first head weight and a decrease by 42% in the total number of plant heads (Migliori *et al.*, 1987). In commercial French artichoke fields, it was transmitted by *Longidorus attenuatus* at a rate of 5–10%. Experimental infection of artichoke seedlings of the hybrid "Bianco tarantino × Nostrano di Ascoli Piceno" induced systemic chlorotic ringspot in the leaves (Rana *et al.*, 1987a).

The globe artichoke strain of the virus (TBRV-A) is more closely related to TBRV-W from beet (Harrison, 1957) (SDI = 1) than to TBRV-BU from potato (Harrison, 1958) or TBRV-Ce from celery (Hollings, 1965) (SDI = 2–3) and is distantly related to AILV-S and AILV-G (SDI = 11 and 12, respectively) (Dodd and Robinson, 1984; Rana *et al.*, 1987a). TBRV-A physicochemical properties conform to those of TBRV strains from other hosts. Particles are isometric ca. 30 nm in diameter, encapsidate two species of ssRNA making up 28% and 38% of M and B particle weight, respectively, and sediment as three components at 53*S* (T), 99*S* (M), and

120*S* (B). M_r of RNAs is 7385 (RNA-1, Acc. No. AY799593) and 4633 (RNA-2, Acc. No. AY157994) and the M_r of CP subunits is 54 kDa. Infected cells show inclusion bodies typical of those induced by nepoviruses, consisting of accumulations of membranous vesicles, endoplasmic reticulum strands, and ribosomes. Virus particles are in groups in the cytoplasm or in rows within tubular structures that are sometimes connected with plasmodesmata (Rana *et al.*, 1987a). In hosts other than artichokes, which have not been tested serologically, TBRV can be detected by DAS-ELISA using commercially available antisera.

Other viruses with nonenveloped isometric particles reported from artichoke are: Artichoke vein banding virus (AVBV), AMCV, CMV, *Broad bean wilt virus* (BBWV), *Pelargonium zonate spot virus* (PZSV), and TSV.

E. Artichoke vein banding virus

AVBV, described by Gallitelli *et al.* (1978, 1984), was found originally in Apulia in the Turkish artichoke cvs Bayrampasa and Sakiz, in which it causes chlorotic discolorations along the leaf veins, and in the Italian cv. Mazzaferrata. Because of its bipartite ssRNA genome, AVBV was originally classified as a tentative species in the genus *Nepovirus* (Harrison and Murant, 1977). More recently, it was transferred to the newly established genus *Cheravirus* (Le Gall *et al.*, 2005b), family *Secoviridae* (Carstens, 2010) with which it shares a major characterizing feature, namely, the presence of three CP subunits with estimated M_r of 22, 24.5, and 27 kDa. Virus particles sediment as three components with coefficients of 56*S* (T), 92*S* (M), and 124*S* (B) have a ssRNA genome made up of two functional species that represent 24% (M) and 37% (B) of the particle weight with an estimated M_r of 2.4×10^6 (RNA-1) and 1.4×10^6 (RNA-2). AVBV is not widespread, has no known vector and a moderately wide experimental host range comprising 20 species in seven dicotyledonous families (Gallitelli *et al.*, 1978). The virus can be maintained in and purified from *C. quinoa* or French bean, while *C. quinoa* and *Chenopodium amaranticolor* are suitable diagnostic species as both react to infection with local lesions followed by severe systemic symptoms. In infected plants, AVBV induces cytopathological changes consisting of cytoplasmic inclusion bodies made up of accumulations of membranes and vesicles, some of which contain finely stranded material resembling nucleic acid. Virus particles occur in the cytoplasm, scattered or in discrete paracrystalline arrays (Gallitelli *et al.*, 1984). The origin of AVBV is unknown, although it may have come to Italy with infected material from Turkey. In the collection plot where the virus was found, it was present in the Turkish varieties Bayrampasa and Sakiz but in none of the neighboring plants belonging to other cultivars.

F. *Artichoke mottled crinkle virus*

AMCV was the first fully characterized artichoke virus (Martelli *et al.*, 1981; Quacquarelli and Martelli, 1966). It is endemic in the Mediterranean area and was recovered from symptomatic plants in southern Italy (Martelli, 1965), Morocco (Fischer and Lockhart, 1974a), Malta (Martelli *et al.*, 1976), and Tunisia (Rana and Cherif, 1981) and was also isolated from symptomless plants in southern Italy and Greece (Rana and Kyriakopoulou, 1982). Symptoms include severe deformation, mottling, and crinkling of the leaves. Growth and yield of symptomatic plants are severely affected, flower heads are distorted and plants may die. If the plant survives the new foliage develops poorly and often shows bright chrome yellow discoloration.

AMCV is a species in the genus *Tombusvirus* (family *Tombusviridae*) (Lommel *et al.*, 2005) serologically closely related to *Petunia asteroid mosaic virus* (SDI = 1). The two viruses, however, occupy different ecological niches (natural hosts and environment), which justifies their retention as different species. Italian, Moroccan, Tunisian, and Maltese isolates are serologically identical, whereas a serological variant was identified in Greece (Rana and Kyriakopoulou, 1982). AMCV persists in the soil from where, like some other tombusviruses, it can infect healthy plants without vector.

AMCV particles sediment as a single component with a coefficient of 132*S* containing ca. 17% of a single-stranded fully sequenced RNA genome, 4789 nt in size (Acc. No. X62493), and have CP subunits with M_r of 41 kDa. The cytopathology of AMCV infection is complex. Among other features, infected cells contain cytopathic structures (Martelli and Russo, 1973) denoted "multivesicular bodies," which are a hallmark of tombusvirus infections (Martelli and Russo, 1984). AMCV supports a satellite RNA with an estimated size of 440 nt that attenuates symptoms induced by the helper virus in *N. benthamiana* (Gallitelli and Hull, 1985).

Diagnosis of AMCV can be achieved by RT-PCR using commercially available primers pair AMCV fwd 5′-ATG GCA ATG GTA AAG AGA AA-3′ and AMCV rev 5′-CTC GGA CTT TCG TCA GGA AGT TTG AA-3′, which yield an amplicon of 1349 bp, and by Dot blot hybridization with DIG-labeled DNA probes. There are no commercially available antisera.

G. *Cucumber mosaic virus*

CMV was recorded by Lisa (1971) in northern Italy from mottled cardoon plants and was successfully transmitted to cardoon seedlings by *Myzus persicae*. Globe artichoke seedlings mechanically inoculated using the cardoon virus isolate remained symptomless. CMV strains belonging to subgroup II (Palukaitis and Garcia-Arenal, 2003; Roossinck *et al.*, 2005) were isolated in France (Migliori *et al.*, 1984a) and in southern Italy (Paradies *et al.*, 2000). In the latter instance, they were in mixed infection with either the potyvirus ArLV or the tospovirus *Tomato spotted wilt virus* (TSWV) in plants showing severe stunting, leaf malformation, and necrosis. Further well-substantiated records of infection by CMV are from Tunisia (Chabbouh and Cherif, 1990) and Spain (Ortega *et al.*, 2005), whereas a report from Slovenia (Baricevich *et al.*, 1995) is uncertain.

A wide range of antisera for CMV detection is commercially available although there are no reports on their use with artichoke. Diagnosis of CMV can also be done by RT-PCR using commercially available primers pair CMV RNA-2–5′ fwd 5′-GTT TAT TTA CAA GAG CGT ACG G-3′ and RNA-2–3′ rev 5′-GGT TCG AA(AG) (AG)(AT)A TAA CCG GG-3′, which yield an amplicon of 637 bp, and by Dot blot hybridization with a DIG-labeled DNA probe. An indication of the CMV subgroup (IA, IB, or II) can be obtained by the restriction pattern yielded by the 637 bp amplicon on digestion with *Mlu*I (Finetti-Sialer *et al.*, 1999).

H. *Broad bean wilt virus*

BBWV was first isolated in southern Italy, in mixed infection with ArLV and, occasionally, CraV and *Bean yellow mosaic virus* (BYMV), from plants of cv. Castellammare showing yellow mottle, mosaic, or line patterns (Russo and Rana, 1978). The virus was further studied by Rana *et al.* (1987b) who, in mechanically inoculated artichoke seedlings of cv. Locale di Mola, reproduced a yellow mottle similar to that seen in the past in Italy in different artichoke cultivars (Gigante, 1951; Martelli and Rana, 1976). This virus also occurs in France (Migliori *et al.*, 1988), Greece (Kyriakopoulou, 1995), and Spain (Ortega *et al.*, 2005).

BBWV is a species in the genus *Fabavirus*, family *Secoviridae* (Carstens, 2010) with a bipartite positive-sense ssRNA genome encapsidated in different particles, made up by two types of protein subunits and sedimenting as T, M, and B. Two BBWV species are recognized, under the name of BBWV-1 and BBWV-2. A comparative biological, physicochemical, serological, and cytopathological characterization of a French (BBWV-FA) and an Italian (BBWV-IA) isolate from artichoke (Migliori *et al.*, 1988) proved the two isolates were serologically more related to each other (SDI = 2–3) than to BBWV-1. BBWV-IA is therefore listed as a strain of BBWV-2 (Le Gall *et al.*, 2005a), has sedimentation coefficients of 37*S* (T), 61*S* (M), 76*S* (B), a genome made up by two RNA molecules 5957 nt (RNA-1, Acc. No. AB051386) and 3593 nt (RNA-2, Acc. No. AB032403) in size, and CP subunits with M_r of 44 and 26 kDa.

In case of mixed infection with ArLV, inoculation of *G. globosa* or *C. quinoa* can be used to separate the two viruses as they are infected only locally by ArLV but systemically by BBWV. Because of aphid transmissibility in a nonpersistent manner, BBWV spreads rapidly in the field and can reach a high incidence, as in France, where infection rates of 40–70% have been recorded in crops of cv. Camus de Bretagne (Migliori *et al.*, 1987). A much lower infection rate was observed in Spain where the virus was reported under the particular name of Artichoke French latent fabavirus (Ortega *et al.*, 2005).

In experimental tests done in France, the most efficient vector of BBWV-FA was the aphid *Capitophorus horni* (Migliori *et al.*, 1988). The same authors showed that BBWV infection reduces the yield remarkably in weight of the first flower head and the number of marketable heads (Migliori *et al.*, 1987).

Diagnosis can be carried out using commercially available antisera against BBWV-1 and BBWV-2.

I. *Pelargonium zonate spot virus*

PZSV is a virus originally described by Gallitelli (1983). It has been characterized molecularly (Finetti-Sialer and Gallitelli, 2003) and classified as the type species of *Anulavirus*, a new genus in the family *Bromoviridae* (Gallitelli *et al.*, 2005). PZSV has quasi-spherical particles 25–35 nm in diameter with a poorly resolved surface. They sediment as tree components with coefficients of 80*S* (T), 90*S* (M), and 118*S* (B), a positive-sense ssRNA genome consisting of three functional species with sizes of 3383 nt (RNA-1, Acc. No. AJ272327), 2435 nt (RNA-2, Acc. No. AJ272328), and 2569 nt (RNA-3, Acc. No. AJ272329) encapsidated in 24 kDa protein subunits. Virions are sensitive to organic solvents and to temperature above 40 °C. The cytopathology is complex differing from that of other members of the family. In infected cells, chloroplasts are severely damaged and nuclei show localized dilations of the envelope that harbor double-membraned vesicles with finely fibrillar material resembling nucleic acid. The vesicles are also present in ground cytoplasm. Virus particles accumulate in the nucleus and cytoplasm in large noncrystalline aggregates or within tubular structures associated with plasmodesmata and cell wall (Castellano and Martelli, 1981).

The virus was isolated in Apulia from artichokes of cv. Romanesco with stunting and chlorotic mottling of the leaves (Rana and Martelli, 1983). It had been recovered previously from tomato and pelargonium in the same area (Martelli and Cirulli, 1969; Quacquarelli and Gallitelli, 1979). An isolate from *Chrysanthemum coronarium* induced yellowish ringspot and line patterns when artificially inoculated to globe artichoke seedlings of cv. Locale di Mola (Rana *et al.*, 1990). PZSV is endemic in Apulia (southern Italy) where it infects *Diplotaxis erucoides* (rocket) latently. It contaminates superficially pollen grains, which are carried to neighboring crops, especially tomato, by thrips thriving in rocket flowers (Vovlas *et al.*, 1989). PZSV is seed-transmitted in *D. erucoides* (Vovlas *et al.*, 1989), *Nicotiana glutinosa* (Gallitelli, 1982), and tomato (Lapidot *et al.*, 2010). Other natural hosts include *Picris echioides* and *Sonchus oleraceus*.

PZSV has experimentally been transmitted to plants of at least 32 species in nine dicotyledonous families. *N. glutinosa* and *Cucurbita pepo* (zucchini squash) are diagnostic species. In *N. glutinosa*, the virus induces chlorotic rings 3–4 days after inoculation followed by severe systemic mosaic with necrosis of margins and downward rolling of the leaves. *C. pepo* reacts to PZSV infection with chlorotic and/or necrotic local lesions, systemic apical necrosis and death. PZSV can be detected by DAS-ELISA, RT-PCR using commercially available primer pair PZSV fwd 5′-ATG CCC CCT AAG AGA CAG-3′ and PZSV rev 5′-AC AGA GGT ATA TAC TCT GC-3′, and Dot blot hybridization using DIG-labeled DNA probes.

J. *Tobacco streak virus*

TSV is a member of the subgroup I of the genus *Ilarvirus* (family *Bromoviridae*). It has particles 26–35 nm in diameter, which sediment as three components with coefficients of 90*S*, 98*S*, and 113*S*, respectively, and encapsidate a positive-sense ssRNA genome consisting of three functional species with sizes of 3491 nt (RNA-1, Acc. No. U80934), 2926 nt (RNA-2, Acc. No. U75538), and 2205 nt (RNA-3, Acc. No. X00435). The viral capsid is made by a single protein subunit with M_r of 26.3 kDa (Roossinck *et al.*, 2005). TSV was reported only once from globe artichoke in Brazil, where it was isolated from stunted plants with malformed leaves (Costa and Tasaka, 1971). This isolate did not differ from other TSV strains from potato, tobacco, and tomato occurring in the same agricultural area and did not have clear-cut harmful effects on the artichoke crop. TSV is transmitted in nature by thrips (*Thrips tabaci*, *Frankliniella occidentalis*) but whether this applies also to artichoke is unknown. Detection is by commercially available ELISA kits.

III. VIRUSES WITH FILAMENTOUS PARTICLES

Of the viruses with filamentous particles that infect globe artichoke, three belong in the family *Potyviridae* (Berger *et al.*, 2005), two in the family *Alphaflexiviridae* (genus *Potexvirus*), three in the family *Betaflexiviridae* (genus *Carlavirus*) (Carstens, 2010), and one in the family *Closteroviridae*, genus *Crinivirus* (Martelli *et al.*, 2005).

A. *Artichoke latent virus*

ArLV was first recovered from symptomless globe artichoke plants by Marrou and Mehani (1964) in Tunisia, then by Fischer and Lockhart (1974b) in Morocco, and by Foddai *et al.* (1977) in Sardinia (insular Italy). In Apulia, ArLV has been found in *Petunia hybrida* in mixed infection with CMV (Di Franco and Gallitelli, 1985). Interestingly, ArLV does not occur in Albania, Yugoslavia, Slovenia, Lebanon, Portugal, and Greece (Baricevich *et al.*, 1995; Rana *et al.*, 1982).

ArLV is a typical potyvirus with particles measuring ca. 12×746 nm sedimenting as a single component with a coefficient of 145*S* and encapsidating a ssRNA with an estimated M_r of 3×10^6. The sequence of a fragment of 500 nt of the ArLV genome is available (Acc. No. X87255). CP subunits have an estimated M_r of 33 kDa (Rana *et al.*, 1982). The virus was transmitted experimentally in the nonpersistent manner to artichoke seedlings by *M. persicae*, *Brachycaudus cardui*, and *Aphis fabae* (Rana *et al.*, 1982). Transmission efficiency in the field was estimated to be high. This was shown by the quick reinfection rate (over 75% in 2 years) registered in a trial carried out with virus-free plants in Sardinia (Foddai *et al.*, 1983b). Conversely, in a recent survey in Apulia virus-free plants of the reflowering type "Brindisino" (see Section VI) showed an estimated reinfection rate between 1% and 2% after 2 years from planting close to a commercial artichoke field. In this field, the ArLV incidence was higher than 80%. The difference between these data may depend either on fluctuations of aphids populations during the two surveys or on the inaccurate estimate of the virus-free status of Sardinian plants (see Section VI).

The host range of ArLV isolates from eight different countries was studied by Rana *et al.* (1982), who also characterized an Italian isolate of the virus, showing that it is serologically unrelated to the following potyviruses: *Bean common mosaic virus*, BYMV, *Clover yellow vein virus*, *Beet mosaic virus*, *Lettuce mosaic virus*, *Zucchini yellow fleck virus*, *Papaya ringspot virus*, *Watermelon mosaic virus*, *Turnip mosaic virus*, *Potato virus Y*, *Soybean mosaic virus*, and *Dasheen mosaic virus*. The virus can be grown in and purified from *N. benthamiana*, while selective diagnostic species can be *G. globosa*, *N. clevelandii*, and *P. vulgaris* (Rana *et al.*, 1982).

Originally, ArLV was thought not to have any damaging effect on infected plants. However, trials conducted in France and Italy on cvs Camus de Bretagne, Spinoso sardo, and Brindisino, and the hybrids Talpiot, H137, and H044 artificially infected with one or two viruses showed that ArLV causes significant decreases of the number and size of marketable heads, premature opening and color breaking of head scales, and shortening of the head stalk. In particular, yield losses varied from 38% to 53% and caused delayed harvesting (Camele *et al.*, 1999; Foddai *et al.*, 1983a,b; Migliori *et al.*, 1987; Rana *et al.*, 1992).

Field control trials for preventing the spread of ArLV in cv. Spinoso sardo showed that treatments with deltametrin and mineral oils lessened the infection from 32% (control) to 7% (treated plots) over a 3-year period (Foddai *et al.*, 1991). Recent studies have shown that accessions of *C. cardunculus* and *C. syriaca* are susceptible to ArLV, whereas an interspecific hybrid between these two species had a high level of resistance (Manzanares *et al.*, 1995). No ArLV antisera are commercially available thus identification can be carried out either by RT-PCR with the following primer pair ArLV fwd 5′-TTG TTC ATA AGG GAG CGC GT-3′ and ArLV rev 5′-CTCAAG CTC TCG AAC TAA CT-3′, which amplify a fragment of 580 bp or by a DIG-labeled DNA probe.

B. *Ranunculus latent virus*

Ranunculus latent virus (RaLV) is a macluravirus (family *Potyviridae*) originally isolated in Liguria (northern Italy) from plants of Persian buttercup (*Ranunculus asiaticus*) varieties showing mosaic and growth disorders (Turina *et al.*, 2006). This virus was serologically undistinguishable from another virus named Cy42 consistently found both in symptomatic and in symptomless artichoke plants grown in Latium (central Italy), Sardinia, and Liguria. Cy42 shows 98% identity (Acc. No. HQ449546) with a fragment of the CI protein, and the whole 6K2, NIa, NIb, and CP of the published nucleotide sequence of RaLV (Acc. No. HQ449550) and 80% identity with the 500 bp sequence of an ArLV genome fragment deposited in GenBank (Acc. No. X87255). Despite the sequence similarity in this region, Cy42 and RaLV were not recognized by either the primer pair or the DIG-labeled DNA probe used for ArLV detection. Therefore, the question is whether Cy42 is a new macluravirus infecting artichoke and if, and to what extent, it is related to ArLV (Ciuffo *et al.*, 2010; M. Turina, personal communication). Anyway, since a reassessment of the taxonomic position of ArLV seems desirable, its complete nucleotide sequence is being determined.

C. *Bean yellow mosaic virus*

BYMV, a member of the genus *Potyvirus*, has filamentous particles ca. 750 nm long, containing ca. 5% of ssRNA 9532 nt in size (Acc. No. NC_003492, D83749, U47033). The virus was initially recorded from globe artichoke in Apulia (Russo and Rana, 1978), then in Greece (Rana and Kyriakopoulou, 1980). Apulian artichokes were also infected by BBWV and ArLV and showed vein yellowing, yellow flecking, and line patterns on the leaves. These symptoms closely resembled those characterizing the etiologically undetermined "mosaic" described by Gigante (1951), and the yellow mottle from Campania and Apulia with which

BBWV is associated (Rana *et al.*, 1987b). In artificial inoculation tests, a BYMV isolate from artichoke cv. Castellammare maintained in broad bean was unable to infect artichoke seedlings of cvs Mazzaferrata and Vert globe systemically but caused erratic local lesions (Russo and Rana, 1978). Maintenance and propagation host species are *Pisum sativum*, *V. faba*, *P. vulgaris*, and *N. clevelandii*, while *C. sativus* can be used as diagnostic species since is not infected by BYMV. The cytopathology of infected *V. faba* conforms to that induced by potyviruses (Russo and Rana, 1978). BYMV detection can be done using commercially available ELISA kits, by RT-PCR with primer pairs BYMV fwd 5′-ACA CAA GCA CAG TTT GAA GC-3′ and BYMV rev 5′-AGA TCA AGC TCA CAC GAG G-3′, which amplify a fragment of 750 bp, and by Dot blot hybridization using a DIG-labeled DNA probe.

D. *Turnip mosaic virus*

TuMV, a potyvirus with an ssRNA genome 9853 nt in size (Acc. No. AF169561), was isolated in Sardinia and Apulia from symptomless artichoke plants of cvs Spinoso sardo (Foddai *et al.*, 1993) and Brindisino (D. Gallitelli, M. Padula, and M. Finetti-Sialer, unpublished information) that were also infected by ArLV. The artichoke isolate of TuMV has particles ca. 730 nm long that were decorated by an antiserum to a German isolate of the virus. The incidence of TuMV infections in Apulia and their economic importance have not been determined but recent surveys suggest the virus is more widespread than originally thought. In *N. benthamiana*, TuMV induces necrotic lesions in the inoculated leaves, which are of diagnostic value. The virus can be detected in artichoke by RT-PCR using primer pair TuMV fwd 5′-AT(TC) CT(GA) TAC AC(GCT) CC (GA) GAG CA-3′ and TuMV rev 5′-GCG CCA CGC AGT GCT G-3′, which amplify a fragment of 473 bp, or by a DIG-labeled DNA probe. ELISA reagents are also available but have so far not been used with artichoke.

E. Artichoke latent virus M and Artichoke latent virus S

Artichoke latent virus M (ArLVM) and ArLVS are tentative species in the genus *Carlavirus* (Adams *et al.*, 2005). ArLVM was found in Apulia in mixed infection with one or two of the following viruses: AILV, AMCV, and ArLV (Di Franco *et al.*, 1989). Virus particles in leaf-dip preparations from naturally infected globe artichoke leaves measured 697 nm in length and contained a single species of RNA with an estimated M_r of 7.5×10^6 and a CP with M_r of 31 kDa. ArLVM was not associated with any particular symptom in the field and was not mechanically transmitted to *C. quinoa*, *C. amaranticolor*, *C. sativus*, *C. pepo*, *C. scolymus*, *G. globosa*, *N. benthamiana*, *N. clevelandii*, *Nicotiana megalosiphon*, *N. tabacum* cvs White Burley and Xanthi, and *P. vulgaris* cv. La Victoire. In immunoelectron microscopy tests, ArLVM was weakly decorated by antisera to *Carnation latent virus* and *Poplar mosaic virus* (PopMV), both members of the genus *Carlavirus*. The cytopathology of globe artichoke cells infected by ArLVM did not differ from that induced by other carlaviruses in their hosts. The cells contained complex cytoplasmic inclusions composed of deranged organelles, lipid droplets, and accumulations of membranes (Di Franco *et al.*, 1989).

ArLVS was identified probably for the first time in symptomless Californian artichoke plants (Costa *et al.*, 1959), although no information was given on virus particle length and cytopathology. Two viruses causing the same symptoms as those induced by ArLVS in *C. amaranticolor*, *N. clevelandii*, and *Zinnia elegans* were later found in symptomless globe artichoke plants in Apulia (Majorana and Rana, 1970b) and Morocco (Fischer and Lockhart, 1974b). The particle lengths of these viruses were 678 and 674 nm, respectively. There is no evidence that a virus found in Brazil in symptomless artichoke plants (Costa and Camargo, 1969; Kitajima *et al.*, 1970) was the same as ArLVS, as infected plants contained the cylindrical inclusions (pinwheels and laminated aggregates) typical of potyviral infections. Artichoke plants from Spain showing severe degeneration were infected by ArLVS (or another carlavirus) and a potyvirus because infected cells had cytopathological changes suggestive of the simultaneous presence of a carlavirus and a potyvirus (Peña-Iglesias and Ayuso-Gonzales, 1972). No antisera to either ArLVM or ArLVS are commercially available.

F. Unnamed putative carlavirus

A virus with filamentous particles 664 nm in length was isolated in Apulia from symptomless globe artichoke plants of cvs Terom and Romanesco. Its biological, physicochemical, and cytopathological properties resembled those of members of the genus *Carlavirus* (Rana *et al.*, 1989). This virus sedimented as a single component with a coefficient of 150*S* and was distinct from ArLVM because of its sap transmissibility to herbaceous hosts, genome size (M_r 7.7 vs. 7.5 kb), CP subunit size (M_r 29 vs. 31 kDa), and particle length (664 vs. 697 nm). Like ArLVM, it was serologically related to PopMV; however, it was distant enough (SDI = 4–5) to support its identification as an artichoke isolate of PopMV (Rana *et al.*, 1989). Since virus characterization is not complete, its taxonomic allocation remains undetermined and it does not appear among the recognized carlaviruses (Adams *et al.*, 2005).

G. *Potato virus X*

Potato virus X (PVX), the type species of the genus *Potexvirus* (Adams *et al.*, 2005), has filamentous particles ca. 515 × 13 nm in size, which sediment as a single component with a coefficient of 117*S*, and contains ca. 6% of an ssRNA 6435 nt in size (Acc. No. D00344). PVX was isolated in Tunisia (Chabbouh, 1989) from artichoke plants showing mosaic, leaf deformation, and a strong decrease in size of the leaf blade. The virus was mechanically transmitted to herbaceous hosts and identified serologically. Cytopathological changes were those expected in cells infected by PVX. Detection can be done serologically by ELISA or by either RT-PCR or Dot blot using a DIG-labeled DNA probe.

H. Artichoke curly dwarf virus

ACDV has filamentous particles 582 × 15 nm in size and is classified as a tentative species in the genus *Potexvirus* (Adams *et al.*, 2005). It was recovered in California (USA) from stunted artichoke plants with distorted leaves, variously extended vein necrosis, delayed flower head development, and low-quality crop. (Leach and Oswald, 1950; Morton, 1961). Mechanically inoculated artichoke and cardoon seedlings reproduced the field disease (Leach and Oswald, 1950). ACDV epidemiology is unknown. Morton (1961) reported ACDV to occur consistently in mixed infections with another virus with longer filamentous particles that may be the same as the putative carlavirus ArLVS.

I. Artichoke degeneration virus

ADV is a virus with filamentous particles which infects globe artichokes in Spain. It was identified in dwarfed plants of cv. Tudela with mottling, curling, and crinkling of the leaves (Peña-Iglesias and Ayuso-Gonzales, 1972). Virus particles from partially purified preparations had an estimated length of 585 nm. The cytopathology of infected cells resembled that induced by potexviruses. Transmission experiments by aphids were unsuccessful. According to Peña-Iglesias and Ayuso-Gonzales (1972), ADV and the putative potexvirus ACDV are not the same. This, however, was not proved experimentally. Since the information on ADV is too limited to allow its classification, the virus was not included among tentative potexvirus species (Adams *et al.*, 2005).

J. *Tomato infectious chlorosis virus*

Tomato infectious chlorosis virus (TICV), genus *Crinivirus*, family *Closteroviridae* (Martelli *et al.*, 2005), was originally detected in California in artichoke, cardoon (*C. cardunculus*), and weeds (*Nicotiana glauca*, *P. echioides*)

(Duffus *et al.*, 1996). Then, the virus was recorded and in several Italian regions (Liguria, Sardinia, Latium, Campania, and Sicily) in artichoke plants with no clear symptoms or showing interveinal yellowing of the leaves. In all instances, plants were infested with high populations of *Trialeurodes vaporariorum* (Davino *et al.*, 2009; Vaira *et al.*, 2002).

TICV has filamentous flexuous particles of two lengths (850 and 900 nm) and an ssRNA genome split into two molecules (RNA-1 and RNA-2), both needed for infectivity and separately encapsidated. RNA-1 consists of 8271 nt (Acc. No. NC_013258) and contains three open reading frames. RNA-2 is 7914 nt in size (Spanish isolate SP5131, Acc. No. FJ542305; Californian isolate CA4, Acc. No. FJ542305) and contains eight open reading frames (Orílio and Navas-Castillo, 2009). Besides artichoke, the virus infects several crops and ornamental plants, including tomato, tomatillo (*Physalis ixocarpa*), potato, lettuce, *Ranunculus* sp., aster, petunia, and zinnia (Smith, 2009). Thus, in the countries where TICV has been recorded in other crops, it can be viewed as a potential threat to artichoke cultivation. The virus is transmitted by *T. vaporariorum* but not by *Bemisia tabaci* (Wisler *et al.*, 1998) and is included in the EPPO Alert list (Smith, 2009) being regarded as an emerging pathogen with a high potential for economic impact. There are no commercially available antisera to TICV. The virus can be detected by Dot blot hybridization using a DIG-labeled DNA probe or by primer pair upstream primer I (5′-ATGAGGTCTTTCACAGTGG-3′) and downstream primer II (5′-GTCCGAAACTGATTGAACC-3′), which yield an amplicon of ca. 700 bp (Li *et al.*, 1998).

K. Filamentous viruses and globe artichoke degeneration disease

Under the name of degeneration of globe artichoke, disorders characterized by foliar mottling, leaf deformation and crinkling, stunting, and decreased yield have been reported from several countries of the Mediterranean region. Common to all, these diseases are the progressiveness of the degenerating condition, which becomes increasingly severe as the crop ages. Diseased plants show also the consistent presence of a filamentous virus (Marrou and Mehani, 1964; Mehani, 1969; Ortega *et al.*, 2005; Peña-Iglesias and Ayuso-Gonzales, 1972; Pochard *et al.*, 1964; Welvaert and Zitouni, 1974). Welvaert and Zitouni (1974) questioned whether artichoke degeneration could be only of virus origin and suggested the possible involvement of bacteria and perhaps other undetermined biotic causes. Since no real advances have been made in proving its nature, artichoke degeneration remains an ill-defined disease. Nevertheless, viruses seem to be involved in its etiology because of the consistent presence of multiple agents of diverse taxonomic position. However, the presence of individual filamentous "latent" viruses reported in the literature as single different potyviruses, carlaviruses, or potexviruses is not always proved by clear experimental evidence and may not be correct.

IV. VIRUSES WITH RIGID ROD-SHAPED PARTICLES

A. *Tobacco mosaic virus*

The presence of TMV (genus *Tobamovirus*), family *Virgaviridae* (Carstens, 2010; Lewandowski, 2005) in symptomless *C. cardunculus* was recorded by Lisa (1971) in Piedmont (Italy). The cardoon isolate of TMV is serologically identical with *Tomato mosaic virus* (ToMV) (Brunt, 1986). Thus, this artichoke virus should be reclassified as ToMV with rigid rod-shaped particles 300 × 18 nm, sedimenting as a single component with coefficient of 190*S*, containing an ssRNA genome with a size of 6383 nt (Acc. No. AF332868), that forms 5% of the particle weight, CP subunits of a single type with M_r of 17.5 kDa. ToMV can be detected by ELISA.

B. *Tobacco rattle virus*

TRV is the type species of the genus *Tobravirus*, family *Virgaviridae* (Carstens, 2010; Robinson, 2005). It was isolated in Brazil (Sao Paulo State) from globe artichoke plants with a disease characterized by bright yellow discoloration of the leaves (Chagas and Silberschmidt, 1972; Chagas *et al.*, 1969). The virus, whose particles were 25 nm thick and 200 and 55 nm in lengths, was transmitted by *Trichodorus christiei* to artichoke seedlings but not by mechanical inoculation (Salomão, 1976). Another TRV isolate was found in France, in mixed infections with ArLV and BBWV in artichoke plants of cv. Camus de Bretagne (Migliori *et al.*, 1985). The French isolate was transmitted by *Trichodorus primitivus* and was recognized by antisera to the following TRV isolates: PRN, CAM, and *Solanum nigrum* (Migliori *et al.*, 1985). *S. nigrum* is a natural source of TRV inoculum since the virus was identified also in this weed sampled from artichoke fields in Sicily. *Paratrichodorus tunisiensis* transmits the Italian isolate (Roca and Rana, 1981). TRV genome consists of two ssRNA species encapsidated separately, RNA-1 (6791 nt, Acc. No. AF166084) and RNA-2 (3855 nt, Acc. No. Z36974) in particles coated by 22.3 kDa protein subunits. TRV infection can be detected by commercially available antisera.

V. VIRUSES WITH ENVELOPED PARTICLES

A. *Tomato spotted wilt virus*

TSWV, a member of the genus *Tospovirus* (Nichol *et al.*, 2005), was first found in Argentinian artichokes with yellowing and withering of the leaves (Gracia and Feldman, 1978), in Australia (Martelli *et al.*, 1981), California (Vilchez *et al.*, 2005), and Spain (Ortega *et al.*, 2005). In southern Italy (Apulia and Basilicata), it was isolated from artichokes of cvs Catanese and Violet de Provence, seed-propagated cultivars, and Nunhems hybrids 6370 and 8546. Infection rates in the field varied from 2% to 3% in Basilicata to ca. 40% in Apulia, where TSWV is now regarded as the most damaging virus of artichoke. Infected plants show severe stunting and deformation, generalized chlorosis and bronzing of the apical leaves, distortion of the head stalk, and necrosis of portions of the inner and outer scales (Camele and Rana, 1999, 2005; Vovlas and Lafortezza, 1994). The virus is transmitted persistently by several thrips species, among which *F. occidentalis* is the most efficient.

TSWV has enveloped roughly spherical particles 70–110 nm in diameter, covered by knob-like surface projections. Virions contain five structural proteins from 29 to 200 kDa in size and three ssRNA segments. RNA L (8897 nt in size, Acc. No. D10066) has a negative polarity, whereas both RNA M (4821 nt in size, Acc. No. S48091) and RNA S (2916 nt in size, Acc. No. D00645) have an ambisense coding strategy (Goldbach and Kuo, 1996). The cytopathology of TSWV infection in globe artichoke was studied by Gracia and Feldman (1978) and Vovlas and Lafortezza (1994).

N. glutinosa and *N. occidentalis* are useful diagnostic species as they react with local necrotic lesions 2–4 days after inoculation, followed by systemic necrotic infection. TSWV can be detected by commercially available antisera and by Dot blot hybridization using a DIG-labeled DNA probe or RT-PCR using the primer pair TSWVNSM fwd 5′-GA (AG) GAA ACA TCT TCC TTT GG-3′ and TSWVNSM rev 5′-CCT CTT CTT CTT CAA CTG ATC-3′, which yield an amplicon of 679 bp.

B. *Cynara* virus

CraV was first observed in degenerated globe artichokes in Spain (Peña-Iglesias and Ayuso-Gonzales, 1972), then in *C. scolymus* and *C. syriaca* in Apulia (Russo *et al.*, 1975). CraV particles are bacilliform, enveloped, measure 170–260 nm (modal length = 243 nm) × 60–65 nm, and localize within dilations of the endoplasmic reticulum, from which they pick up the envelope (Peña-Iglesias and Ayuso-Gonzales, 1972; Rana *et al.*, 1988; Russo *et al.*, 1975). Thus, in principle, CraV could be classified in the genus

Cytorhabdovirus, but the limited knowledge of its properties suggested inclusion among the unassigned species in the family *Rhabdoviridae* (Tordo *et al.*, 2005). CarV has always been found in mixed infections, with an unidentified potyvirus in Spain (Peña-Iglesias and Ayuso-Gonzales, 1972) and, in Apulia, with viruses with filamentous (ArLV or BYMV) and isometric particles (AILV or BBWV) (Russo *et al.*, 1975). The virus, whose vector is unknown, was transmitted mechanically to an experimental host range restricted to plants in the family *Solanaceae* (Rana *et al.*, 1988). *Nicotiana langsdorffii* is a good host for virus maintenance and purification. An antiserum to CraV (Rana *et al.*, 1988) did not decorate Ivy vein clearing virus, another cytoplasmic rhabdovirus (Castellano and Rana, 1981) classified as unassigned species to the family *Rhabdoviridae* (Tordo *et al.*, 2005).

VI. DIAGNOSIS

Mixed infections are common in globe artichoke, thus making symptom-based diagnosis almost impossible. Most of the plants, even when infected by a single virus, do not show distinctive symptoms, either because infections are latent or because varietal reactions, environmental conditions, agronomic practices, and plant age influence symptom expression. Because virus distribution in the host may vary with the season and plant age, the success of any detection technique is strongly dependent on the choice of a proper time for sampling. In our experience (southern Italy), samples collected between early September and mid-November from young leaves of at least 1-year-old plants give the best and most reproducible results (Repetto *et al.*, 1997).

Reaction of herbaceous indicators is rarely of diagnostic value since many of the artichoke-infecting viruses share a common host range and often induce similar symptoms in the same host. Thus, the use of herbaceous indicators is usually restricted to those viruses for which no serological or molecular detection tools are available.

Serology has been used in several instances, either as gel double diffusion tests or as ELISA (Foddai *et al.*, 1992; Migliori *et al.*, 1989; Ortega *et al.*, 2005). Although ELISA has distinctive advantages over immunodiffusion, it is not free of drawbacks when used for artichoke. The rapid oxidation of artichoke sap gives borderline results, especially when the antigen concentration is low, so background readings prevent differentiation of infected from healthy samples. Although ELISA kits for many of the viruses found in artichoke are commercially available; there are few reports of their use for direct detection in artichoke sap. Thus, ELISA is often used for virus identification after transmission to herbaceous hosts.

Electron microscopy (EM) has been used extensively, as some of the artichoke viruses occur in enough high concentration (e.g., AMCV, TMV) in host tissues to be seen in leaf-dip mounts and identified by immune electron microscopy (IEM) (Milne and Luisoni, 1977). ISEM has also been successfully applied to samples containing low amounts of virus (Derrick, 1973). EM-based diagnosis has been further adapted using various decoration techniques and gold-conjugated antibodies. Finally, as cytopathological changes can be specific at the genus or the species level, observation of thin sectioned tissues can prove useful for preliminary identification (Martelli and Russo, 1977, 1984).

Accurate diagnosis combined with rapid and early virus detection can be carried out using nucleic acid-based tests. Cloning of virus nucleic acid and use of nonradioactive detection methods has increased the use of hybridization techniques. Attention is to be paid, however, to the virus looked for, to adapt the molecular reagent (i.e., the probe) and hybridization conditions (i.e., stringency and washings) to its specific detection. For example, all definitive tombusviruses so far described cross-hybridize because of the conserved genome sequences (Martelli *et al.*, 1989), thus hindering a clear-cut identification of individual viruses by molecular hybridization. With AMCV, this obstacle was overcome by the synthesis a virus-specific riboprobe designed on the 3′-end of the CP gene, a genomic region where the lowest degree of sequence homology exists in tombusviruses (Barbarossa *et al.*, 1993).

Dot blotting represents a suitable assay for routine tests and sanitary certification schemes. In a protocol developed in our laboratory (Gallitelli and Saldarelli, 1996; Repetto *et al.*, 1997), artichoke tissues are crushed in an alkaline solution that prevents oxidation, spotted onto a positively charged Nylon membrane, and hybridized with DIG-labeled RNA or DNA probes. Since mixed infections are common in artichoke, specific probes for each individual virus can be mixed in the same hybridization solution (Saldarelli *et al.*, 1996).

For large-scale testing, tissue prints made in the field allow the analysis of multiple samples, thus saving time, space, and transport of infected plant material to the laboratory. Young artichoke leaves are cut transversally, the exposed surface is quickly pressed onto a nylon membrane prewetted with an alkaline solution (i.e., 50 mM NaOH–2.5 mM EDTA) and air dried. Another advantage of Dot blot or tissue print hybridization is that replica filters can be prepared and stored for longtime to be probed with the same or, if needed, a different probe.

Diagnostic tools for artichoke viruses able to identify multiple infections reliably in a single test are desirable. PCR has this potential. Grieco and Gallitelli (1999) have described a method for the independent amplification of ArLV, AILV, and AMCV (i.e., viruses belonging to three different genera) sequences in a six-primer-driven RT-PCR using plant sap.

The simultaneous and specific identification of each virus was achieved with a single test, based on the different size of the amplicon produced by each virus. In this RT-PCR protocol, antisense primers for the synthesis of first strand cDNA using a viral RNA template represented the most critical step for a specific amplification. To ensure strain specificity, the primers were selected within genomic regions that differed among members of the same taxon. Although this procedure has several advantages, it did not overcome the problem of detection in mixed infections. For example, sensitivity level was lower than that reported for other virus–host combinations, since detection limits in ca. 70 ng leaf tissue were 400 fg of AMCV and ArLV RNAs but only 4 pg of AILV RNA. This was perhaps because of the simultaneous presence of primers and templates competing for the same reagents during first strand cDNA synthesis and its following amplification. Another critical point was plant sap concentration. Amplified fragments were seen only when artichoke sap was diluted 10^3-fold before addition to the reaction. With sap diluted 10^2-fold, the reaction did not proceed. Another possible source of problems was the relative concentration of each viral template in naturally infected samples. In mixed infections, different amounts of each virus are likely to occur, especially if the viruses belong to different taxonomic groups. This may selectively drive the reaction for one virus, masking the presence of the others.

Although PCR has the potential for sensitive, specific, and reliable identification of artichoke viruses, it needs downstream analysis (e.g., gel electrophoresis, Southern blotting, sequencing), which makes the technique less appealing for large-scale routine diagnosis. These and other problems have recently been overcome by real-time PCR technologies. Finetti-Sialer *et al.* (2000) have used Scorpion-RT-PCR (Whitcombe *et al.*, 1999) to detect TSWV in naturally infected artichoke plants. In serially diluted plant sap, the method detected as little as 7.5 pg of virus in infected plant tissue and proved to be at least as sensitive as visual estimation of ethidium bromide-stained amplicons separated by gel electrophoresis. The use of different fluorophores for different viruses may allow detection in the same reaction, although this possibility has not yet been explored with artichoke sap.

VII. PRODUCTION OF VIRUS-FREE PLANTS

The damaging effects of virus infections on the quantity and quality of the yield highlight that new artichoke fields must be planted with virus-free propagation material. Its availability would reduce the problem and allow a great increase in productivity. In addition, it would make nursery productions conform to the current EU legislation (Directives 93/61/CEE and 93/62/CEE), which needs that nursery productions originate from virus-free and true-to-type mother plants (Gallitelli and Barba, 2003).

Seed propagation has been regarded as an attractive means to make available virus-free plants. However, most artichoke-infecting viruses are seed and pollen-transmitted and AILV and ArLV have been detected in artichoke seed coats and fully expanded cotyledons (Bottalico *et al.*, 2002). Thus, procedures to detect seed-transmitted viruses must be applied if seed-based nursery productions will be grown.

Some encouraging, but not fully convincing results were obtained with *in vivo* or *in vitro* thermotherapy of infected plants. Exclusion of viruses using *in vitro* culture of shoot tips or young head receptacles (Marras *et al.*, 1982; Pécaut *et al.*, 1985) is less effective than meristem-tip culture (Harbaoui *et al.*, 1982; Peña-Iglesias and Ayuso-Gonzales, 1974, 1982). However, the latter technique results in the loss of earliness (an economically important trait) of reflowering globe artichokes of the "early Mediterranean group," which includes cvs Brindisino, Violet de Provence, and Catanese. Pécaut and Martin (1992, 1993) reported that meristem-tip culture of cultivars of this group yields plants characterized by lateness and deeply divided leaves (denoted "pastel" variants), or producing globular heads (denoted "bull" variants), or showing both traits (denoted "pastel-bull" variants) (Pochard *et al.*, 1969). The frequency of such variants increases with the number of subcultures and can reach 100% (Pécaut and Martin, 1992, 1993). In addition, meristem-tip culture does not guarantee elimination of viruses because some persist through repeated subculturing (Faccioli and Marani, 1998). Recently, it was shown that meristem-tip culture removes ArLV but not AILV. However, AILV can be removed by growing offshoots from infected plants to 38 °C for a minimum 60 days, followed by excision and growth of meristem tips *in vitro* (Acquadro *et al.*, 2010; Papanice *et al.*, 2004). Thus, *in vitro* culture of meristem-tip explants only is not always a dependable method for producing virus-free artichoke plants.

Thermotherapy has been widely used for virus elimination in plants (Mink *et al.*, 1998). The assumption is that it delays virus movement through the plant, therefore inhibiting meristem invasion (Mink *et al.*, 1998), or hinders replication while the virus is degraded, thus resulting in its elimination from shoot tips (Cooper and Walkey, 1978; Kassanis, 1957).

Recently, Wang *et al.* (2008) have shown that *Raspberry bushy dwarf virus* (RBDV) can be removed from raspberry plants by a combination of meristem-tip culture, thermotherapy, and cryotherapy. It has been speculated the extensive RBDV RNA degradation could occur because of RNA silencing, which is enhanced at high temperatures (Chellappan *et al.*, 2005; Qu *et al.*, 2005; Szittya *et al.*, 2003) and likely to be also involved in virus exclusion from meristems (Foster *et al.*, 2002; Mochizuki and Ohki, 2004; Qu *et al.*, 2005; Schwach *et al.*, 2005). Thus, results of Wang *et al.* (2008)

suggested a model by which thermotherapy lessens the part of apical meristem invaded by RBDV by degrading its RNA and reducing the number of virus-infected cells.

As said above, Acquadro *et al.* (2010) have obtained virus-free plants of artichoke cv. Brindisino by a combination of meristem-tip culture and thermotherapy. ArLV infection was removed by meristem-tip culture and AILV by a combination of meristem-tip culture and *in vitro* or *in vivo* heat therapy. Comparing these data with those of Wang *et al.* (2008), Jovel *et al.* (2007), and Siddiqui *et al.* (2008), it can be assumed that RNA silencing as well as other mechanisms may have been involved in removing both viruses. ArLV is a potyvirus, and like other members of this genus, it should code for the RNA-silencing suppressor HC-Pro, which may not be able to interfere with the systemic silencing spread (reviewed by Li and Ding, 2006). Thus, silencing could account for ArLV exclusion from the apical meristem, making it possible to get virus-free regenerants. However, this explanation does not fit AILV, which like other nepoviruses may not have a recognized RNA-silencing suppressor. After a first phase of symptom expression, plants infected by nepoviruses usually undergo a recovery phase in which newly produced leaves are symptomless. Acquadro *et al.* (2010) have proposed a model whereby AILV could escape silencing through a recovery-like mechanism that allows the virus to enter the apical meristem, yielding still infected regenerants. This hypothesis seems supported by the notion that AILV-infected tobacco plants undergo recovery, but recovered tissue still contain infectious virus (Mascia *et al.*, 2009; Santovito *et al.*, 2010). However, AILV may not survive the heat treatment of the first subculture, either because RNA silencing is enhanced or because meristematic cells are damaged by heat or both. So virus-free regenerants can be produced with the next cycle of meristem-tip culture.

An alternative but equally efficient way for producing AILV- and ArLV-free artichoke plants is *in vivo* thermotherapy. With this procedure, artichoke offshoots grown in large pots are acclimatized at 28–30 °C for 1 month before exposure at 38 °C for up to 150 days. Usually all leaves die within the first 2 weeks and meristem tips are collected from the new shoots.

In vitro and *in vivo* heat therapy combined with meristem-tip culture was successful in obtaining virus-free regenerants of artichoke cv. Brindisino that preserved the early-flowering habit and a normal head shape. When visually inspected, sanitized plants showed larger plant size and brighter green foliage and produced leaves and capitula the number of which was 1.5–2-folds that of standard plants of the globe artichoke "Brindisino" type (Fig. 1). The *in vitro* approach is economic in time and space, since heat treatment involves only a few explants, which can be held in a small growth chamber. However, two passages through meristem-tip culture are required, risking off-types to appear. The *in vivo* protocol needs only one round of meristem-tip culture but is more

FIGURE 1 (A) Nursery multiplication of virus-free propagative material of different age of clones of globe artichoke "Brindisino" sanitized by *in vitro* meristem-tip culture and thermotherapy. (B) The fitosanitary status of plants under multiplication is routinely checked by tissue print hybridization with DIG-labeled DNA probes. (C) Virus-free plants of globe artichoke "Brindisino" (central rows) show a larger size and brighter color than standard plant material (on the left) routinely used by farmers. Picture took Nov 2010 in a demonstration field set up by the Apulian phytosanitary extension service. (See Page 7 in Color Section at the back of the book.)

demanding in heated and lighted space, since plants must be kept for long periods at high temperature. Besides, it is less efficient in number of plants recovered after the treatment.

Finally, caution must be paid before declearing as virus-free the regenerants got by any treatment. For example, Acquadro *et al.* (2010) were unable to detect AILV in artichoke shoots after 60 days of heat treatment. The virus, however, reappeared within the following 30 days, suggesting that its presence had been lowered to undetectable levels but not removed. It is therefore necessary to extend the tests for at least 1 year to make sure that regenerants are virus-free.

VIII. CONCLUDING REMARKS

Many of the artichoke-infecting viruses have been characterized molecularly and sensitive tools have been developed for their detection. However, the role of many viruses in the etiology of specific diseases has still not be determined, and whether they can decrease the yield, and to what extent.

Because there is a progressive tendency to move from vegetatively to seed-propagated crops that remain in the field for no longer than a couple of years, it is likely the economic importance of virus infections will decrease in the future. In fact, the high infection rates in new plantations is at first due to the use of offshoots originating from sanitarily uncontrolled mother plants, which may have undergone a superficial visual selection by the farmers. Certified seeds and virus-free germplasm of the traditional early- and late-flowering types should decrease inoculum potential.

Meanwhile, some steps of the infection pathway need to be clarified. Transmissibility of major viruses through true seeds should be studied, incidence and distribution of natural sources of inoculum in different environments should be identified, and the incubation period of single and mixed infections should be determined. This information is essential if any prediction is to be made of the impact of inoculum potential on the rate of reinfection of plants in the field. All this leads to the conclusion that studies on artichoke virology, which are stagnating, except for the interest shown by two or three laboratories in the world, should resume interest and be directed toward practical applications.

REFERENCES

Acquadro, A., Papanice, M., Lanteri, S., Bottalico, G., Portis, E., Campanale, A., Finetti-Sialer, M. M., Mascia, T., Sumerano, P., and Gallitelli, D. (2010). Production and genotyping of virus-free plants in a reflowering globe artichoke varietal type. *Plant Cell Tiss. Org.* **100**:329–337.

Adams, M. J., Accotto, G. P., Agranovsky, A. A., Bar-Joseph, M., Boscia, D., Brunt, A. A., Candresse, T., Coutts, R. H. A., Dolja, V. V., Falk, B. W., Foster, G. D., Gonsalves, D., *et al.* (2005). Family *Flexiviridae*. *In* (C. M. Fauquet, M. A. Mayo, J. Maniloff, U. Desselberger, and L. A. Ball, eds.), Virus Taxonomy. 8th Report of the International Committee on Taxonomy of Viruses, pp. 1089–1124. Elsevier/Academic Press, San Diego, USA.

Avgelis, A. D., and Vovlas, C. (1989). Artichoke yellow ringspot nepovirus naturally infecting cucumber in Crete. *Neth. J. Plant Pathol.* **95**:177–184.

Avgelis, A. D., Katis, N., and Grammatikaki, G. (1992). Broad bean wrinkly seed caused by artichoke yellow ringspot nepovirus. *Ann. Appl. Biol.* **121:**133–142.

Barbarossa, L., Grieco, F., and Gallitelli, D. (1993). Synthesis of in vitro transcripts for the detection of Artichoke motted crinkle tombusvirus in artichoke. *Riv. Patol. Veg.* **3:**1–6.

Baricevich, D., Kus, M., and Pepelnjak, M. (1995). Isolation of IgG immunoglobulins from rabbit ALV-antiserum using protein-A-sefarose and estimation of globe artichoke local cultivar ''Istra'' for the presence of artichoke latent virus (ALV) and cucumber mosaic virus (CMV) in Slovenia *Zbornik Biotechniske Fakultete Univerze Ljubijani, Kmetijstvo* **65:**47–54.

Berger, P. H., Adams, M. J., Barnett, O. W., Brunt, A. A., Hammond, J., Hill, J. H., Jordan, R. L., Kashiwazaki, S., Rybicki, E., Spence, N., Stenger, D. C., Ohki, S. T., *et al.* (2005). Family *Potyviridae*. *In* (C. M. Fauquet, M. A. Mayo, J. Maniloff, U. Desselberger, and L. A. Ball, eds.), Virus Taxonomy. 8th Report of the International Committee on Taxonomy of Viruses, pp. 819–841. Elsevier/Academic Press, San Diego, USA.

Bottalico, G., Padula, M., Campanale, A., Finetti-Sialer, M. M., Saccomanno, F., and Gallitelli, D. (2002). Seed transmission of *Artichoke Italian latent virus* and *Artichoke latent virus* in globe artichoke. *J. Plant Pathol.* **84:**167.

Brown, D. J. F., Kyriakopoulou, P. E., and Robertson, W. M. (1997). Frequency of transmission of artichoke Italian latent nepovirus by *Longidorus fasciatus* (*Nematoda*: *Longidoridae*) from artichoke fields in the Iria and Kandia areas of Argopolis in Northern Peloponnesus, Greece. *Eur. J. Plant Pathol.* **103:**501–506.

Brunt, A. A. (1986). Tomato mosaic virus. *In* ''The Plant Viruses. The Rod-Shaped Plant Viruses'' (M. H. V. Van Regenmortel and H. Fraenkel-Conrat, eds.), pp. 181–204. Plenum Press, New York.

Camele, I., and Rana, G. L. (1999). Rinvenimento di infezioni del *Tospovirus* della bronzatura del pomodoro (TSWV) in due composite coltivate in Basilicata. Atti VII Convegno Nazionale S.I.Pa.V., Piacenza.

Camele, I., and Rana, G. L. (2005). New outbreaks of *Tomato spotted wilt tospovirus* (TSWV) infectious on globe artichoke in Basilicata and Apulia. *Acta Hortic.* **681:**593–595.

Camele, I., Nuzzaci, M., Rana, G. L., and Kyriakopoulou, P. E. (1991). *Papaver rhoeas* L., a host of two pathogenic viruses of cultivated plants. *Petria* **1:**111–115.

Camele, I., Candido, V., Rana, G. L., and Palumbo, M. (1999). Effetti morfologici e produttivi dell'infezione del potyvirus latente del carciofo. *Petria* **9:**43–52.

Carstens, E. B. (2010). Ratification vote on taxonomic proposals to the International Committee on Taxonomy of Viruses (2009). *Arch. Virol.* **155:**133–146.

Castellano, M. A., and Martelli, G. P. (1981). Electron microscopy of *Pelargonium zonate spot virus* in host tissues. *Phytopathol. Mediterr.* **20:**64–71.

Castellano, M. A., and Rana, G. L. (1981). Transmission and ultrastructure of Ivy vein clearing virus infections. *Phytopathol. Mediterr.* **20:**199–202.

Chabbouh, N. (1989). Mise en evidence de quatre virus presents sur l'artichaut en Tunisie. *Ann. INRAT* **62:**3–14.

Chabbouh, N., and Cherif, C. (1990). Outbreaks and new records. Tunisia. Cucumber mosaic virus in artichoke. *FAO Plant Protect. B.* **35:**52–53.

Chagas, C. M., and Silberschmidt, K. M. (1972). Virus da faixa amarela da alcachofra: occorencia, transmissão mecanica e propriedades fisicas. *Biologico* **38:**35–40.

Chagas, C. M., Flores, M., and Carner, J. (1969). Una nova doenca de virus da alcachofra no Estado de São Paulo. *Biologico* **35:**271–274.

Chellappan, P., Vanitharani, R., Ogbe, F., and Fauquet, C. M. (2005). Effect of temperature on geminivirus-induced RNA silencing in plants. *Plant Physiol.* **138:**1828–1841.

Ciccarone, A. (1967). Attuali cognizioni sulle malattie del carciofo. Atti 1° Congresso Internazionale di Studi sul Carciofo, Bari 1967, pp. 181–192.

Ciuffo, M., Lenzi, R., Testa, M., and Turina, M. (2010). Widespread occurrence of Ranunculus latent virus (*Macluravirus, Potyviridae*) in artichoke crops in Italy. *J. Plant Pathol.* **92:** S4–S78.

Cooper, V. C., and Walkey, D. G. A. (1978). Thermal inactivation of cherry leaf roll virus in tissue cultures of Nicotiana rustica raised from seeds and meristem tips. *Ann. Appl. Biol.* **88:**273–278.

Costa, A. S., and Camargo, I. G. B. (1969). Occorência do virus latente da alcachofra en São Paulo. *Rev. Soc. Bras. Fitopatol.* **3:**53–54.

Costa, A. S., and Tasaka, H. (1971). Enfezamento e malformacao foliar de alcachofra inducido pelo virus da necrose branca do fumo. *Biologico* **37:**176–179.

Costa, A. S., Duffus, J. E., Morton, D., Yarwood, C. E., and Bardin, R. (1959). A latent virus of California artichokes. *Phytopathology* **49:**49–53.

Davino, S., Tomassoli, L., Tiberini, A., Mondello, V., and Davino, M. (2009). Outbreak of tomato infectious chlorosis virus in a relevant artichoke producing area of Sicily. *J. Plant Pathol.* **91**(Suppl. 4)**:**S4.57–S4.58.

Derrick, K. S. (1973). Quantitative assay for plant viruses using serologic specific electron microscopy. *Virology* **56:**652–653.

Di Franco, A., and Gallitelli, D. (1985). Petunia hybrida Vilm., ospite naturale in Puglia del virus del mosaico del Cetriolo (CMV) e del virus latente del Carciofo. *Inf. Fitopatol.* **35** (2):45–46.

Di Franco, A., Gallitelli, D., Vovlas, C., and Martelli, G. P. (1989). Partial characterization of Artichoke virus M. *J. Phytopathol.* **127:**265–273.

Dodd, S. M., and Robinson, D. J. (1984). Nucleotide sequence among RNA species of strains of tomato black ring virus and other nepoviruses. *J. Gen. Virol.* **65:**1731–1740.

Duffus, J. E., Liu, H. Y., and Wisler, G. C. (1996). Tomato infectious chlorosisvirus—A new clostero-like virus transmitted by *Trialeurodes vaporariorum*. *Eur. J. Plant Pathol.* **102:**219–226.

Faccioli, G., and Marani, F. (1998). Virus elimination by meristem tip culture and tip micrografting. *In* "Control of Plant Virus Diseases" (A. Hadidi, R. K. Khetarpal, and H. Koganezawa, eds.), pp. 346–380. APS Press, St. Paul, MN, USA.

FAO, 2010: http://faostat.fao.org/site/339/default.aspx

Finetti-Sialer, M., and Gallitelli, D. (2003). Complete nucleotide sequence of Pelargonuim zonate spot virus and its relationships with the family *Bromoviridae*. *J. Gen. Virol.* **84:**3143–3151.

Finetti-Sialer, M., Cillo, F., Barbarossa, L., and Gallitelli, D. (1999). Differentiation of cucumber mosaic virus subgroups by RT-PCR RFLP. *J. Plant Pathol.* **81:**145–148.

Finetti-Sialer, M., Ciancio, A., and Gallitelli, D. (2000). Use of fluorogenic scorpions for fast and sensitive detection of plant viruses. *Bull. OEPP/EPPO Bull.* **30:**437–440.

Fischer, H. V., and Lockhart, B. E. (1974a). Occurrence in Morocco of a virus disease of artichoke related to artichoke mottled crinkle. *Plant Dis. Rep.* **58:**1117–1120.

Fischer, H. V., and Lockhart, B. E. (1974b). A Moroccan isolate of artichoke latent virus. *Plant Dis. Rep.* **58:**1123–1126.

Foddai, A., Marras, F., and Idini, G. (1977). Presenza di un "*Potyvirus*" sul Carciofo (*Cynara scolymus* L.) in Sardegna *Annali della Facoltà di Agraria dell' Università di Sassari* **25:**398–407.

Foddai, A., Corda, P., and Idini, G. (1983a). Influenza del potyvirus latente del carciofo spinoso sardo sulla produttività delle piante in pieno campo. I. Risultati relativi al primo anno di impianto. *Riv. Patol. Veg.* **19:**29–35.

Foddai, A., Corda, P., and Idini, G. (1983b). Influenza del potyvirus latente del carciofo spinoso sardo sulla produttività delle piante in pieno campo. II. Risultati relativi al secondo anno di impianto. *Ann. Fac. Agr. Univ. Sassari* **30:**37–43.

Foddai, A., Marras, F., and Idini, G. (1991). Field control trias for preventing the spread of aphid-borne Artichoke latent virus (ALV) in Artichoke. *Phytopathol. Mediterr.* **30:**1–5.
Foddai, A., Marras, F., and Idini, G. (1992). Il saggio immunoenzimatico (ELISA) per la diagnosi del *Potyvirus* latente del carciofo. *Riv. Patol. Veg. S IV* **18:**103–107.
Foddai, A., Cugusi, M., and Idini, G. (1993). Artichoke (*Cynara scolymus* L.): A new host for turnip mosaic virus. *Phytopathol. Mediterr.* **32:**247–248.
Foster, T. M., Lough, T. J., Emerson, S., Lee, R., Bowman, J. L., Forster, R. L. S., and Lucas, W. J. (2002). A surveillance system regulates selective entry of RNA into the shoot apex. *Plant Cell* **14:**1497–1508.
Gallitelli, D. (1982). Properties of pelargonium zonate spot virus. *Ann. Appl. Biol.* **100:**457–466.
Gallitelli, D. (1983). Pelargonium zonate spot virus.. *CMI/AAB Descr. Plant Viruses*, No. 272.
Gallitelli, D., and Barba, M. (2003). Normativa vivaistica in Italia. *In* Atti Giornate Nazionali di studi sul Carciofo, pp. 47–51. Vivaismo e strategie di sviluppo del Carciofo, Samassi (Ca).
Gallitelli, D., and Hull, R. (1985). Characterization of a satellite RNA associated with Tomato bushy stunt virus and five other tombusviruses. *J. Gen. Virol.* **66:**1533–1543.
Gallitelli, D., and Saldarelli, P. (1996). Molecular identification of phytopathogenic viruses. *In* "Methods in Molecular Biology, Species Diagnostics Protocols: PCR and other Nucleic Acid Methods" (J. P. Clapp, ed.), Vol. 50, pp. 57–79. Humana Press, New Jersey, USA.
Gallitelli, D., Rana, G. L., and Di Franco, A. (1978). Il virus della scolorazione perinervale del carciofo. *Phytopathol. Mediterr.* **17:**1–7.
Gallitelli, D., Rana, G. L., and Di Franco, A. (1984). Artichoke vein bading virus.. *CMI/AAB Descr. Plant Viruses*, No. 285.
Gallitelli, D., Rana, G. L., Vovlas, C., and Martelli, G. P. (2004). Viruses of globe artichoke: An overview. *J. Plant Pathol.* **86:**267–281.
Gallitelli, D., Finetti-Sialer, M., and Martelli, G. P. (2005). *Anulavirus*, a proposed new genus of plant viruses in the family *Bromoviridae*. *Arch. Virol.* **150:**407–411.
Gigante, R. (1951). Il mosaico del carciofo. *Boll. Staz. Pat. Veg. Roma* **7:**177–181.
Goldbach, R., and Kuo, G. (1996). Introduction (tospoviruses and thrips). *Acta Hortic.* **431:**21–26.
Gracia, O., and Feldman, J. M. (1978). Natural infection of artichoke by tomato spotted wilt virus. *Plant. Dis. Rep.* **62:**1076–1077.
Grieco, F., and Gallitelli, D. (1999). Multiplex reverse transcriptase-polymerase chain reaction applied to virus detection in globe artichoke. *J. Phytopathol.* **145:**183–185.
Harbaoui, Y., Smaijn, G., Welvaert, W., and Debergh, P. (1982). Assainissement viral de l'artichaut (*Cynara scolymus*) par la culture in vitro d'apex méristématiques. *Phytopathol. Mediterr.* **21:**15–19.
Harrison, B. D. (1957). Studies on host range, properties and mode of transmission of beet ringspot virus. *Ann. Appl. Biol.* **45:**462–472.
Harrison, B. D. (1958). Relationships between beet ringspot, potato bouquet and tomato black ring viruses. *J. Gen. Microbiol.* **18:**450.
Harrison, B. D., and Murant, A. F. (1977). Nepovirus Group.. *CMI/AAB Descr. Plant Viruses*, No. 185.
Hollings, M. (1965). Some properties of celery yellow vein, a virus serologically related to tomato black ring virus. *Ann. Appl. Biol.* **55:**459–470.
Jankulova, M., Savino, V., Gallitelli, D., Quacquarelli, A., and Martelli, G. P. (1977). Isolation of artichoke Italian latent virus from Grapevine in Bulgaria. Proceedings of the VI Meeting of ICVG Cordova 1976, pp. 143–147.
Jovel, J., Walker, M., and Sanfaçon, H. (2007). Recovery of *Nicotiana benthamiana* plants from a necrotic response induced by a Nepovirus is associated with RNA silencing but not with reduced virus titer. *J. Virol.* **81:**12285–12297.

Kassanis, B. (1957). Effects of changing temperature on plant virus diseases. *Adv. Virus Res.* **4:**221–241.

Kitajima, E. W., Camargo, I. G. B., and Costa, A. S. (1970). Microscopia elettronica de particulas do virus latente da alcachofra e das inclusoes citoplasmicas associadas a sua infeccao. Proceedings of the 1° Corgreso Brasileiro de Microscopia Elettronica, Riberao Preto 1970, pp. 97–98.

Kyriakopoulou, P. E. (1981). Two new virus diseases of tobacco due to Artichoke yellow ringspot virus, alone and in mixed infection with Tobacco mosaic virus. Proceedings of the Panhellenic Congress of Geotechnical Research, Halkidiki, Greece, pp. 120–121.

Kyriakopoulou, P. E. (1995). Artichoke Italian latent virus causes artichoke patchy chlorotic stunting disease. *Ann. Appl. Biol.* **127:**489–497.

Kyriakopoulou, P. E., and Bem, F. P. (1973). Some virus diseases of cultivated plants noticed in Greece in the years 1971 and 1972. Proceedings of First Symposium of Geotechnical Research, B.I., pp. 409–418.

Kyriakopoulou, P. E., and Bem, G. L. (1982). Artichoke Marathon yellow blotch, a new virus disease of artichoke. Benaki Phytopathological Institute Annual Report 1982, pp. 90–91.

Kyriakopoulou, P. E., Rana, G. L., and Roca, F. (1985). Geographic distribution, natural host range, pollen and seed transmissibility of Artichoke yellow ringspot virus. *Annls. Ist. phytopath. Benaki* **14:**139–145.

Lanteri, S., Di Leo, I., Ledda, L., Mameli, M. G., and Portis, E. (2001). RAPD variation within and among populations of globe artichoke (*Cynara scolymus* L.), cv 'Spinoso sardo'. *Plant Breed.* **120:**243–247.

Lapidot, M., Guenoune-Gelbart, D., Leibman, D., Holdengreber, V., Davidovitz, M., Machbash, Z., Klieman-Shoval, S., Cohen, S., and Gal-On, A. (2010). Pelargonium zonate spot virus is transmitted vertically via seed and pollen in tomato. *Phytopathology* **100:**798–804.

Laudanki, F. (1958). Rapport sur les travaux de recherche effectuée en 1956. *Service Botanique et Agronomique de Tunisie.*

Le Gall, O., Iwanami, T., Karasev, A. V., Jones, A. T., Lehto, K., Sanfaçon, H., Wellink, J., Wetezel, T., and Yoshikawa, N. (2005a). Family *Comoviridae. In* (C. M. Fauquet, M. A. Mayo, J. Maniloff, U. Desselberger, and L. A. Ball, eds.), Virus Taxonomy. 8th Report of the International Committee on Taxonomy of Viruses, pp. 807–818. Elsevier/ Academic Press, San Diego, USA.

Le Gall, O., Iwanami, T., Karasev, A. V., Jones, A. T., Lehto, K., Sanfaçon, H., Wellink, J., Wetezel, T., and Yoshikawa, N. (2005b). Genus *Cheravirus. In* (C. M. Fauquet, M. A. Mayo, J. Maniloff, U. Desselberger, and L. A. Ball, eds.), Virus Taxonomy. 8th Report of the International Committee on Taxonomy of Viruses, pp. 803–805. Elsevier/ Academic Press, San Diego, USA.

Leach, L. D., and Oswald, J. W. (1950). Curly dwarf, a virus disease of globe artichoke. *Phytopathology* **40:**967–968.

Lewandowski, D. J. (2005). Genus *Tobamovirus. In* "Virus Taxonomy. 8th Report of the International Committee on Taxonomy of Viruses" (C. M. Fauquet, M. A. Mayo, J. Maniloff, U. Desselberger, and L. A. Ball, eds.), pp. 1009–1011. Elsevier/Academic Press, San Diego, USA.

Li, F., and Ding, S. W. (2006). Virus counterdefense: Diverse strategies for evading the RNA-silencing immunity. *Annu. Rev. Microbiol.* **60:**503–531.

Li, R. H., Wisler, G. C., Liu, H. Y., and Duffus, J. E. (1998). Comparison of diagnostic techniques for detecting tomato infectious chlorosis virus. *Plant Dis.* **82:**84–88.

Lisa, V. (1971). Two viruses naturally infecting cultivated cardoon (*Cynara cardunculus* L.). *Phytopathol. Mediterr.* **10:**231–237.

Lommel, S. A., Martelli, G. P., Rubino, L., and Russo, M. (2005). Family *Tombusviridae. In* "Virus Taxonomy. 8th Report of the International Committee on Taxonomy of Viruses" (C. M. Fauquet, M. A. Mayo, J. Maniloff, U. Desselberger, and L. A. Ball, eds.), pp. 907–936. Elsevier/Academic Press, San Diego, USA.

Majorana, G., and Rana, G. L. (1970a). Un nuovo virus latente isolato da carciofo in Puglia. *Phytopathol. Mediterr.* **9:**193–196.

Majorana, G., and Rana, G. L. (1970b). A latent virus of artichoke belonging to potato virus S group. *Phytopathol. Mediterr.* **9:**200–202.

Maliogka, V. I., Dovas, C. I., Lesemann, D. E., Winter, S., and Katis, N. I. (2006). Molecular identification, reverse transcription-polymerase chain reaction detection, host reactions, and specific cytopathology of Artichoke yellow ringspot virus infecting onion crops. *Phytopathology* **96:**622–629.

Manzanares, M. J., Corre, J., and Hervé, Y. (1995). Evaluation of globe artichoke and related germplasm for resistance to artichoke latent virus. *Euphytica* **84:**219–228.

Marras, F., Foddai, A., and Fiori, M. (1982). Possibilità di risanamento del carciofo da infezioni virali mediante coltura in vitro di apici meristematici. Proceedings of Giornate Fitopatopatologiche 1982, pp. 151–158.

Marrou, J., and Mehani, S. (1964). Etude d'un virus parasite de l'artichaut. *C.R. Acad. Agric. France* **50:**1051–1064.

Martelli, G. P. (1965). L'arricciamento maculato del carciofo. *Phytopathol. Mediterr.* **4:**58–60.

Martelli, G.P, Rana, G. L., and Savino, V. (1977). Artichoke Italian latent virus.. *CMI/AAB Descr. Plant Viruses*, No. 176.

Martelli, G. P., and Cirulli, M. (1969). Le virosi delle piante ortensi in Puglia. III. Una maculatura gialla del pomodoro causata dal virus della necrosi perinervale del tabacco (Tobacco streak virus). *Phytopathol. Mediterr.* **8:**154–156.

Martelli, G. P., and Gallitelli, D. (2008). Viruses of Cynara. *In* Characterization, Diagnosis & Management of Plant Viruses, Vol 1. Industrial Crops, pp. 445–479. Studium Press, Texas, USA.

Martelli, G. P., and Rana, G. L. (1976). Viruses and virus diseases of globe artichoke and cardoon. Nuovi Studi sul Carciofo. Proceedings of the 2° Congresso Internazionale sul Carciofo, Bari 1973, pp. 811–830.

Martelli, G. P., and Russo, M. (1973). Electron microscopy of artichoke mottled crinkle virus in leaves of *Chenopodium quinoa* Willd. *J. Ultrastruct. Res.* **42:**93–107.

Martelli, G. P., and Russo, M. (1977). Rhabdoviruses of plants. *In* "The Atlas of Insect and Plant Viruses" (K. Maramorosch, ed.), pp. 181–213. Academic Press, New York.

Martelli, G. P., and Russo, M. (1984). The use of thin sectioning for visualization and identification of plant viruses. *Methods Virol.* **8:**143–224.

Martelli, G. P., Russo, M., and Rana, G. L. (1976). Occurrence of artichoke mottled crinkle virus in Malta. *Plant Dis. Rep.* **60:**130–133.

Martelli, G. P., Russo, M., and Rana, G. L. (1981). A survey of the virological problems of Cynara species. Proceedings of the 3° Congresso Internazione di Studi sul Carciofo, Bari 1981, pp. 895–927.

Martelli, G.P, Russo, M., and Gallitelli, D. (1989). Tombusvirus group. *CMI/AAB Descr. Plant Viruses*, No. 252.

Martelli, G. P., Agranovsky, A. A., Bar-Joseph, M., Boscia, D., Candresse, T., Coutts, R. H. A., Dolja, V. V., Falk, G. D., Gonsalves, D., Hu, J. S., Jelkmann, W., Karasev, A., *et al.* (2005). Family *Closteroviridae*. *In* (C. M. Fauquet, M. A. Mayo, J. Maniloff, U. Desselberger, and L. A. Ball, eds.), Virus Taxonomy. 8th Report of the International Committee on Taxonomy of Viruses, pp. 1077–1087. Elsevier/Academic Press, San Diego, USA.

Mascia, T., Prigigallo, I., and Gallitelli, D. (2009). *Artichoke Italian latent virus* and RNA silencing: Implications in sanitation schemes of nepovirus infected plants. *J. Plant Pathol.* **91**(Suppl. 4):S4.72–S4.73.

Mayo, M. A., and Jones, A. T. (1999). Nepoviruses (Comoviridae). *In* "Encyclopedia of Virology" (A. Granoff and R. G. Webster, eds.), 2nd edn. pp. 1007–1013. Academic Press, San Diego, USA.

Mehani, S. (1969). La dégénerescence infectieuse des artichauts en Tunisie. *Doc. tech. INRAT* **41**:30.

Migliori, A., Marzin, H., and Rana, G. L. (1984a). Mise en évidence du tomato black ring virus (TBRV) chez l'artichaut en France. *Agronomie* **4**:683–686.

Migliori, A., Lot, H., Pécaut, P., Duteil, M., and Rouzé-Jouan, J. (1984b). Les virus de l'artichaut. I. mise en évidence de trois virus dans les cultures françaises d'artichaut. *Agronomie* **4**:257–268.

Migliori, A., Marzin, H., Legal, V., Homo, E., and Corre, J. (1985). Présence du tobacco rattle virus (TRV) dans les cultures d'artichaut en France. *Agronomie* **5**:549–552.

Migliori, A., Homo, E., Corre, J., Marzin, H., Legal, V., and Curvale, J.-P. (1987). Répartition, fréquence et nuisibilité des virus chez l'artichaut en Bretagne. *PHM Rev. Hortic.* **247**:29–36.

Migliori, A., Rana, G. L., Piazzolla, P., Gourret, J.-P., and Rubinom, L. (1988). Caractérisation d'une nouvelle souche du virus du flétrissement de la fève isolee de l'artichaut en France. *Agronomie* **8**:201–209.

Migliori, A., Fouville, D., and Rana, G. L. (1989). Detection du virus latent de l'artichaut (ALV) et du virus du fletrissement de la feve (BBWV), souche FA, par le test ELISA. *PHM Rev. Hortic.* **299**:59–64.

Milne, R. G., and Luisoni, R. (1977). Rapid immune electron microscopy of virus preparations. *Methods Virol.* **6**:265–281.

Mink, G. I., Wample, R., and Howell, W. E. (1998). Heat treatment of perennial plants to eliminate phytoplasmas, viruses and viroids while maintaining plant survival. *In* "Control of Plant Virus Diseases" (A. Hadidi, R. K. Khetarpal, and H. Koganezawa, eds.), pp. 332–345. APS Press, St. Paul, MN, USA.

Mochizuki, T., and Ohki, S. T. (2004). Shoot meristem tissue of tobacco inoculated with Cucumber mosaic virus is infected with the virus and subsequently recovers from infection by RNA silencing. *J. Gen. Plant Pathol.* **70**:363–366.

Morton, D. J. (1961). Host range and properties of the globe artichoke curly dwarf virus. *Phytopathology* **51**:731–734.

Murant, A. F., Jones, A. T., Martelli, G. P., and Stace-Smith, R. (1996). Nepoviruses: General properties, diseases and virus identification. *In* "The Plant Viruses: Polyhedral Virions and Bipartite RNA Genomes" (B. D. Harrison and A. F. Murant, eds.), pp. 99–137. Plenum Press, New York, USA.

Nichol, S. T., Beaty, B. J., Elliott, R. M., Goldbach, R., Plyusnin, A., Schmaljohn, C. S., and Tesh, R. B. (2005). Genus *Tospovirus*. *In* "Virus Taxonomy. 8th Report of the International Committee on Taxonomy of Viruses" (C. M. Fauquet, M. A. Mayo, J. Maniloff, U. Desselberger, and L. A. Ball, eds.), pp. 712–716. Elsevier/Academic Press, San Diego, USA.

Orílio, A. F., and Navas-Castillo, J. (2009). The complete nucleotide sequence of the RNA2 of the crinivirus tomato infectious chlorosis virus: Isolates from North America and Europe are essentially identical. *Arch. Virol.* **154**:683–687.

Ortega, A. M., Juarez, M., Jordà, M. C., and Armengol, J. (2005). Viral diseases in artichoke crops in Spain. *Acta Hortic.* **681**:611–616.

Palukaitis, P., and Garcia-Arenal, F. (2003). Cucumoviruses. *Adv. Virus Res.* **62**:241–323.

Papanice, M. A., Campanale, A., Bottalico, G., Sumerano, P., and Gallitelli, D. (2004). Produzione di germoplasma risanato di carciofo Brindisino. *Italus Hortus* **11**(5):11–15.

Paradies, F., Finetti-Sialer, M., Di Franco, A., and Gallitelli, D. (2000). First report on the occurrence of cucumber mosaic virus in artichoke in Italy. *J. Plant Pathol.* **82**:244.

Pécaut, P., and Martin, F. (1992). Non-conformity of *in vitro* propagated plants of early Mediterranean varieties of globe artichoke (*Cynara scolymus* L.). *Acta Hortic.* **300**:363–366.

Pécaut, P., and Martin, F. (1993). Variation occurring after natural and *in vitro* multiplication of early Mediterranean cultivars of globe artichoke (*Cynara scolymus* L.). *Agronomie* **13**:909–919.

Pécaut, P. J., Corre, J., Lot, H., and Migliori, A. (1985). Intérêt des plants sains d'artichaut régénérés par la culture *in vitro*. *PHM Rev. Hortic.* **256:**21–26.

Peña-Iglesias, A., and Ayuso-Gonzales, P. (1972). Degeneration of Spanish globe artichoke (*Cynara scolymus* L.) plants. I. Virus isolation, host range, purification and ultrastructure of infected hosts. *An. Inst. Nac. Invest. Agron.* **2:**89–122.

Peña-Iglesias, A., and Ayuso-Gonzales, P. (1974). The elimination of viruses from globe artichoke (Cynara scolymus L.) by etiolated meristem tip culture. Proceedings of the 19th International Horticulture Congress, Varsaw 1974, p. 63.

Peña-Iglesias, A., and Ayuso-Gonzales, P. (1982). The elimination of some globe artichoke viruses by shoot apex culture and in vitro micropropagation of the plant. Proceedings of the 4th Conference of the I.S.H.S. Vegetable Virus Working Group, Wellesbourne 1982, p. 47.

Pochard, E., Mehani, S., and Marrou, J. (1964). Presence d'un virus latent dans plusieurs varieties françoises et Nord-africaines d'Artichaut. *Etud. Virol. App.* **5:**23–29.

Pochard, E., Foury, C., and Chambonnet, D. (1969). Il miglioramento genetico del carciofo. Proceedings of the 1° Congresso internazionale di studi sul carciofo, Bari 1967, pp. 117–143.

Portis, E., Mauromicale, G., Barchi, L., Mayuro, R., and Lanteri, S. (2005). Population structure and genetic variation in autochthonous globe artichoke germplasm from Sicily Island. *Plant Sci.* **168:**1591–1598.

Qu, F., Ye, X. H., Hou, G. C., Sato, S., Clemente, T. E., and Morris, T. J. (2005). RDR6 has a broad-spectrum but temperature-dependent antiviral defense role in *Nicotiana benthamiana*. *J. Virol.* **79:**15209–15217.

Quacquarelli, A., and Gallitelli, D. (1979). Tre virosi del geranio in Puglia. *Phytopathol. Mediterr.* **18:**61–70.

Quacquarelli, A., and Martelli, G. P. (1966). Ricerche sull'agente dell'arricciamento maculato del carciofo. I. Ospiti differenziali e proprietà. Proceedings of the I Congresso dell'Unione Fitopatologica Mediterranea, Bari-Napoli 1966, pp. 168–177.

Quacquarelli, A., Rana, G. L., and Martelli, G. P. (1976). Weed hosts of plant pathogenic viruses in Apulia. *ACS.* **39:**651–653.

Rana, G. L., and Cherif, C. (1981). Occurrence of artichoke mottled crinkle virus in Tunisia. *Phytopathol. Mediterr.* **20:**179–180.

Rana, G. L., and Kyriakopoulou, P. E. (1980). Bean yellow mosaic virus in artichokes in Greece. Proceedings of the 5th Congress of the Mediterranean Phytopathological Union, Patras 1980, pp. 38–40.

Rana, G. L., and Kyriakopoulou, P. E. (1982). Artichoke Italian latent and artichoke mottled crinkle viruses in artichoke in Greece. *Phytopathol. Mediterr.* **22:**46–48.

Rana, G. L., and Martelli, G. P. (1983). Virosi del carciofo. *Italia Agricola* **120**(1)**:**27–38.

Rana, G. L., and Roca, F. (1976). Trasmissione con nematodi del virus. *In* ''latente italiano del carciofo'' (AILV) Proceedings of the 2° Congresso Internazionale di Studi sul Carciofo, Bari 1973, pp. 855–858.

Rana, G. L., Rosciglione, B., and Cannizzaro, G. (1978). La maculatura anulare gialla del cardo e del carciofo. *Phytopathol. Mediterr.* **17:**63–64.

Rana, G. L., Gallitelli, D., Kyriakopoulou, P. E., Russo, M., and Martelli, G. P. (1980). Host range and properties of artichoke yellow ringspot virus. *Ann. Appl. Biol.* **97:**177–185.

Rana, G. L., Russo, M., Gallitelli, D., and Martelli, G. P. (1982). Artichoke latent virus: Characterization, ultrastructure, and geographical distribution. *Ann. Appl. Biol.* **101:**179–188.

Rana, G. L., Kyriakopoulou, P. E., and Martelli, G.P (1983). Artichoke yellow ringspot virus. *CMI/AAB Descr. Plant Viruses*, No. 271.

Rana, G. L., Castrovilli, S., Gallitelli, D., and Kyriakopoulou, P. E. (1985). Studies on two serologically distinct raspberry ringspot virus strains from artichoke. *Phytopathol. Z.* **112:**222–228.

Rana, G. L., Di Franco, A., Piazzolla, P., and Migliori, A. (1987a). Further studies on a tomato black ring virus isolate from artichoke. *J. Phytopathol.* **118:**203–211.

Rana, G. L., Migliori, A., and Ragozzino, A. (1987b). La maculatura gialla del carciofo in Campania e Puglia. *Inf. Fitopatol.* **37**(4)**:**41–43.

Rana, G. L., Di Franco, A., and Galasso, I. (1988). Further studies on *Cynara* rhabdovirus. *J. Phytopathol.* **123:**147–155.

Rana, G. L., Piazzolla, P., Lafortezza, R., and Greco, N. (1989). Characterization of a latent elongated virus from globe artichoke (*Cynara scolymus* L.) in Italy. *J. Phytopathol.* **125:**289–298.

Rana, G. L., Camele, I., and Balducci, V. (1990). Isolamento del virus dell'anulatura concentrica del geranio da crisantemo selvatico in Puglia. *Inf. Fitopatol.* **40**(7–8)**:**59–62.

Rana, G. L., Elia, A., Nuzzaci, M., and Lafortezza, R. (1992). Effect of artichoke latent virus infection on the production of artichoke heads. *J. Phytopathol.* **135:**153–159.

Rana, G. L., Gallitelli, D., Vovlas, C., and Martelli, G. P. (2005). Viruses of globe artichoke: An overview. *Acta Hortic.* **681:**572–575.

Repetto, A., Cadinu, M., Leoni, S., Gallitelli, D., Saldarelli, P., Barbarossa, L., and Grieco, F. (1997). Effetto sulla coltura *in vitro* di apici meristematici sull'ottenimento di piante di carciofo "Spinoso sardo" e "Masedu" esenti da virus *Notiz. Prot. Piante* **7:**189–195.

Robinson, D. J. (2005). Genus *Tobravirus*. *In* "Virus Taxonomy. 8th Report of the International Committee on Taxonomy of Viruses" (C. M. Fauquet, M. A. Mayo, J. Maniloff, U. Desselberger, and L. A. Ball, eds.), pp. 1015–1019. Elsevier/Academic Press, San Diego, USA.

Robinson, D. J., and Clark, J. (1987). Genome sequence homology among strains of raspberry ringspot nepovirus. Report of the Scottish Crop Research Institute 1986, p. 172.

Roca, F., and Rana, G. L. (1981). *Paratrichodorus tunisiensis* (*Nematoda, Trichodoridae*) a new vector of tobacco rattle virus in Italy. *Nem. Mediterr.* **9:**91–101.

Roca, F., Martelli, G. P., Lamberti, F., and Rana, G. L. (1975). Distribution of *Longidorus attenuatus* Hooper in Apulian artichoke fields and its relationship with artichoke Italian latent virus. *Nem. Mediterr.* **3:**91–101.

Roca, F., Rana, G. L., and Kyriakopoulou, P. E. (1982). *Longidorus fasciatus* Roca *et* Lamberti vector of a serologically distinct strain of artichoke Italian latent virus in Greece. *Nem. Mediterr.* **10:**65–69.

Roca, F., Rana, G. L., and Kyriakopoulou, P. E. (1986). Studies on *Longidoridae* (*Nematoda: Dorylaimida*) and raspberry ringspot virus spread in some artichoke fields in Greece. *Nem. Mediterr.* **14:**251–256.

Roossinck, M. J., Bujarski, J., Ding, S. W., Hajimorad, R., Hanad, K., Scott, S., and Tousignant, M. (2005). Genus *Cucumovirus*. *In* "Virus Taxonomy. 8th Report of the International Committee on Taxonomy of Viruses" (C. M. Fauquet, M. A. Mayo, J. Maniloff, U. Desselberger, and L. A. Ball, eds.), pp. 1053–1055. Elsevier/Academic Press, San Diego, USA.

Russo, M., and Rana, G. L. (1978). Occurrence of two legume viruses in Artichoke. *Phytopathol. Mediterr.* **17:**212–216.

Russo, M., Martelli, G. P., and Rana, G. L. (1975). A rhabdovirus of *Cynara* in Italy. *Phytopathol. Z.* **83:**223–231.

Russo, M., Martelli, G. P., Rana, G. L., and Kyriakopoulou, P. E. (1978). The ultrastructure of Artichoke yellow virus infections. *Microbiologica* **1:**81–99.

Saldarelli, P., Barbarossa, L., Grieco, F., and Gallitelli, D. (1996). Digoxigenin-labelled riboprobes applied to phytosanitary certification of tomato in Italy. *Plant Dis.* **80:**1343–1346.

Salomão, T. (1976). Soil transmission of artichoke yellow band virus. Studi sul Carciofo. Proceedings of the 2° Congresso Internazionale sul Carciofo, Bari 1973, pp. 831–854.

Santovito, E., Mascia, T., and Gallitelli, D. (2010). Unravelling the mechanism(s) of *Artichoke Italian Latent Virus* entry in plant Meristems. Proceedings of the 54th Annual Congress of Italian Society of Agricultural Genetics Annual Congress, Matera, Italy, p. 4.58.

Savino, V., Gallitelli, D., Jankulova, M., and Rana, G. L. (1977). A comparison of four isolates of artichoke Italian latent virus (AILV). *Phytopathol. Mediterr.* **16:**41–50.

Schwach, F., Vaistij, F. E., Jones, L., and Baulcombe, D. C. (2005). An RNA-dependent RNA polymerase prevents meristem invasion by potato virus X and is required for the activity but not the production of a systemic silencing signal. *Plant Physiol.* **138:**1842–1852.

Siddiqui, S. A., Sarmiento, C., Kiisma, M., Koivumäki, S., Lemmetty, A., Truve, E., and Lehto, K. (2008). Effects of viral silencing suppressors on tobacco ringspot virus infection on two Nicotiana species. *J. Gen. Virol.* **89:**1502–1508.

Smith, R. E. (1940). Diseases of truck crops. *Calif. Agr. Ext. Ser. Cir.* **119:**112.

Smith, I. M. (2009). Tomato infectious chlorosis virus. *Bull. OEPP/EPPO Bull.* **39:**62–64.

Szittya, G., Silhavy, D., Molnar, A., Havelda, Z., Lovas, A., Lakatos, L., Banfalvi, Z., and Burgyan, J. (2003). Low temperature inhibits RNA silencing-mediated defence by the control of siRNA generation. *EMBO J.* **22:**633–640.

Tordo, N., Benmansour, A., Calisher, C., Dietzgen, R. G., Fang, R. X., Jackson, A. O., Kurath, G., Nadin-Davis, S., Tesh, R. B., and Walker, P. J. (2005). Family *Rhabdoviridae*. *In* "Virus Taxonomy. 8th Report of the International Committee on Taxonomy of Viruses" (C. M. Fauquet, M. A. Mayo, J. Maniloff, U. Desselberger, and L. A. Ball, eds.), pp. 623–644. Elsevier/Academic Press, San Diego, USA.

Turina, M., Ciuffo, M., Lenzi, R., Rostagno, L., Mela, L., Derin, E., and Palmano, S. (2006). Characterization of four viral species belonging to the family Potyviridae isolated from Ranunculus asiaticus. *Phytopathology* **96:**560–566.

Vaira, A. M., Accotto, G. P., Vecchiati, M., and Bragaloni, M. (2002). Tomato infectious chlorosis virus causes leaf yellowing and reddening of tomato in Italy. *Phytoparasitica* **30:**290–294.

Vilchez, M., Paulus, A. O., Moyer, J. W., and Schrader, W. L. (2005). First report of Tomato spotted wilt virus affecting globe artichoke in California, USA. *Acta Hortic.* **681:**607–610.

Vovlas, C., and Lafortezza, R. (1994). Il virus dell'avvizzimento maculato del pomodoro su carciofo in Puglia. *Inf. Fitopatol.* **44**(9)**:**42–44.

Vovlas, C., Martelli, G. P., and Quacquarelli, A. (1971). Le virosi delle piante ortensi in Puglia. VI. Il complesso delle maculature anulari della cicoria. *Phytopathol. Mediterr.* **10:**244–254.

Vovlas, C., Gallitelli, D., and Conti, M. (1989). Preliminary evidence for an unusual mode of transmission in the ecology of pelargonium zonate spot virus. Proceedings of the 4th Plant Virus Epidemiology Workshop, Bari 1989, pp. 205–305.

Wang, Q., Cuellar, W. J., Rajamäki, M. L., Hirata, Y., and Valkonen, J. P. T. (2008). Combined thermotherapy and cryotherapy for efficient virus eradication: Relation of virus distribution, subcellular changes, cell survival and viral RNA degradation in shoot tips. *Mol. Plant Pathol.* **9:**237–250.

Welvaert, W., and Zitouni, B. (1974). Investigation on infectious degeneration of artichoke in Tunisia. Studi sul Carciofo. Proceedings of the 2° Congresso Internazionale sul Carciofo, Bari 1972, pp. 865–875.

Whitcombe, D., Theaker, J., Guy, S. P., Brown, T., and Little, S. (1999). Detection of PCR products using self-probing amplicons and fluorescence. *Nat. Biotechnol.* **17:**804–807.

Wisler, G. C., Duffus, J. E., Liu, H.-Y., and Li, R. H. (1998). Ecology and epidemiology of whitefly-transmitted closteroviruses. *Plant Dis.* **82:**270–280.

CHAPTER 9

Viruses in Sweetpotato

Gad Loebenstein

Contents

Abstract

Sweetpotato in the Mediterranean is mainly grown in Egypt, Spain, Portugal, and Israel. Yields vary from 34 tons/ha in Israel to 7.8 tons/ha in Portugal. As sweetpotatoes are vegetatively propagated, the differences in yields are probably due to the quality in the propagation material, mainly infection by various viruses. The main viruses affecting sweetpotato in Mediterranean countries are

Department of Virology, Agricultural Research Organization, Bet Dagan, Israel

Advances in Virus Research, Volume 84
ISSN 0065-3527, DOI: 10.1016/B978-0-12-394314-9.00009-9

Sweet potato feathery mottle virus potyvirus, *Sweet potato sunken vein virus* (*Sweet potato chlorotic stunt virus*) crinivirus, and their combined infection, causing the sweetpotato disease. Eleven other viruses sporadically reported from Mediterranean countries are also reviewed, as well as possible methods for control.

I. INTRODUCTION

Sweetpotatoes were brought to Europe by Christopher Columbus in 1492, and several kinds were cultivated in Spain by the middle of the sixteenth century, including red, purple, and pale or "white" varieties. Cultivation of sweetpotato was tried unsuccessfully in Belgium in 1576.

Presently, sweetpotato is grown to a certain extent in many of the Mediterranean countries. Yields (tons) in 2008 are summarized in Table I (FAOSTAT, 2008).

As seen from Table I, there is a marked difference in yields between the various countries. Highest yield per hectare is obtained in Israel. These variations could be due to cultivation practices, length of growing period, and very probably to degrees of virus infection. Thus, the high yields in Israel could mainly be the result of using virus-tested planting material (see section III Control).

II. THE VIRUSES

A. *Sweet potato feathery mottle virus*

Sweet potato feathery mottle virus (SPFMV) is an aphid-transmitted potyvirus. Virions are filamentous, not enveloped, usually flexuous, with a modal length of 830–850 nm. The genome consists of single-stranded

TABLE I Sweetpotato production (tons) and Yields (tons/ha) in several Mediterranean countries

Country	Total production	Yields
Egypt	258,983	11.6
Greece	8000	20
Israel	23,316	34.3
Italy	8158	20.1
Morocco	10,895	17.4
Palestinian Territory	9000	29.0
Portugal	26,000	7.8
Spain	25,000	16.7

Adapted from FAOSTAT (2008).

linear RNA, unipartite of 11.6 kb, with a poly(A) 3′-end region (Moyer and Cali, 1985). The complete nucleotide (nt) sequence of an SPFMV severe strain (SPFMV-S) genomic RNA was determined (Sakai *et al.*, 1977). The viral RNA genome (S strain) is 10,820 nts long (GenBank accession D86371) excluding the poly(A) tail and contained one open-reading frame (ORF), potentially encoding 3493 amino acids. Except in the regions of P1 and P3, the polyprotein has a high level of amino acid identity with those of other potyviruses (Sakai *et al.*, 1977).

Many strains (Moyer, 1986), isolates, variants, and serotypes of SPFMV have been reported, but mainly from South America and Africa. By comparing coat protein gene sequences of isolates, it was shown that a Spanish isolate was related to the East African cluster of SPFMV (Kreuze *et al.*, 2000).

Most sweetpotato cultivars infected by SPFMV alone show no or only mild circular spots on their leaves. In controlled experiments, SPFMV infection alone did not reduce yields compared to virus-free controls, while the complex infection with *Sweet potato sunken vein virus* (SPSVV) (synonym *Sweet potato chlorotic stunt virus* (SPCSV)) reduced yields by 50% or more (Milgram *et al.*, 1996). SPFMV is transmitted in a nonpersistent manner by aphids, including *Aphis gossypii, Myzus persicae, Aphis craccivora*, and *Lipaphis erysimi*. Aphid transmissibility is dependent on virus-encoded helper-component proteinase and a DAG triplet in the protein coat. The virus can be transmitted mechanically to various *Ipomoea* spp. as *I. batatas, I. setosa, I. nil, I. incarnata*, and *I. purpurea*.

The virus is transmitted by grafting but not by seed or pollen or by contact between plants. Probably, the major part of SPFMV dissemination is by planting infected cuttings or slips.

The virus can best be diagnosed by grafting on *I. setosa*, causing vein clearing followed by remission, or on *I. incarnata* and *I. nil* inducing systemic vein clearing, vein banding, and ringspots. SPFMV can be diagnosed by ELISA, and antisera are commercially available. However, ELISA reliably detects SPFMV only in leaves with symptoms and when coinfected with SPCSV (Gutiérrez *et al.*, 2003). It is best to sample several leaves from a plant, as the virus seems to be unevenly distributed, especially for meristem-derived plantlets. This has to be followed up by indexing on a susceptible indicator. SPFMV can also be detected by membrane immunobinding (MIBA, also termed NCM-ELISA) and by using a riboprobe (Abad and Moyer, 1992). The latter although being very sensitive, detecting 0.128 pg of RNA, compared with 179 pg of capsid protein, was still somewhat less sensitive than grafting on a sweetpotato indicator (cv. Jewel). The use of a riboprobe also requires a well-equipped laboratory and radioactive materials or digoxigenin for labeling the probe. SPFMV has also been identified by reverse transcription and polymerase chain reaction (RT-PCR), utilizing degenerate genus-specific primers, designed to amplify the variable 5′ terminal region of the potyvirus coat protein gene (Colinet *et al.*, 1998b).

SPFMV can be purified from infected *I. nil* leaf material (Cali and Moyer, 1981), by rather complex purification procedures, and directly from sweetpotatoes (infected also with SPSVV), by a relatively simple method (Cohen *et al.*, 1988). This method yielded 50–100 mg virus/kg tissue, 5–10 times more when compared with the previous methods. This high yield of virus is due to the synergistic interaction between SPFMV and SPSVV, where SPFMV titer attains more than 600-fold increase (Karyeija *et al.*, 2000b). It should be mentioned that virus yields from plants infected by SPFMV alone, without SPSVV, are markedly lower.

Though SPFMV alone generally causes only minor damage, its control is imperative as in combination with other viruses its effect on plant growth and yields may become substantial. The proven approach is to prepare virus-tested plants from meristems. The plants are then grown under insect-proof conditions, and propagation is continued by cuttings. The farmer plants a certain number of vines in the open field and continues to propagate until the stand of the field is complete. The following year he will again start from virus-tested plants. This scheme has been in operation for more than 15 years in Israel, and presently, it is difficult to find SPFMV in the country.

Breeding of SPFMV-resistant plants was initiated by the International Center for Potato (CIP) (Mihovilovich *et al.*, 2000). Several clones that were resistant to SPFMV in CIP's tests were found to be susceptible, when exposed to Israeli (unpublished) and Ugandan isolates (Karyeija *et al.*, 1998). Apparently, strain diversity requires that breeding and selection have to be done in various locations. On the other hand, a substantial number of African sweetpotato landraces have resistance to this virus (Carey *et al.*, 1997).

Another approach was the development of transgenic sweetpotatoes with coat protein (CP) mediated resistance to SPFMV (Okada *et al.*, 2001). Also, CP-mediated resistance has been introduced into several African varieties, and cultivar CPT-560 was evaluated in Kenya in a cooperative project between Monsanto Co., USA and Kenya Agriculture Research Institute (KARI) (Odame *et al.*, 2002). However, these transgenic lines were not resistant to the "complex" infection with SPCSV, causing the Sweet potato virus disease (SPVD) (Wambugu, 1991).

SPFMV has been found in Egypt (IsHak *et al.*, 2003), Israel (Loebenstein *et al.*, 2009), Spain (Valverde *et al.*, 2004), Italy (Parrella *et al.*, 2006), and Syria (Ismail *et al.*, 2006). Very probably, SPFMV also occurs in other Mediterranean countries growing sweetpotato. In Israel, however, it is difficult in the recent years to find SPFMV, due to the use of virus-tested planting material (see above).

B. *Sweet potato sunken vein virus* Genus *Crinivirus* (possible synonym: SPCSV)

Although SPCSV is the name recognized by the International Committee of Taxonomy of Viruses (ICTV), they might be considered as separate viruses for the following reasons: (i) SPCSV and SPSVV differ in their nucleotide and protein sequences by more than 20% (Cuellar *et al.*, 2008); (ii) comparing nucleotide sequences of the Israeli SPSVV with the Ugandan isolate of SPCSV in the heat-shock 70-like protein region, revealed an identity of only 69% (Cuellar *et al.*, 2008); (iii) symptoms induced by SPSVV on *I. setosa* differ markedly from those caused by SPCSV; (iv) infection of sweetpotato by SPSVV alone produced on cv. Georgia Jet mild symptoms consisting of slight yellowing of veins, with some sunken secondary veins on the upper sides of the leaves and swollen veins on their lower sides (at 28–29 °C). Upward rolling of the three to five distal leaves was also observed. Similar symptoms were also obtained on graft-inoculated plants of cv. Papel and Camote Negro (Cohen *et al.*, 1992). The disease could easily be detected in the field, where plants had erect branches, with some upward rolling leaves. In some cultivars, SPCSV isolates cause no symptoms or mild stunting combined with slight yellowing or purpling of old leaves (Gibson *et al.*, 1998; Gutiérrez *et al.*, 2003).

Effects on yields by SPSVV or SPCSV alone are minor or close to a 30% reduction, but in complex infection with SPFMV or other viruses, yield losses of 50% and more are observed (Gutiérrez *et al.*, 2003; Milgram *et al.*, 1996). SPSVV and/or SPCSV are transmitted by the whitefly *Bemisia tabaci* biotype B, *Trialeurodes abutilonea*, and *Bemisia afer* (Gamarra *et al.*, 2010; Ng and Falk, 2006) in a semipersistent manner, requiring at least 1 h for acquisition and infection feeding periods and reaching a maximum after 24 h for both of them. The virus is graft transmissible, but not by mechanical inoculation. The virus was transmitted by whiteflies to *I. setosa*, *Nicotiana clevelandii*, *Nicotiana benthamiana*, and *Amaranthus palmeri* from *I. setosa* and by grafting to other Ipomoeas.

The virus is best being diagnosed on a pair of sweetpotato plants, one healthy, the other infected by SPFMV. On the healthy plants, hardly any symptoms will become apparent, while (if carrying SPFMV) severe symptoms of SPVD will appear. The virus can also be diagnosed by immunosorbent electron microscopy (ISEM) and by ELISA.

SPSVV can be purified from infected *I. nil* or *N. clevelandii*, maintained at 28 °C (Cohen *et al.*, 1992). Normal length of virus particles from purified preparations was 850 nm, the diameter 12 nm, with an open helical structure typical for closteroviruses.

In leaf dip preparations from *I. setosa* infected with an isolate of SPCSV from Nigeria, a modal length of 950 nm was obtained (Winter *et al.*, 1992). The CP of the Israeli and Nigerian isolates had a MW of 34,000 and 29,000, respectively (Cohen *et al.*, 1992; Hoyer *et al.*, 1996). Double-stranded RNA from the Israeli isolate consisted of one major band of 10.5 kbp, and two minor bands of 9.0 and 5.0 kbp. Northern blot analysis of the Nigerian isolate revealed the presence of a large dsRNA with an estimated size of 9.0 kbp and several smaller ones. Several Kenyan cDNA clones revealed an ORF of 774 nts coding for the CP (Hoyer *et al.*, 1996). SPCSV can be serologically divided into two major serotypes, which correlate to two genetically distantly related strains/groups based on the CP and the heat-shock protein 70 homologue (Hsp70h) genes similarities, that is, the East Africa (EA) group and the West Africa (WA) group; the latter also occurs in the Mediterranean (Hoyer *et al.*, 1996; Tairo *et al.*, 2005).

The complete nucleotide sequences of genomic RNA1 (9407 nts) and RNA2 (8223 nts) were determined, revealing that SPCSV possesses the second largest identified positive-strand single-stranded RNA genome among plant viruses after *Citrus tristeza virus* (Kreuze *et al.*, 2002). RNA1 contains two overlapping ORFs that encode the replication module, consisting of the putative papain-like cysteine proteinase, methyltransferase, helicase, and polymerase domains. RNA2 contains the *Closteroviridae* hallmark gene array represented by an Hsp70h, a protein of 50–60 kDa depending on the virus, the major CP, and a divergent copy of the CP. The two genomic RNAs of SPCSV contained nearly identical 208-nt-long three-terminal sequences, and the ORF for a putative small hydrophobic protein was found in SPCSV RNA1. Furthermore, unlike any other plant or animal virus, SPCSV carried an ORF for a putative RNase III-like protein (ORF2 on RNA1). Several subgenomic RNAs (sgRNAs) were detected in SPCSV-infected plants, indicating that the sgRNAs formed from RNA1 accumulated earlier in infection than those of RNA2. Recently, it was shown that the three-proximal part of RNA1 and the partial sequence of the Hsp70h gene of the Israeli SPSVV differ markedly in sequence from most of the SPCSV isolates (Cuellar *et al.*, 2008; see also GenBank accession numbers EU124491 and EU124487EU124491 EU124487). Additionally, the phylogenetic analysis of the partial Hsp70h, CP, and p60 genes of the EA and the WA strains indicate that they may belong to different species in the genus *Crinivirus* (Abad *et al.*, 2007; Tairo *et al.*, 2005).

The disease has been reported from East and Southern Africa, Nigeria, Niger, Indonesia, Israel, Egypt, Spain, Argentina, Brazil, Peru (Gutiérrez *et al.*, 2003), and recently from the USA (Abad *et al.*, 2007).

A Peruvian sweetpotato landrace was transformed with an introduced hairpin construct targeting the replicase-encoding sequences of SPFMV and SPCSV (Kreuze *et al.*, 2008). A high level of resistance to SPCSV was observed in the transgenic plants, which however did not prevent development of SPVD when plant became infected also with SPFMV. Apparently, small amounts of SPFMV and SPCSV are sufficient for causing SPVD.

For control of SPCSV/SPSVV, it is best to start from healthy plant material obtained from meristem tip culture and to test the plants obtained on a pair of sweetpotato plants as described above.

C. *Sweet potato virus disease*

SPVD is caused by the interaction of SPFMV and SPSVV/SPCSV. Characteristic symptoms of the disease include vein clearing, chlorosis, and stunting. The disease was described by Schaefers and Terry (1976) in Nigeria and is the most important virus (complex) disease in EA (Karyeija *et al.*, 1998). The disease was described in Israel by Loebenstein and Harpaz (1960), Spain (Trenado *et al.*, 2007), and occurs probably in Italy (Parrella *et al.*, 2006). During the past decade, sweetpotato plants showing symptoms similar to SPVD have been observed in most areas of Spain (Valverde *et al.*, 2004). It can cause losses over 50%, especially in Uganda and Kenya. In Israel in a 2-year field experiment, yield reductions of 50% were observed in plots planted with SPVD-infected cuttings (Milgram *et al.*, 1996).

As stated above, the optimal approach for controlling SPVD is to supply virus-tested planting material.

D. *Sweet potato mild mottle virus* Genus *Ipomovirus* (synonym: *sweet potato B virus*; Sheffield, 1957)

SPMMV has so far been reported *inter alia* from West- and South Africa, Indonesia, China, Philippines, India, New Zealand, and Egypt.

SPMMV can cause leaf mottling, stunting, and loss of yields. Cultivars differ greatly in their reaction to the virus, some being symptomlessly infected and others apparently immune. The virus is transmitted semi-persistently by *B. tabaci*, by grafting, and by mechanical inoculation. It is not transmitted by seed or by contact between plants. The virus was transmitted to plants in 14 families (Hollings *et al.*, 1976).

Diagnostic species: *Nicotiana tabacum*, *Nicotiana glutinosa*—vein clearing, leaf puckering, mottling, and distortion; *Chenopodium quinoa*—local lesions, not systemic; *I. setosa*—conspicuous systemic vein chlorosis. Diagnosis can be confirmed by serological tests of *I. setosa*. Commercial antisera are available. After 3–4 weeks, new growth is almost symptomless. Sap transmission from sweetpotato to test plants is often difficult. The virus is best maintained in *N. glutinosa*, *N. clevelandii*, and *N. tabacum* and can be assayed quantitatively on *C. quinoa*, where SPMMV induces local lesions.

The virus can be purified from systemically infected *N. tabacum* (Hollings *et al.*, 1976). Virions are flexous rod-shaped particles, 800–950 nm in length, containing 5% RNA and 95% protein. The genome consists of single-stranded RNA. The viral RNA was cloned, and the assembled genomic sequence was 10,818 nts in length with a polyadenylated tract at the 3′-end. The sequence accession code is Z73124. Almost all known potyvirus motifs are present in the polyprotein of SPMMV, except some motifs in the putative helper component and CP, which are incomplete or missing. This may account for its vector relations (Colinet *et al.*, 1998a). The CP has a MW of 37,700.

SPMMV isolates showed a high level of variability with no discrete strain grouping.

Sequences of several SPMMV isolates revealed nucleotide sequence identities of 88.0% and 89.9% or higher for the CP-encoding region and 3-UTR, respectively, while CP amino acid (aa) sequences were 93.0–100% identical. Analysis of the CP-encoding nucleotide sequences did not revealed phylogenetically distinguishable groups of SPMMV isolates. Rather, analysis indicated high genetic variability (Tairo *et al.*, 2005).

A synergism was observed in sweetpotato doubly infected by SPMMV and SPCSV (but not by SPFMV) (Untiveros *et al.*, 2007).

E. *Sweet potato latent virus*

Sweet potato latent virus (SPLV) Genus *Potyvirus* is widespread in China and has been reported also from Egypt (http://research.cip.cgiar.org/typo3/web/uploads/media/Sweetpotato_latent_potyvirus.doc—without published records). SPLV may cause mild chlorosis, but in most cultivars, the infection is symptomless. Symptoms often disappear after infection, but the plants remain infected. Crystal inclusions are observed in the nucleus and pinwheels in the cytoplasm. SPLV isolates from Japan and China were transmitted by the aphid *M. persicae* (Usugi *et al.*, 1991), and the virus can be transmitted by mechanical inoculation and by grafting. It is not transmitted by seed.

Diagnostic species: *N. benthamiana*—systemic mosaic and stunting; *N. clevelandii*—systemic pinprick chlorotic lesions; *C. quinoa*, *Chenopodium amaranticolor*—brown necrotic local lesions, no systemic infection; *I. setosa*—systemic mottle. Diagnosis can be confirmed by serological tests of *I. setosa*.

The virus is best maintained in *N. benthamiana* or *N. clevelandii* and can be assayed on *C. quinoa* or *C. amaranticolor*.

The virus can be purified according to Liao *et al.* (1979). Virus particles are flexuous rods, 750–790 nm in length. The capsid protein has a MW of 36,000. By using MAbs and polyclonal antibodies, some epitopes common to SPLV and SPFMV are found. These can easily be differentiated when potyvirus cross-reactive MAbs are used, indicating a distant relationship (Hammond *et al.*, 1992).

Combining RT-PCR with degenerate oligonucleotide primers derived from the conserved regions of potyviruses, it was possible to identify SPLV, as well as SPFMV and *Sweet potato G virus* (Colinet *et al.*, 1998b), and to differentiate between the two strains of SPLV (Colinet *et al.*, 1997).

Sequence data accession codes are X84011 SPLV (Chinese) mRNA for CP and X84012 SPLV (Taiwan) mRNA for CP. According to Nishiguchi *et al.* (2001), SPLV has 58% homology to SPFMV-S.

The best way to control this virus as well as other viruses infecting sweetpotato is by establishing propagation nurseries derived from virus-tested mother plants.

F. *Sweet potato virus G*

Sweet potato virus G (SPVG) Genus *Potyvirus* is widespread in China (Colinet *et al.*, 1994, 1998b) and was reported in the Mediterranean basin from Egypt (IsHak *et al.*, 2003) and Spain (Trenado *et al.*, 2007).

The virus is transmitted mechanically and by aphids *M. persicae* and *A. gossypii* in a nonpersistent manner (Souto *et al.*, 2003). SPVG causes mottling in *I. nil* and chlorotic spotting in *I. setosa* and *Ipomoea tricolor*. Cylindrical inclusion bodies, which consisted of pinwheels and scrolls, were observed in the cytoplasm of epidermal, mesophyll, and vascular cells of infected *I. nil* and *I. setosa* (Souto *et al.*, 2003). Isolates LSU-1 and -3 obtained from sweetpotato plants in Louisiana, USA (Souto *et al.*, 2003) reacted with MAb PTY-1 (Jordan and Hammond, 1991).

A partial sequence of SPVG (X76944) has been obtained after RT-PCR, showing an identity of around 70% and 80% in the amino acid sequence between the complete and conserved core of the CP of SPFMV, respectively (Colinet *et al.*, 1998b). The SPVG CP has 355 amino acids, while that of SPFMV has 316 amino acids (Colinet *et al.*, 1994). Comparison with CP sequences of known potyviruses indicates that SPVG is a member of the genus *Potyvirus*.

G. *Sweet potato virus 2*

Sweet potato virus 2 (SPV2) is a tentative member of Genus *Potyvirus*. Synonyms: *Sweet potato virus 2*, *Ipomoea vein mosaic virus*, and *Sweet potato virus Y* (Ateka *et al.*, 2004; Moyer *et al.*, 1989; Souto *et al.*, 2003).

The virus was first isolated from sweetpotato plants from Taiwan (Rossel and Thottappilly, 1988) and was found *inter alia* in sweetpotato clones in Portugal (Souto *et al.*, 2003) and Spain (Trenado *et al.*, 2007). The virus has filamentous particles of 850 nm in length and induces cytoplasmic cylindrical inclusions consisting of pinwheels and scrolls (Ateka *et al.*, 2004; Souto *et al.*, 2003). The virus was transmitted nonpersistently by *M. persicae* and mechanically transmitted to several species of genera

Chenopodium, *Datura*, *Nicotiana*, and *Ipomoea*. SPV2 causes vein clearing and leaf distortion on *N. benthamiana*; chlorotic local lesions on *Chenopodium* spp.; and vein mosaic on *I. nil*, *I. setosa*, and *I. tricolor* (Ateka *et al.*, 2007; Souto *et al.*, 2003).

Although SPV2 has been isolated from sweetpotato plants showing mild symptoms of leaf mottle, vein yellowing, and/or ringspots (Rossel and Thottappilly, 1988; Tairo *et al.*, 2006), the significance of this virus to sweetpotato production is not clear as similar symptoms may be caused by other viruses, and sweetpotato cultivars inoculated with SPV2 under greenhouse conditions failed to produce obvious symptoms (Ateka *et al.*, 2004; Souto *et al.*, 2003). However, SPV2 interacts synergistically with SPCSV, increasing symptoms severity on sweetpotato (Kokkinos and Clark, 2006; Tairo *et al.*, 2006), suggesting that SPV2 might be economically important in areas where SPCSV occurs.

It seems that biologically and genetically diverse strains of SPV2 occur. Some differences in test plant reactions and host range appeared to correlate to some extent with the geographic origin and molecular distinctness of the SPV2 isolates (Ateka *et al.*, 2007). Comparison of the CP gene sequences of several isolates revealed nucleotide and amino acid sequences identities ranging from 81% to 99% and from 86% to 99%, respectively. Phylogenetic analysis of sequences distinguished several groups, which partially correlated with the geographic origin of the isolates (Ateka *et al.*, 2007).

H. *Sweet potato mild speckling virus*

Sweet potato mild speckling virus (SPMSV) Genus *Potyvirus* was isolated from plants of cv. Morada in Argentina showing symptoms of "sweetpotato chlorotic dwarf disease" (Di Feo *et al.*, 2000). These plants showed chlorosis, dwarfing, vein clearing, and leaf distortion. The severity of the disease depends on the presence of SPMSV in the complex. Symptoms in plants infected with SPFMV and SPCSV were milder than those in plants infected with the three viruses (Di Feo *et al.*, 2000). SPMSV is synergistic in plants infected with SPCSV but not in plants infected by SPFMV (Untiveros *et al.*, 2008).

SPMSV has been found *inter alia* in Peru, China, and Egypt. SPMSV is transmitted mechanically and by *M. persicae* in a nonpersistent manner. However, it does not react with the MAb PTY-1 that recognizes a cryptotope conserved in most aphid-transmitted potyviruses (Di Feo *et al.*, 2000; Jordan and Hammond, 1991). Its host range is restricted to Convolvulaceae, Chenopodiaceae, and Solanaceae. *I. setosa* and *I. nil* react with vein clearing, blistering, leaf deformation, and mosaic; *N. benthamiana* with vein clearing, and reduction, deformation, and down rolling of leaves;

and *C. quinoa* and *N. tabacum* "Samsun" with local infections (Fuentes *et al.*, 1977). SPMSV can easily be detected by ELISA.

This virus has flexuous particles of ca 800 nm. The CP sequences of SPMSV showed 63% identity with SPFMV, 68–70% with SPLV, 57% with SPVG, and 73% with *Potato virus Y* (*Alvarez et al.*, 1997). Cylindrical inclusion (bundles, laminate aggregates, and pinwheels, neither circles nor scrolls) was observed in the cytoplasm of infected *I. setosa* and *I. batatas* (Nome *et al.*, 2006). This high resemblance favors inclusion of SPMSV within the *Potyvirus* genus.

I. *Sweet potato caulimo-like virus*

Sweet potato caulimo-like virus (SPCaLV) has been found together with SPFMV in widely scattered geographical locations—Madeira, New Zealand, Papua New Guinea (Aritua *et al.*, 2007). The virus was also reported *inter alia* from Kenya, Nigeria, and Egypt.

Sweetpotato plants infected by SPCaLV are usually without symptoms, though a few leaves occasionally have chlorotic or purple spots, probably induced by SPFMV (Atkey and Brunt, 1987). Virus is not transmitted by mechanical inoculation, not by contact between plants or by seed.

Graft inoculation of *I. setosa* seedlings with SPCaLV resulted in appearance of chlorotic flecks along minor veins or circular interveinal spots on several leaves. Such leaves sometimes become completely chlorotic, wilt, and die. Fifty-nanometer caulimo-like particles can be detected in negatively stained sap of *I. setosa* (but not from *I. batatas*) and ca 10 times more by ISEM. In the cytoplasm of infected *I. setosa,* leaf cells, numerous isometric particles, and intracellular inclusions are readily detected. The inclusions are spherical or ovoid, up to 4 μm in diameter with a large central lacuna and usually several smaller peripheral lacunae; they do not contain virions (Atkey and Brunt, 1987).

J. *Ipomoea yellow vein virus*

Ipomoea yellow vein virus (IYVV) Genus *Begomovirus* was first found infecting *Ipomoea indica* plants showing yellow vein symptoms in Spain (Banks *et al.*, 1999) and Italy (Briddon *et al.*, 2005), then found infecting cultivated sweetpotato plants. The complete nucleotide sequence (AJ132548) confirmed its begomovirus nature. Contrary to *Sweet potato leaf curl virus* (SPLCV), IYVV was not transmitted by *B. tabaci* biotype B (or Q and S). Phylogenetic analysis of the three *Ipomoea*-infecting begomoviruses species (SPLCV, *Sweet potato leaf curl Georgia virus* (SPLCGV), and IYVV) recognized by the ICTV revealed that these viruses form a separate cluster that place them apart from all other begomoviruses.

K. *Sweet potato leaf curl virus*

SPLCV a begomovirus was identified in Sicily (Briddon *et al.*, 2005). The virus is monopartite and has a genome organization typical of Old World begomoviruses and distinct from all other begomoviruses.

L. *Ipomoea crinkle leaf curl virus*

In 1992, Cohen *et al.* (1997) observed on *I. setosa*, grafted with scions from sweetpotato cv. Georgia Jet plants, introduced from an unknown source in North America, atypical symptoms of little leaf and crinkle symptoms. Geminate particles were observed in crude sap preparations. The virus was transmitted by *B. tabaci* in a persistent manner and by grafting, but not mechanically. The virus was transmitted to several *Ipomoea* species, including *Ipomoea hederacea*, *Ipomoea trifida*, *I. nil*, *Ipomoea littoralis*, and *I. setosa* and induced symptoms on them, including on some cultivars of *I. batatas*, but not on cv. Georgia Jet. It might be mentioned that SPLCV did not infect *I. hederacea* and symptoms induced by *Ipomoea crinkle leaf curl virus* (ICLCV) on *I. setosa* and *I. nil* differed from those described for SPLCV (Chung *et al.*, 1985). Based on the host range, ICLCV is considered to be distinct from SPLCV, but its exact relationship to other identified begomoviruses remains unclear.

M. *Cucumber mosaic virus*

Cucumber mosaic virus (CMV) Genus *Cucumovirus* is one of the most widespread plant viruses, recorded in more than 190 species, belonging to more than 40 families (Palukaitis and García-Arena, 2003). CMV has been isolated from *I. setifera* (Migliori *et al.*, 1978), and Martin (1962) succeeded in transmitting CMV by mechanical inoculation to *I. nil*, *I. purpurea*, *Ipomoea lacunosa*, and *Ipomoea trichocarpa* but not to *I. batatas* cv. Puerto Rico. Cohen and Loebenstein (1991) failed in transmitting CMV to healthy sweetpotato plants. However, sweetpotatoes carrying the whitefly-transmitted SPSVV can easily be infected by CMV by aphid, mechanical, or graft inoculations (Cohen and Loebenstein, 1991). Untiveros *et al.* (2007) found that CMV was able to infect sweetpotatoes without the assistance of SPCSV. It appears that CMV strains differ in their ability to infect sweetpotato. CMV isolated from cucumber (Cohen and Loebenstein, 1991) or *Arracacia xanthorrhiza* (Untiveros *et al.*, 2007) was able to infect sweetpotato plants when assisted by SPSVV or not, respectively. In some fields in Israel during the 1980s, heavy CMV infections together with SPFMV and SPSVV caused severe yellowing and stunting. Later, when farmers used certified planting material such symptoms were hardly found. Apparently, the presence of another virus

(SPSVV) facilitates replication or translocation of some CMV strains in sweetpotato. It may be that there is a gene-silencing mechanism that inhibits replication of such CMV strains in healthy sweetpotato and is suppressed by SPSVV, allowing CMV to replicate and/or move in the sweetpotato plant.

It is interesting to note that although CMV occurs worldwide, in sweetpotato, it has been reported so far only from Israel, Japan, New Zealand (Fletcher *et al.*, 2000), Spain, WA (Clark and Moyer, 1988), and Egypt. CMV was not found in Kenya (Ateka *et al.*, 2004) and Tanzania (Ndunguru and Kapinga, 2007) even though SPCSV is very widespread; SPCSV strains do not support infection of sweetpotatoes with CMV, while SPSVV is needed to infect sweetpotatoes with CMV.

N. *Sweet potato C-6 virus*

The virus was isolated from sweetpotato Sosa 29, from the Dominican Republic, showing chlorotic spots (Fuentes, 1994). In Louisiana, it was found in most of the black ornamental sweetpotatoes (Blackie, Ace of Spades, and Black Beauty) (Clark and Valverde, 2000). Its host range is restricted to Convolvulaceae. The virus induces on *I. nil* and *I. setosa* fine chlorotic spots and vein clearing but on sweetpotato cv. Paramonguino, Costanero, and Jonathan chlorotic spots. This is followed by yellowing and leaf drop. The virus is transmitted by grafting and by mechanical inoculation (with a low efficiency), using sap from *I. nil* roots but not from leaves. Attempts to transmit the virus with *M. persicae* were unsuccessful. The virus was detected *inter alia* in samples from Peru, Uganda, USA, Egypt, Kenya, South Africa, and New Zealand. C-6 virus gave a mild reaction with Potato virus S (PVS) antiserum.

C-6 virus was purified from infected *I. setosa* plants (Fuentes, 1994). Virus particles are flexuous rods, 750–800 nm in length. No "pinwheels" were observed in infected *I. nil*, but chloroplasts and mitochondria showed hypertrophy.

III. CONTROL

At present, the best way to control virus diseases in sweetpotato is to supply the grower with virus-indexed propagation material. Such programs are operating in Israel and in the Shandong province of China (Gao *et al.*, 2000). In Israel, as a result of planting virus-tested material, yields increased at least by 100%, while in China increases ranged between 22% and 92%. The payoff to the farmer has been high and in Israel use of certified material is common practice, while in China the use of pathogen-free material is being extended. In African countries such programs are operating only on a limited scale because sweetpotatoes are grown mainly as a food security crop and not as a commercial one.

Some cultural/phytosanitation practices may facilitate control of viral diseases. Examples of such cultural practices include selection of disease-free planting material, destroying (roguing) of diseased plants and wild *Ipomoea* spp, especially in young crops, isolating new crops (15–20 m far) from old diseased crops, destroying crop residues, and protecting crops with barriers or intercropping with maize (Gibson and Aritua, 2002).

Breeding programs might be a future answer and such programs are in operation in Uganda, combining SPVD resistance with desirable agronomic traits such as yield, earliness, and acceptable culinary quality (Karyeija *et al.*, 2000a). It will have to be seen if these cultivars will retain their resistance in other places where different strains of virus components of SPVD may be present. Thus, several clones that were resistant to SPFMV in CIP's tests were found to be susceptible, when Israeli (unpublished) and Ugandan isolates were tested (Karyeija *et al.*, 1998).

So far transgenic approaches did not result in cultivars resistant to SPVD. Apparently, the small residues of SPCSV present in the plant were sufficient to cause together with SPFMV the severe SPVD.

As *Ipomoea* spp. are susceptible to most sweetpotato viruses, it is of importance to survey weeds as hosts and potential reservoirs of viruses.

IV. CONCLUDING REMARKS

Sweetpotato in the Mediterranean basin is so far a minor crop, grown mainly for local consumption and in some countries like Israel also for exports to Europe. This has been a developing venture, and exports have been increasing gradually. Yields in Israel are high, averaging about 34 tons/ha. To maintain high and stable high yield, it is necessary to supply the grower with virus-tested planting material every year and to encourage the grower to use this source and not his own planting material.

The most prevalent sweetpotato viruses in the Mediterranean area are SPFMV and SPSVV/SPCSV. When both infect the plant, the severe SPVD is observed. There are a few reports on additional sweetpotato viruses in the Mediterranean basin as *SPVG*, *SPV2*, *IYVV*, *ICLCV*, and CMV. Many of the above reviewed sweetpotato viruses have been reported from Egypt, the major sweetpotato producing country in the Mediterranean in which a diversity of germplasm has been introduced over time. This emphasizes the need for implementing careful quarantine procedures for sweetpotato, as it is done for other important vegetatively propagated crops such as potato.

REFERENCES

Abad, J. A., and Moyer, J. W. (1992). Detection and distribution of sweetpotato feathery mottle virus in sweetpotato by in vitro-transcribed RNA probes (riboprobes), membrane immunobinding assay, and direct blotting. *Phytopathology* **82:**300–305.

Abad, J. A., Parks, E. J., New, S. L., Fuentes, S., Jesper, W., and Moyer, J. W. (2007). First report of *Sweetpotato chlorotic stunt virus*, a component of sweetpotato virus disease, in North Carolina. *Plant Dis.* **91:**327.

Alvarez, V., Ducasse, D. A., Biderbost, E., and Nome, S. F. (1997). Sequencing and characterization of the coat protein and 3′ non-coding region of a new sweetpotato potyvirus. *Arch. Virol.* **142:**1635–1644.

Aritua, V., Bua, B., Barg, E., Vetten, H. J., Adipala, E., and Gibson, R. W. (2007). Incidence of five viruses infecting sweetpotatoes in Uganda; the first evidence of Sweetpotato caulimo-like virus in Africa. *Plant Pathol.* **56:**324–331.

Ateka, E. M., Barg, E., Njeru, R. W., Lesemann, D.-E., and Vetten, H. J. (2004). Further characterization of "sweetpotato virus 2" *Arch. Virol.* **149:**225–239.

Ateka, E. M., Barg, E., Njeru, R. W., Thompson, G., and Vetten, H. J. (2007). Biological and molecular variability among geographically diverse isolates of sweet potato virus 2. *Arch. Virol.* **152:**479–488.

Atkey, P. T., and Brunt, A. A. (1987). Electron microscopy of an isometric caulimo-like virus from sweetpotato (*Ipomoea batatas*). *J. Phytopathol.* **118:**370–376.

Banks, G. K., Bedford, I.d, Beitia, F. J., Rodrigues-Cerezo, E., and Markham, P. G. (1999). A novel Geminivirus of *Ipomoea indica* (*Convolvulaceae*) from Southern Spain. *Plant Dis.* **83:**486.

Briddon, R. W., Bull, S. E., and Bedford, I. D. (2005). Occurrence of *Sweetpotato leaf curl virus* in Sicily. *New Dis. Rep.* **11:**33.

Cali, B. B., and Moyer, J. W. (1981). Purification, serology, and particle morphology of two russet crack strains of sweetpotato feathery mottle virus. *Phytopathology* **71:**302–305.

Carey, E. E., Mwanga, R. O. M., Fuentes, S., Kasule, S., Macharia, C., Gichuki, S. T., and Gibson, R. W. (1997). Sweetpotato viruses in Uganda and Kenya: Results of a survey. Proceedings of the Sixth Triennial Symposium of the International Society Tropical Crops—Africa Branch, pp. 22–28.

Chung, M. L., Liao, C. H., Chen, M. J., and Chiu, R. J. (1985). The isolation, transmission and host range of sweetpotato leaf curl disease agent in Taiwan. *Plant Prot. Bull. (Taiwan)* **27:**333–342.

Clark, C. A., and Moyer, J. W. (1988). Compendium of Sweetpotato Diseases. The American Phytopathological Society, Minnesota, USA, 74 pp.

Clark, C. A., and Valverde, R. A. (2000). Viruses and sweetpotato cultivar decline in Louisiana, USA. *In* (Y. Nakazawa and K. Ishiguro, eds.), Proceedings of International Work-Shop on Sweetpotato Cultivar Decline Study. Kyushu National Agricultural Experiment Station (KNAES), Miyakonjo, Japan, pp. 62–69.

Cohen, J., and Loebenstein, G. (1991). Role of a whitefly-transmitted agent in infection of sweetpotato by cucumber mosaic virus. *Plant Dis.* **75:**291–292.

Cohen, J., Salomon, R., and Loebenstein, G. (1988). An improved method for purification of sweetpotato feathery mottle virus directly from sweetpotato. *Phytopathology* **78:**809–811.

Cohen, J., Franck, A., Vetten, H. J., Lesemann, D. E., and Loebenstein, G. (1992). Purification and properties of closterovirus-like particles associated with a whitefly-transmitted disease of sweetpotato. *Ann. Appl. Biol.* **121:**257–268.

Cohen, J., Milgram, M., Antignus, Y., Pearlsman, M., Lachman, O., and Loebenstein, G. (1997). Ipomoea crinkle leaf curl caused by a whitefly-transmitted gemini-like virus. *Ann. Appl. Biol.* **131:**273–282.

Colinet, D., Kummert, J., and Lepoivre, P. (1994). The complete nucleotide sequences of the coat protein cistron and 3′ non-coding region of a newly-identified potyvirus infecting sweetpotato, as compared to those of sweetpotato feathery mottle virus. *Arch. Virol.* **139:**327–336.

Colinet, D., Kummert, J., and Lepoivre, P. (1997). Evidence for the assignment of two strains of SPLV to the genus Potyvirus based on coat protein and 3′-non-coding region sequence data. *Virus Res.* **49:**91–100.

Colinet, D., Kummert, J., and Lepoivre, P. (1998a). The nucleotide sequence and genome organization of the whitefly transmitted sweetpotato mild mottle virus: A close relationship with members of the family Potyviridae. *Virus Res.* **53:**187–196.

Colinet, D., Nguyen, M., Kummert, J., Lepoivre, P., and Xia, F. Z. (1998b). Differentiation among potyviruses infecting sweetpotato based on genus- and virus-specific reverse transcription polymerase chain reaction. *Plant Dis.* **82:**223–229.

Cuellar, W. J., Tairo, F., Kreuze, J. F., and Valkonen, J. P. T. (2008). Analysis of gene content in sweetpotato chlorotic stunt virus RNA1 reveals the presence of P22 protein RNA silencing suppressor in only few isolates: Implications to viral evolution and synergism. *J. Gen. Virol.* **89:**573–582.

Di Feo, L., Nome, S. F., Biderbost, E., Fuentes, S., and Salazar, L. F. (2000). Etiology of sweetpotato chlorotic dwarf disease in Argentina. *Plant Dis.* **84:**35–39.

FAOSTAT (2008). FAO Statistical Databases. http://apps.fao.org/.

Fletcher, J. D., Lewthwaite, S. L., Fletcher, P. J., and Dannock, J. (2000). Sweetpotato (kumara) virus disease surveys in New Zealand. *In* (Y. Nakazawa and K. Ishiguro, eds.), Proceeding of the International Workshop on Sweetpotato Cultivar Decline Study. Kyushu National Agricultural Experimental Station (KNAES), 8–9 September 2000, Miyakonojo Japan, pp. 42–47.

Fuentes, S. (1994). Preliminary identification of a sweetpotato virus (C-6). *Fitopatologia* **29:**38 (Abstract; in Spanish).

Fuentes, S., Arellano, J., and Meza, M. A. (1977). Preliminary studies of a new virus, C-8, affecting sweetpotato. *Fitopatologia* **32:**9–10 (Abstr. in Spanish).

Gamarra, H. A., Fuentes, S., Morales, F. J., and Barker, I. (2010). *Bemisia afer* sensu lato, a vector of Sweet potato chlorotic stunt virus. *Plant Dis.* **94:**510–514.

Gao, F., Gong, Y. F., and Zhang, P. B. (2000). Production and development of virus-free sweetpotato in China. *Crop Protect.* **19:**105–111.

Gibson, R. W., and Aritua, V. (2002). The perspective of sweetpotato chlorotic stunt virus in sweetpotato production in Africa: A review. *Afr. Crop Sci. J.* **10:**281–310.

Gibson, R. W., Mpembe, I., Alicai, T., Carey, E. E., Mwanga, R. O. M., Seal, S. E., and Vetten, H. J. (1998). Symptoms, aetiology and serological analysis of sweetpotato virus disease in Uganda. *Plant Pathol.* **47:**95–102.

Gutiérrez, D. L., Fuentes, S., and Salazar, L. F. (2003). Sweetpotato virus disease (SPVD): Distribution, incidence, and effect on sweetpotato yield in Peru. *Plant Dis.* **87:**297–302.

Hammond, J., Jordan, R. L., Larsen, R. C., and Moyer, J. W. (1992). Use of monoclonal antisera and monoclonal antibodies to examine serological relationships among three filamentous viruses of sweetpotato. *Phytopathology* **82:**713–717.

Hollings, M., Stone, O. M., and Bock, K. R. (1976). Purification and properties of sweetpotato mild mottle virus, a whitefly-borne virus from sweetpotato (Ipomoea batatas) in East Africa. *Ann. Appl. Biol.* **82:**511–528.

Hoyer, U., Maiss, E., Jelkmann, W., Lesemann, D. E., and Vetten, H. J. (1996). Identification of the coat protein gene of a sweetpotato sunken vein closterovirus isolate from Kenya and evidence for a serological relationship among geographically diverse closterovirus isolates from sweetpotato. *Phytopathology* **86:**744–750.

IsHak, J. A., Krueze, J. F., Johansson, A., Mukasa, S. B., Tairo, F., Abo El-Abbas, F. M., and Valkonen, J. P. T. (2003). Some molecular characteristics of three viruses from SPVD-affected sweetpotato plants in Egypt. *Arch. Virol.* **148:**2449–2460.

Ismail, I. D., Raie, S. U., and Akil, E. (2006). Diagnosis of some sweetpotato viruses using indicator plants and serological tests. *Arab. Near East Plant Prot. Newslett.* **43:**29.

Jordan, R., and Hammond, J. (1991). Comparison and differentiation of potyvirus isolates and identification of strain-, virus-, subgroup-specific and potyvirus group-common epitopes using monoclonal antibodies. *J. Gen. Virol.* **72:**25–36.

Karyeija, R. F., Gibson, R. W., and Valkonen, J. P. T. (1998). Resistance in sweetpotato virus disease (SPVD) in wild East African Ipomoea. *Ann. Appl. Biol.* **133:**39–44.

Karyeija, R. F., Kreuze, J. F., Gibson, R. W., and Valkonen, J. P. T. (2000a). Two serotypes of *Sweetpotato feathery mottle virus* in Uganda and their interaction with resistant sweetpotato cultivars. *Phytopathology* **90:**1250–1255.

Karyeija, R. F., Kreuze, J. F., Gibson, R. W., and Valkonen, J. P. T. (2000b). Synergistic interactions of a potyvirus and a phloem-limited crinivirus in sweetpotato plants. *Virology* **269:**26–36.

Kokkinos, C. D., and Clark, C. A. (2006). Interactions among *Sweetpotato chlorotic stunt virus* and different potyviruses and potyvirus strains infecting sweetpotato in the United States. *Plant Dis.* **90:**1347–1352.

Kreuze, J. F., Karyeija, R. F., Gibson, R. W., and Valkonen, J. P. T. (2000). Comparisons of coat protein gene sequences show that East African isolates of sweetpotato feathery mottle virus form a genetically distinct group. *Arch. Virol.* **145:**567–574.

Kreuze, J. F., Savenkov, E. I., and Valkonen, J. P. T. (2002). Complete genome sequence and analyses of the subgenomic RNAs of sweetpotato chlorotic stunt virus reveal several new features for the genus Crinivirus. *J. Virol.* **76:**9260–9270.

Kreuze, J. F., Samolski, I., Untiveros, M., Cuellar, W. J., Lajo, G., Cipriani, P. G., Ghislain, M., and Valkonen, J. P. T. (2008). RNA silencing mediated resistance to a crinivirus (*Closteroviridae*) in cultivated sweetpotato (*Ipomoea batatas* L.) and development of sweetpotato virus disease following co-infection with a potyvirus.. *Mol. Plant Pathol.* **9:**589–598.

Liao, C. H., Chien, K., Chung, M. L., Chiu, R. J., and Han, Y. H. (1979). A study of a sweetpotato virus disease in Taiwan. I. Sweetpotato yellow spot virus disease. *J. Agric. Res. China* **28:**127–137.

Loebenstein, G., and Harpaz, I. (1960). Virus diseases of sweetpotatoes in Israel. *Phytopathology* **50:**100–104.

Loebenstein, G., Cohen, J., and Dar, Z. (2009). Sweetpotato in Israel. *In* "The Sweetpotato" (G. Loebenstein and G. Thottappilly, eds.), pp. 483–487. Springer Science + Business Media B.V.

Martin, W. J. (1962). Susceptibility of certain Convolvulaceae to internal cork, tobacco ringspot and cucumber mosaic viruses. *Phytopathology* **52:**607–611.

Migliori, A., Marchoux, G., and Quiot, J. B. (1978). Dynamique des populations du virus de la mosaïque du concombre en Guadeloupe. *Ann. Phytopathol.* **10:**455–466.

Mihovilovich, R., Mendoza, H. A., and Salazar, L. F. (2000). Combining ability for resistance to sweetpotato feathery mottle virus. *HortScience* **35:**1319–1320.

Milgram, M., Cohen, J., and Loebenstein, G. (1996). Effects of sweetpotato feathery mottle virus and sweetpotato sunken vein virus on sweetpotato yields and rate of reinfection on virus-free planting material in Israel. *Phytoparasitica* **24:**189–193.

Moyer, J. F. (1986). Variability among strains of sweetpotato feathery mottle virus. *Phytopathology* **76:**1126 (Abstr.).

Moyer, J. W., and Cali, B. B. (1985). Properties of sweetpotato feathery mottle virus RNA and capsid protein. *J. Gen. Virol.* **65:**1185–1189.

Moyer, J. W., Jackson, G. V. H., and Frison, E. A. (eds.) (1989). FAO/IBPGR Technical Guidelines for the Safe Movement of Sweetpotato Germplasm Food and Agriculture Organization of the United Nations/International Board For Plant Genetic Resources, Rome.

Ndunguru, J., and Kapinga, R. (2007). Viruses and virus-like diseases affecting sweetpotato subsistence farming in southern Tanzania. *Afric. J. Agric. Res.* **2:**232–239.
Ng, J. C. K., and Falk, B. W. (2006). Virus-vector interactions mediating nonpersistent and semipersistent transmission of plant viruses. *Annu. Rev. Phytopathol.* **44:**183–212.
Nishiguchi, M., Okada, Y., Sonoda, S., Mori, M., Kimura, T., Hanada, K., Sakai, J., Murata, M., Matsuda, Y., Fukuoka, H., Miyazaki, T., Nakano, M., *et al.* (2001). Sweetpotato feathery mottle virus derived resistance: CP mediated resistance and gene silencing. International Workshop on Sweetpotato Cultivar Decline Study September 8–9, 2000, pp. 120–124. Kyushu National Agricultural Experiment Station, Miyakonojo, Miyazaki.
Nome, C. F., Laguna, I. G., and Nome, S. F. (2006). Cytological alterations produced by Sweetpotato mild speckling virus. *J. Phytopathol.* **154:**504–507.
Odame, H., Kameri-Mbote, P., and Wafula, D. (2002). Innovation and policy process: The case of transgenic sweetpotato in Kenya. *Economics and Political Weekly* **37**(27)**:**2770–2777. (http://www.jstor.org/stable/4412332).
Okada, Y., Saito, A., Nishiguchi, M., Kimura, T., Mori, M., Hanada, K., Sakai, J., Miyazaki, C., Matsuda, Y., and Murata, T. (2001). Virus resistance in transgenic sweetpotato [Ipomoea batatas L. (Lam)] expressing the coat protein gene of sweetpotato feathery mottle virus. *Theor. Appl. Genet.* **103:**743–751.
Palukaitis, P., and García-Arena, l F. (2003). Cucumoviruses. *Adv. Virus Res.* **62:**241–323.
Parrella, G., De Stradis, A., and Giorgini, M. (2006). *Sweet potato feathery mottle virus* is the causal agent of Sweet Potato Virus Disease (SPVD) in Italy. *New Dis. Rep.* **13:**8.
Rossel, H. W., and Thottappilly, G. (1988). Complex virus diseases of sweetpotato. *In* Exploration, Maintenance and Utilization of Sweetpotato Genetic Resources. , Report of 1st Sweetpotato Planning Conference, International Potato Center, Lima, Peru, pp. 291–302.
Sakai, J., Mori, M., Morishita, T., Tanaka, M., Hanada, K., Usugi, T., and Nishigushi, M. (1977). Complete nucleotide sequence and genome organization of sweetpotato feathery mottle virus (S strain) genomic RNA: The large coding region of the P1 gene. *Arch. Virol.* **142:**1553–1562.
Schaefers, G. A., and Terry, E. R. (1976). Insect transmission of sweetpotato agents in Nigeria. *Phytopathology* **66:**642–645.
Sheffield, F. M. L. (1957). Virus diseases of sweetpotato in East Africa. I. Identification of the viruses and their insect vectors. *Phytopathology* **47:**582–590.
Souto, E. R., Sim, J., Chen, J., Valverde, R. A., and Clark, C. A. (2003). Properties of strains of Sweetpotato feathery mottle virus and two newly recognized potyviruses infecting sweetpotato in the United States. *Plant Dis.* **87:**1226–1232.
Tairo, F., Musaka, S. B., Jones, R. A. C., Kullaia, A., Rubaihayo, P. R., and Valkonen, J. P. T. (2005). Unravelling the genetic diversity of the three main viruses involved in Sweetpotato Virus Disease (SPVD), and its practical implications. *Mol. Plant Pathol.* **6:**199–211.
Tairo, F., Jones, R. A. C., and Valkonen, J. P. T. (2006). Potyvirus complexes in sweetpotato: Occurrence in Australia, serological and molecular resolution, and analysis of the Sweetpotato virus 2 (SPV2) component. *Plant Dis.* **90:**1120–1128.
Trenado, H. P., Lozano, G., Valverde, R. A., and Navas-Castillo, J. (2007). First report of *Sweetpotato virus G* and Sweetpotato virus 2 infecting sweetpotato in Spain. *Plant Dis.* **91:**1687 (Abstr.).
Untiveros, M., Fuentes, S., and Salazar, L. F. (2007). Synergistic interaction of sweet potato chlorotic stunt virus (Crinivirus) with Carla-, Cucumo-, Ipomo-, and Potyviruses infecting sweet potato. *Plant Disease* **91:**669–676.
Untiveros, M., Fuentes, S., and Kreuze, J. (2008). Molecular variability of sweetpotato feathery mottle virus and other potyviruses infecting sweetpotato in Peru. *Arch. Virol.* **153:**473–483.
Usugi, T., Nakano, M., Shinkai, A., and Hayashi, T. (1991). Three filamentous viruses isolated from sweetpotato in Japan. *Ann. Phytopathol. Soc. Jpn.* **57:**512–521.

Valverde, R. A., Lozano, G., Navas-Castillo, J., Ramos, A., and Vald's, F. (2004). First report of *Sweet potato chlorotic stunt virus* and *Sweet potato feathery mottle virus* Infecting sweetpotato in Spain. *Plant Dis.* **88:**428.
Wambugu, F. M. (1991). In vitro and epidemiological studies of sweetpotato (*Ipomea batatas*) (L.) Lam. virus diseases in Kenya. University of Bath, PhD Thesis, 271pp.
Winter, S., Purac, A., Leggett, F., Frison, E. A., Rossell, H. W., and Hamilton, R. I. (1992). Partial purification and molecular cloning of a closterovirus from sweetpotato infected with the sweetpotato virus disease complex from Nigeria. *Phytopathology* **82:**869–875.

CHAPTER 10

Viruses of Asparagus

Laura Tomassoli,* Antonio Tiberini,* and **Heinrich-Josef Vetten**[†]

Contents

* Plant Pathology Research Centre, Agricultural Research Council, Rome, Italy
† Julius Kuehn Institute, Federal Research Centre for Cultivated Plants, Institute of Epidemiology and Pathogen Diagnostics, Braunschweig, Germany

Advances in Virus Research, Volume 84
ISSN 0065-3527, DOI: 10.1016/B978-0-12-394314-9.00010-5

Abstract

The current knowledge on viruses infecting asparagus (*Asparagus officinalis*) is reviewed. Over half a century, nine virus species belonging to the genera *Ilarvirus*, *Cucumovirus*, *Nepovirus*, *Tobamovirus*, *Potexvirus*, and *Potyvirus* have been found in this crop. The potyvirus *Asparagus virus 1* (AV1) and the ilarvirus *Asparagus virus 2* (AV2) are widespread and negatively affect the economic life of asparagus crops reducing yield and increasing the susceptibility to biotic and abiotic stress. The main properties and epidemiology of AV1 and AV2 as well as diagnostic techniques for their detection and identification are described. Minor viruses and control are briefly outlined.

I. INTRODUCTION

Asparagus (*Asparagus officinalis*) is a perennial plant of the family Liliaceae. It is native to Mesopotamia, East Mediterranean area, (Knaflewski, 2009) and now naturalized all over the world in temperate and tropical regions. Since antiquity (Brothwell and Brothwell, 1998), asparagus is cultivated as vegetable crop for its succulent stalks (spears) that arise from the crowns from early to late spring. After the last harvest, asparagus stalks are allowed to grow forming a luxuriant bush where stems remain green and function as leaves, while the leaves themselves are reduced to small scales.

Asparagus was actually known for its medicinal properties long before it was considered a food. A Greek physician reported on its capacity to relieve inflammation of the stomach, relaxes the bowels, makes urine, and helps the weak. Today, asparagus is esteemed for being rich in vitamins (i.e., folic acid, vitamin C, and A), minerals (i.e., potassium, calcium, iron, magnesium, selenium, zinc), and dietary fiber (Amaro-López *et al.*, 1998; Sun and Powers, 2007).

The genus *Asparagus* includes 300 species counting wild and ornamental asparagus, but *A. officinalis* is economically the most important species worldwide. Green asparagus and white asparagus (known as spargel) are the products of this species as well as the purple asparagus grown in Italy.

There are about 195,000 ha of asparagus grown worldwide (Benson, 2009), equally divided between white and green, with an increasing trend toward the production of green, due to lower harvesting costs. China (56,000 ha) is the first producing country followed by Peru (26,800 ha) and Germany (22,000 ha). The most important Mediterranean countries are in Europe: Spain (11,000 ha), France (7,000 ha), Italy (6,000 ha), Greece (4,280 ha) (Benson, 2009), and small quantity in Slovenia (Jakše and Maršić, 2008). About 1,000 ha are present in North Africa distributed between

Morocco, Tunisia, and Egypt, while, along the coast of Minor Asia, Turkey is the only country with a recorded production.

Fungi (e.g., *Puccinia asparagi*, *Stemphylium vesicarium*, *Fusarium moniliforme*, *Fusarium oxysporum* f.sp. *asparagi*, and *Botrytis cinerea*) are reported as the most important pathogens causing economical damage in asparagus crop (Blok and Bollen, 1995; Tomassoli *et al.*, 2008b). A bacterial disease, a soft rot caused by *Pseudomonas* and *Erwinia* spp., affects asparagus tips during storage. Two asparagus beetles and an aphid species (*Brachycolus asparagus*) are quite troublesome pests for the crop. In particular, different aphid species, in addition to a direct damage on fern growth, play an important role as virus vector.

Nine viruses have been reported on asparagus, so far. For three of them, named *Asparagus virus 1* (AV1), *Asparagus virus 2* (AV2), and *Asparagus virus 3* (AV3), asparagus is the only natural host. *Tobacco streak virus* (TSV), *Cucumber mosaic virus* (CMV), *Tobacco mosaic virus* (TMV), and three nepoviruses have also been identified in the vegetable crop with different incidence and economical significance. Nevertheless, available evidence suggests that AV1 and AV2 are the most important viruses of asparagus.

II. *ASPARAGUS VIRUS 1*

The first documented presence of a virus in asparagus was in 1960. Hein (1960) used the name AV1 for a viral pathogen that induced local lesions in *Chenopodium* spp. after mechanical inoculation and was transmitted by aphids but not through seeds. Afterward, AV1 was reported in the USA (Mink and Uyeda, 1977), Japan (Fujisawa *et al.*, 1983), and Italy (Bertaccini *et al.*, 1982).

Based on particle morphology, types of inclusion bodies induced by the virus, and mode of aphid transmission, AV1 was assigned to the genus *Potyvirus*, family *Potyviridae* (Fujisawa *et al.*, 1983).

A. Particle morphology and properties

AV1 is characterized by filamentous, flexuous particles measuring 700–880 nm in length and 13 nm in width. Its nonenveloped nucleocapsid exhibits a helical symmetry and an indistinct axial canal. There is one sedimenting component in purified preparations containing particles with a modal length of 750 nm. The ultraviolet adsorption spectrum of purified virions has a maximum at 260 nm and a minimum at 247 nm and an *A*260/280 ratio of 1.24 (Fujisawa *et al.*, 1983) which suggests a nucleic acid content of about 6% (Howell and Mink, 1985). Hein (1969) and Fujisawa *et al.* (1983) reported about slightly different values for the *in vitro* properties of AV1: 58–60 and 50–55 °C for thermal inactivation point (TIP), 10^{-5} and 10^{-3}–2×10^{-4} for dilution end point (DEP), 6 and 8–11 days, for longevity *in vitro* (LIV) at room temperature, respectively. In the cytoplasm of infected cells, AV1 induces the formation of cylindrical inclusions (CI), typical of potyvirus infections, such as pinwheels, scrolls, and laminated aggregates. Based on the CI morphology, potyviruses have been subdivided into four categories: type I, pinwheels and scrolls; type II, pinwheels and laminated aggregates; type III, pinwheels with scrolls and laminated aggregates; type IV, pinwheels with scrolls and short usually curved laminated aggregates (Edwardson, 1992; Edwardson *et al.*, 1984). Ultrastructural studies on CI in AV1-infected leaf tissues assigned an American isolate to type II (Howell and Mink, 1985), whereas German and Italian isolates were assigned to type III (Gröschel and Jank-Ladwig, 1977; Marani *et al.*, 1993).

B. Nucleic acid component

Like all potyviruses, AV1 has a monopartite, linear, positive-sense, and single-stranded RNA genome. No complete AV1 sequence is available from GenBank. The partial sequence of the coat protein (CP) region of an Italian isolate, named AV1-1770, has been published (GenBank accession no. EF576991). Recently, complete genome sequencing of isolate AV1-1770 has been achieved; it was found to be about 10,000 nucleotide (nt) long with a deduced genome organization typical for a potyvirus (L. Tomassoli, unpublished data).

The potyvirus RNA genome contains one large open-reading frame (ORF) that is expressed as a single polyprotein about 350 kDa in size (Riechmann *et al.*, 1992). This is proteolytically processed into 10 polypeptides referred to as P1, helper-component protease (HC-Pro), P3, 6K1, CI, 6K2, viral genome-linked protein of nuclear inclusion protein a (NIa-VPg), proteinase domain of NIa (NIa-Pro), nuclear inclusion protein b (NIb) as RNA-dependent RNA polymerase, and CP (Merits *et al.*, 2002). The 3′-end of the RNA is polyadenylated, the VPg is attached to the 5′-end of the virus genome, and cap-independent translation is mediated by the 5′-untranslated region (UTR) (Hari *et al.*, 1979; Riechmann *et al.*, 1989). Recent studies have shown the presence of an additional short ORF named PIPO (Pretty Interesting *Potyviridae* ORF) embedded within the P3 cistron and expressed as a P3–PIPO fusion product. First identified in *Turnip mosaic virus* (TuMV), this 11th protein has now been identified throughout the family *Potyviridae* and has been shown to be essential for virus replication and cell-to-cell movement (Chung *et al.*, 2008; Wei *et al.*, 2010). There are studies dealing with the mechanism of polyprotein processing, the biochemical and biological functions of the potyvirus gene products. Most of the potyviral proteins are multifunctional, being involved in genome replication (all proteins), RNA binding (P1, CI, VPg-NIa, NIa-Pro, and NIb), cell-to-cell (CI, CP, and P3-PIPO) and systemic movement (HC-Pro, CP), infection process (HC-Pro and VPg-NIa), RNA silencing (HC-Pro), and aphid transmission (HC-Pro and CP) (Mahajan *et al.*, 1996; Merits *et al.*, 2002; Urcuqui-Inchima *et al.*, 2001; Verchot and Carrington, 1995; Wei *et al.*, 2010). P1 and P3 are the most variable protein regions, and studies suggest a role in virus-host interactions and symptom expression (Aleman-Verdaguer *et al.*, 1997; Urcuqui-Inchima *et al.*, 2001).

With regard to the molecular features of AV1, comparison of partial nucleotide sequences of the CI, NIb, and CP regions of AV1-1770 with those of other potyviruses showed identities of 63%, 68%, and 75% with TuMV, respectively (Tiberini *et al.*, 2008). Rabenstein *et al.* (2007) analyzed the CP core region of two German AV1 isolates and observed highest identities to *Narcissus late season yellows virus*.

C. Symptoms and damages

Wherever AV1 was detected in asparagus, no virus-like symptoms have been observed on infected plants. An unspecific decline of the crop, consisting of reduced vigor, stunting, chlorosis and yellowing, wilt, and a shorter economic life of the crop, is often reported and thought to be due to multiple factors including fungal and viral pathogens (Elmer *et al.*, 1996; Grogan and Kimble, 1959; Knaflewski *et al.*, 2008). AV1 has often been isolated from plants affected by a decline, but no symptoms have ever been reproduced in artificially infected asparagus plants. Several epidemiological studies arrived at the conclusion that virus infections negatively affect the profitable life of asparagus crops reducing plant vigor and increasing the susceptibility to abiotic stress and fungal diseases (Jaspers and Falloon, 1999; Jaspers *et al.*, 1999). In particular, AV1 and AV2 infected plants were more often affected by wilt and root rot caused by *Fusarium* spp. (Evans and Stephens, 1989; Knaflewski *et al.*, 2008).

Due to this complexity of asparagus decline, it is difficult to assess the real damage caused by AV1 infection to asparagus quality and yield. Elmer *et al.* (1996) assumed that the effects of AV1 are weaker than those of AV2 and viruses in mixed infections (AV1 plus AV2 or AV1 plus CMV). The latter are known to be responsible for losses in yield by 70%, in number of spears per plant by 39%, and in mean weight per spear by 17% (Evans *et al.*, 1990; Kegler *et al.*, 1991b; Yang, 1979).

D. Propagation hosts and purification

Asparagus is the only known natural host of AV1. At first, *Chenopodium* spp. were shown to be the only experimental hosts suitable for detection and maintenance of the asparagus potyvirus (Hein, 1960, 1969; Mink and

Uyeda, 1977). Then, Fujisawa *et al.* (1983) identified a limited number of new hosts for AV1 by experimentally infecting 6 of 55 herbaceous species with AV1 following mechanical inoculation. The virus produced only local lesions but no systemic infection in *Tetragonia expansa* (Aizoaceae); *Gomphrena globosa* (Amaranthaceae); and *Spinacia oleracea, Chenopodium amaranticolor, Chenopodium quinoa,* and *Chenopodium capitatum* (Chenopodiaceae).

More recently, some isolates from Germany (Rabenstein *et al.*, 2007) and AV1-1770 isolate (Tomassoli *et al.*, 2008a) were able to infect *Nicotiana* spp. In particular, the Italian isolate caused systemic mottle on *Nicotiana benthamiana* and necrotic local spots on *Nicotiana clevelandii, Nicotiana hesperis,* and *Nicotiana occidentalis* (H. J. Vetten, personal communication).

C. quinoa and *N. benthamiana* are good propagation hosts for different AV1 isolates and have been successfully used for virus purification. This was successfully achieved using chloroform clarification, cycles of differential ultracentrifugation, and sucrose density-gradient centrifugation (Fujisawa *et al.*, 1983; Howell and Mink, 1985).

E. Variability and strains

Up to now, two types of AV1 isolates can be distinguished on the basis of their inability (type I) and ability (type II) to cause systemic infections in *N. benthamiana* (Rabenstein *et al.*, 2007; Tomassoli *et al.*, 2008a). With the exception of AV1-1770, the isolates collected in different regions in Italy and those obtained from Delaware (USA) belonged to type I (Tomassoli *et al.*, 2008a).

Attempts to discriminate the two AV1 types by serological means using polyclonal and monoclonal antibodies produced against a German isolate (AV1/VB) were unsuccessful (Rabenstein *et al.*, 2007; H. J. Vetten, unpublished data). Similarly, AV1 isolates from USA were serologically indistinguishable from European and Japanese isolates (Howell and Mink, 1985).

In attempts to correlate the biological properties with genetic differences, the CP and NIb regions of two German AV1 isolates belonging to two biologically identified types were sequenced and, based on differences in the CP N-terminus, these isolates were thought to represent different potyvirus species (Rabenstein *et al.*, 2007).

On the contrary, Tomassoli *et al.* (2008a) compared the CP gene sequences of four asparagus isolates (type I) from Italy with AV1-1770 isolate (type II), two German isolates (type I and type II), and US isolates (type I). Phylogenetic analysis revealed that the two AV1 types share nucleotide sequence identities of 98.8%, indicating they represent the same species according to the taxonomy criteria for species demarcation

(Adams *et al.*, 2005). Further, no correlation between test plant reaction and CP gene sequences was demonstrated.

More recently, sequence analysis of the CP region of some AV1 isolates from Peru and Mexico (Tomassoli *et al.*, 2009) suggested that they form a cluster distinct from Italian, German, and US isolates sharing nucleotide sequence identities of about 97% with them (A. Tiberini, unpublished data).

F. Transmission, epidemiology

Like the vast majority of potyviruses, AV1 is transmitted in a nonpersistent manner by insects of the order Hemiptera, family Aphididae. Experimental aphid transmission tests proved that *Myzus persicae* and *Aphis craccivora* are able to transmit AV1 but not *Aphis gossypii* and *Macrosiphum euphorbiae* (Fujisawa *et al.*, 1983; Howell and Mink, 1985). Evans *et al.* (1990) failed to experimentally transmit AV1 by the European asparagus aphid (*B. asparagus*); therefore, the role of this species in virus spreading needs more investigation. Under natural conditions, however, several aphid species other than *M. persicae* and *A. craccivora* are likely to be efficient vectors of AV1. Two main modes of AV1 spread in the field have been described. A random distribution is the typical pattern for a nonpersistent transmission by aphids irrespective of the virus source being within or outside the planting (Evans *et al.*, 1990; Howell and Mink, 1985; Knaflewski *et al.*, 2008). The observations that 12% of seedlings after only 4 months from transplanting (Bandte *et al.*, 2008) or more than 90% of plants in 2- to 3-year-old plantations (Tomassoli *et al.*, 2008b) became infected by AV1 suggest that AV1 spread occurs very rapidly. Moreover, AV1 infection increases along the rows as the virus is efficiently transmitted during harvest by contaminated knives used for spear cutting (Kegler *et al.*, 1991a; Knaflewski *et al.*, 2008). Root contact transmission is negligible.

Italy is the only Mediterranean country where AV1 presence is documented. Since the report of Bertaccini *et al.* (1982), AV1 has been detected at various incidence levels depending on cultivation areas. In the late 1980s when Bertaccini *et al.* (1990) surveyed asparagus crops in northern Italy, AV1 was less frequent than AV2. Recent surveys in southern Italy suggest that AV1 has become the predominant, if not the only, virus affecting commercial asparagus crops (Tomassoli *et al.*, 2008b, 2009). No significant difference in the variety performance to AV1 infection has been observed (Bandte *et al.*, 2008; Tomassoli *et al.*, 2008b). On the contrary, when *Asparagus maritimus*, a wild species commonly cultivated in southern Italy for its sensory properties, was grown close to highly AV1-infected *A. officinalis*, a very low number of infected plants were found (Tomassoli *et al.*, 2008b).

G. Detection methods

Chenopodium spp. (e.g., *C. quinoa*, *C. amaranticolor*) are good indicator species for AV1 detection (Bandte *et al.*, 2008; Hein, 1960, 1969; Mink and Uyeda, 1977). After 4–12 days postinoculation (dpi), chlorotic or reddish-brown spots, with a more or less diffuse yellow halo, appear on inoculated leaves (Hein, 1960). Later, the affected tissue turns necrotic. Alternatively, *N. benthamiana* may be used to identify the type II isolates of AV1 (Tomassoli *et al.*, 2008a). Sprouts of fully developed stems are the best source for bioassay (Kegler *et al.*, 1991b).

Polyclonal antibodies have been produced by several authors and used for the serological characterization and detection of AV1 in field samples. Antisera have been applied in various tests, such as agar gel double-diffusion test, trapping and decoration in immunoelectron microscopy, different ELISA formats (DAS- and PTA-ELISA), and Western blots (Bandte *et al.*, 2008; Fujisawa *et al.*, 1983; Howell and Mink, 1985; Rabenstein *et al.*, 2007). Recently, a kit for serological detection of AV1 has become commercially available. Serological tests were shown to be more reliable to AV1 detection than biological tests (Kegler *et al.*, 1991b).

A molecular assay is available for AV1 detection and identification. Using AV1-specific primers (AV1-F62: 5′-TCATCGAAAATGCCAAACC-CACG-3′; AV1-R550: 5′-CGAGATACTCGTGGGAAGCCCAC-3′), a one-step RT-PCR amplifies a 511-bp fragment that includes parts of the CP gene and 3′UTR region (Tomassoli *et al.*, 2007).

III. *ASPARAGUS VIRUS 2*

AV2 was first reported from Germany (Hein, 1963; Weissenfels *et al.*, 1976) and later assigned to the ilarvirus group, family *Bromoviridae* by Uyeda and Mink (1981). Based on serological and molecular properties, AV2 belongs to the subgroup 2 of the genus *Ilarvirus* together with *Citrus leaf rugose virus*, *Citrus variegation virus*, *Tulare apple mosaic virus*, and *Elm mottle virus* (Scott *et al.*, 2003; Uyeda and Mink, 1981).

A. Particle morphology and properties

AV2 virions are not enveloped. A virus preparation contains three major particle components differing somewhat in morphology (quasi-isometric to bacilliform), measuring 26, 28, and 32 nm in diameter and having a sedimentation coefficient of 90S ± 2S (top [T] component), 95S ± 2S (middle [M] component), and 104S ± 2S (bottom [B] component), respectively. After three to four cycles of density gradient centrifugation, a fourth heterogeneous component is also obtained containing noninfectious particles of different sizes. Purified virion preparations have the typical ultraviolet absorption profile for a nucleoprotein with an *A*260/280 ratio of 1.31–1.36 (Uyeda and Mink, 1981). Based on data of Fujisawa *et al.* (1983), Mink and Uyeda (1977), and Weissenfels *et al.* (1978), AV2 has the following physical properties, with a TIP, DEP, and LIV at room temperature ranging from 55 to 66 °C, from 10^{-3} to 10^{-4}, and from 48 to 96 h, respectively. Virus infectivity is provided by the three major components of AV2, low infectivity is provided by the combination of the two faster components, while no infection occurs by each component alone (Uyeda and Mink, 1981; Van Vloten-Doting, 1985).

B. Nucleic acid component

Like all ilarviruses, AV2 has a segmented (tripartite) RNA genome that is distributed over the three types of particles: the largest particles (component B) contain one molecule of RNA 1; particles of the M component contain one molecule of RNA 2; and the smallest particles (component T) contain one molecule each of RNA 3 and RNA 4, with the latter being a subgenomic RNA that is derived from RNA 3 (Rafael-Martín and Rivera-Bustamante, 1999; Scott and Zimmerman, 2009). Each RNA segment is linear, single-stranded and of positive polarity. The complete sequences of the three genomic RNAs are available (Table I).

RNA 1 consists of a single ORF (ORF1) that codes for a putative replicase with a calculated molecular mass of 121 kDa. Both methyltransferase and helicase domains are present (Scott and Zimmerman, 2009). RNA 2 contains two ORFs (Scott and Zimmerman, 2009): ORF2a encodes a putative RNA-dependent RNA polymerase (RdRp) with a molecular mass of 91 kDa; ORF2b is proximal to the 3′-terminus and encodes a protein (21 kDa) that, by analogy to cucumoviruses, is thought to be a viral suppressor of RNA silencing (Guo and Ding, 2002). RNA 3 contains ORF3a and ORF3b that are translated into a putative movement protein (MP) with a molecular mass of 31 kDa and coat protein (CP) with

TABLE I GenBank accession numbers of complete genome segment sequences for AV2, RNA segment size and geographic origin

Genome segment	Size (nt)	Isolate	GeneBank accession no.
RNA 1	3431	USA	EU919666; NC011808
RNA 2	2916	USA	EU919667; NC011809
RNA 3	2306	Mexico	X86352; NC011807

a molecular mass of 24 kDa, respectively (Bol, 2005). In particular, CP is encoded by a subgenomic RNA (sgRNA 4) from the minus strand of RNA 3 and is required for genome activation, a property of all ilarviruses and *Alfalfa mosaic virus* (Jaspars, 1999; Rafael-Martín and Rivera-Bustamante, 1999).

C. Symptoms and damages

Wherever AV2 was detected, no obvious macroscopic symptoms, both on ferns and on spears, have been described. Neither symptoms nor differences in growth have been reported for asparagus seedlings raised from infected seed (Uyeda and Mink, 1981). AV2 has often been detected in plants affected by a decline and by root rot fungi (Evans and Stephens, 1989; Jaspers and Falloon, 1999; Jaspers *et al.*, 1999; Knaflewski *et al.*, 2008). There is experimental evidence that virus-infected asparagus roots produce more monosaccharides that favor *F. oxysporum* f.sp. *asparagi* and *Rhizoctonia solani* and stimulate spore germination (Pawlowski *et al.*, 2004).

Yield loss caused by AV2 infection was estimated to be about 25%; it increased to 70% when AV2 was in mixed infection with AV1 (Jaspers, 1996; Yang, 1979). Further, in a 5-year field trial, Jaspers *et al.* (1999) showed that AV2 infection is responsible for a progressive reduction in yield, in diameter size, and number of spears from the third year after infection and that fern stalks could be severely reduced in diameter and height when compared with AV2-free plants. In short, it was demonstrated that marketable yields of an AV2 infected plantation were reduced by 48% and 57% at the fourth and fifth harvests, respectively.

Significant reduction in root initiation, plant regeneration, and survival has been demonstrated also in tissue culture of asparagus (De Vries-Paterson *et al.*, 1992). This technique is widely used for germplasm maintenance, clonal propagation of hybrids, and production of planting material (Fletcher, 1994). For this reason, the micropropagation of asparagus needs to be carried out with AV2-free plants to avoid a severe economic impact from nursery propagation to commercial field production.

D. Propagation hosts and purification

Asparagus is the only known natural host of AV2, so far. At least, 50 herbaceous plant species were shown to be susceptible to AV2 infection by mechanical inoculation (Fujisawa *et al.*, 1983; Phillips and Brunt, 1985; Uyeda and Mink, 1981). The species belong mainly to the following families: Amaranthaceae (e.g., *G. globosa*), Chenopodiaceae (e.g., *Chenopodium* spp.; *Beta vulgaris*), Compositae (e.g., *Zinnia elegans*), Cucurbitaceae (e.g., *Cucumis sativus*), Leguminosae (e.g., *Vicia* spp.), and Solanaceae (e.g., *Nicotiana* spp.; *Capsicum* spp.). For virus propagation and purification, systemically infected *C. quinoa* and *N. tabacum* were mainly used.

Two major purification methods have been reported which are based on chloroform clarification and several cycles of differential ultracentrifugation. The method of Uyeda and Mink (1981) allows a clear separation of the three virus components by sucrose density gradients, even if the procedure was not always reproducible. In contrast, the method of Phillips and Brunt (1985) has consistently given a clean virus preparation that can be used for antiserum production (Giunchedi *et al.*, 1987).

E. Variability and strains

AV2 isolates producing different symptoms on *C. quinoa* and collected from different continents were originally thought to be serologically slightly different but further analysis failed to establish the existence of different serotypes (Fujisawa *et al.*, 1983; Phillips and Brunt, 1985; Uyeda and Mink, 1981). Other characteristics as symptom expression in indicator plants, viral RNA electrophoretic mobility, and CP molecular weight were compared for an American and a Mexican isolate without significant differences (Rafael-Martín and Rivera-Bustamante, 1999).

Since the CP sequence of only one AV2 isolate (from Mexico) has been determined (GenBank accession no. X86352.1) (Rafael-Martín and Rivera-Bustamante, 1999), there is also no molecular evidence for the existence of AV2 genotypes. Further, the CP sequences of AV2 isolates originating from different years and locations in Italy were 98–99% identical to the corresponding RNA 3 sequence of the Mexican isolate (L. Tomassoli, unpublished data).

F. Transmission, epidemiology

AV2 is a seed-borne virus with a demonstrated transmission rate of up to 60% for seeds originating from an infected asparagus plant (Fujisawa *et al.*, 1983; Uyeda and Mink, 1981). Infected seed lots were the major pathway for AV2 spread until tissue-culture techniques have been routinely used for producing virus-free asparagus.

Epidemiological studies were carried out to investigate other ways of field transmission. First, the presence of infective virus particles was demonstrated on pollen grains of infected asparagus plant (Evans and Stephens, 1988), and afterward, it was shown that AV2 contaminated pollen can be transported by wind or insects (honeybees and thrips) to a female asparagus plant (an open-pollinated plant) (Jaspers and Pearson, 1997). As proof, the number of infected female plants was higher (by 69%) than that of male plants. Further, the use of contaminated cutting knives for spear harvesting resulted in rapid spread and high transmission rates of AV2 (Jaspers and Pearson, 1997). The recent report on AV2 spread by root grafting and AV2 transmission by spores of asparagus rust (*P. asparagi*) (Knaflewski *et al.*, 2008) appears insignificant for the epidemiology of AV2. Like other ilarviruses, AV2 is not transmitted by a specific vector, and attempts to transmit AV2 by aphids (*M. persicae*) and *Cuscuta* spp. failed (Fujisawa *et al.*, 1983; Weissenfels *et al.*, 1978).

For long time, AV2 has been considered a major asparagus virus in all asparagus producing countries of the world. Looking at the Mediterranean countries, there are no AV2 reports from Spain and France as the major asparagus producers. The presence of AV2 in asparagus has been ascertained only for Italy (Bertaccini *et al.*, 1982, 1990), the third largest producer of asparagus in the Mediterranean area. Some differences in virus incidence were detected among the cultivars examined (Larac, Violetto d'Alba, and Precoce d'Argenteuil), with infection rates ranging from 30% to 98% except for UC 157 in which AV2 was not found in two different surveys (Bertaccini *et al.*, 1988; Giunchedi *et al.*, 1987). More recently, AV2 was not detected in asparagus plantations (except for an abandoned field) in Sicily where the cultivars UC 157, Grande, and Italo are widely grown (Tomassoli *et al.*, 2008b). In contrast, AV2 was found in Campania (cv Desto), Basilicata (cv Italo), and Lazio (cv Grande) (L. Tomassoli, unpublished data).

G. Detection methods

Since AV2 is sap transmissible, bioassay was the first and, for a long time, the only diagnostic assay for its identification (Hein, 1963; Mink and Uyeda, 1977; Paludan, 1964; Weissenfels *et al.*, 1976). *C. quinoa* is a good indicator plant (Evans *et al.*, 1990; Falloon *et al.*, 1986), showing chlorotic or necrotic local lesions 4–10 dpi and systemic leaf mottling or necrosis and epinasty. Symptom expression varies considerably and seems to depend on environmental conditions (Montasser and Davis, 1987). In addition, AV2 infections in *N. tabacum* (cvs White Burley and Xanthi) produce local ringspots (five to seven dpi) and a systemic mottle with vein necrosis.

Antisera to AV2 have been raised by several researchers, initially for the serological characterization of the virus in agar gel double-diffusion tests (Fujisawa *et al.*, 1983; Giunchedi *et al.*, 1987; Hartung *et al.*, 1985; Phillips and Brunt, 1985; Uyeda and Mink, 1981) and in recent years for AV2 detection and identification by ELISA (Bertaccini *et al.*, 1988). Serological kits for AV2 detection by antigen-coated plate (ACP) and triple-antibody sandwich (TAS) ELISA are commercially available.

Molecular detection by one-step RT-PCR was performed by Roose *et al.* (2002) using specific primers (primer A-forward: GTCTGGTAATGCTATT-GAAGTTAATGGTCG; primer B-reverse: AGCATCTTCCTTTGGAGG-CATCTAAACTCTC) derived from the RNA 3 segment and able to amplify a fragment (1000 bp) that includes the entire CP gene and a part of the 3′UTR region. Degenerate primer sets derived from the RNA 1 and RNA 2 segments of ilarviruses are also available and have successfully amplified two AV2 isolates (Maliogka *et al.*, 2007; Untiveros *et al.*, 2010).

IV. MINOR VIRUSES

A. *Asparagus virus 3*

AV3 was isolated from asparagus plants in Japan (Fujisawa, 1986). So far, The virus has not been reported from countries other than Japan. For this reason, little information on its properties and yield effects are available. Fujisawa (1986) associated the presence of the virus to a faint yellowing on young leaves of field-grown plants, but the symptom was not reproduced on artificially infected asparagus plants.

AV3 has flexuous, filamentous particles containing a positive-sense ssRNA genome encapsidated by a single viral encoded protein. AV3 is mechanically transmissible to 26 of 37 herbaceous plant species tested by Fujisawa (1986). Most of them (i.e., *Chenopodium* spp. and many *Nicotiana* spp.) produce local and systemic symptoms. Attempts to transmit AV3 by aphids (*M. persicae* and *A. gossypii*) failed. Biological assay is currently the only means for AV3 detection. There are no reports on the serological or molecular detection and identification of AV3.

Based on particle shape and size (570–590 nm in length), *in vitro* properties (TIP: 55 °C for 10 min; LIV: 21–23 days at 20 °C; DEP: 2×10^{-4}–10^{-2}), and serological relationships with two potexviruses (*Cactus virus X* and *Narcissus mosaic virus*), AV3 was assigned to potexvirus group already about 20 years ago (Brunt *et al.*, 2000). This taxonomic assignment of AV3 (genus *Potexvirus*, family *Alphaflexiviridae*) has recently been confirmed by full genome sequencing of one AV3 isolate (GenBank accession no. AB304848) (Hashimoto *et al.*, 2008). The complete RNA genome is 6937 nt long, excluding the poly(A) tail. Five ORFs were identified: ORF1 encoding a 182-kDa RdRp with the three conserved domains (methyltransferase, helicase, and polymerase); ORF2–4 encodes the triple gene block (TGB) proteins (28, 13, and 9 kDa, respectively); ORF5 encodes a 26-kDa CP. By phylogenic analysis of all recognized potexvirus species, the closest relatives of AV3 are *Narcissus mosaic virus* and *Alstroemeria virus X*. A potexvirus isolated from Chinese scallion and initially considered a distinct species (Chen *et al.*, 2002) is now regarded as an AV3 strain as they have a complete nucleotide sequence identity of 75% and share CP and RdRp amino acid sequence identities of 87.4% and 80%, respectively.

Since the AV3 outbreak occurred in Japan more than 20 years ago and there have been no additional AV3 reports, the virus may be considered economically insignificant for asparagus production.

B. *Cucumber mosaic virus*

The first occurrence of CMV in asparagus was in Germany (Weissenfels and Schmelzer, 1976), later confirmed by other authors, and often found in mixed infection with AV1 (Bandte *et al.*, 2008; Kegler *et al.*, 1991a; Rabenstein *et al.*, 2007). CMV infected asparagus was also encountered in Britain in 1985 (Phillips and Brunt, 1985), but there have been no reports on the occurrence of CMV in asparagus from Italy (Tomassoli *et al.*, 2008b) or other Mediterranean countries.

CMV is the type member of the genus *Cucumovirus*, family *Bromoviridae* (Palukaitis and García-Arenal, 2003; Palukaitis *et al.*, 1992; Roossinck *et al.*, 1999). CMV causes diseases in more than 1200 plant species in over 100 families worldwide and is responsible for economic losses in vegetable, ornamental, and fruit crops (Palukaitis and García-Arenal, 2003). Studies on the biology, ecology, and genome structure of numerous isolates of CMV showed its high capacity to adapt to new hosts and environments. CMV causes symptoms that vary in type and severity depending on the host and time of infection. On asparagus, no obvious symptoms have been reported.

As CMV can be transmitted by more than 80 aphid species in a nonpersistent manner, it can be assumed that any aphid species when searching for a suitable host plant can efficiently introduce and spread the virus in asparagus plantings. Asparagus isolates of CMV can be readily transmitted by mechanical inoculation to herbaceous indicator plants causing typical local lesions on inoculated leaves of *C. quinoa* and *C. amaranticolor* and systemic mosaic on *N. benthamiana*, *N. clevelandii*, and *Nicotiana glutinosa* (Kegler *et al.*, 1991b; Phillips and Brunt, 1985). CMV is known to be seedborne in at least 20 plant species (Gallitelli, 2000), but there is no information on virus transmission by seed and pollen in asparagus.

CMV has a tripartite, positive-sense, single-stranded RNA genome that is enclosed in small icosahedral virions (29 nm in diameter), with each RNA species being individually encapsidated. The CMV genome is similarly organized and encodes proteins functionally similar to those of many other members of the family *Bromoviridae* (i.e., Ilarvirus, AV2). Strains of CMV support the replication of a satellite RNA (RNA 5 or sat RNA) of 332–334 nt in length. RNA 5 affects disease symptom expression on infected host plants (Kaper *et al.*, 1990).

CMV isolates are divided into subgroups I and II based on serological and molecular properties. Further, subgroup I strains are separated into IA and IB according to symptoms induced on *Vigna unguiculata* and restriction fragment length polymorphism (RFLP) profiles (Roossinck, 2002). Asparagus isolates of CMV have not been characterized in detail (no nucleotide sequences available) and, consequently, not assigned to subgroups.

CMV detection in asparagus can be done by ELISA (Bandte *et al.*, 2008; Kegler *et al.*, 1991b) for which numerous commercial kits are available. Although there are many molecular techniques for the identification and characterization of CMV strains, none of them have been used for asparagus isolates of CMV.

More investigations are needed to establish the economic importance of CMV for asparagus production.

C. *Tobacco streak virus*

Asparagus stunt virus was the name assigned to a virus isolated from stunted asparagus plants in Denmark (Paludan, 1964). Then, the isolate was shown to be related to TSV on the basis of biological, physical, and serological properties (Brunt and Paludan, 1970). Also in California, USA, asparagus was found naturally infected by TSV (Mink and Uyeda, 1977). Despite the fact that there are reports on TSV infections in vegetable and ornamental crops in Europe (Bellardi *et al.*, 2006; Brunt and Paludan, 1970; Piccirillo *et al.*, 1990), TSV has not yet been isolated from asparagus in any country of the Mediterranean area.

TSV was first identified in tobacco and is now known to cause severe diseases in more than 80 plant species (i.e., cotton, soybean, sunflower, tomato) resulting often in heavy crop losses (Jain *et al.*, 2008). TSV infected asparagus plants were stunted and showed chlorotic or brown streaks on stem and browning of needles (Brunt and Paludan, 1970), but there are no data about virus effects on yield and spear quality.

Some differences were observed in biological properties between the asparagus isolate and TSV isolates from other plant species. In fact, the asparagus isolate did not produce symptoms on *G. globosa* and the local lesions induced on *C. quinoa* by the asparagus isolate were not necrotic when compared to those of the type isolate (Mink and Uyeda, 1977).

TSV is not seed transmissible in asparagus, but it can be spread by thrips-mediated pollen transfer since *Frankliniella occidentalis* and *Thrips tabaci* occur at high population densities in asparagus plantations in summer (Cockfield and Beers, 2008; Jones, 2005).

TSV is the type member of the genus *Ilarvirus* and the sole member of subgroup 1. TSV is not serologically related to AV2, and as they differ in their nucleotide sequences, they have been assigned to different subgroups of the genus *Ilarvirus* (Fauquet *et al.*, 2005). Various diagnostic methods (e.g., RT-PCR and hybridization assays) can be reliably used for TSV identification. This also includes ELISA for which several commercial kits are available.

TSV currently does not seem to pose a risk for asparagus cultivation.

D. Other viruses

Apart from the viral species described above, four more viruses have been reported to infect asparagus. Tobamovirus-like particles were isolated from asparagus in Italy (Faccioli, 1965) and in Germany (Kegler *et al.*, 1991b). The virus has not been identified until now. Faccioli (1965) reported that the Italian isolate infected certain indicator plants following mechanical inoculation and observed rod-like particles measuring 314 nm in length by electron microscopy.

Lastly, three nepoviruses (*Arabis mosaic virus*, *Strawberry latent ringspot virus*, and *Tomato black ring virus*) were reported to be transmitted by nematodes to asparagus plants growing at a site that had previously been cultivated with infected strawberry (Posnette, 1969). However, no more field surveys or epidemiological studies have been carried out.

V. CONTROL METHODS: PRESENT AND FUTURE

Seeds, seedlings, and crowns play an important role for establishing new asparagus plantations. High-quality planting material that has been certified to be virus-free is the key management tool for a long economic life (10–12 years) and profitable production of this crop. Recent surveys carried out in countries where a high incidence of AV2 had been recorded one or two decades ago (Bandte *et al.*, 2008; Bertaccini *et al.*, 1990; Kegler *et al.*, 1991a; Tomassoli *et al.*, 2008b) demonstrated that the production of AV2-free seedlings by *in vitro* culture propagation significantly reduced the presence of this ilarvirus. The seed-borne nature of AV2 prompted seed companies to apply a seed sanitation program while less attention was given to AV1 whose incidence rapidly increased worldwide. Therefore, breeding programs and tissue-culture activities have to improve their screening tests for AV1 detection and elimination.

This sanitation practice may be more efficient if the new asparagus plantings are established in areas where AV1 inoculum pressure is low or absent, as demonstrated in Sicily, Italy (Tomassoli *et al.*, 2008b). If external virus sources (weeds and adjacent infected crops) as well as the level of virus inoculum inside the crop are minor, it has been demonstrated that the rate of secondary virus spread during the growing season remains low with beneficial effects for the commercial production (Rennie and Cockerell, 2006). Further, the absence of internal sources of virus inoculum could significantly reduce the transmission of AV1 and AV2 by means of contaminated knives used for spear harvesting and by other means of contact transmission. Nevertheless, a simple knife wiping may be an important hygiene measure for avoiding or limiting virus spread during spear harvesting (Knaflewski *et al.*, 2008).

AV1 is significantly spread by aphids. These vectors acquire AV1 from infected plants by probing and in the same way transmit it to other healthy asparagus plants. Insecticides are usually inefficient in controlling viruses that are transmitted by aphids in a nonpersistent manner and can even result in a higher number of infections (Perring *et al.*, 1999).

No resistant asparagus genotypes have been identified (Bandte *et al.*, 2008; Bertaccini *et al.*, 1990; Jaspers and Falloon, 1996; Tomassoli *et al.*, 2008b). Recently, field surveys indicated that *A. maritimus* (syn: *Asparagus amarus, Asparagus scaber*) might be a tolerant species since the AV1 incidence was always very low in all of the *A. maritimus* surveyed cultivations in areas with a high incidence of AV1 in traditional cultivars (Tomassoli *et al.*, 2008b). Some wild asparagus species (*Asparagus acutifolius, Asparagus albus,* and *Asparagus stipularis*) have also been tested for viruses and were found to be free of AV1 and AV2 (L. Tomassoli, unpublished data).

In view of the aforementioned information, much effort needs to be invested in breeding programs that aim at producing new commercial hybrids capable of escaping AV1 infection. Unfortunately, normal diploid *A. officinalis* genotypes ($2n=20$ chromosomes) appeared sexually compatible only with *A. maritimus* but not with other wild species. Therefore, bridge crosses are necessary to overcome the sexual incompatibility with the majority of the wild species. Research on fungal resistance is currently in progress (Falavigna, 2009). It would be desirable that these new hybrids are also checked for virus resistance.

Until now, sowing healthy seed, transplanting healthy seedlings and crowns, using recommended cultural practices that minimize AV1 and AV2 spread, should remain the mainstays of successful virus management in asparagus.

REFERENCES

Adams, M. J., Antoniw, J. F., and Fauquet, C. M. (2005). Molecular criteria for genus and species discrimination within the family Potyviridae. *Arch. Virol.* **150:**459–479.

Aleman-Verdaguer, M. E., Goudou-Urbino, C., Dubern, J., Beachy, R. N., and Fauquet, C. (1997). Analysis of the sequence diversity of the P1, HC, P3, NIb and CP genomic regions of several yam mosaic potyvirus isolates: Implications for the intraspecies molecular diversity of potyviruses. *J. Gen. Virol.* **78:**1253–1264.

Amaro-López, M. A., Zurera-Cosano, G., and Moreno-Rojas, R. (1998). Trends and nutritional significance of mineral content in fresh white asparagus spears. *Int. J. Food Sci. Nutr.* **49:**353–363.

Bandte, M., Grubits, E., von Bargen, S., Rabenstein, F., Weber, D., Uwihs, F., and Büttner, C. (2008). Field study on the occurrence of virus infections in asparagus (Asparagus officinalis L.) in North Germany. "Ernähren uns in der Zukunft Energiepflanzen", 63rd ALVA-Tagung in Raumberg-Gumpenstein, Austria, pp. 97–99.

Bellardi, M. G., Sghedoni, L., and Bertaccini, A. (2006). *Tobacco streak virus* infecting *Buxus sempervirens*. *Acta Hortic.* **722:**229–234.

Benson, B. L. (2009). Update of the world's asparagus production area, spear utilization and production periods. Abstract XII International Asparagus Symposium, Lima, Peru, p. 86.

Bertaccini, A., Marani, F., Martini, L., and Ventura, A. M. (1982). Le virosi dell'asparago nell'Italia settentrionale epidemiologia e possibilità di prevenzione. *Atti Gior. Fitopal* (suppl.1):27–33.

Bertaccini, A., Macri, S., and Poggi Pollini, C. (1988). Diagnosis of asparagus virus 2 with ELISA method. *Inf. Fitopat.* **38:**39–41.

Bertaccini, A., Giunchedi, L., and Poggi Pollini, C. (1990). Survey on asparagus virus diseases in Italy. *Acta Hortic.* **271:**279–283.

Blok, W. J., and Bollen, G. J. (1995). Fungi on roots and stem bases of asparagus in the Netherlands: Species and pathogenicity. *Eur. J. Plant Pathol.* **10:**15–24.

Bol, J. F. (2005). Replication of alfamo- and ilarviruses: Role of the coat protein. *Ann. Rev. Phytopathol.* **43:**39–62.

Brothwell, D. D. R., and Brothwell, P. (1998). Food in Antiquity: A Survey of the Diet of Early Peoples. JHU Press, Baltimore, Maryland, p. 283.

Brunt, A. A., and Paludan, N. (1970). The serological relationship between "asparagus stunt virus" and tobacco streak virus *Phytopathol. Z.* **69:**277–282.

Brunt, A. A., Foster, G. D., Morozov, S. Y., and Zavriev, S. K. (2000). Genus potexvirus. *In* "Virus Taxonomy. Seventh Report of the International Committee on Taxonomy of Viruses" (M. H. V. van Regenmortel, C. M. Fauquet, D. H. L. Bishop, E. B. Castens, M. K. Estes, S. M. Lemon, J. Maniloff, M. A. Mayo, D. J. McGeoch, C. R. Pringle, and R. B. Wickner, eds.), pp. 975–981. Academic Press, San Diego, CA.

Chen, J., Zheng, H. Y., Chen, J. P., and Adams, M. J. (2002). Characterisation of a potyvirus and a potexvirus from Chinese scallion. *Arch. Virol.* **147:**683–693.

Chung, B. Y., Miller, W. A., Atkins, J. F., and Firth, A. E. (2008). An overlapping essential gene in the Potyviridae. *Proc. Natl. Acad. Sci. USA* **105:**5897–5902.

Cockfield, S. D., and Beers, E. H. (2008). Management of dandelion to supplement control of western flower thrips (Thysanoptera: Thripidae) in apple orchards. *J. Entomol. Soc. B. C.* **105:**89–96.

De Vries-Paterson, R. M., Evans, T. A., and Stephens, C. T. (1992). The effect of asparagus virus infection on asparagus tissue culture. *Plant Cell Tissue Organ Cult.* **31:**31–35.

Edwardson, J. R. (1992). Inclusion bodies. *Arch. Virol.* Suppl **5:**25–30.

Edwardson, J. R., Christie, R. G., and Ko, N. J. (1984). Potyvirus cylindrical inclusions—Subdivision-IV. *Phytopathology* **74:**1111–1114.

Elmer, W. H., Johnson, D. A., and Mink, G. I. (1996). Epidemiology and management of the diseases causal to asparagus decline. *Plant Dis.* **80:**117–125.

Evans, T. A., and Stephens, C. T. (1988). Association of asparagus virus II with pollen from infected asparagus (*Asparagus officinalis*). *Plant Dis.* **72:**195–198.

Evans, T. A., and Stephens, C. T. (1989). Increased susceptibility to Fusarium crown and root rot in virus-infected asparagus. *Phytopathology* **79:**253–258.

Evans, T. A., De Vries-Paterson, R. M., Wacker, T. L., and Stephens, C. T. (1990). Epidemiology of asparagus viruses in Michigan asparagus. *Acta Hortic.* **271:**285–290.

Faccioli, G. (1965). Further researches on a virus isolated from Asparagus officinalis L. *Phytopathol. Medit.* **4:**163–167.

Falavigna, A. (2009). Recent progress of asparagus breeding in Italy. Abstract International Asparagus Symposium, Lima, Peru, p. 50.

Falloon, P. G., Falloon, L. M., and Grogan, R. G. (1986). Survey of California asparagus for asparagus virus I, asparagus virus II and tobacco streak virus. *Plant Dis.* **70:**103–105.

Fauquet, C., Mayo, M. A., Maniloff, J., Desselberger, U., and Ball, L. A. (2005). Virus taxonomy: Classification and nomenclature of viruses. Eighth Report of the International Committee on Taxonomy of Viruses. Academic Press, Inc, London, UK, p. 1259.

Fletcher, P. J. (1994). In vitro long-term storage of asparagus. *N. Z. J. Crop Hort. Sci.* **22:**351–359.
Fujisawa, I. (1986). Asparagus virus III: A new member of potexvirus from asparagus. *Ann. Phytopathol. Soc. Jap.* **52:**193–200.
Fujisawa, I., Goto, T., Tsuchizaki, T., and Iizuka, N. (1983). Host range and some properties of asparagus virus I isolated from *Asparagus officinalis* in Japan. *Ann. Phytopathol. Soc. Jap.* **49:**299–307.
Gallitelli, D. (2000). The ecology of Cucumber mosaic virus and sustainable agriculture. *Virus Res.* **71:**9–21.
Giunchedi, L., Poggi Pollini, C., Bertaccini, A., and Marani, F. (1987). Purification and some properties of an Italian asparagus virus 2 isolate. *Phytopathol. Medit.* **26:**117–120.
Grogan, R. G., and Kimble, K. A. (1959). The association of fusarium wilt with the asparagus decline and replant problem in California. *Phytopathology* **49:**122–125.
Gröschel, H., and Jank-Ladwig, R. (1977). Pinwheel detection in local lesions of *Chenopodium quinoa* upon infection by asparagus virus 1 (in German). *Phytopathol. Z.* **88:**180–183.
Guo, H. S., and Ding, S. W. (2002). A viral protein inhibits the long range signaling activity of the gene silencing signal. *EMBO J.* **21:**398–407.
Hari, V., Siegel, A., Rozek, D., and Timberlake, W. E. (1979). The RNA of tobacco etch virus contains poly(A). *Virology* **92:**568–571.
Hartung, A. C., Evans, T. A., and Stephens, C. T. (1985). Occurrence of asparagus virus II in commercial asparagus fields in Michigan. *Plant Dis.* **69:**501–504.
Hashimoto, M., Ozeki, J., Komatsu, K., Senshu, H., Kagiwada, S., Mori, T., Yamaji, Y., and Namba, S. (2008). Complete nucleotide sequence of asparagus virus 3. *Arch. Virol.* **153:**219–221.
Hein, A. (1960). On the occurrence of a virosis of Asparagus. *Z. PflKrankh.* **67:**217–219.
Hein, A. (1963). Virosen am Spargel. *Mitt. Biol. Bundesanst. Land-Forstwirtsch. Berlin-Dahlem.* **108:**70–74.
Hein, A. (1969). On virus diseases of asparagus. Asparagus virus I. *Z. PflKrankh.* **76:**395–406.
Howell, W. E., and Mink, G. I. (1985). Properties of asparagus virus 1 isolated from Washington State asparagus. *Plant Dis.* **69:**1044–1046.
Jain, R. K., Vemana, K., and Sudeep, B. (2008). Tobacco streak virus—An emerging virus in vegetable crops. *In* "Characterization, Diagnosis and Management of Plant Viruses: Vegetable and Pulse Crop" (G. P. Rao, P. L. Kumar, and R. J. Holguin-Peña, eds.), Vol. 3, pp. 203–212. Studium Press LLC, USA.
Jakše, M., and Maršić, N. K. (2008). Asparagus production in Slovenia. *Acta Hortic.* **776:**351–356.
Jaspars, E. M. (1999). Genome activation in alfamo- and ilarviruses. *Arch. Virol.* **144:**843–863.
Jaspers, M. V. (1996). Effect of asparagus virus 2 on yield of *Asparagus officinalis. Acta Hortic.* **415:**383–386.
Jaspers, M. V., and Falloon, P. G. (1996). Asparagus virus 2: A contributing factor in asparagus decline. *Acta Hortic.* **479:**263–270.
Jaspers, M. V., and Pearson, M. N. (1997). Transmission of asparagus virus 2 in an asparagus crop. Proceedings 50th New Zealand Plant Protection Society Conference, pp. 84–88.
Jaspers, M. V., Falloon, P. G., and Pearson, M. N. (1999). Long-term effects of asparagus virus 2 infection on growth and productivity in asparagus. *Ann. Appl. Biol.* **135:**379–384.
Jones, D. R. (2005). Plant viruses transmitted by thrips. *Eur. J. Plant Pathol.* **113:**119–157.
Kaper, J. M., Tousignant, M. E., and Geletka, L. M. (1990). Cucumber-mosaic-virus-associated RNA-5. XII. Symptom-modulating effect is codetermined by the helper virus satellite replication support function. *Res. Virol.* **141:**487–503.
Kegler, H., Schmidt, H. B., Wolterstorff, B., Reinhardt, I., Weber, I., and Proll, E. (1991a). Spread of viruses in asparagus fields. *Arch. Phytopathol. Plant Prot.* **27:**251–258.

Kegler, H., Wolterstorff, B., Reinhardt, I., Richter, J., Weber, I., and Meyer, U. (1991b). Contribution to the virus testing of asparagus plants. *Arch. Phytopathol. Plant Prot.* **27**:353–360.

Knaflewski, M. (2009). History of asparagus cultivation and International asparagus. Abstract XII International Asparagus Symposium, Lima, Peru, p. 20.

Knaflewski, M., Fiedorow, Z., and Pawlowski, A. (2008). Viral diseases and their impact on asparagus performance and yield. *Acta Hortic.* **776**:191–197.

Mahajan, S., Dolja, V. V., and Carrington, J. C. (1996). Roles of the sequence encoding tobacco etch virus capsid protein in genome amplification: Requirements for the translation process and a cis-active element. *J. Virol.* **70**:4370–4379.

Maliogka, V. I., Dovas, C. I., and Katis, N. I. (2007). Demarcation of ilarviruses based on the phylogeny of RNA2-encoded RdRp and a generic ramped annealing RT-PCR. *Arch. Virol.* **152**:1687–1698.

Marani, F., Bertaccini, A., Rabiti, A. L., and Benni, A. (1993). Cytopathology induced by an Italian asparagus virus 1 isolate. *Adv. Hort. Sci.* **7**:15–17.

Merits, A., Rajamäki, M. L., Lindholm, P., Runeberg-Roos, P., Kekarainen, T., Puustinen, P., Mäkeläinen, K., Valkonen, P. T., and Saarma, M. (2002). Proteolytic processing of potyviral proteins and polyprotein processing intermediates in insect and plant cells. *J. Gen. Virol.* **83**:1211–1221.

Mink, G. L., and Uyeda, I. (1977). Three mechanically-transmissible viruses isolated from asparagus in Washington. *Plant Dis. Rep.* **61**:398–401.

Montasser, M. S., and Davis, R. F. (1987). Survey for asparagus viruses I and II in new Jersey. *Plant Dis.* **71**:497–499.

Paludan, N. (1964). Virus disease of *Asparagus officinalis*. *Manedsovers. Plantesygd.* **407**:11–16.

Palukaitis, P., and García-Arenal, F. (2003). Cucumoviruses. *Adv. Virus Res.* **62**:241–323.

Palukaitis, P., Roossinck, M. J., Dietzgen, R. G., and Francki, R. I. B. (1992). Cucumber mosaic virus. *Adv. Virus Res.* **41**:281–348.

Pawlowski, A., Fiedorow, Z., Golehniak, B., and Weber, Z. (2004). The influence of exudates and sap from asparagus (Asparagus officinalis) roots infected by Asparagus virus 2 (AV2) on the growth of soil fungi pathogenic to the plants. *Phytopathol. Pol.* **33**:31–40.

Perring, T. M., Gruenhagen, N. M., and Farrar, C. A. (1999). Management of plant viral diseases through chimical control of insect vectors. *Annu. Rev. Entomol.* **44**:457–481.

Phillips, S., and Brunt, A. A. (1985). Occurrence of cucumber mosaic virus and asparagus virus II in asparagus (*Asparagus officinalis* L. var. *officinalis*) in Britain. *Plant Pathol.* **34**:440–442.

Piccirillo, P., Diana, G., and Barba, M. (1990). Il virus della striatura del tabacco (TSV) isolato su tabacco in Campania. *Inf. Fitopatol.* **40**:57–59.

Posnette, A. F. (1969). Nematode transmitted viruses in asparagus. *J. Hort. Sci.* **44**:403–406.

Rabenstein, F., Schubert, J., and Habekuß, A. (2007). Identification and differentiation of viruses on asparagus in Germany. Plant Virus Epidemiology Symposium, 15–19 October 07, ICRISAT, India.

Rafael-Martín, M., and Rivera-Bustamante, R. F. (1999). Molecular characterization of the RNA3 of asparagus virus 2. *Arch. Virol.* **144**:185–192.

Rennie, W. J., and Cockerell, V. (2006). Seed-borne diseases. *In* "The Epidemiology of Plant Diseases" (B. M. Cooke, D. Gareth Jones, and B. Kaye, eds.), pp. 357–372. Springer, Dordrecht, the Netherlands.

Riechmann, J. L., Lain, S., and Garcia, J. A. (1989). The genome-linked protein and 5′ end RNA sequence of plum pox potyvirus. *J. Gen. Virol.* **70**:2785–2789.

Riechmann, J. L., Lain, S., and Garcia, J. A. (1992). Highlights and prospects of potyvirus molecular biology. *J. Gen. Virol.* **73**:1–16.

Roose, M. L., Stone, N. K., Mathews, D., and Dodds, J. A. (2002). RT-PCR detection of Asparagus 2 Ilarvirus. *Acta Hortic.* **589**:357–363.

Roossinck, M. J. (2002). Evolutionary history of *Cucumber mosaic virus* deduced by phylogenetic analyses. *J. Virol.* **76:**3382–3387.
Roossinck, M. J., Bujorski, J., Ding, S. W., Hajimorad, R., Hanada, K., Scott, S., and Tousignant, M. (1999). Family Bromoviridae. *In* "Virus Taxonomy—Seventh Report of the International Committee on Taxonomy of Viruses" (M. H. V. van Regenmortel, C. M. Fauquet, D. H. L. Bishop, E. B. Castens, M. K. Estes, S. M. Lemon, J. Maniloff, M. A. Mayo, D. J. McGeoch, C. R. Pringle, and R. B. Wickner, eds.), pp. 923–935. Academic Press, San Diego, CA.
Scott, S. W., and Zimmerman, M. T. (2009). The nucleotide sequences of the RNA1 and RNA2 of asparagus virus 2 show a close relationship to citrus variegation virus. *Arch. Virol.* **154:**719–722.
Scott, S. W., Zimmerman, M. T., and Ge, X. (2003). Viruses in subgroup 2 of the genus *Ilarvirus* share properties at the molecular level. *Arch. Virol.* **148:**2063–2075.
Sun, T., and Powers, J. R. (2007). Chapter 12—Antioxidants and antioxidant activities of vegetables. *In* "Antioxidant Measurement and Applications, ACS Symposium Series", 956, pp. 160–183.
Tiberini, A., Tomassoli, L., and Vetten, H. J. (2008). Asparagus virus 1: Biological, serological and molecular properties of a virus of the genus Potyvirus. Abstract IWGLVV Conference, Ljubljana, Slovenia, p. 29.
Tomassoli, L., Zaccaria, A., and Tiberini, A. (2007). Use of one-step RT-PCR for detection of Asparagus virus 1. *J. Plant Pathol.* **89:**413–415.
Tomassoli, L., Tiberini, A., Zaccaria, A., and Vetten, J. H. (2008a). Molecular and biological studies of Asparagus virus 1 (genus Potyvirus). *J. Plant Pathol.* **90**(Suppl. 2)**:**437.
Tomassoli, L., Zaccaria, A., Valentino, A., and Tamietti, G. (2008b). First investigations on the diseases affecting green asparagus crops in Sicily. *Col. Prot.* **37:**83–90.
Tomassoli, L., Tiberini, A., Zaccaria, A., and Falavigna, A. (2009). Prevalence of Asparagus virus 1 in asparagus commercial crops. Abstract XII International Asparagus Symposium, Lima, Peru.
Untiveros, M., Perez-Egusquiza, Z., and Clover, G. (2010). PCR assay for the detection of members of the genus *Ilarvirus* and family *Bromoviridae*. *J. Virol. Methods* **165:**97–104.
Urcuqui-Inchima, S., Haenni, A. L., and Bernardi, F. (2001). Potyvirus proteins: A wealth of functions. *Virus Res.* **74:**157–175.
Uyeda, I., and Mink, G. I. (1981). Properties of asparagus virus II, a new member of the Ilarvirus Group. *Phytopathology* **71:**1264–1269.
Van Vloten-Doting, L. (1985). Coat protein is required for infectivity of tobacco streak virus: Biological equivalence of the coat proteins of tobacco streak and alfalfa mosaic viruses. *Virology* **65:**215–225.
Verchot, J., and Carrington, J. C. (1995). Debilitation of plant potyvirus infectivity by P1 proteinase-inactivating mutations and restoration by second site modifications. *J. Virol.* **69:**1582–1590.
Wei, T., Zhang, C., Hong, J., Xiong, R., Kasschau, K., Zhou, X., Carrington, J., and Wang, A. (2010). Formation of complexes at plasmodesmata for potyvirus intercellular movement is mediated by viral protein P3N-PIPO. *PLoS Pathog.* **6**(6)**:**e1000962. doi: 10.1371/journal.ppat.1000962.
Weissenfels, M., and Schmelzer, K. (1976). Untersuchungen uber das Schadausmass durch Viren am Spargel (*Asparagus officinalis L.*). *Arch. Phytopathol. Pflanzenschutz* **12:**67–73.
Weissenfels, M., Schimidt, H. B., and Schmelzer, K. (1976). Partikelgestalt des Spargel-Virus 2 (Asparagus virus 2). *Arch. Phytopathol. Pflanzenschutz* **12:**63–64.
Weissenfels, M., Schmelzer, K., and Schmidt, H. B. (1978). A contribution to the characterization of asparagus virus 2. *Zentralbl. Bakteriol. Naturwiss.* **133:**65–79.
Yang, H. J. (1979). Early effects of viruses on the growth and productivity of asparagus plants. *Hort. Sci.* **14:**734–735.

CHAPTER 11

Virus Diseases of Peas, Beans, and Faba Bean in the Mediterranean Region

Khaled Makkouk,* **Hanu Pappu,**[†] and
Safaa G. Kumari[‡]

Contents

Abstract

In the Mediterranean region, pea, bean, and faba bean production is affected by around 17 major viruses. These viruses do not have the same ecology and consequently require a variety of different preventive measures to control them. Some of these viruses have a narrow host range, such as *Faba bean necrotic yellows virus* (FBNYV), and others, such as *Alfalfa mosaic virus* (AMV) and *Cucumber mosaic virus* (CMV), a very wide host range. Such features are important when identifying sources of virus inoculum in a region, and the vectors can transmit viruses from natural reservoirs to the crop plants. Some of these viruses are seed borne and, consequently, can be disseminated long distances through infected

* National Council for Scientific Research, Beirut, Lebanon
[†] Washington State University, Pullman, Washington, USA
[‡] International Center for Agricultural Research in the Dry Areas (ICARDA), P.O. Box 5466, Aleppo, Syria

Advances in Virus Research, Volume 84
ISSN 0065-3527, DOI: 10.1016/B978-0-12-394314-9.00011-7

seeds. Crop losses caused by these viruses are variable, depending on the sensitivity and susceptibility of the crop to infection. Host resistance genes have been identified for some of these viruses, but in others, such as FBNYV, no resistance genes in faba bean have been identified yet. Significant progress was made in developing precise methods for the identification of these viruses, and new virus problems are being identified every year. This chapter is not intended to be a review for pea, bean, and faba bean viruses, but rather focuses on the major viruses which affect these crops in the Mediterranean basin with focus on the progress made over the past two decades.

I. INTRODUCTION

Pea (*Pisum sativum* L.), bean (*Phaseolus vulgaris* L.), and faba bean (*Vicia faba* L.) are among the most important food legumes in the Mediterranean region. These crops are infected naturally with many viruses, and the number of these viruses continues to increase (Bos *et al.*, 1988; Kumari and Makkouk, 2007). The impact of infection with these viruses on crop yield varies significantly depending on the individual virus, the crop, and the location where infection occurs. In this chapter, we have attempted to review the work done on the economically most important viruses that attack these crops in the Mediterranean region, their ecology and epidemiology, transmission, sensitive assays available for their detection, and appropriate measures for their control.

II. IMPORTANCE

Peas are reported to be infected with at least 30 viruses worldwide. The most important viruses which infect this crop in the Mediterranean region are *Bean leafroll virus* (BLRV), *Bean yellow mosaic virus* (BYMV), *Pea seed-borne mosaic virus* (PSbMV), and *Pea enation mosaic virus* (PEMV)-*1* and -*2*.

Beans are reported to be attacked by around 20 major viruses worldwide, but only 11 viruses have been reported on this crop in the Mediterranean region; the most important among them are *Bean common mosaic virus* (BCMV), *Bean common mosaic necrosis virus* (BCMNV), BYMV, *Alfalfa mosaic virus* (AMV), and *Cucumber mosaic virus* (CMV).

Faba bean is naturally infected by around 50 viruses worldwide (Bos *et al.*, 1988; Cockbain, 1983; Makkouk *et al.*, 2003), and the number continues to increase. Fortunately, only few are of major economic importance in the Mediterranean region. Around 20 viruses have so far been identified as affecting faba bean in the Mediterranean region (Bos *et al.*, 1988; Kumari and Makkouk, 2007; Makkouk *et al.*, 2003). Those with major economic importance are: *Faba bean necrotic yellows virus* (FBNYV), BLRV, BYMV, and *Broad bean mottle virus* (BBMV). Most of these viruses cause losses to the crops they infect; the amount of damage depends on the extent of virus spread and the susceptibility of genotypes planted. Yield loss could vary from almost none to a complete crop failure, as observed for FBNYV on faba bean in Middle Egypt during the 1991/1992 (Makkouk *et al.*, 1994) and 1997/1998 growing seasons.

III. PRODUCTION

In the Mediterranean region, pea production reached around 4.85 million tons, bean production around 18.2 million tons, and faba bean production around 3.7 million tons (FAO statistics, 2009). These figures represent both dry and green production for peas and beans and only dry grain for faba bean, even though in some Mediterranean countries, faba bean is also consumed as green vegetable. In the Mediterranean region, France and Spain are the major pea producing countries; Egypt and Turkey are the major bean producing countries; and Egypt, Morocco, Spain, and Italy are the major faba bean producing countries. Details of production in each of the Mediterranean countries are summarized in Table I. Accurate crop loss data due to virus diseases in the Mediterranean region are very limited, and the available data cover mainly incidence of infection in the field and potential yield losses determined in infection experiments.

IV. THE MAIN VIRUS DISEASES

A. Virus diseases of pea

1. *Bean yellow mosaic virus* (genus *Potyvirus*, family *Potyviridae*)

Distribution in the Mediterranean countries. BYMV is distributed worldwide including several Mediterranean countries. It is reported on peas in Syria (Alkhalaf *et al.*, 2010), Egypt (Eldin *et al.*, 1981), Italy (Faccioli *et al.*, 1990), and Libya (Zidan *et al.*, 2002).

Symptoms and strains. Commonly referred to as pea mosaic, BYMV-associated symptoms in peas include mosaic and veinal chlorosis and the intensity varies with the strain. Sometimes necrosis may be seen in stems and veins.

Transmission. BYMV is transmitted by several species of aphids in a nonpersistent manner. The virus can be acquired by aphids from other hosts such as alfalfa, red clover, white clover, and vetch. The main vector species are *Acyrthosiphon pisum* (Harris), *Aphis fabae* (Scopoli), *Aphis*

TABLE I Area harvested and production of pea, bean, and faba bean in the Mediterranean countries (FAO statistics, 2009)

Region	Area harvested (ha)					Production (tons)				
	Beans (dry)	Beans (green)	Peas (dry)	Peas (green)	Faba bean (dry)	Beans (dry)	Beans (green)	Peas (dry)	Peas (green)	Faba bean (dry)
World	25,211,468	960,272	6,158,809	1,164,145	2,507,855	19,723,330	6,814,403	10,379,890	9,168,666	4,096,682
Algeria	1616	8918	8487	28,724	32,278	1159	45,096	5969	102,971	36,495
Egypt	35,000	25,000	50	30,000	86,519	100,000	300,000	90	295,000	295,182
Libya	320[a]	M[b]	3250	1200	1000	1000[a]	M[b]	5200	6000	1500
Morocco	18,000[a]	8140[a]	37,900[a]	17,000	181,900[a]	13,000[a]	182,180[a]	15,570[a]	118,000	a108,680
Tunisia	170[a]	410[a]	11,000	3600[a]	58,000	100[a]	2200[a]	10,000	15,000[a]	60,000
Cyprus	170	120	M[b]	72	500	223	2450	M[b]	750	410
Israel	M[b]	700	300[a]	3200[a]	3700	M[b]	8969	300[a]	12,940[a]	18,429
Lebanon	167[a]	1420[a]	950[a]	880[a]	175[a]	200[a]	14,200[a]	2400[a]	5200[a]	200[a]
Syria	637	3547	4127	1970	17,477	1312	32,498	6848	14,014	37,782
Turkey	97,451	72,000	1224	14,000	9383	181,205	603,653	3604	95,046	21,150
Albania	14,000	587	M[b]	281	150	23,000	6902	M[b]	2572	200
Bosnia and Herzegovina	9497	500[a]	1425	M[b]	M[b]	14,906	1800[a]	3201	M[b]	M[b]
Croatia	1947	967	372	793	M[b]	2460	9329	955	4671	M[b]
France	2500	7000	100,230[a]	30,000	88,000	6300	48,000	446,850[a]	405,000	437,300
Greece	8000	6300	450	900	1800	12,800	65,000	640	5500	4000
Italy	6400	19,100	10,800	11,500	56,100	12,800	175,200	28,700	76,000	97,800
Malta	150	M[b]	M[b]	30	280	300	M[b]	M[b]	135	800
Montenegro	M[b]	675[a]	M[b]	137[a]	M[b]	M[b]	1066[a]	M[b]	258[a]	M[b]
Slovenia	355	661	648	73	24	700	3368	1420	280	34
Spain	7600	12,500	163,800	12,500	19,000	12,500	163,700	143,600	81,700	23,200
Mediterranean region	203,980	168,545	345,013	156,860	556,286	383,965	1,665,611	675,347	1,241,037	1,143,162

[a] Data of 2008.
[b] M = missing data (not available).

gossypii (Glover), *Aulacorthum solani* (Kaltenbach), *Brevicoryne brassicae* L., *Myzus persicae* (Sulzer), and *Rhopalosiphum maidis* (Fitch). The virus is also transmitted through seed of most temperate pulses, including faba beans, field peas, lentils, and lupins and through seed of a number of forage legumes and clovers (Table II).

Particle morphology and properties. The particles are usually flexuous, ca. 750 nm long and 12–15 nm wide. Particles are helically constructed (pitch 3.4 nm), but their axial canal is obscure. The virus particles sediment as one component ca. 151S (varying from 140 to 166S), which is composed of 5.5–6% nucleic acid and 94–94.5% of one type of protein. Serological specificity is located particularly in the N-terminal part of the coat protein molecule which is readily lost during virus purification and storage (Shukla *et al.*, 1988).

Genome structure and organization. The nucleic acid is unipartite, single-stranded, positive-sense RNA of MW ca. 6000 kDa and a genome size ca. 10 kb. The coat protein gene of a number of BYMV strains/isolates is 819 nucleotides (nt) long and encodes the 273 amino acid residues of the coat protein (Tracy *et al.*, 1992) with a MW of 30.9 kDa (Hammond and Hammond, 1989).

Propagation hosts and purification. The host range of BYMV is wide and not limited to *Fabaceae*. The virus is reported to infect nearly 200 species in 14 families. Temperate pulse hosts include chickpeas, faba beans, field peas, lentils, and lupins. Temperate legume pasture hosts include lathyrus, lucerne, vetch and medic and clover species. BYMV has a number of other pulse hosts, including soybeans, peanuts, and French beans. It also infects ornamental hosts, the most common being gladiolus species. BYMV can be purified following the procedure described by Bos (1970). A purified preparation of the virus has an A_{260}:A_{280} ratio of 1.21.

Detection methods available. Monoclonal (Jordan and Hammond, 1991) and polyclonal (Makkouk *et al.*, 1988c) antibodies were produced for BYMV detection by ELISA and primers for detection by RT-PCR (Bariana *et al.*, 1994; Ortiz *et al.*, 2006; Uga, 2005) (Table III).

Control methods. Several BYMV-resistant pea cultivars and breeding lines were developed using a single recessive gene, *mo*. Due to the non-persistent transmission by aphids, chemical control of aphid populations offers little virus control.

2. *Bean leafroll virus* (genus *Luteovirus*, family *Luteoviridae*)

Distribution in the Mediterranean countries. BLRV is reported to infect pea in Italy (Larsen and Webster, 1999) and France (Leclercq-Le Quillec and Maury, 1998).

Symptoms and strains. BLRV, first described in faba bean, causes serious disease in peas. BLRV infection results in yellowing of leaves and hence the disease was also referred to as "top yellows" and was initially considered to be caused by pea leafroll virus. BLRV-induced symptoms include chlorosis, upward rolling/curling of leaves, and stunting. Infection may result in brittle leaves. Yield is affected due to misshapen and/or poorly filled pods.

Information on particle morphology. Genome structure and organization (Domier *et al.*, 2002; Vemulapati *et al.*, 2011), propagation hosts, transmission, and detection methods are described under Section IV.C.2.

Control methods. Aphids play an important role in BLRV outbreaks. Control tactics include monitoring and managing aphid populations and growing resistant cultivars.

For more information on BLRV check the Section IV.C.2.

3. *Pea seed-borne mosaic virus* (genus *Potyvirus*, family *Potyviridae*)

Distribution in the Mediterranean countries. PSbMV is reported to infect peas in Egypt (Abdelmaksoud *et al.*, 2000), France (Leclercq-Le Quillec and Maury, 1998), Libya (Zidan *et al.*, 1997), Syria (Makkouk *et al.*, 1993b), and Turkey (Fidan and Yorganci, 1989).

Symptoms and strains. A wide range of symptoms are associated with PSbMV infection in pea. Symptoms include mosaic, vein clearing, leaf rolling, chlorosis, and stunting. If the infection was the result of seed-borne infection, symptoms tend to be more severe. Pods may fail to set or result in deformed pods. Virus-infected seed may show some abnormalities but is not a reliable indicator of virus presence. Plants dually infected with PSbMV and PEMV may show more severe symptoms than those infected with PSbMV alone.

Transmission. PSbMV is transmitted by aphids in a nonpersistent manner by more than 20 different aphid species and through seed. The pea aphid (*A. pisum*), green peach aphid (*M. persicae*), and cotton aphid (*A. gossypii*) are known aphid vectors of PSbMV. The virus could be introduced to new locations/regions through infected seed and could get established if the aphid vector is present (Table II).

Particle morphology and properties. Virus particles are flexuous rods about 770 nm in length.

Genome structure and organization. The PSbMV genome consists of a positive-sense single-stranded RNA of about 9.9 kb and codes for a polyprotein of approximately 334 kDa in size. This polyprotein is cleaved proteolytically into various functional proteins by virally coded proteases.

Propagation hosts and purification. Pea is a good propagation host. *V. faba* is a systemic host. Local lesion hosts include *Chenopodium amaranticolor* Cost. & Reyn. and *C. quinoa* Willd where virus infection results in necrotic local lesions and chlorotic local lesions, respectively. PSbMV can be purified following the procedure of Alconero *et al.* (1986). Purified virus preparation had an A_{260}:A_{280} of 1.14–1.18. PSbMV has filamentous particles ca. 770 nm long and 12 nm wide which sediment as one component ca. 154S and contain a single positive-sense strand RNA of ca. 10 kb.

Detection methods available. Serological (ELISA) using polyclonal (Makkouk *et al*., 1993b) and nucleic acid (PCR)-based methods were developed and are useful in virus detection and characterization.

Control methods. Growing resistant cultivars offer the best strategy for reducing the impact of PSbMV. Rate of seed transmission varies with the cultivar and virus strain. Genetic resistance to seed transmission provides the best option for reducing the virus spread and its impact. Planting virus-tested, clean seed also reduces the spread of the virus and its subsequent establishment especially in areas where the virus was not previously reported, since infected seed could be the primary means of virus introduction. Managing aphid vectors could offer some protection by reducing the aphid-mediated secondary spread from the initial disease loci that resulted from infected seed.

4. *Pea enation mosaic virus* (PEMV-1, genus *Enamovirus*, family *Luteoviridae*; PEMV-2, genus *Umbravirus*)

Distribution in the Mediterranean countries. PEMV is reported to infect peas in France (Leclercq-Le Quillec and Maury, 1998), Italy (Rosciglione and Cannizzaro, 1975), and Spain (Tornos *et al*., 2008).

Symptoms and strains. PEMV mainly infects members of the family Leguminosae including *Anthillus, Astrogalus, Cicer, Glycine, Lens, Lotus, Lupinus, Medicago, Melilotus, Phaseolus, Pisum, Trifolium*, and *Vicia* (Gonsalves and Shepherd, 1972; Hagedorn *et al*., 1964; Hull and Lane, 1973; Izadpanah and Shepherd, 1966a; Mahmood and Peters, 1973). In peas, PEMV causes downward curling of the leaves, chlorotic and translucent spots on the leaves, and vein clearing. As the disease progresses, malformations, epinasty, severe stunting, and rugosity may be observed on the leaves. Enations (hyperplastic outgrowths) are associated with the veins, develop underneath the leaf lamina and hence the name, pea enation. The pods are distorted, covered with warts, and bear few peas. The first reference to the ''mosaic disease of the sweet pea'' was made by Taubenhaus (1914). The term ''enation'' and further description and characterization of the disease were provided by Osborn (1935). The virus is mechanically as well as aphid transmissible. The unique set of properties of PEMV led Harrison *et al*. (1971) to assign the virus to a monotypic group, later given the name *Enamovirus* (De Zoeten and Demler, 1995). At that time, PEMV was thought to have a single-stranded, bipartite genome with two different RNAs separately encapsidated in morphologically distinct particles.

Transmission. Several aphid species were reported to transmit PEMV (Osborn, 1938; Nault, 1967, 1975). Of these, the pea aphid (*A. pisum*) and the green peach aphid (*M. persicae*) are the most common aphid species reported to transmit PEMV. PEMV is transmitted by aphids in a circulative, nonpropagative manner (Demler *et al.*, 1996) (Table II).

Particle morphology and properties. The virus particles are isometric, ca. 27 nm in diameter with an ssRNA genome of 6 kb.

Genome structure and organization. PEMV has a bipartite genome comprising two autonomously replicating and taxonomically unrelated positive-sense RNAs. The large RNA or RNA 1 (5706 nt; PEMV-1) has nucleotide and translated amino acid sequence similarities with beet western yellows–potato leafroll poleroviruses subgroup (Demler and de Zoeten, 1991), while the smaller RNA or RNA 2 (4253 nt; PEMV-2) belongs to the genus *Umbravirus*. Recently, the genome structure and organization of PEMV-1 and PEMV-2 isolates from the pacific northwestern region of the United States were determined (Vemulapati *et al.*, 2010, 2011). The RNA 2 can infect the whole plant by itself, but due to lack of coat protein, it depends on RNA 1 for structural and vector transmission functions. In turn, RNA 2 provides functions related to systemic movement and mechanical transmission lacking in RNA 1 (Demler *et al.*, 1997). This symbiotic association between RNA 1 and RNA 2 successfully establishes a systemic infection in susceptible hosts. PEMV-1 genomic RNA contains five open reading frames (ORFs). ORF-1 encodes a 34-kDa protein of unknown function. ORF-2 overlaps the ORF-1 in a unique reading frame and encodes an 84-kDa protein that contains protease-like motifs, which might be involved in posttranslational processing of virus translation products. ORF-3 was found to encode the RNA-dependent RNA polymerase (RdRp) and contain a helicase-like motif typical of RdRp. It is thought to be expressed by a frameshift fusion of the ORF-2 and ORF-3 products. ORF-4 codes for a 21-kDa coat protein (CP), which is separated from ORF-3 by an intergenic region spanning 189 bases. ORF-4 is immediately followed by ORF-5, which codes for a 33-kDa protein and is thought to be an aphid transmission subunit of the virus (Demler and de Zoeten, 1991).

Propagation hosts and purification. Virus propagation and purification of PEMV is done using *P. sativum* (Progress No. 9 and cv. 8221). In peas, the highest virus titer is usually reached 10–14 days post inoculation, when the plants are grown at 18–22 °C. Nonleguminous hosts such as *Nicotiana clevelandii* A. Gray, *N. benthamiana*, *N. tabacum* L. cv. White Burley, and *Gomphrena globosa* L. were found to be susceptible. Purification from other hosts has been reported (French *et al.*, 1973; Izadpanah and Shepherd, 1966b; Motoyoshi and Hull, 1974).

Detection methods available. Molecular and serological methods are available for the detection of PEMV in plant hosts as well as insect vectors. ELISA-based detection (D'Arcy *et al.*, 1989b) and RT-PCR (Chomič *et al.*, 2010; Ortiz *et al.*, 2005) were reported for the detection of luteoviruses. Though RT-PCR is more sensitive than ELISA, it may not be practical for testing a large number of samples in situations such as screening of germplasm, indexing, and surveys. Polyclonal antisera developed by conventional methods often resulted in serological cross-reactions among luteoviruses (Waterhouse *et al.*, 1988). The development of monoclonal antibodies for detecting different members of the family *Luteoviridae* was reported (D'Arcy *et al.*, 1989b).

Control. Managing vector populations and alternate/reservoir hosts offer some control. Resistance toward PEMV infection has been identified in dry land legume crops such as pea (Schroeder and Barton, 1958) and lentil (Aydin *et al.*, 1987). Pathogen-derived resistance and RNAi/host-mediated gene silencing strategies have potential for introducing resistance to PEMV (Sudarshana *et al.*, 2007).

5. Other viruses

Other viruses that were reported and are considered to be of less economically important are BCMV and CMV.

B. Virus diseases of bean

1. *Bean common mosaic virus* and *Bean common mosaic necrosis virus* (genus *Potyvirus*, family *Potyviridae*)

These two viruses cause what is known as the common mosaic disease of bean. They cause significant yield loss in bean which can reach 80% (Galvez and Morales, 1989; Morales, 2003).

Distribution in the Mediterranean countries. Both viruses commonly occur in all Mediterranean countries that grow common bean with BCMV more widely distributed than BCMNV (Acikgoz and Citir, 1986; Azzam and Makkouk, 1985; CAB, 1974; Habib *et al.*, 1981; Halupecki *et al.*, 2003; Kheder, 2002; Lisa, 2000; Lockhart and Fischer, 1974; Ravnikar *et al.*, 1996; Saiz *et al.*, 1995; Yilmaz *et al.*, 2002).

Symptoms and strains. Eight major BCMV and three BCMNV strains have been reported (Drijfhout *et al.*, 1978; Morales, 2003). Blackeye cowpea mosaic virus, a virus known to infect *P. vulgaris* for many years, is now classified as a strain of BCMV (Fauquet *et al.*, 2005). The BCMV and BCMNV strains cause two main types of symptoms, common mosaic, often associated with leaf malformation, and "black root," characterized by systemic necrosis and plant death (Drijfhout *et al.*, 1978).

Transmission. Both BCMV and BCMNV are mechanically transmitted and by several aphid species in the nonpersistent manner, especially by *A. pisum*, *A. fabae*, and *M. persicae* (Kennedy *et al.*, 1962). Seed transmission, around 35% on average, is reported for both viruses (Morales, 2003) and is an important factor for initial crop infection (Table II).

Particle morphology and properties. Both viruses have flexuous, filamentous particles ca. 750 nm long and 12–15 nm wide, with sedimentation coefficient 154–158S. Purified virus preparation has an A_{260}:A_{280} ratio of 1.12–1.27. Buoyant density in CsCl is 1.31–1.32 g/m^3 (Morales and Bos, 1988).

Genome structure and organization. The structure and organization of BCMV and BCMNV genomes is similar to that of BYMV described under Section IV.C.3.

Propagation hosts and purification. All common bean cultivars that produce mild mosaic symptoms are suitable propagation and maintenance hosts. Highly purified virus preparations for both viruses can be obtained by following the procedure reported by Morales and Bos (1988).

Detection methods available. Monoclonal and polyclonal antibodies were produced for BCMV and BCMNV detection by ELISA (Morales, 1979; Wang *et al.*, 1982, 1984) and primers for detection by RT-PCR (Xu and Hampton, 1996) (Table III).

Control. The most effective control measure against both BCMV and BCMNV is the cultivation of common bean resistant varieties. In the absence of resistant cultivars, the use of virus-free common bean seed minimizes the risk of outbreaks.

2. *Bean yellow mosaic virus* (genus *Potyvirus*, family *Potyviridae*)

Distribution in the Mediterranean countries. BYMV has been reported on common bean in Turkey (Acikgoz and Citir, 1986), Croatia (Halupecki *et al.*, 2003), Egypt (Habib *et al.*, 1981), Italy (Lisa, 2000), Lebanon (Azzam and Makkouk, 1985), Libya (Shagrun, 1973), and Spain (Saiz *et al.*, 1993).

Symptoms. Although symptoms produced in response to BYMV infection may vary depending on time of infection, bean variety, and virus strain, symptoms generally include crinkling, downward cupping, yellow mottling, and dead areas along the veins of infected leaves. Death of vine tips and new leaves may occur on pole and half-runner bean types. Vines may die back several feet, thereby destroying the bean plants. Plants affected with root and stem rot, manganese toxicity, and bacterial blight may show some symptoms similar to those of the virus, so laboratory diagnosis is often advisable.

Control. BYMV is carried to beans by aphids. These insects pick up the virus mainly from red or white clover or other legume weed hosts growing in the same field or in nearby fields, including earlier plantings of beans. The following suggestions may reduce the damage caused by BYMV: (1) Avoid planting beans near clover or other legumes, (2) Destroy legume weeds along the borders or other areas in the field, (3) Plant a barrier of sweet corn or other tall-growing crop upwind of the beans, (4) Plant bush-type beans if possible. It appears that Kentucky Wonder Pole beans and White Half-Runner beans are more susceptible to the virus, and (5) Avoid planting successive crops near each other.

For more information on BYMV, check the Section IV.A.1.

3. *Alfalfa mosaic virus* (genus *Alfamovirus*, family *Bromoviridae*)

Distribution in the Mediterranean countries. AMV has a worldwide distribution. It has been reported on common bean in Italy (Lisa, 2000). The virus incidence reported in most bean fields is low, and it is not considered at present of economical importance on beans.

Symptoms and strains. AMV has a very wide host range infecting at least 697 species in 167 genera of 71 families (Edwardson and Christie, 1997), and it is reported to naturally infect bean and faba bean. Lucerne (*Medicago sativa* L.) is probably the main source of the virus in many countries, but other perennial legumes, including red and white clovers (*Trifolium pratense* L. and *T. repens* L.), are also known sources (Hull, 1969). Symptoms induced by AMV are generally in the form of systemic mosaic or mottle. Under some conditions, symptoms could disappear or become extremely mild. Mechanical inoculation of many strains to *P. vulgaris* will give necrotic local lesions which can be used as an assay of the virus.

Transmission. AMV is transmitted by sap and aphids, such as *A. pisum*, *Aphis craccivora*, *A. fabae*, and *M. persicae* in the nonpersistent manner. Many aphid species have been reported as vectors (Edwardson and Christie, 1997) (Table II).

Particle morphology and properties. AMV has bacilliform particles which sediment as four major components, having a constant diameter of 18 nm and varying from 30 to 57 nm in length, depending on the nucleic acid species encapsidated.

Genome structure and organization. In most strains, each particle contains one species of positive single-stranded RNA, and their size varies from 1.0 to 3.4 kb. Virions are readily separated into components by sucrose density gradient centrifugation RNA-1, -2, and -3 act as mRNAs. The coat protein ORF is expressed from a subgenomic RNA that is usually encapsidated.

Propagation hosts and purification. AMV can be purified following the procedure of van Vloten-Doting and Jaspars (1972). Purified virus preparation had an A_{260}:A_{280} ratio of 1.7–1.8.

Detection methods available. AMV can be detected by biological methods (Bailiss and Offei, 1990), serological methods (e.g., ELISA and TBIA) (Bailiss and Offei, 1990; Makkouk and Kumari, 1996) using polyclonal (Makkouk *et al.*, 1987) and monoclonal (Hajimorad *et al.*, 1990) antibodies, or molecular methods (RT-PCR) (Bariana *et al.*, 1994). For routine testing, serological methods are usually sufficient. The virus is easy to purify (Hull, 1969; Makkouk *et al.*, 1987), and it is a reasonably good antigen with strains of differing pathogenicity, and from different geographic origins, being closely serologically related (Hull, 1969). For strain differentiation, biological methods involving host range and symptom expression in indicator species are most widely used. Monoclonal antibodies can

TABLE II The main features of viruses reported to naturally infect pea, bean, and faba bean in Mediterranean countries

Virus	Crop infected[a]	Manner of transmission[b]	Virus particle: shape[c] and Size (nm) diameter/length	Occurrence in Mediterranean countries[d]	References
Alfalfa mosaic virus (AMV, genus *Alfamovirus*, family *Bromoviridae*)	B, FB, P	Sa, Se[e], Aphids NP[f]	Ba, 30–57	It (B, P), E (FB), Li (FB), Mo (FB), Le (FB), Sy (FB)	Jaspars and Bos (1980), Lisa (2000), Kumari and Makkouk (2007), Faccioli *et al*. (1990)
Bean common mosaic virus (BCMV) and *Bean common mosaic necrosis virus* (BCMNV) (genus *Potyvirus*, family *Potyviridae*)	B	Sa, Se, Aphids NP	Fi, 750	Le (B), E (B), It (B), Mo (B), Tr (B), Cr (B), Sl (B), Sp (B)	Morales and Bos (1988), Azzam and Makkouk (1985), Lockhart and Fischer (1974), Kheder (2002), Ravnikar *et al*. (1996), Yilmaz *et al*. (2002), Halupecki *et al*. (2003), Lisa (2000), Saiz *et al*. (1995)
Bean leafroll virus (BLRV, genus *Luteovirus*, family *Luteoviridae*)	FB, P	Aphids P[f]	Is, 27	Ag (Cp), E (FB), Mo (FB), Tn (FB), Le (FB), Sy (FB), Sp (FB), Tr (Cp), It (P)	Ashby (1984), Fortass and Bos (1991), Makkouk *et al*. (1994, 1998), Najar *et al*. (2000), Ortiz *et al*. (2005), Kumari and Makkouk (2007), Larsen and Webster (1999)
Bean yellow mosaic virus (BYMV, genus *Potyvirus*, family *Potyviridae*)	B, FB, P	Sa, Se, Aphids NP	Fi, 750	Ag (Cp), Le (FB, B), E (B, FB, P), G (FB), Is (FB), It (B, FB, P), Cr (B), Sp (B, FB), Li (B, FB, P), Mo (FB), Sy (FB, P), Tn (FB), Tr (B, FB)	Acikgoz and Citir (1986), Bos (1970), Gamal-Eldin *et al*. (1982), Fortass and Bos (1991), Habib *et al*. (1981), Halupecki *et al*. (2003), Kurçman (1977), Lisa (2000), Makkouk *et al*. (1982, 1988b, 1994), Mouhanna *et al*. (1994), Najar *et al*. (2000),

						Nienhaus and Saad (1967), Nitzany and Cohen (1964), Rana and Kyriakopoulou (1981), Saiz *et al.* (1993), Shagrun (1973), Vovlas and Russo (1978), Younis *et al.* (1992), Kumari and Makkouk (2007), Azzam and Makkouk (1985), Zidan *et al.* (2002), Eldin *et al.* (1981), Faccioli *et al.* (1990)
Beet western yellows virus (BWYV, genus *Polerovirus*, family *Luteoviridae*)	FB	Aphids P	Is, 26		Ag (Cp), E (FB), Mo (Cp), Tn (FB), Le (Cp), Sy (FB, P), Tr (Cp), Sp (FB)	Duffus (1972), Kumari and Makkouk (2007), Fresno *et al.* (1997), Fegla *et al.* (2009), Haj Kassem *et al.* (2001)
Broad bean mottle virus (BBMV, genus *Bromovirus*, family *Bromoviridae*)	FB, B	Sa, Se, Beetles	Is, 26		Ag (FB), E (FB), Mo (FB, B), Sy (FB, P), Tn (FB)	Fortass and Diallo (1993), Fortass and Bos (1991), Gibbs (1972), Makkouk *et al.* (1988b), Najar *et al.* (2000), Zagh and Ferault (1980), Hassan *et al.* (1999)
Broad bean stain virus (BBSV, genus *Comovirus*, family *Comoviridae*)	FB	Sa, Se, Beetles	Is, 25		E (FB), Le (FB), Mo (FB), Tn (FB), Sy (FB, P), Tr (Le)	Gibbs and Smith (1979), Kumari and Makkouk (2007), Haj Kassem *et al.* (2001)
Broad bean true mosaic virus (BBTMV, genus *Comovirus*, family *Comoviridae*)	FB	Sa, Se, Beetles	Is, 25		E (FB), Mo (FB), Tn (FB), Le (FB), Sy (FB)	Gibbs and Paul (1970), Kumari and Makkouk (2007)
Broad bean wilt virus (BBWV, genus *Fabavirus* family *Comoviridae*)	B, FB	Aphids NP	Is, 25		It (B, FB), E (FB), Mo (FB), Tn (FB), Le (FB), Sy (FB)	Lisa (2000), Kumari and Makkouk (2007), Rosciglione and Cannizzaro (1977), Makkouk *et al.* (1990)

(*continued*)

TABLE II *(continued)*

Virus	Crop infected[a]	Manner of transmission[b]	Virus particle: shape[c] and Size (nm) diameter/length	Occurrence in Mediterranean countries[d]	References
Chickpea chlorotic dwarf virus (CpCDV, genus *Mastrevirus*, family *Geminiviridae*)	FB	Leafhoppers	Ge, 15×25	E (FB), Sy (Cp), Tr (CP)	Horn *et al.* (1993), Kumari and Makkouk (2007)
Clover yellow vein virus (ClYVV, genus *Potyvirus*, family *Potyviridae*)	B	Sa, Se, Aphids NP	Fi, 750	It (B), F (P), Sp (FB, B)	Lisa (2000), Leclercq-Le Quillec and Maury (1998), Ortiz *et al.* (2009)
Cucumber mosaic virus (CMV, genus *Cucumovirus*, family *Bromoviridae*)	B, FB, P	Sa, Se, Aphids NP	Is, 28	It (B, P), E (B, FB), F (B), G (B), Li (FB), Sl (B), Tn (FB), Sy (FB, P), Mo (Cp), Le (B), Sp (FB, B)	Francki *et al.* (1979), Ravnikar *et al.* (1996), Marchoux *et al.* (1997), Lisa (2000), Varveri and Boutsika (1999), Mazyad *et al.* (1974), Kumari and Makkouk (2007), Azzam and Makkouk (1985), Fresno *et al.* (1997), Bos and Maat (1974), Faccioli *et al.* (1990), Haj Kassem *et al.* (2001)
Faba bean necrotic yellows virus (FBNYV, genus *Nanovirus*, family *Nanoviridae*)	FB	Aphids P	Is, 18	Ag (Cp), E (FB), Le (FB), Li (FB, P), Mo (FB), Sp (FB), Sy (FB, B, P), Tn (FB), Tr (FB)	Babin *et al.* (2000), El-Amri (1999), Fadel et *al.* (2005), Katul *et al.* (1993), Makkouk *et al.* (1994), Najar *et al.* (2000), Franz *et al.* (1995), Kumari and Makkouk (2007), Zidan *et al.* (2002), Haj Kassem *et al.* (2001)

Pea early-browning virus (PEBV, genus *Tobravirus*)	FB, P	Sa, Se, Nematodes	Short rod 105–215	Ag (FB), Li (FB), Mo (FB), It (FB)	Harrison (1973), Bos *et al*. (1993), Mahir *et al*. (1992), Lockhart and Fischer (1976), Russo *et al*. (1984)
Pea enation mosaic virus-1 (PEMV-1, genus *Enamovirus*, family *Luteoviridae*) and *Pea enation mosaic virus-2* (PEMV-2, genus *Umbravirus*)	FB, P	Sa, Se, Aphids P	Is, 28	E (FB), Mo (FB), Tn (FB), Sy (FB, P), F (P), It (FB, P), Sp (P, FB)	Peters (1982), Kumari and Makkouk (2007), Leclercq-Le Quillec and Maury (1998), Rosciglione and Cannizzaro (1975), Tornos *et al*. (2008), Haj Kassem *et al*. (2001)
Peanut stunt virus (PSV, genus *Cucumovirus*, family *Bromoviridae*)	B	Sa, Se, Aphids NP	Is, 28	Mo (B), It (B), F (B)	Fischer and Lockhart (1978), Lisa (2000), Douine and Devergne (1978)
Pea seed-borne mosaic virus (PSbMV, genus *Potyvirus*, family *Potyviridae*)	FB, P	Sa, Se, Aphids NP	Fi, 770	Ag (Cp), E (FB, P), Le (FB), Li (FB, P), Mo (FB), Sy (FB, P), Tn (FB), Tr (P), F (P)	Hampton and Mink (1975), Makkouk *et al*. (1993b), Kumari and Makkouk (2007), Leclercq-Le Quillec and Maury (1998), Haj Kassem *et al*. (2001), Makkouk *et al*. (1993b), Zidan *et al*. (1997), Fidan and Yorganci (1989), Abdelmaksoud *et al*. (2000)
Southern bean mosaic virus (SBMV, genus *Sobemovirus*)	B	Sa, Se, Beetles	Is, 30	Mo (B)	Tremaine (1983), Segundo *et al*. (2004)
Soybean dwarf virus (SbDV, genus *Luetovirus*, family *Luteoviridae*)	FB	Aphids P	Is, 28	Tn (FB), Sy (Cp, Le)	Kumari and Makkouk (2007)

(*continued*)

TABLE II *(continued)*

Virus	Crop infected[a]	Manner of transmission[b]	Virus particle: shape[c] and Size (nm) diameter/length	Occurrence in Mediterranean countries[d]	References
Tomato spotted wilt virus (TSWV, genus *Tospovirus*, family *Bunyaviridae*)	FB, B	Sa, Se, Thrips	Is, 70–90	It (B), Sp (FB), F (FB)	Ie (1970), Lisa (2000), Fresno *et al.* (1997), Marchoux *et al.* (1991)
Tomato yellow leaf curl virus (TYLCV, genus *Begomovirus*, family *Geminiviridae*)	B	Whiteflies	Ge, 25 × 13	Tn (B), G (B), It (B), Sp (B)	Chouchane *et al.* (2007), Papayiannis *et al.* (2007), Davino *et al.* (2007), Navas-Castillo *et al.* (1999)

[a] B, beans; FB, faba bean; P, pea; Le, lentil; Cp, chickpea.
[b] Sa, sap; Se, seeds.
[c] Ba, bacilliform; Fi, filamentous; Ge, gemini; Is, isometric.
[d] Ag, Algeria; Ab, Albania; BH, Bosnia-Herzegovina; Cr, Croatia; Cy, Cyprus; E, Egypt; F, France; G, Greece; Is, Israel; It, Italy; Le, Lebanon; Li, Libya; Ma, Malta; Mn, Montenegro; Mo, Morocco; Sl, Slovenia; Sp, Spain; Sy, Syria; Tn, Tunisia; Tr, Turkey.
[e] Seed transmission of the virus.
[f] NP, nonpersistent transmission; P, persistent transmission; SP, semipersistent transmission.

also distinguish between strains (Hajimorad *et al.*, 1990), and there should be potential for RT-PCR, using strain-specific primers (Table III).

Control. AMV does not cause serious losses on beans, and no work on control has been reported. However, and since AMV is both seed and aphid transmitted, a range of general control measures can be applied: (i) use of virus-free seed, (ii) control of virus spread from overwintering hosts by spatial separation, and (iii) aphid vector control. Insecticides have limited use as the virus is nonpersistently transmitted by aphids. No source of resistance to the virus in bean has been reported.

4. *Cucumber mosaic virus* (genus *Cucumovirus*, family *Bromoviridae*)

Distribution in the Mediterranean countries. CMV is distributed worldwide. In recent years, it has increasingly been reported as the causal agent in disease epidemics of major crops including *P. vulgaris* throughout the world, especially in the tropics (Palukaitis *et al.*, 1992) and southern Europe (Gallitelli, 2000). CMV has been reported on common bean in many Mediterranean countries, such as France (Marchoux *et al.*, 1997), Greece (Varveri and Boutsika, 1999), Italy (Lisa, 2000), Egypt (Mazyad *et al.*, 1974), Lebanon (Azzam and Makkouk, 1985), and Slovenia (Ravnikar *et al.*, 1996).

Symptoms and strains. CMV has a wide host range and infects more than 800 species of both monocotyledonous and dicotyledonous plants from over 85 families including *P. vulgaris* (Palukaitis *et al.*, 1992). Several CMV strains can induce different symptoms in common bean, ranging from mild mosaic to severe plant malformation. Yield losses vary from 5% to 75% depending on the cultivar, age of infection, virus trains, and environmental conditions (Bird *et al.*, 1974).

Transmission. Over 80 species of aphids can transmit CMV in a nonpersistent manner (Gallitelli, 2000; Palukaitis *et al.*, 1992). The most common aphid vectors are *M. persicae* and *A. gossypii*. The virus is stylet borne and all instars of aphids can act as efficient vectors. The transmission specificity of CMV is determined by its coat protein (Chen and Francki, 1990). Two domains of the coat protein at amino acid positions 129 and 168 are involved in aphid transmissibility (Palukaitis *et al.*, 1992) (Table II).

Other known routes of virus transmission are by mechanical inoculation and through seed. Over 10% seed transmission has been recorded in *P. vulgaris* a property which is coded by RNA 1 (Hampton and Francki, 1992).

Particle morphology and properties. Particles of CMV are isometric, approximately 29 nm in diameter, and each particle contains 180 protein subunits, but their arrangements are not easily visible by electron microscopy. The virus coat protein has a molecular weight of 25 kDa.

Genome structure and organization. The viral genome consists of three linear, positive-sense, single-stranded RNA molecules with estimated sizes of RNA1 = 3389 nt, RNA2 = 3035 nt, and RNA3 = 2197 nt. Genome organization is similar to that of AMV.

Propagation hosts and purification. CMV can be purified readily in significant quantities by the purification method described by Lot *et al.* (1972), as modified by Peden and Symons (1973). A modified purification protocol where the infected leaves are treated with chloroform vapor prior to differential centrifugation has been described (Krstic *et al.*, 1997). A purified preparation of the virus has an A_{260}:A_{280} ratio of 1.7.

Detection methods available. When infected with CMV, the diagnostic hosts *C. amaranticolor* and *C. quinoa* produce chlorotic lesions, *Cucumis sativus* L. produces systemic mosaic, and *Vigna unguiculata* (L.) Walp produces necrotic local lesions. Host range can be used to differentiate CMV from other viruses but is not useful in determining the subgroup of any CMV isolate. A rapid immunofilter paper assay was described, which can detect CMV in leaf extracts within 30 s (Ohki and Kameya-Iwaki, 1996) and which is now commercially available. CMV can be detected by ELISA (Elliott *et al.*, 1996) using polyclonal and monoclonal (Wahyuni *et al.*, 1992) antibodies and by RT-PCR (Bariana *et al.*, 1994; Uga, 2005) (Table III).

Control. Control measures for CMV are mainly preventive. Conventional methods of virus control are difficult to apply due to the wide host range of CMV which infects many weeds that can act as virus reservoirs and infect crops in adjacent fields. Developing new crop varieties resistant to CMV, either by conventional breeding methods or by gene technology, is gaining momentum. Since CMV is transmitted by over 80 aphid species, the diverse behavior of various aphid vectors can greatly reduce the impact of insecticide sprays, which are more effective when the insects are a direct "pest" rather than a "vector." A more sustainable approach will be the prevention of aphids reaching the crops. This can be achieved by planting barrier crops, applying sticky traps, or covering the ground with an aphid deterrent material like aluminum foil boards. Spraying the crops with mineral oil is sometimes effective.

5. Other viruses

There are other viruses of less economic importance reported to infect common bean in the Mediterranean region such as *Broad bean wilt virus-1*, *Clover yellow vein virus*, *Peanut stunt virus*, *Southern bean mosaic virus*, BBMV, and *Tomato yellow leaf curl virus* (Chouchane *et al.*, 2007; Fischer and Lockhart, 1978; Fortass and Diallo, 1993; Lisa, 2000; Papayiannis *et al.*, 2007; Segundo *et al.*, 2004).

C. Virus diseases of faba bean

1. *Faba bean necrotic yellows virus* (genus *Nanovirus*, family *Nanoviridae*)

Distribution in the Mediterranean countries. FBNYV was first isolated from faba bean near Lattakia, Syria (Katul *et al.*, 1993). The virus can be very damaging in those years in which an epidemic occurs, as happened in

Middle Egypt during 1992, 1997, and 1998 and in Tunisia in 2001. In Middle Egypt, losses due to FBNYV infection reached 80–90% during 1992 (Makkouk *et al.*, 1994). In Mediterranean countries, FBNYV is considered to be the virus most damaging to faba bean. The virus is reported to occur on faba bean in Egypt, Lebanon, Libya, Morocco, Spain, Syria, Tunisia, and Turkey (Babin *et al.*, 2000; El-Amri, 1999; Fadel e*t al.*, 2005; Katul *et al*, 1993; Makkouk *et al.*, 1994; Najar *et al.*, 2000).

Symptoms and strains. One-week-old faba bean plants show retarded growth as early as 5 days after inoculation. At 2 weeks after infection, the plants are usually severely stunted. The leaves become thick and brittle and show interveinal chlorotic blotches, which begin at the leaf margins. Young leaves remain very small and are cupped upward, whereas older leaves are rolled downward. New shoots, leaves, and flowers develop poorly. About 3–4 weeks after infection, interveinal chlorosis usually turns necrotic and infected plants die within 5–7 weeks after infection. Similar symptoms were also observed in other susceptible host plants. Some infected *Trifolium* and *Medicago* species, however, develop leaf reddening instead of, or in addition to, chlorosis (Katul *et al.*, 1993). FBNYV is taxonomically very closely related to *Milk vetch dwarf virus* (MDV), and less so to *Subterranean clover stunt virus*. All three viruses cause similar symptoms in legume species (Franz *et al.*, 1996; Sano *et al.*, 1998; Timchenko *et al.*, 2000; Vetten *et al.*, 2005).

Transmission. Three aphid species, *A. pisum*, *A. craccivora*, and *A. fabae*, are reported as vectors of FBNYV and transmit it in a circulative non-propagative manner (Franz *et al.*, 1998; Makkouk *et al.*, 1998; Ortiz *et al.*, 2006). The efficiency of transmission by aphids was found to be high for the first two species and very low for *A. fabae* (Franz *et al.*, 1995, 1998; Katul *et al.*, 1993). Aphids require long acquisition and inoculation feeding periods to become efficient vectors, and FBNYV persists in the aphids for almost their entire life (Franz *et al.*, 1998). As with all phloem-limited viruses, FBNYV is not known to be transmitted by seed or mechanical means. Since all its hosts are propagated by true seed, the only method of natural spread of FBNYV is by aphid vectors (Table II).

Particle morphology and properties. The average diameter of FBNYV isometric particles is 18 nm. They have an angular outline, but without obvious substructure (Katul *et al.*, 1993).

Genome structure and organization. The genome is circular ssDNA. Ten DNA components have been isolated from virus preparations; individual molecules, each of which appears to be encapsidated in a separate particle, are approximately 1 kb in size. The 10 circular DNA components (C1–C10) found associated with the FBNYV genome potentially encode four distinct replication-associated proteins (Rep) and six non-Rep proteins (Katul *et al.*, 1995a, 1997, 1998). MDV reacts strongly with FBNYV antiserum (Katul *et al.*, 1993), and both MDV and FBNYV share amino acid sequence identities of 83% and 97% in their capsid (CP) and master replication initiator (M-Rep) proteins (Sano *et al.*, 1998; Timchenko *et al.*, 2000).

Propagation hosts and purification. FBNYV has a relatively narrow host range, mostly restricted to leguminous species. Faba bean is the main natural host, but other legume crops such as chickpea, lentil, dry bean, pea, and cowpea are also natural hosts of FBNYV (Franz *et al.*, 1995; Horn *et al.*, 1995; Katul *et al.*, 1993; Makkouk *et al.*, 1992). The virus also occurs naturally in the wild legume species *Lathyrus* sp., *Medicago* sp., *Melilotus* sp., *Tetragonolobus* sp., *Trifolium* sp., *Vicia* sp., as well as in perennial species of the genus *Onobrychis* and *M. sativa*. The virus also occurs naturally in some nonleguminous species including *Amaranthus blitoides* S. Watson, *A. retroflexus* L., and *A. gracilis* L. [*A. viridis*] (Franz *et al.*, 1997; Mouhanna *et al.*, 1994).

The virus has been successfully purified from faba bean by Katul *et al.* (1993). A purified preparation of the virus has an A_{260}:A_{280} ratio of 1.4.

Detection methods available. The virus can be diagnosed in an infected tissue using ELISA or tissue-blot immunoassay (TBIA) using either polyclonal (Katul *et al.*, 1993; Kumari *et al.*, 2001) or monoclonal antibodies (Franz *et al.*, 1996) and by dot-blot hybridization (Franz *et al.*, 1996; Katul *et al.*, 1993, 1995b). These tests are essential to differentiate FBNYV from viruses in the family *Luteoviridae*, which induce similar symptoms in legume species such as BLRV, BWYV, and SbDV, or viruses of the family *Geminiviridae*, such as *Chickpea chlorotic dwarf virus* (CpCDV). Primers for FBNYV detection by PCR are available (Kumari *et al.*, 2010; Shamloul *et al.*, 1999) (Table III).

Control. The most effective measure to control FBNYV in faba bean crop is an integrated virus management approach consisting of (i) planting at an appropriate time, to avoid peak numbers of viruliferous aphids coming from susceptible summer legume crops or wild hosts, (ii) well-timed application (either once or twice) of an aphicide, (iii) seed treatment with Imidacloprid before planting (Makkouk and Kumari, 2001), (iv) roguing of infected plants early in the season, and (v) use of an appropriate seeding rate. Such measures are essential to reach an economic yield in regions where FBNYV epidemics are likely to occur. In addition, genotypes resistant to this virus have been reported in lentil only (Makkouk *et al.*, 2001). Several laboratories attempted to produce transgenic faba beans resistant to FBNYV infection, but such product is not yet available.

2. *Bean leafroll virus* (genus *Luteovirus*, family *Luteoviridae*)

Distribution in the Mediterranean countries. BLRV is the most important virus known to infect faba bean. It was first isolated from faba bean by Quantz and Volk (1954) in Germany. BLRV and the other luteoviruses can have a marked effect on yield. Yield losses of 50–90% have been reported in Europe (Schmutterer and Thottappilly, 1972). The virus is reported to

TABLE III Primers available for PCR detection of pea, bean, and faba bean viruses

Virus	Genome	Sequence of primer pairs 5′–3′	Amplified fragment (bp)	Reference
AMV	ssRNA	CGTCAGCTTTCGTCGAACA GCCGTCGCGCATGGTAAT	288	Bariana *et al.* (1994)
BCMV		ACCACGCTGCAGCTAAAGAGAACA AATCTAGATGATATCATACTCTCTA	1456	Xu and Hampton (1996)
BLRV	ssRNA	TCCAGCAATCTTGGCATCTC GAAGATCAAGCCAGGTTCA	391	Ortiz *et al.* (2005)
		AAAGAGGTTCTACAGGCCAC GATCAAGTTCCTCGCAGAAC	440	Kumari *et al.* (2006)
BYMV	ssRNA	GGTTTGGCYAGRTATGCTTTTG GAGAATTTAAAGACGGATA	240	Bariana *et al.* (1994)
		CAGTTTATTATGCAGCGG GTTATCATCAATCTTCCTGC	644	Uga (2005)
		CAAGGTGAGTGGACAATGATGG GAGAGAATGATACACATACTGAA	525	Ortiz *et al.* (2006)
BWYV	ssRNA	ATGAATACGGTCGTGGGTAC GATAGTTGAGGAAAGGGAGTTG	429	Kumari *et al.* (2006)
CMV	ssRNA	CGAGTCATGGACAAATCTGAATCAA AGYCCTTCCGAAGAAAYCTAGGAGA	879–881	Uga (2005)
		GCCGTAAGCTGGATGGACAA TATGATAAGAAGCTTGTTTCGCG	482–501	Wylie *et al.* (1993)
		GGCGAATTCGAGCTCGCCGTAAGCTGGATGGAC CTCGAATTCGGATCCGCTTCTCCGCGAG	920	Abdullahi *et al.* (2001)
		TATGATAAGAAGCTTGTTTCGCGCA TTTTAGCCGTAAGCTGGATGGACAACCC	500	Bariana *et al.* (1994)
		GTTTATTTACAAGAGCGTACGG GGTTCGAAAGTATAACCGGG	650	Sclavounos *et al.* (2006)

(continued)

TABLE III *(continued)*

Virus	Genome	Sequence of primer pairs 5′–3′	Amplified fragment (bp)	Reference
		AGTGACTTCAGGCAGT GCTTGTTTCGCGCATTCA	436	Davino *et al*. (2005)
FBNYV	ssDNA	TACAGCTGTCTTTGCTTCCT CGCGGAGTAATTAAATCAAAT	666	Kumari *et al*. (2010)
		ACATCGAAGAGCAGTATCTGG ACGTTGTCGTTTTCACCTTGG	487	Shamloul *et al*. (1999)
		TTTCCCGCTTCGCTAAGTTTAA ACACCCTCCTTGGAACTGGTATAA	931	Shamloul *et al*. (1999)
		CATTTCGGATGAACATCTGGG ATGAACTATCAAGCGATGGAG	1002	Shamloul *et al*. (1999)
PEMV	ssRNA	GAGGGTGCCACCACGACTAC TGAAAATTAGATAAGGAAAACCCAAG	114	Skaf *et al*. (2000)
PSbMV	ssRNA	GATTTCTTCGTTGTTTGTT CTTGAGTGCTGGCGTGGTT	494	Phan *et al*. (1997)
		GCTCTAGACTCGAGGGGAARTCRAAAGCTAAAAC GTCCTAGAGCTTGCGCAATWGGATTGTA	654	Phan *et al*. (1997)
		TACATCTAGATTACATGGCTCTCATTCCGAGAAG CAAACGCGTGACGAAACCAAGGATGATGAAAG	888	Roberts *et al*. (2003)
		TACATCTAGATTACATGGCTCTCATTCCGAGAAG GGTTGCTCGAGGGTGATGAGACCAAAGATGAAAG	958	Roberts *et al*. (2003)
SbDV	ssRNA	AGGCCAAGGCGGCTAAGAG AAGTTGCCTGGCTGCAGGAG	440	Kumari *et al*. (2006)

occur on faba bean in Egypt, Morocco, Tunisia, Lebanon, Syria, and Spain (Fortass and Bos, 1991; Makkouk *et al.*, 1994, 1998; Najar *et al.*, 2000; Ortiz *et al.*, 2005).

Symptoms and strains. Infection with BLRV is restricted to legume species including faba bean, lentil, pea, and chickpea in many parts of the world (Bos *et al.*, 1988; Makkouk *et al.*, 2003). The main symptoms produced by the virus are interveinal chlorosis, yellowing, stunting, leaf rolling, reddening and thickening of the leaves, suppression of flowering, and pod setting. In faba bean, old leaves become leathery. Because the symptoms produced by BLRV infection can vary slightly among the species infected, the common disease names vary based on the crop species involved. Hence, the disease names "pea leafroll," "pea yellow top," "chickpea stunt," and "lentil yellows" reflect the main symptoms produced in these species in response to infection with BLRV.

Transmission. BLRV is transmitted only by aphids in a persistent, non-propagative manner. *A. pisum*, *A. craccivora*, *A. fabae*, and *M. persicae* are the most common aphid species reported to transmit BLRV (Cockbain, 1983; Kaiser, 1973). The virus is retained when the vector molts, does not multiply in the vector, is not transmitted by mechanical inoculation, is transmitted by grafting, and is not transmitted by seed or pollen (Table II).

Particle morphology and properties. The virus particles are isometric, ca. 27 nm in diameter with an ssRNA genome of 6 kb.

Genome structure and organization. So far, only one complete sequence of BLRV (accession number AF441393) is available in the GenBank database. The BLRV genome consists of five ORFs typical of other members of the genus *Luteovirus* (Domier *et al.*, 2002). ORF-1 encodes a protein of 42 kDa and is overlapped by ORF-2 by 15 nt. ORF-2 encodes a protein of 62 kDa. ORF-1 and ORF-2 have been predicted to encode a tombusvirus-like RdRp, where ORF-2 is predicted to be expressed as a translational fusion product of ORF-1 through a frameshift, which is promoted by a "slippery" sequence (GGUUUUU) at the junction of ORF-1 and ORF-2. ORF-3 encodes a 22-kDa coat protein and is separated from ORF-2 by an intergenic region. ORF-4 and ORF-5 are predicted to encode proteins of 16 and 59 kDa, respectively (Domier *et al.*, 2002).

Propagation hosts and purification. BLRV can be purified following the procedure described by D'Arcy *et al.* (1989a). A purified preparation of the virus has an A_{260}:A_{280} ratio of 1.83.

Detection methods available. BLRV can be diagnosed in an infected tissue by ELISA or TBIA, using either polyclonal (Ashby and Huttinga, 1979; D'Arcy *et al.*, 1989a) or monoclonal antibodies, and by RT-PCR (Kumari *et al.*, 2006; Ortiz *et al.*, 2005) (Table III).

Control. The spread of BLRV within crops can sometimes be reduced by aphicides (Schmutterer and Thottappilly, 1972), seed treatment with Imidacloprid before planting (Makkouk and Kumari, 2001), and manipulation of sowing date (Johnstone and Rapley, 1979). Genotypes resistant to BLRV have been reported in faba bean (Makkouk *et al.*, 2002) and lentil (Makkouk *et al.*, 2001).

3. *Bean yellow mosaic virus* (genus *Potyvirus*, family *Potyviridae*)
Distribution in the Mediterranean countries. BYMV was reported in faba bean by Boning (1927) in Germany. The number of fields infected with BYMV can vary greatly among locations. A high incidence, up to 100%, has been noted in some regions of Egypt, Iraq, and Sudan. Two of these countries (Egypt and Sudan) are known for their relatively warm winters, which favor increased aphid populations and movement. BYMV is reported to occur on faba bean in Egypt, Greece, Israel, Italy, Lebanon, Libya, Morocco, Spain, Syria, Tunisia, and Turkey (Fortass and Bos, 1991; Gamal-Eldin *et al.*, 1982; Kurçman, 1977; Makkouk *et al.*, 1982, 1988b; Mouhanna *et al.*, 1994; Najar *et al.*, 2000; Nienhaus and Saad, 1967; Nitzany and Cohen, 1964; Rana and Kyriakopoulou, 1981; Saiz *et al.*, 1993; Shagrun, 1973; Vovlas and Russo, 1978; Younis *et al.*, 1992).

Symptoms and strains. The symptoms that result from BYMV infection on faba bean are variable and are greatly affected by the strain of the virus and the genotype of the plant involved. Symptoms produced by the virus include mosaic, mottling, green vein banding, and chlorosis. Severe strains of the virus can produce stem and tip necrosis and early death in some faba bean genotypes. Kaiser (1973) found that pods of faba beans, infected with BYMV, occasionally developed necrotic ring spotting and discolored seeds.

Transmission. Sasaya *et al.* (1993) detected BYMV in seeds of 15 of the 17 faba bean cultivars tested, and in 45 of 56 faba bean lines from 14 countries, with a range of seed transmission rates of 0.1–2.4% (Kaiser, 1973). The virus was also detected in seed of 10 of 18 commercial faba bean samples tested during 1989–1992, the range of seed infection being 0–9.2%, as determined by grow out tests of the harvested seeds (Sasaya *et al.*, 1993) (Table II).

Detection methods available. ELISA, immunosorbent EM TBIA, and dot immunobinding assay were used to detect BYMV in faba bean seeds (Makkouk *et al.*, 1993a; Raizada *et al.*, 1991). RT-PCR has also been advocated for seed testing (Bariana *et al.*, 1994) (Table III).

Control. Measures to control BYMV include manipulation of sowing date (Abu Salih *et al.*, 1973), spraying with mineral oils (Proeseler *et al.*, 1976), soil mulching with reflective polyethylene sheets (Tachibana, 1981), and ensuring that crops are not grown in the vicinity of known overwintering disease sources. In regions where seed-borne infection is a problem, it is clearly advisable to use virus-free seeds as much as possible. In addition, BYMV-resistant faba bean genotypes were selected in Canada (Ghad and Bernier, 1984) and Syria (Makkouk and Kumari, 1995b).

For more information on BYMV, please check the Section IV.A.1.

4. ***Broad bean mottle virus* (genus *Bromovirus*, family *Bromoviridae*)**

Distribution in the Mediterranean countries. BBMV was first described in faba bean by Bawden *et al.* (1951). A faba bean grain yield loss of 35–55%, depending on time of BBMV infection, has been reported (Makkouk *et al.*, 1988a,b). The virus has been found to reach high incidence levels in faba bean fields in Morocco, Sudan, and Tunisia (Fortass and Bos, 1991; Makkouk *et al.*, 1988a,b; Najar *et al.*, 2000). In addition, BBMV is known to occur in Syria, Egypt, and Algeria (Najar *et al.*, 2000; Zagh and Ferault, 1980). A detailed survey in Morocco revealed its occurrence in 56% of the fields inspected, with a maximum incidence of 33% recorded in one field (Fortass and Bos, 1991). The virus is beetle and seed transmitted and may cause serious losses.

Symptoms and strains. The symptoms produced in faba bean are, mainly, mottling, marbling, or diffuse mosaic, which is often associated with leaf malformation and sometimes with plant stunting. Some faba bean genotypes may show necrosis in response to BBMV infection. BBMV can affect seed quality by causing necrosis and shriveling of the seed.

Transmission. Virus is transmitted by mechanical inoculation and grafting. Seed transmission is suspected and has been reported but at low rate (0.1–1.4%) especially when the virus occurs in mixed infection with BYMV (Makkouk *et al.*, 1988b). Virus is transmitted by arthropods, by insects of the order Coleoptera: *Acalymma trivittata* Mannerheim, *Apion arrogans* Wencher, *A. radiolus* Kirby, *A. vorax* Hbst, *Colaspis flavida* Say, *Diabrotica undecimpunctata* Mannerheim, *Hypera variabilis* Herbst, *Pachytychius strumarius* Gyll, *Sitina lineatus* var. *viridifrons* Motsch, *Sitona crinite* Herbst, *S. limosa* Rossi, *S. lineatus* L., *Spodoptera exigua* Hübner, and *Smicronyx cyaneus* Gyll (Ahmed and Eisa, 1999; Borges and Louro, 1974; Cockbain, 1983; Fortass and Diallo, 1993; Makkouk and Kumari, 1989, 1995a; Walters and Surin, 1973) (Table II).

Particle morphology and properties. The virus has isometric particles ca. 26 nm in diameter.

Genome structure and organization. The virus particles have a multiple ssRNA genome which is ca. 8.25 kb in size. The genomic RNAs 1 and 2 each encode a single large ORF (ORF-1 and ORF-2) and RNA 3 encodes a 5′ protein, the movement protein, and the coat protein (CP) which is translated from a subgenomic RNA.

Propagation hosts and purification. For a long time, BBMV attracted academic interest because of its structural characteristics; it reaches high concentration in infected cells and is an easy virus to purify. The virus was successfully purified as described by Lane (1974). A purified preparation of the virus has an A_{260}:A_{280} ratio of 5.4.

Detection methods available. BBMV can be diagnosed in infected tissue by using ELISA or TBIA, using polyclonal antisera (Makkouk *et al.*, 1987, 1988a).

Control. The use of healthy seed and beetle control reduces the spread of BBMV. No faba bean cultivars resistant to BBMV infection are available.

5. Other viruses

In addition to the viruses mentioned above, few other viruses were reported to cause damage to faba bean in specific countries and limited areas, such as *Chickpea chlorotic stunt virus* (genus *Polerovirus*, family *Luteoviridae*), *Beet western yellows virus* (BWYV, genus *Polerovirus*, family *Luteoviridae*), *Broad bean stain virus* (genus *Comovirus*, family *Comoviridae*), *Pea seed-borne mosaic virus* (genus *Potyvirus*, family *Potyviridae*), and *Pea early-browning virus* (genus *Tobravirus*).

V. CONCLUDING REMARKS

When some viruses covered in this chapter are not reported to infect pea, bean, or faba bean in some Mediterranean countries, it does not mean that in fact they do not naturally infect these crops in these countries, but rather not enough effort was done to detect them. Since diagnostic tools to handle large number of samples are now available, periodic surveys to detect potential viruses, including new ones, are essential to monitor their presence in all agricultural ecologies of the Mediterranean region.

Continuous research on viruses which attack legume vegetables in the Mediterranean region is needed, as it is evident that in this region scientists and farmers alike find themselves confronted with new virus problems every year. Virus diseases of leguminous vegetables, similar to other vegetables, should be looked at as a dynamic system in a state of flux, where equilibrium has been disrupted by agricultural and other human activities.

Host plant resistance and cultural practices seem to be the most promising control approaches welcomed by growers, and significant effort was made to identify sources of resistance and understanding resistance mechanisms to many major viruses. However, more effort is still needed to develop cultivars combining such resistance with other desirable traits.

More than 50% of the viruses that infect pea, bean, and faba bean are seed borne and often these are also disseminated by flying vectors. This means in regions where vector activity is high during the growing season, virus epidemics are likely to occur even when rate of virus-infected seeds is low. Consequently, providing farmers with healthy seeds is essential in reducing virus damage in these crops.

Experience has shown that single control measures are often not effective in reducing virus spread in field situations, and combining a number of components in an integrated disease management package leads to a better control. These are usually applied before, at planting time, or during early growth stage of the crop. However, when virus disease pressure is low, farmers are inclined not to apply these measures to reduce cost. Consequently, the development of prediction models to forecast damaging outbreaks is essential to guide growers to apply control measures only when needed. Such prediction models are not available in many Mediterranean countries.

REFERENCES

Abdelmaksoud, H. M., Gamal Eldin, A. S., Ibrahim, S. M., and Sallam, A. A. (2000). Nucleotide sequence of the coat protein gene of pea seed borne mosaic Potyviridae (Egyptian strain). *In* "Abstract book of Seventh Arab Congress of Plant Protection, Amman, Jordan, 22-26 October, 2000" (A. Katbeh Bader and H. S. Hasan, eds.), p. 349. Arab Society for Plant Protection, Beirut, Lebanon.

Abdullahi, I., Ikotun, T., Winter, S., Thottappilly, G., and Atiri, G. I. (2001). Investigation on seed transmission of cucumber mosaic virus in cowpea. *Afr. Crop Sci. Soc.* **9**:677–684.

Abu Salih, H. S., Ishag, H. M., and Siddig, S. A. (1973). Effects of sowing date on incidence of Sudanese broad bean mosaic virus in, and yield of, *Vicia faba*. *Ann. Appl. Biol.* **74**:371–378.

Acikgoz, S., and Citir, A. (1986). Incidence, epidemiology and identification of viruses on *Phaseolus vulgaris* L. in Erzincan plain in Turkey. *J.Turk. Phytopathol.* **15**(2):61–76.

Ahmed, A. H., and Eisa, E. B. (1999). Transmission of broad bean mottle virus by the larvae of *Spodoptera exigua*. *FABIS Newsl.* **28**:30–31.

Alconero, R., Provvidenti, R., and Gonsalves, D. (1986). Three pea seedborne mosaic virus pathotypes from pea and lentil germ plasm. *Plant Dis.* **70**:783–786.

Alkhalaf, M., Kumari, S. G., Haj Kasem, A., Makkouk, K. M., and Al-Chaabi, S. (2010). *Bean yellow mosaic virus* on cool-season food legumes and weeds: Distribution and its effect on faba bean yield and control in Syria. *Arab J. Plant Prot.* **28**:38–47.

Ashby, J. W. (1984). Bean leaf roll virus. *In CMI/AAB Descriptions of Plant Viruses No. 286*, p. 4.

Ashby, J. W., and Huttinga, H. (1979). Purification and some properties of pea leafroll virus. *Neth. J. Plant Pathol.* **85**:113–123.

Aydin, H., Muehlbauer, F. J., and Kaiser, W. J. (1987). *Pea enation mosaic virus* resistance in lentil (Lens culinaris). *Plant Dis.* **71**:635–638.

Azzam, O. I., and Makkouk, K. M. (1985). A survey of viruses affecting dry been and cowpea in Lebanon. *Arab J. Plant Prot.* **3**:76–80.

Babin, M., Ortiz, V., Castro, S., and Romero, J. (2000). First detection of *Faba bean necrotic yellow virus* in Spain. *Plant Dis.* **84**:707.

Bailiss, K. W., and Offei, S. K. (1990). Alfalfa mosaic virus in lucerne seed during seed maturation and storage, and in seedlings. *Plant Pathol.* **39**:539–547.

Bariana, H. S., Shannon, A. L., Chu, P. W. G., and Waterhouse, P. M. (1994). Detection of five seedborne legume viruses in one sensitive multiplex polymerase chain reaction test. *Phytopathology* **84**:1201–1205.

Bawden, F. C., Chaudhuri, R. P., and Kassanis, B. (1951). Some properties of broad bean mottle virus. *Ann. Appl. Biol.* **38**:774–784.

Bird, J., Sanchez, J., Rodrigues, R. L., Cortes-Monllor, A., and Kaiser, W. J. (1974). A mosaic of beans (*Phaseolus vulgaris* L.) caused by a stain of common cucumber mosaic virus. *J. Agric. Univ. P.R.* **58**:151–161.

Boning, K. (1927). Die Mosaikkrankheit der Ackerbohne (*Vicia faba*). *Forsch. Geb. Pflzkrkh. Immunität Pflanzenr* **4**:43–111.

Borges, M. D., and Louro, O. (1974). A biting insect as a vector of broad bean mottle virus? *Agronomia Lusitania* **36**:215–216.

Bos, L. (1970). Bean yellow mosaic virus. *CMI/AAB Descriptions of Plant Viruses No. 40.*

Bos, L., and Maat, D. Z. (1974). A strain of cucumber mosaic virus, seed-transmitted in beans. *Neth. J. Plant Pathol.* **80:**113–123.

Bos, L., Hampton, R. O., and Makkouk, K. M. (1988). Viruses and virus diseases of pea, lentil, faba bean and chickpea. *In* "World Crops: Cool Season Food Legumes" (R. J. Summerfield, ed.), pp. 591–615. Kluwer Academic Publishers, Dordrecht, The Netherlands.

Bos, L., Mahir, M. A.-M., and Makkouk, K. M. (1993). Some properties of pea early-browning tobravirus from faba bean (*Vicia faba* L.) in Libya. *Phytopathol. Mediterr.* **32:**7–13.

Chen, B., and Francki, R. I. B. (1990). Cucumovirus transmission by the aphid Myzus persicae is determined solely by the viral coat protein. *J. Gen. Virol.* **71:**939–944.

Chomič, A., Pearson, M. N., Clover, G. R. G., Farreyrol, K., Saul, D., Hampton, J. G., and Armstrong, K. F. (2010). A generic RT-PCR assay for the detection of *Luteoviridae. Plant Pathol.* **59:**429–442.

Chouchane, S. G., Gorsane, F., Nakhla, M. K., Maxwell, D. P., Marrakchi, M., and Fakhfakh, H. (2007). First report of tomato yellow leaf curl virus-Israel species infecting tomato, pepper and bean in Tunisia. *J. Phytopathol.* **155:**236–240.

Cockbain, A. J. (1983). Viruses and virus-like disease of *Vicia faba* L. *In* "The Faba Bean (Vicia faba L.)" (P. D. Hebblethwaite, ed.), pp. 421–461. Butterworths, London, UK.

D'Arcy, C. J., Martin, R. R., and Spiegel, S. (1989a). A comparative study of luteovirus purification methods. *Can. J. Plant Pathol.* **11:**251–255.

D'Arcy, C. J., Torrance, L., and Martin, R. R. (1989b). Discrimination among luteoviruses and their strains by monoclonal antibodies and identification of common epitopes. *Phytopathology* **79:**869–873.

Davino, S., Bellardi, M. G., Di Bella, M., Davino, M., and Bertaccini, A. (2005). Characterization of a *Cucumber mosaic virus* isolate infecting *Mandevilla sanderi* (Hemsl.) Woodson. *Phytopathol. Mediterr.* **44:**220–225.

Davino, S., Salamone, S., Iacono, G., Accotto, G. P., and Davino, M. (2007). First report of *Tomato yellow leaf curl virus* on bean in Italy in greenhouse. *Informatore Fitopatologico* **57:**47–49.

De Zoeten, G. A., and Demler, S. A. (1995). Virus Taxonomy, Classification and Nomenclature of viruses. *Sixth Report of the International Committee on Taxonomy of Viruses.*

Demler, S. A., and de Zoeten, G. A. (1991). The nucleotide sequence and luteovirus-like nature of RNA 1 of an aphid non-transmissible strain of pea enation mosaic virus. *J. Gen. Virol.* **72:**1819–1834.

Demler, S. A., de Zoeten, G. A., Adam, G., and Harris, K. F. (1996). Pea enation mosaic enamovirus: Properties and aphid transmission. *In* "The Plant Viruses, Volume 5, Polyhedral Virions and Bipartite RNA Genomes" (B. D. Harrison and A. F. Murant, eds.), pp. 303–344. Plenum Press, New York.

Demler, S. A., Rucker-Feeny, D. G., Skaf, J. S., and de Zoeten, G. A. (1997). Expression and suppression of circulative aphid transmission in *Pea enation mosaic virus. J. Gen. Virol.* **78:**511–523.

Domier, L. L., Mc Coppin, N. K., Larsen, R. C., and D'Arcy, C. J. (2002). Nucleotide sequence shows that Bean leafroll virus has a Luteovirus-like genome organization. *J. Gen. Virol.* **83:**1791–1798.

Douine, L., and Devergne, J. C. (1978). Isolation of peanut stunt virus in France. *Ann. Phytopathol.* **10:**79–92.

Drijfhout, E., Silbernagel, M. J., and Burke, D. W. (1978). Differentiation of strains of bean common mosaic virus. *Neth. J. Plant Pathol.* **84:**13–26.

Duffus, J. E. (1972). Beet western yellows virus. *CMI/AAB Descriptions of Plant Viruses No. 89.*

Edwardson, J. R., and Christie, R. G. (1997). Viruses Infecting Peppers and Other Solanaceous Crops. Vol. 1. Monograph 18-1. University of Florida Agricultural Experiment Station, Florida.

El-Amri, A. (1999). Identification and repartition of faba bean necrotic yellows virus (FBNYV) in Morocco. *Al Awamia* **99:**19–26.

Eldin, A. S. G., El-Kady, M. A., and El-Amrety, A. A. (1981). Pea mosaic virus (PMV) strain of bean yellow mosaic virus isolated from pea. *Egypt. J. Phytopathol.* **13:**23–28.

Elliott, M. S., Zettler, F. W., Zimmerman, M. T., Barnett, O. W., Jr., and LeGrande, M. D. (1996). Problems with interpretation of serological assays in a virus survey of orchid species from Puerto Rico, Ecuador, and Florida. *Plant Dis.* **80:**1160–1164.

Faccioli, G., Stefanelli, D., and Nascetti, D. (1990). Spread of viruses in pea crops. *Informatore Fitopatologico* **40:**57–58.

Fadel, S., Khalil, J., and Shagrun, M. (2005). First record of *Faba bean necrotic yellows virus* and a Luteovirus in faba bean crop (*Vicia faba* L.) in Libya. *Arab J. Plant Prot.* **23:**132.

FAO (Food and Agricultural Organization of the United Nation) (2009). Statistical FAO-STAT, Food and Agricultural Organization of the United Nation. Website: http://faostat.fao.org/

Fauquet, C. M., Mayo, M. A., Maniloff, J., Desselberger, U., and Ball, L. A. (2005). Virus Taxonomy: Classification and Nomenclature of Viruses. *Eighth Report of the International Committee on Taxonomy of Viruses.* Elsevier Academic Press. p. 1259.

Fegla, G., El-Sayed, W., El-Faham, Y., and Kawanna, M. (2009). Biological, serological and molecular detection of the most dominant viruses affecting faba bean in Northern Egypt. *Arab J. Plant Prot.* **27**(Special Issue; Supplement):E-80.

Fidan, U., and Yorganci, U. (1989). Investigations on the detection and seed transmission of the virus diseases occurring on the pulse crops in Aegean Region. I. The identification of viruses infecting pulse crops in Aegean Ragion. *J. Turk. Phytopathol.* **18:**93–105.

Fischer, H. U., and Lockhart, B. E. L. (1978). Host range and properties of peanut stunt virus from Morocco. *Phytopathology* **68:**289–293.

Fortass, M., and Bos, L. (1991). Survey of faba bean (*Vicia faba* L.) for viruses in Morocco. *Neth. J. Plant Pathol.* **97:**369–380.

Fortass, M., and Diallo, S. (1993). Broad bean mottle bromovirus in Morocco; curculionid vectors, and natural occurrence in food legumes other than faba bean (*Vicia faba* L.). *Neth. J. Plant Pathol.* **99:**219–226.

Francki, R. I. B., Mossop, D. W., and Hatta, T. (1979). Cucumber mosaic virus. *CMI/AAB Descriptions of Plant Viruses No. 213.*

Franz, A., Makkouk, K. M., and Vetten, H. J. (1995). Faba bean necrotic yellows virus naturally infects *Phaseolus* Bean and cowpea in the coastal area of Syria. *J. Phytopathol.* **143:**319–320.

Franz, A., Makkouk, K. M., Katul, L., and Vetten, H. J. (1996). Monoclonal antibodies for the detection and differentiation of faba bean necrotic yellows virus isolates. *Ann. Appl. Biol.* **128:**255–268.

Franz, A., Makkouk, K. M., and Vetten, H. J. (1997). Host range of faba bean necrotic yellows virus and potential yield loss in infected faba bean. *Phytopathol. Mediterr.* **36:**94–103.

Franz, A., Makkouk, K. M., and Vetten, H. J. (1998). Acquisition, retention and transmission of faba bean necrotic yellows virus by two of its aphid vectors, *Aphis craccivora* (Koch) and *Acyrthosiphon pisum* (Harris). *J. Phytopathol.* **146:**347–355.

French, J. V., Bath, J. E., Tsai, J. H., and Thottappilly, G. (1973). Purification of *Pea enation virus* from its vector, *Acyrthosiphon pisum*, and aphid transmission characteristics. *Virology* **51:**78–84.

Fresno, J., Castro, S., Babin, M., Carazo, G., Molina, A., de Blas, C., and Romero, J. (1997). Virus diseases of broad bean in Spain. *Plant Dis.* **81:**112.

Gallitelli, D. (2000). The ecology of cucumber mosaic virus and sustainable agriculture. *Virus Res.* **71:**9–21.

Galvez, G. W., and Morales, F. J. (1989). Aphid-transmitted viruses. *In* "Bean Production Problems in the Tropics" (H. F. Schwartz and M. A. Pastor-Corrales, eds.), pp. 333–362. Centro Internacional de Agricultura Tropical (CIAT), Cali, Colombia.

Gamal-Eldin, A. S., El-Amrety, A. A., Mazyad, H. M., and Rizkallah, L. R. (1982). Effect of bean yellow mosaic and broad bean wilt viruses on broad bean yield. *Agric. Res. Rev. (Egypt)* **60:**195–204.

Ghad, I. P. S., and Bernier, C. C. (1984). Resistance in faba bean (*Vicia faba*) to bean yellow mosaic virus. *Plant Dis.* **68:**109–111.

Gibbs, A. J. (1972). Broad bean mottle virus. *CMI/AAB Descriptions of Plant Viruses No. 101.*

Gibbs, A. J., and Paul, H. L. (1970). Echtes ackerbohnemosaik-virus. *CMI/AAB Descriptions of Plant Viruses No. 20.*

Gibbs, A. J., and Smith, H. G. (1979). Broad bean stain virus. *CMI/AAB Descriptions of Plant Viruses No. 29.*

Gonsalves, D., and Shepherd, R. J. (1972). Biological and physical properties of the two nucleoprotein components of pea enation mosaic and their associated nucleic acids. *Virology* **48:**709–723.

Habib, S. A., El-Atta, O. K., El-Hammady, M., and Awad, M. (1981). Interaction between bean common mosaic virus and bean yellow mosaic virus in relation to morphological characters of bean plants (*Phaseolus vulgaris*). *Res. Bull. Fac. Agric. Ain Shams Univ.* **1606:** 1–14.

Hagedorn, D. J., Layne, R. E. C., and Ruppell, E. G. (1964). Host range of *Pea enation mosaic virus* and use of *Chenopodium album* Willd. as a local lesion host. *Phytopathology* **64:**843–852.

Haj Kassem, A. A., Makkouk, K. M., and Attar, N. (2001). Viruses on cultivated forage legume in Syria. *Arab J. Plant Prot.* **19:**73–79.

Hajimorad, M. R., Dietzgen, R. G., and Francki, R. I. (1990). Differentiation and antigenic characterization of closely related alfalfa mosaic virus strains with monoclonal antibodies. *J. Gen. Virol.* **71:**2809–2816.

Halupecki, E., Cvjetkovic´, B., Ban, D., and Borošic´, J. (2003). Detection of viruses in selected lines of dwarf dry bean (Phaseolus vulgaris var. nanus Martens) using ELISA-test. Lectures and papers presented at the 6th Slovenian Conference on Plant Protection, Zreče, 4-6 March 2003.

Hammond, J., and Hammond, R. W. (1989). Molecular cloning, sequencing and expression in *Escherichia coli* of the bean yellow mosaic virus coat protein gene. *J. Gen. Virol.* **70:** 1961–1974.

Hampton, R. O., and Francki, R. I. B. (1992). RNA-1 dependent seed transmissibility of cucumber mosaic virus in Phaseolus vulgaris. *Phytopathology* **82:**127–130.

Hampton, R. O., and Mink, G. I. (1975). Pea seed-borne mosaic virus. *CMI/AAB Descriptions of Plant Viruses No. 146.*

Harrison, B. D. (1973). Pea early-browning virus. *CMI/AAB Descriptions of Plant Viruses No. 120.*

Harrison, B. D., Finch, J. T., Gibbs, A. J., Hollings, M., Shepherd, R. J., Velenta, V., and Wetter, C. (1971). Sixteen groups of plant viruses. *Virology* **45:**356–363.

Hassan, H. T., Makkouk, K. M., and Haj Kassem, A. A. (1999). Viral disease on cultivated legume crops in Al-Ghab Plain, Syria. *Arab J. Plant Prot.* **17:**17–21.

Horn, N. M., Reddy, S. V., Roberts, I. M., and Reddy, D. V. R. (1993). Chickpea chlorotic dwarf virus, a new leafhopper-transmitted geminivirus of chickpea in India. *Ann. Appl. Biol.* **122:**467–479.

Horn, N. M., Makkouk, K. M., Kumari, S. G., van den Heuvel, J. F. J. M., and Reddy, D. V. R. (1995). Survey of chickpea (*Cicer arietinum* L.) for chickpea stunt disease and associated viruses in Syria, Turkey and Lebanon. *Phytopathol. Mediterr.* **34:**192–198.

Hull, R. (1969). Alfalfa mosaic virus. *Advan. Virus Res.* **15:**265–433.

Hull, R., and Lane, L. C. (1973). The unusual nature of the components of a strain of pea enation mosaic. *Virology* **55:**l–13.

Ie, T. S. (1970). Tomato spotted wilt virus. *CMI/AAB Descriptions of Plant Viruses No. 39.*

CAB International, Nosworth Way, Wallingford, Oxforshire, OX10 8DE, UK (1974). Bean common mosaic virus. *Distribution Map of Plant Dis., 1974 No.* October (edition 3).

Izadpanah, K., and Shepherd, R. J. (1966a). Galactia sp as a local lesion host for *Pea enation mosaic virus*. *Phytopathology* **56:**458–459.

Izadpanah, K., and Shepherd, R. J. (1966b). Purification and properties of *Pea enation mosaic virus*. *Virology* **28:**463–476.

Jaspars, E. M. J., and Bos, L. (1980). Alfalfa mosaic virus. *CMI/AAB Descriptions of Plant Viruses No. 229.*

Johnstone, G. R., and Rapley, P. E. L. (1979). The effect of time of sowing on the incidence of subterranean clover red leaf virus infection in broad bean (*Vicia faba*). *Ann. Appl. Biol.* **99:**135–141.

Jordan, R., and Hammond, J. (1991). Comparison and differentiation of potyvirus isolates and identification of strain-, virus-, subgroup-specific and potyvirus group-common epitopes using monoclonal antibodies. *J. Gen. Virol.* **72:**25–36.

Kaiser, W. J. (1973). Biology of bean yellow mosaic and pea leaf roll viruses affecting *Vicia faba* in Iran. *Phytopath. Z.* **78:**253–263.

Katul, L., Vetten, H. J., Maiss, E., Makkouk, K. M., Lesemann, D. E., and Casper, R. (1993). Charecteristics and serology of virus-like particles associated with faba bean necrotic yellows. *Ann. Appl. Biol.* **123:**629–647.

Katul, L., Maiss, E., and Vetten, H. J. (1995a). Sequence analysis of a *Faba bean necrotic yellows virus* DNA component containing a putative replicase gene. *J. Gen. Virol.* **76:**475–479.

Katul, L., Vetten, H. J., Lesemann, D. E., Maiss, E., and Makkouk, K. M. (1995b). Diagnostic methods for the detection of faba bean necrotic yellows virus, a circular ssDNA virus. *EPPO Bull.* **25:**329–336.

Katul, L., Maiss, E., Morozov, S. Y., and Vetten, H. J. (1997). Analysis of six DNA components of the faba bean necrotic yellows virus genome and their structural affinity to related plant virus genomes. *Virology* **233:**247–259.

Katul, K., Timchenko, T., Gronenborn, B., and Vetten, H. J. (1998). Ten distinct circular ssDNA components, four of which encode putative replication-associated proteins, are associated with the faba bean necrotic yellows virus genome. *J. Gen. Virol.* **79:**3101–3109.

Kennedy, J. S., Day, M. F., and Eastop, V. F. (1962). A Conspectus of Aphids as Vectors of Plant Viruses. Commonwealth Institute of Entomology, London, 114pp.

Kheder, M. A. (2002). Isolation and identification of bean common mosaic virus and its effect on ultrastructure of infected *Phaseolus vulgaris* L. leaves. *Agric. Bot. Dept. Fac. Agric. Zagazig Univ. Zagazig (Egypt)* **40:**1473–1485.

Krstic, B., Tosic, M., Stojanovic, G., and Vico, I. (1997). Purification of some plant viruses by the improved chloroform-evaporation method. *Zastita Bilja* **48:**239–244.

Kumari, S. G., and Makkouk, K. M. (2007). Virus diseases of faba bean (*Vicia faba* L.) in Asia and Africa. *Plant Viruses* **1:**93–105.

Kumari, S. G., Makkouk, K. M., Katul, L., and Vetten, H. J. (2001). Polyclonal antibodies to the bacterially expressed coat protein of *Faba bean necrotic yellows virus*. *J. Phytopathol.* **149:**543–550.

Kumari, S. G., Rodoni, B., Hlaing Loh, M., Makkouk, K. M., Freeman, A., and van Leur, J. (2006). Distribution, identification and characterization of Luteoviruses affecting food legumes in Asia and North Africa. Proceeding of 12th Mediterranean Phytopathological Congress, 11-15 June 2006, Rhodes Island, Greece, pp. 412–416, 590pp.

Kumari, S. G., Rodoni, B., Vetten, H. J., Freeman, A., van Leur, J., Loh, M., Shiying, B., and Xiaoming, W. (2010). Detection and partial characterization of *Milk vetch dwarf virus* in faba bean (*Vicia faba* L.) in Yunnan Province, China. *J. Phytopathol.* **158:**35–39.

Kurçman, S. (1977). Determination of virus diseases on cultural plants in Turkey. *J. Turkish Phytopath.* **6:**27–48.

Lane, L. C. (1974). The Bromovirus group. *Adv. Virus Res.* **19:**151–220.

Larsen, R. C., and Webster, D. M. (1999). First report of bean leaf roll luteovirus infecting pea in Italy. *Plant Dis.* **83:**399.
Leclercq-Le Quillec, F., and Maury, Y. (1998). Désordres provoqués par les virus sur pois protéagineux (Disorders caused by viruses on pea crops). *Phytoma* **502:**41–43.
Lisa, V. (2000). Viruses of *Phaseolus* bean in Italy. *Italus Hortus* **7:**51–54.
Lockhart, B. E. L., and Fischer, H. U. (1974). Chronic infection by seedborne bean common mosaic virus in Morocco. *Plant Dis. Rep.* **58:**307–308.
Lockhart, B. E. L., and Fischer, H. U. (1976). Some properties of an isolate of pea early-browning virus occuring in Morocco. *Phytopathology* **66:**1391–1394.
Lot, H., Marrou, J., Quiot, J. B., and Esvan, C. (1972). Contribution à l'étude du virus de la mosaïque du concombre (CMV). I. Méthode de purification rapide du virus. *Ann. Phytopathol.* **4:**25–38.
Mahir, M. A.-M., Fortass, M., and Bos, L. (1992). Identification and properties of a deviant isolate of the broad bean yellow band serotype of pea early-btrowning virus from faba bean (*Vicia faba*) in Algeria. *Neth. J. Plant Pathol.* **98:**237–252.
Mahmood, K., and Peters, D. (1973). Purification of *Pea enation mosaic virus* and the infectivity of its components. *Neth. J. Plant Pathol.* **79:**138–147.
Makkouk, K. M., and Kumari, S. G. (1989). *Apion arrogans*, a weevil vector of broad bean mottle virus. *FABIS Newsl.* **25:**26–27.
Makkouk, K. M., and Kumari, S. G. (1995a). Transmission of broad bean stain comovirus and broad bean mottle bromovirus by weevils in Syria. *J. Plant Dis. Prot.* **102:**136–139.
Makkouk, K. M., and Kumari, S. G. (1995b). Screening and selection of faba bean (*Vicia faba* L.) germplasm for resistance to bean yellow mosaic potyvirus. *J. Plant Dis. Prot.* **102:**461–466.
Makkouk, K. M., and Kumari, S. G. (1996). Detection of ten viruses by the tissue-blot immunoassay (TBIA). *Arab J. Plant Prot.* **14:**3–9.
Makkouk, K. M., and Kumari, S. G. (2001). Reduction of spread of three persistently aphid-transmitted viruses affecting legume crops by seed-treatment with Imidacloprid (Gaucho®). *Crop Prot.* **20:**433–437.
Makkouk, K. M., Lesemann, D. E., and Haddad, N. A. (1982). Bean yellow mosaic virus from broad bean in Lebanon: Incidence, host range, purification, and serological properties. *J. Plant Dis. Prot.* **89:**59–66.
Makkouk, K. M., Katul, L., and Rizkallah, A. (1987). Electrophoretic separation: An alternative simple procedure for the purification of broad bean mottle and alfalfa mosaic viruses. *FABIS Newsl.* **19:**12–14.
Makkouk, K. M., Bos, L., Rizkallah, A., Azzam, O. I., and Katul, L. (1988a). Broad bean mottle virus: Identification, serology, host range and occurrence on faba bean (*Vicia faba*) in West Asia and North Africa. *Neth. J. Plant Pathol.* **94:**195–212.
Makkouk, K. M., Bos, L., Azzam, O. I., Kumari, S. G., and Rizkallah, A. (1988b). Survey of viruses affecting faba bean in six Arab countries. *Arab J. Plant Prot.* **6:**53–61.
Makkouk, K. M., Katul, L., and Rizkallah, A. (1988c). A simple procedure for the purification and antiserum production of bean yellow mosaic virus. *J. Phytopathol.* **122:**89–93.
Makkouk, K. M., Kumari, S. G., and Bos, L. (1990). Broad bean wilt virus: Host range, purification, serology, transmission characteristics, and occurrence in faba bean in West Asia and North Africa. *Neth. J. Plant Pathol.* **96:**291–300.
Makkouk, K. M., Kumari, S. G., and Al-Daoud, R. (1992). Survey of viruses affecting lentil (*Lens culinaris*) in Syria. *Phytopathol. Mediterr.* **31:**188–190.
Makkouk, K. M., Hsu, H. T., and Kumari, S. G. (1993a). Detection of three plant viruses by dot-blot and tissue-blot immunoassays using chemiluminescent and chromogenic substrates. *J. Phytopathol.* **139:**97–102.
Makkouk, K. M., Kumari, S. G., and Bos, L. (1993b). Pea seed-borne mosaic virus: Occurrence in faba bean (*Vicia faba* L.) and lentil (*Lens culinaris* Med.) in West Asia and North Africa,

and further information on host range, purification, serology, and transmission characteristics. *Neth. J. Plant Pathol.* **99:**115–124.

Makkouk, K. M., Rizkallah, L., Madkour, M., El-Sherbeeny, M., Kumari, S. G., Amriti, A. W., and Solh, M. B. (1994). Survey of faba bean (*Vicia faba* L.) for viruses in Egypt. *Phytopathol. Mediterr.* **33:**207–211.

Makkouk, K. M., Vetten, H. J., Katul, L., Franz, A., and Madkour, M. A. (1998). Epidemiology and control of faba bean necrotic yellows virus. *In* "Plant Virus Disease Control" (A. Hadidi, R. K. Khetarpal, and H. Koganezawa, eds.), pp. 534–540. APS Press, The American Phytopathological Society, St. Paul, Minnesota, USA.

Makkouk, K. M., Kumari, S., Sarker, A., and Erskine, W. (2001). Registration of six lentil germplasm lines with combined resistance to viruses. *Crop Sci.* **41:**931–932.

Makkouk, K. M., Kumari, S. G., and van Leur, J. A. G. (2002). Screening and selection of faba bean (*Vicia faba* L.) germplasm resistant to *Bean leafroll virus*. *Aust. J. Agric. Res.* **53:**1077–1082.

Makkouk, K. M., Kumari, S. G., Hughes, J. A., Muniyappa, V., and Kulkarni, N. K. (2003). Other legumes: Faba bean, chickpea, lentil, pigeonpea, mungbean, blackgram, lima bean, horegram, bambara groundnut and winged bean. *In* "Virus and Virus-like Diseases of Major Crops in Developing Countries" (G. Loebenstein and G. Thottappilly, eds.), pp. 447–476. Kluwer Academic Publishers, Dordrecht, The Netherlands.

Marchoux, G., Gébre-Selassié, K., and Villevieille, M. (1991). Detection of tomato spotted wilt virus and transmission by *Frankliniella occidentalis* in France. *Plant Pathol.* **40:**347–351.

Marchoux, G., Quiot, J. B., and Devergne, J. C. (1997). Characterization of an isolate of cucumber mosaic virus transmitted by bean (*Phaseolus vulgaris* L.) seeds. *Ann. Phytopathol.* **9:**421–434.

Mazyad, H. M., El-Hammady, M., and Sabak, A. (1974). Occurrence of cucumber mosaic virus in bean plants in Egypt. First Congress of Egypt Phytopathological Society, Cairo, Egypt.

Morales, F. J. (1979). Purification and serology of bean common mosaic virus. *Turrialba* **29:**320–324.

Morales, F. J. (2003). Common bean. *In* "Virus and Virus-Like Diseases of Major Crops in Developing Countries" (G. Loebenstein and G. Thottappilly, eds.), pp. 425–445. Kluwer Acdemic Publishers, The Netherlands.

Morales, F. J., and Bos, L. (1988). Bean common mosaic virus. *Association of Applies Biologists (AAB) Descriptions of Plant Viruses No. 337*. p 6.

Motoyoshi, F., and Hull, R. (1974). The infection of tobacco protoplasts with *Pea enation mosaic virus*. *J. Gen. Virol.* **24:**89–99.

Mouhanna, A. M., Makkouk, K. M., and Ismail, I. D. (1994). Survey of virus disease of wild and cultivated legumes in the coastal region of Syria. *Arab J. Plant Prot.* **12:**12–19.

Najar, A., Makkouk, K. M., Boudhir, H., Kumari, S. G., Zarouk, R., Bessai, R., and Ben Othman, F. (2000). Viral diseases of cultivated legume and cereal crops in Tunisia. *Phytopathol. Mediterr.* **39:**423–432.

Nault, L. R. (1967). Inoculation of *Pea enation mosaic virus* by the green peach, potato, and foxglove aphids. *J. Econ. Entomol.* **60:**1586–1587.

Nault, L. R. (1975). Tests for transmission of pea enation mosaic virus by oligophagous mustard and grain aphids. *Phytopathology* **65:**496–497.

Navas-Castillo, J., Sánchez-Campos, S., Díaz, J. A., Sáez-Alonso, E., and Moriones, E. (1999). Tomato yellow leaf curl virus-Is causes a novel disease of common bean and severe epidemics in tomato in Spain. *Plant Dis.* **83:**29–32.

Nienhaus, F., and Saad, A. T. (1967). First report on plant virus diseases in Lebanon, Jordan and Syria. *Z. PflKrankh. PflPath. PflSchutz* **74:**459–471.

Nitzany, F. E., and Cohen, S. (1964). Virus affecting broad beans in Israel. *Phytopathol. Mediterr.* **3:**1–8.

Ohki, S. T., and Kameya-Iwaki, M. (1996). Simplifying of the rapid immunofilter paper assay for faster detection of plant viruses: Simplified RIPA. *Ann. Phytopathol. Soc. Jpn.* **62:** 240–242.

Ortiz, V., Castro, S., and Romero, J. (2005). Optimization of RT-PCR for the detection of Bean leaf roll virus in plant host and insect vectors. *J. Phytopathol.* **153:**68–72.

Ortiz, V., Navarro, E., Castro, S., Carazo, G., and Romero, J. (2006). Incidence and transmission of *Faba bean necrotic yellows* virus (FBNYV) in Spain. *Span. J. Agric. Res.* **4:**255–260.

Ortiz, V., Castro, S., and Romero, J. (2009). First report of *Clover yellow vein virus* in grain legumes in Spain. *Plant Dis.* **93:**106.

Osborn, H. T. (1935). Incubation of pea mosaic in the aphid, *Macrosiphum pisi. Phytopathology* **25:**160–177.

Osborn, H. T. (1938). Studies on pea virus 1. *Phytopathology* **28:**923–936.

Palukaitis, P., Roossinck, M. J., Dietzgen, R. G., and Francki, R. I. B. (1992). Cucumber mosaic virus. *Adv. Virus Res.* **41:**281–348.

Papayiannis, L. C., Paraskevopoulos, A., and Katis, N. I. (2007). First report of *Tomato yellow leaf curl virus* infecting common bean (*Phaseolus vulgaris*) in Greece. *Plant Dis.* **91:**465.

Peden, K. W. C., and Symons, R. H. (1973). Cucumber mosaic virus contains a functionally divided genome. *Virology* **53:**487–492.

Peters, D. (1982). Pea enation mosaic virus. *CMI/AAB Descriptions of Plant Viruses No. 257.*

Phan, T. T. H., Khetarpal, R. K., Le, T. A. H., and Maury, Y. (1997). Comparison of immunocapture PCR and ELISA in quality control of pea seed for pea seedborne mosaic potyvirus. *In* "Seed Health Testing: Progress Towards the 21st Century" (J. D. Hutchins and J. C. Reeves, eds.), pp. 193–199. National Institute of Agricultural Botany, Cambridge, UK.

Proeseler, G., Fritzsche, R., and Schimanski, B. (1976). Laboratory trials with insecticides, mineral oil and combinations of them for the reduction of aphid transmission of non-persistent viruses. *Arch. Phytopathol. Pflanzenschutz* **12:**19–26.

Quantz, L., and Volk, J. (1954). Die Blattrollkrankheit der Ackerbohne und Erbse, eine neue Viruskrankheit bei Leguminosen. *NachrBl. dt. PflSchtzdienst* **6:**177–182.

Raizada, R. K., Aslam, M., and Singh, B. P. (1991). Immunological detection of bean yellow mosaic virus in seeds of faba bean (*Vicia faba* L.). *Indian J. Virol.* **7:**179–183.

Rana, G. L., and Kyriakopoulou, P. E. (1981). Bean yellow mosaic virus in artichokes in Greece. Proceedings of the Fifth Congress of the Mediterranean Phytopathological Union, Patras, Greece, 21-27 September 1980. Hellenic Phytopathological Society. Athens Greece, pp. 38–40.

Ravnikar, M., Grum, M., Mavrič, I., and Camloh, M. (1996). Determination and elimination of seedborne bacteria and viruses of bean (Phaseolus vulgaris L.). Novi izzivi v poljedelstvu '96. Zbornik simpozija, Ljubljana, Slovenia, 9-10 December 1996, pp. 195–199.

Roberts, I. M., Wang, D., Thomas, C. L., and Maule, A. J. (2003). Pea seed-borne mosaic virus seed transmission exploits novel symplastic pathways to infect the pea embryo and is, in part, dependent upon chance. *Protoplasma* **222:**31–43.

Rosciglione, B., and Cannizzaro, G. (1975). Record of pea enation mosaic virus in Sicily. *Phytopathol. Mediterr.* **14:**34–36.

Rosciglione, B., and Cannizzaro, G. (1977). Identification of broad bean wilt virus on Linosa island. *Phytopathol. Mediterr.* **16:**140–142.

Russo, M., Gallitelli, D., Vovlas, C., and Savino, V. (1984). Properties of broad bean yellow band virus, a possible new tobravirus. *Ann. Appl. Biol.* **105:**223–230.

Saiz, M., Castro, S., Carazo, G., Romero, J., and de Blas, C. (1993). First report of bean yellow mosaic virus in Spain. *Plant Dis.* **77:**429.

Saiz, M., de Blas, C., Carazo, G., Fresno, J., Romero, J., and Castro, S. (1995). Incidence and characterization of bean common mosaic virus isolates in Spanish bean fields. *Plant Dis.* **79:**79–81.

Sano, Y., Wada, W., Hashimoto, Y., Matsumoto, T., and Kojima, M. (1998). Sequences of ten circular ssDNA components associated with the Milk vetch dwarf virus genome. *J. Gen. Virol.* **79:**3111–3118.

Sasaya, T., Iwasaki, M., and Yamamoto, T. (1993). Seed transmission of bean yellow mosaic virus in broad bean (*Vicia faba*). *Ann. Phytopathol. Soc. Jpn.* **59:**559–562.

Schmutterer, H., and Thottappilly, G. (1972). Zur wirtschaftlichen Bedeutung und Ausbreitung des Erbsenblattrollvirus im Ackerbohnenbestand sowie zur chemischen Bekampfung der Vektoren. *Z. PflanzenKr. Pflanzenschutz* **79:**478–484.

Schroeder, W. T., and Barton, D. W. (1958). The nature and inheritance of resistance to the *Pea enation mosaic virus* in garden pea, *Pisum sativum L.*. *Phytopathology* **48:**628.

Sclavounos, A. P., Voloudakis, A. E., Arabatzis, Ch., and Kyriakopoulou, P. R. (2006). A severe Hellenic CMV tomato isolate: Symptom variability in tobacco, characterization and discrimination of variants. *Eur. J. Plant Pathol.* **15:**163–172.

Segundo, E., Gil-Salas, F. M., Janssen, D., Martin, G., Cuadrado, I. M., and Remah, A. (2004). First report of *Southern bean mosaic virus* infecting French bean in Morocco. *Plant Dis.* **88:**1162.

Shagrun, M. (1973). Bean yellow mosaic virus on broad bean plants in Libya. 1. Identification of the causal agent. *Libyan J. Agric.* **11:**33–38.

Shamloul, A. M., Hadidi, A., Madkour, M. A., and Makkouk, K. M. (1999). Sensitive detection of banana bunchy top and faba bean necrotic yellows viruses from infected leaves, in vitro tissue cultures, and viruliferous aphids using polymerase chain reaction. *Can. J. Plant Pathol.* **21:**326–337.

Shukla, D. D., Strike, P. M., Tracy, S. L., Gough, K. H., and Ward, C. W. (1988). The N and C termini of the coat proteins of potyviruses are surface-located and the N terminus contains the major virus-specific epitopes. *J. Gen. Virol.* **69:**1497–1508.

Skaf, J. S., Schultz, M. H., Hirata, H., and de Zoeten, G. A. (2000). Mutational evidence that the VPg is involved in the replication and not the movement of *Pea enation mosaic virus*-1. *J. Gen. Virol.* **81:**1103–1109.

Sudarshana, M. R., Roy, G., and Falk, B. W. (2007). Methods for engineering resistance to plant viruses. *Methods Mol. Biol.* **354:**183–195.

Tachibana, Y. (1981). Control of aphid-borne viruses in faba bean by mulching with silver polyethylene film. *FABIS Newsl.* **3:**56.

Taubenhaus, J. J. (1914). The diseases of the sweet pea. *Delaware Agric. Exp. Stat. Bull.* **106:**62–69.

Timchenko, T., Katul, L., Sano, Y., de Kouchkovsky, F., Vetten, H. J., and Gronenborn, B. (2000). The master Rep concept in nanovirus replication: Identification of missing genome components and potential for natural genetic reassortment. *Virology* **274:**189–195.

Tornos, T., Cebrián, M. C., Córdoba-Sellés, M. C., Alfaro-Fernández, A., Herrera-Vásquez, J. A., Font, M. I., and Jorda, M. C. (2008). First report of *Pea enation mosaic virus* infecting pea and broad bean in Spain. *Plant Dis.* **92:**1469.

Tracy, S. L., Frenkel, M. J., Gough, K. H., Hanna, P. J., and Shukla, D. D. (1992). Bean yellow mosaic, clover yellow vein, and pea mosaic are distinct potyviruses: Evidence from coat protein gene sequences and molecular hybridization involving the 3′ non-coding regions. *Arch. Virol.* **122:**249–261.

Tremaine, J. H. (1983). Southern bean mosaic virus. *CMI/AAB Descriptions of Plant Viruses No. 274.*

Uga, H. (2005). Use of crude sap for one-step RT-PCR-based assays of Bean yellow mosaic virus and the utility of this protocol for various plant-virus combinations. *J. Gen. Plant Pathol.* **71:**86–89.

van Vloten-Doting, L., and Jaspars, E. M. J. (1972). The uncoating of alfalfa mosaic virus by its own RNA. *Virology* **48:**699–708.

Varveri, C., and Boutsika, L. (1999). Characterization of cucumber mosaic cucumovirus isolates in Greece. *Plant Pathol.* **48:**95–100.

Vemulapati, B., Druffel, K. L., Eigenbrode, S., Karasev, A., and Pappu, H. R. (2010). Molecular characterization of *Pea enation mosaic virus* (genus *Enamovirus*) and *Bean leaf roll virus* (genus *Luteovirus*) from the Pacific Northwestern USA. *Arch. Virol.* **155:**1713–1715.

Vemulapati, B., Druffel, K. L., Eigenbrode, S. D., Karasev, A., and Pappu, H. R. (2011). Genomic characterization of *Pea enation mosaic virus*-2 from the Pacific Northwestern USA. *Arch. Virol.* **156:**1897–1900.

Vetten, H. J., Chu, P. W. G., Dale, J. L., Harding, R., Hu, J., Katul, L., Kojima, M., Randles, J. W., Sano, Y., and Thomas, J. E. (2005). Nanoviridae. *In* "Virus Taxonomy. VIIIth Report of the ICTV" (C. M. Fauquet, M. A. Mayo, J. Maniloff, U. Desselberger, and L. A. Ball, eds.), pp. 343–352. Elsevier/Academic Press, London.

Vovlas, C., and Russo, M. (1978). Virus disease of vegetable crops in Apulia. XXII. Bean yellow mosaic virus in broad bean. *Phytopath. Medit.* **17:**201–204.

Wahyuni, W. S., Dietzgen, R. G., Hanada, K., and Francki, R. I. B. (1992). Serological and biological variation between and within subgroup I and II strains of cucumber mosaic virus. *Plant Pathol.* **41:**282–297.

Walters, H. J., and Surin, P. (1973). Transmission and host range studies of broad bean mottle virus. *Plant Dis. Rep.* **57:**833–836.

Wang, W. Y., Mink, G. I., and Silbernagel, M. J. (1982). Comparison of direct and indirect enzyme-linked immunosorbet assay (ELISA) in the detection of bean common mosaic virus strains (Abstr). *Phytopathology* **72:**954.

Wang, W. Y., Mink, G. I., Silbernagel, M. J., and Davis, W. C. (1984). Production of hybridoma lines secreting specific antibodies to bean common mosaic virus (BCMV) strains (Abstr.). *Phytopathology* **74:**1142.

Waterhouse, P. M., Gildow, F. E., and Johnstone, G. R. (1988). Luteovirus Group, Descriptions of Plant Viruses. Commonwealth Mycological Institute and Association of Applied Biologists, Kew, UK, No. 339.

Wylie, S., Wilson, C. R., Jones, R. A. C., and Jones, M. G. K. (1993). A polymerase chain reaction assay for cucumber mosaic virus in lupin seeds. *Aust. J. Agric. Res.* **44:**41–51.

Xu, L., and Hampton, R. O. (1996). Molecular detection of *Bean common mosaic* and *Bean common mosaic necrosis potyviruses* and pathogroups. *Arch. Virol.* **141:**1961–1977.

Yilmaz, N. D. K., Gümüş, M., and Erkan, S. (2002). Studies on determination of virus diseases in the seeds of bean from Tokat Province. *Ege Üniv. Ziraat Fak. Derg.* **39:**49–55.

Younis, H. A., Shagrun, M., and Khalil, J. (1992). Isolation of bean yellow mosaic virus from broad bean plants in Libya. *Libyan J. Agric.* **13:**165–170.

Zagh, S., and Ferault, A. C. (1980). A broad bean virus diseases occurring in Algeria. *Ann. Phytopathol.* **12:**153–159.

Zidan, F., Khalil, J., and Shagrun, M. (1997). Survey and identification of pea viruses in the western region of Libya. (W. Khoury and B. Bayaa, eds.), *In* "Abstract book of Sixth Arab Congress of Plant Protection", Beirut, Lebanon, October 27-31, 1997, p. 188.

Zidan, F., Khalil, J., and Shagrun, M. (2002). Survey and identification of pea viruses in Libya. *Arab J. Plant Prot.* **20:**154–156.

CHAPTER 12

Tospoviruses in the Mediterranean Area

Massimo Turina,*[*] Luciana Tavella,[†] and
Marina Ciuffo[*]

Contents

[*] Istituto di Virologia Vegetale-CNR, Strada delle Cacce, Torino, Italy
[†] DIVAPRA Entomologia e Zoologia applicate all'Ambiente, University of Torino, Via L. da Vinci, Grugliasco (TO), Italy

Advances in Virus Research, Volume 84
ISSN 0065-3527, DOI: 10.1016/B978-0-12-394314-9.00012-9

Abstract

Tospoviruses are among the most serious threats to vegetable crops in the Mediterranean basin. Tospovirus introduction, spread, and the diseases these viruses cause have been traced by epidemiological case studies. Recent research has centered on the close relationship between tospoviruses and their arthropod vectors (species of the *Thripidae* family). Here, we review several specific features of tospovirus–thrips associations in the Mediterranean. Since the introduction of *Frankliniella occidentalis* in Europe, *Tomato spotted wilt virus* (TSWV) has become one of the limiting factors for vegetable crops such as tomato, pepper, and lettuce. An increasing problem is the emergence of TSWV resistance-breaking strains that overcome the resistance genes in pepper and tomato. *F. occidentalis* is also a vector of *Impatiens necrotic spot virus*, which was first observed in the Mediterranean basin in the 1980s. Its importance as a cause of vegetable crop diseases is limited to occasional incidence in pepper and tomato fields. A recent introduction is *Iris yellow spot virus*, transmitted by the onion thrips *Thrips tabaci*, in onion and leek crops. Control measures in vegetable crops specific to Mediterranean conditions were examined in the context of their epidemiological features and tospovirus species which could pose a future potential risk for vegetable crops in the Mediterranean were discussed.

I. INTRODUCTION

The genus *Tospovirus* (family *Bunyaviridae*) comprises plant viruses transmitted by thrips (order *Thysanoptera*, family *Thripidae*). All members share a common structure: virus particles quasi-spherical in shape, from 80 to 120 nm in diameter, and surrounded by an envelope of a host-derived membrane in which two glycoproteins, Gn and Gc, are embedded. Within the virion, there is a tripartite negative-sense (with two ambisense segments) RNA genome with each segment independently packaged by a number of copies of nucleoprotein (N) and a low number of L protein molecules (the viral RNA-dependent RNA polymerase, RdRP; Fig. 1).

As the production of vegetable crops is vital to the Mediterranean agriculture, viral diseases caused by tospoviruses have wide-reaching implications across the entire region. Tospoviruses are one of the most important groups of plant viruses responsible for vegetable crop diseases in the Mediterranean, and among the tospovirus species present in the Mediterranean basin, *Tomato spotted wilt virus* (TSWV) ranks first in importance because of the severe diseases caused in tomato and pepper. Further, given the specific conditions for vegetable production in the Mediterranean, major efforts have gone into controlling tospovirus

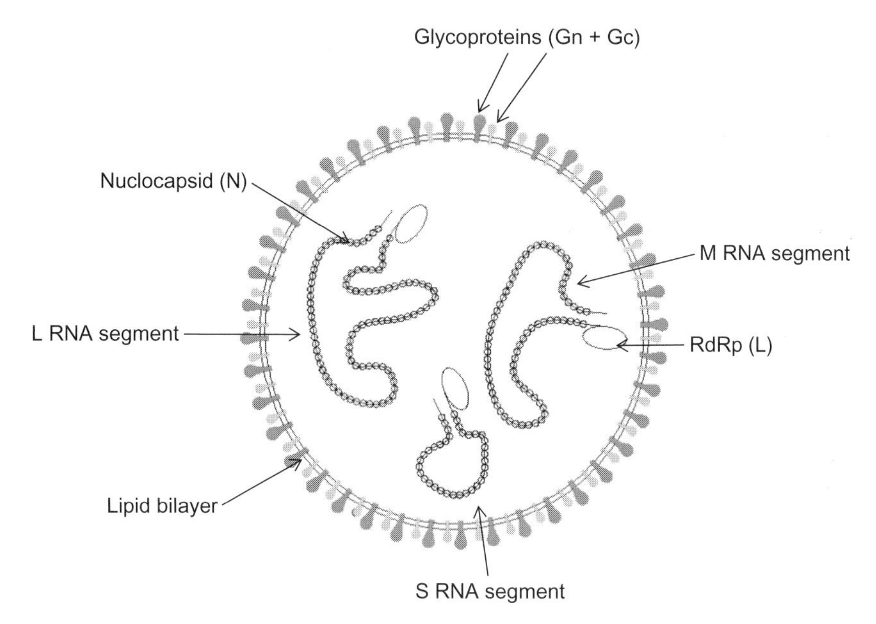

FIGURE 1 Diagram of tospovirus virions with the host-derived membrane and the three RNA segments encapsidated by the nucleocapsid (N) protein.

diseases, since even small epidemic episodes can impact the economic cost-effectiveness of these crops. Also, it is not uncommon to find TSWV infections even in tomato and pepper cultivars containing TSWV resistance genes, probably due to the exclusive dependence on these for TSWV control (Aramburu and Marti, 2003; Ciuffo *et al.*, 2005; Margaria *et al.*, 2004; Roggero *et al.*, 2002a) rather than on a more complex integrated pest management (IPM) strategy.

Impatiens necrotic spot virus (INSV) has been present in the Mediterranean for a number of years, but its economic impact in vegetable crops is less than that in ornamental crops. *Iris yellow spot virus* (IYSV) is an emerging problem. It was reported in Israel in 1998 (Gera *et al.*, 1998), but since ca. 2000, it has spread to other Mediterranean countries where the incidence in bulbous crops such as onion is high. A fourth tospovirus species, *Polygonum ringspot virus* (PolRSV), has been commonly found in Italy on the weed, *Polygonum dumetorum*, but so far it has not been reported on economically important crops (Ciuffo *et al.*, 2008). *Chrysanthemum stem necrosis virus* (CSNV) so far has been limited to occasional interceptions (Jones, 2005; Ravnikar *et al.*, 2003); its endemic presence in the Mediterranean has not yet been convincingly demonstrated.

Here, we present an update on the state of tospoviruses in the Mediterranean basin, broadly outlining the more general features of the genus and each virus species present in the area, with particular focus on their molecular biology, epidemiological features, relationships with their thrips vectors, and general control measures. For details of the diseases tospoviruses cause in specific vegetable crops, the reader is referred to other chapters in this volume.

II. RECENT ADVANCES IN TOSPOVIRUS RESEARCH

Comprehensive reviews covering tospovirus taxonomy and molecular biology have been previously published (Adkins, 2000; Goldbach and Peters, 1996; Pappu, 2008; Pappu *et al.*, 2009; Tsompana and Moyer, 2008). An excellent review giving a global perspective on tospovirus epidemiology (Pappu *et al.*, 2009) describes the 19 virus species characterized till 2009. Two possible new tospovirus species have been characterized since then: a tospovirus from Colombia affecting alstroemeria (Hassani-Mehraban *et al.*, 2010) and a new tospovirus on tomato from Thailand, named *Tomato necrotic ringspot virus* (Seepiban *et al.*, 2011).

The primary obstacle to tospovirus molecular biology research is the lack of a reverse genetic system. Such reverse genetic systems were developed for other genera in the *Bunyaviridae* family (Blakqori and Weber, 2005; Flick and Pettersson, 2001; Flick *et al.*, 2003a,b; Habjan *et al.*, 2008; Ogawa *et al.*, 2007). Still, a great deal is known about tospovirus molecular biology and determinants affecting their interactions with their plant and thrips hosts. All tospovirus species characterized so far have a conserved genome organization comprising three genomic segments (Fig. 2) classified as small (S), medium (M), and large (L). The S segment is ambisense and encodes the NSs and N proteins expressed through two subgenomic RNAs capped by a "cap snatch" mechanism (Kormelink *et al.*, 1992; van Knippenberg *et al.*, 2005). The NSs protein was discovered to be a potent-silencing suppressor in plants (Takeda *et al.*, 2002). The NSs protein was also shown to act as a silencing suppressor in tick cells (Garcia *et al.*, 2006), but there is no definitive evidence as yet that it interferes with the silencing pathway in thrips, since NSs may also interfere with other antiviral pathways in the insect. NSs is thought to be a plant-silencing suppressor for all tospoviruses. The N protein is the nucleocapsid, which encapsidates each tospovirus genomic RNA.

The M segment is also ambisense, encoding the glycoprotein precursor (Gn+Gc) and the NSm protein. NSm has been shown to be a movement protein, allowing infections to spread cell to cell within the host plants. After initial characterization as a tubule-forming cell-to-cell movement protein, NSm attracted renewed attention through a heterologous expression system that allowed the dissection of functional domains (Lewandowski and Adkins, 2005; Li *et al.*, 2009). The L segment is a negative-sense RNA encoding only the RdRP (Goldbach and Peters, 1996).

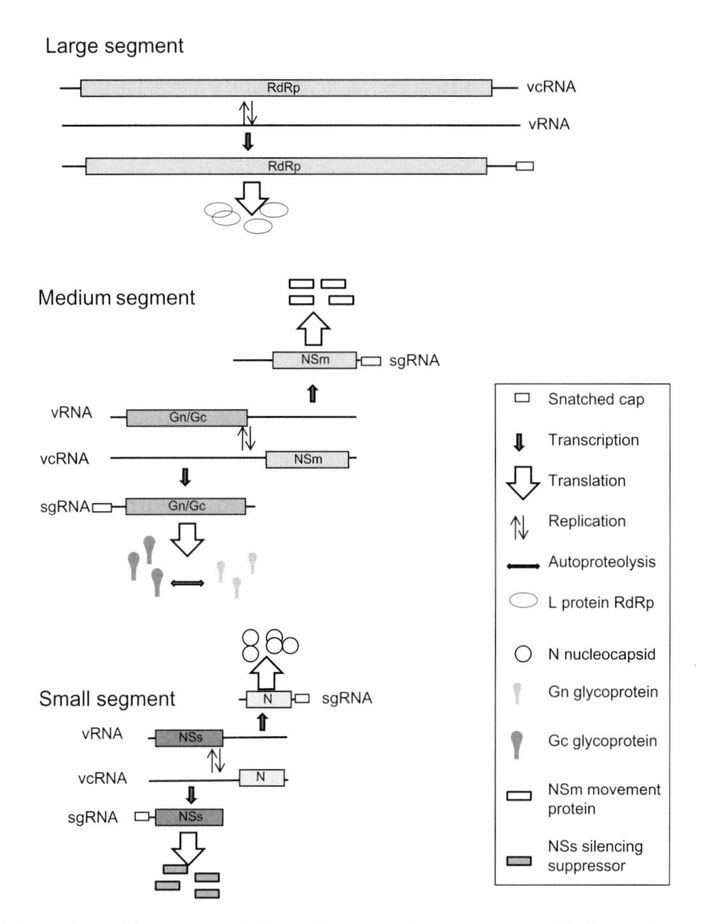

FIGURE 2 Schematic representation of a tospovirus genome with the main steps for protein expression, subgenomic (sg) transcription, and genome replication cycle. V, encapsidated genomic segment; VC, complementary strand to virus encapsidated genomic strand.

The bulk of tospovirus research in recent years has dealt with attempting to understand molecular determinants affecting interactions between tospoviruses, their host plants and their thrips vectors. An intriguing finding was that the preference of thrips for tospovirus-infected plants derives from a reduced antiherbivore defense in infected plants (Belliure *et al.*, 2005), whereas a direct effect of the virus on the thrips remains to be established. The tospovirus genetic determinants for thrips transmission have also been investigated (Sin *et al.*, 2005; Whitfield *et al.*, 2005, 2008). These data show that the Gn protein is a major determinant, likely binding to thrips receptors for uptake of the tospovirus and thrips infection. A possible practical implication of one of these experiments showed that supplying excess Gn protein to the thrips midgut can inhibit TSWV transmission by a competent thrips population, presumably by competing for binding sites on thrips gut receptors (German, 2009; Whitfield *et al.*, 2008). Much less progress has been made toward the identification of a receptor in the thrips–tospovirus interaction after initial promising results (Bandla *et al.*, 1998; Kikkert *et al.*, 1998; Medeiros *et al.*, 2000).

A recent groundbreaking study has brought the "omics" approach to the thrips–tospovirus interaction: a cDNA library of thrips mRNA has been characterized (Rotenberg and Whitfield, 2010), and future studies will likely take advantage of this library, particularly for elucidating the thrips antiviral pathways possibly activated during tospovirus infection.

III. TOSPOVIRUS DIAGNOSIS

Generally, the diagnosis of tospoviruses does not pose specific challenges, particularly for the tospovirus species occurring in the Mediterranean region. Tospoviruses cause systemic infections, and they accumulate abundantly in various tissues and organs. Only in the case of IYSV in onion should particular care be taken to properly select symptomatic areas, since infection in this plant is often localized (Tomassoli *et al.*, 2009). Excellent commercial-specific serological tests are available making rapid and accurate diagnosis relatively easy. Some problems might still be encountered, but so far these concern mostly viruses occurring in Asia (Chen *et al.*, 2010), where monoclonal antibodies are still necessary to distinguish different tospovirus species. One case where the taxonomic unit is not yet well defined, and soon might involve the Mediterranean, is that of *Tomato yellow ring virus* (TYRV), also known as *Tomato yellow ring fruit virus*, where, based on specific ecological niches, two species are proposed, although a polyclonal antiserum cannot distinguish them. Given that the two ecological niches are based only on host specificity, the "strain" status probably better describes this specific case (Hassani-Mehraban *et al.*, 2007). Serological analysis largely relies on DAS-ELISA, and squash blots have been used to detect TSWV in thrips (Aramburu *et al.*, 1996). Recent progress in field diagnosis was made with the use of immunostrips lateral flow devices (Roggero *et al.*, 2002b).

Molecular assays are also widely used in tospovirus diagnostics. Initially, cDNAs or riboprobes were used for detection by Dot blot (Huguenot *et al.*, 1990; Rice *et al.*, 1990; Ronco *et al.*, 1989). A standard RT-PCR method was first developed in 1994 (Mumford *et al.*, 1994), and around the same time, immunocapture-RT-PCR was developed in other laboratories (Nolasco *et al.*, 1993). An RT-PCR protocol was adapted to detect TSWV in single thrips (Mason *et al.*, 2003), and a qRT-PCR protocol using TaqMan™ chemistry was also developed for tospoviruses (Roberts *et al.*, 2000).

Particular effort was placed on generating general tospovirus tests, and Chu *et al.* (2001) showed that at least one pair of primers can be used for this purpose. A multiplex RT-PCR was recently developed for the detection of five different tospovirus species of interest for Asian countries (Kuwabara *et al.*, 2010).

In recent years, the tendency to shift from standard RT-PCR to qRT-PCR has not been widely applied for diagnosing tospoviruses in plants, but qRT-PCR was proven to be a major step forward in the quantification of tospovirus infection in single thrips (Boonham *et al.*, 2002).

IV. THRIPS VECTORS OF TOSPOVIRUS IN THE MEDITERRANEAN BASIN

Tospoviruses are transmitted in a persistent and propagative manner exclusively by thrips (*Thysanoptera*: *Thripidae*) (Whitfield *et al.*, 2005). The order *Thysanoptera* comprises more than 6000 known species (Mound and Morris, 2007), several of which are important plant pests and responsible for direct damage to many economically important crops. Fourteen thrips species now identified as tospovirus vectors belong to five genera *Frankliniella*, *Thrips*, *Scirtothrips*, *Ceratothripoides*, and *Dictyothrips* (*Thrysanoptera*: *Thripidae*) (Ciuffo *et al.*, 2010; Jones, 2005; Pappu *et al.*, 2009; Premachandra *et al.*, 2005; Tsompana and Moyer, 2008; Whitfield *et al.*, 2005). The thrips–tospovirus transmission relation is highly specific and unique in that first-instar larvae acquire the virus from infected plants. Upon ingestion, virions travel into the midgut where the virus replicates during a temperature-related incubation period and then infects muscle cells and salivary glands. Adults (sometimes already second-instar larvae) can transmit the virus by injecting saliva into plant tissues (Moritz *et al.*, 2004; Whitfield *et al.*, 2005). The thrips–tospovirus relationship is peculiar in that, these insect-infecting viruses during their evolution adapted to the plant kingdom and became major plant pathogens. This evolutionary pathway was followed by only two other virus families (*Rhabdoviridae* and *Reoviridae*), while most other insect-transmitted plant viruses do not replicate in the vector and are not phylogenetically related to insect-infecting viruses (Hull, 2002).

Currently, 14 thrips species are reported to be tospovirus vectors worldwide: *Frankliniella occidentalis* Pergande, *Thrips tabaci* Lindeman, and *F. schultzei* Trybom are the most frequently reported in addition to *Ceratothripoides claratris* Shumsher, *F. bispinosa* Morgan, *F. cephalica* Crawford, *F. fusca* Hinds, *F. gemina* Bagnall, *F. intonsa* Trybom, *F. zucchini* Nakahara & Monteiro, *Thrips palmi* Karny, *T. setosus* Moulton, *Scirtothrips dorsalis* Hood, and *Dictyothrips betae* Uzel (Ciuffo *et al.*, 2010; de Borbon *et al.*, 2006; Jones, 2005; Ohnishi *et al.*, 2006; Premachandra *et al.*, 2005;

Whitfield *et al.*, 2005). Among these species, the following six have been recorded in the Mediterranean: *F. occidentalis* (western flower thrips), *F. intonsa*, *F. schultzei* (cotton bud or common blossom thrips), *T. tabaci* (onion thrips), *T. palmi* (melon thrips), and *D. betae*. However, only *F. intonsa*, *T. tabaci*, and *D. betae* are natives of the Palaearctic region, whereas the others are exotic species though they can be now considered virtually cosmopolitan because of their worldwide dispersion through increased international trade (Vierbergen, 1995).

Thrips have likely been spread worldwide due to global commercial distribution of plants, aided in their dispersal by their small size and cryptic habits (e.g., eggs within leaf tissues, larvae in the flowers or between the inner leaves, pupae in the soil). These properties, as well as their high reproduction rate and polyphagy, make thrips difficult to control (Morse and Hoddle, 2006). Most thrips are generally polyphagous and can reproduce and develop on several host plants, crop and noncrop herbaceous and woody plants, feeding on parenchyma cells and pollen, especially the flower-living *Frankliniella* spp. By their feeding, they can cause direct economic damage, such as discoloration, silvering, deformity, scarring on leaves, flowers, and fruits, in vegetable, flower, and fruit crops (Childers, 1997).

Presently, the important thrips vectors of TSWV and IYSV in the Mediterranean basin are *F. occidentalis* and *T. tabaci*. The western flower thrips *F. occidentalis*, native of North America's west coast, is now the most widespread vector thrips worldwide (Jones, 2005). It first appeared in Europe in 1983 when it was found in ornamental nurseries in the Netherlands (Kirk and Terry, 2003). Since then, it has spread throughout the continent, colonizing crops in the Mediterranean in the late 1980s (Kirk and Terry, 2003). Thriving in Mediterranean climatic conditions, it can have continuous generations both in greenhouses and in outdoors, slowing down its development when temperatures fall. Direct damage due to feeding, and oviposition, has been reported on numerous ornamental, fruit (including grapes), and vegetable crops in the open field and greenhouses.

The first reported vector of TSWV was *T. tabaci*, which was long believed to be the main thrips vector (Pittman, 1927; Sakimura, 1962). Later studies showed, however, that not all TSWV isolates were transmitted by this species and that only arrhenotokous but not thelytokous populations were associated with tobacco infections of TSWV. Presently, the potential of *T. tabaci* populations to transmit tospoviruses is variable; this has been linked to differences in reproductive strategy (thelytokous and arrhenotokous populations) and/or host preference (Chatzivassiliou *et al.*, 2002; Wijkamp *et al.*, 1995), as well as differences in virus isolates (Tedeschi *et al.*, 2001). Recent evidence suggests that vector capability is an inheritable recessive trait (Cabrera-La Rosa and Kennedy, 2007).

Several populations of *T. tabaci* were found to be poor vectors of TSWV and unable to transmit other tospoviruses (Chatzivassiliou *et al.*, 2002; Tedeschi *et al.*, 2001; Wijkamp *et al.*, 1995). In contrast, *T. tabaci* is the only known vector of IYSV (Cortes *et al.*, 1998; Kritzman *et al.*, 2001; Nagata *et al.*, 1999).

From its eastern Mediterranean origin, *T. tabaci* has spread worldwide and is now cosmopolitan (Jones, 2005). Its populations can be made up of both sexes, or males are very scarce or absent and reproduction occurs by thelytokous parthenogenesis. However, the sex ratio varies widely over the species' geographic distribution and does not appear to be correlated with longitude, as had been previously thought (Kendall and Capinera, 1990). *T. tabaci* is highly polyphagous; it has been reported on vegetable and fruit crops (Jones, 2005; Tommasini and Maini, 1995) but shows a preference for plants of the *Alliaceae* family (onion, leek, garlic), on which its populations are usually composed of both females and males, unlike the all-female populations collected on other crops, such as sweet pepper, in the same geographic area (Bosco and Tavella, 2010; Vierbergen and Ester, 2000). The presence and abundance of male thrips in onion populations could be crucial because it might be correlated with tospovirus transmission efficiency, as was found for TSWV (Chatzivassiliou *et al.*, 2002; Wijkamp *et al.*, 1995).

Among other thrips vectors, *F. schultzei* and *T. palmi* are noted for their efficiency in tospovirus transmission, but mostly in regions outside the Mediterranean area. *F. schultzei* has a pantropical distribution; its origin remains unclear: it is generally considered to be from South America, although it might also have originated from Africa (Hoddle *et al.*, 2008). The common blossom thrips is a polyphagous species, mainly living in flowers and on plants involved in international trade (Jones, 2005). In the Mediterranean, it is found in Israel, Egypt, Morocco, and mainland Spain (Anon, 1999). *T. palmi* is believed to be from Southeast Asia, and it has spread both widely and rapidly in tropical and subtropical regions (Jones, 2005). The melon thrips has often been intercepted in international trade in Europe, especially in northern Europe where it can survive and overwinter in protected environments. Although past outbreaks were eradicated, in 2004 it was detected in the flowers of kiwi fruit in Portugal (Cannon *et al.*, 2007). Nevertheless, the occurrence of *T. palmi* must be monitored carefully because it is a major vector of several tospoviruses, different from those transmitted by *F. occidentalis* (Pappu *et al.*, 2009). In contrast, the Palaearctic *F. intonsa* is commonly found throughout Europe and Asia; it is polyphagous and normally occurs together with other thrips species in flowers due to its high dependence on pollen (Jones, 2005; Palmer *et al.*, 1989). Its presence in the Mediterranean has been documented by several field surveys, especially on cotton, tobacco, strawberry, vegetables (Cosmi *et al.*, 2003; Deligeorgidis *et al.*, 2002;

Martini *et al.*, 2009). Despite its dispersion, its competence for transmission was assessed only in one laboratory trial (Wijkamp *et al.*, 1995). Unlike other thrips species, not much is known about *D. betae*, newly identified as the vector of PolRSV (Ciuffo *et al.*, 2010). It has been reported in many parts of the eastern Palaearctic region (Strassen zur, 2007) but has been rarely found and is poorly investigated because of its limited economic impact. Its presence has been mainly recorded on weeds, among which *Polygonum convolvulus* and *P. dumetorum* are hosts of this new tospovirus (Ciuffo *et al.*, 2008).

V. TOSPOVIRUS SPECIES AFFECTING VEGETABLE CROPS IN THE MEDITERRANEAN BASIN

A. *Tomato spotted wilt virus*

TSWV is by far the economically most important tospovirus in the Mediterranean basin, particularly because it is one of the major threats (or the major threat in some specific situations) for tomato and pepper crops. But given its wide host range, TSWV is also a threat to other horticultural crops such as artichoke (Gallitelli *et al.*, 2004; Testa *et al.*, 2008; Vovlas and Lafortezza, 1994), lettuce (Moreno *et al.*, 2004), celery and basil (Gallo *et al.*, 1995), eggplant (Betti, 1992; Parisi *et al.*, 1998), and chicory (Vovlas and Lafortezza, 1992). TSWV was described as causing disease at the beginning of the twentieth century (Brittlebank, 1919), and in Europe, it was first described in England in 1932 (Smith, 1932). Effective control of its vector (mainly *T. tabaci* at the time) resulted in a decline in TSWV-caused diseases in western Europe, whereas TSWV has remained present in eastern Europe and in northern Greece on tobacco (Tsakiridis and Gooding, 1972). TSWV began to be a major threat to European horticulture in the 1980s, when a new and more efficient vector, the Nearctic western flower thrips *F. occidentalis*, extended its reproductive area to the northern and southern hemispheres from its center of origin, the western United States (Smith *et al.*, 1992). In the Mediterranean, this thrips species was first reported in northern Italy in 1987 (Rampinini, 1987), after its initial introduction in northern Europe. Since the introduction of this new vector, TSWV has become a major threat, causing severe yield losses in many vegetable and ornamental crops all over the world.

TSWV infection is characterized by severe symptoms on the majority of the plant species it infects. On tomato, it causes leaf bronzing, small brown flecks, stunting, and dieback of growing tips; ringspots often appear on green fruit, which turn yellow on mature fruits. In pepper, TSWV causes severe stunting of young plants, and chlorotic mosaic or yellow flecking of the leaves; infections of mature plants cause chlorotic line patterns with necrotic spots. Necrotic spots are also present on pepper fruits, often also displaying ring patterns. In lettuce, TSWV can cause general yellowing, with necrotic spotting, and distortion and necrosis of the heart leaves. Further, a distinguishing feature is the large number of species it can infect: from ca. 150 reported in 1968 (Best, 1968), to 650 in 1994 (Goldbach and Peters, 1994) to about 1100 in 2003 (Parrella *et al.*, 2003). Its epidemiology is complicated by the innumerable infected weeds that serve as a reservoir for primary infection.

TSWV is likely established in all Mediterranean countries, although information about specific situations in some areas of North African countries is not exhaustive. In the Mediterranean, TSWV became a problem in horticultural and ornamental crops first in Italy (Lisa *et al.*, 1990; Vaira *et al.*, 1993), Spain (Jordà, 1993) and France (Marchoux *et al.*, 1991), then in Portugal (Louro, 1996), Slovenia (Mavric and Ravnikar, 2001), Albania (Çota and Merkuri, 2004), Greece (Chatzivassiliou *et al.*, 1996, 2000), Turkey (Yardimci and Kilic, 2009; Yurtmen *et al.*, 1999), Israel (Antignus *et al.*, 1994, 1997), Jordan (Anfoka *et al.*, 2006), Egypt (El-Wahab *et al.*, 2008), and Tunisia (Ben Moussa *et al.*, 2000, 2005). Its presence in Montenegro was also recently ascertained (Zindovic *et al.*, 2011).

TSWV isolates have been collected in various areas around the world, and sequences of some of the protein encoding regions are deposited in public databases. A thorough study of the TSWV population structure and of the driving forces of TSWV evolution was carried out by Tsompana *et al.* (2005). Based on phylogenetic analysis, the study suggested a geographical structuring of TSWV population possibly due to founder effects; with rare exceptions likely due to geographic exchanges of isolates (gene flow was shown between some subpopulations), five subpopulations were detected in 41 isolates characterized in their N gene region. No intragenic recombination was detected in the population. The study also showed positive selective pressure on specific sites.

As for the Mediterranean, a different phylogenetic analysis in the NSm coding region showed two well-differentiated populations of distinct isolates in a small area in Apulia in southern Italy (D-type and A-type) (Sialer *et al.*, 2002).

Sequences of the N gene of many isolates from the Mediterranean region are now in the databases. Since no differences were shown between different coding regions of the tospovirus genome for determining population structure, we performed a phylogenetic analysis of the Mediterranean isolates based on the N gene sequence (Fig. 3). Our analysis showed that these isolates group uniformly in one of five subpopulations in some countries (France, Spain, Bulgaria), whereas in other countries, the isolates clustered into two or three distinct subpopulations as in Italy, confirming the complexity of the interchanges of isolates in the area. Noteworthy is that an Egyptian isolate present in the database

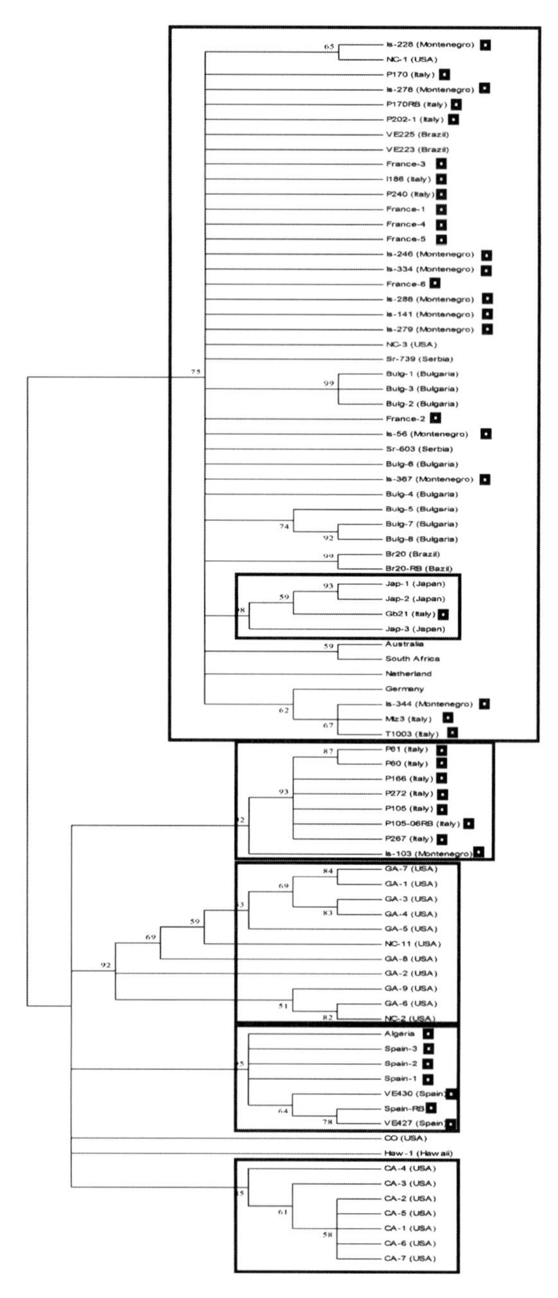

FIGURE 3 Consensus phylogenetic tree of *Tomato spotted wilt virus* isolates from nucleocapsid (N) sequences obtained from the GenBank. Isolates from the Mediterranean basin are marked by a square dot. The tree was drawn using the neighbor-joining method, but similar topologies were obtained with other methods. The major clades supported by bootstrap values over 60 are included in squares. The name of the each isolate is that associated to the sequence in the GenBank database.

(GenBank accession no. DQ479968), although labeled TSWV, is by far the most distant from all the others, with only 83% identity in the N region at the protein level. Further characterization of the isolate could shed more light on this potentially new TSWV strain.

B. *Impatiens necrotic spot virus*

INSV is the second member of the genus *Tospovirus* to have been characterized and distinguished from TSWV in the late 1980s (Law and Moyer, 1990). In Europe, the virus was first identified in the Netherlands on New Guinea impatiens (isolate NL-07) (De Avila *et al.*, 1992). Later, its presence was reported throughout Mediterranean countries: the Italian Riviera (Vaira *et al.*, 1993), Slovenia (Mavric and Ravnikar, 2001), France (Marchoux *et al.*, 1991), Spain (Lavina and Battle, 1994), Portugal (Louro, 1996), and Israel (Gera *et al.*, 1999).

Presently, INSV is widely dispersed owing to the distribution of its vector *F. occidentalis*. Its host range is second only to TSWV among the tospoviruses and includes, besides ornamental plant species, many horticultural plants. Among these, natural infection in some economically important vegetable crops has been described. Sweet pepper infection by INSV was reported in the Netherlands in 1998 (Verhoeven and Roenhorst, 1998): symptoms on plants were described only on fruits, which may be explained by the preference of thrips for flowers instead of leaves. The study also reported INSV in spinach and pepino, but no symptom descriptions were given.

In the Mediterranean, natural infection in some horticultural plants in both field and protected crops was reported (Vicchi *et al.*, 1999): in lettuce, INSV caused necrosis on leaves along veins, simple and concentric chlorotic rings, and malformation and necrosis of youngest leaves. In cucumber, it caused chlorotic rings and severe deformation only on fruits, but infection seemed to be localized only on symptomatic fruits without true systemic invasion. Indeed, the virus was detected only in symptomatic fruits and not on leaves or asymptomatic fruits on the same plant. In pepper, deformation on apical leaves and chlorotic/necrotic areas were observed (Roggero *et al.*, 1999). INSV was also shown to cause only local infection on pepper in Slovenia (Mavric and Ravnikar, 2001). Mixed infections with TSWV were detected in some hosts, such as in tomato in Apulia in Italy (Finetti Sialer and Gallitelli, 2000). High temperature (constant 33 °C) blocks the movement of INSV in *Capsicum* spp. (Roggero *et al.*, 1999) but does not alter its infectivity.

A number of studies on INSV investigated whether infected weed species could be sources of infection for ornamental and horticultural crops. In Japan, for example, the susceptibility of 32 weed species in different families was evaluated: common weed species, such as *Stellaria media*, *Portulaca oleracea*, *Cerastium glomeratum*, *Cardamine scutata*, and *Veronica arvensis*, were found susceptible to INSV (Okuda *et al.*, 2010).

For many years, the western flower thrips *F. occidentalis* was the only thrips reported to be able to transmit INSV with an efficiency between 80% and 90% (Deangelis *et al.*, 1993, 1994; Wijkamp and Peters, 1993). In 2004, however, it was demonstrated that *F. intonsa* was another INSV vector, though its efficiency was far lower than that of *F. occidentalis* (males 18%, females 4% vs. about 80% for *F. occidentalis*) (Sakurai *et al.*, 2004). More recently, *F. fusca* was also demonstrated to be an INSV vector (Naidu *et al.*, 2001). *T. tabaci* was confirmed not to be a vector for INSV in Italian thrips populations (Tedeschi *et al.*, 2001). Nevertheless, direct analysis on insects showed the presence of the virus, indicating that a given vector's inability to transmit it is not related to acquisition but instead to failure of the virus to invade the vector's specific organs (e.g., salivary glands).

C. *Iris yellow spot virus*

IYSV was the last among the various tospovirus species to be reported in the Mediterranean region. It causes economic losses in a wide variety of monocotyledonous plants, particularly on onions (Gent *et al.*, 2006; Pappu *et al.*, 2009). IYSV is transmitted by the onion thrips *T. tabaci* (Chatzivassiliou *et al.*, 1999; Kritzman *et al.*, 2001; Nagata *et al.*, 1999). The discovery of this virus has at least four independent sources, although its phylogeny suggests a possible Middle Eastern-European origin. Initial reports of IYSV came from the Treasure Valley in northwestern United States (Hall *et al.*, 1993), Brazil (Pozzer *et al.*, 1999), the Netherlands (Cortes *et al.*, 1998), and Israel (Gera *et al.*, 1998; Kritzman *et al.*, 2000). IYSV is now present worldwide (Pappu *et al.*, 2009). In Europe, after the initial report in the Netherlands (Cortes *et al.*, 1998), IYSV epidemics in onions and leeks have also been confirmed in Slovenia (Mavric and Ravnikar, 2000), Spain (Cordoba-Selles *et al.*, 2005, 2007), Greece (Chatzivassiliou *et al.*, 2009), and Serbia (Bulajic *et al.*, 2009). In Italy, IYSV has only recently been reported and characterized (Tomassoli *et al.*, 2009), although its presence was detected before (Cosmi *et al.*, 2003). Onion infection with IYSV has also been reported in France and Germany (Huchette *et al.*, 2008; Leinhos *et al.*, 2007), but molecular features of these isolates are not yet available in the databases. The S, M, and L segments of the virus have been sequenced, and the N coding region of isolates has been used for studying the diversity of IYSV populations in various regions of the world (Bag *et al.*, 2010; Nischwitz *et al.*, 2007; Pappu *et al.*, 2006; Smith *et al.*, 2006).

Two recent studies included isolates from the Mediterranean (Bulajic *et al.*, 2009; Tomassoli *et al.*, 2009) and both showed that the IYSV isolates group in different clades. The Slovenian leek isolate constitutes one clade;

a second clade includes isolates from the Netherlands, Israel, Japan, and Australia; a third clade includes mostly isolates from the western United States, Chile, Guatemala, one isolate from Serbia and one from Brazil; a fourth clade includes isolates from Peru and Georgia (United States); a fifth statistically well-supported clade includes isolates from Serbia, Spain, and Italy, with the Italian isolates distinguished into two different subclades. As mentioned, the isolates from the Palaearctic region are the most diverse (they group in four different clades) and the fact that the clade with the most distant isolates is the one collected from leek in Slovenia (although verification of this sequence is needed) supports the hypothesis that the Mediterranean basin may be the origin of this virus. Molecular diversity seems to correspond to biological diversity, although direct comparison among isolates is arduous because of the difficulties in mechanical transmission. Symptoms on diseased onion plants are often localized, diamond-shaped lesions. Symptomless infections were also found to occur in Serbia and the Netherlands (Bulajic *et al.*, 2009), pointing out the likelihood of different strains in Europe. Evidence for serological diversity among IYSV isolates was also shown (Tomassoli *et al.*, 2009).

There are no clear estimates of disease severity or economic damage caused by IYSV on vegetable crops in the Mediterranean basin. More epidemiological data are needed to establish an integrated disease management (IDM) approach for this pathogen. No seed transmission of the virus has been observed (Bulajic *et al.*, 2009), and a direct role of weeds as a source of IYSV infection in onion field is yet to be proved (Gent *et al.*, 2006). However, proximity with other ornamental crops, hosts for the virus, was thought to be a contributing factor to IYSV infection in vegetable crops (Kritzman *et al.*, 2001). Several weeds were found to be naturally infected with IYSV (Evans *et al.*, 2009a,b). Volunteer-infected onions and overwintering IYSV-infected thrips diapausing or quiescent in the soil may play a role in primary infection in onion fields (Gent *et al.*, 2006). The same authors also highlighted differences in the epidemiology of IYSV versus TSWV. IYSV has only one thrips species as vector and its natural host range is mostly restricted to *Liliaceae* and *Alliaceae*, whereas TSWV is transmitted by a number of thrips species and its natural host range covers more than 50 families. Further, *T. tabaci* has an ability to rapidly develop large populations in onion crops. This implies potential secondary spread during the growing season and the possibility of interfering with virus spread through thrips control using insecticide. This possibility is limited in TSWV-infected vegetable crops, where secondary spread is minimal, and most of the infection comes from infected weeds or neighboring infected crops (Culbreath *et al.*, 2003; Gitaitis *et al.*, 1998). Management options for onion bulb or seed production have been reviewed (Gent *et al.*, 2006), but no specific studies have investigated their application in the Mediterranean.

VI. EPIDEMIOLOGY AND CONTROL MEASURES FOR TOSPOVIRUSES

A. Tospovirus epidemiology

Tospoviruses are generally regarded as having specific epidemiological features influenced by the relationship with their insect vectors. All tospoviruses replicate in their vectors, and individual thrips are capable of transmitting the virus acquired only during the first two larval stages. These two features imply that the vector is also a virus host and that there is a certain time lapse (2–3 weeks) between acquisition and the ability to transmit the virus. A viruliferous vector then requires probes of 5 min in order to transmit the virus to a plant host. A third feature shared by the majority of the economically important tospoviruses is their wide natural host range, which overlaps with the fairly wide host range of the main thrips vector species in the Mediterranean. Further, from an epidemiological point of view, it is essential to establish for each plant species whether the thrips can simply feed or can complete their whole life cycle, including reproduction. No tospovirus has been shown convincingly to be transmitted through seed. Therefore, for each specific horticultural crop, disease epidemiology is influenced by virus-related variables and biological features of the insect in addition to specific virus–plant host interactions. Since very few studies to date have investigated the specificity of tospovirus infection on vegetable crops in the Mediterranean, information needs to be culled from studies carried out mainly in the United States and Australia, where the epidemiology of the important tospoviruses has been studied.

Other sources of infection are volunteer plants or nearby infected ornamental or vegetable crops, as well as infected weeds. The identification of reservoirs in specific outbreaks has been the goal of numerous studies, wherein each geographical and agronomic situation has its specific reservoir hosts (Bautista *et al.*, 1996; Bitterlich and MacDonald, 1993; Groves *et al.*, 2002; Latham and Jones, 1997). In the Mediterranean region, the question of reservoir hosts was addressed in Italy (van Os *et al.*, 1993), Turkey (Arli-Sokmen *et al.*, 2005), and Greece (Chatzivassiliou *et al.*, 2001, 2007).

One of the most controversial questions about the epidemiology of tospovirus epidemics is whether the spread within the field is only from primary infection or whether secondary spread occurs in the field. The issue has been tackled in a number of different specific situations, and the answer depends essentially on crop duration (a tendency to polycyclic epidemics only applies to longer lasting crops), whether the specific crop is a host for the vector, and whether a sufficient number of weed or volunteer host plants are present within the crop. In lettuce, for example, all infections were primary (Bald, 1937; Wilson, 1998), whereas secondary spread was found on tobacco in Greece (Thresh, 1983). In a number of other systems, infection was predominantly primary, with limited secondary spread (Camann *et al.*, 1995; Coutts *et al.*, 2004; Gitaitis *et al.*, 1998). But in the favorable climatic conditions of the Mediterranean, thrips vectors have continuous generations on the crops, resulting in a greater occurrence of secondary spread, as suggested from a study in sweet pepper crops in greenhouses in the Italian Riviera (Tavella *et al.*, 1997). This problem deserves further investigation given its implications for the possible efficacy of insecticide interventions against thrips.

B. Control measures: General principles

Because of difficulties with chemical control of thrips populations, an IDM approach is key to containing tospovirus epidemics in vegetable crops. Among the specific studies in the Mediterranean, few were conducted in open fields (Fanigliulo *et al.*, 2009; Garcia *et al.*, 1997). As some of the conditions are similar, here we will only summarize common control strategies employed in IDM and refer the reader to specific papers on tospovirus control in tomato (Cho *et al.*, 1989; Momol *et al.*, 2004; Riley and Pappu, 2000), pepper (Garcia *et al.*, 1997; Reitz et al., 2003), onion (Gent *et al.*, 2006), and lettuce (Coutts *et al.*, 2004; Yudin *et al.*, 1990). For a theoretical discussion and a general IDM approach to tospoviruses, the reader is referred to relatively recent and comprehensive reviews (Jones, 2004, 2006; Pappu *et al.*, 2009).

The following outline mentions the main possible interventions, taking into account that some are well suited to open field situations, whereas others are limited to nurseries or greenhouses; all three situations are present in vegetable crops in the Mediterranean.

The aim of the most common phytosanitary intervention is to minimize the primary source of inoculum by eliminating cultural residues at the end of the crop cycle and by isolating vegetable crops from ornamentals, weed, or volunteer plant control. Roguing can also be effective since most tospovirus–host combinations have very obvious symptoms. The use of virus-free seedlings is also recommended. To this purpose, drenches of healthy transplants with systemic neonicotinoid insecticides just before transplanting may suppress the onset of epidemics (Coutts and Jones, 2005), particularly in short-duration crops like lettuce. Agronomic measures to be employed include scheduling planting date to avoid thrips population peaks, use of silver mulches to diminish thrips landing rates, use thrips-proof netting in nurseries and greenhouses, and when possible, interfering with wind transport by employing windbreaks. Finally, allowing for susceptible crop and weed-free periods in wide areas is also a suggested cultural practice, although very difficult to deploy in the Mediterranean area.

These phytosanitary and agronomic measures are chiefly preventive and therefore can only limit, but not control, tospovirus epidemics; some are particularly difficult to carry out owing to the conditions specific to the Mediterranean that hamper regional interventions. Other measures as the use of crops carrying resistance genes for tospoviruses, chemical control with insecticides, and biocontrol of thrips should be considered. In this context, a risk assessment for interventions, such as that successfully adopted for groundnut in the United States (Brown *et al.*, 2005), would be particularly useful for each vegetable crop for which tospoviruses are a limiting factor.

C. Resistance genes to tospoviruses and tospovirus resistance-breaking strains

The best way to control tospoviruses is the use of commercial varieties carrying resistance genes introgressed generally from closely related species. Most seed companies now list in their catalogs different varieties generally termed "tolerant" to TSWV infection. These usually carry the *Sw-5* gene (tomato) or the *Tsw* gene (pepper).

An initial screen of *Lycopersicon esculentum* and *L. pimpinellifolium* identified a range of dominant and recessive alleles, which, however, could not be used, since the resistance proved to be strain specific (Finlay, 1953). Screening of other *Lycopersicon* species, such as *L. hirsutum* or *L. chilense*, identified resistances controlled by polygenic systems or resistance that could not be introgressed in *L. esculentum* (Rosello *et al.*, 1998). In surveys of *L. peruvianum* genetic resources, a number of accessions showed resistance to TSWV and to other closely related tospoviruses, and it is in this species that the *Sw-5* gene was found (Boiteux and Giordano, 1993; Paterson *et al.*, 1989). Molecular markers have been developed to assist in the breeding process, and SCAR locus-specific codominant markers recently became available (Dianese *et al.*, 2010). The *Sw-5* gene was cloned (Brommonschenkel and Tanksley, 1997; Brommonschenkel *et al.*, 2000) and characterized as belonging to the class of the NBS-LRR resistance gene.

Another resistance gene in tomato, named *Sw-6*, derived from line UPV 32 (Rosello *et al.*, 1998, 2001) proved to be isolate specific and not useful in breeding programs. From *L. chilense* accession LA1938, the *Sw-7* resistance gene was derived (Canady *et al.*, 2001; Scott *et al.*, 2005). A recent screening of two *L. peruvianum* collections revealed other possible sources of resistance, although initial attempts at introgressing long-term durable resistance other than *Sw-5* failed (Gordillo *et al.*, 2008).

In pepper, resistance to TSWV was identified in accessions PI159236, PI152225, CNPH-27, and 7204 (Black *et al.*, 1991; Boiteux and De Avila, 1994; Stevens *et al.*, 1992). This resistance to TSWV is monogenic, dominant, and located in the same locus in all the above-mentioned accessions (Boiteux, 1995; Boiteux and De Avila, 1994; Moury *et al.*, 1997). The resistance gene, named *Tsw*, was mapped to the distal portion of chromosome 10 (Jahn *et al.*, 2000). In a more recent screening, ECU-973, an accession of *Capsicum chinense*, was shown to carry a resistance phenotypically similar to that of *Tsw* (Cebolla-Cornejo *et al.*, 2003).

The resistance gene *Tsw* is widely used in the majority of commercial pepper varieties as the unique source of TSWV resistance. However, high temperatures and inoculation at an early stage of plant development results in breakdown of resistance. The mechanism of resistance is at the cell-to-cell movement level, and hypersensitive response observed in the original accession of *C. chinense* is not the major component of resistance (Margaria *et al.*, 2007). Further, the hypersensitive response and dehiscence of the inoculated leaf observed in the *C. chinense*-resistant accessions did not occur in *Capsicum annuum* commercial cultivars, where the resistance gene was introgressed.

TSWV also can occasionally inflict severe damage on potato crops, causing necrotic areas in infected tubers (Abad *et al.*, 2005; Al-Shahwan *et al.*, 1997; Pourrahim *et al.*, 2001). A screening for resistance did not show any variability of the cultivars tested when inoculation by thrips was performed in open field conditions. However, mechanical inoculation showed consistent differences among cultivars in the efficiency of translocation from infected plant to tuber and from infected tuber to progeny plants, two distinctive steps in TSWV epidemiology in this crop (Wilson, 2001).

Resistance-breaking (RB) strains for *Tsw* and *Sw-5* genes were reported initially during cultivar selections. The occurrence of RB strains in open field conditions was reported in Spain (Aramburu and Marti, 2003; Margaria *et al.*, 2004) and Italy (Ciuffo *et al.*, 2005; Roggero *et al.*, 2002a). Further, it was recently shown that mixed infection of TSWV with tomato criniviruses can also break the *Sw-5* resistance (Garcia-Cano *et al.*, 2006). An initial molecular characterization of an RB strain on tomato carrying *Sw-5* was performed through reassortment, and the TSWV M segment genomic RNA was shown to carry the avirulence factor (Hoffmann *et al.*, 2001). When the M segment of RB isolates from Spain were compared to wild-type isolates from the same region, two mutations in the NSm region (C118Y and T120N) were consistently observed (Lopez *et al.*, 2011). Unfortunately, few RB strains from other countries were included in this study from which a possible "marker" character for RB mutation in TSWV overcoming *Sw-5* might have been derived.

In research on isolates breaking the *Tsw* resistance genes in pepper, an initial study using reassortant isolates mapped the viral genetic determinants to the S segment (Jahn *et al.*, 2000). A study on TSWV isolates from Italy and Spain field epidemics and from RB strains obtained in controlled conditions showed that different mutations in the NSs gene are associated with the RB phenotype (Margaria *et al.*, 2007). RB isolates expressing truncated forms of the NSs protein were characterized, showing that NSs is responsible for maintaining infection in newly emerging leaves (Margaria *et al.*, 2007).

A further possibility of field resistance against tospoviruses is "resistance" to thrips. A source of resistance to thrips was found in pepper, and it is now being introgressed in commercial pepper varieties. Its indirect effect on TSWV incidence in open field was postulated based on lower reproduction rate and lower preference (Fery and Schalk, 1991; Maris *et al.*, 2003a,b, 2004; Peters *et al.*, 2007). Onion and leek varieties resistant to thrips have also been investigated (Brar *et al.*, 1993; Coudriet *et al.*, 1979).

Numerous studies have dealt with transgenic resistance to tospovirus in various vegetable crops using pathogen-derived resistance (Accotto *et al.*, 2005; Gubba *et al.*, 2002; Herrero *et al.*, 2000; Kim *et al.*, 1994; Pang *et al.*, 1996; Pappu, 1997; Yang *et al.*, 2004), but this technology awaits regulatory approval in Europe and acceptance among consumers is still low.

D. Thrips control

Reduction of external virus sources and prevention of the development of vector populations are key to effective control in an integrated management program (Jones, 2004). The use of insecticides remains a mainstay in the management of TSWV spread, but the intensity of insecticide treatment could lead to insecticide resistance for both *F. occidentalis* (Bielza, 2008; Brodsgaard, 1989) and *T. tabaci* (Foster *et al.*, 2010; Martin *et al.*, 2003; McIntyre Allen *et al.*, 2005). Alternative pest management approaches such as intercropping (Theunissen and Schelling, 1997) or the evaluation of biologically active plant volatiles against *T. tabaci* (Koschier *et al.*, 2002) could be useful.

The combined use of UV-reflective mulch, acibenzolar-*S*-methyl, and insecticides was highly effective in reducing TSWV incidence in tomatoes in Florida (Momol *et al.*, 2004) and southern Italy (Fanigliulo *et al.*, 2009). The positive effect of UV-reflective mulch to manage TSWV on crisp lettuce in Spain was also assessed (Diaz *et al.*, 2006), while the positive impact of straw mulch on *T. tabaci* populations on onion was assessed in New York (Larentzaki *et al.*, 2008).

Biotic agents, such as predatory mites and bugs, pathogenic fungi, and nematodes, integrated with other physical and cultural tactics (i.e., screens, traps, nursery hygiene, resistant cultivars, and agro-ecosystem manipulation), have been successful in thrips control inside greenhouses (Jacobson, 1997; Sanchez and Lacasa, 2006). For example, species of the genus *Orius* (*Hemiptera*: *Anthocoridae*) are well known as predators able to control pest outbreaks in different crops. Some Nearctic species have been widely investigated because of their efficiency in controlling thrips (Funderburk *et al.*, 2000; Sabelis and van Rijn, 1997), and they are produced by commercial insectaries and largely used in IPM programs. After the accidental introduction of *F. occidentalis* in Europe, some Palaearctic species were discovered to be effective predators of *F. occidentalis* and of other thrips on horticultural crops and to be better adapted to the environmental conditions than the Nearctic species (Riudavets, 1995; Tavella *et al.*, 1994, 2000; van de Veire and Degheele, 1992). Several native Mediterranean species of *Orius* are now well known as thrips control agents including *O. laevigatus* (Fieber) in southern France on protected strawberry (Villevieille and Millot, 1991), in Greece on both weed plants and pepper (Barbetaki *et al.*, 1999), in Egypt on corn and cotton (Tawfik and Ata, 1973), in Spain and Italy on pepper (Lacasa *et al.*, 1996; Tavella *et al.*, 1991, 1994; Vacante and Tropea Garzia, 1993); *O. niger* Wolff in Greece on different cultivated plants (Barbetaki *et al.*, 1999; Lykoyressis, 1993), in Italy on strawberry, pepper, basil, green bean, and lettuce (Bosco *et al.*, 2008; Tavella *et al.*, 1994); *O. albidipennis* Reuter in Egypt on corn and cotton (Zaki, 1989), and in south-western Spain (Lacasa *et al.*, 1996).

VII. POTENTIAL FUTURE THREATS OF TOSPOVIRUSES TO VEGETABLE CROPS IN THE MEDITERRANEAN BASIN

The Mediterranean vegetable crops where tospoviruses create an economically relevant problem are still limited for the time being. Elsewhere in the world, however, other tospovirus species are indeed important, and other vegetable crops, including *Cucurbitaceae* and *Fabaceae*, are also affected by tospoviruses. Therefore, although the Mediterranean in general is affected by tospoviruses, the situation could worsen considerably if new tospovirus species, present in Asia and the Americas, were introduced in the Mediterranean. Within this scenario, the question of potential thrips vectors gains utmost importance: one example of a foreseeable immediate threat is the number of species of tospoviruses transmitted by *T. palmi*, and among these, the recently characterized TNSV on tomato in Thailand. The vector species is not yet endemic to the Mediterranean but it has been occasionally intercepted (see previous section).

So far, cucurbitaceous crops in the Mediterranean are not challenged by tospoviruses and sporadic detection of TSWV and INSV never led to a serious economic problem. This situation could change in the near future. A recently characterized new tospovirus species found in melon crops in Mexico, named *Melon severe mosaic virus* (Ciuffo *et al.*, 2009), could pose a threat, although the specific thrips vector species for this virus is still unknown.

Another interesting case, which is now only a botanical curiosity, is the widespread occurrence in Italy of PolRSV. This tospovirus has a potential host range extending to solanaceous hosts, but is naturally limited to buckwheat by its strict association with its thrips vector *D. betae* (Ciuffo *et al.*, 2008, 2010). So far, only two species in the family *Polygonaceae* have been reported as PolRSV hosts. Nevertheless, the virus could endanger tomato crops if *D. betae* adapts to solanaceous hosts or if PolRSV adapts so as to be transmitted by other thrips species such as *F. occidentalis* or *T. tabaci*.

Another potential threat to watch out for is CSNV, a tospovirus species intercepted occasionally in Europe and found in England before it was eradicated (Mumford *et al.*, 2003; Verhoeven *et al.*, 1996). Unclear is whether its presence in Slovenia is a case of ''interception'' or it is indeed present in the area (Ravnikar *et al.*, 2003).

Another possible threat, given Iran's proximity to the Mediterranean, comes from the recently characterized tospoviruses present there (TYRV and TFYRV), which infect tomato and potato (Hassani-Mehraban *et al.*, 2005; Winter *et al.*, 2006) and are transmitted by *T. tabaci*, a vector endemic to the Mediterranean.

The risk of importing new tospoviruses (and/or their thrips vectors) to the Mediterranean is particularly high, not only through the exchange of fresh market produce with other areas of the world but also through the exchange of germplasm. Continuous updating of the virus alert list is therefore mandatory. Far more important over the long term is preventing the exchange of infected materials. This could be achieved by improving the diagnostics and the new disease characterization capabilities in those developing countries where the climatic and economic conditions are well suited for carrying out vegetable crop breeding programs.

VIII. CONCLUSIONS

The tospovirologist community is undergoing a generational change, but this shift does not appear to hamper the knowledge or advancements in tospovirus research. New technological platforms allow somewhat for circumventing the lack of a reverse genetic system for this group of viruses. Nevertheless, we think that further attempts at developing a reverse genetic system should be a part of the research agenda. The insect–virus relationships have been examined in other model systems. We believe that the study of TSWV–thrips interactions will lead to novel discoveries. In this respect, although for several reasons thrips are not as amenable as fruit flies to genetic studies, sequencing their genome would be an invaluable tool for studying their interactions with tospoviruses, opening new inroads into pest control with the identification of specific targets in the tospovirus–thrips interaction. A thrips genome-sequencing project awaits adequate funding and leadership, but increasingly cheaper sequencing costs and advances in bioinformatics are bringing it within reach. The search for a specific receptor molecule, for which there is a growing body of indirect proof, will gain momentum from the use of new proteomic approaches once the thrips genome sequence becomes available.

Within the Mediterranean, commercial trade between the countries along its northern and southern rim is growing. In addition to the historically close ties with Israel, commercial and technological exchanges with other Middle Eastern and North African countries have increased. Accordingly, cooperative projects for the development of facilities and knowledge transfer in plant disease monitoring in these countries are being carried out through European Union and international programs. Such investment is vital for assisting these countries in implementing effective pest and disease control strategies and reducing the risk of new potential threats to vegetable crops in the Mediterranean.

ACKNOWLEDGMENTS

We are grateful to Hanu Pappu, Bryce Falk, and Lara Bosco for reading the chapter.

REFERENCES

Abad, J. A., Moyer, J. W., Kennedy, G. G., Holmes, G. A., and Cubeta, M. A. (2005). *Tomato spotted wilt virus* on potato in eastern North Carolina. *Am. J. Potato Res.* **82:**255–261.

Accotto, G. P., Nervo, G., Acciarri, N., Tavella, L., Vecchiati, M., Schiavi, M., Mason, G., and Vaira, A. M. (2005). Field evaluation of tomato hybrids engineered with *Tomato spotted wilt virus* sequences for virus resistance, agronomic performance, and pollen-mediated transgene flow. *Phytopathology* **95:**800–807.

Adkins, S. (2000). *Tomato spotted wilt virus*—Positive steps towards negative success. *Mol. Plant Pathol.* **1:**151–157.

Al-Shahwan, I. M., Abdalla, O. A., and Al-Saleh, M. A. (1997). Viruses in the northern potato-producing regions of Saudi Arabia. *Plant Pathol.* **46:**91–94.

Anfoka, G. H., Abhary, M., and Stevens, M. R. (2006). Occurrence of *Tomato spotted wilt virus* (TSWV) in Jordan. *OEPP/EPPO Bull.* **36:**517–522.

Anon(1999). *Frankliniella schultzei* (Trybom) . *CABI/EPPO Distribution Maps of Plant Pests No. 598.* CAB International, Wallingford, UK.

Antignus, Y., Gera, A., Pearlsman, M., Ben-Joseph, R., Ganaim, N., Raccah, B., and Cohen, S. (1994). *Tomato spotted wilt*, a new disease of flowers and vegetables in Israel. *Phytoparasitica* **22:**86–87.

Antignus, Y., Lapidot, M., Ganaim, N., Cohen, J., Lachman, O., Pearlsman, M., Raccah, B., and Gera, A. (1997). Biological and molecular characterization of *Tomato spotted wilt virus* in Israel. *Phytoparasitica* **25:**319–330.

Aramburu, J., and Marti, M. (2003). The occurrence in north-east Spain of a variant of *Tomato spotted wilt virus* (TSWV) that breaks resistance in tomato (*Lycopersicon esculentum*) containing the Sw-5 gene. *Plant Pathol.* **52:**407.

Aramburu, J., Riudavets, J., Arno, J., Lavina, A., and Moriones, E. (1996). Rapid serological detection of *Tomato spotted wilt virus* in individual thrips by squash-blot assay for use in epidemiological studies. *Plant Pathol.* **45:**367–374.

Arli-Sokmen, M., Menna, H., Sevik, M. A., and Ecevit, O. (2005). Occurrence of viruses in field-grown pepper crops and some of their reservoir weed hosts in Samsun, Turkey. *Phytoparasitica* **33:**347–358.

Bag, S., Druffel, K. L., and Pappu, H. R. (2010). Structure and genome organization of the large RNA of Iris yellow spot virus (genus Tospovirus, family Bunyaviridae). *Arch. Virol.* **155:**275–279.

Bald, J. G. (1937). Investigations on ''spotted wilt'' of tomatoes. III. Infection in field plots *Council Sci. Ind. Res. Aust. Bull.* **106:**32.

Bandla, M. D., Campbell, L. R., Ullman, D. E., and Sherwood, J. L. (1998). Interaction of *Tomato spotted wilt tospovirus* (TSWV) glycoproteins with a thrips midgut protein, a potential cellular receptor for TSWV. *Phytopathology* **88:**98–104.

Barbetaki, A., Lykouressis, D., and Perdikis, D. (1999). Predatory species of the genus *Orius* recorded in fields of vegetable crops in Greece. *IOBC/WPRS Bull.* **22**(5)**:**97–101.

Bautista, R., Mau, R. F. L., Cho, J. J., and Custer, D. (1996). Thrips, tospovirus and host-plant associations in Hawaiian farm ecosystem: Prospects for reducing yield losses. *Acta Hort.* **341:**477–482.

Belliure, B., Janssen, A., Maris, P. C., Peters, D., and Sabelis, M. W. (2005). Herbivore arthropods benefit from vectoring plant viruses. *Ecol. Lett.* **8:**70–79.

Ben Moussa, A., Makni, M., and Marrakchi, M. (2000). Identification of the principal viruses infecting tomato crops in Tunisia. *OEPP/EPPO Bull.* **30:**293–296.

Ben Moussa, A., Marrakchi, M., and Makni, M. (2005). Characterisation of Tospovirus in vegetable crops in Tunisia. *Infect. Genet. Evol.* **5:**312–322.

Best, R. J. (1968). Tomato spotted wilt virus. *Adv. Virus Res.* **13:**65–146.

Betti, L. (1992). Tomato spotted wilt tospovirus on eggplant in Sicily. *Phytopathol. Mediterr.* **31:**119–120.

Bielza, P. (2008). Perspective—Insecticide resistance management strategies against the western flower thrips, *Frankliniella occidentalis*. *Pest Manag. Sci.* **64:**1131–1138.

Bitterlich, I., and MacDonald, L. S. (1993). The prevalence of *Tomato spotted wilt virus* in weeds and crop plants in south western British Columbia. *Can. Plant Dis. Surv.* **72:**137–142.

Black, L. L., Hobbs, H. A., and Gatti, J. M. (1991). *Tomato spotted wilt virus*-resistance in Capsicum chinense PI 152225 and 159236. *Plant Dis.* **75:**863.

Blakqori, G., and Weber, F. (2005). Efficient cDNA-based rescue of La Crosse bunyaviruses expressing or lacking the nonstructural protein NSs. *J. Virol.* **79:**10420–10428.

Boiteux, L. S. (1995). Allelic relationships between genes for resistance to tomato spotted wilt tospovirus in Capsicum chinense. *Theor. Appl. Genet.* **90:**146–149.

Boiteux, L. S., and De Avila, A. C. (1994). Inheritance of a resistance specific to Tomato spotted wilt tospovirus in *Capsicum chinense* PI-159236. *Euphytica* **75:**139–142.

Boiteux, L. S., and Giordano, L. D. (1993). Genetic-basis of resistance against 2 tospovirus species in tomato (*Lycopersicon esculentum*). *Euphytica* **71:**151–154.

Boonham, N., Smith, P., Walsh, K., Tame, J., Morris, J., Spence, N., Bennison, J., and Barker, I. (2002). The detection of *Tomato spotted wilt virus* (TSWV) in individual thrips using real time fluorescent RT-PCR (TaqMan). *J. Virol. Methods* **101:**37–48.

Bosco, L., and Tavella, L. (2010). Population dynamics and integrated pest management of *Thrips tabaci* on leek under field conditions in northwest Italy. *Entomol. Exp. Appl.* **135:**276–287.

Bosco, L., Giacometto, E., and Tavella, L. (2008). Colonization and predation of thrips (Thysanoptera: Thripidae) by *Orius* spp. (Heteroptera: Anthocoridae) in sweet pepper greenhouses in Northwest Italy. *Biol. Control* **44:**331–340.

Brar, K. S., Sidhu, A. S., and Chadha, M. L. (1993). Screening onion varieties for resistance to *Thrips tabaci* Lind. and *Helicoverpa armigera* (Hubner). *J. Insect Sci.* **6:**123–124.

Brittlebank, C. C. (1919). Tomato diseases. *J. Vic. Dept. Agric.* **17:**231–235.

Brodsgaard, H. F. (1989). *Frankliniella occidentalis* (Thysanoptera; Thripidae)—A new pest in Danish glasshouses: A review. *Tidsskr Planteavl* **93:**83–91.

Brommonschenkel, S. H., and Tanksley, S. D. (1997). Map-based cloning of the tomato genomic region that spans the Sw-5 tospovirus resistance gene in tomato. *Mol. Gen. Genet.* **256:**121–126.

Brommonschenkel, S. H., Frary, A., and Tanksley, S. D. (2000). The broad-spectrum tospovirus resistance gene Sw-5 of tomato is a homolog of the root-knot nematode resistance gene Mi. *Mol. Plant Microbe Interact.* **13:**1130–1138.

Brown, S. L., Culbreath, A. K., Todd, J. W., Gorbet, D. W., Baldwin, J. A., and Beasley, J. P. (2005). Development of a method of risk assessment to facilitate integrated management of spotted wilt of peanut. *Plant Dis.* **89:**348–356.

Bulajic, A., Djekic, I., Jovic, J., Krnjajic, S., Vucurovic, A., and Krstic, B. (2009). Incidence and Distribution of *Iris yellow spot virus* on onion in Serbia. *Plant Dis.* **93:**976–982.

Cabrera-La Rosa, J. C., and Kennedy, G. G. (2007). *Thrips tabaci* and tomato spotted wilt virus: Inheritance of vector competence. *Entomol. Exp. Appl.* **124:**161–166.

Camann, M. A., Culbreath, A. K., Pickering, J., Todd, J. W., and Demski, J. W. (1995). Spatial and temporal patterns of spotted wilt epidemics in peanut. *Phytopathology* **85:**879–885.

Canady, M. A., Stevens, M. R., Barineau, M. S., and Scott, J. W. (2001). *Tomato spotted wilt virus* (TSWV) resistance in tomato derived from *Lycopersicon chilense* Dun. LA 1938. *Euphytica* **117:**19–25.

Cannon, R. J. C., Matthews, L., and Collins, D. W. (2007). A review of the pest status and control options for *Thrips palmi*. *Crop Prot.* **26:**1089–1098.

Cebolla-Cornejo, J., Soler, S., Gomar, B., Soria, D. M., and Nuez, F. (2003). Screening Capsicum germplasm for resistance to tomato spotted wilt virus (TSWV). *Ann. Appl. Biol.* **143:**143–152.

Chatzivassiliou, E. K., Livieratos, I. C., Katis, N., Avegelis, A., and Lykouressis, D. (1996). Occurrence of tomato spotted wilt virus in vegetables and ornamentals in Greece. *Acta Hort.* **431:**44–50.

Chatzivassiliou, E. K., Nagata, T., Katis, N. I., and Peters, D. (1999). Transmission of tomato spotted wilt tospovirus by *Thrips tabaci* populations originating from leek. *Plant Pathol.* **48:**700–706.

Chatzivassiliou, E. K., Weekes, R., Morris, J., Wood, K. R., Barker, I., and Katis, N. I. (2000). *Tomato spotted wilt virus* (TSWV) in Greece: Its incidence following the expansion of *Frankliniella occidentalis*, and characterisation of isolates collected from various hosts. *Ann. Appl. Biol.* **137:**127–134.

Chatzivassiliou, E. K., Boubourakas, I., Drossos, E., Eleftherohorinos, I., Jenser, G., Peters, D., and Katis, N. I. (2001). Weeds in greenhouses and tobacco fields are differentially infected by Tomato spotted wilt virus and infested by its vector species. *Plant Dis.* **85:**40–46.

Chatzivassiliou, E. K., Peters, D., and Katis, N. I. (2002). The efficiency by which *Thrips tabaci* populations transmit Tomato spotted wilt virus depends on their host preference and reproductive strategy. *Phytopathology* **92:**603–609.

Chatzivassiliou, E. K., Peters, D., and Katis, N. I. (2007). The role of weeds in the spread of Tomato spotted wilt virus by *Thrips tabaci* (Thysanoptera: Thripidae) in tobacco crops. *J. Phytopathol.* **155:**699–705.

Chatzivassiliou, E. K., Giavachtsia, V., Mehraban, A. H., Hoedjes, K., and Peters, D. (2009). Identification and Incidence of Iris yellow spot virus, a new pathogen in onion and leek in Greece. *Plant Dis.* **93:**761.

Chen, T. C., Lu, Y. Y., Cheng, Y. H., Li, J. T., Yeh, Y. C., Kang, Y. C., Chang, C. P., Huang, L. H., Peng, J. C., and Yeh, S. D. (2010). Serological relationship between Melon yellow spot virus and Watermelon silver mottle virus and differential detection of the two viruses in cucurbits. *Arch. Virol.* **155:**1085–1095.

Childers, C. C. (1997). Feeding and oviposition injuries to plants. *In* ''Thrips as Crop Pests'' (T. Lewis, ed.), p. 505. CABI, Wallingford, UK.

Cho, J. J., Mau, R. F. L., German, T. L., Hartman, R. W., Yudin, L. S., Gonsalves, D., and Provvidenti, R. (1989). A multidisciplinary approach to management of *Tomato spotted wilt virus*. *Plant Dis.* **73:**375–383.

Chu, F. H., Chao, C. H., Chung, M. H., Chen, C. C., and Yeh, S. D. (2001). Completion of the genome sequence of Watermelon silver mottle virus and utilization of degenerate primers for detecting tospoviruses in five serogroups. *Phytopathology* **91:**361–368.

Ciuffo, M., Finetti-Sialer, M. M., Gallitelli, D., and Turina, M. (2005). First report in Italy of a resistance-breaking strain of *Tomato spotted wilt virus* infecting tomato cultivars carrying the Sw5 resistance gene. *Plant Pathol.* **54:**564.

Ciuffo, M., Tavella, L., Pacifico, D., Masenga, V., and Turina, M. (2008). A member of a new Tospovirus species isolated in Italy from wild buckwheat (*Polygonum convolvulus*). *Arch. Virol.* **153:**2059–2068.

Ciuffo, M., Kurowski, C., Vivoda, E., Copes, B., Masenga, V., Falk, B. W., and Turina, M. (2009). A new tospovirus sp in cucurbit crops in Mexico. *Plant Dis.* **93:**467–474.

Ciuffo, M., Mautino, G. C., Bosco, L., Turina, M., and Tavella, L. (2010). Identification of Dictyothrips betae as the vector of Polygonum ring spot virus. *Ann. Appl. Biol.* **157:**299–307.

Cordoba-Selles, C., Martınez-Priego, L., Munoz-Gomez, R., and Jordá-Gutierrez, C. (2005). Iris yellow spot virus: A new onion disease in Spain. *Plant Dis.* **89:**1243.

Cordoba-Selles, C., Cebrian-Mico, C., Alfaro-Fernandez, A., Munoz-Yerbes, M. J., and Jorda-Gutierrez, C. (2007). First report of Iris yellow spot virus in commercial leek (*Allium porrum*) in Spain. *Plant Dis.* **91:**1365.

Cortes, I., Livieratos, I. C., Derks, A., Peters, D., and Kormelink, R. (1998). Molecular and serological characterization of Iris yellow spot virus, a new and distinct tospovirus species. *Phytopathology* **88:**1276–1282.

Cosmi, T., Marchesini, E., and Martini, G. (2003). Presenza e diffusione di Tospovirus e di tripidi vettori in Veneto. *Inf. Agrar.* **59:**69–72 (in Italian).

Çota, E., and Merkuri, J. (2004). Introduction of *Frankliniella occidentalis* and occurrence of Tomato spotted wilt tospovirus in Albania. *EPPO Bull.* **34:**421–422.

Coudriet, D. L., Kishaba, A. N., McCreight, J., and Bohn, W. G. (1979). Varietal resistance in onions to thrips (Thysanoptera: Thripidae). *J. Econ. Entomol.* **72:**614–615.

Coutts, B. A., and Jones, R. A. C. (2005). Suppressing spread of *Tomato spotted wilt virus* by drenching infected source or healthy recipient plants with neonicotinoid insecticides to control thrips vectors. *Ann. Appl. Biol.* **146:**95–103.

Coutts, B. A., Thomas-Carroll, M. L., and Jones, R. A. C. (2004). Patterns of spread of *Tomato spotted wilt virus* in field crops of lettuce and pepper: Spatial dynamics and validation of control measures. *Ann. Appl. Biol.* **145:**231–245.

Culbreath, J. K., Todd, J. W., and Brown, S. L. (2003). Epidemiology and management of *Tomato spotted wilt virus* in peanut. *Annu. Rev. Phytopathol.* **41:**53–75.

De Avila, A. C., Dehaan, P., Kitajima, E. W., Kormelink, R., Resende, R. D., Goldbach, R. W., and Peters, D. (1992). Characterization of a distinct isolate of *Tomato spotted wilt virus* (TSWV) from impatiens sp in the Netherlands. *J. Phytopathol.* **134:**133–151.

de Borbon, C. M., Gracia, O., and Piccolo, R. (2006). Relationships between tospovirus incidence and thrips populations on tomato in Mendoza, Argentina. *J. Phytopathol.* **154:**93–99.

Deangelis, J. D., Sether, D. M., and Rossignol, P. A. (1993). Survival, development, and reproduction in western flower thrips (Thysanoptera, Thripidae) exposed to *Impatiens necrotic spot virus*. *Environ. Entomol.* **22:**1308–1312.

Deangelis, J. D., Sether, D. M., and Rossignol, P. A. (1994). Transmission of *Impatiens necrotic spot virus* in peppermint by western flower thrips (Thysanoptera, Thripidae). *J. Econ. Entomol.* **87:**197–201.

Deligeorgidis, P. N., Athanassiou, C. G., and Kavallieratos, N. G. (2002). Seasonal abundance, spatial distribution and sampling indices of thrip populations on cotton; a 4-year survey from central Greece. *J. Appl. Entomol.* **126:**343–348.

Dianese, E. C., de Fonseca, M. E. N., Goldbach, R., Kormelink, R., Inoue-Nagata, A. K., Resende, R. O., and Boiteux, L. S. (2010). Development of a locus-specific, co-dominant SCAR marker for assisted-selection of the Sw-5 (Tospovirus resistance) gene cluster in a wide range of tomato accessions. *Mol. Breed.* **25:**133–142.

Diaz, B. M., Biurrun, R., Moreno, A., Nebreda, M., and Fereres, A. (2006). Impact of ultraviolet-blocking plastic films on insect vectors of virus diseases infesting crisp lettuce. *Hort. Sci.* **41:**711–716.

El-Wahab, A., El-Sheikh, A. S., and Elnagar, M. A. K. (2008). First record in Egypt of thrips *Frankliniella occidentalis* and Impatiens necrotic spot tospovirus. International Conference 2008 Diversifying Crop Protection, La Grande Motte, France.

Evans, C. K., Bag, S., Frank, E., Reeve, J., Ransom, C., Drost, D., and Pappu, H. R. (2009a). Natural infection of Iris yellow spot virus in Twoscale saltbush (Atriplex micrantha) growing in Utah. *Plant Dis.* **93:**430.

Evans, C. K., Bag, S., Frank, E., Reeve, J., Ransom, C., Drost, D., and Pappu, H. R. (2009b). Green foxtail (Setaria viridis), a naturally infected grass host of Iris yellow spot virus in Utah. *Plant Dis.* **93:**670.

Fanigliulo, A., Comes, S., Pacella, R., Crescenzi, A., Momol, M. T., Olson, S. M., Reitz, S., Saygili, H., Sahin, F., and Aysan, Y. (2009). Integrated management of viral diseases in field-grown tomatoes in Southern Italy. *Acta Hort.* **808:**387–391.

Fery, R. L., and Schalk, J. M. (1991). Resistance in pepper (*Capsicum annuum*) to western flower thrips *Frankliniella occidentalis* (Pergande). *Hort. Sci.* **26:**1073–1074.

Finetti Sialer, M. M., and Gallitelli, D. (2000). The occurrence of *Impatiens necrotic spot virus* and *Tomato spotted wilt virus* in mixed infection in tomato. *J. Plant Pathol.* **82:**244.

Finlay, K. W. (1953). Inheritance of spotted wilt resistance in tomato. II. Five genes controlling spotted wilt resistance in four tomato types. *Aust. J. Biol. Sci.* **6:**153–163.

Flick, R., and Pettersson, R. F. (2001). Reverse genetics system for Uukuniemi virus (Bunyaviridae): RNA polymerase I-catalyzed expression of chimeric viral RNAs. *J. Virol.* **75:**1643–1655.

Flick, K., Hooper, J. W., Schmaljohn, C. S., Pettersson, R. F., Feldmann, H., and Flick, R. (2003a). Rescue of Hantaan virus minigenomes. *Virology* **306:**219–224.

Flick, R., Flick, K., Feldmann, H., and Elgh, F. (2003b). Reverse genetics for Crimean-Congo hemorrhagic fever virus. *J. Virol.* **77:**5997–6006.

Foster, S. P., Gorman, K., and Denholm, I. (2010). English field samples of *Thrips tabaci* show strong and ubiquitous resistance to deltamethrin. *Pest Manag. Sci.* **66:**861–864.

Funderburk, J. E., Stavisky, J., and Olson, S. M. (2000). Predation of *Frankliniella occidentalis* (Thysanoptera: Thripidae) in field peppers by *Orius insidiosus* (Hemiptera: Anthocoridae). *Environ. Entomol.* **29:**376–382.

Gallitelli, D., Rana, G. L., Vovlas, C., and Martelli, G. P. (2004). Viruses of globe artichoke: An overview. *J. Plant Pathol.* **86:**267–281.

Gallo, S., Gotta, P., and Lisa, V. (1995). Tomato spotted wilt virus. *Piemonte Agric.* **19** (Suppl 1):1–8 (in Italian).

Garcia, F., GreatRex, R. M., and Gomez, J. (1997). Development of integrated crop management systems for sweet peppers in southern Spain. *IOBC/WPRS Bull.* **20**(4):8–15.

Garcia, S., Billecocq, A., Crance, J. M., Prins, M., Garin, D., and Bouloy, M. (2006). Viral suppressors of RNA interference impair RNA silencing induced by a Semliki Forest virus replicon in tick cells. *J. Gen. Virol.* **87:**1985–1989.

Garcia-Cano, E., Resende, R. O., Fernandez-Munoz, R., and Moriones, E. (2006). Synergistic interaction between Tomato chlorosis virus and Tomato spotted wilt virus results in breakdown of resistance in tomato. *Phytopathology* **96:**1263–1269.

Gent, D. H., du Toit, L. J., Fichtner, S. F., Mohan, S. K., Pappu, H. R., and Schwartz, H. F. (2006). Iris yellow spot virus: An emerging threat to onion bulb and seed production. *Plant Dis.* **90:**1468–1480.

Gera, A., Cohen, J., Salomon, R., and Raccah, B. (1998). Iris yellow spot tospovirus detected in onion (Allium cepa) in Israel. *Plant Dis.* **82:**127.

Gera, A., Kritzman, A., Cohen, J., and Raccah, B. (1999). First report of Impatiens necrotic spot tospovirus (INSV) in Israel. *Plant Dis.* **83:**587.

German, T. (2009). Exploiting vector specificity to inhibit tospovirus transmission. *Phytopathology* **99:**S153.

Gitaitis, R. D., Dowler, C. C., and Chalfant, R. B. (1998). Epidemiology of tomato spotted wilt in pepper and tomato in southern Georgia. *Plant Dis.* **82:**752–756.

Goldbach, R., and Peters, D. (1994). Possible causes of the emergence of tospovirus diseases. *Semin. Virol.* **5:**113–120.

Goldbach, R., and Peters, D. (1996). Molecular and biological aspects of tospoviruses. *In* "The Bunyaviridae" (R. M. Elliot, ed.), p. 129. Plenum Press, New York.

Gordillo, L. F., Stevens, M. R., Millard, M. A., and Geary, B. (2008). Screening two *Lycopersicon peruvianum* collections for resistance to *Tomato spotted wilt virus*. *Plant Dis.* **92:**694–704.

Groves, R. L., Walgenbach, J. F., Moyer, J. W., and Kennedy, G. G. (2002). The role of weed hosts and tobacco thrips, *Frankliniella fusca*, in the epidemiology of *Tomato spotted wilt virus*. *Plant Dis.* **86:**573–582.

Gubba, A., Gonsalves, C., Stevens, M. R., Tricoli, D. M., and Gonsalves, D. (2002). Combining transgenic and natural resistance to obtain broad resistance to tospovirus infection in tomato (*Lycopersicon esculentum* Mill.). *Mol. Breed.* **9:**13–23.

Habjan, M., Penski, N., Spiegel, M., and Weber, F. (2008). T7 RNA polymerase-dependent and -independent systems for cDNA-based rescue of Rift Valley fever virus. *J. Gen. Virol.* **89:**2157–2166.

Hall, J. M., Mohan, K., Knott, E. A., and Moyer, J. W. (1993). Tospoviruses associated with scape blight of onion (*Allium cepa*) seed crops in Idaho. *Plant Dis.* **77:**952.

Hassani-Mehraban, A., Saaijer, J., Peters, D., Goldbach, R., and Kormelink, R. (2005). A new tomato-infecting tospovirus from Iran. *Phytopathology* **95:**852–858.

Hassani-Mehraban, A., Saaijer, J., Peters, D., Goldbach, R., and Kormelink, R. (2007). Molecular and biological comparison of two Tomato yellow ring virus (TYRV) isolates: Challenging the Tospovirus species concept. *Arch. Virol.* **152:**85–96.

Hassani-Mehraban, A., Botermans, M., Verhoeven, J. T. J., Meekes, E., Saaijer, J., Peters, D., Goldbach, R., and Kormelink, R. (2010). A distinct tospovirus causing necrotic streak on Alstroemeria sp in Colombia. *Arch. Virol.* **155:**423–428.

Herrero, S., Culbreath, A. K., Csinos, A. S., Pappu, H. R., Rufty, R. C., and Daub, M. E. (2000). Nucleocapsid gene-mediated transgenic resistance provides protection against Tomato spotted wilt virus epidemics in the field. *Phytopathology* **90:**139–147.

Hoddle, M. S., Mound, L. A., and Paris, D. L. (2008). Thrips of California. CBIT Publishing, Queensland.

Hoffmann, K., Qiu, W. P., and Moyer, J. W. (2001). Overcoming host- and pathogen-mediated resistance in tomato and tobacco maps to the M RNA of *Tomato spotted wilt virus*. *Mol. Plant Microbe Interact.* **14:**242–249.

Huchette, O., Bellamy, C., Filomenko, R., Pouleau, B., Seddas, S., and Pappu, H. R. (2008). Iris yellow spot virus on shallot and onion in France. *Online. Plant Health Progress.* doi: 10.1094/PHP-2008-0610-01-BR.

Huguenot, C., Vandendobbelsteen, G., Dehaan, P., Wagemakers, C. A. M., Drost, G. A., Osterhaus, A., and Peters, D. (1990). Detection of tomato spotted wilt virus using monoclonal-antibodies and riboprobes. *Arch. Virol.* **110:**47–62.

Hull, R. (2002). Transmission I: By invertebrate, nematodes and fungi. *Matthews' Plant Virology*. Academic Press, San Diego, USA, p. 485.

Jacobson, R. J. (1997). Integrated pest management (IPM) in glasshouses. *In* "Thrips as Crop Pests" (T. Lewis, ed.), p. 639. CABI, Wallingford, UK.

Jahn, M., Paran, I., Hoffmann, K., Radwanski, E. R., Livingstone, K. D., Grube, R. C., Aftergoot, E., Lapidot, M., and Moyer, J. (2000). Genetic mapping of the Tsw locus for resistance to the Tospovirus Tomato spotted wilt virus in Capsicum spp. and its relationship to the Sw-5 gene for resistance to the same pathogen in tomato. *Mol. Plant Microbe Interact.* **13:**673–682.

Jones, R. A. C. (2004). Using epidemiological information to develop effective integrated virus disease management strategies. *Virus Res.* **100:**5–30.

Jones, D. R. (2005). Plant viruses transmitted by thrips. *Eur. J. Plant Pathol.* **113:**119–157.

Jones, R. A. C. (2006). Control of plant virus diseases. *Adv. Virus Res.* **67:**205–244.

Jordà, C. (1993). Nuevas virosis de mayor incidencia en cultivos horticolas. *Phytoma España* **50:**7–13.

Kendall, D. M., and Capinera, J. L. (1990). Geographic and temporal variation in the sex-ratio of onion thrips (Thysanoptera, Thripidae). *Southwest. Entomol.* **15:**80–88.

Kikkert, M., Meurs, C., van de Wetering, F., Dorfmuller, S., Peters, D., Kormelink, R., and Goldbach, R. (1998). Binding of *Tomato spotted wilt virus* to a 94-kDa thrips protein. *Phytopathology* **88:**6369.

Kim, J. W., Sun, S. S. M., and German, T. L. (1994). Disease resistance in tobacco and tomato plants transformed with the *Tomato spotted wilt virus* nucleocapsid gene. *Plant Dis.* **78:**615–621.

Kirk, W. D. J., and Terry, L. I. (2003). The spread of western flower thrips *Frankliniella occidentalis* (Pergande). *Agric. For. Entomol.* **5:**301–310.

Kormelink, R., Vanpoelwijk, F., Peters, D., and Goldbach, R. (1992). Nonviral heterogeneous sequences at the 5′ ends of *Tomato spotted wilt virus* messenger-RNAs. *J. Gen. Virol.* **73:**2125–2128.

Koschier, E. H., Sedy, K. A., and Novak, J. (2002). Influence of plant volatiles on feeding damage caused by the onion thrips *Thrips tabaci*. *Crop Prot.* **21:**419–425.

Kritzman, A., Beckelman, H., Alexandrov, S., Cohen, J., Lampel, M., Zeidan, M., Raccah, B., and Gera, A. (2000). Lisianthus leaf necrosis: A new disease of lisianthus caused by Iris yellow spot virus. *Plant Dis.* **84:**1185–1189.

Kritzman, A., Lampel, M., Raccah, B., and Gera, A. (2001). Distribution and transmission of Iris yellow spot virus. *Plant Dis.* **85:**838–842.

Kuwabara, K., Yokoi, N., Ohki, T., and Tsuda, S. (2010). Improved multiplex reverse transcription-polymerase chain reaction to detect and identify five tospovirus species simultaneously. *J. Gen. Plant Pathol.* **76:**273–277.

Lacasa, A., Contreras, J., Sanchez, J. A., Lorca, M., and Garcia, F. (1996). Ecology and natural enemies of *Frankliniella occidentalis* (Pergande, 1895) in South-east Spain. *Folia Entomol. Hung.* **57:**67–74.

Larentzaki, E., Plate, J., Nault, B. A., and Shelton, A. M. (2008). Impact of straw mulch on populations of onion thrips (Thysanoptera: Thripidae) in onion. *J. Econ. Entomol.* **101:**1317–1324.

Latham, L. J., and Jones, R. A. C. (1997). Occurrence of tomato spotted wilt tospovirus in native flora, weeds and horticultural crops. *Aust. J. Agric. Res.* **48:**359369.

Lavina, A., and Battle, A. (1994). First report of *Impatiens necrotic spot virus* in *Asplenium nidus-avis* in Spain. *Plant Dis.* **78:**316.

Law, M. D., and Moyer, J. W. (1990). A tomato spotted wilt-like virus with a serologically distinct n-protein. *J. Gen. Virol.* **71:**933–938.

Leinhos, G., Muller, J., Heupel, M., and Krauthausen, H. J. (2007). Iris yellow spot virus an Bund- und Speisezwiebeln-erster Nachweis in Deutschland. *Nachrichtenbl. Deut. Pflanzenschutzd.* **59:**310–313.

Lewandowski, D. J., and Adkins, S. (2005). The tubule-forming NSm protein from Tomato spotted wilt virus complements cell-to-cell and long-distance movement of Tobacco mosaic virus hybrids. *Virology* **342:**26–37.

Li, W., Lewandowski, D. J., Hilf, M. E., and Adkins, S. (2009). Identification of domains of the Tomato spotted wilt virus NSm protein involved in tubule formation, movement and symptomatology. *Virology* **390:**110121.

Lisa, V., Vaira, A. M., Milne, R. G., Luisoni, E., and Rapetti, S. (1990). *Tomato spotted wilt virus* in cinque specie coltivate in Liguria. *Inf. Fitopatol.* **40:**34–41 (in Italian).

Lopez, C., Aramburu, J., Galipienso, L., Soler, S., Nuez, F., and Rubio, L. (2011). Evolutionary analysis of tomato Sw-5 resistance-breaking isolates of *Tomato spotted wilt virus*. *J. Gen. Virol.* **92:**210–215.

Louro, D. (1996). Detection and identification of *Tomato spotted wilt virus* and impatiens necrotic spot virus in Portugal. *Acta Hort.* **431:**99–105.

Lykoyressis, D. P. (1993). The occurrence of the polyphagous predator *Orius niger* (Wolff) (Hemiptera: Anthocoridae) in Greece. *Entomol. Hell.* **11:**43–44.

Marchoux, G., Gebreselassie, K., and Villevieille, M. (1991). Detection of tomato spotted wilt virus and transmission by *Frankliniella occidentalis* in France. *Plant Pathol.* **40:**347–351.

Margaria, P., Ciuffo, M., and Turina, M. (2004). Resistance breaking strain of Tomato spotted wilt virus (Tospovirus; Bunyaviridae) on resistant pepper cultivars in Almeria, Spain. *Plant Pathol.* **53:**795.

Margaria, P., Ciuffo, M., Pacifico, D., and Turina, M. (2007). Evidence that the nonstructural protein of Tomato spotted wilt virus is the avirulence determinant in the interaction with resistant pepper carrying the Tsw gene. *Mol. Plant Microbe Interact.* **20:**547–558.

Maris, P. C., Joosten, N. N., Goldbach, R. W., and Peters, D. (2003a). Restricted spread of Tomato spotted wilt virus in thrips-resistant pepper. *Phytopathology* **93:**1223–1227.

Maris, P. C., Joosten, N. N., Peters, D., and Goldbach, R. W. (2003b). Thrips resistance in pepper and its consequences for the acquisition and inoculation of *Tomato spotted wilt virus* by the western flower thrips. *Phytopathology* **93:**96–101.

Maris, P. C., Joosten, N. N., Goldbach, R. W., and Peters, D. (2004). Decreased preference and reproduction, and increased mortality of *Frankliniella occidentalis* on thrips-resistant pepper plants. *Entomol. Exp. Appl.* **113:**149–155.

Martin, N. A., Workman, P. J., and Butler, R. C. (2003). Insecticide resistance in onion thrips (*Thrips tabaci*) (Thysanoptera: Thripidae). *New Zealand J. Crop Hort. Sci.* **31:**99–106.

Martini, G., Cosmi, T., and Zorzi, M. (2009). Potenziale pericolosità di TSWV su tabacco in Italia. *Inf. Agrar.* **65:**57–58 (in Italian).

Mason, G., Roggero, P., and Tavella, L. (2003). Detection of *Tomato spotted wilt virus* in its vector *Frankliniella occidentalis* by reverse transcription-polymerase chain reaction. *J. Virol. Methods* **109:**69–73.

Mavric, I., and Ravnikar, M. (2000). Iris yellow spot tospovirus in Slovenia. *In* ''5th Congress of the European Foundation for Plant Pathology: Biodiversity in Plant Pathology'' (A. Catara, G. Albanese, and M. Catara, eds.), p. 223. Taormina-Giardini Naxos, Italy.

Mavric, I., and Ravnikar, M. (2001). First report of *Tomato spotted wilt virus* and *Impatiens necrotic spot virus* in Slovenia. *Plant Dis.* **85:**1288.

McIntyre Allen, J. K., Scott-Dupree, C. D., Tolman, J. H., and Harris, R. C. (2005). Resistance of *Thrips tabaci* to pyrethroid and organophosphorus insecticides in Ontario, Canada. *Pest Manag. Sci.* **61:**809–815.

Medeiros, R. B., Ullman, D. E., Sherwood, J. L., and German, T. L. (2000). Immunoprecipitation of a 50-kDa protein: A candidate receptor component for tomato spotted wilt tospovirus (Bunyaviridae) in its main vector, Frankliniella occidentalis. *Virus Res.* **67:**109–118.

Momol, M. T., Olson, S. M., Funderburk, J. E., and Stavisky, J. (2004). Integrated management of Tomato spotted wilt on field-grown tomatoes. *Plant Dis.* **88:**882–890.

Moreno, A., de Blas, C., Biurrun, R., Nebreda, M., Palacios, I., Duque, M., and Fereres, A. (2004). The incidence and distribution of viruses infecting lettuce, cultivated Brassica and associated natural vegetation in Spain. *Ann. Appl. Biol.* **144:**339–346.

Moritz, G., Kumm, S., and Mound, L. (2004). Tospovirus transmission depends on thrips ontogeny. *Virus Res.* **100:**143–149.

Morse, J. G., and Hoddle, M. S. (2006). Invasion biology of thrips. *Annu. Rev. Entomol.* **51:**67–89.

Mound, L. A., and Morris, D. C. (2007). The insect order Tysanoptera: Classification versus systematics. *Zootaxa* **1668:**395–411.

Moury, B., Palloix, A., Selassie, K. G., and Marchoux, G. (1997). Hypersensitive resistance to tomato spotted wilt virus in three *Capsicum* chinense accessions is controlled by a single gene and is overcome by virulent strains. *Euphytica* **94:**45–52.

Mumford, R. A., Barker, I., and Wood, K. R. (1994). The detection of tomato spotted wilt virus using the polymerase chain-reaction. *J. Virol. Methods* **46:**303–311.

Mumford, R. A., Jarvis, B., Morris, J., and Blockley, A. (2003). First report of Chrysanthemum stem necrosis virus (CSNV) in the UK. *Plant Pathol.* **52:**779.

Nagata, T., Almedia, A. L., Resende, R., and De Avila, C. (1999). The identification of the vector species of iris yellow spot tospovirus occurring in onion in Brazil. *Plant Dis.* **83:**399.

Naidu, R. A., Deom, C. M., and Sherwood, J. L. (2001). First report of *Frankliniella fusca* as a vector of Impatiens necrotic spot tospovirus. *Plant Dis.* **85:**1211.

Nischwitz, C., Pappu, H. R., Mullis, S. W., Sparks, A. N., Langston, D. R., Csinos, A. S., and Gitaitis, R. D. (2007). Phylogenetic analysis of Iris yellow spot virus isolates from onion (*Allium cepa*) in Georgia (USA) and Peru. *J. Phytopathol.* **155:**531–535.

Nolasco, G., Deblas, C., Torres, V., and Ponz, F. (1993). A method combining immunocapture and pcr amplification in a microtiter plate for the detection of plant-viruses and subviral pathogens. *J. Virol. Methods* **45:**201–218.

Ogawa, Y., Sugiura, K., Kato, K., Tohya, Y., and Akashi, H. (2007). Rescue of Akabane virus (family Bunyaviridae) entirely from cloned cDNAs by using RNA polymerase I. *J. Gen. Virol.* **88:**3385–3390.

Ohnishi, J., Katsuzaki, H., Tsuda, S., Sakurai, T., Akutsu, K., and Murai, T. (2006). *Frankliniella cephalica*, a new vector for Tomato spotted wilt virus. *Plant Dis.* **90:**685.

Okuda, M., Fuji, S., Okuda, S., Sako, K., and Iwanami, T. (2010). Evaluation of the potential of thirty two weed species as infection sources of impatiens necrotic spot virus. *J. Plant Pathol.* **92:**357–361.

Palmer, J. M., Mound, L. A., and du Heame, G. J. (1989). CIE Guides to Insects of Importance to Man. 2. Thysanoptera. CAB International Institute of Entomology, London, UK.

Pang, S. Z., Jan, F. J., Carney, K., Stout, J., Tricoli, D. M., Quemada, H. D., and Gonsalves, D. (1996). Post-transcriptional transgene silencing and consequent tospovirus resistance in transgenic lettuce are affected by transgene dosage and plant development. *Plant J.* **9:**899–909.

Pappu, H. R. (1997). Managing tospoviruses through biotechnology: Progress and prospects. *Biotechnol. Dev. Monitor.* **32:**14–17.

Pappu, H. R. (2008). Tomato spotted wilt virus. *In* ''Encyclopedia of Virology'' (B. W. J. Mahy and M. H. V. Van Regenmorte, eds.), p. 133. Elsevier Ltd., Oxford, UK.
Pappu, H. R., du Toit, L. J., Schwartz, H. F., and Mohan, S. K. (2006). Sequence diversity of the nucleoprotein gene of *Iris yellow spot virus* (genus Tospovirus, family Bunyaviridae) isolates from the western region of the United States. *Arch. Virol.* **151:**1015–1023.
Pappu, H. R., Jones, R. A. C., and Jain, R. K. (2009). Global status of tospovirus epidemics in diverse cropping systems: Successes achieved and challenges ahead. *Virus Res.* **141:**219–236.
Parisi, B., Grassi, G., Roggero, P., and Masenga, V. (1998). Tospovirus (TSWV) su melanzana in coltura protetta. *Inf. Agrar.* **54**(23)**:**45–48 (in Italian).
Parrella, G., Gognalons, P., Gebre-Selassie, K., Vovlas, C., and Marchoux, G. (2003). An update of the host range of tomato spotted wilt virus. *J. Plant Pathol.* **8:**227–264.
Paterson, R. G., Scott, S. J., and Gergerich, R. J. (1989). Resistance in two Lycopersicon species to an Arkansas isolate of tomato spotted wilt virus. *Euphytica* **43:**173–178.
Peters, D., Maris, P. N., Joosten, N., and Goldbach, R. W. (2007). Impeded spread of Tomato spotted wilt virus to pepper plants less preferred by *Frankliniella occidentalis. J. Insect Sci.* **7:** Abs. N 26.
Pittman, H. A. (1927). Spotted wilt of tomatoes. Preliminary note concerning the transmission of the ''spotted wilt'' of tomatoes by an insect vector (*Thrips tabaci* Lind.) *J. Council Sci. Ind. Res.* **1:**74–77.
Pourrahim, R. S., Farzadfar, S., Moini, A. A., Shahraeen, N., and Ahoonmanesh, A. (2001). First report of *Tomato spotted wilt virus* on potatoes in Iran. *Plant Dis.* **85:**442.
Pozzer, L., Bezerra, I. C., Kormelink, R., Prins, M., Peters, D., Resende, R. D., and De Avila, A. C. (1999). Characterization of a tospovirus isolate of iris yellow spot virus associated with a disease in onion fields in Brazil. *Plant Dis.* **83:**345–350.
Premachandra, W., Borgemeister, C., Maiss, E., Knierim, D., and Poehling, H. M. (2005). *Ceratothripoides claratris,* a new vector of a capsicum chlorosis virus isolate infecting tomato in Thailand. *Phytopathology* **95:**659–663.
Rampinini, G. (1987). Un nuovo parassita della Saintpaulia: *Frankliniella occidentalis. Clamer Informa* **1:**20–23 (in Italian).
Ravnikar, M., Vozelj, N., Marvic, I., Svigelj, S. D., Zupancic, M., and Petrovic, N. (2003). Detection of Chrysanthemum stem necrosis virus and Tomato spotted wilt virus in chrysanthemum. 8th International Congress of Plant Pathology, Christchurch, New Zealand, 2–7 February 2003.
Reitz, S. R., Yearby, E. L., Funderburk, J. E., Stavisky, J., Momol, M. T., and Olson, S. M. (2003). Integrated management tactics for *Frankliniella* thrips (Thysanoptera: Thripidae) in field-grown pepper. *J. Econ. Entomol.* **96:**1201–1214.
Rice, D. J., German, T. L., Mau, R. F. L., and Fujimoto, F. M. (1990). Dot blot detection of tomato spotted wilt virus-RNA in plant and thrips tissues by cDNA clones. *Plant Dis.* **74:**274–276.
Riley, D. G., and Pappu, H. R. (2000). Evaluation of tactics for management of thrips-vectored Tomato spotted wilt virus in tomato. *Plant Dis.* **84:**847–852.
Riudavets, J. (1995). Predators of *Thrips tabaci*. *In* ''Biological control of thrips pests'' (J. van Lenteren, ed.), p. 47. Wageningen Agricultural University Papers, Wageningen, The Netherlands.
Roberts, C. A., Dietzgen, R. G., Heelan, L. A., and MacLean, D. J. (2000). Real-time RT-PCR fluorescent detection of tomato spotted wilt virus. *J. Virol. Methods* **88:**1–8.
Roggero, P., Dellavalle, G., Ciuffo, M., and Pennazio, S. (1999). Effects of temperature on infection in *Capsicum* sp. and *Nicotiana benthamiana* by impatiens necrotic spot tospovirus. *Eur. J. Plant Pathol.* **105:**509–512.
Roggero, P., Masenga, V., and Tavella, L. (2002a). Field isolates of *Tomato spotted wilt virus* overcoming resistance in pepper and their spread to other hosts in Italy. *Plant Dis.* **86:**950–954.

Roggero, P., Salomone, A., and Gotta, P. (2002b). La diagnosi dei Tospovirus. *Inf. Fitopatol.* **52:**17–21 (in Italian).

Ronco, A. E., Dal Bo, E., Ghiringhelli, P. D., Medrano, C., Romanowski, V., Sarachu, A. N., and Grau, O. (1989). Cloned cDNA probes for the detection of *Tomato spotted wilt virus. Phytopathology* **79:**1309–1313.

Rosello, S., Diez, M. J., and Nuez, F. (1998). Genetics of tomato spotted wilt virus resistance coming from Lycopersicon peruvianum. *Eur. J. Plant Pathol.* **104:**499–509.

Rosello, S., Ricarte, B., Diez, M. J., and Nuez, F. (2001). Resistance to Tomato spotted wilt virus introgressed from Lycopersicon peruvianum in line UPV 1 may be allelic to Sw-5 and can be used to enhance the resistance of hybrids cultivars. *Euphytica* **119:**357–367.

Rotenberg, D., and Whitfield, A. E. (2010). Analysis of expressed sequence tags for *Frankliniella occidentalis,* the western flower thrips. *Insect Mol. Biol.* **19:**537–551.

Sabelis, M. W., and van Rijn, P. C. J. (1997). Predation by insects and mites. *In* "Thrips as Crop Pests" (T. Lewis, ed.), p. 259. CABI, Wallingford, UK.

Sakimura, K. (1962). The present status of thrips-borne viruses. *In* "Biological Transmission of Disease Agents" (K. Maramorosch, ed.), pp. 33–40. Academic Press, New York, NY.

Sakurai, T., Inoue, T., and Tsuda, S. (2004). Distinct efficiencies of Impatiens necrotic spot virus transmission by five thrips vector species (Thysanoptera: Thripidae) of tospoviruses in Japan. *Appl. Entomol. Zool.* **39:**71–78.

Sanchez, J. A., and Lacasa, A. (2006). A biological pest control story. *IOBC/WPRS Bull.* **29** (4)**:**19–24.

Scott, J. W., Stevens, M. R., and Olson, S. M. (2005). An alternative source of resistance to *Tomato spotted wilt virus. Tomato Genet. Coop. Rep.* **55:**40–41.

Seepiban, C., Gajanandana, O., Attathom, T., and Attathom, S. (2011). Tomato necrotic ringspot virus, a new tospovirus isolated in Thailand. *Arch. Virol.* **156:**263–274.

Sialer, M. M. F., Lanave, C., Padula, M., Vovlas, C., and Gallitelli, D. (2002). Occurrence of two distinct *Tomato spotted wilt virus* subgroups in southern Italy. *J. Plant Pathol.* **84:**145–152.

Sin, S. H., McNulty, B. C., Kennedy, G. G., and Moyer, J. W. (2005). Viral genetic determinants for thrips transmission of Tomato spotted wilt virus. *Proc. Natl. Acad. Sci. U.S.A.* **102:**5168–5173.

Smith, K. M. (1932). Studies on plant virus diseases. XI. Further experiments with a ringspot virus. Its identification of spotted wilt of the tomato. *Ann. Appl. Biol.* **19:**305–330.

Smith, I. M., McNamara, D. G., Scott, P. R., and Harris, K. M. (1992). Frankliniella occidentalis. *In* "Quarantine Pests for Europe" (CAB/EPPO, ed.), p. 145. University Press, Cambridge.

Smith, T. N., Jones, R. A. C., and Wylie, S. J. (2006). Genetic diversity of the nucleocapsid gene of Iris yellow spot virus. *Australas. Plant Pathol.* **35:**359–362.

Stevens, M. R., Scott, S. J., and Gergerich, R. C. (1992). Inheritance of a gene for resistance to tomato spotted wilt virus (TSWV) from *Lycopersicon* species for resistance to tomato spotted wilt virus (TSWV). *Euphytica* **80:**79–84.

Strassen zur, R. (2007). Fauna Europaea: Thysanoptera Thripidae. Fauna Europaea. http://www.faunaeur.org Online version 1.3.

Takeda, A., Sugiyama, K., Nagano, H., Mori, M., Kaido, M., Mise, K., Tsuda, S., and Okuno, T. (2002). Identification of a novel RNA silencing suppressor, NSs protein of Tomato spotted wilt virus. *FEBS Lett.* **532:**75–79.

Tavella, L., Arzone, A., and Alma, A. (1991). Researches on *Orius laevigatus* (Fieb.), a predator of *Frankliniella occidentalis* (Perg.) in greenhouses. A preliminary note. *IOBC/WPRS Bull.* **14**(5)**:**65–72.

Tavella, L., Alma, A., and Arzone, A. (1994). Attività predatrice di *Orius* spp. (Anthocoridae) su *Frankliniella occidentalis* (Perg.) (Thripidae) in coltura protetta di peperone. *Inf. Fitopatol.* **44**(1)**:**40–43 (in Italian).

Tavella, L., Alma, A., Conti, A., Arzone, A., Roggero, P., Ramasso, E., Dellavalle, G., and Lisa, V. (1997). Tripidi e TSWV nelle serre di peperone in Liguria. *Colture Protette* **26**(7–8)**:**79–83 (in Italian). ISSN 0390-0444.

Tavella, L., Tedeschi, R., Arzone, A., and Alma, A. (2000). Predatory activity of two *Orius* species on the western flower thrips in protected pepper crops (Ligurian Riviera, Italy). *IOBC/WPRS Bull.* **23**(1):231.

Tawfik, M. F. S., and Ata, A. M. (1973). The life-history of *Orius laevigatus* (Fieber) (Hemiptera: Anthocoridae). *Bull. Soc. ent. Egypte* **57:**145–151.

Tedeschi, R., Ciuffo, M., Mason, G., Roggero, P., and Tavella, L. (2001). Transmissibility of four tospoviruses by a thelytokous population of *Thrips tabaci* from Liguria, northwestern Italy. *Phytoparasitica* **29:**37–45.

Testa, M., Marras, P. M., Turina, M., and Ciuffo, M. (2008). Presenza del virus dell'avvizzimento maculato del pomodoro (TSWV) su carciofo in Sardegna. *Protezione delle colture* **2:**34–36 (in Italian). ISSN 2279-7602.

Theunissen, J., and Schelling, G. (1997). Damage threshold for *Thrips tabaci* (Thysanoptera: Thripidae) in monocropped and intercropped leek. *Eur. J. Entomol.* **94:**253–261.

Thresh, J. M. (1983). Progress curves of plant virus disease. *Adv. Appl. Biol.* **8:**1–85.

Tomassoli, L., Tiberini, A., Masenga, V., Vicchi, V., and Turina, M. (2009). Characterization of Iris yellow spot virus isolates from onion crops in northern Italy. *J. Plant Pathol.* **91:**733–739.

Tommasini, M. G., and Maini, S. (1995). *Frankliniella occidentalis* and other thrips harmful to vegetable and ornamental crops in Europe. *In* ''Biological Control of Thrips Pests'' (J. van Lenteren, ed.), p. 1. Wageningen Agricultural University Papers, Wageningen, The Netherlands.

Tsakiridis, J. P., and Gooding, G. V. (1972). *Tomato spotted wilt virus* in Greece. *Phytopathol. Mediterr.* **11:**42–47.

Tsompana, M., and Moyer, J. W. (2008). Tospoviruses. *In* ''Encyclopedia of Virology'' (B. W. J. Mahy and M. H. V. Van Regenmortel, eds.), 3rd edn. p. 157. Elsevier, Oxford UK.

Tsompana, M., Abad, J., Purugganan, M., and Moyer, J. W. (2005). The molecular population genetics of the Tomato spotted wilt virus (TSWV) genome. *Mol. Ecol.* **14:**53–66.

Vacante, V., and Tropea Garzia, G. (1993). Prime osservazioni sulla risposta funzionale di *Orius laevigatus* (Fieber) nel controllo di *Frankliniella occidentalis* (Pergande) su peperone in serra fredda. *Colture Protette* **22**(1):33–36 (in Italian).

Vaira, A. M., Roggero, P., Luisoni, E., Masenga, V., Milne, R. G., and Lisa, V. (1993). Characterization of 2 tospoviruses in Italy—Tomato spotted wilt and impatiens necrotic spot. *Plant Pathol.* **42:**530–542.

van de Veire, M., and Degheele, D. (1992). Biological control of the western flower thrips, *Frankliniella occidentalis* (Pergande) (Thysanoptera: Thripidae), in glasshouse sweet peppers with *Orius* spp. (Hemiptera: Anthocoridae). A comparative study between *O. niger* (Wolff) and *O. insidiosus* (Say). *Biocontrol Sci. Technol.* **2:**281–283.

van Knippenberg, I., Lamine, M., Goldbach, R., and Kormelink, R. (2005). *Tomato spotted wilt virus* transcriptase in vitro displays a preference for cap donors with multiple base complementarity to the viral template. *Virology* **335:**122–130.

van Os, B., Stancanelli, G., Mela, L., and Lisa, V. (1993). Ruolo delle piante spontanee ed infestanti nell'epidemiologia di Tospovirus in Liguria. *Inf. Fitopatol.* **43**(10)**:**40–44 (in Italian).

Verhoeven, T. J., and Roenhorst, J. W. (1998). Occurrence of tospoviruses in the Netherlands. (D. Peters and R. Goldbach, eds.), *In* ''Fourth International Symposium on Tospoviruses and Thrips in Floral and Vegetable Crops'', Graduate Schools of Experimental Plant Sciences and Production Ecology, Wageningen, The Netherlands, p. 77.

Verhoeven, J. T. J., Roenhorst, J. W., Cotes, I., and Peters, D. (1996). Detection of a novel tospovirus in chrysanthemum. *Acta Hort.* **432:**44–51.

Vicchi, V., Fini, P., and Cardoni, M. (1999). Presence of impatiens necrotic spot tospovirus (INSV) on vegetable crops in Emilia-Romagna region. *Inf. Fitopatol.* **49**(4)**:**52–55 (in Italian).

Vierbergen, G. (1995). International movement, detection, and quarantine of Thysanoptera pests. *In* "Thrips Biology and Management" (B. L. Parker, M. Skinner, and T. Lewis, eds.), p. 119. Plenum, New York.

Vierbergen, G., and Ester, A. (2000). Natural enemies and sex ratio of *Allium porrum* in the Netherlands. Mededelingen van de Faculteit Landbouwkundige en Toegepaste Biologische Wetenschappen, Universiteit Gent, p. 335.

Villevieille, M., and Millot, P. (1991). Lutte biologique contre *Orius laevigatus* sur fraisier. *OILB/WPRS Bull.* **14**(5):57.

Vovlas, C., and Lafortezza, R. (1992). Infezioni dal virus dell'avvizzimento maculato del pomodoro (TSWV) su coltivazioni di cicoria. *Inf. Agrar.* **21**:63–64.

Vovlas, C., and Lafortezza, R. (1994). Il virus dell'avvizzimento maculato del pomodoro su carciofo in Puglia. *Inf. Fitopatol.* **44**:42–44 (in Italian).

Whitfield, A. E., Ullman, D. E., and German, T. L. (2005). Tospovirus-thrips interactions. *Annu. Rev. Phytopathol.* **43**:459–489.

Whitfield, A. E., Kumar, N. K. K., Rotenberg, D., Ullman, D. E., Wyman, E. A., Zietlow, C., Willis, D. K., and German, T. L. (2008). A soluble form of the *Tomato spotted wilt virus* (TSWV) glycoprotein G_N(G_N-S) inhibits transmission of TSWV by *Frankliniella occidentalis*. *Phytopathology* **98**:45–50.

Wijkamp, I., and Peters, D. (1993). Determination of the median latent period of two tospovirus in *Frankliniella occidentalis*, using a novel leaf disk assay. *Phytopathology* **83**:986–991.

Wijkamp, I., Almarza, N., Goldbach, R., and Peters, D. (1995). Distinct levels of specificity in thrips transmission of tospoviruses. *Phytopathology* **85**:1069–1074.

Wilson, C. R. (1998). Incidence of weed reservoirs and vectors of tomato spotted wilt tospovirus on southern Tasmanian lettuce farms. *Plant Pathol.* **47**:171–176.

Wilson, C. R. (2001). Resistance to infection and translocation of Tomato spotted wilt virus in potatoes. *Plant Pathol.* **50**:402–410.

Winter, S., Shahraeen, N., Koerbler, M., and Lesemann, D. E. (2006). Characterization of tomato fruit yellow ring virus: A new Tospovirus species infecting tomato in Iran. *Plant Pathol.* **55**:287.

Yang, H., Ozias-Akins, P., Culbreath, A. K., Gorbet, D. W., Weeks, J. R., Mandal, B., and Pappu, H. R. (2004). Field evaluation of Tomato spotted wilt virus resistance in transgenic peanut. *Plant Dis.* **88**:259–264.

Yardimci, N., and Kilic, H. C. (2009). Tomato Spotted Wilt Virus in vegetable growing areas in the West Mediterranean Region of Turkey. *Afr. J. Biotechnol.* **8**:4539–4541.

Yudin, L. S., Tabashnik, B. E., Cho, J. J., and Mitchell, W. C. (1990). Disease-prediction and economic-models for managing *Tomato spotted wilt virus*-disease in lettuce. *Plant Dis.* **74**:211–2016.

Yurtmen, M., Guldur, M. E., and Yilmaz, M. A. (1999). *Tomato spotted wilt virus* on peppers in Icel province of Turkey. *Petria* **9**:342–343.

Zaki, F. N. (1989). Rearing of two predators, *Orius albidipennis* (Reut.) and *Orius laevigatus* (Fieber) (Hem., Anthocoridae) on some insect larvae. *J. Appl. Entomol.* **107**:107–109.

Zindovic, J., Bulajic, A., Krstic, B., Ciuffo, M., Margaria, P., and Turina, M. (2011). First report of Tomato spotted wilt virus on pepper in Montenegro. *Plant Dis.* **95**:882.

CHAPTER 13

Cucumber Mosaic Virus

Mireille Jacquemond

Contents

INRA, UR407 Pathologie Végétale, Domaine Saint Maurice, Montfavet, France

Advances in Virus Research, Volume 84
ISSN 0065-3527, DOI: 10.1016/B978-0-12-394314-9.00013-0

Abstract

Cucumber mosaic virus (CMV) is an important virus because of its agricultural impact in the Mediterranean Basin and worldwide, and also as a model for understanding plant–virus interactions. This review focuses on those areas where most progress has been made over the past decade in our understanding of CMV. Clearly, a deep understanding of the role of the recently described CMV *2b* gene in suppression of host RNA silencing and viral virulence is the most important discovery. These findings have had an impact well beyond the virus itself, as the *2b* gene is an important tool in the studies of eukaryotic gene regulation. Protein 2b was shown to be involved in most of the steps of the virus cycle and to interfere with several basal host defenses. Progress has also been made concerning the mechanisms of virus replication and movement. However, only a few host proteins that interact with viral proteins have been identified, making this an area of research where major efforts are still needed. Another area where major advances have been made is CMV population genetics, where contrasting results were obtained. On the one hand, CMV was shown to be prone to recombination and to show high genetic diversity based on sequence data of different isolates. On the other hand, populations did not exhibit high genetic variability either within plants, or even in a field and the nearby wild plants. The situation was partially clarified with the finding that severe bottlenecks occur during both virus movement within a plant and transmission between plants. Finally, novel studies were undertaken to elucidate mechanisms leading to selection in virus population, according to the host or its environment, opening a new research area in plant–virus coevolution.

I. INTRODUCTION

Cucumber mosaic virus (CMV) is responsible for important agronomic losses in many crops worldwide, probably has one of the broadest host range among plant viruses, and shows a high degree of diversity, as revealed by a large number of isolates differing in both biological and molecular properties. In addition, features facilitating experimental manipulation make CMV an important model for research. These include mechanical transmission; strong accumulation in infected hosts, which allows easy purification from small samples; and infectious cDNAs available for several different strains for a reverse genetic approach. Reviews of current knowledge of the virus have been published approximately every 10 years since the beginning of the 1980s. The last was published by Palukaitis and García-Arenal in 2003 and will be taken as a baseline for this chapter. In particular, only results published after their review will be referenced here, and the reader interested in details about previously published data is invited to consult the previous review. The main striking advances made since concern (i) the role of the CMV 2b protein in plant gene silencing suppression and overall virus virulence, (ii) an increased clarity of molecular typing and grouping of natural isolates, and (iii) new insights into the effects of the selective forces exerted on viral populations.

CMV is the type member of the genus *Cucumovirus*, which also includes *Tomato aspermy virus* (TAV) and *Peanut stunt virus* (PSV), with the newly described *Gayfeather mild mottle virus* (GMMV) as a putative member (Adams *et al.*, 2009). In contrast to TAV and PSV, which have a quite restricted host range, CMV infects a large number of species. In 1991, Edwardson and Christie reported 1241 host species within 101 plant families, including monocots and dicots, and many new hosts have been described since. The host range includes all types of cultures (for food and feed products, ornamentals, etc.), as well as numerous wild species, which are important for the year-round persistence of the virus in the open field.

II. GENOME ORGANIZATION AND EXPRESSION

A. The viral genome

CMV virions are icosahedral particles 29 nm in diameter and are composed of 180 subunits of a single capsid protein (CP) and 18% RNA. The genome is composed of three single-stranded positive-sense RNAs, named 1–3 in order of decreasing size (Fig. 1). RNA1 is monocistronic and codes for protein 1a, which possesses a putative methyltransferase domain in its N-terminal part and a helicase motif in the C-terminal part. RNA2 encodes the large 2a protein, which possesses the GDD motif typical for an RNA-dependent RNA polymerase (RdRp), and the small 2b protein, which is expressed from an open reading frame (ORF) overlapping the 3′-terminal part of ORF 2a (+1 frame shift). The 2b protein interferes with the host RNA interfering (RNAi) pathway. Bicistronic RNA3 encodes the 3a or movement protein (MP) of the virus, and the CP. Although the first ORF of each bicistronic RNA is expressed from the genomic RNA, the second ORFs are expressed from the subgenomic RNA4A (protein 2b) and RNA4 (CP) (Fig. 1). The total length of each RNA can differ slightly according to the strain or strain grouping, but

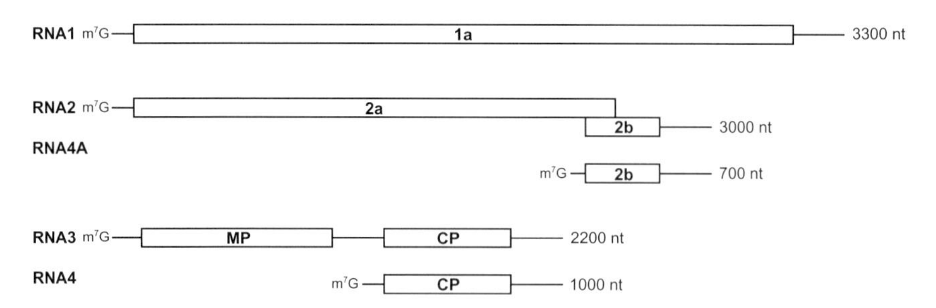

FIGURE 1 CMV genomic organization. The number of nucleotides (nt) correspond to approximate sizes. ORFs are indicated by boxes and named according to the proteins they code for. MP, movement protein; CP, capsid protein. Draw not to scale.

each ORF of different strains has a similar size, except for ORF 2a and ORF 2b from subgroup II, which are smaller than their counterparts in subgroup I (Table III). Each RNA has a cap structure at its 5′ end, and can adopt a tRNA-like structure at its 3′-hydroxylated end. In addition, the approximately 150 3′-terminal nucleotides (nt) are highly conserved among the different RNAs of a strain.

B. Small encapsidated RNAs

Small RNAs can also be encapsidated in CMV particles, of which two, RNA5 and satellite RNA, are well characterized. RNA5 is present not only in subgroup II CMV strains but also in TAV. It is approximately 300 nt long and consists of a mixture of the 3′-termini of RNAs 2 and 3. Contrary to the genomic and subgenomic RNAs, RNA5 is not capped and no polypeptides associated with its presence have been observed. The 20 or so 5′-terminal nucleotides of RNA5, composing Box-1, are highly conserved between subgroup II CMV and TAV, but Box-1 is not present in subgroup I CMV. Box-1 is also present in the unrelated *Beet necrotic yellow vein virus* in the genus *Benyvirus*, where it has been suggested that it could be part of a subgenomic promoter. Thompson *et al.* (2008) identified four stem-loop structures surrounding Box-1 in CMV. Mutations in these structures or deletion of Box-1 affected RNA5 production but had no effects on virus accumulation or symptom development. De Wispelaere and Rao (2009) further showed that production of RNA5 is independent of virus multiplication. Taken together, these results suggest that RNA5 probably arises by specific cleavage of RNAs 2, 3, and 4, rather than by transcription from them.

CMV satellite RNAs are small noncoding RNAs that have almost no sequence similarity to the viral genome but are fully dependent on functions encoded by the CMV genome for replication, encapsidation, dissemination, etc. Special attention was drawn to satellite RNA at the beginning of the 1980s, when it was demonstrated that even if some satellite RNAs were responsible for the development of a lethal disease in tomato, nearly all caused a remarkable decrease in virus accumulation and symptoms severity in all other hosts tested, acting thus as a true parasite of the virus. Epidemics of satellite-RNA-containing isolates have been erratic. After having caused particularly important damage in tomato crops in France, Spain, or Italy in the 1980s and 1990s, satellite RNA has not been implicated in further CMV outbreaks. Its origin remains mysterious, even though it has been shown that a satellite RNA can emerge following successive passages of satellite-RNA-free isolates in some hosts (particularly tobacco) (Jacquemond and Tepfer, 1998) and even off a cDNA-derived inoculum (Hajimorad *et al.*, 2009).

C. Defective RNAs

Finally, it should be noted that defective RNA3 has occasionally been identified in some viral progenies. These RNAs have deletions of a few hundred nucleotides in either the *MP* or the *CP* gene, or both. Their origin is unclear, as they occurred under different experimental conditions: in tobacco back-inoculated with a progeny from RNA transcripts (Kaplan *et al.*, 2004), in a natural isolate following multiplication on tobacco (López *et al.*, 2007), or in a preparation of purified virus of the well-characterized Y strain (Takeshita *et al.*, 2008). In no case, however, were these RNAs observed directly in a natural CMV isolate, nor has any role in the virus cycle been demonstrated, making a natural origin totally speculative.

III. THE VIRUS CYCLE

A. Virus replication

Upon entering the cell, virus particles are uncoated and genomic RNAs are translated for production of viral proteins, of which 1a and 2a are involved in the replication complex in association with host proteins. Replication occurs on the tonoplast, where both viral proteins have been localized. A model has been proposed for replication of *Brome mosaic virus* (BMV), the type member of the *Bromoviridae* family, to which CMV belongs, and which has the same genetic organization. In the BMV model, protein 1a is addressed to the tonoplast, and then protein 2a is recruited. Additional data obtained with CMV suggest that the C-terminal part of protein 1a interacts with the N-terminal region of protein 2a. Phosphorylation of protein 2a prevents association and probably reserves protein 2a for functions in which 1a protein is not involved. Two tonoplast intrinsic proteins

(TIPs) from *Arabidopsis thaliana*, TIP1 and TIP2, were shown to interact *in vitro* with protein 1a, but not with protein 2a, and to colocalize in transfected protoplasts, suggesting that they could have a role in either protein 1a anchoring and/or virus replication (Kim *et al.*, 2006). More recently, the same group showed that protein Tsip1 from tobacco can interact not only with protein 1a but also with protein 2a, alone or in combination, both *in vitro* and *in planta* on the tonoplast of infected cells. Tsip1 also regulates virus replication (Huh *et al.*, 2011), and the zinc-finger-like domain of Tsip1 appears to be necessary for interaction with the viral proteins. The exact roles of all these proteins in addressing and/or anchoring proteins 1a and 2a to the tonoplast, and in the replication complex, still need additional studies.

The first step of replication corresponds to the synthesis of minus (−) strand RNA, which is in turn used for production of plus (+) strands. These (+) strands will then have three functions: (i) mRNA for translation, (ii) template for further transcription, and (iii) production of virions. In addition, it is noteworthy that replication is asymmetric and that double-stranded RNA (dsRNA) accumulates in the infected cells. Although dsRNA can be exploited for virus diagnosis, their role is still unclear: do they play a role in the virus cycle, or are they simply a storage form of excess RNAs? Asymmetry of replication is considered to be common for positive-stranded RNA viruses. In the case of CMV, (−) strand accumulation reaches a plateau soon after infection, while (+) strand accumulation continues to increase and can reach a level nearly 100-fold that of (−) strands (Seo *et al.*, 2009). These authors showed that, while both proteins 1a and 2a are required for synthesis of the (−) strand, protein 2a alone can produce (+) strands from a (−) template of either the genomic or the subgenomic RNAs (Seo *et al.*, 2009). This possibility could account not only for the presence of free protein 2a in the cytoplasm but also for the higher proportion of (+) strands in infected cells. Also, phosphorylation of protein 2a, by preventing its association with protein 1a, could induce the switch from (−) to (+) strand synthesis or, alternatively, the switch from transcription to translation of the (+) strands. On the other hand, encapsidation sequesters (+) strands and thus prevents transcription or translation. Finally, optimized virus growth rate will result from a balance between transcription and translation and will be influenced by encapsidation and, probably, by virus movement.

B. Cell-to-cell movement

Following replication in the initially infected cells, the virus migrates to adjacent cells through the plasmodesmata. Although all the viral proteins may be involved in movement, the 3a protein encoded by RNA3 is considered to be the MP because it possesses the main characteristics of

MPs: (i) localization to plasmodesmata; (ii) the ability to increase the plasmodesmal size exclusion limit (SEL); (iii) promoting trafficking not only of RNA but also of itself, through plasmodesmata; (iv) binding to single-stranded RNAs without sequence specificity and cooperative binding to CMV viral RNAs through a domain located toward its C-terminal region (positions 174–233) or two additional smaller domains in the central part of the molecule; and (v) ability to act *in trans* and to complement a CMV isolate deficient for virus movement, when expressed in transgenic plants.

Increase of the plasmodesmal SEL has been associated with the ability of the MP to interact with and sever F-actin filaments (Su *et al.*, 2010). Targeting to and trafficking through plasmodesmata has been associated with the central part of the MP, a region rich in cysteine and histidine residues and which acts as a zinc-binding domain *in vitro* (Sasaki *et al.*, 2006). Particularly, two amino acids proved to be determinant for plasmodesmata targeting and cell-to-cell movement, as well as zinc-binding ability. A third amino acid played a role only in cell-to-cell movement, suggesting that other factors than the zinc-binding activity are required for efficient trafficking into and/or out of the cell (Sasaki *et al.*, 2006).

MP and the viral RNA can form ribonucleoprotein complexes, which have been visualized by atomic force microscopy (AFM) (Andreev *et al.*, 2004; Kim *et al.*, 2004). These structures are called "beads-on-a-string," because of their similarity to the similar structures first described for TMV (*Tobacco mosaic virus*, genus *Tobamovirus*; Kiselyova *et al.*, 2001), and show thicker regions corresponding to coated RNA separated by thinner ones, approximately 15 nm-long, that are partially sensitive to RNase treatment. Although these complexes are essential for movement, it seems that it is not strictly in a form associated with the MP that the viral RNAs move through the plasmodesmata.

The CMV MP protein also generates tubules in protoplasts, which could be related to virus movement, although there are no reports of the presence of such tubules in infected tissues, and a modified MP unable to form tubules could still promote virus movement. Cytopathological studies were carried out in our laboratory, in tobacco plants infected with the CMV R strain (B. Delécolle, unpublished results). In several sections of one sample, tubules containing particle-like structures, approximately 15 nm in diameter, were observed in several plasmodesmata of infected cells at or near the infection front (Fig. 2). These particle-like structures are much smaller than CMV virions and may be related to cell-to-cell movement of CMV. Additional immunolabeling experiments would be required in order to establish their exact composition.

Although the virus does not seem to migrate as virions, CP is necessary for cell-to-cell movement, even though no direct interactions between the two proteins have been observed *in vitro*. Studies involving recombinants

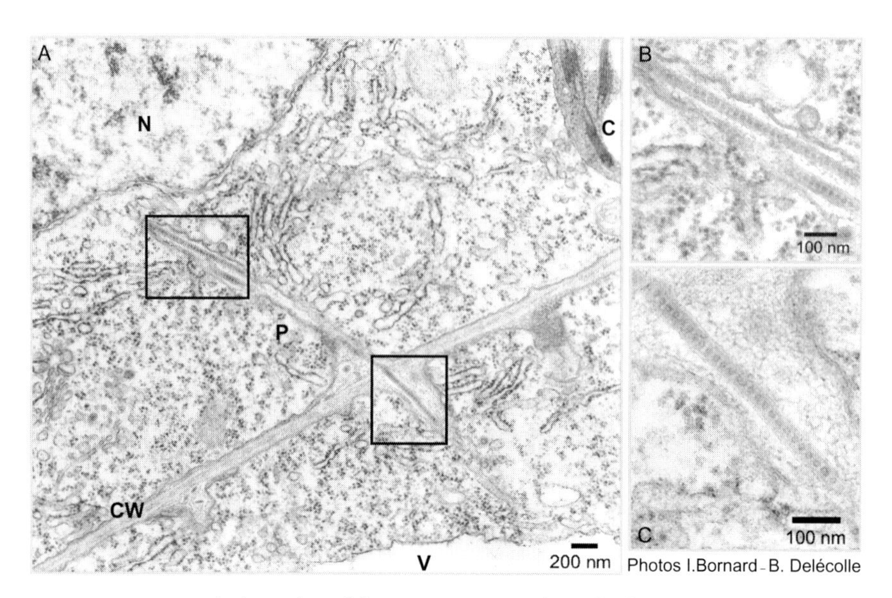

FIGURE 2 Cytopathological modifications in CMV-infected tobacco cells (A). Panels (B) and (C) are enlarged images of the areas enclosed by black boxes in panel (A). C, chloroplast; CW, cell wall; N, nucleus; P, plasmodesmata; V, vacuole.

made between CMV and TAV RNA3 showed that efficient movement requires compatibility between the 29 C-terminal amino acids of the MP and the C-terminal two-thirds of the CP (Salánki *et al.*, 2004). Llamas *et al.* (2006), using different recombinants, further revealed a more complex situation, as interactions between different regions of the CP appeared to condition MP–CP compatibility. Moreover, Kim *et al.* (2004) also showed that cell-to-cell movement was still possible for a CMV isolate in which the 33 C-terminal amino acids of the MP were deleted, although its systemic movement was impaired. This CP-independent MP variant displayed a higher affinity for viral RNA than the wild-type MP. It has been proposed that CP determines a conformational state of MP, allowing it to associate to viral RNAs and/or promote cell-to-cell movement (Blackman *et al.*, 1998). Curiously, this interaction is relevant only for CMV, as in a CMV isolate in which the MP was replaced by the MP of a virus that does not need CP for cell-to-cell movement (*Cymbidium ringspot virus*, genus *Tombusvirus*), movement was unimpaired (Palukaitis and García-Arenal, 2003).

CMV MP belongs to the "30K superfamily" of MPs, which also includes MPs of other members of family *Bromoviridae* (in genera *Alfamo-*, *Bromo-*, *Cucumo-*, and *Ilarvirus*) as well as members of *Como-* and *Tobamovirus*. BMV and CPMV (*Cowpea mosaic virus*, *Comovirus*) migrate as virus particles within tubules, while TMV migrates as ribonucleic complexes. It has been shown that, depending on the CP characteristics, *Alfalfa mosaic virus* (AMV, *Alfamovirus*) can migrate as particles in a CP-dependent manner or as ribonucleic complexes in a CP-independent one (Sánchez-Navarro *et al.*, 2006). CMV and AMV shares several common properties regarding cell-to-cell movement: (i) their MPs are able to form tubules, (ii) the CP is required for movement, (iii) movement supposes an interaction between the CP and the C-terminal part of the MP, but this part is not essential (Sánchez-Navarro *et al.*, 2006). Like the MP of BMV, CPMV, and PNRSV (*Prunus necrotic ringspot virus, Ilarvirus*), the CMV MP was able to complement a modified AMV RNA3, deleted for its own MP, only if the AMV MP C-terminal was maintained, suggesting that these heterologous MPs were unable to interact with AMV CP. On the other hand, the C-terminal AMV MP region was not required for efficient complementation of movement by TMV-MP (Sánchez-Navarro *et al.*, 2006). Moreover, in all cases, virus particles were not required for movement. These results demonstrate that AMV could use either of the two main transport mechanisms, suggesting a common functionality of all the MPs of the superfamily.

The two proteins encoded by RNA2 are also involved in cell-to-cell movement, at least in certain hosts. Two mutations in the MP and the 2a protein were shown to account for reduced movement (both local and systemic) in squash, acting independently but having an additive effect when simultaneously present (Choi *et al.*, 2005). In addition, *in vitro* interactions between the MP and 2a protein were also reported (Hwang *et al.*, 2005). Later, deletion mutants of both proteins indicated that the N-terminal 21 amino acids and the region surrounding the GDD motif of the polymerase interact strongly with the MP, while a serine at position 14 in the MP is critical for this interaction (Hwang *et al.*, 2007). To what extent does this interaction determine cell-to-cell movement ability, as the N-terminal part of the MP is not essential? Since the amino acid at position 14 corresponds to a putative phosphorylation site, does that mean that MP phosphorylation plays a role in interaction with the polymerase or, as shown for TMV, in MP localization in the cell? These questions will be answered by determining the form under which the virus migrates, either from cell-to-cell or systemically, and what host proteins are involved in these processes.

Earlier work demonstrated the role of the 2b protein in virus movement. Deletion or interruption of the *2b* ORF generally results in less efficient or altered local movement of CMV (Soards *et al.*, 2002), cucumovirus reassortants (Shi *et al.*, 2003), and PSV (Netsu *et al.*, 2008). However, as this protein is involved not only in nearly all the steps of the virus cycle but also in suppressing the RNAi-mediated defense mechanism of the plant, an indirect role in movement through RNAi suppression cannot be excluded.

C. Long-distance movement

Overall, there is less information concerning long-distance virus movement than cell-to-cell movement. Viruses use the phloem-assimilate flow pathway for systemic invasion of their hosts, and this was shown to be the case for CMV in cucumber plants (Moreno *et al.*, 2004). Long-distance movement supposes three distinct steps: loading into the sieve elements, movement within them, and then unloading from these elements, all of which can constitute as many barriers to invasion.

Studies of different combinations of strains—or their mutants—and host species show that all the viral proteins can have effects on long-distance movement, although the CP, MP, and 2b proteins seem to be the major actors. Even if some variants deficient for encapsidation, or characterized by unstable particles (Ng *et al.*, 2005; Pierrugues *et al.*, 2007), can migrate systemically, the CP is essential for efficient long-distance movement and also has host-specific determinants for this function. Its roles in long- and short-distance movements are clearly distinct. CMV migrates to the different organs through sieve tubes as viral particles (Requena *et al.*, 2006). These authors also showed that particles interact with a protein present in the phloem exudate of infected cucumbers, which is a homolog of phloem protein 1 (PP1) of pumpkin. PP1 has plasmodesmal-gating ability and translocates with the phloem stream. It could act either by protecting the viral RNAs against the ribonucleases present in the phloem or by facilitating the movement of ribonucleic complexes to or through the sieve tubes (Requena *et al.*, 2006).

The main question regarding loading of the virus into the sieve elements concerns the exact structure of what is translocated and thus where encapsidation occurs. There are several lines of evidence that the MP is also involved in long-distance virus movement. As mentioned above, MP mutants in which the 33 C-terminal region is deleted cannot invade tobacco plants systemically (Kim *et al.*, 2004), and the MP has been clearly associated with the ability of some strains to systemically infect wild soybean (Hong *et al.*, 2007). Amino acids at positions 51 and 240 have been shown to condition the ability to systemically infect cucurbits. Moreover, in both infected nontransgenic plants and noninfected transgenic plants expressing MP, the protein accumulates in all types of plasmodesmata, including those connecting the companion cells and the sieve elements, where MP reaches the highest amounts (Blackman *et al.*, 1998). In these experiments, viral particles were visualized in the cytoplasm of companion cells and in the sieve element, where they form aggregates often embedded in a discrete membrane, but not in the branched plasmodesmata connecting these cells. From these results, Blackman *et al.* (1998) proposed that the virus moves into sieve elements as an RNA–protein complex, thanks to the MP, and viral particles are then assembled there. To our knowledge, no additional data have been published since. Similarly, exit from the phloem has not been extensively studied. As demonstrated for only a few cases, viruses are thought to exit phloem to reach the sink tissues almost entirely from major veins (for review, see Pallás *et al.*, 2011).

The 2b protein conditions long-distance migration in a host-specific manner. Deletion of the *2b* gene prevented systemic infection in cucumber, squash, pepper, and tomato, but not in several *Nicotiana* sp. or *A. thaliana* plants, even if systemic spread and virus accumulation could be reduced in these hosts (Ding *et al.*, 1995; Lewsey *et al.*, 2009; Mascia *et al.*, 2010; Masiri *et al.*, 2011; Soards *et al.*, 2002; Wang *et al.*, 2004b). Virus movement deficiency resulting from deletion of the *2b* gene could be complemented in transgenic plants expressing the protein or when it was expressed by a virus vector (Lewsey *et al.*, 2009; Wang *et al.*, 2004b), suggesting that interference with movement is due to this protein and not the overlapping C-terminal part of protein 2a.

RNA1 has been shown to affect the rate of systemic infection in squash and to control systemic infection in lily (Yamaguchi *et al.*, 2005a). In a yeast two-hybrid system using a cDNA library from tobacco, one protein, named Tcoi1, was shown to interact with protein 1a (Kim *et al.*, 2008). Tcoi1, located in the cytosol, was previously unknown and was shown to contain a putative methyltransferase domain. Interaction with protein 1a was confirmed not only by an *in vitro* GST pulldown approach, but also *in vivo* in *Arabidopsis* protoplasts, using the BRET (Bioluminescence Resonance Energy Transfer) technique and co-immunoprecipitation experiments. These authors showed that the two proteins interact through their methyltransferase domains. Also, Tcoi1 methylated the helicase and the methyltransferase domains of protein 1a *in vitro*. Systemic accumulation of the virus was enhanced in plant expressing Tcoi1 (but not local accumulation), while it was reduced in plants expressing a *Tcoi1* antisense gene. By return, CMV infection induced increased expression of the *Tcoi1* gene. Taken together, these results indicate that Tcoi1 modulates the replication and/or the spread of the virus, although it remains to determine if it is through its capacity to methylate protein 1a (Kim *et al.*, 2008).

Meristematic tissues are generally thought to escape virus invasion, and regeneration of plants from shoot apical meristems (SAMs) is used as a sanitary control measure to eliminate CMV, for example, in *Vanilla planifolia* (Retheesh and Bhat, 2010). CMV, however, can be transiently detected in SAMs of infected tobacco plants, in either the tunica or the corpus or both, 6–8 days after infection (Mochizuki and Ohki, 2004). The meristematic cells then seem to recover, as the virus is no longer detected 14 or 18 days after infection. By comparing two strains having differing abilities to infect SAMs, these authors further showed that the amino acid at position 129 of the CP is determinant for SAM infection (Mochizuki and

Ohki, 2005). This amino acid is implicated in many other functions in plant–virus infection, such as symptom expression, aphid transmissibility, and movement (both local and systemic). As a role for CP in movement has been observed in a number of hosts, including several cucurbits, maize, *Tetragonia expansa*, and *Physalis floridana* (references in Kobori *et al.*, 2002; Palukaitis and García-Arenal, 2003), it can be supposed that efficiency of invasion of the SAM in tobacco is also related to this function. Finally, the 2b protein was also shown to determine SAM invasion through its ability to suppress RNA silencing (Sunpapao *et al.*, 2009). More precisely, by using a modified strain in which the *2b* gene is not expressed (without altering the 2a protein), these authors observed that the variant was absent from the SAM, while the wild-type strain was successfully detected, and that small viral RNAs accumulated earlier in the SAM of plants infected with the variant. Thus, in this case, the role of the 2b protein in virus movement is associated to its silencing suppressor activity, rather than to a direct role of the protein itself, as previously suggested. Similarly, one can wonder if the role of proteins 1a and 2a in movement may also be indirect, a consequence of lower virus titer when these proteins have been modified.

IV. DISSEMINATION IN NATURE

In the field, CMV is transmitted by aphids in a nonpersistent manner and also through seeds in some, but not all, host species. A general question concerning CMV epidemics is the role of reservoir hosts that allow the virus to survive between successive crops; this will also be considered below.

A. Dissemination by aphids

It is now generally agreed that transmission of viruses by aphids follows two main modes, distinguished by whether the viruses are circulative or noncirculative in the vector (Ng and Perry, 2004). Transmission of CMV is in a nonpersistent manner, previously also called stylet-borne transmission, and corresponds to the noncirculative mode. It is characterized by short acquisition and inoculation times (from seconds to minutes), without retention in the insect. Aphids remain viruliferous for short periods and lose the ability to transmit virus following feeding on a plant or even following a simple feeding probe.

CMV is transmitted by a large number of aphid species; *Myzus persicae* and *Aphis gossypii* are probably among the most important in vegetable crops and have been extensively used in laboratory experiments. There are a few studies considering at the same time viral epidemics and aphid populations in the open field. This was done recently by Gildow *et al.* (2008), in snap bean crops in the United States. Among the 25 aphid species collected in several fields, they identified six as the major contributors to aphid populations (*Rhopalosiphum maidis*, *A. gossypii*, *Therioaphis trifolii*, *A. fabae*, *A. glycines*, and *Acyrthosiphon pisum* by the order of decreasing prevalence). Twelve species were compared for their ability to transmit CMV in group testing experiments. Four of them proved to be the most efficient vectors (*A. gossypii*, *A. glycines*, *A. pisum*, and *T. trifolii*), which are also among the most prevalent in the open field (Gildow *et al.*, 2008).

The relationship between transmission efficiency and virus accumulation in the source plant remains somewhat unclear. Earlier studies showed that these parameters are not related above a certain virus accumulation level. This is consistent with the severe bottleneck occurring during aphid transmission (Ali *et al.*, 2006; Betancourt *et al.*, 2008; see below). Below this level, transmission efficiency depends on virus accumulation. This is well illustrated by the transmissibility of satellite-RNA-containing isolates. As mentioned before, satellite RNA drastically reduces replication of the viral genome, particularly in solanaceous hosts, and such plants are poor sources for aphids (Escriu *et al.*, 2000). In such cases, the density of the vector population will determine the success of virus epidemics. Models have been developed in order to explain the contrasted epidemics of the three phenotypic diseases CMV can induce in tomato crops (Escriu *et al.*, 2003). Tomatoes infected by CMV can develop mosaic and fern-leaf symptoms (satellite-RNA-free isolates), attenuated symptoms (isolates containing a nonnecrogenic satellite RNA), or lethal necrosis (isolates containing a necrogenic satellite RNA). Satellite RNAs are not widespread in natural isolates, except precisely during epidemics of necrosis in tomato crops. Although a satellite-RNA-free isolate can be transmitted from a source containing a nonnecrogenic satellite RNA, this has never been observed for necrogenic isolates (Escriu *et al.*, 2000). Based on plant biomass, necrogenic isolates appeared to be the most virulent, followed by isolates inducing fern-leaf symptoms (Escriu *et al.*, 2003). Considering estimated transmissibility, ranking of isolates was as follows: satellite-RNA-free > isolates containing nonnecrogenic satellite RNA > necrogenic isolates. A coinfection model, including competition between isolates, predicted the invasion by necrogenic isolates under higher aphid densities (Escriu *et al.*, 2003) and could explain the dramatic necrogenic epidemics that occurred in several Mediterranean countries in the 1980s and 1990s.

Studies of strains showing very poor transmission efficiency by aphids demonstrated that the CP is the sole viral determinant of transmissibility. Several, and sometimes different, amino acid positions were shown to interfere with transmission efficiency by either *A. gossypii* or *M. persicae* in laboratory experiments. Some of them, such as position 129, are exposed on the outer surface of the protein (for which a high-resolution atomic structure is available), while others (such as position 162) are on the inner face. These positions could be essential for either interaction with components of the aphid stylet or particle stability. Correlation between virion stability and transmission efficiency was established for position 162, but not 129 (Ng *et al.*, 2005). Position 161 was also implicated in both virion instability and loss of aphid transmission in another study involving different CMV strains and reassortants (Pierrugues *et al.*, 2007), probably by affecting surface charge or structure.

During their experiments, Escriu *et al.* (2000) observed that aphids took about twice as long to probe tomato leaves infected with a necrogenic isolate as to probe leaves infected with a nonnecrogenic one (either containing or not a nonnecrogenic satellite RNA). This suggested that different isolates could cause different appetence of the aphids for the same infected host. In fact, pathogen infection can modify host phenotype and the host's interactions with other organisms. This is the case of CMV infection in squash, where it has been shown that population density of *A. gossypii* appeared to be less important on CMV-infected plants than on healthy ones and winged aphids less frequently colonized infected plants (Mauck *et al.*, 2010a). On the other hand, infection greatly increased attractiveness of the plants for both *A. gossypii* and *M. persicae*, by inducing increased emissions of a mixture of volatiles by the plants (Mauck *et al.*, 2010a). So CMV infection attracts aphids, which then disperse rapidly, increasing thus the efficiency of virus dissemination. In addition, these authors also showed that CMV infection induced negative changes in the behavior and population density of the squash bug *Anasa tristis* (Mauck *et al.*, 2010b), showing that virus infection can have broader implications for ecosystem function.

B. Dissemination through seeds

Transmission through seeds by either cultivated crops or wild species is an important parameter for the development of epidemics, as plants grown from infected seeds constitute a primary source, in the crops or at their proximity, from which the virus can be efficiently disseminated by aphids. Transmission of CMV through seeds has been described for several species, based on biological or serological techniques (Palukaitis *et al.*, 1992), and more recently by RT-PCR detection of the virus. Testing ungerminated seeds for the presence of the virus is pertinent from an agronomic point of view but can lead to overestimation of seed contamination, due to contamination of the seed coat, even if, in the case of CMV, it is generally admitted that the virus is too unstable to remain viable as an external contaminant of the seeds. True transmission of CMV through seeds, that is, by the embryo, has been clearly established for bean, spinach (Yang *et al.*, 1997), lentil (Makkouk and Attar, 2003), lupin

(O'Keefe *et al.*, 2007), and more recently, pepper (Ali and Kobayashi, 2010). Detection of the virus in seed lots or in plants germinated from seeds has also been documented for common bean, *Vigna unguiculata* and *V. radiata* (Abdullahi *et al.*, 2001; Babovic *et al.*, 1997; Bashir and Hampton, 1996; Bhattiprolu, 1991; Chalam *et al.*, 2008; Lahoz *et al.*, 1994); pea and faba bean (Latham and Jones, 2001); chickpea, vetch, and several clovers (Latham *et al.*, 2001); tomato (Park and Cha, 2002); alfalfa (Jones, 2004a); butterfly pea (Odedara *et al.*, 2007); and hull-less oil pumpkin (Tóbiás *et al.*, 2008). Transmission rates are generally quite low (below 2.5%), although sufficient to successfully initiate epidemics, but higher rates were reported for lentil (up to 9.5%), tomato (8%), spinach (15%), and cowpea (21%).

The importance of using healthy seeds for controlling CMV epidemics is illustrated by the strategy developed in Australia at the end of the 1980s. CMV epidemics can cause important losses in lupin, and seed-infected lupin plants are the main primary sources from which infection spreads as large patches, while neither pastures nor weed hosts play an essential role as reservoirs (Jones, 2005). Because of the economic impact of the disease, a commercial testing service was created in 1988 to help farmers avoid sowing lupin seeds with CMV infection levels that could compromise the crop. Thresholds of 0.5% and 0.1% of seed infection were determined for zones of moderate or high risk, respectively, and were effective in reducing yield losses in most years. Both aphid populations and CMV outbreaks were forecasted, based upon climatology data for estimating aphid population development and the level of seed infection by the virus (Thackray *et al.*, 2004). The model was validated by 14 years of field data and incorporated into a decision support system for farmers.

In most of the studies dealing with CMV transmission from seeds collected on inoculated and infected plants, the presence of the virus was primarily based on symptom development by the germinated plants. It is both striking and worrisome that plants infected through their seed can be asymptomatic, as observed for common bean (Lahoz *et al.*, 1994) and undoubtedly demonstrated recently for pepper (Ali and Kobayashi, 2010). These authors observed high transmission rates (95–100%) in dry seeds extracted from fruits of peppers infected with the Fny strain of CMV, which otherwise induces a severe mosaic on pepper. Mean rates of infection of 60% and 30% were observed for seed coat and embryos, respectively. Moreover, when the virus was estimated by grow-out tests, the rate of seed transmission ranged from 10% to 14%, and all the plants remained asymptomatic. It would be interesting to test the ability of such plants to serve as virus sources for aphid-mediated transmissions. More generally, this result suggests that the transmission efficiency for several host–virus combinations for which the results were based on symptom development only should be revisited.

C. Reservoir hosts

CMV has been described in many wild and weedy plants in nature, where it generally induces a latent infection, as the plants develop no apparent symptoms. These plants are thought to constitute the primary external sources of infection for cultivated crops and, in turn, to be contaminated from cultivated crops and further serve as successional/seasonal survival hosts for the virus between crops, even more so as some of them can transmit the virus through the seed. Also, they constitute reservoirs for aphids. For these reasons, removal of wild plants is one of the key components of integrated virus disease management for economically important crops (Jones, 2004b; Lecoq and Pitrat, 1983). However, comparisons of virus populations in cultivated and wild species remain rare. In a 3-year study, the presence of CMV and its incidence in vegetation was surveyed in several types of plant populations: (i) weeds between fields under cultivation, (ii) edges delimiting fields, (iii) pasture wastelands, and (iv) melon crops (Sacristán *et al.*, 2004). CMV was detected in 40 species, including nearly all the most frequent species, and CMV incidence was positively correlated with the number of plants but not with the area of individuals. CMV population dynamics in melons was not related to that in weeds at the same period or in the months previous to the crop. However, population genetic and biological structure appeared to be the same in both melons and weeds, showing migration events between habitats (Sacristán *et al.*, 2004).

A key unanswered question concerns the distance of migration of the inoculum from the sources to the primary infection foci in the crop. For a virus transmitted in a nonpersistent manner, this distance is expected to be relatively short, although aphid migration can vary considerably depending on climatic factors, particularly dominant winds. Detection of the virus in aphids is not easy to achieve, even with very sensitive methods such as RT-PCR. Moreover, detection of the virus in an aphid does not mean that transmission would occur. Increasing the distance of sampling in weeds from cultivated crops to approximately 1 km resulted in a lower incidence of the virus in weeds, suggesting that distance of migration was of this scale (Sacristán *et al.*, 2004). In a study involving another nonpersistently transmitted virus (WMV, *Watermelon mosaic virus*, genus *Potyvirus*), Joannon *et al.* (2010) also estimated that the average distance of virus dissemination was low, less than a few kilometers.

V. HOST RESPONSE

A. Symptom development in susceptible hosts

Most of the infected hosts develop systemic mosaic symptoms, more or less severe, depending on the host genotype and the strain, and all viral proteins have been shown to be involved in symptom development.

Despite a lot of work aimed at determining the biochemical changes occurring in infected plants (at the protein or RNA level), the mechanisms underlying symptom development are still poorly understood. New insights were brought very recently; when taking advantage of the powerful tools available today such as high-throughput sequencing, two groups elucidated the mechanisms involved in induction of bright yellow chlorosis in tobacco by a satellite RNA (Shimura *et al.*, 2011; Smith *et al.*, 2011). Y-Sat RNA (satellite RNA of Y strain) has a 22-nt sequence complementary to the mRNA of tobacco magnesium protophorphyrin chelatase unit I (*ChlI*), a gene essential in chlorophyll synthesis. Both studies reported that the small RNAs derived from Y-sat RNA directed development of chlorosis in tobacco by downregulating the *ChlI* gene through RNA silencing (see below). Expression in tobacco of a gene mutated in the 22-nt stretch of complementarity abolished induction of chlorosis, and other *Nicotiana* species, which do not develop yellow chlorosis when Y-Sat RNA is associated to the virus, showed a point mutation in the 22-nt target sequence (Smith *et al.*, 2011). Conversely, when Y-Sat RNA was modified in order to correspond to the complementary sequence of the *ChlI* gene of *A. thaliana* or tomato, these hosts developed chlorosis following infection (Shimura *et al.*, 2011). A new satellite RNA was described very recently that induces a yellow chlorosis on pepper (as does Y-Sat RNA since pepper *ChlI* also has complementarity to this satellite). However, the genetic determinant for chlorosis mapped to another region of the satellite (Choi *et al.*, 2011), illustrating the difficulty in generalizing a result obtained with a strain–host combination.

Necrotic symptoms can occur in either just the inoculated leaves or as a systemic syndrome, which can in some cases lead to the death of the plants. Necrotic lesions in the inoculated leaves restrict the virus to a few cells around infection sites and correspond to a defensive hypersensitive reaction (HR) response. The HR was associated with protein 1a, or 2a or CP, depending on the host–strain combination (Palukaitis and García-Arenal, 2003). Looking at the plant side, the biochemistry of the HR is well documented (for a review, see Palukaitis and Carr, 2008). Moreover, the hormone spermine was recently shown to also act as a signal for the HR defensive response to CMV in *A. thaliana* (Mitsuya *et al.*, 2009). Hypersensitive hosts belong to different families, although plants in the *Fabacea* (beans) and *Chenopodiacea* (*Chenopodium* sp.) are the most common, and are used experimentally for biological cloning of CMV isolates.

Systemic necrosis is described in a few cases and is exemplified by the Fulton strain of CMV, isolated from tobacco in the United States by R. W. Fulton in 1953. The genetic determinant for necrosis of a biologically cloned isolate of this strain (Ns) was recently mapped to one amino acid of the 1a protein (Divéki *et al.*, 2004). Analysis of different mutants and modeling the partial structure of the region surrounding this residue suggested that it is involved in a basic-neutral amphiphilic α-helix, whose integrity is essential for virus replication (Salánki *et al.*, 2007). However, none of the mutants was able to induce necrosis as strong as that of the wild-type strain (Salánki *et al.*, 2007).

Systemic necrosis has been more recently described in *Nicotiana glutinosa* infected by strain Cb7, a IB strain isolated from tomato in China, and this reaction was linked to the region corresponding to the *2b* gene, since introduction of this gene in the Fny strain resulted in the development of top necrosis by infected *N. glutinosa* plants (Du *et al.*, 2007a). It was further demonstrated that amino acid 55 of the 2b protein is the major determinant for necrosis (Du *et al.*, 2008). However, because several other well-characterized CMV strains possess the same residue as Cb7 at position 55 and do not induce necrosis in *N. glutinosa*, these authors identified two other conserved positions in the 2a/2b overlapping region, which could also account for necrosis induction. Because the two proteins could not be individually mutated at these positions, a new construction resulting in premature termination of protein 2a was tested. Partial necrosis occurred, confirming an additional role of the C-terminus of protein 2a in necrosis induction (Du *et al.*, 2008).

Necrotic lesions also occur, in both inoculated and noninoculated upper leaves, in several ecotypes of *A. thaliana*, following inoculation with strain HL, isolated from lily. Biological properties of reassortants created between this strain and strain Y indicated that RNA2 is the genetic determinant for systemic necrosis (Inaba *et al.*, 2011). Moreover, formation of necrotic lesions was associated with increased levels of hydrogen peroxide. Because catalase (CAT) is involved in the regulation of the reactive oxygen species in the cells, and because *A. thaliana* mutants deficient for two catalases display necrosis, these authors examined putative interactions between CAT proteins of *A. thaliana* and protein 2b, known to be a determinant for virulence (see below) and necrosis induction as previously presented. 2b protein interacted with CAT3 (but not the two other catalases of *A. thaliana*) in yeast two-hybrid assays and this ability was associated with its 40 C-terminal amino acids. They also showed that protein 2b recruited the cytoplasmic CAT3 in the nucleus of transfected *A. thaliana* protoplasts. In addition, transgenic plants expressing the HL-2b protein developed necrotic spots, produced hydrogen peroxide, and showed an important reduction of catalase activity. Finally, overexpression of the *Cat3* gene in *A. thaliana* resulted in increased catalase activity and considerably delayed development of necrosis after inoculation. Overall, this work allowed identification of a host component interacting with 2b protein, resulting in the modification of the hydrogen peroxide levels and catalase activity in the cell and development of necrosis (Inaba *et al.*, 2011).

A fourth case of development of systemic necrosis corresponds to tomato plants infected with isolates containing a necrogenic satellite RNA. In this case, cell necrosis has the same characteristics as programmed cell death and corresponds to high accumulation of the (−) strand of the satellite RNA. In particular, the development of necrosis corresponds to the induction of several defense responses, including synthesis of PR (pathogenesis related) proteins, and here also, accumulation of hydrogen peroxide and increase in catalase (CAT2) levels (Xu *et al.*, 2003). A more complete transcriptomic analysis further showed that, among the numerous genes that were deregulated upon necrosis, genes involved in ethylene synthesis and signaling played an important role. Inoculation of ethylene-insensitive mutants or chemical treatment inhibiting ethylene action confirmed the role of ethylene in cell death (Irian *et al.*, 2007).

B. Coinfections and synergism

Coinfection modulates symptom development in two opposite ways. Coinfection by CMV and a virus from another genus (*Crinivirus, Potexvirus, Potyvirus, Tobamovirus*) can worsen the disease (synergy), particularly in cucurbits or solanaceous hosts, and generally correlates with an increase in CMV accumulation only. Synergy can also occur in plants that are tolerant to one of the two viruses or, finally, by restoring a deficient function of one of the two viruses (see Palukaitis and García-Arenal, 2003). However, in cucumbers resistant to CMV, coinfection with ZYMV (*Zucchini yellow mosaic virus, Potyvirus*) resulted in increased CMV accumulation, but without disease worsening (Wang *et al.*, 2004a). The main advance made in recent years concerns the mechanism underlying synergism, and particularly the role of protein 2b via its silencing suppressor activity (see below). Malformation of young leaves of tobacco plants coinfected with CMV and TMV was also observed when transgenic plants expressing protein 2b were infected with TMV (Siddiqui *et al.*, 2011). Moreover, these authors also showed that the synergistic effect resulted from a joint effect of the two viral silencing suppressors (protein 2b and TMV replicase). Similarly, tomato plants exhibited worsened symptoms when coinfected with PVY (*Potato virus Y, Potyvirus*) and CMV. This synergy was accompanied by an increase in CMV level and a decrease in PVY level, essentially through the action of CMV 2b protein (Mascia *et al.*, 2010). When the *2b* gene was deleted in CMV, the virus became unable to move systemically, but its defect could be compensated when plants were coinfected with PVY (Mascia *et al.*, 2010). However, it is not obvious that this assistance relates to synergism, because in a similar situation in squash, ZYMV could not complement the same CMV mutant for systemic movement (Wang *et al.*, 2004b). Conversely, CMV protein 2b could complement PVY for restricted movement in tobacco. In this host, necrotic PVY strains accumulated poorly in younger leaves, which did not develop necrotic symptoms. Coinfection with CMV resulted in uniform distribution and enhanced accumulation of PVY only when the viral *2b* gene was present (Ryang *et al.*, 2004). Another evidence that the viral silencing suppressors play an essential role in synergism came from the work of Fukuzawa *et al.* (2010), which showed that PVY HC-Pro abolished cycling of CMV in *Nicotiana benthamiana*.

On the other hand, although coinfection of two CMV strains can occur, resulting in an additive effect of the symptoms induced by each strain, infection with a CMV strain can result in protection against a second infection by a related strain (cross-protection). The mechanism behind cross-protection is not well known, and either RNA silencing or competition between the two strains have been proposed. Consistent with this second hypothesis is that when plants are inoculated with an artificial quadripartite CMV composed of one strain plus an additional RNA3 from another strain, the heterologous RNA3 accumulated poorly in both inoculated leaves and noninoculated upper ones (Takeshita *et al.*, 2004a). Moreover, coinoculation of both strains in cowpea resulted in exclusive distribution in all leaves (Takeshita *et al.*, 2004b). In another study, a CMV mutant in which the *2b* gene was deleted could protect tobacco and *N. benthamiana* against infection by the same strain, and even a less closely related strain from another subgroup, suggesting that it is likely that RNA silencing did not play a role in cross-protection (Ziebell *et al.*, 2007).

C. Basal RNA-mediated defenses

Plants can deploy several basal defense mechanisms (for a review, see Palukaitis and Carr, 2008). Particularly, salicylic acid (SA) and jasmonic acid (JA) are important signaling molecules in plant defense responses; their activation generally induces expression of *PR* genes, and they have been shown to play antagonistic roles in plant defense. However, only RNA-mediated resistance will be considered here, as many advances have been made in this area in the past decade, with studies involving CMV having played a major role.

In eukaryotes, regulation of endogenous gene expression as well as protection against foreign genes, invasive elements, or repeated sequences is directed by RNA silencing. We will focus here on posttranscriptional gene silencing (PTGS), which corresponds to RNA-mediated processes that downregulate gene expression, because they constitute the defense mechanism directed against infection by RNA viruses. Indeed, studies on plant viruses played an important role in our understanding of the mechanisms that underlie PTGS (for a review, see Voinnet, 2005). Although diverse pathways account for RNA silencing, they all rely on a set of common reactions and start with the production of small RNA duplexes, 21- to 24-nt long, named micro RNAs (miRNAs) or small interfering RNAs (siRNAs), depending on the pathway. These small RNAs are produced by digestion of dsRNAs or imperfect double-stranded hairpins in single-stranded RNAs, by RNase III-type enzymes named DICER-LIKE (DCL). Following strand dissociation, one is incorporated into the RISC complex (RNA-induced silencing complex) and directs this complex to degrade RNAs that are complementary in sequence. The core component of this complex is a protein Argonaute (AGO), which has an endonucleolytic cleavage activity and degrades the target RNA in a sequence-specific manner. Small RNAs can also be copied by cellular RNA-dependent RNA polymerase (RDR) to produce new dsRNAs, from which a secondary gene silencing cycle occurs. The small RNAs also constitute the mobile signals allowing silencing to spread throughout the plant (Dunoyer *et al.*, 2010).

Knowledge of the enzymes involved in RNA silencing in plants is primarily from studies on *A. thaliana*, where there are four DCL genes, ten AGO genes and six RDR genes. Loss-of-function mutants for each gene, as well as multiple mutants, have been used to determine which enzymes were involved in the different PTGS pathways. It was shown that the CMV genome is mainly targeted by DCL4, generating 21-nt-long siRNAs (Bouché *et al.*, 2006). DCL2 could substitute for DCL4 in *dcl4* mutants, giving rise to a 22-nt siRNA species. Alternatively, DCL3 generated 24-nt siRNAs in double *dcl2 dcl4* mutants (Bouché *et al.*, 2006). This illustrated the hierarchical functioning of DICER proteins in RNA silencing. Although it was initially thought that siRNAs derived essentially from the dsRNA forms that are typical for virus infection, it seems that hairpins in single-stranded RNAs are also targets for DICER proteins. This has been shown for CMV satellite RNA. Sequencing diverse cloned siRNAs produced in *A. thaliana* infected with a satellite-RNA-containing strain showed that 43% of the siRNAs derived from the satellite and were 21 or 22 nt long. They were generated mainly by the action of DCL4, resulting in the production of the 21-nt species (Du *et al.*, 2007b). Interestingly, these authors noted an asymmetry in strand polarity, with a majority of siRNAs deriving from the (+) strand (62%), and also that siRNAs of inversed polarities did not always match perfectly on each side of the sequence. Moreover, even if all the satellite sequence could be recovered, some regions were more represented than others and corresponded to T-shaped hairpins in the molecule (Du *et al.*, 2007b).

Among the 10 argonaute proteins of *A. thaliana*, only two are thought to direct CMV genome cleavage. AGO1 was shown to bind CMV siRNAs, as did AGO5 (Takeda *et al.*, 2008), to recruit siRNAs *in vivo* (Zhang *et al.*, 2006) and to be upregulated upon CMV infection (Cillo *et al.*, 2009; Zhang *et al.*, 2006). Also, *ago1* mutants appeared to be hypersusceptible to CMV infection (Morel *et al.*, 2002), suggesting that AGO1 is essential for protection against CMV. Recently, *ago2* mutants were also shown to be hypersusceptible to CMV and the protein was also shown to be induced by CMV infection, demonstrating its role in antiviral defense in plants (Harvey *et al.*, 2011).

As mentioned above, novel small RNA duplexes need to be produced by a host RDR for induction of secondary PTGS and efficient protection. Inoculation of different *rdr* mutants with either CMV or a mutated virus in which the *2b* gene was deleted indicated that RDR1 directs production of CMV secondary siRNAs, although this could be observed only in the absence of the 2b protein (Diaz-Pendon *et al.*, 2007). An in-depth study of virus infection and production of siRNAs of each genome segment in different mutants deficient for RDR1, RDR6, or both, still using a CMV mutant that does not express the 2b protein, indicated that both RDR1 and 6 are required for viral immunity (Wang *et al.*, 2010). While wild-type *A. thaliana* develops no symptoms when infected with the mutated CMV, *A. thaliana rdr1 rdr6* mutants developed symptoms and accumulated over 10-fold more virus. They also displayed a nearly twofold reduction in the level of siRNAs of all three genome segments. Ratios of siRNAs/genomic RNAs in *rdr* mutants were 20-fold lower than in wild-type plants, indicating that RNA silencing in *A. thaliana* is consistently associated with RDR amplification of siRNAs. Interestingly, they also demonstrated a specialization of the RDR proteins; RDR1 directed the production of siRNAs of the 5′-terminal part of the viral RNAs, while RDR6 preferentially targeted the 3′-terminal part (Wang *et al.*, 2010). These authors also showed that satellite RNA escaped secondary PTGS by being not targeted by host RDR, resulting in an optimized replication and providing an explanation for its unusually high accumulation in infected plants.

D. 2b Protein is a virulence factor and counteracts host defense

Protein 2b is localized primarily in the nucleus of infected cells, and in *A. thaliana*, this localization is dependent on an interaction with host protein AtKAPα, a karyopherin-like protein involved in protein localization to the nucleus (Wang *et al.*, 2004b). Indeed, two nuclear localization signals (NLS), rich in arginine residues, are present at positions 22–27 and 33–36 (amino acid positions) of protein 2b. Deletion of both signals was required for protein delocalization. In addition, protein 2b possesses a putative phosphorylation site at positions 39–43, whose presence was also associated with nuclear localization (Wang *et al.*, 2004b). Alignment of several 2b sequences showed that it is the most variable protein of the CMV genome, even if the C-terminal sequence of approximately 17 amino acids is conserved (Lewsey *et al.*, 2009). TAV 2b protein is able to bind small double- and single-stranded RNAs (Rashid *et al.*, 2008) and also forms dimers *in vitro* (Chen *et al.*, 2008). Similarly, CMV 2b protein can bind small RNA duplexes and longer dsRNAs to a lesser extent, *in vitro* (Goto *et al.*, 2007). Also, CMV 2b protein was detected as monomers in the nuclei-enriched fraction from infected plants, but as dimers in the cytoplasm- and membrane-enriched fraction (Lewsey *et al.*, 2009).

The first demonstration that the 2b protein is a virulence factor came from the work of Ding *et al.* (1995), who showed that deletion of the *2b* gene in CMV resulted in an asymptomatic infection in *N. glutinosa*, accompanied by loss of systemic infection in cucumber. Further work involving other hosts or CMV strains confirmed that deletion (or premature termination or untranslatable versions) of the *2b* gene resulted in asymptomatic infections, or at least considerable attenuation of the symptoms, showing that protein 2b was not required for systemic movement in these hosts but was an essential determinant for symptom induction (Hou *et al.*, 2011; Lewsey *et al.*, 2009; Netsu *et al.*, 2008; Soards *et al.*, 2002; Wang *et al.*, 2004b). Virus accumulation in the inoculated leaves was reduced to a greater or lesser extent. In one study with protoplasts, the mutant replicated as efficiently as the wild-type strain. Moreover, even if its local movement profile was modified within the different types of cells, it remained efficient for this function (Soards *et al.*, 2002). All studies reported a lower efficiency of systemic movement, resulting in a decrease in accumulation of the virus in the upper noninoculated leaves. Nevertheless, virus accumulation in these leaves permitted to conclude that it was not involved in the asymptomatic phenotype of the plants.

On the other hand, the virulence of protein 2b was found to be directly related to the severity of the strain it arose from and its subgrouping within CMV strains. This was shown through two distinct approaches. First, constitutive expression of protein 2b in transgenic plants showed that these plants could develop symptom-like developmental anomalies. Particularly, plants expressing the *2b* gene from subgroups IA or IB generally developed symptoms (Lewsey *et al.*, 2007, 2009, Praveen *et al.*, 2008; Zhang *et al.*, 2006), while plants expressing the gene from subgroup II strains had a normal phenotype (Lewsey *et al.*, 2007; Siddiqui *et al.*, 2008). In the second approach, hybrid viruses between strains from different subgroups were created by exchanging their *2b* genes. This led to all possible scenarios: attenuation of the gravity of the disease of the receptor strain when a subgroup II 2b was introduced into a IA strain (Cillo *et al.*, 2009), development of the symptoms typical for the donor strain (hybrids between subgroups IA and IB) (Du *et al.*, 2007a), or finally, hypervirulence (subgroup I 2b in a II strain) (Shi *et al.*, 2002). This last situation also occurred when TAV 2b was introduced into the subgroup II strain Q (Ding *et al.*, 1996; Shi *et al.*, 2003). Determinants for virulence were mapped on the 2b sequence of subgroup IA strain Fny (Lewsey *et al.*, 2009). The two NLS and the N-terminal region of the protein were shown to be the main determinants. Moreover, the putative phosphorylation site and the C-terminal region were also involved to a lesser extent. Also, depending on the host, protein 2b appeared not to be the only determinant of symptomatology, suggesting that protein 2b could synergize symptom induction by other viral factors (Lewsey *et al.*, 2009). In most cases, however, the nature of the symptoms did not correlate with virus accumulation.

The second main characteristic of protein 2b is that it counteracts the basal defenses of the host based on RNA silencing. In plants in which a *gfp* gene (coding a green fluorescent protein) was silenced, expression of the 2b protein resulted in fluorescence emission typical for *gfp* gene expression. Protein 2b appeared to prevent long-distance transmission of the systemic silencing signals (Brigneti *et al.*, 1998). Protein 2b was further shown to also suppress silencing in a protoplast system (Qi *et al.*, 2004). The protein suppresses RNA silencing through different and probably complementary actions, at different steps of the silencing pathway.

Protein 2b interferes strongly with accumulation of siRNAs. In wild-type *A. thaliana* plants, CMV infection resulted in drastically reduced accumulation of 21-, 22-, and 24-nt siRNA species, compared to infection with the same strain lacking the *2b* gene, although virus accumulation did not differ (Diaz-Pendon *et al.*, 2007). Defects of the mutant CMV were corrected in *A. thaliana* mutants with deficient *dcl* or *rdr* genes (Diaz-Pendon *et al.*, 2007; Lewsey *et al.*, 2010). A mutation at position 46 of the 2b protein resulted in an attenuated strain and also in a weakened ability to suppress RNA silencing in petunia (Goto *et al.*, 2007). These authors also showed that the mutated protein was not only no longer localized in the nucleus but also bound small RNAs poorly *in vitro*, compared to the wild-type strain. They proposed that an efficient RNA-binding activity was required for silencing suppression (Goto *et al.*, 2007). The same group further showed that a modified 2b protein, lacking the C-terminal one-third, was still able to bind small RNAs *in vivo*. Moreover, 2b protein facilitated accumulation of siRNAs in the nuclei of infected tobacco protoplasts (Kanazawa *et al.*, 2011). The crystal structure of TAV 2b protein bound to a 21-nt dsRNA was established (Chen *et al.*, 2008). Protein 2b formed dimers in solution, but addition of siRNA duplexes resulted in formation of tetramer structures, composed of four molecules of protein and two molecules of RNA. Under its dimeric form, protein 2b recognized siRNAs by a pair of hook-like structures. Different mutations in the protein decreased both RNA-binding affinity and suppression of RNA silencing, suggesting that protein 2b interferes with RNA silencing by sequestering siRNAs (Chen *et al.*, 2008). That indeed RNA binding is indispensable for the suppressor silencing function was confirmed more recently (González *et al.*, 2010). Interaction of protein 2b with the host miRNA pathway is also documented. All studies dealt with genes involved in plant development and morphogenesis, as well as genes coding for protein of the silencing pathway. Although different levels of interference were observed depending on host species, strains, and overall environmental conditions, all these studies indicated a true interference, with subgroup II protein 2b being less efficient (Cillo *et al.*, 2009; Lang *et al.*, 2011; Lewsey *et al.*, 2007; Zhang *et al.*, 2006).

Transient coexpression of proteins 2b and AGO1, or crosses between transgenic *A. thaliana* plants expressing each protein, followed by specific immunoprecipitation assays, indicated that the two proteins interacted *in vivo*, and *in vitro* interaction was also demonstrated (Zhang *et al.*, 2006). Moreover, these authors also showed that protein 2b inhibited AGO1 slicer activity *in vitro*. Thus, protein 2b would suppress RNA silencing in part by blocking AGO1. A novel study confirmed the interaction and showed that protein 2b interacted also with itself and with protein AGO4, mainly in the nucleus of infected cells (González *et al.*, 2010). Analysis of different 2b mutants indicated that deletion of each NLS or both, and of the putative phosphorylation site, did not modify the protein–protein interactions but resulted in the loss of binding affinity with a 21-nt small RNA and loss of suppression of RNA silencing (González *et al.*, 2010).

Virulence and silencing suppressor activity were compared for protein 2b of subgroup I and II strains. Both strains reached similar levels in systemic leaves of *N. benthamiana*. However, the subgroup II protein 2b induced a less virulent phenotype and also suppressed RNA silencing less efficiently. The two proteins differ mainly in three regions (N-terminal part and middle of the molecule), and particularly, sequence alignment indicated a 10-amino-acid deletion in the subgroup II protein 2b at positions 62–71. This region was shown to be a key determinant of suppressor activity (Ye *et al.*, 2009). Thus, reduced virulence and reduced ability to suppress RNA silencing are associated in subgroup II strains but are independent, as different regions of the molecule account for these characteristics.

Coherent with the RNA silencing suppressor activity of its protein 2b, CMV can suppress PTGS-based resistance to another virus in transgenic plants if CMV inoculation precedes that with the target virus (Mitter *et al.*, 2003; Simón-Mateo *et al.*, 2003). In transgenic tobacco plants expressing a dsRNA derived from PVY, however, suppression occurred only in leaves that had emerged when CMV was inoculated. Moreover, plants reverted to susceptibility only transiently, since they recovered from infection by the target virus and remained immune to further infection (Mitter *et al.*, 2003). Thus, inhibition of PTGS by protein 2b was not efficient enough to restore a fully susceptible phenotype. More striking was the observation that protein 2b could counter a resistance conferred by a potyviral gene (PVY-HcPro) against an unrelated virus (TRV, *Tobacco rattle virus, Tobravirus*), when expressed by this virus (Shams-Bakhsh *et al.*, 2007). Here also, plants recovered from infection, as is commonly observed for this virus–host combination. Thus, typically, in these two studies, different mechanisms were involved, with protein 2b interfering with some, but not all, of them.

RNA silencing is not the sole host-defense mechanism triggered by protein 2b, as this protein also allows CMV to escape SA-induced resistance in *N. glutinosa* (Ji and Ding, 2001). The mechanisms involved in

SA-mediated resistance seem to be host dependent, making it difficult to apply findings on one host to other host species. So in tobacco and *A. thaliana*, SA-induced resistance corresponds to an inhibition of virus systemic movement, while in squash, it is local movement that is inhibited (Mayers *et al.*, 2005). Nevertheless, major alterations in SA and JA signaling pathways were associated with protein 2b in *A. thaliana* (Lewsey *et al.*, 2010). Interestingly, increase of SA in infected plants did not occur when protein 2b was not expressed, even if only a few SA-regulated genes were inhibited in its presence. In contrast, the level of JA was unchanged, but most of the genes regulated by JA were inhibited (Lewsey *et al.*, 2010). It is probable that several other changes in plant metabolism due to protein 2b will be discovered in the future. It is also striking that protein 2b can function as a suppressor of silencing in other organisms, such as the green alga *Chlamydomonas reinhartii*, providing useful new tools for studying RNA silencing in this organism (Ahn *et al.*, 2010).

E. Genes conferring resistance to CMV

1. Natural resistance genes

There are only a few natural CMV-resistance genes in either cultivated crops or related wild species (Table I) and most of them were presented by Palukaitis and García-Arenal (2003). Since then, attempts to identify resistances in tomato or different *Solanum* species yielded unfortunately disappointing results, as only lower susceptibility or, in a very few cases, tolerance were identified (Akhtar *et al.*, 2010; Cillo *et al.*, 2007). Moreover, breeding for resistance in tomato is often hampered by difficulties in obtaining viable and/or fertile hybrids between tomato and its wild relatives.

On the other hand, some single resistance genes have been characterized, mainly in *A. thaliana*. *RCY1* belongs to the NBS-LRR gene family and was identified in *A. thaliana* ecotype C24. Its expression results in an HR toward the Y strain of the virus, elicited by the CP, and expression of host-defense genes *PR-1* and *PR-5*. Resistance and production of PR proteins were compromised in mutants in which accumulation of SA was blocked, while they were not affected in mutants without the JA signaling pathway (Takahashi *et al.*, 2004). Moreover, plants in which both the JA and SA pathways were downregulated appeared more resistant to the virus. Takahashi and collaborators proposed that expression of the resistance gene was modulated through antagonistic interactions between the two signaling pathways. Single mutations in each of the main domains of *RCY1* (CC, NBS, or RRR) resulted in loss of resistance and the virus spreading systemically, but impacted differently the development of necrotic lesions (Sekine *et al.*, 2006). Also, transgenic expression of *RCY1* in susceptible Col0 ecotype conferred resistance to the Y strain (Sekine *et al.*, 2008). In these plants, the strength of resistance correlated with the

TABLE I Host resistance characters to CMV

Plant	Resistance type	Resistance gene(s)	Protein	Avirulence factor	References
Bean	Extreme resistance	*RT4-4*	TIR-NBS-LRR	Protein 2a	Seo *et al.* (2006)
Cucumber	Replication + no symptoms	Two recessive genes	–	–	Wang *et al.* (2004a)
Melon	Infection	Three recessive genes	–	–	Pitrat (2002)
	Infection	*cmv* (seven QTLs)	–	–	Pitrat (2002)
	Infection	*cmv1*	–	–	Essafi *et al.* (2009)
	Aphid transmission	*Vat*	CC-NBS-LRR	–	Pitrat (2002)
Pumpkin	No symptoms	*Cmv*	–	–	Kang *et al.* (2005)
Squash	No systemic symptoms	Two recessive genes	–	–	Kang *et al.* (2005)
Lactuca saligna	Systemic movement	*Rsv*	–	RNAs 2 and 3	Kang *et al.* (2005)
Pepper	Infection	Two QTLs	–	–	Palukaitis and García-Arenal (2003)
	Replication	Four QTLs	–	–	Palukaitis and García-Arenal (2003)
	Systemic movement	Four QTLs	–	–	Palukaitis and García-Arenal (2003)
	Systemic movement	*Cmr1*	–	–	Kang *et al.* (2010)
	Movement	Not determined	–	3′ Part of RNA2	Zhang *et al.* (2011)
Soybean	Systemic movement	Three QTLs	–	*2b* Gene, RNA3-3′ half	Ohnishi *et al.* (2011)

(continued)

TABLE I *(continued)*

Plant	Resistance type	Resistance gene(s)	Protein	Avirulence factor	References
Cowpea	Hypersensitive response	*Cry*	–	Protein 2a	Palukaitis and García-Arenal (2003)
A. thaliana	Hypersensitive response	*RCY1*	CC-NBS-LRR	CP	Sekine *et al.* (2006)
	Systemic movement	*ssi2*	–	–	Sekine *et al.* (2004)
	Replication	*cum1*	eIF4E1	–	Yoshii *et al.* (2004)
	Replication	*cum2*	eIF4G	–	Yoshii *et al.* (2004)

expression level of the transgene, resulting in extreme resistance in the highest expressors (Sekine *et al.*, 2008). Also in *A. thaliana*, a recessive mutant of the *SSI2* gene, which is involved in lipid synthesis, had increased PR-1 and SA accumulation, and CMV was restricted to the inoculated leaves (Sekine *et al.*, 2004). However, this resistance proved to be partial, as not all the plants were resistant and mediated in an SA-independent manner (Sekine *et al.*, 2004). Finally, the recessive genes *cum1* and *cum2*, which were earlier shown to reduce CMV accumulation in plants, proved to correspond to initiation factors eIF4E1 and eIF4G, respectively, and to act by affecting the efficiency of translation of protein 3a in a manner dependent on the 5′ noncoding sequence (Yoshii *et al.*, 2004).

Using an innovative strategy for cloning randomly NBS-LRR genes that are upregulated during infection, Seo *et al.* (2006) identified one such gene in common bean, which they named *RT4-4*, which directed a resistance-like response to CMV (systemic necrosis) when expressed transgenically in *N. benthamiana*. Interestingly, this resistance was effective against several CMV strains coming from tomato or pepper, independent of their subgrouping, but not against a strain isolated from common bean. In the common bean from which *RT4-4* was cloned, the pepper and tomato isolates remained restricted to the inoculated leaves without inducing a HR, while the bean isolate infected the plants systemically. Thus, the resistance gene of a legume can function in a solanaceous genetic background, even if it induces a different resistant phenotype. Protein 2a was shown to be the elicitor of RT4-4, and amino acid 631, a residue already known to be involved in induction of an HR in cowpea, proved to be determinant in RT4-4 recognition of protein 2a (Seo *et al.*, 2006).

Polygenic resistances directed against CMV are present in several *Capsicum* sp., where they control either the initial steps of infection or virus replication, or even virus systemic movement, but no major resistant genes were identified until recently. A single dominant gene named *Cmr1* was characterized in an Asian pepper cultivar. This gene conferred resistance to systemic movement of some, but not all, CMV strains. It was localized on linkage group 2 of the pepper genome, at a position syntenic to the *Tm-1* gene which confers resistance to *Tomato mosaic virus* (ToMV, *Tobamovirus*) in tomato (Kang *et al.*, 2010). Also, a resistance of a chili pepper line, the determinism of which is unknown, was recently shown to confer a high resistance against several strains of CMV, whatever their subgrouping (Zhang *et al.*, 2011). The virulence factor (virulence is used here as the ability to infect a resistant genotype) of strains overcoming this resistance mapped to the 3′ part of RNA2, encompassing the end of protein 2a, the whole *2b* gene, and the 3′ noncoding region. However, the ability to overcome the resistance depended on the virus genetic background, suggesting that other parts of the genome might also be involved (Zhang *et al.*, 2011).

Resistance characters that block systemic movement were suspected to exist in some wild and cultivated soybean cultivars, based on certain specific strain–host genotype interactions. This resistance was dissected very recently, as were the viral determinants for virulence (Ohnishi *et al.*, 2011). These authors discovered a complex situation in both sides, as at least three QTLs were shown to control resistance, while the entire *2b* gene and the 5′ noncoding region of RNA3, and potentially the 40 N-terminal amino acids of protein 3a from a virulent strain, were required to overcome the resistance. Such multigenic resistances are thought to be more durable than resistance conferred by a single gene, as the virus would have to accumulate several mutations to adapt and overcome the resistance. Although commonly hypothesized, the higher durability of a polygenic resistance has been shown in a few cases, for PVY in tobacco or pepper (Acosta-Leal and Xiong, 2008; Palloix *et al.*, 2010).

2. Pathogen-derived resistance

The first transgenic plants expressing a virus-derived CMV-resistance gene are now a quarter of a century old. Since then, that a virus-derived transgene can confer resistance to the target virus has been validated for several viral genera (including DNA viruses) and several host species (for a review, see Prins *et al.*, 2008). CMV is one of the viruses for which the largest range of transgenes, either derived from the viral genome or not, have been tested (reviewed by Morroni *et al.*, 2008). *CP* genes proved to be efficient in many cases and conferred a broad spectrum of resistance. This resistance is probably mediated by the protein. Genes derived from the viral replicase 2a protein also proved to be effective, although the resistance could be broken by some virus strains. The mechanisms underlying this resistance have not been elucidated, even if some data suggest that the protein is here also involved (Morroni *et al.*, 2008). The constructs that are probably the most promising contain inverted repeats of CMV sequence, leading to production of dsRNA upon expression in plants and activation of PTGS against the target sequence. The first constructs corresponded to quite large regions (700–1500 nt) of the *CP* or the *2a* genes and could confer high levels of resistance (Hu *et al.*, 2011; Morroni *et al.*, 2008). In such plants, production of siRNAs was an absolute prerequisite for resistance (Dalakouros *et al.*, 2011). Shorter constructs, resulting in the production of 21-nt-long viral hairpins from the *2b* gene, or the precursor of a plant miRNA modified in order to contain the viral sequence also proved to be effective, although, as expected, the resistance was specific of the donor strain (Qu *et al.*, 2007). Ideally, and in order to reach an as broad as possible resistance spectrum, it would be necessary to express simultaneously hairpins corresponding to the three main subgroup strains.

VI. CMV IN THE MEDITERRANEAN BASIN

CMV is distributed all around the Mediterranean basin, where it has been detected in many crops including olive, the emblematic crop of the region (Martelli, 1999). Globally, Libya is the only country where CMV has not been described, probably because no surveys were done there. There is a vast literature describing CMV on vegetable crops in the Mediterranean basin and it would be impossible to cite all. For convenience, we choose to cite the oldest papers giving evidence for the presence of the virus in the main vegetable crops: cucurbits, eggplant, pepper, and tomato (Table II).

Methods for detecting the virus have evolved considerably since the early studies, at the beginning of the 1960s. Initially, CMV was characterized, thanks to the disease symptom phenotype of a small number of differential hosts, accompanied, when possible, by electron microscopy observations, and in some cases, by aphid transmission assays. Serology was included at the beginning of the 1970s, essentially through the use of an agar immunodiffusion test, which did not need any special equipment, but consumed large amounts of serum. ELISA was then adopted by most laboratories and is probably still the most used technique, for which polyclonal reagents, as well as monoclonal reagents specific for each of the two main subgroups (I and II), are commercialized today. ELISA remains the most convenient technique for CMV diagnosis—at least for vegetable—because of its ease of implementation and lower costs compared to molecular techniques and also because CMV generally reaches high levels in the infected plants. On the other hand, such high levels have the disadvantage of creating false-positive data if special care is not taken during preparation of crude extracts (all techniques) or washing steps following incubation of these extracts in ELISA procedures.

Molecular techniques, generally targeting the *CP* gene, were developed in order to better characterize virus populations. Several pairs of primers allowing specific amplification of the two subgroups have been described. Analysis is achieved by sequencing or by restriction length polymorphism of the amplified RT-PCR products. This last technique has, however, to be used carefully for confident subgrouping. Particularly, it is necessary to consider several enzymes, cutting specifically each subgroup-derived fragment or both, in order to alleviate the problem of partial digestion of PCR products and to reduce the impact of loss of a restriction site by mutation. Finally, as stated below in Section VII.A, a confident genotyping of any CMV isolate requires analysis of all three genomic segments, as reassortants or recombinants can occur in nature.

The first important epidemiological studies were carried out in France at the end of the 1970s in the main vegetable crops of the region that are cultivated twice a year (melon, marrow, bean, tomato, and pepper).

TABLE II Presence of CMV in the Mediterranean basin

Country	Hosts	Detection methods	References
Albania	Cucumber, eggplant, pepper, squash, tomato	Molecular hybridization	Finetti-Sialer *et al.* (2005)
Algeria	Tomato	?	Badr (1989)
Bosnia	Pepper	BT	Subasic *et al.* (1990)
Cyprus	Tomato	BT, S	Ioannou (1985)
Croatia	Spinach, beet	BT, dsRNAs, S	Krajacic and Juretic (1993)
Egypt	Pepper	S	Abdel-Salam *et al.* (1989)
France	Cucumber, eggplant, melon, pepper, tobacco, tomato, squash	BT	Messiaen *et al.* (1963)
Greece	Squash, artichoke	BT, EM, IT, S	Kyriakopoulou and Bem (1982)
Iran	Cucumber, melon, squash	BT, EM, IT, S	Rahimian and Izadpanah (1978)
Israel	Cucumber, eggplant, marrow, muskmelon, pepper, tomato	BT, IT	Nitzany and Wilkinson (1961)
Italy	Tomato, pepper, squash, eggplant, spinach	BT, EM, S	Faccioli (1972) and references therein
Jordan	Tomato	BT, S	Al-Musa and Mansour (1983)
	Cucurbits	BT, S	Al-Musa (1989)
Lebanon	Cucumber, squash	ELISA	Katul and Makkouk (1987)
Morocco	Pepper	BT, EM, IT, S	Lockhart and Fisher (1976)
Serbia	Cucurbits	BT, ELISA	Dukić *et al.* (2006)
Slovenia	Cucurbits, pepper, tomato	ELISA	Vozelj *et al.* (2003)
Spain	Cucumber, eggplant, melon, pepper, tomato	BT, EM, S	Garcia-Luque *et al.* (1983)
Syria	Cucumber, squash	ELISA	Katul and Makkouk (1987)
Tunisia	Pepper	?	Lange and Hammi (1977)
	Tomato	ELISA	Ben Moussa *et al.* (2000)
	Cucurbits	ELISA	Mnari-Hattab *et al.* (2008)
Turkey	Cucurbits	BT, EM	Nogay and Yorganc (1984, 1985)
	Pepper, tomato, watermelon	BT, EM, S	Ylmaz and Davis (1985)

BT, biological test; ELISA, enzyme-linked-immunosorbent assay; EM, electron microscopy; IT, insect (aphid) transmission; S, other serological methods.

In a 3-year survey, melon crops appeared to be quickly infected by CMV, while the prevalence of the virus in the other crops varied according to the species, the year, or the season during a given year (Quiot *et al.*, 1979a). The two main subgroups, I and II, were found in both cultivated plants and weeds growing near or within the crops, but coinfection occurred only rarely (1.2%) (Quiot *et al.*, 1979b,c). In both cases, some hosts appeared to suit better one subgroup of strains. For example, cucurbits (and purslane) were preferentially infected with subgroup I strains, while tomato (and madder) were preferentially infected with subgroup II strains. The more recent work of Sacristán *et al.* (2004) showed, however, that the population genetics in both cultivated plants and weeds were similar (see Section IV.C).

During these early surveys, careful consideration was given to tomato lethal necrosis, a new syndrome associated with CMV that had caused serious damage in tomato crops since a few years in France, where previously CMV was known for inducing mosaic and/or fern-leaf symptoms. It was observed that necrosis epidemics occurred only under high aphid population densities, which fits well with the models developed later by Escriu *et al.* (2003). Although necrosis developed as patches in the crops, fern-leaf symptoms developed randomly. Also, the fern-leaf disease did not protect against necrosis, and even some plants with fern-leaf symptoms developed necrosis later (Quiot *et al.*, 1979d). Necrotic epidemics were then observed in Spain (1986) and Italy (1988), and the necrogenic satellite RNA associated with the disease was characterized. Surveys done in these countries during the 1990s revealed, however, a more complex situation in tomato crops. Two new syndromes were observed in 1989: curling and stunting of tomato plants, which were also associated with a satellite RNA (Grieco *et al.*, 1992; Jordá *et al.*, 1992), and fruit necrosis (also called internal browning), which was due to the helper virus itself, even if a satellite RNA, of the nonnecrogenic type, could be present (Crescenzi *et al.*, 1993). In addition, satellite RNA that did not modulate the helper-strain-induced disease was also described. Finally, satellite RNAs fell into four groups according to the symptoms developed by the infected tomatoes: necrogenic satellite RNA, "ameliorative" satellite RNA, "curling-stunting" satellite RNA, satellite RNA without any observable associated symptomatology. The main differences between the situations in Spain and Italy lie in the genetic characteristics of the helper viral isolates. In Italy, lethal necrosis was more frequently associated with CMV strains belonging to subgroup II, while such strains were rare in Eastern and Central Spain where tomato necrosis outbreaks were associated to subgroup I strains, which were later assigned to subgroup IB (Bonnet *et al.*, 2005) (see below). On the other hand, in Italy, development of "curling-stunting" epidemics was associated with the introduction of subgroup

IB isolates, which constituted the most represented subgroup in tomato crops by the end of the 1990s (for a review, see Gallitelli, 2000). The various types of satellite RNA were associated with subgroup I strains in tomato crops in Greece where severe CMV outbreaks occurred during the 1990s (Varveri and Boutsika, 1999). Similarly, the severe necrogenic outbreaks observed in Croatia in 1989 were associated with subgroup I strains and satellite RNA (Skorić *et al.*, 1996). So, it is evident that satellite-RNA-bearing CMV isolates were widespread in the Mediterranean basin over a 20-year period, where they caused new and damaging diseases, but only in tomato crops. Since then, satellite RNAs seem to have disappeared or to be only marginally present, and concomitantly the prevalence of subgroup IB strains has decreased considerably. Although the presence and dissemination of a satellite RNA is strongly associated with tomato disease outbreaks, satellite RNAs were also regularly detected in isolates from other crops than tomato, and even in weeds, although apparently at a low frequency. It is noteworthy that satellite RNA showed a very high genetic diversity due to both mutations and recombination, resulting in heterogeneous and genetically unstructured populations (Alonso-Prados et al., 1998; Aranda et al., 1993, 1997; Grieco et al., 1997), in contrast to the metapopulation structure of its helper virus. Because of these properties, of the unusual biological activity of satellite RNAs and also their unknown origin, reemergence of severe outbreaks of satellite-RNA-containing isolates cannot be excluded, particularly in a context of increasing aphid populations due to climate change and reduction of the use of insecticides in crops.

VII. VIRUS DIVERSITY AND EVOLUTION

A. Phylogeny

Initially, CMV strains were classified into two subgroups, I and II, based on converging biological, serological, and physical criteria. Strains belonging to subgroup I were also called heat-resistant, while strains belonging to subgroup II were called heat-sensitive and were also distinguished from the former by inducing typical etch symptoms on tobacco plants. While within-subgroup nucleotide similarity is high, between-subgroup similarity is around 70–75%, which constitutes a uniquely high degree of sequence diversity within a plant virus species; it could be argued that the two subgroups should be considered separate species. Subgroup I strains were further divided into subgroups IA and IB, based on nucleotide variation in the 5′ noncoding region of RNA3 (Roossinck *et al.*, 1999). Phylogenetic analysis of the CP ORF confirmed this new grouping, and suggested three distinct major events in CMV evolution. A first radiation would have given rise to subgroup II strains; a second would have led to subgroup IB, from which subgroup IA would have derived in a third event (Roossinck *et al.*, 1999).

Until recently, isolate typing was based essentially on the sequence of the *CP* gene. However, a more complete analysis with the full sequence of 15 strains showed that the trees constructed with proteins other than the CP did not completely support the grouping previously proposed for CP and suggested that recombination, or at least reassortment, might have played an important role in CMV evolution (Roossinck, 2002). A similar approach was used with the nucleotide sequence of the 38 fully sequenced isolates available in the databases (Table III), completed by the sequences of the three reference strains of our laboratory (I17F, Vir, and R). Trees were constructed using the neighbor-joining method and 1000 bootstrap replicates, and rooted with PSV strain P. Each genomic segment was treated separately in a first approach (Fig. 3A, B, and C for RNA1, 2, and 3, respectively). The three trees are robust, since bootstrap values close to 100 were observed at the branch points defining the species and subgroups. For all three RNAs, a first branching resulted in clear clustering of group II strains. A second one separated out two strains, BX and PHz, which were collected on *Pinellia ternata* in China, and placed in a new subgroup III by Liu *et al.* (2009). The next branching of the RNA2 tree distinguished unambiguously subgroups IA and IB. Subgroup IA appears more compact, although some strains clustered separately within each subgroup. The branching distinguishing these two subgroups appeared more complex for RNAs 1 and 3, some strains constituting intermediate clusters (strains Ca, CS, D8, and SD for RNA1 and strains SD, Ca, and CS for RNA3). Careful examination of each isolate identified three reassortants: D8 whose RNA1 clusters in subgroup IA, while the two other genomic RNAs cluster in subgroup IB; Tsh and PF, both isolated from tomato in China and Japan, respectively, and already described as natural reassortants between subgroups I and II (Chen *et al.*, 2007; Maoka *et al.*, 2010). In a second approach, we constructed a tree corresponding to the whole genome by concatenating the sequences of the three genomic segments of each isolate (Fig. 3D). In some way, this tree clarified the grouping, as it immediately allowed identification of the reassortants, which clustered separately. Particularly, isolate Tsh whose RNAs 1 and 3 belong to subgroup II and RNA2 to subgroup IA was closer to subgroup II, and isolate PF composed of RNAs 1 and 2 from subgroup IA and RNA3 from subgroup II appeared closer to subgroup I. The reassortant D8 clustered with SD, in an intermediate position between subgroups IA and IB. Additional studies using programs devoted to localizing recombination events are needed to clarify the position of isolates at the boundaries of subgroups IA and IB. This tree also indicates a high genetic diversity within each subgroup, even if it is generally

TABLE III Name, origin, and accession number of the CMV strains

Name	Host species	Country	Accession numbers (1, 2, 3)	RNA lengths (1, 2, 3) (nt)	ORF 1a (aa)	ORF 2a (aa)	ORF 2b (aa)	ORF MP (aa)	ORF CP (aa)
BX	*Pinellia ternata*	China	DQ399548, DQ399549, DQ399550	3336, 3037, 2179	993	858	110	279	218
Ca	*Arachis hypogea*	China	AY429434, AY429433, AY429432	3356, 3045, 2219	993	858	111	279	218
Cah1	*Canna* sp.	China	FJ268744, FJ268745, FJ268746	3356, 3045, 2220	993	858	111	279	218
Cb7	Tomato	China	EF216866, DQ785470, EF216867	3356, 3045, 2218	993	858	111	279	218
CS	*Arachis hypogea*	China	AY429435, AY429436, AY429437	3356, 3045, 2212	993	858	111	279	217
CTL	*Brassica chinensis*	China	EF213023, EF213024, EF213025	3357, 3047, 2217	993	858	111	278	218
D8	*Raphanus sativus*	Japan	AB179764, AB179765, AB004781	3380, 3046, 2218	993	858	111	279	218
Fny	?	United States	D00356, D00355, D10538	3357, 3050, 2216	993	857	110	279	218
IA	?	Indonesia	AB042292, AB042293, AB042294	3358, 3036, 2214	993	857	111	279	218
I17F	Tomato	France	Y18137 (RNA3)	3360, 3048, 2216	993	857	110	279	218
Ixora	Tomato	Philippines	U20220, U20218, U20219	3361, 3060, 2216	993	857	134	279	218
Legume	Cowpea	Japan	D16403, D16406, D16405	3359, 3047, 2213	993	857	110	279	218
Li		Korea		3396, 3064, 2228	993	857	110	279	218

	Lilium tsingtauense		AB506795, AB506796, AB506797						
LiCB	Lily	Korea	AB506798, AB506799, AB506800	3396, 3062, 2228	993	857	110	279	218
LS	Lettuce	United States	AF416899, AF416900, AF127976	3390, 3038, 2198	992	840	100	279	218
LY	*Lupinus angustifolius*	Australia	AF198101, AF198102, AF198103	3391, 3038, 2203	992	840	100	279	218
Ly2	*Lilium longiflorum*	Korea	AJ535913, AJ535914, AJ296154	3400, 3062, 2226	993	857	110	279	218
Mf	?	Korea	AJ276479, AJ276480, AJ276481	3357, 3053, 2214	993	857	112	279	218
New Delhi	Tomato	India	GU111227, GU111228, GU111229	3358, 3042, 2214	993	857	110	283	218
Ns	Tobacco	United States	AJ580953, AJ511989, AJ511990	3366, 3052, 2216	993	857	110	279	218
NT9	?	Taiwan	D28778, D28779, D28780	3358, 3042, 2214	993	857	110	283	218
Pepo	Squash	Japan	AB124834, AB124835, AF103991	3360, 3052, 2216	993	857	110	279	218
PF	Tomato	Japan	AB368499, AB368500, AB368501	3364, 3049, 2206	993	857	110	279	218
Phy	?	China	DQ402477, DQ412731, DQ412732	3356, 3048, 2220	993	857	111	279	218
PHz	*Pinellia ternata*	China	EU723568, EU723570, EU723569	3356, 3047, 2190	993	857	110	279	218
PI-1	Tomato	Spain	AM183114, AM183115, AM183116	3361, 3043, 2217	993	857	110	283	218
Q	?	Australia	X02733, X00985, M21464	3389, 3035, 2197	991	839	100	279	218
R	*Ranunculus*	France	Y18138 (RNA3)	3392, 3038, 2206	992	840	100	279	218

(continued)

TABLE III (*continued*)

Name	Host species	Country	Accession numbers (1, 2, 3)	RNA lengths (1, 2, 3) (nt)	ORF 1a (aa)	ORF 2a (aa)	ORF 2b (aa)	ORF MP (aa)	ORF CP (aa)
Rb	*Rubdeckia hirta*	Korea	GU327363, GU327364, GU327365	3362, 3049, 2214	993	857	110	279	218
Ri-8	Tomato	Spain	AM183117, AM183118, AM183119	3360, 3049, 2192	993	857	110	279	218
Rs	*Raphanus sativus*	Hungary	AJ511988, AJ517801, AJ517802	3359, 3052, 2216	993	857	110	279	218
SD	?	China	AF071551, D86330, AB008777	3379, 3048, 2219	993	857	111	279	218
Tagetes	*Tagetes erecta*	China	EU665000, EU665001, EU665002	3394, 3039, 2204	993	840	100	279	218
Tfn	Tomato	Italy	Y16924, Y16925, Y16926	3365, 3042, 2220	993	857	110	283	218
TN	Tomato	Japan	AB176849, AB176848, AB176847	3397, 3039, 2205	992	840	100	279	218
Trk7	Clover	Hungary	AJ007933, AJ007934, L15336	3391, 3039, 2209	992	840	100	279	218
Tsh	Tomato	China	EF202595, EF202596, EF202597	3394, 3047, 2206	993	857	110	279	218
Vir	Pepper	Italy	Not deposited	3358, 3054, 2218	993	858	111	279	218
Y	Tobacco	Japan	D12537, D12538, D12499	3361, 3051, 2217	993	857	110	279	218
Z1	Squash	Korea	GU327366, GU327367, GU327368	3359, 3050, 2215	993	858	110	279	218
42CM	Cucumber	Japan	AB368496, AB368497, AB368498	3371, 3050, 2191	993	857	110	279	218

considered that subgroup IA is more homogeneous than subgroup IB, which is not a monophyletic group. Taken as a whole, these results show that analysis of at least a part of each genomic segment is necessary for confident genotyping of CMV isolates.

B. Natural CMV populations

There are rather few recent papers relating specific surveys for CMV in vegetable crops in the Mediterranean basin. One of the possible explanations is that CMV is a quite old and well-known disease, which does not attract particular attention. However, CMV appears in general virus

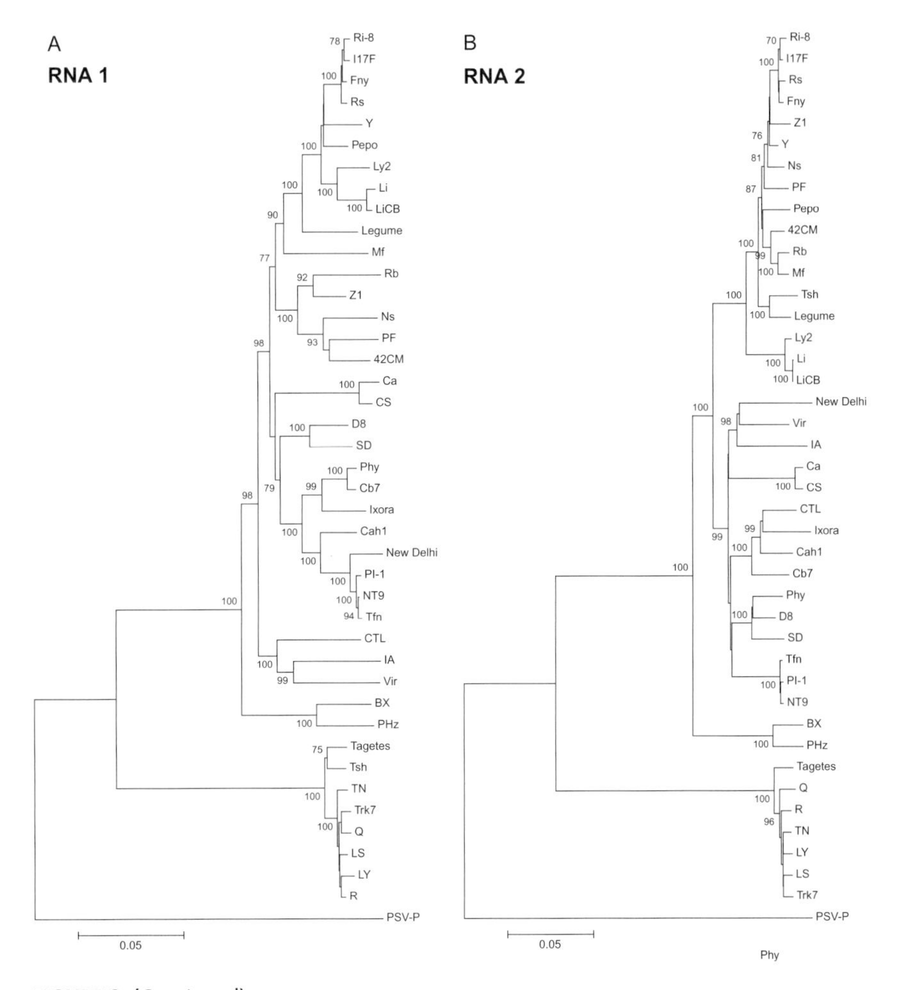

FIGURE 3 (Continued)

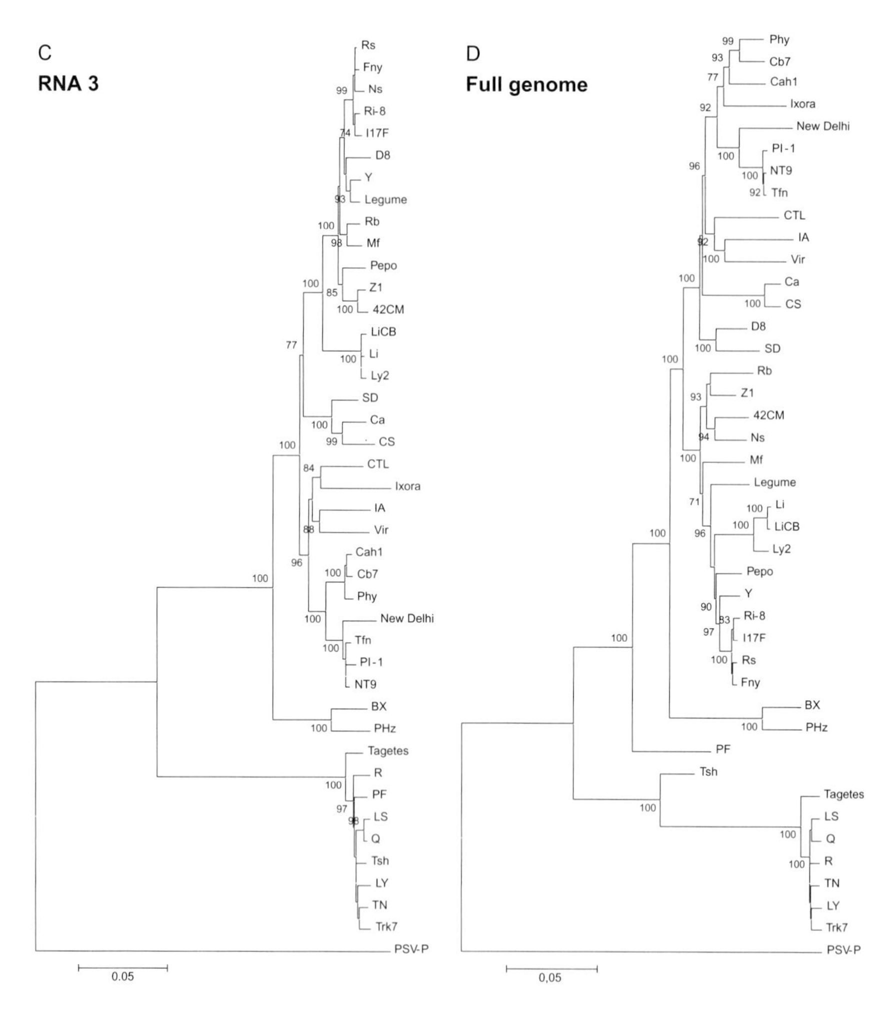

FIGURE 3 Neighbor-joining rooted trees of each CMV genomic segment (figure A, B, and C for RNA1, 2, and 3, respectively) or the whole genomic sequence resulting from concatenation (D), constructed with 1000 bootstrap replicates. Bootstrap values lower than 70 are not indicated.

surveys conducted for specific crops (for more details, see Chapters 2–4, 7 and 11). A survey made in 2007 in tomato crops in eastern Spain indicated that CMV had a minor incidence compared to other viruses, and particularly PVY or ToMV, whose incidence has increased because of the cultivation of old but susceptible local varieties, which are preferred by the consumers and sold at higher prices (Soler *et al.*, 2010).

There is an impressive amount of sequence data for the CMV *CP* gene in the databases; however, most sequences concern Asiatic isolates, particularly China, Japan, and India, potentially inducing a bias in subgroup prevalence worldwide. Despite this, it is evident that subgroups II and IA have a worldwide distribution, while subgroup IB is mainly present in Asia. There are only a few reports of the presence of subgroup IB in other countries, such as Spain (see below, Bonnet *et al.*, 2005; Fraile *et al.*, 1997), Italy (strains Tfn and Vir were collected in South and North Italy, respectively, at the end of the 1980s) (Gallitelli, 2000), and Greece (Sclavounos *et al.*, 2006), for what concerns the Mediterranean basin. A few subgroup IB isolates have also been described in the United States (Lin *et al.*, 2003) and Brazil (Eiras *et al.*, 2004).

Extensive genetic characterization of CMV isolates in vegetable crops has been carried out only in Spain and California. In California, studies were carried out in order to evaluate the potential role of transgenic squash plants expressing a CMV *CP* gene in emergence of new virus strains that could arise by recombination between the transgene and an infecting CMV. This study thus considered both samples collected in fields of resistant plants (conventional or transgenic resistance) in 1999 (63), as well as older isolates collected in various crops 5–14 years before (18) (Lin *et al.*, 2003). These isolates fell into five pathotypes, according to their ability to infect susceptible or resistant cultivars. Typing by SSCP (single-strand conformation polymorphism) analysis of a cDNA amplified from the *CP* gene and the 3′ noncoding region differentiated 14 groups. Finally, sequencing certain isolates showed that all isolates belonged to subgroup IA, except two belonging to subgroup IB. No relationship was found between the year, the crop, the geographic location, or the type of resistance gene and the genetic characteristics of the virus. Sequencing parts of all other ORFs further completed this work. Phylogenetic analyses showed that one of the isolates was indeed a reassortant between subgroups IA (RNA3) and IB (RNAs 1 and 2) (Lin *et al.*, 2004). In addition, reassortment was also thought to have occurred within subgroup IA, as isolates from this group clustered differently according to the region under analysis, a situation similar to what can be observed in Fig. 3 when comparing the trees constructed for the three genomic RNAs.

In Spain, extensive surveys were done in different vegetable crops from 1989 to 2002 in different production areas, and isolates were analyzed by ribonuclease protection assays (RPA) and partial sequencing of all three genomic RNAs (Bonnet *et al.*, 2005; Fraile *et al.*, 1997). The results illustrated an evolving situation, with a transitional period around 1995. Until 1995, the three main subgroups—IA, IB, and II—were present in the three regions under study where they represented 22%, 39%, and 15% of the entire sample set, respectively. Subgroup II strains were mainly found in Léon, in northwestern Spain. The remaining isolates represented mixed infections with subgroups IA and IB (16% of subgroup I population) or either reassortants (4%) or recombinants between these subgroups (7%) (Fraile *et al.*, 1997). Only four among the six possible reassortants were observed. The recombinants identified corresponded to the exchange of the two RNA3 ORFs, and no recombinants in the two other genomic segments were found. Thus, the presence of reassortants and recombinants was not random, suggesting that some combinations were not favored under natural infection in the field. Sequencing a region of each RNA was done for some samples and confirmed the typing done by RPA. Comparison of the sequence diversity at nonsynonymous and synonymous sites of all subgroup I isolates, or within subgroups IA and IB, revealed that RNA1 was under higher evolutionary constraints during the divergence between the two subgroups than the two other RNAs. No correlation was observed between genetic composition and host, location, or year of sampling (Fraile *et al.*, 1997). Isolates collected from 1996 showed that subgroup IB, which was previously mainly present in tomato crops in eastern Spain, has disappeared, at the same time as CMV outbreaks in this crop. Reassortants represented the same proportion of the population, except that reassortants between subgroups I and II were also found. Recombinant isolates were observed in a higher proportion (17%), which corresponded to an increased prevalence of a recombinant RNA3 of IB-IA type, suggesting that this recombinant benefited from a selective advantage in nature compared to others. The genetic composition of isolates collected on cultivated plants or weeds did not differ (Bonnet *et al.*, 2005). Subgroup IB reemerged in 2003 in tomato crops in northeastern Spain, where it altered fruit quality without inducing symptoms in leaves of the plants (Aramburu *et al.*, 2007). The sequence of one isolate was established and showed high sequence identity (99%) with the Tfn strain isolated in 1989 in Italy (Campania) or isolates previously found in the region of Barcelona in 1992, suggesting a common origin.

C. Host adaptation

There are several lines of evidence for host adaptation in virus populations. The first example is the results obtained for a satellite RNA by Moriones *et al.* (1991). They observed that when Ix-sat RNA (satellite RNA from the Ixora strain of CMV) was amplified in squash or tobacco, the progeny showed the same sequence as the parent preparation. However, when this satellite RNA was amplified on tomato, a U $\rightarrow$ C mutation occurred at position 102 and was maintained during successive passages in this host. Interestingly, when this variant was passed again from tomato to squash, reversion toward the initial residue occurred at position 102.

There are at least two species—lily and soybean—for which adaptation probably occurred, as isolates from these hosts show a restricted host range. Common CMV strains generally do not infect lily plants systemically. Curiously, isolates collected on lily in Japan and Korea cannot infect tobacco, different cucurbits, and even tomato (Lee *et al.*, 2007; Masuta *et al.*, 2002; Ryu *et al.*, 2002). These isolates cluster separately within subgroup IA (Masuta *et al.*, 2002; Table III and Fig. 3), and their RNA3 has been proposed to be a recombinant having an IB-type MP ORF and an IA-type CP ORF (Bonnet *et al.*, 2005). To what extent adaptation to lily is associated with the recombinant nature of RNA3 remains to be determined. CMV was also isolated from lily in France and Poland, and based on the sequence of the *CP* gene, these isolates were shown to cluster with Asian lily isolates (Berniak *et al.*, 2010; Chen *et al.*, 2001). Moreover, subgroup II isolates were also identified in Poland.

CMV isolates from lily have another singularity, as they are not able to assist replication of satellite RNA, a situation rare for CMV strains. RNA1 has been shown to be responsible for this inability (Yamaguchi *et al.*, 2005b), while it was also shown to determine the ability to infect lily plants (Yamaguchi *et al.*, 2005a). Although the regions involved in the two properties are different, it is tempting to associate loss of the ability to infect different hosts of the virus and to replicate a satellite RNA, two properties common to most CMV strains, and adaptation to lily.

Similarly, isolates from soybean, a species in which the virus is efficiently seed transmitted, are described as having a narrow host range, as they cannot infect cucumber (Hong *et al.*, 2007). Phylogenetic analyses of RNA3 from several isolates collected on either cultivated or wild soybeans indicated that they clustered separately within subgroup IB (Hong *et al.*, 2003). Comparison of the nonsynonymous and synonymous substitutions in the MP and CP ORFs with the corresponding values for subgroup IA strains indicated that the MP of the soybean isolates was under high evolutionary constraint and could indicate adaptation to soybean, as this gene is involved in systemic movement of the virus in this host (Hong *et al.*, 2007).

We encountered a similar situation with CMV isolates collected on rosemary plants in southeastern France in 2000 and 2010. Among the 20 or so diagnostic CMV hosts that were tested (including several cucurbit and solanaceous hosts), only *N. benthamiana* and *Nicotiana clevelandii* could be infected systemically. Partial sequence data on the three genome segments suggest that these isolates constitute a new CMV subgroup (M. Jacquemond, unpublished results). Research is currently underway in the laboratory to further characterize these isolates.

The interesting question that arises regarding the narrow host range isolates is what could be the advantage, if any, for a virus to be limited to one host or to a limited number of hosts in nature. It may be significant that for the three hosts in question, annual aphid-mediated infection is not required, either because the plants are perennial (rosemary, lily) or there is efficient virus transmission through seeds. This could make the virus less dependent on the ability to infect a broad range of alternate host species in the surrounding environment.

D. Role of recombination in CMV evolution

Even if reassortant genotypes can occur, they are most often at a low frequency in viral populations. Similarly, the number of reassortants among the strains that have been sequenced entirely is low (3 of 41). This suggests that, in mixed-infected plants, reassortants are at a selective disadvantage. This hypothesis was tested in plants infected with a mixture of subgroup IA and IB strains (four different paired combinations), using a local lesion host of CMV (*Chenopodium quinoa*) for cloning individual infectious units. Single local lesions were then inoculated to tobacco plants and the genomic composition of the progeny in each tobacco plant was characterized by RPA (Escriu *et al.*, 2007). In *C. quinoa* infected with a combination of two strains, all the expected genotypes (two parentals and six reassortants) were recovered, although with frequencies greatly different from the frequency expected for random reassortment (0.125). The frequency of the parental IA strain was 0.2, that of the IB strain was 0.08, and genotypes with RNA3 from subgroup IA were a majority (0.78). In inoculated leaves of tobacco plants, the situation was similar, although frequency of the genotypes having an IA RNA3 was even greater (0.97). In systemically infected leaves, only four genotypes were recovered, among which the parental IA strain was the majority (0.83). Finally, no recombinants were identified (Escriu *et al.*, 2007). These results clearly show that selection operated at the beginning of the infection process (local lesions in *C. quinoa*) was even more severe during colonization of the inoculated leaf (cell-to-cell movement in tobacco leaves) and increased further during systemic movement, as only half of the possible genotypes were recovered and one of them was dominant. They also support the concept of coadaptation of genes within the viral genome and illustrate the role of epistasis in virus fitness, which is expected to be strong for a pathogen where all proteins are multifunctional (Sanjuán *et al.*, 2004). Similarly, only one reassortant of PSV has been described (Hu and Ghabrial, 1998).

Evidence for recombination within a CMV ORF in nature is rare. It has been described only in two or three cases during the surveys in Spain, among the 159 isolates studied (Bonnet *et al.*, 2005), despite a high frequency of mixed infections, which are the prerequisite for recombination to occur. *In vitro* recombinants created between different strains, within each ORF, are, however, generally viable, although their fitness was not often evaluated. However, in one case, when recombinants were made between subgroup IA and II strains in the CP ORF, 2 of 13 recombinants were not viable, four displayed important deficiencies (movement, replication, or aphid transmission), and neither of the two viable recombinants that were tested could outcompete the parental IA strain (Pierrugues *et al.*, 2007). It is thus probable that recombinants are generally less fit and are thus counterselected in nature. On the other hand, recombination in RNA3 by exchange of the two ORFs was more frequent in both weeds and cultivated crops in Spain, or in lily isolates in Asia (Bonnet *et al.*, 2005), and has also been described for an isolate of PSV (Kiss *et al.*, 2008), suggesting that the cost associated with this type of recombination is lower.

Recombinants in the noncoding regions of the genome could be thought to be more frequent as these regions seem less constrained, even if they include the genomic promoters, and as reassortants between different strains of CMV or even different cucumovirus are viable, suggesting a weak specificity of these promoters. Moreover, *in silico* studies showed that the 3′ noncoding region of CMV subgroup II is more closely related to that of TAV than to its counterpart in CMV subgroup I and that subgroup II termini probably arose from recombination between CMV and TAV (Thompson and Tepfer, 2009). Nevertheless, here also, recombination in this region has been observed in only a few isolates, and associated with adaptation of CMV to *Alstromeria* (Chen *et al.*, 2002). The recombination events were found in both RNAs 2 and 3, and corresponded to a duplication of part of the 3′ noncoding region. By contrast, recombination has been more frequently observed under experimental conditions, particularly in reassortants created between CMV and TAV. This was firstly described for a reassortant composed of RNAs 1 and 2 of CMV and RNA3 of TAV (Fernández-Cuartero *et al.*, 1994). Recombination in this reassortant led to acquisition of the 3′-terminal part of the noncoding region of CMV RNA2 by TAV RNA3. Moreover, this recombinant proved to be more fit than the parental reassortant or even the CMV strain. A similar situation was observed for reassortants/recombinants between CMV strains (Pierrugues *et al.*, 2007). A second example of recombination between CMV and TAV was described in the progeny of a complex inoculum containing TAV RNAs 1 and 2 and CMV RNAs 2 and 3. The recombinant described was composed of TAV RNA1, CMV RNA2, CMV RNA2 in which the 3′-terminal part was substituted by the one of TAV RNA2 and, finally, CMV RNA3 (Masuta *et al.*, 1998). Sequencing the 3′-termini of RNAs 3 and 4 further showed that, here also, recombination had occurred in some molecules which brought the 3′-termini of either TAV RNA1 or RNA2 (Suzuki *et al.*, 2003). Finally, a similar recombination event was identified in RNA3 progeny in plants infected with CMV RNA1, CMV RNA2 bearing the 2b of TAV, and TAV RNA3, following serial passages (Shi *et al.*, 2008). This recombinant was selected by TAV 2b protein and was maintained when coinoculated with the two parental strains (Shi *et al.*, 2009). A more recent work showed that the N-terminal 12 amino acids of the protein were sufficient for its selection (Shi and Palukaitis, 2011). In all these cases, recombination aimed at conferring to the heterologous RNA, the terminal part of the RNAs involved in the replication complex, which may have increased its replication efficiency.

Studies on intra- and intermolecular recombination increased greatly when questions about the potential risks associated with transgenic plants expressing virus-derived genes arose, soon after the first demonstration that expression of such genes could confer resistance against the virus. Template-switching of the replicase complex from an RNA transcribed from a transgene and infecting viral RNAs was first observed in transgenic plants expressing a full-length copy of RNA1, following inoculation with RNAs 2 and 3 (Canto *et al.*, 2001). In three out of the five plants that became infected, a recombination event was identified at the 3′ end of RNA1, resulting in the replacement of a more or less important part of this region by part of the 3′ end of either RNA2 or RNA3. These events occurred in 5% of the cloned RT-PCR products of the progeny and were also observed at a similar frequency in nontransgenic plants infected with RNA transcripts (Canto *et al.*, 2001). When studying the risk associated with recombination in transgenic plants, situation in nontransgenic plants infected by two strains (the strain donor of gene for transformation and the infecting strain) is taken as a baseline. If the same population of recombinants is observed in the two situations, then it can be considered that this type of transgenesis does not constitute a specific risk for the environment. A map of the crossover sites for recombination in RNA3 progeny, in plants doubly infected by CMV and TAV, was first established (de Wispelaere *et al.*, 2005). Recombination occurred in 9.6% of the plants and 28 different sites of precise homologous recombination could be identified, mostly in large blocks of sequence identity between TAV and CMV RNA 3. While recombination was observed in the noncoding regions as well as in the MP ORF, no recombinant was observed in the CP ORF (de Wispelaere *et al.*, 2005). Interestingly, similar recombination events were observed during *in vitro* reverse transcription of a mixture of the two strains when using an RNase-H-free reverse transcriptase (Fernandez-Delmond *et al.*, 2004). By doing so with a mixture of two CMV strains, several recombinants were created, including in the *CP* gene. The recombinant RNA3 population was then characterized in transgenic plants expressing a CMV *CP* gene and the 3′ noncoding region, following infection with either another CMV strain or TAV, without selection pressure in favor of recombination (Turturo *et al.*, 2008). In TAV-infected plants, recombination occurred mainly in the 3′ noncoding region, as already shown in doubly infected plants. In CMV-infected plants, a major recombination event was identified in the *CP* gene (Turturo *et al.*, 2008), although this recombinant had been shown to be nonviable (Pierrugues *et al.*, 2007). Populations of recombinants in infected transgenic plants and coinfected nontransgenic plants were equivalent, suggesting thus that such transgenic plants would not be at the origin of an emerging new virus in nature (Turturo *et al.*, 2008). When plants were infected under conditions that favor recombination with the transgenic RNA, using a disabled RNA3 in the inoculum, the previous recombinant was still present, but was not predominant. Rather, most of the recombinants occurred in the 3′ noncoding region and were of aberrant type. Finally, nearly half of the population corresponded to intragenomic recombinants between RNA3 and RNA1 (Morroni *et al.*, 2009).

All the work done with infected transgenic plants or nontransgenic doubly infected plants clearly identified a hotspot for recombination corresponding precisely to the beginning of Box-1 of RNA5. The recombinants previously mentioned in the progeny of diverse CMV–TAV reassortants, as well as the recombinant isolate from *Alstroemeria*, correspond also to this zone. Box-1 and RNA5 production were indeed shown to direct production of nonhomologous recombinants (de Wispelaere and Rao, 2009). Also, the 2b protein is involved in intermolecular recombination between RNAs 1 or 2 and RNA3, directing both the crossover site and the rate of recombination in CMV–TAV reassortant/recombinant. It has been proposed to act through its RNA-binding activity and to facilitate movement of selected recombinants (Shi *et al.*, 2008). High accumulation of both RNA5 and 2b was observed for a reassortant/recombinant TAV/CMV, promoting its systemic movement (Asaoka *et al.*, 2010).

E. Role of genetic drift in CMV evolution

Despite a huge replication rate without repairing mechanisms, viral genetic diversity within a plant appears low, suggesting that severe population bottlenecks occur during infection and plant invasion. The kinetics of plant infection and colonization were investigated at the beginning of plant virology, but took advantage of molecular tools only recently (for review, see García-Arenal and Fraile, 2011). Bottlenecks associated with virus movement (local and systemic) are evaluated by comparing the genetic composition of the progeny in plants inoculated with two or more viral genotypes. Then mathematical modeling is developed to estimate the number of viral entities constituting the founders in both inoculated and systemically infected leaves. This approach was firstly developed for TMV in tobacco, where it was shown that founder numbers were in all cases small, on the order of units (Sacristán *et al.*, 2003). Similarly, the genetic composition of a population initially composed of 12 marked CMV mutants was monitored in inoculated and systemically infected leaves of tobacco plants (Li and Roossinck, 2004). All mutants could be detected in the inoculated leaves, but a considerable decrease was observed in systemically infected leaves, with an average of only seven mutants being detected. Interestingly, movement of the different mutants occurred unevenly, as different parts of a leaf showed different genetic compositions (Li and Roossinck, 2004). The same mutant population was tested on other hosts, namely, tomato, pepper, squash, and *N. benthamiana* (Ali and Roossinck, 2010). Here also, the 12 mutants were detected in the inoculated leaves of all hosts, indicating that they replicated and moved from cell to cell efficiently. Analysis of systemically infected leaves indicated a significant decrease in their number, confirming the bottleneck in systemic movement in these hosts also. Comparison of the genetic composition in the first systemically infected leaves (emerging at the time of inoculation) and in leaves which developed later (not yet formed at the time of inoculation) showed an even greater decrease in the number of mutants that were detected, resulting in numbers as low as one to two in the three cultivated hosts and around four for *N. benthamiana*. In addition, some differences could be observed in the genetic composition in different hosts, suggesting forces driven by the host genotype (Ali and Roossinck, 2010).

The occurrence of bottlenecks was also examined during aphid transmission of the virus, once it was proposed that individual aphids transmit a few number of particles of another nonpersistently transmitted virus (PVY, Moury *et al.*, 2004). This was even more pertinent for CMV, which requires three particles to be simultaneously transmitted to establish an infection. The 12 previously mentioned mutants were used to inoculate cotyledons of squash, and these cotyledons (where all mutants replicated and moved) were used as virus sources for single *A. gossypii* or *M. persicae*. Efficiencies of acquisition and transmission were evaluated by detecting the mutants in either the aphid or the test plant (Ali *et al.*, 2006). An average of three mutants was transmitted per aphid, whatever the aphid species. Composition of the progeny in test plants indicated that transmission of each mutant was stochastic, although some mutants were better transmitted by an aphid species than the other. Most of the mutants were detected in aphids by quantitative RT-PCR, suggesting that the severe bottleneck during these experiments occurred during transmission (inoculation), but not during acquisition of the virus (Ali *et al.*, 2006). In another study, the effective number of founders initiating an infection following aphid transmission from plants infected with two CMV genotypes was determined, using a modified version of the model previously developed by Sacristán *et al.* (2003) (Betancourt *et al.*, 2008). It appeared that the number of founders was small, between one to two, in accordance with the number of mutants found to be transmitted by one aphid (Ali *et al.*, 2006), but also with the results obtained with PVY (Moury *et al.*, 2004). These recent findings are essential when studying virus evolution, as they suggest that random genetic drift during transmission or movement is probably a major force exerted on viral population.

VIII. VIRUS AND HOST FITNESS

The impact of virus infection on host fitness is poorly documented for plants. It is generally considered that viruses reduce host fitness, but this mainly results from observations of symptom severity in crops, and in fact mostly from estimates of crop losses. That virus infection could be to the advantage of its host has, however, been shown recently, as infection of several plant species by CMV improved their tolerance to abiotic stresses such as drought, and even freezing in the case of beets (Xu *et al.*, 2008). Also, it is assumed that viruses must not kill their hosts in order to survive and disseminate in nature, and indeed host death occurs rarely. Thus, evaluating host fitness cannot be based on host mortality, as classically done for bacterial or animal pathogens. Studies on host fitness were developed only recently on a few virus–host combinations, including CMV, and were based on plant morbidity (biomass, height, or weight of the plants; number and viability of the seeds, etc.). Fitness of the virus, on the other hand, is generally evaluated by its accumulation in infected hosts or by its ability to compete when different strains or species are compared.

In studies involving several CMV strains representing the two subgroups and a set of *A. thaliana* accessions representing most of the natural genetic variation of this species, Fernando García-Arenal's group brought important new insights in plant–virus relations and coevolution. Particularly, they observed that the quantitative aspects of viral accumulation and its effects on host fitness (growth and fecundity) were fully dependent on the virus genotype–host genotype combination (Pagán *et al.*, 2007). Except for certain specific combinations, virus virulence (used here as effects of virus infection on host fitness) was not correlated with virus accumulation, thus not supporting the trade-off hypothesis for the evolution of virulence. Studies were also carried out in order to evaluate to what extent plants can modify their life-history traits in response to virus infection, and particularly if reproduction was favored over growth when plants were infected. Virus infection resulted in decreases of both growth and fecundity of the plants. Host genotype differences explained most of the observed variance. Although all genotypes delayed flowering when infected, only genotypes with a longer life cycle increased their reproductive structures and progeny (Pagán *et al.*, 2008). Moreover, population density was also shown to play an important role in plant fitness in an infection situation. Particularly, for those genotypes that allocate more resources to vegetative growth than to reproduction, growth was affected by infection only when population density was increased, that is, when individuals had to compete for natural resources (Pagán *et al.*, 2009). Overall, these results illustrate the multiplicity of requirements for studying host–virus interactions and also constitute a warning against general conclusions drawn from a single host–virus combination under fixed experimental conditions.

Generalist pathogens such as CMV are transmitted and dispersed very efficiently. According to evolutionary theory, the generalist strategy is expected to penalize virus fitness, as maximal fitness cannot be achieved in all host species. This raises the questions of whether some hosts could select the virus, or alternatively, whether the virus could adapt to different hosts. If the virus adapts to its hosts, then successive passages of an isolate would result in increased virulence and/or increased virus accumulation. Using isolates collected on distantly related hosts, bean, cucumber, and tomato, Sacristán *et al.* (2005) compared virus fitness (infectivity and virus accumulation) and virulence (impact on plant growth) before and after 10 passages through their host of origin or the two other hosts. In the least permissive host, bean, they observed host-dependent infectivity, as only isolates from bean infected this host efficiently. Also in this host, virus accumulation increased after 10 successive passages, suggesting improvement of this component of virus fitness. However, as this was not observed when the two other hosts were infected with bean isolates, it is unlikely that this gain of accumulation would reflect host adaptation. Significant differences in virus accumulation were observed in the different hosts. On the whole, 10 passages did not induce significant differences in virus accumulation, suggesting that the isolates were already at their maximal fitness in their host of origin. Infection with any of the isolates, in all three host species resulted in decreased host fitness (estimated by plant dry weight), although to different extents depending on the host–isolate combination. So, as already demonstrated on *A. thaliana*, virulence is not correlated to virus accumulation in these three hosts (Sacristán et al., 2005).

The possibility of virus selection by the host was investigated in 21 wild species, for four viruses transmitted by aphids in a nonpersistent manner (AMV, CMV, WMW) or in a circulative manner (BWYV, *Beet western yellows virus*, genus *Polerovirus*) or by thrips (TSWV, *Tomato spotted wilt virus*, genus *Tospovirus*) (Malpica *et al.*, 2006). Nearly half of the host species were not randomly infected by the different viruses, indicating host selectivity, and most of them exhibited a high infection level. Similarly, three viruses, including CMV, appeared also to be nonrandomly distributed in the wild. This suggests that different viruses specialize on different hosts. So, despite its broad host range, CMV does not infect all potential hosts it has the opportunity to encounter in nature. The second important finding of this work was that, when coinfection occurred (10–30% of the plants infected depending on host species), it resulted mostly in positive association between the viruses (Malpica *et al.*, 2006). This has important implications for virus dissemination and evolution, as coinfection can favor emergence of new viruses through recombination or can facilitate virus dispersion through complementation.

IX. CONCLUDING REMARKS

The properties and functioning of protein 2b is doubtless the area of CMV research where the most progress has been made in the past decade, as it has largely exceeded research on the virus itself. Plant viral suppressors of RNA silencing such as 2b have been particularly useful in exploring fine mechanisms of gene regulation, putting plant virology at the forefront of fundamental research. However, the role of protein 2b in several steps of the virus cycle, and particularly as a pathogenicity determinant, still requires a major research effort. More generally, progress in unraveling the interactions between viral proteins or viral RNAs and host components has not advanced to the same degree during the same period. Even if several interactions between viral and host proteins have been demonstrated, their number remains low and their significance is not always well understood. This situation has two possible explanations. The first is our lack of knowledge of plant genomes, except of course for certain model plants. However, not only are those plants not well suited for all viruses, but also many of their genes have still unknown function. Second, the considerable burden of the experimental approaches for such studies, such as screening libraries for *in vitro* interactions in yeast hybrid systems or transcriptomic assays, often results in the adoption of *a priori* and restrictive approaches, such as focusing on candidate genes. For all steps of the virus cycle, the lack of knowledge of the host components involved in virus–host interactions clearly constitutes a blocking point. An additional problem is that results obtained in one host species or family cannot be extended to other hosts or families. This is well illustrated by studies on virus movement, where clearly contrasting results have been obtained with cucurbit and solanaceous hosts, probably due to different characteristics of their plasmodesmata and flow pathway. Thus, further improvement of our knowledge in plant virology is strongly associated with that of enhanced understanding of host physiology and metabolism.

The reverse genetic approach has been, continues to be, and will remain a remarkable tool for associating specific properties to a genome region, to a single protein, or even to a single amino acid or nucleotide residue. However, there are situations where interpretation of the results is difficult and even misleading. First, some hybrid genomes are not viable, and others can adapt to become viable. In the former case, studying the properties of converse pairs of recombinants, which is a prerequisite for certainty of analysis, can be compromised (Llamas *et al.*, 2006). In the latter, compensatory mutations can occur in the target region or in another region of the genome (Thompson *et al.*, 2006). Ideally, verification of the full genome of the progeny viruses is needed when using a reverse genetic approach, but this is never done because it would be too time consuming and too expensive. Also, as all virus genes are multifunctional, epistatic or additive effects can be expected to occur. If so, the function of one gene can be masked or enhanced, depending on the genetic background. This is exemplified by PVY, for which it has been demonstrated that ability to overcome a monogenic recessive resistance in pepper, through mutation in its virulence factor (protein VPg), was considerably increased in a recombinant virus expressing the cylindrical inclusion protein of another strain (Montarry *et al.*, 2011). Quantitative genetics has not been considered until now in plant virology. But obviously, this approach would be required in situations where more than one viral gene is involved, either directly or indirectly, in the function(s) being studied. Multipartite viruses such as CMV may well have advantages when using a reverse genetic approach, as smaller genome segments are easier to manipulate experimentally.

Virus populations in naturally infected hosts behave as metapopulations, with founder and extinction events. Even if coinfection occurs regularly in nature, virus diversity in a cultivated crop (and even more so in a single plant) is rather low, with only a few genotypes being dominant. Also, virus diversity is similar in cultivated plants and nearby wild plants. On the other hand, there is experimental evidence that CMV is prone to recombination and most of the reassortants and recombinants created *in vitro* are viable. One of the important advances made these past years has been to reconcile these apparently contradictory observations. First, there are strong bottlenecks in virus populations, either during within-plant movement or during between-plant transmission. In both cases, the number of founders is low, being one or a few units only. Second, for a recombination event to occur, the plants must be coinfected with two strains. This situation is not very frequent for strains belonging to subgroups I and II, either in space or in time. Concerning subgroup IB strains, they are not disseminated worldwide and thus cohabit with subgroup IA mainly in Asian countries. Moreover, extensive surveys are scarce. Also, the two parental genotypes need to be in the same cell, a situation that is also unlikely to occur frequently because of segregation of virus genotypes in coinfected plants. Finally, even if viable, most of the recombinants tested so far seem less fit than parental strains. So even if in principle a large set of reassortants or recombinants could develop in a single plant, their emergence will be strictly dependent on a selective advantage relative to the parental strains. The same reasoning can be applied to mutations, the second main selective force exerted on virus population, which has been less studied for CMV. Thus, taking into account genetic drift exerted on virus populations at many steps of the virus cycle, increased fitness of any variant appears to be the key determinant of successful emergence in nature. This also results in isolates having an optimized fitness in their host of origin, with a poor potential for improvement following repeated multiplication of the virus under experimental conditions.

Even if our knowledge of virus–plant interactions has increased, methods for controlling virus epidemics have not evolved. Cultivation of resistant varieties remains the only efficient way to control CMV. Unfortunately, natural resistances are scarce for most CMV host crops. Use of transgenic plants expressing virus-derived genes would thus be of considerable interest, as several types of transgenes proved to be very effective in conferring resistance to the virus. Moreover, a considerable effort has been made for assessing the safety of transgenes bearing CMV-derived sequences, making it possible to design transgenes satisfying both efficacy and biosafety criteria. Despite this, the only CMV-resistant crop available commercially is a transgenic squash that is grown to a limited extent in the United States. Moreover, efforts to design and test new transgenes have decreased considerably these past years. Disapproval of transgenic products by consumers, particularly in Europe, played an important role at two levels. First, this could impose real constraints on international trade for countries where cultivation of transgenic varieties is a current practice. Second, it also reduced the sources of funding for laboratory research, which, in all industrialized countries, is increasingly dependent on external contracts. It should also be noted that fighting viruses in vegetable crops is no longer the main priority of private breeders, at least for viruses such as CMV, because their economical impact in industrialized countries is too low. This situation is regrettable for two main reasons. First, a decreased research effort will impact both our ability to operate if conditions become more favorable to GMOs and our ability to carry out the additional studies that are needed concerning the efficacy and durability of these resistances in the open field. Further, what is secondary for industrialized countries is essential for developing ones, where the need for conferring tolerance to both biotic and abiotic stresses is vital.

ACKNOWLEDGMENT

I sincerely thank Mark Tepfer for enriching discussions about every part of this review and for correcting the English.

REFERENCES

Abdel-Salam, A. M., Kararah, M. A., Nakhla, M. K., and Ibrahim, L. M. (1989). Purification of a severe isolate of *Cucumber mosaic virus* (CMV) isolated from nurseries growing pepper in Egypt. *Egypt. J. Phytopathol.* **21**:85–99.

Abdullahi, I., Ikotun, T., Winter, S., Thottapilly, G., and Atiri, G. I. (2001). Investigation on seed transmission of *Cucumber mosaic virus* in cowpea. *Afr. Crop Sci. J.* **9**:677–684.

Acosta-Leal, R., and Xiong, Z. (2008). Complementary functions of two recessive R-genes determine resistance durability of tobacco 'Virgin A mutant' (VAM) to *Potato virus Y*. *Virology* **379:**275–283.

Adams, I. P., Glover, R. H., Monger, W. A., Mumford, R., Jackeviciene, E., Navalinskiene, M., Samuitiene, M., and Boonham, N. (2009). Next-generation sequencing and metagenomic analysis: A universal diagnostic tool in plant virology. *Mol. Plant Pathol.* **10:**537–545.

Ahn, J.-W., Yin, C.-J., Liu, J. R., and Jeong, W.-J. (2010). *Cucumber mosaic virus* 2b protein inhibits RNA silencing pathways in green alga *Chlamydomonas reinhardtii*. *Plant Cell Rep.* **29:**967–975.

Akhtar, K. P., Saleem, M. Y., Asghar, M., Ahmad, M., and Sarwar, N. (2010). Resistance of *Solanum* species to *Cucumber mosaic virus* subgroup IA and its vector Myzus persicae. *Eur. J. Plant Pathol.* **128:**435–450.

Ali, A., and Kobayashi, M. (2010). Seed transmission of *Cucumber mosaic virus* in pepper. *J. Virol. Methods* **163:**234–237.

Ali, A., Li, H., Schneider, W. L., Sherman, D. J., Gray, S., Smith, D., and Roossinck, M. J. (2006). Analysis of genetic bottlenecks during horizontal transmission of *Cucumber mosaic virus*. *J. Virol.* **80:**8345–8350.

Ali, A., and Roossinck, M. J. (2010). Genetic bottlenecks during systemic movement of *Cucumber mosaic virus* vary in different host plants. *Virology* **404:**279–283.

Al-Musa, A. M. (1989). Oversummering hosts for some cucurbits viruses in the Jordan valley. *J. Phytopathol.* **127:**49–54.

Al-Musa, A., and Mansour, A. (1983). Plant viruses affecting tomatoes in Jordan. Identification and prevalence. *Pythopath. Z.* **106:**186–190.

Alonso-Prados, J., Aranda, M. A., Malpica, J. M., García-Arenal, F., and Fraile, A. (1998). Satellite RNA of cucumber mosaic cucumovirus spreads epidemically in natural populations of its helper virus. *Phytopathology* **88:**520–524.

Andreev, I. A., Kim, S. H., Kalinina, N. O., Rakitina, D. V., Fitzgerald, A. G., Palukaitis, P., and Taliansky, M. E. (2004). Molecular interactions between a plant virus movement protein and RNA: Force spectroscopy investigation. *J. Mol. Biol.* **339:**1041–1047.

Aramburu, J., Galipienso, L., and López, C. (2007). Reappearance of *Cucumber mosaic virus* isolates belonging to subgroup IB in tomato plants in north-eastern Spain. *J. Phytopathol.* **155:**513–518.

Aranda, M. A., Fraile, A., and García-Arenal, F. (1993). Genetic variability and evolution of the satellite RNA of *Cucumber mosaic virus* during natural epidemics. *J. Virol.* **67:**5896–5901.

Aranda, M. A., Fraile, A., Dopazo, J., Malpica, J. M., and García-Arenal, F. (1997). Contribution of mutation and RNA recombination to the evolution of a plant pathogenic RNA. *J. Mol. Evol.* **44:**81–88.

Asaoka, R., Shimura, H., Arai, M., and Masuta, C. (2010). A progeny from a cucumovirus pseudorecombinant evolved to gain the ability to accumulate its RNA-silencing suppressor leading to systemic infection in tobacco. *Mol. Plant Microbe. Interact.* **23:**332–339.

Babovic, M., Bulajic, A., Delibasic, G., Milijic, S., and Todorovic, D. (1997). Role of bean seed in transmitting *Bean common mosaic virus* and *Cucumber mosaic virus*. *Acta Hort.* **462:**253–258.

Badr, A. B. (1989). Prevalence of TMV and CMV in tomatoes grown in commercial greenhouse and field. *Ann. Agric. Sci. (Cairo)* **34:**713–721.

Bashir, M., and Hampton, R. O. (1996). Detection and identification of seed-borne viruses from cowpea (*Vigna unguiculata* (L.) Walp.) germplasm. *Plant Pathol.* **45:**54–58.

Ben Moussa, A., Makni, M., and Marrakchi, M. (2000). Identification of the principal viruses infecting tomato crops in Tunisia. *EPPO B.* **30:**293–296.

Berniak, H., Kamińska, M., and Malinowski, T. (2010). *Cucumber mosaic virus* groups IA and II are represented among isolates from naturally infected lilies. *Eur. J. Plant Pathol.* **127:**305–309.

Betancourt, M., Fereres, A., Fraile, A., and García-Arenal, F. (2008). Estimation of the effective number of founders that initiate an infection after aphid transmission of a multipartite plant virus. *J. Virol.* **82:**12416–12421.

Bhattiprolu, S. L. (1991). Seed-borne *Cucumber mosaic virus* infecting French bean (*Phaseolus vulgaris*) in India. *Indian J. Virol.* **7:**67–76.

Blackman, L. M., Boevink, P., Santa Cruz, S., Palukaitis, P., and Oparka, K. J. (1998). The movement protein of *Cucumber mosaic virus* traffics into sieve elements in minor veins of *Nicotiana clevelandii. Plant Cell* **10:**525–537.

Bonnet, J., Fraile, A., Sacristán, S., Malpica, J. M., and García-Arenal, F. (2005). Role of recombination in the evolution of natural populations of *Cucumber mosaic virus*, a tripartite RNA plant virus. *Virology* **332:**359–368.

Bouché, N., Lauressergues, D., Gasciolli, V., and Vaucheret, H. (2006). An antagonistic function for *Arabidopsis* DCL2 in development and a new function for DCL4 in generating viral siRNAs. *EMBO J.* **25:**3347–3356.

Brigneti, G., Voinnet, O., Li, W.-X., Ji, L.-H., Ding, S.-W., and Baulcombe, D. C. (1998). Viral pathogenicity determinants are suppressors of transgene silencing in *Nicotiana benthamiana. EMBO J.* **17:**6739–6746.

Canto, T., Choi, S. K., and Palukaitis, P. (2001). A subpopulation of RNA1 of *Cucumber mosaic virus* contains 3′ termini originating from RNAs 2 or 3. *J. Gen. Virol.* **82:**941–945.

Chalam, V. C., Parakh, D. B., Khetarpal, R. K., Maurya, A. K., Anju, J., and Singh, S. (2008). Interception of seed-transmitted viruses in cowpea and mungbean germplasm imported during 2003. *Indian J. Virol.* **19:**12–16.

Chen, Y. K., Derks, A. F. L. M., Langeveld, S., Goldbach, R., and Prins, M. (2001). High sequence conservation among *Cucumber mosaic virus* isolates from lily. *Arch. Virol.* **146:**1631–1636.

Chen, Y.-K., Goldbach, R., and Prins, M. (2002). Inter- and intramolecular recombinations in the *Cucumber mosaic virus* genome related to adaptation to *Alstroemeria. J. Virol.* **76:**4119–4124.

Chen, Y., Chen, J., Zhang, H., Tang, X., and Du, Z. (2007). Molecular evidence and sequence analysis of a natural reassortant between *Cucumber mosaic virus* subgroup IA and II strains. *Virus Genes* **35:**405–413.

Chen, H.-Y., Yang, J., Lin, C., and Yuan, Y. A. (2008). Structural basis for RNA-silencing suppression by *Tomato aspermy virus* protein 2b. *EMBO Rep.* **9:**754–760.

Choi, S. K., Palukaitis, P., Min, B. E., Lee, M. Y., Choi, J. K., and Ryu, K. H. (2005). *Cucumber mosaic virus* 2a polymerase and 3a movement proteins independently affect both virus movement and the timing of symptom development in zucchini squash. *J. Gen. Virol.* **86:**1213–1222.

Choi, S.-K., Jeon, Y.-W., Yoon, J.-Y., and Choi, J.-K. (2011). Characterization of a satellite RNA of *Cucumber mosaic virus* that induces chlorosis in *Capsicum annuum. Virus Genes* **43:**111–119.

Cillo, F., Pasciuto, M. M., De Giovanni, C., Finetti-Sialer, M. M., Ricciardi, L., and Gallitelli, D. (2007). Response of tomato and its wild relatives in the genus *Solanum* to *Cucumber mosaic virus* and satellite combinations. *J. Gen. Virol.* **88:**3166–3176.

Cillo, F., Mascia, T., Pasciuto, M. M., and Gallitelli, D. (2009). Differential effects of mild and severe *Cucumber mosaic virus* strains in the perturbation of microRNA-regulated gene expression in tomato map to the 3′ sequence of RNA2. *Mol. Plant Microbe. Interact.* **22:**1239–1249.

Crescenzi, A., Barbarossa, L., Gallitelli, D., and Martelli, G. P. (1993). Cucumber mosaic cucumovirus populations in Italy under natural epidemic conditions and after a satellite-mediated protection test. *Plant Dis.* **77:**28–33.

Dalakouros, A., Tzanopoulou, M., Tsagris, M., Wassenegger, M., and Kalantidis, K. (2011). Hairpin transcription does not necessarily lead to efficient triggering of the RNAi pathway. *Transgenic Res.* **20:**293–304.

De Wispelaere, M., and Rao, A. L. N. (2009). Production of *Cucumber mosaic virus* RNA5 and its role in RNA recombination. *Virology* **384:**179–191.
De Wispelaere, M., Gaubert, S., Trouilloud, S., Belin, C., and Tepfer, M. (2005). A map of the diversity of RNA3 recombinants appearing in plants infected with *Cucumber mosaic virus* and *Tomato aspermy virus*. *Virology* **331:**117–127.
Diaz-Pendon, J. A., Li, F., Li, W.-X., and Ding, S.-W. (2007). Suppression of antiviral silencing by *Cucumber mosaic virus* 2b protein in *Arabidopsis* is associated with drastically reduced accumulation of three classes of small interfering RNAs. *Plant Cell* **19:**2053–2063.
Ding, S.-W., Li, W.-X., and Symons, R. H. (1995). A novel naturally occurring hybrid gene encoded by a plant RNA virus facilitates long distance virus movement. *EMBO J.* **14:**5762–5772.
Ding, S.-W., Shi, B.-J., Li, W.-X., and Symons, R. H. (1996). An interspecies hybrid RNA virus is significantly more virulent than either parental virus. *Proc. Natl. Acad. Sci. U.S.A.* **93:**7470–7474.
Divéki, Z., Salánki, K., and Balázs, E. (2004). The necrotic pathotype of the *Cucumber mosaic virus* (CMV) Ns strain is solely determined by amino acid 461 of the 1a protein. *Mol. Plant Microbe. Interact.* **8:**837–845.
Du, Z.-Y., Chen, F.-F., Liao, Q.-S., Zhang, H.-R., Chen, Y.-F., and Chen, J.-S. (2007a). 2b ORFs encoded by subgroup IB strains of *Cucumber mosaic virus* induce differential virulence on *Nicotiana* species. *J. Gen. Virol.* **88:**2596–2604.
Du, Q.-S., Duan, C.-G., Zhang, Z.-H., Fang, Y.-Y., Fang, R.-X., Xie, Q., and Guo, H.-S. (2007b). DCL4 targets *Cucumber mosaic virus* satellite RNA at novel secondary structures. *J. Virol.* **81:**9142–9151.
Du, Z., Chen, F., Zhao, Z., Liao, Q., Palukaitis, P., and Chen, J. (2008). The 2b protein and the C-terminus of the 2a protein of Cucumber mosaic virus subgroup I strains both play a role in viral accumulation and induction of symptoms. *Virology* **380:**363–370.
Dukić, N., Krstić, B., Vico, I., Berenji, J., and Duduk, B. (2006). First report of *Zucchini yellow mosaic virus, Watermelon mosaic virus,* and *Cucumber mosaic virus* in bottlegourd (*Lagenaria siceraria*) in Serbia. *Plant Dis.* **90:**380.
Dunoyer, P., Schott, G., Himber, C., Meyer, D., Takeda, A., Carrington, J. C., and Voinnet, O. (2010). Small RNA duplexes function as mobile silencing signals between plant cells. *Science* **328:**912–916.
Edwardson, J. R., and Christie, R. G. (1991). Cucumoviruses. *In* CRC Handbook of Viruses Infecting Legumes, pp. 293–319. CRC Press, Boca Raton, FL.
Eiras, M., Boari, A. J., Colariccio, A., Chaves, A. L. R., Briones, M. R. S., Figueira, A. R., and Harakava, R. (2004). Characterization of isolates of the cucumovirus *Cucumber mosaic virus* present in Brazil. *J. Plant Pathol.* **86:**61–69.
Escriu, F., Perry, K. L., and García-Arenal, F. (2000). Transmissibility of *Cucumber mosaic virus* by *Aphis gossypii* correlates with viral accumulation and is affected by the presence of its satellite RNA. *Phytopathology* **90:**1068–1072.
Escriu, F., Fraile, A., and García-Arenal, F. (2003). The evolution of virulence in a plant virus. *Evolution* **57:**755–765.
Escriu, F., Fraile, A., and García-Arenal, F. (2007). Constraints to genetic exchange support gene coadaptation in a tripartite RNA virus. *PLoS Pathog.* **3:**67–74.
Essafi, A., Díaz-Pendón, J. A., Moriones, E., Monforte, A. J., Garcia-Mas, J., and Martín-Hernández, A. M. (2009). Dissection of the oligogenic resistance to *Cucumber mosaic virus* in the melon accession PI 161375. *Theor. Appl. Genet.* **118:**275–284.
Faccioli, G. (1972). Un ceppo di virus del cetriolo isolato da spinaccio in Emilia. *Phytopathol. Mediterr.* **11:**67–70.
Fernández-Cuartero, B., Burgyán, J., Aranda, M. A., Salánki, K., Moriones, E., and García-Arenal, F. (1994). Increase in the relative fitness of a plant virus RNA associated with its recombinant nature. *Virology* **203:**373–377.

Fernandez-Delmond, I., Pierrugues, O., de Wispelaere, M., Guilbaud, L., Gaubert, S., Divéki, Z., Godon, C., Tepfer, M., and Jacquemond, M. (2004). A novel strategy for creating recombinant infectious RNA virus genomes. *J. Virol. Methods* **121:**247–257.

Finetti-Sialer, M., Mërkuri, J., Tauro, G., Myrta, A., and Gallitelli, D. (2005). Viruses of vegetable crops in Albania. *EPPO B.* **35:**491–495.

Fraile, A., Alonso-Prados, J. L., Aranda, M. A., Bernal, J. J., Malpica, J. M., and García-Arenal, F. (1997). Genetic exchange by recombination or reassortant is infrequent in natural populations of a tripartite RNA plant virus. *J. Virol.* **71:**934–940.

Fukuzawa, N., Itchoda, N., Ishihara, T., Goto, K., Masuta, C., and Matsumura, T. (2010). HC-Pro, a potyvirus RNA silencing suppressor, cancels cycling of *Cucumber mosaic virus* in *Nicotiana benthamiana* plants. *Virus Genes* **40:**440–446.

Gallitelli, D. (2000). The ecology of *Cucumber mosaic virus* and sustainable agriculture. *Virus Res.* **71:**9–21.

García-Arenal, F., and Fraile, A. (2011). Population dynamics and genetics of plant infection by viruses. *In* "Recent Advances in Plant Virology" (C. Caranta, M. A. Aranda, M. Tepfer, and J. J. Lopez-Moya Eds, eds.), pp. 263–281. Caister Academic Press, UK.

Garcia-Luque, I., Diaz-Ruiz, J. R., Rubio-Huertos, M., and Kaper, J. M. (1983). *Cucumovirus* survey in Spanish economically important crops. *Phytopathol. Mediterr.* **22:**127–132.

Gildow, F. E., Shah, D. A., Sackett, W. M., Butzler, T., Nault, B. A., and Fleischer, S. J. (2008). Transmission efficiency of *Cucumber mosaic virus* by aphids associated with virus epidemics in snap bean. *Phytopathology* **98:**1233–1241.

González, I., Martínez, L., Rakitina, D. V., Lewsey, M. G., Atencio, F. A., Llave, C., Kalinina, N. O., Carr, J. P., Palukaitis, P., and Canto, T. (2010). *Cucumber mosaic virus* 2b protein subcellular targets and interactions: Their significance to RNA silencing suppressor activity. *Mol. Plant Microbe. Interact.* **23:**294–303.

Goto, K., Kobori, T., Kosaka, Y., Natsuaki, T., and Masuta, C. (2007). Characterization of silencing suppressor 2b of *Cucumber mosaic virus* based on examination of its small RNA-binding abilities. *Plant Cell Physiol.* **48:**1050–1060.

Grieco, F., Cillo, F., Barbarossa, L., and Gallitelli, D. (1992). Nucleotide sequence of a *Cucumber mosaic virus* satellite RNA associated with a tomato top stunt. *Nucleic Acids Res.* **20:**6733.

Grieco, F., Lanave, C., and Gallitelli, D. (1997). Evolutionary dynamics of *Cucumber mosaic virus* satellite RNA during natural epidemics in Italy. *Virology* **229:**166–174.

Hajimorad, M. R., Ghabrial, S. A., and Roossinck, M. (2009). De novo emergence of a novel satellite RNA of *Cucumber mosaic virus* following serial passages of the virus derived from RNA transcripts. *Arch. Virol.* **154:**137–140.

Harvey, J. J. W., Lewsey, M. G., Patel, K., Westwood, J., Heimstädt, S., Carr, J. P., and Baulcombe, D. C. (2011). An antiviral defense role of AGO2 in plants. *PLoS One* **6:**e14639.

Hong, J. S., Masuta, C., Nakano, M., Abe, J., and Ueyda, I. (2003). Adaptation of *Cucumber mosaic virus* soybean strains (SSVs) to cultivated and wild soybeans. *Theor. Appl. Genet.* **107:**49–53.

Hong, J. S., Ohnishi, S., Masuta, C., Choi, J. K., and Ryu, K. H. (2007). Infection of soybean by *Cucumber mosaic virus* as determined by viral movement protein. *Arch. Virol.* **152:**321–328.

Hou, W.-N., Duan, C.-G., Fang, R.-X., Zhou, X.-Y., and Guo, H.-S. (2011). Satellite RNA reduces expression of the 2b suppressor protein resulting in the attenuation of symptoms caused by *Cucumber mosaic virus* infection. *Mol. Plant Pathol.* **12:**595–605.

Hu, C. C., and Ghabrial, S. A. (1998). Molecular evidence that strain BV-15 of peanut stuntcucumovirus is a reassortant between subgroup I and II strains. *Phytopathology* **88:**92–97.

Hu, Q., Niu, Y., Zhang, K., Liu, Y., and Zhou, X. (2011). Virus-derived transgenesexpressing hairpin RNA give immunity to *Tobacco mosaic virus* and *Cucumber mosaic virus*. *Virol. J.* **8:**41.

Huh, S. U., Kim, M. J., Ham, B.-K., and Paek, K.-E. (2011). A zinc finger protein Tsip1 controls *Cucumber mosaic virus* infection by interacting with the replication complex on vacuolar membranes of the tobacco plant. *New phytol* **191:**746–762. doi: 10.1111/j.1469-8137.2011.03717.x.

Hwang, M. S., Kim, S. H., Lee, J. H., Bae, J. M., Paek, K. H., and Park, Y. I. (2005). Evidence for interaction between the 2a polymerase protein and the 3a movement protein of *Cucumber mosaic virus*. *J. Gen. Virol.* **86:**3171–3177.

Hwang, M. S., Kim, K. N., Lee, J. H., and Park, Y. I. (2007). Identification of amino acid sequences determining interaction between the cucumber mosaic virus-encoded 2a polymerase and 3a movement proteins. *J. Gen. Virol.* **88:**3445–3451.

Ioannou, N. (1985). Yellow leaf curl and other virus diseases of tomato in Cyprus. *Plant Pathol.* **34:**428–434.

Inaba, J.-I., Kim, B. M., Shimura, H., and Masuta, C. (2011). Virus-induced necrosis is a consequence of direct protein-protein interaction between a viral RNA-silencing suppressor and a host catalase. *Plant Physiol.* **156:**2026–2036.

Irian, S., Xu, P., Dai, X., Zhao, P. X., and Roossinck, M. J. (2007). Regulation of a virus-induced lethal disease in tomato revealed by LongSAGE analysis. *Mol. Plant Microbe. Interact.* **20:**1477–1488.

Jacquemond, M., and Tepfer, M. (1998). Satellite RNA-mediated resistance to plant viruses: Are the ecological risks well assessed? *In* "Plant Virus Disease Control" (A. Hadidi, R. K. Kheterpal, and H. Koganezawa Eds, eds.), pp. 94–120. APS Press, MN, USA.

Ji, L.-H., and Ding, S.-W. (2001). The suppressor of transgene RNA silencing encoded by *Cucumber mosaic virus* interferes with salicylic acid-mediated virus resistance. *Mol. Plant Microbe. Interact.* **14:**715–724.

Joannon, B., Lavigne, C., Lecoq, H., and Desbiez, C. (2010). Barriers to gene flow between emerging populations of *Watermelon mosaic virus* in southeastern France. *Phytopathology* **100:**1373–1379.

Jones, R. A. C. (2004a). Occurrence of virus infection in seed stocks and 3-year-old pastures of lucerne (*Medicage sativa*). *Aust. J. Agr. Res.* **55:**757–764.

Jones, R. A. C. (2004b). Using epidemiological information to develop effective integrated virus disease management strategies. *Virus Res.* **100:**5–30.

Jones, R. A. C. (2005). Patterns of spread of two non-persistently aphid-borne viruses in lupin stands under four different infection scenarios. *Ann. Appl. Biol.* **146:**337–350.

Jordá, C., Alfaro, A., Aranda, M. A., Moriones, E., and García-Arenal, F. (1992). Epidemic of *Cucumber mosaic virus* plus satellite RNA in tomatoes in Eastern Spain. *Plant Dis.* **76:**363–366.

Kanazawa, A., Inaba, J.-I., Shimura, H., Otagaki, S., Tsukahara, S., Matsuzawa, A., Kim, B. M., Goto, K., and Masuta, C. (2011). Virus-mediated efficient induction of epigenetic modifications of endogenous genes with phenotypic changes in plants. *Plant J.* **65:**156–168.

Kang, B.-C., Yeam, I., and Jahn, M. M. (2005). Genetics of plant virus resistance. *Ann. Rev. Phytopathol.* **43:**18.1–18.41.

Kang, W.-H., Hoang, N. H., Yang, H.-B., Kwon, J.-K., Jo, S.-H., Seo, J.-K., Kim, K.-H., Choi, D., and Kang, B.-C. (2010). Molecular mapping and characterization of a single dominant gene controlling CMV resistance in peppers (*Capsicum annuum* L.). *Theor. Appl. Genet.* **120:**1587–1596.

Kaplan, I. B., Lee, K.-C., Canto, T., Wong, S.-K., and Palukaitis, P. (2004). Host-specific encapsidation of a defective RNA3 of *Cucumber mosaic virus*. *J. Gen. Virol.* **85:**3757–3763.

Katul, L., and Makkouk, K. M. (1987). Occurrence and serological relatedness of five cucurbit potyviruses in Lebanon and Syria. *Phytopathol. Mediterr.* **26:**36–42.

Kim, S. H., Kalinina, N. O., Andreev, I., Ryabov, E. V., Fitzgerald, A. G., Taliansky, M. E., and Palukaitis, P. (2004). The C-terminal 33 amino acids of the *Cucumber mosaic virus* 3a protein affect virus movement, RNA binding and inhibition of infection and translation. *J. Gen. Virol.* **85:**221–230.

Kim, M. J., Kim, H. R., and Paek, K.-H. (2006). *Arabidopsis* tonoplast proteins TIP1 and TIP2 interact with the *Cucumber mosaic virus* 1a replication protein. *J. Gen. Virol.* **87:**3425–3431.

Kim, M. J., Huh, S. U., Ham, B.-K., and Paek, K.-H. (2008). A novel methyltransferase methylates *Cucumber mosaic virus* 1a protein and promotes systemic spread. *J. Virol.* **82:**4823–4833.

Kiselyova, O. I., Yaminski, I. V., Karger, E. M., Frolova, O. Y., Dorokhov, Y. L., and Atabekov, J. G. (2001). Visualization by atomic force microscopy of *Tobacco mosaic virus* movement protein-RNA complexes formed *in vitro. J. Gen. Virol.* **82:**1503–1508.

Kiss, L., Sebestyén, E., László, E., Salamon, P., Bálazs, E., and Salánki, K. (2008). Nucleotide sequence analysis of *Peanut stunt virus* Rp strain suggests the role of homologous recombination in cucumovirus evolution. *Arch. Virol.* **153:**1373–1377.

Kobori, T., Miyagawa, M., Nishioka, K., Ohki, S. T., and Osaki, T. (2002). Amino acid 129 of *Cucumber mosaic virus* coat protein determines local symptom expression and systemic movement in *Tetragonia expansa, Momordica charantia* and *Physalis floridana. J. Gen. Plant Pathol.* **68:**81–88.

Krajacic, M., and Juretic, N. (1993). Natural infection of Swiss chard by *Cucumber mosaic virus* in Croatia. *Petria* **3:**93–98.

Kyriakopoulou, P. E., and Bem, F. (1982). *Cucumber mosaic virus* on *Cucurbitaceae* and other crops in Greece. *Ann. Inst. Phytopath. Benaki* **13:**151–162.

Lahoz, E., Piccirillo, P., and Ragozzino, A. (1994). Studies on the seed transmissibility of three isolates of CMV (*Cucumber mosaic virus*) in different cultivars of common bean. *Petria* **4:**117–122.

Lang, Q., Jin, C. Z., Lai, L., Feng, J., Chen, S., and Chen, J. (2011). Tobacco microRNAs prediction and their expression infected with *Cucumber mosaic virus* and Potato virus X. *Mol. Biol. Rep.* **38:**1523–1531.

Lange, E., and Hammi, M. (1977). Essais sur l'efficacité des pulvérisations d'huile contre les viroses transmises par des pucerons au piment en Tunisie. *Phytopathol. Mediterr.* **16:**18–21.

Latham, L. J., and Jones, R. A. C. (2001). Incidence of virus infection in experimental plots, commercial crops, and seed stocks of cool season crop legumes. *Aust. J. Agr. Res.* **52:**397–413.

Latham, L. J., Jones, R. A. C., and McKirdy, S. J. (2001). Cucumber mosaic cucumovirus infection of cool-season crop, annual pasture, and forage legumes: Susceptibility, sensitivity, and seed transmission. *Aust. J. Agr. Res.* **52:**683–697.

Lecoq, H., and Pitrat, M. (1983). Field experiments on the integrated control of aphid-borne viruses in muskmelon. *In* "Plant Virus Epidemiology. The Spread and Control of Insect-Borne Viruses" (R. T. Plumb and J. M. Thresh, eds.), pp. 169–176. Blackwell Scientific Publications, Oxford, UK.

Lee, J.-A., Choi, S. K., Yoon, J. Y., Hong, J. S., Ryu, K. H., Lee, S. Y., and Choi, J. K. (2007). Variation in the pathogenicity of lily isolates of *Cucumber mosaic virus. Plant Pathol. J.* **23:**251–259.

Lewsey, M., Robertson, F. C., Canto, T., Palukaitis, P., and Carr, J. P. (2007). Selective targeting of miRNA-regulated plant development by a viral counter-silencing protein. *Plant J.* **50:**240–252.

Lewsey, M., Surette, M., Robertson, F. C., Ziebell, H., Choi, S. H., Ryu, K. H., Canto, T., Palukaitis, P., Payne, T., Walsh, J. A., and Carr, J. P. (2009). The role of the *Cucumber mosaic virus* 2b protein in viral movement and symptom induction. *Mol. Plant Microbe. Interact.* **6:**642–654.

Lewsey, M. G., Murphy, A. M., MacLean, D., Dalchau, N., Westwood, J. H., Macaulay, K., Bennett, M. H., Moulin, M., Hanke, D. E., Powell, G., Smith, A. G., and Carr, J. P. (2010). Disruption of two defensive signaling pathways by a viral RNA silencing suppressor. *Mol. Plant Microbe. Interact.* **23:**835–845.

Li, H., and Roossinck, J. M. (2004). Genetic bottlenecks reduce population variation in an experimental RNA virus population. *J. Virol.* **78:**10852–10857.

Lin, H.-X., Rubio, L., Smythe, A., Jiminez, M., and Falk, B. W. (2003). Genetic diversity and biological variation among California isolates of *Cucumber mosaic virus. J. Gen. Virol.* **84:**249–258.

Lin, H.-X., Rubio, L., Smythe, A. B., and Falk, B. W. (2004). Molecular population genetics of *Cucumber mosaic virus* in California: Evidence for founder effects and reassortment. *J. Virol.* **78:**6666–6675.

Liu, Y. Y., YU, S. L., Lan, Y. F., Zhang, C. L., Hou, S. S., Li, X. D., Chen, X. Z., and Zhu, X. P. (2009). Molecular variability of five *Cucumber mosaic virus* isolates from China. *Acta Virol.* **53:**89–97.

Llamas, S., Moreno, I. M., and García-Arenal, F. (2006). Analysis of the viability of coat-protein hybrids between *Cucumber mosaic virus* and *Tomato aspermy virus. J. Gen. Virol.* **87:**2085–2088.

Lockhart, B. E. L., and Fisher, H. U. (1976). *Cucumber mosaic virus* infections of pepper in Morocco. *Plant Dis. Rep.* **60:**262–264.

López, C., Aramburu, J., Galipienso, L., and Nuez, F. (2007). Characterisation of several heterogeneous species of defective RNAs derived from RNA3 of *Cucumber mosaic virus. Arch. Virol.* **152:**621–627.

Makkouk, K. M., and Attar, N. (2003). Seed transmission of *Cucumber mosaic virus* and *Alfalfa mosaic virus* in lentil seeds. *Arab J. Plant Prot.* **21:**49–52.

Malpica, J. M., Sacristán, S., Fraile, A., and García-Arenal, F. (2006). Association and host selectivity in multi-host pathogens. *PLoS One* **1:**e41.

Maoka, T., Hayano, Y. S., Iwasaki, M., Yoshida, K., and Masuta, C. (2010). Mixed infection in tomato to ensure frequent generation of a natural reassortant between two subgroups of *Cucumber mosaic virus. Virus Genes* **40:**148–150.

Martelli, G. P. (1999). Infectious diseases and certification of olive: An overview. *EPPO B.* **29:**127–133.

Mascia, T., Cillo, F., Fanelli, V., Finetti-Sialer, M. M., De Stradis, A., Palukaitis, P., and Gallitelli, D. (2010). Characterization of the interactions between *Cucumber mosaic virus* and *Potato virus Y* in mixed infections in tomato. *Mol. Plant Microbe. Interact.* **23:**1514–1524.

Masiri, J., Velasquez, N. V., and Murphy, J. F. (2011). *Cucumber mosaic virus* 2b-deficient mutant causes limited, asymptomatic infection of bell pepper. *Plant Dis.* **95:**331–336.

Masuta, C., Ueda, S., Suzuki, M., and Uyeda, I. (1998). Evolution of a quadripartite hybrid virus by interspecific exchange and recombination between replicase components of two related tripartite RNA viruses. *Proc. Natl. Acad. Sci.* **95:**10487–10942.

Masuta, C., Seshimo, Y., Mukohara, M., Jung, H. J., Ueda, S., Ryu, K. H., and Choi, J. K. (2002). Evolutionary characterization of two lily isolates of *Cucumber mosaic virus*isolated in Japan and Korea. *J. Gen. Plant Pathol.* **68:**163–168.

Mauck, K. E., De Moraes, C. M., and Mescher, M. C. (2010a). Deceptive chemical signals induced by a plant virus attract insect vectors to inferior hosts. *Proc. Natl. Acad. Sci.* **107:**3600–3605.

Mauck, K. E., De Moraes, C. M., and Mescher, M. C. (2010b). Effects of *Cucumber mosaic virus* infection on vector and non-vector herbivores of squash. *Commun. Integr. Biol.* **3:**579–582.

Mayers, C. N., Lee, K.-C., Moore, C. A., Wong, S.-M., and Carr, J. P. (2005). Salicylic acid-induced resistance to *Cucumber mosaic virus* in squash and *Arabidopsis thaliana*: Contrasting mechanisms of induction and antiviral action. *Mol. Plant Microbe. Interact.* **18:**428–434.

Messiaen, C. M., Maison, P., and Migliori, A. (1963). Le virus I du concombre dans le Sud-est de la France. *Phytopathol. Mediterr.* **11:**251–260.

Mitsuya, Y., Takahashi, Y., Berberich, T., Miyazaki, A., Matsumura, H., Takahashi, H., Terauchi, R., and Kusano, T. (2009). Spermine signaling plays a significant role in the defense response of *Arabidopsis thaliana* to *Cucumber mosaic virus*. *J. Plant Physiol.* **166:**626–643.

Mitter, N., Sulistyowati, E., and Dietzgen, R. G. (2003). *Cucumber mosaic virus* infection transiently breaks dsRNA-induced transgenic immunity to *Potato virus Y* in tobacco. *Mol. Plant Microbe. Interact.* **16:**936–944.

Mnari-Hattab, M., Jebari, H., and Zouba, A. (2008). Identification et distribution des virus responsables de mosaïques chez les cucurbitacées en Tunisie. *EPPO B.* **38:**497–506.

Mochizuki, T., and Ohki, S. T. (2004). Shoot meristem tissue of tobacco inoculated with *Cucumber mosaic virus* is infected with the virus and subsequently recovers from infection by RNA silencing. *J. Gen. Plant Pathol.* **70:**363–366.

Mochizuki, T., and Ohki, S. T. (2005). Amino acid 129 in the coat protein of *Cucumber mosaic virus* primarily determines invasion of the shoot apical meristem of tobacco plants. *J. Gen. Plant Pathol.* **71:**326–332.

Montarry, J., Doumayrou, J., Simon, S., and Moury, B. (2011). Genetic background matters: A plant-virus gene-for-gene interaction is strongly influenced by genetic contexts. *Mol. Plant Pathol.* **12:**911–920. doi: 10.1111/j.1364-3703.2011.00724.x.

Morel, J.-B., Godon, C., Mourrain, P., Béclin, C., Boutet, S., Feuerbach, F., Proux, F., and Vaucheret, H. (2002). Fertile hypomorphic ARGONAUTE (*ago1*) mutants impaired in post-transcriptional gene silencing and virus resistance. *Plant Cell* **14:**629–639.

Moreno, I. M., Thompson, J. R., and García-Arenal, F. (2004). Analysis of the systemic colonization of cucumber plants by *Cucumber green mottle mosaic virus*. *J. Gen. Virol.* **85:**749–759.

Moriones, E., Fraile, A., and García-Arenal, F. (1991). Host-associated selection of sequence variants from a satellite RNA of *Cucumber mosaic virus*. *Virology* **184:**465–468.

Morroni, M., Thompson, J. R., and Tepfer, M. (2008). Twenty years of transgenic plants resistant to *Cucumber mosaic virus*. *Mol. Plant Microbe. Interact.* **21:**675–684.

Morroni, M., Thompson, J. R., and Tepfer, M. (2009). Analysis of recombination between viral RNAs and transgene mRNA under conditions of high selection pressure in favour of recombinants. *J. Gen. Virol.* **90:**2798–2807.

Moury, B., Fabre, F., and Senoussi, R. (2004). Estimation of the number of virus particles transmitted by an insect vector. *Proc. Natl. Acad. Sci. U.S.A.* **104:**17891–17896.

Netsu, O., Hiratsuka, K., Kuwata, S., Hibi, T., Ugaki, M., and Suzuki, M. (2008). *Peanut stunt virus* 2b cistron plays a role in viral local and systemic accumulation and virulence in *Nicotiana benthamiana*. *Arch. Virol.* **153:**1731–1735.

Ng, J. C. K., and Perry, K. L. (2004). Transmission of plant viruses by aphid vectors. *Mol. Plant Pathol.* **5:**501–511.

Ng, J. C. K., Josefsson, C., Clark, A. J., Franz, A. W. E., and Perry, K. L. (2005). Virion stability and aphid vector transmissibility of *Cucumber mosaic virus* mutants. *Virology* **332:**397–405.

Nitzany, F. E., and Wilkinson, R. E. (1961). The identification of *Cucumber mosaic virus* from different hosts in Israel. *Phytopathol. Mediterr.* **1:**71–76.

Nogay, A., and Yorganc, U. (1984). Investigations on the identification, seed transmission and host range of viruses infecting the culture plants in the *Cucurbitaceae* in Marmara region. 1. The identification of viruses infecting cucurbits in Marmara region. *J. Turkish Phytopathol.* **13:**9–27.

Nogay, A., and Yorganc, U. (1985). Investigations on the identification, seed transmission and host range of viruses infecting the culture plants in the *Cucurbitaceae* in Marmara region. 2. The seed transmissibilities and cucurbit hosts of CMV and WMV-2 isolated from the culture plants in the *Cucurbitaceae*. *J. Turkish Phytopathol.* **14:**9–16.

O'Keefe, D. C., Berryman, D. I., Coutts, B. A., and Jones, R. A. C. (2007). Lack of seed coat contamination with *Cucumber mosaic virus* in lupin permits reliable, large-scale detection of seed transmission in seed samples. *Plant Dis.* **91:**504–508.

Odedara, O. O., Hughes, J. D., and Ayo-John, E. I. (2007). Diagnosis, occurrence and seed transmission studies of viruses infecting four *Centrosema* species in Nigeria. *Trop. Sci.* **47**:244–252.

Ohnishi, S., Echizenya, I., Yoshimoto, E., Boumin, K., Inukai, T., and Masuta, C. (2011). Multigenic system controlling viral systemic infection determined by the interactions between *Cucumber mosaic virus* genes and quantitative trait loci of soybeans cultivars. *Phytopathology* **101**:575–582.

Pagán, I., Alonso-Blanco, C., and García-Arenal, F. (2007). The relationship of within-host multiplication and virulence in a plant-virus system. *PLoS One* **2**:e786.

Pagán, I., Alonso-Blanco, C., and García-Arenal, F. (2008). Host responses in life-history traits and tolerance to virus infection in *Arabidopsis thaliana*. *PLoS One* **4**:e1000124.

Pagán, I., Alonso-Blanco, C., and García-Arenal, F. (2009). Differential tolerance to direct and indirect density-dependent costs of viral infection in *Arabidopsis thaliana*. *PLoS One* **5**: e1000531.

Pallás, V., Genovés, A., Sánchez-Pina, M. A., and Navarro, J. A. (2011). Systemic movement of viruses via the plant phloem. *In* "Recent Advances in Plant Virology" (C. Caranta, M. A. Aranda, M. Tepfer, and J. J. Lopez-Moya, eds.), pp. 75–101. Caister Academic Press, UK.

Palloix, A., Ayme, V., and Moury, B. (2010). Durability of plant major resistance genes to pathogens depends on the genetic background, experimental evidence and consequences for breeding strategies. *New Phytol.* **183**:190–199.

Palukaitis, P., and Carr, J. P. (2008). Plant resistance responses to viruses. *J. Plant Pathol.* **90**:153–171.

Palukaitis, P., and García-Arenal, F. (2003). Cucumoviruses. *Adv. Virus Res.* **62**:242–323.

Palukaitis, P., Roossinck, M. J., Dietzgen, R. G., and Francki, R. I. B. (1992). Cucumber mosaic virus. *Adv. Virus Res.* **41**:281–348.

Park, K. H., and Cha, B. J. (2002). Detection of TMV, ToMV and CMV from tomato seeds and plants. *Res. Plant Dis.* **8**:101–106.

Pierrugues, O., Guilbaud, L., Fernandez-Delmond, I., Fabre, F., Tepfer, M., and Jacquemond, M. (2007). Biological properties and relative fitness of inter-subgroup cucumber mosaic virus RNA3 recombinants produced *in vitro*. *J. Gen. Virol.* **88**:2852–2861.

Pitrat, M. (2002). Gene list for melon. *Cucurbit Genet. Coop. Rep.* **25**:76–93.

Praveen, S., Mangrauthia, S. K., Singh, P., and Mishra, A. K. (2008). Behavior of RNAi suppressor protein 2b of *Cucumber mosaic virus* in planta in presence and absence of the virus. *Virus Genes* **37**:96–102.

Prins, M., Laimer, M., Noris, E., Schubert, J., Wassenegger, M., and Tepfer, M. (2008). Strategies for antiviral resistance in transgenic plants. *Mol. Plant Pathol.* **9**:73–83.

Qi, Y., Zhong, X., Itaya, A., and Ding, B. (2004). Dissecting RNA silencing in protoplasts uncovers novel effects of viral suppressors on the silencing pathway at the cellular level. *Nucleic Acids Res.* **32**:e179.

Qu, J., Ye, J., and Fang, R. (2007). Artificial microRNA-mediated virus resistance in plants. *J. Virol.* **81**:6690–6699.

Quiot, J. B., Douine, L., and Gébré Sélassié, K. (1979a). Fréquence des principales viroses identifiées dans une exploitation maraîchère du Sud-Est de la France. *Ann. Phytopathol.* **11**:283–290.

Quiot, J. B., Devergne, J. C., Cardin, L., Verbrugghe, M., Marchoux, G., and Labonne, G. (1979b). Ecologie et épidémiologie du virus de la mosaïque du concombre dans le Sud-Est de la France VII. Répartition de deux types de populations virales dans des cultures sensibles. *Ann. Phytopathol.* **11**:359–374.

Quiot, J. B., Devergne, G. M., Cardin, L., and Douine, L. (1979c). Ecologie et épidémiologie du virus de la mosaïque du concombre dans le Sud-Est de la France. VI. Conservation de deux types de populations virales dans les plantes sauvages. *Ann. Phytopathol.* **11:**349–358.

Quiot, J. B., Leroux, J. P., Labonne, G., and Renoust, M. (1979d). Epidémiologie de la maladie filiforme et de la nécrose de la tomate provoquées par le virus de la mosaïque du concombre dans le Sud-Est de la France. *Ann. Phytopathol.* **11:**393–408.

Rahimian, H., and Izadpanah, K. (1978). Identity and prevalence of mosaic-inducing cucurbits viruses in Shiraz, Iran. *Phytopath. Z.* **92:**305–312.

Rashid, U. J., Hoffmann, J., Brutschy, B., Piehler, J., and Chen, C.-H. (2008). Multiple targets for suppression of RNA interference by *Tomato aspermy virus* protein 2B. *Biochemistry* **47:**12655–12657.

Requena, A., Simón-Buela, L., Salcedo, G., and García-Arenal, F. (2006). Potential involvement of a cucumber homolog of phloem protein 1 in the long-distance movement of *Cucumber mosaic virus* particles. *Mol. Plant Microbe. Interact.* **19:**734–746.

Retheesh, S. T., and Bhat, A. I. (2010). Simultaneous elimination of *Cucumber mosaic virus* and *Cymbidium mosaic virus* infecting *Vanilla planifolia* through the meristem culture. *Crop Prot.* **29:**1214–1217.

Roossinck, M. J. (2002). Evolutionary history of *Cucumber mosaic virus* deduced by phylogenetic analyses. *J. Virol.* **76:**3382–3387.

Roossinck, M. J., Zhang, L., and Hellwald, K.-H. (1999). Rearrangements in the 5′ nontranslated region and phylogenetic analyses of *Cucumber mosaic virus* RNA3 indicate radial evolution of three subgroups. *J. Virol.* **73:**6752–6758.

Ryang, B.-S., Kobori, T., Matsumoto, T., Kosaka, Y., and Ohki, S. T. (2004). *Cucumber mosaic virus* 2b protein compensates for restricted spread of *Potato virus Y* in doubly infected tobacco. *J. Gen. Virol.* **85:**3405–3414.

Ryu, K. H., Park, H. W., and Choi, J. K. (2002). Characterization and sequence analysis of a lily isolate of *Cucumber mosaic virus* from *Lilium tsingtauense. Plant Pathol. J* **18:**85–92.

Sacristán, S., Malpica, J. M., Fraile, A., and García-Arenal, F. (2003). Estimation of population bottlenecks during systemic movement of *Tobacco mosaic virus* in tobacco plants. *J. Virol.* **77:**9906–9911.

Sacristán, S., Fraile, A., and García-Arenal, F. (2004). Population dynamics of *Cucumber mosaic virus* in melon crops and in weeds in Central Spain. *Phytopathology* **94:**992–998.

Sacristán, S., Fraile, A., Malpica, J. M., and García-Arenal, F. (2005). An analysis of host adaptation and its relationship with virulence in *Cucumber mosaic virus. Phytopathology* **95:**827–833.

Salánki, K., Gellért, A., Huppert, E., Náray-Szabó, G., and Balázs, E. (2004). Compatibility of the movement protein and the coat protein of cucumoviruses is required for cell-to-cell movement. *J. Gen. Virol.* **85:**1039–1048.

Salánki, K., Gellért, A., Náray-Szabó, G., and Balázs, E. (2007). Modeling-based characterization of the elicitor function of amino acid 461 of *Cucumber mosaic virus* 1a protein in the hypersensitive response. *Virology* **358:**109–118.

Sánchez-Navarro, J. A., Herranz, M. C., and Pallás, V. (2006). Cell-to-cell movement of *Alfalfa mosaic virus* can be mediated by the movement proteins of ilar-, bromo-, cucumo-, tobamo- and comoviruses and does not require virion formation. *Virology* **346:**66–73.

Sanjuán, R., Moya, A., and Elena, S. F. (2004). The contribution of epistasis to the architecture of fitness in an RNA virus. *Proc. Natl. Acad. Sci. U.S.A.* **26:**15376–15379.

Sasaki, N., Park, J.-W., Maule, A. J., and Nelson, R. S. (2006). The cysteine-histidine-rich region of the movement protein of *Cucumber mosaic virus* contributes to plasmodesmata targeting, zinc binding and pathogenesis. *Virology* **349:**396–408.

Sclavounos, A. P., Voloudakis, A. E., Arabatzis, Ch., and Kyriakopoulou, P. E. (2006). A severe Hellenic CMV tomato isolate: Symptom variability in tobacco, characterization and discrimination of variants. *Eur. J. Plant Pathol.* **115:**163–172.

Sekine, K.-T., Nandi, A., Ishihara, T., Hase, S., Ikegami, M., Shah, J., and Takahashi, H. (2004). Enhanced resistance to *Cucumber mosaic virus* in the *Arabidopsis thaliana ssi2* mutant is mediated via an SA-independent mechanism. *Mol. Plant Microbe. Interact.* **17:**623–632.

Sekine, K.-T., Ishihara, T., Hase, S., Kusano, T., Shah, J., and Takahashi, K. (2006). Single amino acid alterations in *Arabidopsis thaliana* RCY1 compromise resistance to *Cucumber mosaic virus,* but differentially suppress hypersensitive response-like cell death. *Plant Mol. Biol.* **62:**669–682.

Sekine, K.-T., Kawakami, S., Hase, S., Kubota, M., Ichinose, Y., Shah, J., Kang, H.-G., Klessig, D. F., and Takahashi, H. (2008). High level expression of a virus resistance gene, *RCY1,* confers extreme resistance to *Cucumber mosaic virus* in *Arabidopsis thaliana. Mol. Plant Microbe. Interact.* **21:**1398–1407.

Seo, Y.-S., Rojas, M. R., Lee, J.-Y., Lee, S.-W., Jeon, J.-S., Ronald, P., Lucas, W. J., and Gilbertson, R. L. (2006). A viral resistance gene from common bean functions across plant families and is up-regulated in a non-virus-specific manner. *Proc. Natl. Acad. Sci. U.S.A.* **103:**11856–11861.

Seo, J.-K., Kwon, S.-J., Choi, H.-S., and Kim, K.-Y. (2009). Evidence for alternate states of *Cucumber mosaic virus* replicase assembly in positive- and negative-strand RNA synthesis. *Virology* **383:**248–260.

Shams-Bakhsh, M., Canto, T., and Palukaitis, P. (2007). Enhanced resistance and neutralization of defense responses by suppressors of RNA silencing. *Virus Res.* **130:**103–109.

Shi, B.-J., and Palukaitis, P. (2011). The N-terminal 12 amino acids of Tomato aspermy virus 2b protein function in infection and recombination. *J. Gen. Virol.* **92:**2706–2710. doi: 10.1099/vir.0.035071-0.

Shi, B.-J., Palukaitis, P., and Symons, R. H. (2002). Differential virulence by strains of *Cucumber mosaic virus* is mediated by the *2b* gene. *Mol. Plant Microbe. Interact.* **15:**947–965.

Shi, B.-J., Miller, J., Symons, R. H., and Palukaitis, P. (2003). The 2b protein of cucumoviruses has a role in promoting the cell-to-cell movement of pseudorecombinant viruses. *Mol. Plant Microbe. Interact.* **16:**261–267.

Shi, B.-J., Symons, R. H., and Palukaitis, P. (2008). The cucumovirus 2b gene drives selection of inter-viral recombinants affecting the crossover site, the acceptor RNA and the rate of selection. *Nucleic Acids Res.* **36:**1057–1071.

Shi, B.-J., Symons, R. H., and Palukaitis, P. (2009). Stability and competitiveness of interviral recombinant RNAs derived from a chimeric cucumovirus. *Virus Res.* **140:**216–221.

Shimura, H., Pantaleo, V., Ishihara, T., Myojo, N., Inaba, J.-I., Sueda, K., Burgyán, J., and Masuta, C. (2011). A viral satellite RNA induces yellow symptoms on tobacco by targeting a gene involved in chlorophyll biosynthesis using the RNA silencing machinery. *PLoS Pathog.* **7:**e1002021.

Siddiqui, S. A., Sarmiento, C., Truve, E., Lehto, H., and Lehto, K. (2008). Phenotypes and functional effects caused by various viral RNA silencing suppressors in transgenic *Nicotiana benthamiana* and *N. tabacum*. *Mol. Plant Microbe. Interact.* **21:**178–187.

Siddiqui, S. A., Valkonen, J. P. T., Rajamäki, M.-L., and Lehto, K. (2011). The 2b silencing suppressor of a mild strain of *Cucumber mosaic virus* alone is sufficient for synergistic interaction with *Tobacco mosaic virus* and induction of severe leaf malformation in 2b-transgenic tobacco plants. *Mol. Plant Microbe. Interact.* **24:**685–693.

Simón-Mateo, C., López-Moya, J. J., Guo, H. S., González, E., and García, J. A. (2003). Suppressor activity of potyviral and cucumoviral infections in potyvirus-induced transgene silencing. *J. Gen. Virol.* **84:**2877–2883.

Skorić, D., Krajačić, M., Barbarossa, L., Cillo, F., Grieco, F., Sarić, A., and Gallitelli, D. (1996). Occurrence of cucumber mosaic cucumovirus with satellite RNA in lethal necrosis affected tomatoes in Croatia. *J. Phytopathol.* **144:**543–549.

Smith, N. A., Eamens, A. L., and Wang, M.-B. (2011). Viral small interfering RNAs target host genes to mediate disease symptoms in plants. *PLoS Pathog.* **7:**e1002022.

Soards, A. J., Murphy, A. M., Palukaitis, P., and Carr, J. P. (2002). Virulence and differential local and systemic spread of *Cucumber mosaic virus*in tobacco are affected by the CMV 2b protein. *Mol. Plant Microbe. Interact.* **15:**647–653.

Soler, S., Prohens, J., López, C., Aramburu, J., Galipienso, L., and Nuez, F. (2010). Viruses infecting tomato in València, Spain: Occurrence, distribution and effect of seed origin. *J. Phytopathol.* **158:**797–805.

Su, S., Liu, Z., Chen, C., Zhang, Y., Wang, X., Zhu, L., Miao, L., Wang, X.-C., and Yuan, M. (2010). *Cucumber mosaic virus* movement protein severes actin filaments to increase the plasmodesmal size exclusion limit in tobacco. *Plant Cell* **22:**1373–1387.

Subasic, D., Rusak, G., and Milicic, D. (1990). *Cucumber mosaic virus* from pepper (*Capsicum annuum* L.) in Bosnia and Herzegovina. *Acta Bot. Croat.* **49:**7–12.

Sunpapao, A., Nakai, T., Dong, F., Mochizuki, T., and Ohki, S. T. (2009). The 2b protein of *Cucumber mosaic virus* is essential for viral infection of the shoot apical meristem and for efficient invasion of leaf primordia in infected tobacco plants. *J. Gen. Virol.* **90:**3015–3021.

Suzuki, M., Hibi, T., and Masuta, C. (2003). RNA recombination between cucumoviruses: Possible role of predicted stem-loop structures and an internal subgenomic promoter-like motif. *Virology* **306:**77–86.

Takahashi, H., Kanayama, Y., Zheng, M. S., Kusano, T., Hase, S., Ikegami, M., and Shah, J. (2004). Antagonistic interactions between the SA and JA signalling pathways in *Arabidopsis* modulate espression of defense genes and gene-for-gene resistance to *Cucumber mosaic virus*. *Plant Cell Physiol.* **45:**803–809.

Takeda, A., Iwasaki, S., Watanabe, T., Usumi, M., and Watanabe, Y. (2008). The mechanism selecting the guide strand from small RNA duplexes is different among argonaute proteins. *Plant Cell Physiol.* **49:**493–500.

Takeshita, M., Kikuhara, K., Kuwata, S., Furuya, N., and Takanami, Y. (2004a). Competition between wild-type virus and a reassortant from subgroups I and II of CMV and activation of antiviral responses in cowpea. *Arch. Virol.* **149:**1851–1857.

Takeshita, M., Shigemune, N., Kikuhara, K., Furuya, N., and Takanami, Y. (2004b). Spatial analysis for exclusive interactions between subgroups I and II of *Cucumber mosaic virus* in cowpea. *Virology* **328:**45–51.

Takeshita, M., Matsuo, Y., Yoshikawa, T., Suzuki, M., Furuya, N., Tsuchiya, K., and Takanami, Y. (2008). Characterisation of a defective RNA derived from RNA3 of the Y strain of *Cucumber mosaic virus*. *Arch. Virol.* **153:**579–583.

Thackray, D. J., Diggle, A. J., Berlandier, F. A., and Jones, R. A. C. (2004). Forecasting aphid outbreaks and epidemics of *Cucumber mosaic virus* in lupin crops in a Mediterranean-type environment. *Virus Res.* **100:**67–82.

Thompson, J. R., and Tepfer, M. (2009). The 3′ untranslated region of *Cucumber mosaic virus* (CMV) subgroup II RNA3 arose by interspecific recombination between CMV and *Tomato aspermy virus*. *J. Gen. Virol.* **90:**2293–2298.

Thompson, J. R., Doun, S., and Perry, K. L. (2006). Compensatory capsid protein mutations in *Cucumber mosaic virus* confer systemic infectivity in squash (*Cucurbita pepo*). *J. Virol.* **80:**7740–7743.

Thompson, J. R., Buratti, E., de Wispelaere, M., and Tepfer, M. (2008). Structural and functional characterization of the 5′ region of subgenomic RNA5 of *Cucumber mosaic virus*. *J. Gen. Virol.* **89:**1729–1738.

Tóbiás, I., Szabó, B., Salánki, K., Sári, L., Kuhlmann, H., and Palkovics, L. (2008). Seedborne transmission of *Zucchini yellow mosaic virus* and *Cucumber mosaic virus* in Styrian Hulless group of *Cucurbita pepo*. Cucurbitaceae 2008 Proceedings of the IXth EUCARPIA Meeting on Genetics and Breeding of Cucurbitaceae, pp. 189–197.

Turturo, C., Friscina, A., Gaubert, S., Jacquemond, M., Thompson, J. R., and Tepfer, M. (2008). Evaluation of potential risks associated with recombination in transgenic plants expressing viral sequences. *J. Gen. Virol.* **89:**327–335.

Varveri, C., and Boutsika, K. (1999). Characterization of cucumber mosaic cucumovirus isolates in Greece. *Plant Pathol.* **48:**95–100.

Voinnet, O. (2005). Induction and suppression of RNA silencing: Insights from viral infection. *Nat. Rev. Genet.* **6:**206–221.

Vozelj, N., Petrovic, N., Novak, M. P., Tusek, M., Mavric, I., and Ravnikar, M. (2003). The most frequent viruses on selected ornamental plants and vegetables in Slovenia. Zbornik predavanj in referatov 6. Slovenskega Posvetovanje o Varstvu Rastlin, Zrece, Slovenije, 4-6 marec 2003, pp. 300–304.

Wang, Y., Lee, K. C., Gaba, V., Wong, S. M., Palukaitis, P., and Gal-On, A. (2004a). Breakage of resistance to *Cucumber mosaic virus* by co-infection with *Zucchini yellow mosaic virus*: Enhancement of CMV accumulation independent of symptom expression. *Arch. Virol.* **149:**379–396.

Wang, Y., Tzfira, T., Gaba, V., Citovsky, V., Palukaitis, P., and Gal-On, A. (2004b). Functional analysis of the *Cucumber mosaic virus* 2b protein: Pathogenicity and nuclear localization. *J. Gen. Virol.* **85:**3135–3147.

Wang, X.-B., Wu, Q., Ito, T., Cillo, F., Li, W.-X., Chen, X., Yu, J.-L., and Ding, S.-W. (2010). RNAi-mediated viral immunity requires amplification of virus-derived siRNAs in *Arabidopsis thaliana. Proc. Natl. Acad. Sci. U.S.A.* **107:**484–489.

Xu, P., Blancaflor, B., and Roossinck, M. J. (2003). In spite of induced multiple defense responses, tomato plants infected with *Cucumber mosaic virus* and D satellite RNA succumb to systemic necrosis. *Mol. Plant Microbe. Interact.* **16:**467–476.

Xu, P., Chen, F., Mannas, J. P., Feldman, T., Sumner, L. W., and Roossinck, M. J. (2008). Virus infection improves drought tolerance. *New Phytol.* **180:**911–921.

Yamaguchi, N., Seshimo, Y., and Masuta, C. (2005a). Mapping the sequence domain for systemic infection in edible lily on the viral genome of *Cucumber mosaic virus. J. Gen. Plant Pathol.* **71:**373–376.

Yamaguchi, N., Seshimo, Y., and Masuta, C. (2005b). Genetic mapping of the compatibility between a lily isolate of *Cucumber mosaic virus* and a satellite RNA. *J. Gen. Virol.* **86:**2359–2369.

Yang, Y., Kim, K. S., and Anderson, E. J. (1997). Seed transmission of *Cucumber mosaic virus* in spinach. *Phytopathology* **87:**924–931.

Ye, J., Qu, J., Zhang, J.-F., Geng, Y.-F., and Fang, R.-X. (2009). A critical domain of the *Cucumber mosaic virus* 2b protein for RNA silencing suppressor activity. *FEBS Lett.* **583:**101–106.

Ylmaz, M. A., and Davis, R. F. (1985). Identification of viruses infecting vegetable crops along the Mediterranean Sea coast in Turkey. *J. Turkish Phytopath.* **14:**1–8.

Yoshii, M., Nishikiori, M., Tomita, K., Yoshioka, N., Kozuka, R., Naito, S., and Ishikawa, M. (2004). The *Arabidopsis Cucumovirus multiplication 1* and *2* loci encode translation initiation factors 4E and 4G. *J. Virol.* **78:**6102–6111.

Zhang, X., Yuan, Y.-R., Pei, Y., Lin, S.-S., Tuschl, T., Patel, D. J., and Chua, N.-H. (2006). *Cucumber mosaic virus*-encoded 2b suppressor inhibits *Arabidopsis* argonaute1 cleavage activity to counter plant defense. *Gene Dev.* **20:**3255–3268.

Zhang, D., Tan, X., Willingmann, P., Adam, G., and Heinze, C. (2011). Problems encountered with the selection of *Cucumber mosaic virus* (CMV) isolates for resistance breeding programs. *J. Phytopathol.* **159:**621–629.

Ziebell, H., Payne, T., Berry, J. O., Walsh, J. A., and Carr, J. P. (2007). A *Cucumber mosaic virus* mutant lacking the 2b counter-defense protein gene provides protection against wild-type strains. *J. Gen. Virol.* **88:**2862–2871.

CHAPTER 14

Pepino Mosaic Virus and *Tomato Torrado Virus*: Two Emerging Viruses Affecting Tomato Crops in the Mediterranean Basin

Pedro Gómez,[*,1] **Raquel N. Sempere,**[†] and **Miguel A. Aranda**[*]

Contents

* Centro de Edafología y Biología Aplicada del Segura (CEBAS), Consejo Superior de Investigaciones Científicas (CSIC), Campus Universitario de Espinardo, Espinardo, Murcia, Spain
† Bioprodin SL, Campus de Espinardo s/n, Espinardo, Murcia, Spain
1 Current address: School of Biosciences, University of Exeter, Penryn, United Kingdom

Advances in Virus Research, Volume 84
ISSN 0065-3527, DOI: 10.1016/B978-0-12-394314-9.00014-2

Abstract The molecular biology, epidemiology, and evolutionary dynamics of *Pepino mosaic virus* (PepMV) are much better understood than those of *Tomato torrado virus* (ToTV). The earliest descriptions of PepMV suggest a recent jump from nontomato species (e.g., pepino; *Solanum muricatum*) to tomato (*Solanum lycopersicum*). Its stability in contaminated plant tissues, its transmission through seeds, and the global trade of tomato seeds and fruits may have facilitated the global spread of PepMV. Stability and seed transmission also probably account for the devastating epidemics caused by already-established PepMV strains, although additional contributing factors may include the efficient transmission of PepMV by contact and the often-inconspicuous symptoms in vegetative tomato tissues. The genetic variability of PepMV is likely to have promoted the first phase of emergence (i.e., the species jump) and it continues to play an important role as the virus becomes more pervasive, progressing from regional outbreaks to pandemics. In contrast, the long-term progression of ToTV outbreaks is not yet clear and this may reflect factors such as the limited accumulation of the virus in infected plants, which has been shown to be approximately two orders of magnitude less than PepMV. The efficient dispersion of ToTV may therefore depend on dense populations of its principal vectors, *Bemisia tabaci* and *Trialeurodes vaporariorum*, as has been proposed for the necrogenic satellite RNA of *Cucumber mosaic virus*.

I. INTRODUCTION

A. What is an emerging infectious disease?

Emerging infectious diseases (EIDs) are historically one of the most important problems in agriculture. Examples of EIDs with significant socioeconomical and political consequences include the epidemics of late blight in European potato crops in the mid-nineteenth century and more recently the epidemics of cassava mosaic virus in Africa in the late twentieth century. EIDs can be defined as diseases whose incidence is increasing following an outbreak in a new host population or whose incidence is increasing in an existing host population reflecting long-term changes in underlying epidemiology (Woolhouse, 2002; Woolhouse *et al.*, 2005). Therefore, EIDs include (i) new diseases caused by a previously unknown pathogen or a new strain of a known pathogen; (ii) known diseases that have moved into a new geographic area or host population; and (iii) known diseases that have increased in incidence due to a new transmission mode, changes in the environment or new virulence mutations. The majority of recent EIDs in plants are caused by viruses (Anderson *et al.*, 2004). Generally, they have a negative impact on food security due to their adverse effects on food production and quality (Vurro *et al.*, 2010). They are also extremely difficult to combat due to the scarcity of effective countermeasures.

B. What determines the emergence of a disease?

The emergence of a viral disease involves three major phases: (i) introduction of the viral pathogen into a new host species (species jump); (ii) subsequent adaptation of the viral pathogen to its new host; and (iii) epidemiological spread of the well-adapted viral pathogen in populations of the new host species, bringing about outbreaks, epidemics, and pandemics (Anderson *et al.*, 2004; Domingo, 2010). The genetic plasticity of virus populations is necessary for the emergence of a viral disease, particularly during phases (i) and (ii). Critical aspects include mutation and recombination in the virus genome, fitness tradeoffs across hosts, genetic relatedness between reservoir and naïve hosts, and molecular virus–host interactions.

The environment in which host and virus populations evolve provides an additional layer of complexity and determines the likelihood that a potentially pathogenic virus comes into contact with a new host (this includes long- and short-range virus dispersal). The environment in this context must take into account spatial structuring, virus persistence, vector availability and genetic characteristics, transmission efficiency, and agronomic practices. For example, agricultural regions with intensive cultivation practices and overlapping seasons are thought to pose a substantial risk of disease emergence, particularly in the context of global trade. Finally, environmental changes may have a significant influence in virus emergence, especially global climate change (Canto *et al.*, 2009). Broader ecological and socioeconomic factors therefore need to be taken into consideration.

C. PepMV and ToTV as emerging tomato crop pathogens in the Mediterranean Basin

The Mediterranean basin is a temperate climate region characterized by environmental conditions that make it highly suitable for the production of fruits and vegetables. Tomato (*Solanum lycopersicum*) crops in the Mediterranean basin are economically valuable to growers and provide an important source of hard currency in some countries. Tomato is consumed as fresh fruit and also processed on an industrial scale. Turkey, Egypt, Italy, Spain, Greece, and Morocco are therefore among the 20 largest tomato producers and exporters in the world, yielding up to 13 million tons of tomato fruits in 2008 (FAO, 2009).

Emerging viral pathogens in tomato crops have been recently reviewed (Hanssen *et al.*, 2010b). Eleven emerging viruses were described, including several reemerging viruses that largely disappeared when resistant cultivars were introduced. Despite the wealth of data on tomato infectious diseases, little is known about the epidemiology and ecology of these emerging pathogens. It is therefore necessary to study the population genetics and ecology of tomato viruses, their hosts and vectors, and other factors influencing the emergence of viral diseases in new areas or host populations, as this will lead to the establishment of programs for disease identification, prevention, and control.

In this chapter, we review the current status of *Pepino mosaic virus* (PepMV) and *Tomato torrado virus* (ToTV) epidemics. Interestingly, both viruses have recently been described as emerging tomato pathogens, but our perception is that the diseases caused by these viruses appear to have different long-term epidemiological outcomes.

II. PEPINO MOSAIC VIRUS

A. Genome organization

PepMV belongs to the genus *Potexvirus* (family: *Flexiviridae*) and, like other members of this genus, has virions that are nonenveloped flexuous rods approximately 508 nm in length (Adams *et al.*, 2004; Jones *et al.*, 1980). The PepMV genome consists of a positive-sense, single-stranded RNA molecule, approximately 6.4 kb in length, containing five open reading frames (ORFs) flanked by two untranslated regions (UTRs). At the 3′ end of the genomic RNA (gRNA) there is a poly(A) tail (Fig. 1; Aguilar *et al.*, 2002; Cotillon *et al.*, 2002; Mumford and Metcalfe, 2001). ORF1 encodes the putative viral RNA-dependent RNA polymerase (RdRp), which has three well-conserved domains found in the replicases of the other potexviruses: (i) the putative methyltransferase domain, (ii) the NTPase/helicase domain containing NTP-binding motifs, and (iii) the RdRp domain. ORFs 2, 3, and 4 encode the triple gene block (TGB) proteins TGBp1, TGBp2, and TGBp3, which are essential for virus movement (Morozov and Solovyev, 2003; Verchot-Lubicz, 2005). *Potato virus X* TGBp1 is a multifunctional protein that induces plasmodesmatal gating, moves from cell to cell, has ATPase and RNA-helicase activities, binds viral RNAs, and acts as suppressor of RNA silencing (Lough *et al.*, 2001; Verchot-Lubicz *et al.*, 2007). ORF5 encodes the coat protein (CP) which, in addition to its structural role, is required for cell-to-cell and long-distance movement (Mathioudakis *et al.*, 2012; Sempere *et al.*, 2011).

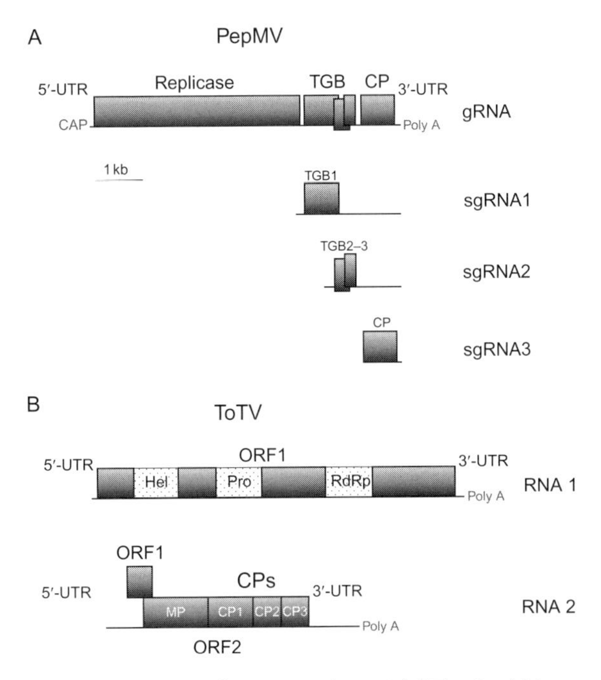

FIGURE 1 Genome organization of PepMV and ToTV. (A) The PepMV genome comprises a single, positive-sense, ~6400-nt RNA strand containing five open reading frames encoding the putative viral polymerase (RdRp), the triple gene block proteins (TGBp1, TGBp2, and TGBp3), the coat protein (CP), and two short untranslated regions (UTRs) flanking the coding regions; there is a poly(A) tail at the 3′ end of the genomic RNA. Genes are expressed from genomic and subgenomic RNAs. (B) The ToTV genome is divided into two positive-sense RNA molecules of about 7800 bp (RNA1) and 5400 bp (RNA2). RNA1 contains one open reading frame encoding a polyprotein with motifs representing RNA-dependent RNA polymerase (RdRp), protease cofactor (Pro-Co), and helicase (He). RNA2 contains two open reading frames encoding the putative movement protein (MP) and three coat proteins (CPs). Two UTRs flank the coding regions with a poly (A) tail at the 3′ end of the genomic RNAs 1 and 2.

Gene expression is thought to be similar to that in other potexviruses, that is, the viral replicase is expressed from the gRNA whereas the TGB proteins are expressed from subgenomic RNAs 1 and 2 and the CP from sgRNA 3 (Fig. 1; Verchot *et al.*, 1998; Aguilar *et al.*, 2002).

B. PepMV epidemics: Phenotypic diversity

PepMV was initially identified in Peru, in 1974, as the agent responsible for a previously uncharacterized disease affecting pepino (*Solanum muricatum*) (Jones *et al.*, 1980). The virus was first isolated from tomato in the Netherlands, in 1999 (van der Vlugt *et al.*, 2000). PepMV then spread rapidly in tomato crops throughout the Northern hemisphere and it is now considered one of the most important epidemic viruses in agriculture. It is thought that PepMV entered Europe from South America, giving rise to multiple, geographically dispersed outbreaks between 2000 and 2002 (Table I). In the past few years, the global status of PepMV epidemics has not changed significantly (Fig. 2 and Table I). PepMV still persists in many temperate regions in America, Africa, Asia, and Europe (Hanssen *et al.*, 2008; Hasiow *et al.*, 2008; Ling, 2007; Pagan *et al.*, 2006; Soler *et al.*, 2005; Spence *et al.*, 2006), despite the implementation of permanent surveillance and prevention programs. PepMV has been reported in 19 European countries, remaining a severe economic threat in the rest of Europe, and has been included in the European Plant Protection Organization alert list (EPPO, 2009).

PepMV infections in tomato cause a wide range of symptoms in leaves, including mild and severe mosaics, bubbling and laminal distortions, and, as recently reported, necrosis (Hasiów-Jaroszewska *et al.*, 2010, 2011). Affected plants are often stunted (Hanssen *et al.*, 2008, 2009; Pagan *et al.*, 2006). However, the main impact of PepMV is on fruit quality (Fig. 3A; Hanssen *et al.*, 2009), although it does not appear to affect the yield (Spence *et al.*, 2006). Fruit symptoms can arise with or without symptoms in the rest of the plant, and symptom expression is dependent on the cultivar (Fakhro *et al.*, 2010), lighting and/or temperature within glasshouses (our unpublished results), and on the PepMV isolate (Hanssen *et al.*, 2011).

Different PepMV isolates sampled from commercial tomato crops have been biologically characterized under experimental conditions, and often there appears to be no strict correlation between the disease phenotype in the field and symptoms observed in the laboratory (Hanssen *et al.*, 2008, 2009, 2010a). The establishment of a correlation between virus genotype and disease phenotype seems to be particularly complex for PepMV, probably reflecting the numerous factors that influence the expression of symptoms, but also difficulties in obtaining pure cultures of this virus. Local-lesion passages are not possible for cloning due to the absence of a local lesion host, so PepMV infectious cDNA clones appear to be the best available option. Infectious clones have been obtained only very recently so there are few publications describing their use (Hasiów-Jaroszewska *et al.*, 2009; Sempere *et al.*, 2011). Site-directed mutagenesis and other cDNA manipulations would enable the identification of regions in the PepMV genome responsible for different biological properties (Hasiów-Jaroszewska & Borodynko, 2012).

C. Host range and potential reservoirs

PepMV appears to have an economically significant impact only in tomato crops grown under intensive cultivation practices, although the virus can infect pepino, aubergine (*Solanum melongena*), potato (*Solanum tuberosum*),

TABLE I PepMV and ToTV outbreaks and current status of epidemics

Virus	Country		Situation[a]	Region[a]	Year[b]	Comment	References
PepMV	Europe	Austria	Present, few records	Vienna, Oberösterreich, and Steiermark	2006	Eradication in progress	EPPO, RS 2002/092; 2007/06
		Belgium	Present, few records	Several regions	2002	Under protection. Several surveys have been conducted	EPPO, RS 2001/098; 2003/132; Hanssen *et al.* (2008)
		Bulgaria	Present, few records	–	2004	Eradication in progress	EPPO, RS 2004/076
		Czech Republic	Present	South Moravian	2008	Eradication in progress	EPPO, RS 2008/144
		Denmark	Present, few records	–	2002	No finds in 2003	EPPO, RS 2002/092; 2003/132
		Finland	Present, few records	–	2003	Eradication in progress	EPPO, RS 2001/088; 2003/074
		France	Present, unpublished	Center and Bretagne (first descriptions), currently present in the major tomato production areas	2000	Thought to be eradicated, but now occasionally detected	EPPO, RS 2003/132; Verdin (personal communication)
		Germany	Present, few records	Hessen, Thüringen, Hamburg, and Sachsen	2001	Eradication in progress	EPPO, RS 2000/171; 2001/041; 2001/128; 2002/092; 2003/002; 2003/132

(*continued*)

TABLE I *(continued)*

Virus	Country	Situation[a]	Region[a]	Year[b]	Comment	References
	Greece	Present	Aitoloakarnania Prefecture	2010	First record. Detected on glasshouse cherry tomatoes	Efthimiou *et al.* (2011)
	Hungary	Present, few records	Center	2002	Under official control	EPPO, RS 2003/132; 2004/167
	Ireland	Absent	–	2002	Found once and eliminated	EPPO, RS 2003/132
	Italy	Present, limited distribution	Sardinia Sicilia	2001 2005	Detected on glasshouse tomatoes	EPPO, RS 2005/072; 2007/080; Davino *et al.* (2006); Parrella and Crescenzi (2005); Roggero *et al.* (2001); Salomone and Roggero (2002)
	Netherlands	Present, limited distribution	–	1999	Under official control	EPPO, RS 2000/003; 2001/041; 2003/132
	Norway	Absent	Stavanger	2001	Found once. Eradicated	EPPO, RS 2003/132
	Poland	Present, few records	Slupia Wielka Wielkopolska	2002	After destroying the affected plants, the virus was detected again in 2005	EPPO, RS 2003/043; RS 2007/107; Hasiow *et al.* (2008); Pospieszny and Borodynko (2006)
	Slovakia	Present, few records	–	2004	Eradication in progress	EPPO, RS 2004/008; 2004/107

	Spain	Present, limited distribution	Andalucia, Cataluña, Galicia, Comunidad Valenciana, Murcia, and Canary Islands	2000	Under protection. Several surveys are in progress	EPPO, RS 2000/132; 2001/142; 2001/159; 2003/132; 2005/050; 2007/079; 2007/137; 2009/172; Alfaro-Fernández *et al.* (2008a, 2009b); Gómez *et al.* (2009a,b); Jordá *et al.* (2000); Pagan *et al.* (2006)
	Sweden	Present, few records	–	2001	A survey in 2004 found no new infections	EPPO, RS 2002/092; 2004/148
	Switzerland	Present, limited distribution	Ticino and Zurich	2004	No outbreaks reported in 2007	EPPO, RS 2006/56; 2007/129; Stäubli (2005)
	United Kingdom	Present, few records	–	1999	No outbreaks reported in 2005. Under official control	EPPO, RS 2000/003; 2001/041; 2002/092; 2003/132; 2004/146; Spence *et al.* (2006); Wright and Mumford (1999); Sansford and Jones (2001)
Asia	Syria	Present	Latakia	2008	–	EPPO, RS 2011/011; Fakhro *et al.* (2010)
	China	Interception only	–	2002	No surveys conducted	Zhang *et al.* (2003)
Africa	Morocco	Interception only	–	2001	No surveys conducted	Anon (2002)
North America	USA	Present, limited distribution	Alabama, Arizona, California, Colorado, Florida, Minnesota, Oklahoma, and Texas	2001	Under official control	EPPO, RS 2001/058; 2003/067; Ling *et al.* (2008)

(*continued*)

TABLE I *(continued)*

Virus	Country		Situation[a]	Region[a]	Year[b]	Comment	References
		Canada	Present, few records	British Colombia, and Ontairo	2000	Under official control	EPPO, RS 2001/158; Ling *et al.* (2008)
	South America	Ecuador	Present	Along the Pacific coast	2000	Found in wild *Lycopersicon*	EPPO, RS 2005/144; 2009/011; Soler *et al.* (2005)
		Chile		Metropolitana and north	2001	Limited distribution	EPPO, RS 2003/112; 2005/093; 2005/144; Muñoz *et al.* (2002)
		Peru		Center and south	2000	Widespread	EPPO, RS 2000/003; 2002/56; Soler *et al.* (2002)
ToTV	Europe	Italy	Transient	Sicilia	2009	All infected plants were destroyed	EPPO, RS 2011/011; Davino *et al.* (2010)
		France	Present, few records	Perpignan	2008	–	EPPO, RS 2011/011; Verdin *et al.* (2009)
		Hungary	Absent	Szeged, Öcsöd, and Csongrád	2007	Eradicated	EPPO, RS 2008/129; 2011/011; Alfaro-Fernández *et al.* (2009a)
		Poland	Absent	Wielkopolska	2003	Eradicated	EPPO, RS 2007/174; Pospieszny *et al.* (2007, 2010)

	Spain	Present	Andalucia, Cataluña, Comunidad Valenciana, Murcia, and Canary Islands	2001	Glasshouse tomatoes. Low incidence	EPPO, RS 2007/128; 2010/84; 2011/011; Alfaro-Fernández *et al.* (2006, 2007a,b, 2008a,b, 2010c,d); Gómez *et al.* (2010); Jordá *et al.* (2003); Verbeek *et al.* (2007)
Central America	Panama	Present	Coclé, Herrera, Los Santos, and Veraguas	2008	Field tomatoes	EPPO, RS 2009/030; Herrera-Vasquez *et al.* (2009)
Oceania	South Australia	Present, few records	Adelaide	2008	Glasshouse tomatoes. Under eradication	EPPO, RS 2009/031; Gambley *et al.* (2010)

[a] Situation and geographical records of PepMV and ToTV according to the EPPO reporting service (RS) on quarantine pests.
[b] Year when virus was reported.

FIGURE 2 The geographic distribution of PepMV and ToTV. The map illustrates the countries where the presence of PepMV (filled red) and ToTV (filled green) have been reported, and the areas where both viruses have been eradicated (empty circles). (See Page 8 in Color Section at the back of the book.)

FIGURE 3 PepMV and ToTV symptoms on tomato plants. (A) Tomato fruits showing the discoloration caused by PepMV. (B) Leaf symptoms induced by ToTV in a field plant. (See Page 8 in Color Section at the back of the book.)

tobacco (*Nicotiana tabacum*), and basil (*Ocimum basilicum*) (Gómez *et al.*, 2009b; Jones *et al.*, 1980; Salomone and Roggero, 2002). In addition, garlic (*Allium sativum*) and broad bean (*Vicia faba*) have recently been identified as PepMV hosts (Fakhro *et al.*, 2011). Surveys of common weed species growing around fields or glasshouses with PepMV-infected tomato plants have shown that PepMV can infect *Amaranthus* spp., *Bassia scoparia*, *Calystegia sepium*, *Chenopodium murale*, *Convolvulus althaeoides*, *Convolvulus arvensis*, *Conyza albida*, *Coronopus* spp., *Datura inoxia*, *Diplotaxis erucoides*, *Echium creticum*, *Echium humile*, *Heliotropium europaeum*, *Malva parviflora*, *Moricandia arvensis*, *Nicotiana glauca*, *Onopordum* spp., *Piptatherum multiflorum*, *Plantago afra*, *Rumex* spp., *Sisymbrium irio*, *Solanum nigrum*, *Solanum luteum*, *Sonchus oleraceus*, *Sonchus tenerrimus*, and *Taraxacum vulgare* (Córdoba *et al.*, 2004; Hasiów-Jaroszewska *et al.*, 2010; Jordá *et al.*, 2001) plants. These alternative hosts may act as PepMV reservoirs and/or virus sources. PepMV can also infect wild *Solanum* species, such as *S. chilense*, *S. chmielewskii*, *S. hirsutum*, *S. parviflorum*, *S. peruvianum*, and *S. pimpinellifolium* (Davino *et al.*, 2008; Soler *et al.*, 2002, 2005). Interestingly, recent data suggest that *Nicotiana glutinosa* and *N. tabacum* plants can be infected by PepMV but only after simultaneous inoculation with isolates Sp13 (EU type) and PS5 (CH2 type) (Gómez *et al.*, 2009b). Therefore, mixed infections can influence PepMV host range. More detailed studies are required to determine the role of alternative hosts in PepMV epidemics.

D. Long-range and short-range dispersal: Control measures

Long-range PepMV dispersal probably occurs through contaminated seeds (Córdoba-Sellés *et al.*, 2007), although tests with plants artificially infected with the EU and CH2 genotypes suggest that transmission occurs at a low efficiency (0.005–2%) (Hanssen *et al.*, 2010b). The seed-borne virus is localized in the seed coat not in the embryo (Ling, 2008). It has been suggested that isolates of the EU type might have an advantage over the CH2 type in seed transmission, which may explain the earlier establishment of the former isolates in European countries (Hanssen *et al.*, 2010c). This transmission mode gives PepMV a high potential for long-distance dissemination, being a principal source of primary infections, either through international seed trade or by seedlings obtained from a contaminated tomato seed source (Ling, 2010). Long-range dispersal is also likely to occur via tomato fruits, since PepMV has been detected in tomato export regions and infected lots have been intercepted by customs officials (Anon., 2002, 2003). The crossborder spread of PepMV is a matter of grave concern because the primary transfer route to noninfected production areas is still uncertain, requiring further research.

The mechanical transmission of PepMV is very efficient (Jones *et al.*, 1980). The virus therefore spreads rapidly and infects most plants if it is introduced into a greenhouse used for tomato cultivation. Other factors that promote short-range dispersal include the mild symptoms that occur in vegetative tissues, therefore delaying the identification of infected plants, and the frequent hands-on activities needed for protected crops (e.g., planting, pruning, tying, grafting, deleafing, spraying, and fruit picking). Growers must follow exhaustive hygiene measures to avoid transmission, especially because PepMV can spread without symptoms. PepMV can also be transmitted with an efficiency of 8% by the fungus *Olpidium virulentus* (Alfaro-Fernández *et al.*, 2010b). The high density of zoospores from this fungus in drainage water may increase PepMV transmission by irrigation or the recirculation of contaminated nutrient solution in a closed hydroponic system (Schwarz *et al.*, 2010). At least two studies have shown that PepMV can be vectored by bumble bees (Lacasa *et al.*, 2003; Shipp *et al.*, 2008). These studies indicated an enhanced risk of PepMV infection during pollination, as the virus may be transmitted between plants on insect body parts or on pollen grains.

The success of PepMV may also reflect its stability and persistence, which are similar to those of *Tobacco mosaic virus* (Córdoba-Sellés *et al.*, 2007). Elimination of the virus from contaminated tools, plant material, and seeds is therefore essential for PepMV control. Both chemical and heat treatments have been shown to reduce PepMV transmission rates from seed to seedling, for example, dry heating tomato seeds at 72–80 °C for 48–72 h, or disinfecting them with 0.5–1.0% sodium hypochlorite or 10% trisodium phosphate will eradicate the virus without affecting seed germination (Córdoba-Sellés *et al.*, 2007; Ling, 2010). A break in the tomato-growing season during which the soil and greenhouse are exposed to solarization can also help to control the virus. Crossprotection can also reduce virus accumulation and limit the severity of aggressive PepMV isolates (Hanssen *et al.*, 2010a; Schenk *et al.*, 2010).

There are no commercial tomato cultivars with genetic resistance to PepMV, so this control strategy is currently unavailable (Gómez *et al.*, 2009a). Tomato germplasm collections have been screened to identify sources of PepMV resistance, but most accessions were found to be susceptible (Ling and Scott, 2007; Soler-Aleixandre *et al.*, 2007). Moderate resistance to PepMV-EU was found in some accessions of the wild species *S. chilense* and *S. peruvianum* (Soler-Aleixandre *et al.*, 2007) but some sources may at least show strain-specific PepMV resistance (Ling and Scott, 2007). The accession that appears most promising in this respect is LA1731 from *S. habrochaites*, which is resistant to PepMV-EU, US1, and CH2 isolates (Ling and Scott, 2007). These forms of resistance appear to be polygenic in nature and many technical challenges would need to be overcome before these traits could be introduced into commercial cultivars.

E. Genetic variability and evolution

The sequencing of different PepMV isolates has identified four main genotypes: the original Peruvian genotype (LP), the EU genotype that infects European tomato crops, a North American genotype (US1), and a Chilean genotype (CH2) (Hanssen and Thomma, 2010). Phylogenetic analysis has revealed two main clusters, one containing the EU and LP genotypes and the other containing the more recently described US and CH2 genotypes, suggesting they have arisen by divergent evolution (Fig. 4). The LP genotype is represented by the original Peruvian isolate (LP2001), which shares 95% overall nucleotide sequence identity with EU isolates (Ling *et al.*, 2008; Lopez *et al.*, 2005; Pagan *et al.*, 2006). However, it is considered distinct from EU isolates because it does not cause any symptoms in tomato (Lopez *et al.*, 2005). The EU genotype encompasses several European tomato isolates that share >98% nucleotide sequence identity (Aguilar *et al.*, 2002; Cotillon *et al.*, 2002; Mumford and Metcalfe, 2001; Verhoeven *et al.*, 2003). The US1 genotype (and its relative, US2) corresponds to two unique isolates from tomato characterized in the United States, which share ~80% nucleotide sequence identity to each other and to the EU genotypes (Maroon-Lango *et al.*, 2005). The CH2

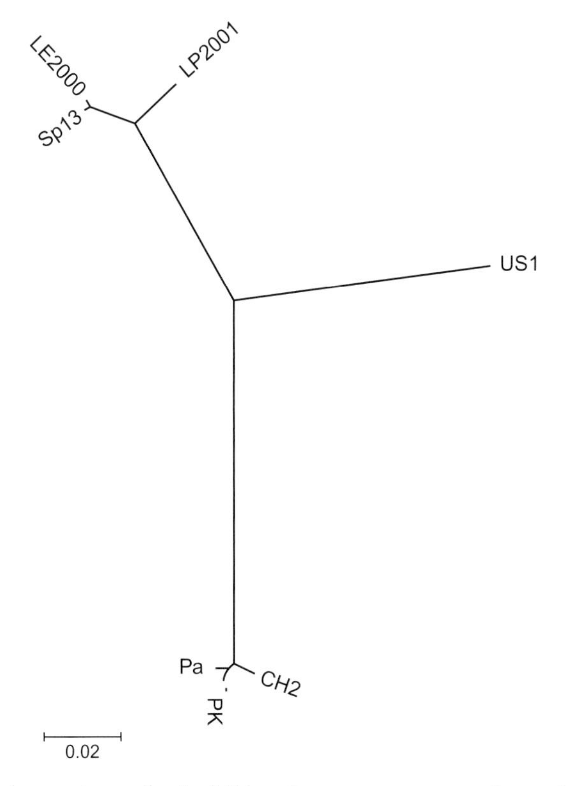

FIGURE 4 Phylogenetic tree for the full-length genome sequence of seven PepMV isolates. The tree was built using the neighbor-joining method with 1000 bootstrap replicates in MEGA 4. LE2000 and Sp13 are Spanish isolates (accession numbers AJ600359 and NC004067, respectively), LP2000 is a Peruvian isolate (accession number AJ606361), US1 is an american isolate (accession number AY509926), CH2 is a Chilean isolate (accession number DQ000985) and Pa and PK are Polish isolates (accession numbers FJ612601 and EF408821, respectively)

genotype was isolated from infected tomato seeds in Chile and is more closely related to the US2 genotype (91% identity) than US1 or EU (about 80% in each case) (Ling, 2007). Interestingly, sequence variation between CH2 and US2 was not evenly distributed throughout the genome, and *in silico* analysis suggested that US2 arose through the recombination of ancestors similar to US1 and CH2. Similarities in the biological and molecular properties of CH2 and the US isolates also suggest that both types may share a common origin. More recently, Polish isolates have been identified with an unusual phenotype, that is, the induction of severe necrosis on tomato plants and local necrotic lesions on thorn-apple (*D. inoxia*) (Hasiów-Jaroszewska *et al.*, 2010). The necrotic Polish isolates (PepMV-PK and -Pa) appear to cluster with the CH2 genotype (Fig. 4), although necrotic isolates have also been identified in the EU group (Schenk *et al.*, 2010).

During early PepMV epidemics in Europe, PepMV populations were shown to be genetically highly homogeneous in Spain (Pagan *et al.*, 2006), England (Mumford and Metcalfe, 2001), and France (Cotillon *et al.*, 2002). Isolates obtained at the time were predominantly the EU genotype, suggesting a common origin (Verhoeven *et al.*, 2003). Since 2004, the situation has changed and other strains have been cited as emerging and spreading in European tomato production greenhouses. For example, PepMV-CH2 isolates have spread and have become dominant in Spain (Alfaro-Fernández *et al.*, 2009b; Gómez *et al.*, 2009b), Belgium (Hanssen *et al.*, 2008), and Poland (Hasiów-Jaroszewska *et al.*, 2010), as well as in North America (Ling *et al.*, 2008). The PepMV-CH2 genotype is often found together with PepMV-EU in mixed infections (Gómez *et al.*, 2009b).

The genetic variability and evolutionary dynamics of Spanish PepMV populations have been studied with detail (Gómez *et al.*, 2011; Pagan *et al.*, 2006). Spanish PepMV populations sampled between 2000 and 2004 were found to be genetically homogeneous, spatially and temporally unstructured, and mostly representative of the EU genotype (Pagan *et al.*, 2006). Later, hybridization analysis showed that PepMV populations sampled in the Murcia region of south-eastern Spain between 2005 and 2008 comprised a mixture of cocirculating genotypes, PepMV-CH2 and PepMV-EU. Although the CH2 isolates predominated, the EU isolates were not displaced but persisted in mixed infections (Fig. 5). A similar situation appears to have arisen in Belgian greenhouse tomato crops (Hanssen *et al.*, 2008). Combining a population genetics approach with the *in planta* analysis of virus strains, Gómez *et al.* (2009b) showed that PepMV virulence and host range contributed to the observed epidemiological patterns. Phylogenetic analysis of the TGB and CP coding regions from 50 isolates confirmed the existence of two well-defined clusters in Murcia, both of them under purifying selection. Inoculations of two cloned isolates from each cluster showed that mixed infections could extend the PepMV host range. In tomato, the CH2 isolate had a higher fitness than the EU isolate.

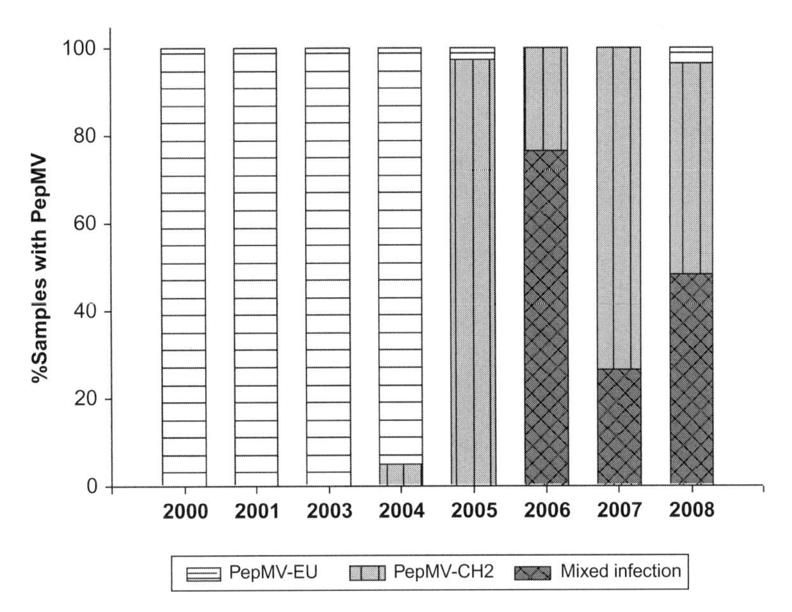

FIGURE 5 Detection of PepMV in samples from symptomatic tomato plants. Detection of PepMV-EU and/or PepMV-CH2 in infected samples of commercial tomato crops in the Murcia region over a period of 8 years. Data from 2000 to 2004 are from Pagan *et al.* (2006).

Mixed infections in tomato did not appear to affect the accumulation of the EU isolate, but the accumulation of the CH2 isolate was reduced, alleviating the symptoms normally induced by the CH2 isolate in single infections.

The rate of PepMV molecular evolution, estimated by sequencing isolates collected in south-eastern Spain, was found to be 2.527×10^{-3} substitutions/site/year (Gómez *et al.*, 2012). This value lies within the range reported for ssRNA viruses, but is an order of magnitude higher than the rates reported recently for other plant RNA viruses, suggesting that PepMV evolves faster than other viruses, possibly due to specific epidemiological factors (Duffy and Holmes, 2008; Fargette *et al.*, 2008; Gibbs *et al.*, 2008; Simmons *et al.*, 2008). This may reflect greater opportunities for horizontal transmission in host populations, which increase contact rates and thus the effective virus population size, resulting in a higher rate of substitutions (Duffy *et al.*, 2008). The coalescence process for the effective number of PepMV infections ($N_e\tau$), illustrated by the Bayesian skyline plot (Fig. 6), suggested that $N_e\tau$ remained constant for a lengthy period and after a small dip (2003–2005) there was a dramatic epidemic expansion, coincident with the onset of PepMV-CH2 infections in south-eastern Spain. This clearly shows that genetic and ecological

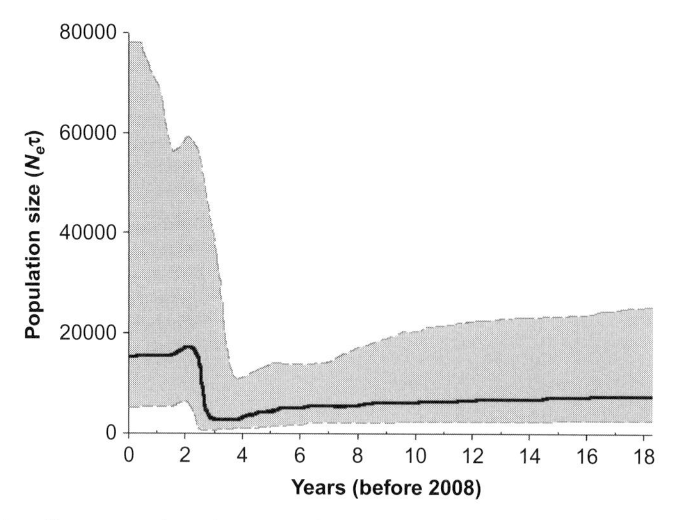

FIGURE 6 Effective number of PepMV infections. Bayesian skyline plot ($m=10$) showing the evolution of the effective number of PepMV infections. $N_e\tau$ represents the effective number population size times the generation time (in years). The thick line is the median estimate, and the dashed gray lines show the 95% HPD limits. The plot shows a sharp increase in the effective number of infections around 2005 (Gómez *et al.*, 2012).

interactions among different viral strains can modulate the evolutionary dynamics of PepMV and determine its epidemiological profile.

III. TORRADOVIRUSES

A. Genome organization and species diversity

The agent responsible for Torrado disease was identified in 2007 as a picornavirus-like bipartite plant RNA virus and was named *Tomato torrado virus* (Verbeek *et al.*, 2007). The recent discovery of ToTV prompted the establishment of the new genus *Torradovirus* in the family *Secoviridae* (order *Picornavirales*) (Sanfancon *et al.*, 2009). ToTV shares many features with other *Secoviridae*, including a similar virion structure and genome organization. Additional torradoviruses with similar properties to the type member have been described and characterized more recently, such as *Tomato marchitez virus* (ToMarV), its isolate *Tomato apex necrosis virus* (ToANV), and *Tomato chocolate spot virus* (ToCSV). These viruses have been identified in tomato crops with distinct necrotic symptoms on leaves and fruits in Mexico and Guatemala, respectively (Batuman *et al.*, 2010; Turina *et al.*, 2007; Verbeek *et al.*, 2008).

Torradoviruses have small icosahedral particles (28–30 nm in diameter) and genomes comprising two single-stranded positive-sense RNA molecules flanked by UTRs (Fig. 1). Both UTRs vary in size among the different torradovirus species with the 5′ UTR ranging from 106 to 140 nt and the 3′ UTR ranging from 628 to 1210 nt (Verbeek *et al.*, 2008; Batuman *et al.*, 2010). RNA 1 is 7.5–7.8 kb in length and contains a single ORF with the characteristic functional domains of the protease cofactor (Pro-Co), helicase, protease, and RdRp at the C-terminus. RNA 2 is 4–5.5 kb in length and contains two overlapping ORFs. Protein putatively coded by ORF1 has no homology to other viral proteins nor any known functional domains (Verbeek *et al.*, 2007; Budziszewska *et al.*, 2008), whereas ORF2 encodes a polyprotein containing the domains for three CPs and a putative movement protein (Verbeek *et al.*, 2007). The CPs are named Vp35, Vp26, and Vp23 according to their predicted molecular weights. The putative torradovirus movement protein has no homology with *Comoviridae* movement proteins, although it does contain a conserved LxxPxL motif (Verbeek *et al.*, 2007).

ToTV, ToMarV, and ToCSV are now thought to represent different species in the genus *Torradovirus* based on the criteria proposed by Sanfancon *et al.* (2009). On average, they present less than 75% amino acid sequence identity in the CPs and less than 80% amino acid sequence identity in the proteinase-polymerase region. ToMarV and ToANV are highly similar, with 99% amino acid sequence identity for the proteinase-polymerase region and 89–98% for the CPs, indicating they are likely to be different isolates of the same species.

B. ToTV epidemics

Torrado disease was first reported in spring 2001 affecting protected tomato crops in the Murcia region of south-eastern Spain (Alfaro-Fernández *et al.*, 2006; Jordá *et al.*, 2003). Initially, the disease was restricted to a small geographical area, but outbreaks were later reported elsewhere in Europe (Italy, France, Spain, Hungary, and Poland), as well as Central America and Australia (Fig. 2 and Table I). ToTV has been eradicated only in Hungary and Poland, where all infected tomato plants were destroyed. In the rest of Europe, ToTV epidemics can be described as transient, and ToTV is cited in only a few records (Table I). Nevertheless, ToTV is included in the European Plant Protection Organization alert list because of the commercial importance of tomato and the uncertain economic impact of Torrado disease (EPPO, 2009).

Torrado disease symptoms in tomato plants begin with the appearance of chlorotic spots at the base of the leaflets, and these later develop into necrotic spots and/or conspicuous shot-holes, giving the leaves a burned appearance (Fig. 3B; Gómez *et al.*, 2010). Fruits can also suffer necrotic lesions and cracking, making them unmarketable. ToTV appears to have a more severe impact in commercial glasshouse tomato crops than under experimentally controlled conditions, and this has been proposed to be the consequence of mixed infections with PepMV (Alfaro-Fernández *et al.*, 2010c,d; Verbeek *et al.*, 2007). However, no evidence was found for any association between ToTV, PepMV, and *Tomato chlorosis virus* in tomato samples with Torrado disease symptoms of varying severity (Gómez *et al.*, 2010). These results suggest that environmental factors may play an important role in the reproducibility and manifestation of Torrado disease symptoms.

C. Host range and potential reservoirs

Tomato is a natural host for ToTV, but under experimental conditions the virus can also systemically infect aubergine (*S. melongena*) and pepper (*Capsisum annuum*), which are also economically valuable crops (Amari *et al.*, 2008). Several weeds and additional experimental hosts can also be infected by ToTV, including *Amaranthus* spp., *Atriplex* spp., *Chenopodium ambrosioides*, *Chenopodium* spp., *Halogetum sativus*, *Malva* spp., *N. benthamiana*, *N. clevelandii*, *N. glauca*, *N. glutinosa*, *N. occidentalis*, *N. rustica*, *N. tabacum* cv. Samsun, *N. tabacum* cv. Xanthi, *Physalis floridana*, *Polygonum* spp., *Senebiera didyma*, *S. nigrum*, and *Spergularia* spp. (Alfaro-Fernández *et al.*, 2008b; Amari *et al.*, 2008). As for PepMV, these potential alternative hosts might act as reservoirs and/or sources of ToTV infection during epidemics.

D. Long-range and short-range dispersal: Control measures

ToTV can be transmitted by whiteflies of the species *Bemisia tabaci* and *Trialeurodes vaporariorum* although the mode of transmission is currently unknown (Amari *et al.*, 2008; Pospieszny *et al.*, 2007). Notably, these vectors may disseminate ToTV over great distances and may therefore spread ToTV in tomato crops over vast geographical areas. The control and prevention of Torrado disease must therefore involve protecting tomato plants from whiteflies, using physical barriers and/or insecticides. The recent discovery of ToTV means there is a lack of large-scale epidemiological data and its prevalence and impact on tomato crops is therefore unknown. In terms of the evolution of ToTV epidemics, our unpublished data suggest that ToTV is becoming less prevalent in the Murcia region of Spain.

The control of ToTV using genetic resistance is an exciting possibility because a homozygous locus on tomato chromosome 4 appears to confer resistance to the virus (Maris *et al.*, 2008). Breeding programs are now in place to incorporate ToTV resistance into commercial tomato cultivars. Indeed, our own unpublished observations indicate that susceptibility to

ToTV among tomato accessions is an exception rather than a rule, and it is tempting to speculate that ToTV epidemics are, at least in part, a consequence of the widespread use of susceptible cultivars.

E. Genetic variability and evolution

Little is known about the genetic diversity of ToTV isolates and the evolutionary dynamics of virus populations. A comparative analysis of the complete genome sequences of 22 ToTV isolates (including isolates from Spain, Poland, and Hungary) revealed 97–100% identity for all genes. The most conserved region was RdRp, followed by Pro-Co, and the highest differences were found in the RNA 2 ORFs, mainly in Vp23. Phylogenetic analysis of the Pro-Co, RdRp, movement protein, Vp23, and Vp35 genes showed that all ToTV isolates cluster together in a single monophyletic group (Alfaro-Fernández *et al.*, 2010a).

IV. SOME CONCLUSIONS AND MANY HYPOTHESES

There is much more information available about the molecular biology, epidemiology, and evolutionary dynamics of PepMV compared to ToTV. Early records of PepMV suggest that it jumped recently to tomato from nontomato species such as pepino, probably in regions such as South America where adapted hosts such as pepino coexist with tomato, providing opportunities for transmission. The subsequent pandemic spread of PepMV was probably facilitated by the stability of the virus and therefore, its persistence in contaminated plant material, its transmission through seeds, and the global trade of tomato seeds and fruits. Once well-adapted tomato PepMV strains have entered a new region, successful epidemic infections are facilitated again by the stability of the virus and the seed transmission route, but also the efficient contact-based transmission and the inconspicuous symptoms, which prevent the prompt identification of infected plants. The genetic diversity of PepMV probably favored its emergence by promoting the jump from adapted to nonadapted hosts and has subsequently favored the evolution of strains adapted to tomato. As an example, the PepMV-CH2 genotype recently identified in south-eastern Spain has increased in prevalence since 2005 probably because it is fitter than the original PepMV-EU genotype. Indeed, this biological advantage could explain the increase in the effective number of PepMV infections observed from 2005, as well as its rapid spread and persistence from one season to the next. Interactions among PepMV strains add a further layer of complexity and also have a significant effect on PepMV evolutionary dynamics.

Although there is no doubt that the early PepMV outbreaks are now evolving into pandemics, the long-term outcome of ToTV outbreaks is less clear. Our perception is that the incidence of ToTV in Spain, where it was first described, has declined sharply over the past 2 years. If our preliminary observations were confirmed, it would be interesting to determine the factors responsible for the dramatic difference between ToTV and PepMV epidemiology. Although only speculation is possible for the time being, it is interesting to note that ToTV accumulates to much lower levels in infected plants (two orders of magnitude lower) than PepMV (Gómez *et al.*, 2010). Under these circumstances, perhaps the efficient dispersion of ToTV is only possible if there is a very high population density of its vectors, as has been proposed for the necrogenic satellite RNA of *Cucumber mosaic virus*, which affected tomatoes in the same region (Alonso-Prados *et al.*, 1998). In addition, our unpublished results suggest that susceptibility to ToTV is infrequent among tomato accessions. This may explain the limited impact of ToTV outbreaks and also provides the necessary tools to introduce disease control measures using sources of natural resistance. It is tempting to speculate that the emergence of ToTV on tomato reflects the increased use of susceptible tomato cultivars, which may result from the pressure on plant breeders and pathologists to introgress quality and productivity traits that perhaps are linked to a locus or loci associated with susceptibility to ToTV. Further studies are required to test this hypothesis.

There is a perception that the emergence and reemergence of viral diseases are occurring with greater frequency compared to previous decades. The cases of PepMV and ToTV are two recent examples but there are many others. In the context of global trade and climate change, there is an urgent need to establish global plant health systems for surveillance, diagnosis, integrated research, communication, and technology transfer to address these problems. Significant recent technological achievements (e.g., next-generation sequencing) can greatly facilitate this process by allowing the rapid, high-throughput analysis of viral isolates, and the tracing of emerging strains. Our perception as scientists is that the major limitation preventing the establishment of such a global plant health system is the lack of political will to create a stable, global funding base.

ACKNOWLEDGMENTS

P. G. was supported by a "Juan de la Cierva" postdoctoral contract from Ministerio de Ciencia e Innovación (Spain). Financial support was provided by grants PET2008_0239 and EUI2009-04009, both from Ministerio de Ciencia e Innovación (Spain). We thank Dr. Richard M. Twyman (richard@writescience.com) for editorial assistance.

REFERENCES

Adams, M. J., Antoniw, J. F., Bar-Joseph, M., Brunt, A. A., Candresse, T., Foster, G. D., Martelli, G. P., Milne, R. G., Zavriev, S. K., and Fauquet, C. M. (2004). The new plant virus family Flexiviridae and assessment of molecular criteria for species demarcation. *Arch. Virol.* **149:**1045–1060.

Aguilar, J. M., Hernandez-Gallarod, M. D., Cenis, J. L., Lacasa, A., and Aranda, M. A. (2002). Complete sequence of the Pepino mosaic virus RNA genome. *Arch. Virol.* **147:**2009–2015.

Alfaro-Fernández, A., Córdoba-Sellés, C., Cebrian, M. C., Font, I., Juarez, M., Medina, V., Lacasa, A., Sanchez-Navarro, J. A., Pallas, V., and Jordá, C. (2006). Necrosis del tomate: "Torrao" o cribado *Boletín de Sanidad Vegetal, Plagas* **32**(1)**:**545–562.

Alfaro-Fernández, A., Córdoba-Sellés, C., Cebrian, M. C., Espino, A., Martfn, R., and Jordá, C. (2007a). First report of tomato torrado virus in tomato in the Canary Islands, Spain. *Plant Dis.* **91:**1060.

Alfaro-Fernández, A., Córdoba-Sellés, C., Cebrian, M. C., Font, I., Juarez, M., Medina, V., Lacasa, A., Sanchez-Navarro, J. A., Pallás, V., and Jordá, C. (2007b). Avances en el estudio del "Torrao" o cribado del tomate *Boletín de Sanidad Vegetal, Plagas* **33**(1)**:**99–109.

Alfaro-Fernández, A., Cebrian, M. C., Córdoba-Sellés, C., Herrera-Vasquez, J. A., and Jordá, C. (2008a). First report of the US1 strain of Pepino mosaic virus in tomato in the Canary Islands, Spain. *Plant Dis.* **92:**1590.

Alfaro-Fernández, A., Córdoba-Sellés, C., Cebrian, M. C., Herrera-Vasquez, J. A., Sanchez-Navarro, J. A., Juarez, M., Espino, A., Martin, R., and Jordá, C. (2008b). First report of tomato torrado virus on weed hosts in Spain. *Plant Dis.* **92:**831.

Alfaro-Fernández, A., Bese, G., Córdoba-Sellés, C., Cebrian, M. C., Herrera-Vasquez, J. A., Forray, A., and Jordá, C. (2009a). First report of tomato torrado virus infecting tomato in Hungary. *Plant Dis.* **93:**554.

Alfaro-Fernández, A., Sánchez-Navarro, J. A., del Carmen Cebrián, M., del Carmen Córdoba-Sellés, M., Pallás, V., and Jordá, C. (2009b). Simultaneous detection and identification of Pepino mosaic virus (PepMV) isolates by multiplex one-step RT-PCR. *Eur. J. Plant Pathol.* **125:**143–158.

Alfaro-Fernández, A., Cebrian, M. C., Herrera-Vasquez, J. A., Córdoba-Sellés, M. C., Sanchez-Navarro, J. A., and Jordá, C. (2010a). Molecular variability of Spanish and Hungarian isolates of tomato torrado virus. *Plant Pathol.* **59:**785–793.

Alfaro-Fernández, A., Córdoba-Sellés, M. D., Herrera-Vasquez, J. A., Cebrian, M. D., and Jordá, C. (2010b). Transmission of Pepino mosaic virus by the fungal vector Olpidium virulentus. *J. Phytopathol.* **158:**217–226.

Alfaro-Fernández, A., Córdoba-Sellés, M. D., Juarez, M., Herrera-Vasquez, J. A., Sanchez-Navarro, J. A., Cebrian, M. D., Font, M. I., and Jordá, C. (2010c). Occurrence and geographical distribution of the "Torrado" disease in Spain *J. Phytopathol.* **158:**457–469.

Alfaro-Fernández, A., Medina, V., Córdoba-Sellés, M. C., Font, M. I., Jornet, J., Cebrian, M. C., and Jordá, C. (2010d). Ultrastructural aspects of tomato leaves infected by Tomato torrado virus (ToTV) and co-infected by other viruses. *Plant Pathol.* **59:**231–239.

Alonso-Prados, J. L., Aranda, M. A., Malpica, J. M., Garcia-Arenal, F., and Fraile, A. (1998). Satellite RNA of cucumber mosaic cucumovirus spreads epidemically in natural populations of its helper virus. *Phytopathology* **88:**520–524.

Amari, K., Gonzalez-Ibeas, D., Gómez, P., Sempere, R. N., Sanchez-Pina, M. A., Aranda, M. A., Diaz-Pendon, J. A., Navas-Castillo, J., Moriones, E., Blanca, J., Hernandez-Gallardo, M. D., and Anastasio, G. (2008). Tomato torrado virus is transmitted by Bemisia tabaci and infects pepper and eggplant in addition to tomato. *Plant Dis.* **92:**1139.

Anderson, P. K., Cunningham, A. A., Patel, N. G., Morales, F. J., Epstein, P. R., and Daszak, P. (2004). Emerging infectious diseases of plants: Pathogen pollution, climate change and agrotechnology drivers. *Trends Ecol. Evol.* **19:**535–544.

Anon. (2002). Update of the situation of Pepino mosaic potexvirus-Supplement. *EPPO Working Party on Phytosanitary Regulations*. Document 02/9495.
Anon. (2003). New pest situations-Pepino mosaic potexvirus. *EPPO Working Party on Phytosanitary Regulations*. Document 03/10351.
Batuman, O., Kuo, Y. W., Palmieri, M., Rojas, M. R., and Gilbertson, R. L. (2010). Tomato chocolate spot virus, a member of a new torradovirus species that causes a necrosis-associated disease of tomato in Guatemala. *Arch. Virol.* **155:**857–869.
Budziszewska, M., Obrepalska-Steplowska, A., Wieczorek, P., and Pospieszny, H. (2008). The nucleotide sequence of a polish isolate of Tomato torrado virus. *Virus Genes* **37:**400–406.
Canto, T., Aranda, M. A., and Fereres, A. (2009). Climate change effects on physiology and population processes of hosts and vectors that influence the spread of hemipteran-borne plant viruses. *Glob. Change Biol.* **15:**1884–1894.
Córdoba, M. C., Martinez-Priego, L., and Jordá, C. (2004). New natural hosts of Pepino mosaic virus in Spain. *Plant Dis.* **88:**906.
Córdoba-Sellés, M. C., Garcia-Rández, A., Alfaro-Fernández, A., and Jordá-Gutierrez, C. (2007). Seed transmission of Pepino mosaic virus and efficacy of tomato seed disinfection treatments. *Plant Dis.* **91:**1250–1254.
Cotillon, A. C., Girard, M., and Ducouret, S. (2002). Complete nucleotide sequence of the genomic RNA of a French isolate of Pepino mosaic virus (PepMV). *Arch. Virol.* **147:**2231–2238.
Davino, S., Bellardi, M., Agosteo, G., Iacono, G., and Davino, M. (2006). Characterization of a strain of Pepino mosaic virus found in Sicily. *Plant Pathol.* **88**(Suppl. 3)**:**S31–S63.
Davino, S., Davino, M., Bellardi, M. G., and Agosteo, G. E. (2008). Pepino mosaic virus and Tomato chlorosis virus causing mixed infection in protected tomato crops in Sicily. *Phytopathol. Mediterr.* **47:**35–41.
Davino, S., Bivona, L., Iacono, G., and Davino, M. (2010). First report of tomato torrado virus infecting tomato in Italy. *Plant Dis.* **94:**1172.
Domingo, E. (2010). Mechanisms of viral emergence. *Vet. Res.* **41:**38.
Duffy, S., and Holmes, E. C. (2008). Phylogenetic evidence for rapid rates of molecular evolution in the single-stranded DNA Begomovirus tomato yellow leaf curl virus. *J. Virol.* **82:**957–965.
Duffy, S., Shackelton, L., and Holmes, E. (2008). Rates of evolutionary change in viruses: Patterns and determinants. *Nat. Rev. Genet.* **9:**267–276.
Efthimiou, K. E., Gatsios, A. P., Aretakis, K. C., Papayiannis, L. C., and Katis, N. I. (2011). First report of Pepino mosaic virus infecting Greenhouse Cherry tomatoes in Greece. *Plant Dis.* **95:**78.
EPPO (2009). EPPO Alert List-Viruses. Pepino mosaic potexvirus—A new virus of tomato introduced into Europe. www.eppo.org/QUARANTINE/Alert_List/alert_list.htm.
Fakhro, A., Von Bargen, S., Bandte, M., and Buttner, C. (2010). Pepino mosaic virus, a first report of a virus infecting tomato in Syria. *Phytopathol. Mediterr.* **49:**99–101.
Fakhro, A., von Bargen, S., Bandte, M., Büttner, C., Franken, P., and Schwarz, D. (2011). Susceptibility of different plant species and tomato cultivars to two isolates of Pepino mosaic virus. *Eur. J. Plant Pathol.* **129:**579–590.
FAO (2009). *Food and Agriculture Organization*. http://faostat.fao.org/.
Fargette, D., Pinel, A., Rakotomalala, M., Sangu, E., Traore, O., Sereme, D., Sorho, F., Issaka, S., Hebrard, E., Sere, Y., Kanyeka, Z., and Konate, G. (2008). Rice yellow mottle virus, an RNA plant virus, evolves as rapidly as most RNA animal viruses. *J. Virol.* **82:**3584–3589.
Gambley, C. F., Thomas, J. E., Persley, D. M., and Hall, B. H. (2010). First report of tomato torrado virus on tomato from Australia. *Plant Dis.* **94:**486.
Gibbs, A. J., Ohshima, K., Phillips, M. J., and Gibbs, M. J. (2008). The prehistory of potyviruses: Their initial radiation was during the dawn of agriculture. *PLoS One* **3:**e2523.

Gómez, P., Rodríguez-Hernández, A. M., Moury, B., and Aranda, M. A. (2009a). Genetic resistance for the sustainable control of plant virus diseases: Breeding, mechanisms and durability. *Eur. J. Plant Pathol.* **125:**1–22.

Gómez, P., Sempere, R. N., Elena, S. F., and Aranda, M. A. (2009b). Mixed infections of Pepino mosaic virus strains modulate the evolutionary dynamics of this emergent virus. *J. Virol.* **83:**12378–12387.

Gómez, P., Sempere, R. N., Amari, K., Gómez-Aix, C., and Aranda, M. A. (2010). Epidemics of Tomato torrado virus, Pepino mosaic virus and Tomato chlorosis virus in tomato crops: Do mixed infections contribute to torrado disease epidemiology? *Ann. Appl. Biol.* **156:**401–410.

Gómez, P., Sempere, R. N., Aranda, M. A., and Elena S. F. (2012). Phylodynamics of Pepino Mosaic Virus in Spain (submitted).

Hanssen, I. M., and Thomma, B. (2010). Pepino mosaic virus: A successful pathogen that rapidly evolved from emerging to endemic in tomato crops. *Mol. Plant Pathol* **11:**179–189.

Hanssen, I. M., Paeleman, A., Wittemans, L., Goen, K., Lievens, B., Bragard, C., Vanachter, A., and Thomma, B. (2008). Genetic characterization of Pepino mosaic virus isolates from Belgian greenhouse tomatoes reveals genetic recombination. *Eur. J. Plant Pathol.* **121:**131–146.

Hanssen, I. M., Paeleman, A., Vandewoestijne, E., Bergen, L. V., Bragard, C., Lievens, B., Vanachter, A. C. R. C., and Thomma, B. P. H. J. (2009). Pepino mosaic virus isolates and differential symptomatology in tomato. *Plant Pathol.* **58:**450–460.

Hanssen, I. M., Gutierrez-Aguirre, I., Paeleman, A., Goen, K., Wittemans, L., Lievens, B., Vanachter, A., Ravnikar, M., and Thomma, B. (2010a). Cross-protection or enhanced symptom display in greenhouse tomato co-infected with different Pepino mosaic virus isolates. *Plant Pathol.* **59:**13–21.

Hanssen, I. M., Lapidot, M., and Thomma, B. P. H. J. (2010b). Emerging viral diseases of tomato crops. *Mol. Plant Microbe Interact.* **23:**539–548.

Hanssen, I. M., Mumford, R., Blystad, D.-R., Cortez, I., Hasiow-Jaroszewska, B., Hristova, D., Pagan, I., Pereira, A.-M., Peters, J., Pospieszny, H., Ravnikar, M., Stijger, I., *et al.* (2010c). Seed transmission of Pepino mosaic virus in tomato. *Eur. J. Plant Pathol.* **126:**145–152.

Hanssen, I. M., van Esse, H. P., Ballester, A. R., Hogewoning, S. W., Parra, N. O., Paeleman, A., Lievens, B., Bovy, A. G., and Thomma, B. P. H. J. (2011). Differential tomato transcriptomic responses induced by Pepino mosaic virus isolates with differential aggressiveness. *Plant Phys.* **156:**301–318.

Hasiow, B., Borodynko, N., and Pospieszny, H. (2008). Complete genomic RNA sequence of the Polish Pepino mosaic virus isolate belonging to the US2 strain. *Virus Genes* **36:**209–214.

Hasiow-Jaroszewska, B., and Borodynko, N. (2012). Characterization of the necrosis determinant of the European genotype of Pepino mosaic virus by site-specific mutagenesis of an infectious cDNA clone. *Arch. Virol.* **157:**337–341.

Hasiów-Jaroszewska, B., Borodynko, N., and Pospieszny, H. (2009). Infectious RNA transcripts derived from cloned cDNA of a pepino mosaic virus isolate. *Arch. Virol.* **154:**853–856.

Hasiów-Jaroszewska, B., Borodynko, N., Jackowiak, P., Figlerowicz, M., and Pospieszny, H. (2010). Pepino mosaic virus—A pathogen of tomato crops in Poland: Biology, evolution and diagnostics. *J. Plant Protect. Res.* **50:**470–476.

Hasiów-Jaroszewska, B., Borodynko, N., Jackowiak, P., Figlerowicz, M., and Pospieszny, H. (2011). Single mutation converts mild pathotype of the Pepino mosaic virus into necrotic one. *Virus Res.* **159:**57–61.

Herrera-Vasquez, J. A., Alfaro-Fernandez, A., Codoba-Sells, M. C., Cebrian, M. C., Font, M. I., and Jordá, C. (2009). First report of tomato torrado virus infecting tomato in single and mixed infections with cucumber mosaic virus in Panama. *Plant Dis.* **93:**198.

Jones, R. A. C., Koenig, R., and Lesemann, D. E. (1980). Pepino mosaic virus, a new potexvirus from pepino (*Solanum muricantum*). *Ann. Appl. Biol.* **94:**61–68.

Jordá, C., Lázaro, A., Font, I., Lacasa, A., Guerrero, M., and Cano, A. (2000). Nueva enfermeded en el tomate. *Phytoma* **119:**23–28.
Jordá, C., Lázaro, Pérez A., Martinez, P. V., and Lacasa, A. (2001). First report of Pepino mosaic virus on natural hosts. *Plant Dis.* **85:**1292.
Jordá, C., Martínez, L., Córdoba, M., Martínez, O., Juárez, M., Font, I., Lacasa, A., Guerrero, M., Cano, A., Monserrat, A., Barceló, N., and Alcázar, A. (2003). El ''cribado'' o ''torrao'', una nueva enfermedad del cultivo del tomate? *Phytoma* **152:**130–136.
Lacasa, A., Guerrero, M. M., Hita, I., Martinez, M. A., Jordá, C., Bielza, P., Contreras, J., Alcazar, A., and Cano, A. (2003). Implication of bumble bees (Bombus spp.) on Pepino mosaic virus (PepMV) spread on tomato crops. *Plagas* **29:**393–403.
Ling, K. S. (2007). Molecular characterization of two Pepino mosaic virus variants from imported tomato seed reveals high levels of sequence identity between Chilean and US isolates. *Virus Genes* **34:**1–8.
Ling, K. S. (2008). Pepino mosaic virus on tomato seed: Virus location and mechanical transmission. *Plant Dis.* **92:**1701–1705.
Ling, K. S. (2010). Effectiveness of chemo- and thermotherapeutic treatments on Pepino mosaic virus in tomato seed. *Plant Dis.* **94:**325–328.
Ling, K. S., and Scott, J. W. (2007). Sources of resistance to Pepino mosaic virus in tomato accessions. *Plant Dis.* **91:**749–753.
Ling, K.-S., Wintermantel, W. M., and Bledsoe, M. (2008). Genetic composition of Pepino mosaic virus population in North American Greenhouse tomatoes. *Plant Dis.* **92:**1683–1688.
Lopez, C., Soler, S., and Nuez, F. (2005). Comparison of the complete sequences of three different isolates of Pepino mosaic virus: Size variability of the TGBp3 protein between tomato and L. peruvianum isolates. *Arch. Virol.* **150:**619–627.
Lough, T. J., Emerson, S. J., Lucas, W. J., and Forster, R. L. S. (2001). Trans-complementation of long-distance movement of *White clover mosaic virus* triple gene block (TGB) mutants: Phloem-associated movement of TGBp1. *Virology* **288:**18–28.
Maris, P. C., De Haan, A. A., Barten, J. H. M., Van Den Heuvel, J. F., and Van Den Heuvel, J. F. J. M. (2008). Plants having tomato torrato virus resistance. Patent application US2009188007.
Maroon-Lango, C. J., Guaragna, M. A., Jordan, R. L., Hammond, J., Bandla, M., and Marquardt, S. K. (2005). Two unique US isolates of Pepino mosaic virus from a limited source of pooled tomato tissue are distinct from a third (European-like) US isolate. *Arch. Virol.* **150:**1187–1201.
Mathioudakis, M. M., Veiga, R., Ghita, M., Tsikou, D., Medina, V., Canto, T., Makris, A. M., and Livieratos, I. C. (2012). Pepino mosaic virus capsid protein interacts with a tomato heat shock protein cognate 70. *Virus Res.* **163:**28–39.
Morozov, S. Y., and Solovyev, A. G. (2003). Triple gene block: Modular design of a multifunctional machine for plant virus movement. *J. Gen. Virol.* **84:**1351–1366.
Mumford, R. A., and Metcalfe, E. J. (2001). The partial sequencing of the genomic RNA of a UK isolate of Pepino mosaic virus and the comparison of the coat protein sequence with other isolates from Europe and Peru. *Arch. Virol.* **146:**2455–2460.
Muñoz, M., Bustos, A., Cabrera, M., and López, L. (2002). Prospección del Pepino mosaic virus (PepMV) en cultivos de tomate. Abstract of a paper presented at the XII Congreso Nacional de Fitopatología, 10-01/04, Puerto Varas - X Región - Chile.
Pagan, I., Córdoba-Sellés, M., Martinez-Priego, L., Fraile, A., Malpica, J., Jordá, C., and Garcia-Arenal, F. (2006). Genetic structure of the population of Pepino mosaic virus infecting tomato crops in Spain. *Phytopathology* **96:**274–279.
Parrella, G., and Crescenzi, A. (2005). The present status of tomato viruses in Italy. Proceedings of the 1st International Symposium on Tomato Diseases, pp. 37–42.
Pospieszny, H., and Borodynko, N. (2006). New polish isolate of Pepino mosaic virus highly distinct from European tomato, Peruvian, and US2 strains. *Plant Dis.* **90:**1106.

Pospieszny, H., Borodynko, N., Obrepalska-Steplowska, A., and Hasiow, B. (2007). The first report of tomato torrado virus in Poland. *Plant Dis.* **91:**1364.

Pospieszny, H., Budziszewska, M., Hasiow-Jaroszewska, B., Obrepalska-Steplowska, A., and Borodynko, N. (2010). Biological and molecular characterization of Polish isolates of tomato torrado virus. *J. Phytopathol.* **158:**56–62.

Roggero, P., Masenga, V., Lenz, R., Coghe, F., Ena, S., and Winter, S. (2001). First report of Pepino mosaic virus in tomato in Italy. *Plant Pathol.* **50:**798.

Salomone, A., and Roggero, P. (2002). Host range, seed transmission and detection by ELISA and lateral flow of an Italian isolate of Pepino mosaic virus. *Plant Pathol.* **84:**65–68.

Sanfancon, H., Wellink, J., Le Gall, O., Karasev, A., van der Vlugt, R., and Wetzel, T. (2009). Secoviridae: A proposed family of plant viruses within the order Picornavirales that combines the families Sequiviridae and Comoviridae, the unassigned genera Cheravirus and Sadwavirus, and the proposed genus Torradovirus. *Arch. Virol.* **154:**899–907.

Sansford, C. E., and Jones, D. R. (2001). Pest Risk Analysis of Pepino Mosaic Virus. Central Science Laboratory, Sand Hutton, York, p. 15.

Schenk, M. F., Hamelink, R., van der Vlugt, R. A. A., Vermunt, A. M. W., Kaarsenmaker, R. C., and Stijger, I. (2010). The use of attenuated isolates of Pepino mosaic virus for cross-protection. *Eur. Plant Pathol.* **127:**249–261.

Schwarz, D., Beuch, U., Bandte, M., Fakhro, A., Buttner, C., and Obermeier, C. (2010). Spread and interaction of Pepino mosaic virus (PepMV) and Pythium aphanidermatum in a closed nutrient solution recirculation system: Effects on tomato growth and yield. *Plant Pathol.* **59:**443–452.

Sempere, R. N., Gómez, P., Truniger, V., and Aranda, M. A. (2011). Development of expression vectors based on pepino mosaic virus. *Plant Methods* **7:**6.

Shipp, J. L., Buitenhuis, R., Stobbs, L., Wang, K., Kim, W. S., and Ferguson, G. (2008). Vectoring of Pepino mosaic virus by bumble-bees in tomato greenhouses. *Ann. Appl. Biol.* **153:**149–155.

Simmons, H. E., Holmes, E. C., and Stephenson, A. G. (2008). Rapid evolutionary dynamics of zucchini yellow mosaic virus. *J. Gen. Virol.* **89:**1081–1085.

Soler, S., Prohens, J., Diez, M. J., and Nuez, F. (2002). Natural occurrence of Pepino mosaic virus in Lycopersicon species in central and southern Peru. *J. Phytopathol.* **150:**49–53.

Soler, S., Lopez, C., and Nuez, F. (2005). Natural occurrence of viruses in Lycopersicon spp in Ecuador. *Plant Dis.* **89:**1244.

Soler-Aleixandre, S., Lopez, C., Cebolia-Cornejo, J., and Nuez, F. (2007). Sources of resistance to Pepino mosaic virus (PepMV) in tomato. *HortScience* **42:**40–45.

Spence, N. J., Basham, J., Mumford, R. A., Hayman, G., Edmondson, R., and Jones, D. R. (2006). Effect of Pepino mosaic virus on the yield and quality of glasshouse-grown tomatoes in the UK. *Plant Pathol.* **55:**595–606.

Stäubli, A. (2005). Faits marquants dans la recherche 2004. Nouvelle virose sur tomate. *Revue suisse Viticult. Arboricult. Horticult.* **37**(3)**:**150.

Turina, M., Ricker, M. D., Lenzi, R., Masenga, V., and Ciuffo, M. (2007). A severe disease of tomato in the Culiacan area (Sinaloa, Mexico) is caused by a new picorna-like viral species. *Plant Dis.* **91:**932–941.

van der Vlugt, R. A., Stijger, C. M., Verhoeven, J. T. J., and Lesemann, D. E. (2000). First report of Pepino mosaic virus on tomato. *Plant Dis.* **84:**103.

Verbeek, M., Dullemans, A. M., van den Heuvel, J., Maris, P. C., and van der Vlugt, R. A. A. (2007). Identification and characterisation of tomato torrado virus, a new plant picorna-like virus from tomato. *Arch. Virol.* **152:**881–890.

Verbeek, M., Dullemans, A. M., van den Heuvel, J., Maris, P. C., and van der Vlugt, R. A. A. (2008). Tomato marchitez virus, a new plant picorna-like virus from tomato related to tomato torrado virus. *Arch. Virol.* **153:**127–134.

Verchot, J., Angell, S. M., and Baulcombe, D. C. (1998). In vivo translation of the triple gene block of *Potato Virus X* requires two subgenomic mRNAs. *J. Virol.* **72:**8316–8320.
Verchot-Lubicz, J. (2005). A new cell-to-cell transport model for *Potexviruses. Mol. Plant-Microbe Int.* **18:**283–290.
Verchot-Lubicz, J., Ye, C. M., and Bamunusinghe, D. (2007). Molecular biology of potexviruses: Recent advances. *J. Gen. Virol.* **88:**1643–1655.
Verdin, E., Gognalons, P., Wipf-Scheibel, C., Bornard, I., Ridray, G., Schoen, L., and Lecoq, H. (2009). First report of tomato torrado virus in tomato crops in France. *Plant Dis.* **93:**1352.
Verhoeven, J. T. J., van der Vlugt, R. A. A., and Roenhorst, J. W. (2003). High similarity between isolates of pepino mosaic virus suggests a common origin. *Eur. J. Plant Pathol.* **109:**419–425.
Vurro, M., Bonciani, B., and Vannacci, G. (2010). Emerging infectious diseases of crop plants in developing countries: Impact on agriculture and socio-economic consequences. *Food Security* **2:**113–132.
Woolhouse, M. E. (2002). Population biology of emerging and re-emerging pathogens. *Trends Microbiol.* **10:**S3–S7.
Woolhouse, M. E., Haydon, D. T., and Antia, R. (2005). Emerging pathogens: The epidemiology and evolution of species jumps. *Trends Ecol. Evol.* **20:**238–244.
Wright, D., and Mumford R. (1999). Pepino mosaic virus (PepMV). First records in tomato in the United Kingdom. *Plant Dis. Notice,* **89** (unpublished).
Zhang, Y., Shen, Z. J., Zhong, J., Lu, X. L., Cheng, G., and Li, R. D. (2003). Preliminary characterization of Pepino mosaic virus Shanghai isolate (PepMV-Sh) and its detection with ELISA. *Acta Agr Shanghai* **19:**90–92.

CHAPTER 15

Control Methods of Virus Diseases in the Mediterranean Basin

Yehezkel Antignus

Contents

Abstract

Viral pathogens form an important group of obligatory parasites of plants. About 977 plant viruses have been described and classified in 14 families and 70 genera. This group of pathogens has complex interactions with their host plants and vectors due to their integration in the molecular mechanisms of living cells, interfering with our ability to manage the malfunctions of virus infected plants by curing means. These constraints led to the perception that the best protection from virus diseases is by prevention.

Department of Phytopathology, Virology Unit, ARO, The Volcani Center, Bet Dagan, Israel

Advances in Virus Research, Volume 84
ISSN 0065-3527, DOI: 10.1016/B978-0-12-394314-9.00015-4

Many cultural procedures used for virus control are aimed at eradicating or altering one or more of the primary participants in the transmission process (vector, virus source plants, and the crop) or preventing their coming together. Part of these control measures were devised to reduce to a minimum, the number of inoculative vector individuals that are active in the crop or interfere with the transmission process at any of its phases, thereby arresting virus spread. Advances in plant virology and a better understanding of plant vector interactions provide strategies based on the formation of mechanical and optical barriers that interfere with the ability of the viral pathogen or its vector to reach the plant and initiate an epidemic.

I. INTRODUCTION

The losses caused by plant pathogens to many crops throughout the world and the need to develop a sustainable agriculture have led to the concept of Integrated Pest Management (IPM). IPM is a management system that emphasizes using information to make pest management decisions and integrating those decisions into ecologically and economically sound system. The components of the IPM system include biological, environmental, and economic monitoring, such as predictive models and economic thresholds, in addition to a wide selection of genetic, biological, cultural, and when necessary chemical control measures (Dent, 1994). The present review is focused on the description of cultural strategies to impede virus spread: (a) Formation of ''physical barriers'' designed to mechanically block the contact between the plant and the viral inoculum and (b) formation of ''optical barriers'' that interfere with the vision mechanism of insect vectors, in a way that affects their flight orientation and landing behavior.

II. PHYTOSANITATION: A GENERAL TOOL TO COMBAT VIRUS DISEASES

The polyphagous nature of insect vectors (Greathead, 1986) increases the risk of virus spread from infected cultivated plants and weeds that serve as natural virus reservoirs. Eradication of potential host plants is relevant in isolated areas, especially in arid regions, where natural vegetation is poor. The phytosanitation approach was successfully implemented by imposing a crop-free period of several weeks, thus decreasing the vector (whiteflies, aphids) populations as well as the inoculum sources (Hilje *et al.*, 2001; Ucko *et al.*, 1998).

III. MANAGEMENT OF SOIL-BORNE VIRUSES

Most soil-borne virus diseases are spread by nematodes or fungi (Hull, 2002). This group of viruses may be dispersed efficiently both in soil-grown crops and in hydroponics or other soil-less systems. Another group of soil-borne viruses are nonvectored viruses like tobamoviruses that are mechanically transmitted without the assistance of a known vector. This group of viruses is characterized by high stability and long persistence in soil and drainage water (Broadbent, 1965; Lewandowski, 2000; Mandahar, 1990). Presumably, the transmission process of these viruses is associated with wounding and abrasion inflicted on root system of transplanted plants and systemic movement of virus particles through the phloem system (Simon-Buela and Garcia-Arenal, 1999). This hypothesis is supported by the finding that undisturbed seedlings grown in virus-infested soils without being transplanted shows low, erratic, or no infection (Antignus *et al.*, 2005a; Broadbent, 1965).

A. Management by planting into "intermediating media"

Transfer of cucumber plants from speedling-type polystyrene trays or polystyrene trays with a vacuum-forming insert, into pots filled with perlite infested with *Cucumber fruit mottle mosaic virus*, resulted in high disease incidence. However, when the plants were removed from the trays into condensed peat cells or into perlite sleeves, prior to their introduction into the infested medium, infection with the virus was completely prevented. The results support the hypothesis that tobamovirus invasion into the root cells is through wounds inflicted through the disturbance of the root system during transplanting. Unwounded, newly formed roots that penetrated into the infested medium were not invaded by the virus. The use of condensed peat trays as well as perlite sleeves can serve as an efficient means to block soil-borne epidemics of tobamoviruses in infested soils (Antignus *et al.*, 2005a). However, this practice should be combined with sanitation of the greenhouse space to eliminate virus transmission via the plant foliage.

This procedure was proved highly efficient to protect pepper crops in commercial greenhouses from root infection by *Pepper mild mottle virus* (PMMoV). Pepper seedlings planted into shallow depressions filled with virus free tuff, perlite, or compost were protected from infection with the virus compared with those planted directly into infested greenhouse soil (Yehezkel Antignus, unpublished, Plate 1B).

B. Management by grafting

By using resistant rootstocks, it is possible to produce a plant that is protected from soil-borne pathogens. Much of the driving force behind the recent push for vegetable grafting worldwide is due to the phase out

PLATE 1 (A) Protection of greenhouse tomato crops against tomato yellow leaf curl virus (TYLCV) by UV-blocking polyethylene films. *Left*: Protected, un infected tomato plants grown in a commercial greenhouse covered with a UV-absorbing polyethylene film (IR-Veradim®, Ginegar Plastic, Israel). *Right*: Tomato plants grown in a commercial greenhouse covered with an ordinary polyethylene film, heavily infected by TYLCV. Both greenhouses were planted with the same variety at the same time and stood few meters apart. (B) The use of virus free intermediate medium (perlite) to protect pepper from the soil-borne pepper mild mottle tobamovirus (PMMV). The perlite is applied to depressions in soil with a diameter of approximately 15 cm. Seedlings are inserted into the perlite that protect the wounded root system from contact with the viral inoculum in soil. (C) Protection of pepper crops by colored shadow nets from aphid-borne nonpersistent viruses. *Left*: Pepper crop grown in a "walk-in" tunnel covered with pearl shading net efficiently protected against the aphid-borne cucumber mosaic virus. *Right*: Pepper crop grown under a conventional black shading net 100% infected with CMV. Both tunnels were planted with the same variety in the same time and stood 1 m apart.

of methyl bromide. A wide use of the grafting technology started in Japan where 60% of vegetables like cucumbers, watermelon, tomato, and eggplants are grafted (Lee, 1994; Oda, 1999). The use of the grafting technology may help to decrease the use of pesticides that negatively affect the public health and the environment (Besri, 2000; Oda, 1999).

Squash rootstocks were used successfully to protect melons from the watermelon strain of *Melon necrotic spot virus* (Antignus, unpublished). Pepper rootstocks carrying the L^4 allele of the gene for resistance against PMMoV successfully protected susceptible pepper scions from infection through virus-infested soil in a commercial greenhouse (Antignus, unpublished). However, when the grafted plants were subjected to foliage inoculation they collapsed due to the hypersensitive resistance response of the rootstock.

IV. MANAGEMENT OF AIR-BORNE VIRUSES

The major group of plant viruses are air-borne, transmitted in nature by members of the subphylum *Insecta* that feed on plants, for example, 192 aphid species are known as vectors of 275 plant viruses (Hull, 2002). Whiteflies and thrips are also known as major virus vectors that are responsible for the spread of virus diseases that may infect economically important crops. The simplest way to prevent contact between these viruliferous vectors and the crop is by forming a physical barrier to prevent mechanically the vector from reaching the crop. In general, any material that blocks insect invasion on the basis of its hole size can form an efficient barrier. Among the materials that are being used, are polypropylene (Agryl®), a lightweight material with a texture that allows a certain air flow and fine mesh nets (50 mesh) that were first implemented in Israeli greenhouses to protect tomato crops from *Tomato yellow leaf curl virus* (TYLCV; Berlinger *et al.*, 1991; Cohen and Berlinger, 1986). A different strategy is the use of barrier cropping that act by different mechanisms as an impediment to the spread of aphid-borne nonpersistent viruses (Fereres, 2000; Hooks and Fereres, 2006). Studies on vision behavior of virus insect vectors opened the way to the introduction of efficient tools to protect open field as well as greenhouse crops from the spread of virus pandemics.

A. Virus management in the open field

1. Virus management by mechanical barriers

Agryl covers are used in Israel to protect zucchini (*Cucurbita pepo*) and melon (*Cucumis melo*) crops from squash leaf curl geminivirus (SLCV) and cucurbit yellow stunt disorder virus (CYSDV) respectively, both vectored by *Bemisia tabaci*. A serious disadvantage of both Agryl and 50 mesh screens is the excess of heat that is generated due to the poor ventilation. In regions with hot climates greenhouses should be equipped with roof openings that allow the hot air to flow out of the structure.

Crops grown in the open field can be protected efficiently from virus infection by either floating row covers that lay over the plants without any support, or by the use of "low tunnels" where the protecting cover rests on wire arches. In both the cases, the protection is time limited, because the cover should be removed for cultural practices or bee pollination (Hilje *et al.*, 2001). However, delay in infection may mitigate effects of yield reduction (Antignus *et al.*, 2004; Cohen and Berlinger, 1986).

2. Management of nonpersistent viruses by oil sprays

Aphids are important vectors of many persistent and nonpersistent viruses worldwide. The peach aphid, *Myzus persicae* (Sulzer), and the potato aphid, *Macrosiphum euphorbiae* (Thomas), alone are able to transmit above 40 viruses to different crops (Blackman and Eastop, 1984).

The nonpersistent mode of virus transmission is characterized by a low vector virus specificity, a very short acquisition and inoculation access periods, and a brief retention time. This high-speed transmission process prevents an efficient control of virus spread by insecticide applications, which require a relative long time period to produce a killing effect (Loebenstein and Raccah, 1980). Spraying weekly with mineral oil has been shown to efficiently reduce nonpersistent virus transmission in both laboratory and field studies (Loebenstein and Raccah, 1980; Loebenstein *et al.*, 1964; Powell, 1992; Simons *et al.*, 1977; Wood, 1962).

The mechanism underlying the inhibition of virus transmission is not yet completely understood. It has been suggested that mineral oil acts by interfering with the binding or the release of virus particles from the insect mouthparts or by delaying stylet penetration (Loebenstein and Raccah, 1980; Powell, 1991; Simons *et al.*, 1977). Mineral oil film on treated plant leaves was suggested as a factor disrupting virus retention on the aphids' stylet (Wang and Pirone, 1996). It has also been suggested to affect stylet penetration behavior, possibly via interference with the sensory structures of the aphid labium, stylets, or epipharynx (Harris, 1977, 1983). A delay in the initiation of a first penetration on oil coated leaves has been also reported (Simons *et al.*, 1977; Wyman, 1971).

Oil sprays are known as enhancers of semiochemicals release, by plants which are assumed to play a major role in insect–plant interaction. Oil application may change the chemical profile of the plant odor by masking or adding a repellent effect, thus modifying host-selection behavior of the insect vectors (Cen *et al.*, 2002). In laboratory experiments with *M. euphorbiae*, it was shown that mineral oil treatment induced a lack of attractiveness of the host plant, which lasted at least 24 h, while the feeding behavior on treated plants was modified throughout 1 week posttreatment. Based on that, it was assumed that modification of host-selection behavior by mineral oil may be partly responsible for the inhibitory effect on virus transmission by aphids (Ameline *et al.*, 2009). A reduction in the occurrence of stylet punctures of epidermal cells membranes was shown by Powell (1991, 1992). Ameline *et al.* (2009, 2010) showed that *M. euphorbiae* increased its xylem sap consumption on treated plants. In additional experiments, these workers have demonstrated that nymph survival was significantly reduced on oil-treated plants, suggesting either a direct toxicity of the treated plant or an indirect effect.

Studies on the effect of various oil application methods on the spread of potato virus Y (PVY) and the phytotoxicity they might cause, have shown that control of spread, is largely depended on the concentration of the oil, and to a lesser extent, on its delivery rate (Boiteau and Singh, 1982). Tests did not show any significant effect of the spray pressure. There was no significant difference between eight commercial oil formulations where in all, control of PVY by oil sprays varied between 35% and 64% (Boiteau and Singh, 1982). Effectiveness of oil formulations can vary according to their own physical properties (Zitter and Ozaki, 1978). de Wijs (1980) observed that oil was more effective when applied at a viscosity 66–150 Sus. Similarly, Zitter and Simons (1980) found the range 60–120 Sus was more effective than lighter or heavier oils.

No significant foliar phytotoxicity or yield reduction resulted from applications of oil, except when a combination of high oil concentration (3% water emulsion) and high rate of application (2240 L/ha) was used. Fungicides mixed with oil or applied immediately after oil produced foliar phytotoxicity. Less phytotoxicity occurred when they were applied 24 h later than oil. Although the relatively high concentration of adjuvant in some of the formulations (17% vs. 3%) did not increase their effectiveness, the adjuvant alone (Triton B-1956) did decrease the spread of PVY (Boiteau and Singh, 1982). This result is contrary to the observations of Vanderveken (1968), who had found that surfactants used in commercial preparations had no inhibitory effect. Since the use of oil sprays is a preventative measure, application requirements are quite different from insecticides (Reagan *et al.*, 1979). Application methods must be chosen to provide maximum coverage of all leaf area exposed to aphids. The effectiveness of treatments serve as an indirect measure of this coverage.

The degree of PVY control and the persistence of oil deposits are better with the higher oil concentrations. Oil emulsions of 1.5% provided consistent reduction in PVY spread, whereas 1% sprays were not as reliable. Tests on the frequency of oil sprays were inconclusive since no difference could be shown between controls and treatments. The literature on the subject is controversial. For example, Shands (1977) noted that six weekly applications of oil gave better control of PVY than three but Bradley *et al.* (1966) found little difference between their effectiveness. The trend is for an enhanced control with an increase in the number of applications (e.g., 2). It was shown that oil/pyrethroid combined sprays in seed potato production, for example, SC811oil and cypermethrin, performed well and improved control efficiency in the presence of a high infection pressure (Bell, 1989). This synergistic effect may be associated with the knock-down effect of pyrethroids that amplifies the disruption of the virus transmission process. High temperatures which are common during late spring and during summer at the eastern and southern parts of the Mediterranean basin limit the use of concentrated oil emulsions due their increased phytotoxic effects under high temperatures. The ability of oil sprays to protect against nonpersistent viruses is limited in cases where the crop is under a pressure of large populations of viruliferous aphid vectors.

3. Virus management by barrier cropping

The use of barrier plants is a management tool based on secondary plants used within or bordering a primary crop for the purpose of disease control. Barrier cropping is a kind of "ecological engineering" that is designed to manipulate the habitat in a way that interferes with insect flight and landing behaviors, resulting in a reduced virus disease incidence.

The use of living plant barriers has been most successful with nonpersistent aphid transmitted viruses (Fereres, 2000; Hooks and Fereres, 2006). Several mechanisms were suggested to explain how barrier plants may affect the spread of nonpersistently aphid transmitted viruses: the virus sink hypothesis, the defense by forming a physical barrier, the host plant camouflaging or the vector trapping by attraction (Hooks and Fereres, 2006).

Some successful results were reported when barrier cropping was used to control whitefly transmitted viruses in tomato (Al-Musa, 1982) and cassava (Fargette and Fauquet, 1988).

Low-growing plants used as "living mulches," for example, coriander (*Coriandrum sativum*) have been shown to be effective in reducing the number of incoming whitefly adults, delaying virus dissemination and decreasing viral disease severity (Hilje *et al.*, 2001).

4. Virus management by "optical barriers"

The spectral sensitivities of insects to both the UV and visible ranges of the light spectrum have been extensively investigated (Bertholf, 1931, 1934; Coombe, 1981, 1982; Goldsmith, 1994; Lubbock 1882; Menzel, 1979; Mound, 1962; Scherer and Kolb, 1987; Vaishampayan *et al.*, 1975a,b). Light manipulation is therefore a powerful tool to interfere with the "vision behavior" of insect vectors in a way that decreases their ability to identify the target plants and to perform a successful virus transmission. The understanding of the spectral sensitivities of virus insect vectors is a basic requirement for the development of innovative virus control tactics.

i. Phototactic action spectrum for whiteflies Mound (1962) correlated the reaction to UV of *B. tabaci* Gennadius to the induction of migratory behavior, and showed that yellow wavelengths induced vegetative behavior which may be part of a natural host-selection mechanism. Coombe (1981) found that the greenhouse whitefly took off more readily and walked faster when exposed to 400 nm radiation than under 500 nm. He confirmed Mound's suggestion that the two types of radiation are complementary, with an apparent balance between migratory behavior induced by UV wavelengths and the landing reaction controlled by yellow wavelengths. These results were later supported by the finding of Antignus *et al.* (1996, 2001a,b, 2004) who showed that filtration of UV light in the range of 280–380 hindered the ability of whiteflies to disperse inside tunnels covered with UV-absorbing films.

ii. Phototactic action spectrum for aphids For flying aphids, it has been suggested that the primary function of color vision lies in distinguishing plants from the sky. During flight, aphids respond strongly to visual stimuli (Kring, 1972) and locate host plants by contrasting the soil background with the green color of plant foliage (Kennedy *et al.*, 1959, 1961). Eastop (1955) suggested that the aphid's sensitivity to color may be related to its host range for any given species. It was proposed that the spread of nonpersistent viruses starts first at the crop edges because viruliferous aphids entering a field tend to land on the margins due to the contrasting colors between the soil background and the plant canopy (Irwin *et al.*, 2000). Our recent studies (Chyzik *et al.*, 2003) have shown that the flight activity of alate aphids (*M. persicae* Sulzer) in a UV-deficient environment was dramatically reduced. Moreover, the propagation rate of aphids grown under filtered light lacking the range of wavelengths between 280 and 380 nm was 1.5–2 times lower than under normal solar irradiation. These findings may reflect wavelength effects on insect physiology, leading to reduced motoric activity (Chyzik *et al.*, 2003).

iii. Phototactic action spectrum for thrips All anthophilous thrips are attracted to colors that match flowers, that is, UV, white, blue, and yellow, but few are attracted to green, red, and black. Reflectance of UV-A wavelengths (320–400 nm) is an important determinant of whether thrips land on a host. If UV reflectance is very high, anthophilous thrips in contrast to grass-feeding thrips are repelled from the surface of attractive colors (Terry, 1997). However, based on electroretinograms of *Frankliniella occidentalis* Pergande exposed to flashes of light ranging from 365 to 620 nm, it was found that there were two peaks of efficiency, one at 365 nm and another in the green–yellow region around 540 nm. Based on these findings, it was suggested that flower thrips have two types of photoreceptors: one sensitive to UV wavelengths, and the other sensitive to green–yellow wavelengths. There is no physiological evidence for a third photopigment sensitive to blue wavelengths (Matteson and Terry, 1992).

5. Virus management by colored soil mulches

The use of soil mulches to protect tomato plants from infestation by whiteflies was reported by Avidov (1956) who used sawdust or white-wash spray to mulch the crop seed beds. Similar results were obtained by straw mulches that markedly reduced whitefly population and delayed the spread of *Cucumber vein yellowing virus* and TYLCV vectored by the whiteflies (Cohen, 1982). Cohen and Melamed-Madjar (1978) also tested yellow, aluminum, and blue polyethylene film, demonstrating the high efficiency of the yellow polyethylene in delaying infection of tomatoes by TYLCV. Reflective surface covering was also used to repel aphids as a tool to control virus dissemination (Kring, 1964). Aluminum foil and gray plastic mulches effectively protected peppers from infection with *Cucumber mosaic virus* (CMV) and PVY by repelling winged forms of *M. persicae* Sulzer. Infection rates in the mulched plots at harvest reached only 4.5–6% compared with 45% in the unmulched control (Loebenstein and Raccah, 1980; Loebenstein *et al.*, 1975). Similar protection effects against whiteflies, aphids, and their vectored viruses were reported later by others (Csizinszky *et al.*, 1995, 1997; Summers *et al.*, 2005; Suwwan *et al.*, 1988). Aphids repellency was also obtained by the use of transparent polyethylene mulches. Field observations indicate a significant reduction in the spread of the aphid-borne, nonpersistent *Maize dwarf mosaic virus* in sweet corn fields during spring, in the Jordan valley, Israel. The transparent polyethylene mulches were used to increase soil temperature for early yields and protection against virus spread was a bonus (Antignus, unpublished). Lower numbers of whiteflies and aphids as well as virus incidence were found in cantaloupe (*C. melo*) grown over transparent mulch in Mexico (Orozco-Santos *et al.*, 1995). The repellency effect, generated by soil mulches is time limited and can work efficiently only during a relatively short period before the plant foliage is covering the entire plastic surface. However, in many cases, infection delay results in reducing the damage level (Plate 2B and C). To overcome this, limit-reflective whitewashes were sprayed over the plant's canopy. Weekly sprays of potato crops with 15% whitewashes reduced the incidence of the persistently aphid-borne potato leaf roll virus up to 61% compared to the untreated control (Marco, 1986).

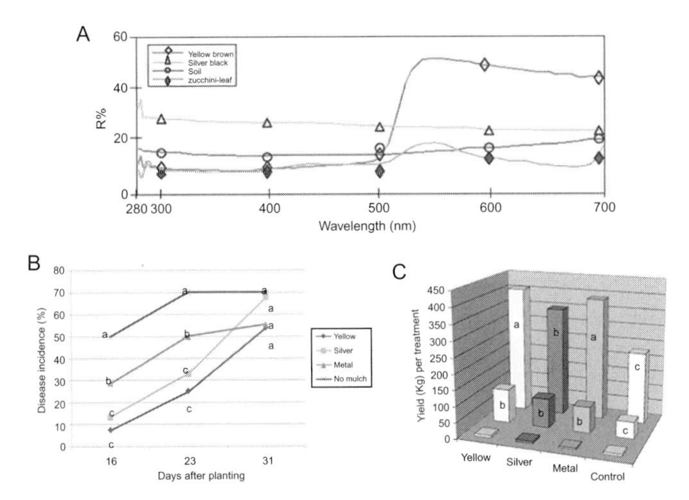

PLATE 2 (A) Spectrophotometric analysis of light reflection (R%) from colored polyethylene films, soil surface, and leaf surface (◇). A reflection peak at 540 nm (50 R%) measured from yellow mulching polyethylene film (Ginegar Plastic, Israel) (△). A constant reflection level at 30 R% at both the UV and visible light ranges measured from silver gray mulching polyethylene film (Ginegar Plastic, Israel) (○). A constant reflection level at ca. 18 R% at both the UV and visible light ranges measured from bare soil, (◆) A reflection peak at 540 nm (19 R%) measured from zucchini leaf. The contrast between the plant canopy and soil is highest when plants are grown on bare soil. Plant image is less visible for landing insects when plants are grown on yellow or silver gray mulches. (B) Delay in the spread of squash leaf curl virus (SLCV) infection in zucchini crops grown on different polyethylene mulches. A disease incidence of 70% was found 20 days after planting when plants grew on bare soil while 25%, 32%, and 50% disease incidence was recorded in zucchini plots grown on yellow, silver gray, and metal polyethylene mulching films respectively. (C) Protection effect of colored soil mulches on the yield of zucchini crops. SLCV disease delay by reflective mulches resulted in a double increase of high grade fruit yield compared to the unmulched control.

In the past, the protection effect of yellow soil mulches was explained by a combination of two factors: the yellow color that attracts the white-flies and the heat emission from the plastic surface that kills the insects (Cohen, 1982). A different explanation was proposed based on a set of experiments in which several colored polyethylene mulches were compared for their protection effect. Yellow, silver, and metal polyethylene mulches were tested for their ability to protect zucchini plants from the spread of *Squash leaf curl virus* (SLCV). Two weeks after planting, disease incidence was lowest (10–20%) in plants grown over yellow and silver mulches compared to 50% disease incidence in the unmulched plots (Plate 2B). The landing of whiteflies on plants grown over silver and yellow soil mulches was five- to sevenfold lower compared with the landing on plants that grew over bare soil. No landing of whiteflies was observed on any of the tested plastic mulches. Our experiments indicate that large yellow sheets failed to attract whiteflies in contrast to the high attraction efficiency of small size yellow traps (Antignus *et al.*, 2005b). The failure of large yellow surfaces to attract whiteflies was formerly reported by Cohen (1982) who demonstrated that sticky polyethylene sheets located vertically, 70 cm above ground failed to trap whiteflies. It is assumed that attraction to yellow plates occurs only when the size of the plate is small enough to allow the insect eye to perceive the yellow color on a background formed by the soil, providing a proper contrast between the yellow plate and the bare soil (Antignus *et al.*, 2005b).

To get a better insight into the protection mechanism by polyethylene soil mulches, a spectrophotometric analysis was conducted, comparing the spectrum and level of light reflection from yellow and silver polyethylene mulches as well as from the soil surface and the plant foliage. This examination has shown that the soil surface is reflecting light at a uniform low level along the entire light spectrum between 300 and 700 nm, while plant leaves have a distinct reflection peak at 540 nm (Plate 2A). Under these circumstances, the contrast between the soil background and the plant canopy is maximal, enabling insects to detect the crop for landing. In cases where the background of the plant is formed by yellow or silver mulches the amount of the reflected light in the visible range is considerably higher compared to the reflection of the soil and plant canopy (Plate 2A). The resulting poor contrast that is formed due to the plastics reflection is interfering with the ability of the insect to detect the plant image and to perceive a landing signal (Antignus *et al.*, 2004, 2005b).

B. Virus management in protected crops

1. Greenhouse protection by UV-absorbing films

Vegetable crops are efficiently protected from insect pests, and viral diseases transmitted by them, when grown in "walk-in" tunnels or greenhouses covered either with UV-absorbing polyethylene films or with UV-absorbing 50-mesh nets (Antignus, 2001; Antignus *et al.*, 1996, 1998). These covers act as filters that eliminate the majority of the UV portion of the light spectrum between 280–380 nm. This light filtration has been shown to significantly reduce the infestation of crops by a wide range of insect pests, including whiteflies, aphids, thrips, and leaf miners (Antignus *et al.*, 1996, 1998). Tomatoes and cucumbers grown under UV-absorbing greenhouse covers were highly protected against whitefly-borne viruses such as TYLCV (Plate 1A) and CYSDV (Antignus *et al.*, 1996; Kumar and Poehling, 2006; Monci *et al.*, 2002). These results suggest that the elimination of the UV portion of the light spectrum is interfering with the "UV vision" of insects, and as a consequence may affect their ability to orient themselves to the crop (Antignus, 2001; Antignus and Ben-Yakir, 2004; Antignus *et al.*, 2000, 2001a,b; Raviv and Antignus, 2004). Later on similar observations were made in different geographical zones of the world confirming our results (Costa and Robb, 1999; Costa *et al.*, 2002; Doukas, 2002; Doukas and Payne, 2007a,b; Kumar and Poehling, 2006; Monci, *et al.*, 2002; Mutwiwa *et al.*, 2005; Rapisarda and Tropea Garzia, 2002).

i. Parameters affecting the efficiency of protection The degree of the UV blocking by a plastic film is an important determinant of protection efficiency (Doukas, 2002). Polyvinyl chloride (PVC) films, which are highly efficient UV blockers, gave significantly better protection against insect pests than standard UV-absorbing polyethylene films (Antignus *et al.*, 1996).

ii. The effect on natural enemies Kajita (1986) found that parasitism of whiteflies by *Encarsia formosa* was the same under both standard and UV-blocking films. In choice experiments, significantly more (two to three times) specimens of *E. formosa* were trapped under a standard film than under UV-blocking films. It seems that the parasitoids, like their hosts, preferred an environment with high-UV radiations. However, when they have no choice, they can perform well in a UV-deficient environment (Doukas, 2002; Doukas and Payne, 2007b). The effect of UV-absorbing plastic sheets on the host location ability of three commercially available parasitoids *Aphidius colemani* Viereck (Hymenoptera: Braconidae), *Diglyphus isaea* Walker (Hymenoptera: Eulophidae) and *Eretmocerus mundus* Mercet (Hymenoptera: Aphelinidae)—was tested in the laboratory and in field trials. The parasitoids' preference between natural light and UV-filtered light was tested under laboratory conditions in a Y-shaped pipe system. In these experiments, around 90% of the tested insects of all three species were strongly attracted to non-UV-filtered light. In field trials parasitoid's ability to locate a host-infested plant from a distance (approximately 10 m) was tested. Host location by *A. colemani* and *D. isaea* expressed by parasitization rates and was not affected by greenhouse covering plastic type (regular vs. UV-absorbing plastic). *E. mundus*, on the other hand, was unable to locate the host-infested plant when the latter was placed in the center of the UV-absorbing plastic covered greenhouses. When the host-infested plants were located in the corners of the greenhouses and the wasps were released at the center, the parasitization rates were lower under the UV-absorbing plastic than under the regular plastic covered greenhouses. UV-absorbing plastic sheets and screens can be used concurrently with *D. isaea* and *A. colemani*, without interruption to their host location ability. *E. mundus* should probably be introduced in multiple release points, or as close as possible to the *B. tabaci* infested plants, in order to facilitate its host location process (Chiel, *et al.*, 2006).

iii. The effect on pollinators Bumblebees (*Bombus terrestris* Linnaeus) are important pollinators of angiosperms. The pollination of tomato plant flowers requires the agitation of flower anther cones to enable an efficient pollination process. Bumblebees are widely used in tomato greenhouses for this purpose (Kevan *et al.*, 1991).

Studies carried out under laboratory conditions have shown that bumblebees perceive when ultraviolet radiation is either removed or added to an illumination source, and are capable of using their visual system to forage efficiently in a UV-deficient environment; thus their forage efficiency is not affected by the type of greenhouse covering (Dyer and Chittka, 2004).

A delay in the hive start up of the bumblebee *B. terrestris* (Bio-Bee, Ltd., Israel) was observed in experimental mini greenhouses covered with UV-blocking films (Steinberg *et al.*, 1997). Later, this problem was solved by placing the hives near the greenhouse walls, where they were exposed to unfiltered light (Antignus, unpublished). In a field study, no significant differences were found in bumblebee activity or in the numbers of flowers visited, under standard or UV-blocking films (Antignus and Ben-Yakir, 2004). Studies in commercial tomato greenhouses have demonstrated that biomass and size of hives were not significantly affected, whether the greenhouses were covered with standard or UV-blocking films (Antignus and Ben-Yakir, 2004; Hefez *et al.*, 1999; Seker, 1999).

iv. The putative mechanism of action The migration of some insects in the environment is driven by the UV radiation (Coombe, 1981; Kevan *et al.*, 1991). Observations carried out in commercial greenhouses where roofs were covered alternately by regular and UV-absorbing films, indicated that insect and virus epidemics were confined to plants growing under roof arches covered with a nonabsorbing films. This unique phenomenon was designated as the "two compartments effect" (Antignus, unpublished). It is suggested that when insects have an option to select between a UV-deficient environment and an environment with a normal level of UV radiation, their choice is to move toward zones with normal levels of UV light. A greenhouse with a UV-absorbing roof forms a UV-deficient compartment while the space outside the greenhouse is a UV-rich compartment. Insects that approach the greenhouse wall from the external environment are exhibiting a positive UV-phototoxic behavior and are diverted away from the UV-deficient greenhouse walls at the moment that contact with UV irradiation is lost near the greenhouse walls. The protection effect of UV-absorbing films is also associated with the reduction of insects flight activity in a UV-deficient environment (Antignus and Ben-Yakir, 2004; Antignus *et al.*, 2001a; Chyzik *et al.*, 2003). Under these circumstances the efficiency of virus transmission is reduced.

C. Protection by colored shade nets

Photoselective shade netting is designed to selectively screen various light spectral components of the solar radiation and/or transform direct light into diffused light. These spectral manipulations are utilized to promote desired physiological responses in ornamental plants, vegetables, and fruit trees (Shahak *et al.*, 2008). Recently, it was shown that pepper crops grown under pearl and yellow ChromatiNets® were protected against the aphid-borne nonpersistent viruses, PVY and CMV (Plate 1C) (Antignus *et al.*, 2009; Shahak *et al.*, 2008). Similar results were reported by Cohen (1981) who used white coarse nets to get a highly efficient protection of pepper crops from aphids and their vectored viruses—CMV and PVY. Although the shade nets permit passage of insect pests, the infestation levels by aphids and whiteflies in "walk-in" tunnels covered with the yellow and pearl nets were consistently lower than in tunnels covered by the red and black nets. The incidence of CMV in pepper grown under the black net was 35–89%, while disease incidence under the colored nets was 2–10-fold lower. The incidence of necrotic PVY in tomatoes grown under the black net was 42%, while two- to threefold less infected plants were recorded under the yellow and pearl nets. TYLCV incidence under the red and black nets was 15–50%, while two- to four fold lower infection rate was found under the pearl and yellow nets. We suggest that the putative protection mechanism of the pearl and yellow nets is a result of a combination of insect repellence by light reflection and the take off respond of the insects, induced by the nonproductive probing attempts on the plastic nets.

D. Protection and monitoring by colored sticky traps

Mound (1962) suggested that *B. tabaci* is attracted by two groups of wavelengths of transmitted light, the blue/ultraviolet and yellow parts of the spectrum. He correlated the reaction to ultraviolet to the induction of migratory behavior, whereas yellow radiation induces landing behavior, which may be part of the host-selection mechanism. The attraction of whiteflies to yellow has been utilized as an important instrument in sampling and monitoring of whiteflies populations (e.g., Gerling and Horowitz, 1984). This vision cue of whiteflies as well as other insect pests was also utilized as a useful, often overlooked tool for management of pest populations in greenhouses (Van de Veire and Vacante, 1984). The so-called yellow-sticky cards are practically used for mass trapping of winged aphids, whiteflies, thrips, leafminers, fungus gnats, and shore flies. However, beneficial insects such as the whitefly parasitoid, *E. formosa*, can also be caught at times. Yellow sticky traps in various forms can catch large numbers of adult whiteflies aphids and thrips

(Natwick *et al.*, 2007). Large yellow-sticky boards or tapes are used in ''hot spots'' at a rate of about one per plant. Alternatively, reams of yellow, sticky tape can be draped between posts along plant rows.

REFERENCES

Al-Musa, A. (1982). Incidence, economic importance, and control of tomato yellow leaf curl in Jordan. *Plant Dis.* **66:**561–563.

Ameline, A., Couty, A., Martoub, M., and Giordanengo, P. (2009). Effect of mineral oil application on the orientation and feeding behavior of *Macrosiphum euphorbiae* (Homoptera: Aphidae). *Entomol. Exp. Appl.* **135:**77–84.

Ameline, A., Couty, A., Martoub, M., Giordanengo, S., and Giordanengo, P. (2010). Modification of *Macrosiphum euphorbiae* colonization behaviour and reproduction on potato plants treated by mineral oil. *Acta Entomologica Sinica* **52:**617–623.

Antignus, Y. (2001). Manipulation of wavelength-dependent behaviour of insects: An IPM tool to impede insects and restrict epidemics of insect-borne viruses. *Virus Res.* **71:**213–220.

Antignus, Y., and Ben-Yakir, D. (2004). Greenhouse photoselective cladding materials serve as an IPM tool to control the spread of insect pests and their vectored viruses. *In* ''Insect Pest Management'' (R. Horowitz and Y. Ishaya, eds.), pp. 319–333. Springer, Berlin.

Antignus, Y., Mor, N., Ben-Joseph, R., Lapidot, M., and Cohen, S. (1996). UV-absorbing plastic sheets protect crops from insect pests and from virus diseases vectored by insects. *Environ. Entomol.* **25:**919–924.

Antignus, Y., Lapidot, M., Hadar, D., Messika, Y., and Cohen, S. (1998). UV absorbing screens serve as optical barriers to protect vegetable crops from virus diseases and insect pests. *J. Econ. Entomol.* **91:**1401–1405.

Antignus, Y., Lapidot, M., and Cohen, S. (2000). Interference with UV vision of insects: An IPM tool to impede epidemics of insect pests and insect associated virus diseases. *In* ''Virus-Insect-Plant Interactions'' (K. F. Harris, J. E. Duffus, and O. P. Smith, eds.), pp. 331–347. Academic Press, San Diego.

Antignus, Y., Nestel, D., Cohen, S., and Lapidot, M. (2001a). Ultraviolet-deficient greenhouse environment affects whitefly attraction and flight behaviour. *Environ. Entomol.* **30:**394–399.

Antignus, Y., Lapidot, M., and Cohen, S. (2001b). Interference with Ultraviolet vision of insects to impede insect pests and insect-borne plant viruses. *In* ''Virus-Insect-Plant Interactions'' (K. S. Harris, ed.), pp. 331–350. Academic Press, New York.

Antignus, Y., Lachman, O., Pearlsman, M., Koren, A., Matan, E., Tregerman, M., Ucko, O., Messika, Y., Omer, S., and Unis, H. (2004). Development of an IPM system to reduce the damage of squash leaf curl begomovirus in zucchini squash crops. Abstract Compendium, 2nd European Whitefly Symposium, Cavtat, Croatia.

Antignus, Y., Lachman, O., and Pearlsman, M. (2005a). Light manipulation by soil mulches protect crops from the spread of Begomoviruses. Abst. of the IX Int. Pl. Virus Epidem. Symp., Lima, Peru.

Antignus, Y., Lachman, O., Pearlsman, M., and Koren, A. (2005b). Containment of cucumber fruit mottle mosaic *Tobamovirus* (CFMMV) infection through roots by planting into a virus-free intermediating medium. *Phytoparasitica* **33:**85–87.

Antignus, Y., Ben-Yakir, D., Offir, Y., Messika, Y., Dombrovsky, A., Chen, M., Ganot, L., Yehezkel, H., Ganz, S., and Shahak, Y. (2009). Colored shade nets form optical barrier protecting pepper and tomato crops against aphid-borne non-persistent viruses. *Sade Va'Yerek* **12:**60–62 (in Hebrew).

Avidov, Z. (1956). Bionomics of the tobacco whitefly (*Bemisia tabaci* Gennad.) in Israel. *Ktavim* **7**:25–41.

Bell, A. C. (1989). Use of oil and pyrethroid sprays to inhibit the spread of potato virus Y^n in the field. *Crop Protect.* **8**:37–39.

Berlinger, M. J., Mordechai, S., Lipper, A., Piper, A., Katz, Y., and Levav, Y. (1991). Protection of greenhouses against *Bemisia tabaci* the whitefly by plastic nets. *Hassadeh* **71**:1579–1583 (In Hebrew).

Bertholf, L. M. (1931). The distribution of stimulative efficiency in the ultraviolet spectrum for the honeybee. *J. Agric. Res.* **43**:703–713.

Bertholf, L. M. (1934). The extent of the spectrum for Drosophila and the distribution of stimulative efficiency in it. *Z. Vergl. Physiol.* **18**:32–64.

Besri, M. (2000). Tomatoes in Morocco: Integrated pest management and grafted plants. *In* Case Studies on Alternatives to Methyl Bromide. Technologies with low environment impact, UNEP, pp. 14–17, Devision of technology, industry and economics, Ozone action program.

Blackman, R. L., and Eastop, V. F. (1984). Aphids on The World's Crop: An Identification and Information Guide. John Wiley and Sons, New York, p. 466.

Boiteau, G., and Singh, R. P. (1982). Evaluation of mineral oil for reduction of virus Y spread in potatoes. *Am. Potato J.* **59**:253–262.

Bradley, R. H. E., Moore, C. A., and Pond, D. D. (1966). Spread of potato virus Y curtailed by oil. *Nature* **209**:1370–1371.

Broadbent, L. (1965). The epidemiology of tomato mosaic VIII Virus infection through tomato roots. *Ann. Appl. Biol.* **55**:57–66.

Cen, Y. J., Tian, M. Y., Pang, X. F., and Rae, D. J. (2002). Repellency, anti feeding effect and toxicity of an horticultural mineral oil against citrus red mite. *In* "Spray Oils—Beyond 2000 Sustainable Pest and Disease Management" (G. A. C. Beatie, ed.), pp. 134–141. University Of Western Sydney, Australia.

Chiel, E., Messika, Y., Steinberg, S., and Antignus, Y. (2006). The effect of UV-absorbing plastic sheet on the attraction and host location ability of three parasitoids: *Aphidius colemani, Diglyphus isaea* and *Eretmocerus mundus*. *Biocontrol* **51**:65–78.

Chyzik, R., Dobrinin, S., and Antignus, Y. (2003). The Effect of a UV-deficient environment on the biology and flight activity of *Myzus persicae* and its Hymenopterous parasite *Aphidius matricariae*. *Phytoparasitica* **31**:467–477.

Cohen, S. (1981). Reducing the spread of aphid-transmitted viruses in peppers by coarse-net cover. *Phytoparasitica* **9**:69–76.

Cohen, S. (1982). Control of whitefly vectors of viruses by color mulches. *In* "Pathogens, Vectors and Plant Diseases, Approaches to Control" (K. F. Harris and K. Maramorosch, eds.), pp. 45–56. Academic Press, New York.

Cohen, S., and Berlinger, M. J. (1986). Transmission and cultural control of whitefly borne viruses. *Agric. Ecosys. Environ.* **17**:89–97.

Cohen, S., and Melamed-Madjar, V. (1978). Prevention by soil mulching of spread of tomato yellow leaf curl virus transmitted by *Bemisia tabaci* (Gennadius) (Hemiptera: Aleyrodidae) in Israel. *Bull. Entomol. Res.* **68**:465–470.

Coombe, P. E. (1981). Wavelength behavior of the whitefly *Trialeurodes vaporariorum* (Homoptera: Aleyrodidae). *J. Comp. Physiol.* **144**:83–90.

Coombe, P. E. (1982). Visual behavior of the greenhouse whitefly, *Trialeurodes vaporariorum*. *Physiol. Entomol.* **7**:243–251.

Costa, H. S., and Robb, K. L. (1999). Effects of ultraviolet—Absorbing greenhouse plastic films on flight behavior of *Bemisia argentifolii* (Homoptera: Aleyrodidae) and *Frankliniella occidentalis* (Thysanoptera: Thripidae). *J. Econ. Entomol.* **92**:557–562.

Costa, H. S., Robb, K. L., and Wilen, C. A. (2002). Field trials measuring the effects of ultraviolet-absorbing greenhouse plastic films on insect populations. *J. Econ. Entomol.* **95**:113–120.

Csizinszky, A. A., Schuster, D. J., and Kring, J. B. (1995). Color mulches influence yield and insect pest populations in tomatoes. *J. Am. Soc. Hortic. Sci.* **120:**778–784.

Csizinszky, A. A., Schuster, D. J., and Kring, J. B. (1997). Evaluation of color mulches and oil sprays for yield and for the control of silverleaf whitefly, *Bemisia Argentifolii* (Bellows and Perring) on tomatoes. *Crop Prot.* **16:**475–481.

de Wijs, J. J. (1980). The characteristics of mineral oils in relation to their inhibitory activity on the aphid transmission of potato virus Y. *Neth. J. Plant Pathol.* **86:**291–300.

Dent, D. (1994). Principles of Integrated pest management. *In* "Integrated Pest Management" (D. Dent, ed.), pp. 8–46. Chapman & Hall, London.

Doukas, D. (2002). Impact of spectral cladding materials on the behaviour of glasshouse whitefly *Trialeurodes vaporariorum* and *Encarsia formosa*, its hymenopteran parasitoid. British Crop Protection Conference, Pests and Diseases 2002, Brighton, pp. 773–776.

Doukas, D., and Payne, C. (2007a). The use of Ultraviolet-blocking films in insect pest management in the UK; effects on naturally occuring arthropod pest and natural enemy populations in a protected cucumber crop. *Ann. Appl. Biol.* **151:**221–231.

Doukas, D., and Payne, C. (2007b). Greenhouse whitefly (Homoptera: Aleyrodidae) dispersal under different UV-light environments. *J. Econ. Entomol.* **100:**380–397.

Dyer, A. G., and Chittka, L. (2004). Bumblebee search time without ultraviolet light. *J. Exp. Biol.* **207:**1683–1688.

Eastop, V. F. (1955). Selection of aphid species by different kinds of traps. *Nature (Lond)* **176:**936.

Fargette, D., and Fauquet, C. (1988). A preliminary study on the influence of intercropping maize and cassava on the spread of African cassava mosaic virus by whiteflies. *Aspects Appl. Biol.* **17:**195–202.

Fereres, A. (2000). Barrier crops as a cultural measure of non-persistenly transmitted aphid-borne viruses. *Virus Res.* **71:**221–231.

Gerling, D., and Horowitz, A. R. (1984). Yellow traps for evaluating the population levels and dispersal patterns of *Bemisia tabaci* (Gennadius) (Homoptera: Aleyrodidae). *Ann. Entomol. Soc. Am.* **77:**753–759.

Goldsmith, T. H. (1994). Ultraviolet receptors and color vision: Evolutionary implications and dissonance of paradigms. *Virus Res.* **34:**1479–1487.

Greathead, A. H. (1986). Host plants. *In* "*Bemisia tabaci*—A Literature Survey" (M. J. W. Cock, ed.), pp. 17–26. CAB International Institute of Biological Control, Silwood Park, UK.

Harris, K. F. (1977). An ingestion-egestion hypothesis of non-circulative virus transmission. *In* "Aphids as Virus Vectors" (K. F. Harris and K. Maramorosch, eds.), pp. 165–220. Academic Press, New York.

Harris, K. F. (1983). Sternorrhynchous vectors of plant viruses: Virus-vector interactions and transmission mechanisms. *Adv. Virus Res.* **28:**113–140.

Hefez, A., Izikovitch, D., and Dag, A. (1999). Effects of UV-absorbing films on the activity of pollinators (honey bees and bumble bees). A report to the chief scientist. Israeli Ministry of Agriculture, Bet-Dagan ID code 891-0117-96.

Hilje, L., Costa, H. S., and Stansly, P. A. (2001). Cultural practices for managing *Bemisia tabaci* and associated viral diseases. *Crop Prot.* **20:**801–812.

Hooks, C. R. R., and Fereres, A. (2006). Protecting crops from non-persistently aphid-transmitted viruses: A review on the use of barrier plants as a management tool. *Virus Res.* **120:**1–16.

Hull, R. (2002). Matthews' Plant Virology. Academic Press, San Diego, CA, p. 1001.

Irwin, M. E., Ruesink, W. G., Isard, S. A., and Kampmeier, G. E. (2000). Mitigating epidemics caused by non-persistently transmitted aphid-borne viruses: The role of plant environment. *Virus Res.* **71:**185–211.

Kajita, H. (1986). Parasitism of the greenhouse whitefly, *Trialeurodes vaporariorum* (Westwood) (Homoptera: Aleurodidae) by *Encarsia formosa* (Hymenoptera: Aphelinidae) in a greenhouse covered with near-ultraviolet absorbing vinyl film. *Proc. Plant Protect. Kyushu* **32:**155–157.

Kennedy, J. S., Booth, C. O., and Kershaw, W. J. S. (1959). Host finding by aphids in the field II. *Aphis fabae* Scop. (Gynoparae) and *Brevicoryne brassicae* L. with a reappraisal of the role of host finding behaviour in virus spread. *Ann. Appl. Biol.* **47:**424–444.

Kennedy, J. S., Booth, C. O., and Kershaw, W. J. S. (1961). Host finding by aphids in the field III. Visual attraction. *Ann. Appl. Biol.* **49:**1–21.

Kevan, P. G., Straver, W. A., Offer, O., and Laverty, T. M. (1991). Pollination of greenhouse tomatoes by bumblebees in Ontario. *Proc. Entomol. Soc. Ont.* **122:**15–19.

Kring, J. B. (1964). New ways to repel aphids. *Front. Plant Sci.* **17:**6–7.

Kring, J. B. (1972). Flight behaviour of aphids. *Annu. Rev. Entomol.* **17:**461–492.

Kumar, P., and Poehling, H. M. (2006). UV- blocking plastic films and nets influence vectors and virus transmission on greenhouse tomatoes in the humid tropics. *Environ. Entomol.* **35:**1069–1082.

Lee, J. M. (1994). Cultivation of grafted vegetables 1. Current status, grafting methods and benefits. *HortScience* **29:**235–239.

Lewandowski, D. (2000). Genus Tobamovirus. *In* "Virus Taxonomy, seventh report of the international committee of taxonomy of viruses" (M. H. V. van Regenmortel, C. M. Fauquet, D. H. L. Bishop, E. B. Carstens, M. K. Estes, S. M. Lemon, J. Maniloff, M. A. Mayo, D. J. McGeoch, C. R. Pringle, and R. B. Wickner, eds.), pp. 889–894. Academic Press, San Diego, CA.

Loebenstein, G., and Raccah, B. (1980). Control of non-persistently transmitted aphid-borne viruses. *Phytoparasitica* **8:**221–235.

Loebenstein, G., Alper, M., and Deutsch, M. (1964). Preventing aphid-spread cucumber mosaic virus with oils. *Phytopathology* **54:**960–962.

Loebenstein, G., Alper, M., Levy, S., and Menagem, E. (1975). Protecting peppers from aphid-borne viruses with aluminium foil or plastic mulch. *Phytoparasitica* **3:**43–53.

Lubbock, J. (1882). Ants, Bees and Wasps: A Record of Observations on the Habits of the Social Hymenoptera. Kegan Paul, Trench, Trubner, London (New edition, based on the 17th, 1929).

Mandahar, C. L. (1990). Virus transmission. *In* "Plant Viruses" (C. L. Mandahar, ed.), Vol. II, pp. 205–242. CRC Press, Boca Raton, FL.

Marco, S. (1986). Incidence of aphid transmitted virus infections reduced by whitewashes sprays on plants. *Phytopathology* **76:**1344–1348.

Matteson, N., and Terry, L. I. (1992). Response to colour by male and female *Frankliniella occidentalis*. *Entomol. Exp. Appl.* **63:**187–201.

Menzel, R. (1979). Spectral sensitivity and colour vision in invertebrates. *In* "Handbook of Sensory Physiology" (H. Autrum, ed.), Vol. VII/6A, pp. 503–580. Springer, Heidelberg.

Monci, F., Garcia-Andres, S., Sanchez, F., and Moriones, E. (2002). Tomato yellow leaf curl disease control with UV-blocking plastic covers in commercial plastichouses of southern spain. Presented at the XXVIth Int. Hort. Cong. & Exhibit., Toronto Canada, 11–17 August 2002.

Mound, L. A. (1962). Studies on the olfaction and colour sensitivity of *Bemisia tabaci* (Genn.) (Homoptera, Aleurodidae). *Entomol. Exp. Appl.* **5:**99–104.

Mutwiwa, U. N., Brogemeister, C., Von Elsner, B., and Tantu, H. (2005). Effects of UV-absorbing plastic films on greenhouse whitefly (Homoptera: Aleyrodidae). *J. Econ. Entomol.* **98:**1221–1228.

Natwick, E. T., Byers, J. A., Chu, C., Lopez, M., and Henneberry, T. J. (2007). Early detection and mass trapping of *Frankliniella occidentalis* and *Thrips tabaci* in vegetable crops. *Southwest. Entomol.* **32:**229–238.

Oda, M. (1999). Grafting of vegetables to improve greenhouse production. http://www.agnet.org/library/eb/480.

Orozco-Santos, M., Perez-Zamora, O., and Lopez-Arriaga, O. (1995). Effect of transparent mulch on insect populations, virus diseases, soil temperature, and yield of cantaloup in a tropical region. *N. Z. J. Crop Hortic. Sci.* **23:**199–204.
Powell, G. (1991). Cell membrane punctures during epidermal penetration by aphids: Consequences for the transmission of two potyviruses. *Ann. Appl. Biol.* **119:**313–321.
Powell, G. (1992). The effect of mineral oil on stylet activities and potato virus Y transmission by aphids. *Entomol. Exp. Appl.* **63:**237–242.
Rapisarda, C., and Tropea Garzia, G. (2002). Tomato yellow leaf curl Sardinia virus and its vector *Bemisia tabaci* in Sicilia (Italy): Present status and control. *OEPP/EPPO Bull.* **32:**25–29.
Raviv, M., and Antignus, Y. (2004). UV Radiation effects on pathogens and insect pests of greenhouse-grown crops. *Photochem. Photobiol.* **79:**219–226.
Reagan, T. E., Goodine, G. V., Jr., and Kennedy, G. G. (1979). Evaluation of insecticides and oil for suppression of aphid-borne viruses in tobacco. *J. Econ. Entomol.* **72:**538–540.
Scherer, C., and Kolb, G. (1987). Behavioral experiments on the visual processing of color stimuli in *Pieris brassicae* L. (Lepidoptera). *J. Comp. Physiol.* **160:**647–656.
Seker, I. (1999). The use of UV-blocking films to reduce insect pests damage in tomato greenhouses and the effect of UV-filtration on the pollination activity of bumble bees. *Gan Sadeh VaMeshek* **12:**55–59 (in Hebrew).
Shahak, Y., Gal, E., Offir, Y., and Ben-Yakir, D. (2008). Photoselective shade netting integrated with greenhouse technologies for improved performance of vegetable and ornamental crops. *Acta Hortic.* **797:**75–80.
Shands, W. A. (1977). Control of aphid-borne potato virus Y in potatoes with oil emulsions. *Am. Potato J.* **54:**179–187.
Simon- Buela, L., and Garcia-Arenal, F. (1999). Virus particles of cucumber green mottle mosaic tobamovirus move systemically in the phloem of infected cucumber plants. *Mol. Plant Microbe Interact.* **12:**112–118.
Simons, J.N, McLean, D. L., and Kinsey, M. G. (1977). Effects of mineral oil on probingbehaviour and transmission of stylet-borne viruses by *Myzus persicae*. *J. Econ. Entomol.* **70:**309–315.
Steinberg, S., Prag, H., Gouldman, D., Antignus, Y., Pressman, E., Asenheim, D., Moreno, Y., and Schnitzer, M. (1997). The effect of ultraviolet-absorbing plastic sheets on pollination of greenhouse tomatoes by bumblebees. Proc. Int. Cong. for Plastics in Agric. (CIPA), Tel-Aviv Israel.
Summers, C. G., Michell, J. P., and Stapleton, J. J. (2005). Mulches reduce aphid-borne viruses and whiteflies in cantaloupe. *California Agric.* **59:**90–94.
Suwwan, M. A., Akkawi, M., Al-Musa, M. A., and Mansour, A. (1988). Tomato performance and incidence of tomato yellow leaf curl (TYLC) virus as affected by type of mulch. *Sci. Hortic.* **37:**39–45.
Terry, L. I. (1997). Host selection, communication and reproductive behaviour. *In* ''Thrips as Crop Pests'' (T. Lewis, ed.), pp. 65–118. CAB International, Oxon (UK), New York (USA).
Ucko, O., Cohen, S., and Ben-Joseph, R. (1998). Prevention of virus epidemics by a crop free period in the Arava region of Israel. *Phytoparasitica* **26:**313–321.
Vaishampayan, S. M., Kogan, M., Waldbauer, G. P., and Woolley, J. T. (1975a). Spectral specific responses in the visual behaviour of the greenhouse whitefly, *Trialeurodes vaporariorum* (Homoptera: Aleyrodidae). *Entomol. Exp. Appl.* **18:**344–356.
Vaishampayan, S. M., Waldbauer, G. P., and Kogan, M. (1975b). Visual and olfactory responses in orientation to plants by the greenhouse whitefly *Trialeurodes vaporariorum* (Homoptera: Aleyrodidae). *Entomol. Exp. Appl.* **18:**412–422.
Van de Veire, M., and Vacante, V. (1984). Greenhouse whitefly control through the combined use of color attraction system with the parasite wasp *Encarsia formosa* (Hym. Aphelinidae). *Entomophaga* **29:**303–310.

Vanderveken, J. (1968). Effects of mineral oils and lipids on aphid transmission of beet mosaic and beet yellows viruses. *Virology* **34:**807–809.

Wang, R. Y., and Pirone, T. P. (1996). Mineral oil interferes with retention of tobacco etch potyvirus in the stylets of *Myzus persicae*. *Phytopathology* **86:**820–823.

Wood, F. A. (1962). Aphid transmission of potato virus Y inhibited by oils. *Virology* **18:**327–329.

Wyman, J. (1971). Use of oils and other materials in the reduction of aphid-transmitted plant viruses. Dissertation, University of Wisconsin.

Zitter, T. A., and Ozaki, H. Y. (1978). Aphid-borne vegetable viruses controlled with oil sprays. *Proc. Fla. State Hortic. Soc.* **91:**287–289.

Zitter, T. A., and Simons, J. N. (1980). Management of viruses by alteration of vector efficiency and by cultural practices. *Ann. Rev. Phytopath.* **18:**289–310.

INDEX

Note: Page numbers followed by "*f*" indicate figures, and "*t*" indicate tables.

D

E

F

G

I

N

O

P

Z

Figure 1, Inge M. Hanssen and Moshe Lapidot (See Page 34 of this Volume)

Figure 1, Hervé Lecoq and Cécile Desbiez (See Page 74 of this Volume)

Number of mutations required for resistance breakdown

3

2

1

6015G
6058G
6058A
6069A
6072U
6075A
6069G + 6070A + 6075A
6069G + 6075A
8424G

Resistance gene	$pvr2^{3}$	$\mathbf{pvr2^{1}}$	$\mathbf{pvr2^{2}}$	**Pvr4**
Durability	poor	average	very high	very high

Figure 3, Benoît Moury and Eric Verdin (See Page 135 of this Volume)

Figure 1, Aranzazu Moreno and Alberto Fereres (See Page 254 of this Volume)

Figure 2, Aranzazu Moreno and Alberto Fereres (See Page 255 of this Volume)

Figure 3, Aranzazu Moreno and Alberto Fereres (See Page 265 of this Volume)

Figure 4, Aranzazu Moreno and Alberto Fereres (See Page 268 of this Volume)

Figure 5, Aranzazu Moreno and Alberto Fereres (See Page 268 of this Volume)

Figure 6, Aranzazu Moreno and Alberto Fereres (See Page 272 of this Volume)